T0134286

Solid State Physics

By identifying unifying concepts across solid state physics, this text covers theory in an accessible way to provide graduate students with the basis for making quantitative calculations and an intuitive understanding of effects. Each chapter focuses on a different set of theoretical tools, using examples from specific systems and demonstrating practical applications to real experimental topics. Advanced theoretical methods including group theory, many-body theory, and phase transitions are introduced in an accessible way, and the quasiparticle concept is developed early, with discussion of the properties and interactions of electrons and holes, excitons, phonons, photons, and polaritons. New to this edition are sections on graphene, surface states, photoemission spectroscopy, two-dimensional spectroscopy, transistor device physics, thermoelectricity, metamaterials, spintronics, exciton-polaritons, and flux quantization in superconductors. Exercises are provided to help put knowledge into practice, with a solutions manual for instructors available online, and appendices review the basic math methods used in the book. A complete set of the symmetry tables used in group theory (presented in Chapter 6) is available at www.cambridge.org/snoke.

David W. Snoke is a Professor at the University of Pittsburgh where he leads a research group studying quantum many-body effects in semiconductor systems. In 2007, his group was one of the first to observe Bose-Einstein condensation of polaritons. He is a Fellow of the American Physical Society.

Solid State Physics

Essential Concepts

Second Edition

DAVID W. SNOKE

University of Pittsburgh

CAMBRIDGE UNIVERSITY PRESS

CAMBRIDGE
UNIVERSITY PRESS

University Printing House, Cambridge CB2 8BS, United Kingdom

One Liberty Plaza, 20th Floor, New York, NY 10006, USA

477 Williamstown Road, Port Melbourne, VIC 3207, Australia

314–321, 3rd Floor, Plot 3, Splendor Forum, Jasola District Centre, New Delhi – 110025, India

79 Anson Road, #06–04/06, Singapore 079906

Cambridge University Press is part of the University of Cambridge.

It furthers the University's mission by disseminating knowledge in the pursuit of education, learning, and research at the highest international levels of excellence.

www.cambridge.org
Information on this title: www.cambridge.org/9781107191983
DOI: 10.1017/9781108123815

First published 2020

Printed in the United Kingdom by TJ International Ltd, Padstow Cornwall

A catalogue record for this publication is available from the British Library.

ISBN 978-1-107-19198-3 Hardback

Additional resources for this publication at www.cambridge.org/snoke

There is beauty even in the solids.

I tell you, if these were silent, even the rocks would cry out!
— Luke 19:40

For his invisible attributes, namely, his eternal power and divine nature, have been clearly perceived, ever since the creation of the world, in the things that have been made.
— Romans 1:20

Contents

Preface

Imagine teaching a physics course on classical mechanics in which the syllabus is organized around a survey of every type of solid shape and every type of mechanical device. Or imagine teaching thermodynamics by surveying all of the phenomenology of steam engines, rockets, heating systems, and such things. Not only would that be tedious, much of the beauty of the unifying theories would be lost. Or imagine teaching a course on electrodynamics which begins with a lengthy discussion of all the faltering attempts to describe electricity and magnetism before Maxwell. Thankfully, we don't do this in most courses in physics. Instead, we present the main elements of the unifying theories, and use a few of the specific applied and historical cases as examples of working out the theory.

Yet in solid state physics courses, many educators seem to feel a need to survey every type of solid and every significant development in phenomenology. Students are left with the impression that solid state physics has no unifying, elegant theories and is just a grab bag of various effects. Nothing could be further from the truth. There are many unifying concepts in solid state physics. But any book on solid state physics that focuses on unifying concepts must leave out some of the many specialized topics that crowd books on the subject.

This book centers on essential theoretical concepts in all types of solid state physics, using examples from specific systems with real units and numbers. Each chapter focuses on a different set of theoretical tools. "Solid state" physics is particularly intended here, because "condensed matter" physics includes liquids and gases, and this book does not include in-depth discussions of those states. These are covered amply, for example, by Chaikin and Lubensky.[1]

Some books attempt to survey the phenomenology of the entire field, but solid state physics is now too large for any book to do a meaningful survey of all the important effects. The survey approach is also generally unsatisfying for the student. Teaching condensed matter physics by surveying the properties of various materials loses the essential beauty of the topic. On the other hand, some books on condensed matter physics deal only with "toy models," never giving the skills to calculate real-world numbers.

Researchers in the field seem to be split in regard to the importance of the advanced topics of group theory and many-body theory. Some solid state physicists say that all of solid state physics starts with group theory, while others dismiss it entirely – I would guess that well over half of academic researchers in the field have never studied group theory at all. As I discuss in Chapter 1, the existence of electron bands does not depend crucially on

[1] P.M. Chaikin and T.C. Lubensky, *Principles of Condensed Matter Physics* (Cambridge University Press, 2000).

symmetry properties, although the symmetry theory provides a wide variety of tools to use for systems that approximate certain symmetries.

In the same way, there is a divide on many-body theory. Experimentalists tend to avoid the subject altogether, while theorists start with it. This leads to an "impedance mismatch" when experimentalists and theorists talk to each other. In Chapter 8 of this book, I introduce the elements of many-body theory which will allow experimentalists to cross this divide without taking years of theoretical courses, and which will serve as an introduction to students planning to go deeper into these methods. It may be a surprise to some people that there are actually several different diagrammatic approaches, including the Rayleigh–Schrödinger theory common in optics circles, the Feynmann diagrammatic method, and the Matsubara imaginary-time method. All three are surveyed in Chapter 8, with a discussion of their connections.

While group theory and many-body theory may come across as high-level topics to some, others may be surprised to see "engineering" topics such as semiconductor devices, stress and strain matrices, and optics included. While some experimentalists skip group theory and many-body theory in their education, too many theorists skip these basic topics in their training. Understanding the details of these methods is crucial for understanding many of the experiments on fundamental phenomena, as well as applications in the modern world.

In this book, I have tried to focus on unifying and fundamental theories. This raises the question: Does solid state physics really involve fundamental physics? Are there really any important questions at stake? Many physics students think that astrophysics and particle physics address fundamental questions, but solid state physics doesn't. Perhaps this is because of the way we teach it. Astrophysics and particle physics courses tend to focus much more on unifying, grand questions, especially at the introductory level, while solid state physics courses often focus on a grab bag of various phenomena. If we can get past the listing of material properties, solid state physics does deal with fascinating questions.

One deep philosophical issue is the question of "reductionism" versus "emergent behavior." Since the time of Aristotle and Democritus, philosophers have debated whether matter can be reduced to "basic building blocks" or if it is infinitely divisible. For the past two centuries, many scientists have tended to assume that Democritus was right – that all matter is built from a few indivisible building blocks, and once we understand these, we can deduce all other behavior of matter from the laws of these underlying building blocks. In the past few decades, many solid state physicists, such as Robert Laughlin, have vociferously rejected this view.[2] They would argue that possibly every quantum particle is divisible, but it doesn't matter for our understanding of the essential properties of things.

At one time, people thought atoms were indivisible, but it was found they are made of subatomic particles. Then people thought subatomic particles were indivisible, but it was found that at least some of them are made of smaller particles such as quarks. Are quarks indivisible? Many physicists believe there is at least one level lower. As the distance scale gets smaller, the energy cost gets higher. This debate came to a head in the 1980s when the high-energy physics community proposed to spend billions of dollars on the

---

[2] R. Laughlin, *A Different Universe* (Basic Books, 2005).

Superconducting Supercollider in Texas, far more than the total budget of all other physics in the USA, and some solid state physicists such as Rustum Roy opposed it. In the anti-reductionist view, it is pointless to keep searching for one final list of all particles and forces.

Those who hold to the anti-reductionist view often point to the concept of "renormalization" in condensed matter physics. This is a very general concept. Essentially, it means that we can redefine a system at a higher level, ignoring the component parts from which it is made. Then we can work entirely at the higher level, ignoring the underlying complexities. The properties at this higher level depend only on a few basic properties of the system, which could arise from any number of different microscopic properties.

There are two versions of this. The first is many-body renormalization, introduced in Chapter 2 of this book and developed further in Chapter 8. In this theory, the ground state of a system is defined as the "vacuum," and excitations out of this state are "quasiparticles" with properties very different from the particles making up the underlying ground state. These quasiparticles then become the new particles of interest, and can themselves make up a new vacuum ground state with additional excitations. As discussed in Chapter 11, this process can be continued to any number of higher levels.

A second type of renormalization is that of renormalization groups, introduced in Chapter 10. In this approach, the essential properties of a system can be described using subsets of the whole, in which properties are averaged. From this a whole field of theory on **universality** has been developed, in which certain properties of systems can be predicted based on just a few attributes of the underlying system, without reference to the microscopic details.

Another deep topic that comes up in solid state physics is the foundations of statistical mechanics. There was enormous controversy at the founding of the field, and much of this controversy was simply swept under the rug in later years, and there is still philosophical debate.[3] The fundamental questions of statistical mechanics arise especially when we deal with nonequilibrium systems, a major topic of solid state physics. In Chapter 4, I present the quantum mechanical basis of irreversible behavior, which involves the concept of "dephasing" which arises in later chapters, especially Chapter 9.

This connects to another important philosophical question, the "measurement" problem of quantum mechanics, that is, what leads to "collapse" of the wave function and what constitutes a measurement. In both quantum statistical mechanics and quantum collapse, we have irreversible behavior arising from an underlying system which is essentially reversible. Is there a connection? The essential paradoxes of quantum mechanics all arise in the context of condensed matter, and going to subatomic particles does not help at all in the resolution of the paradoxes, nor raise new paradoxes.

One of the deepest issues of our day is the question of emergent phenomena. Is life as we know it essentially a generalization of condensed matter physics, in which structure arises entirely from simple interactions at the microscopic level, or do we need entirely new ways of thinking when approaching biophysics, with concepts such as feedback, systems

---

[3] See, e.g., Harvey Brown, "One and for all: the curious role of probability in the Past Hypothesis," in *The Quantum Foundations of Statistical Mechanics*, D. Bedingham, O. Maroney, and C. Timpson (eds.) (Oxford University Press, 2017).

engineering, and transmission and processing of information?[4] Phase transitions are often viewed as examples of order coming out of disorder, through the process known as spontaneous symmetry breaking (introduced in Chapters 10 and 11 of this book). The effects that come about in solid state physics due to phase transitions can be dramatic, but we are a long way from extrapolating these to an explanation of the origin of life.

This book does not survey the rapidly evolving field of topological effects in condensed matter physics, except briefly at the end of Chapters 2 and 9. We have yet to create a canon of the truly essential phenomena, though it is already possible to list the various topology classes.[5] A discussion of surface states, which arise in many examples of topological effects, is presented at the end of Chapter 1.

Many people contributed to improving this book. I would like to thank in particular Dan Boyanovsky, David Citrin, Hrvoye Petek, Chris Smallwood, and Zoltan Vörös for critical reading of parts of this manuscript. I would also like to thank my wife Sandra for many years of warm support and encouragement.

David Snoke
Pittsburgh, 2019

[4] See, e.g., A.D. Lander, "A calculus of purpose," *PLoS Biology 2*, e164 (2004).
[5] See A.P. Schnyder, S. Ryu, A. Furusaki, and A.W.W. Ludwig, "Classification of topological insulators and superconductors in three spatial dimensions," *Physical Reviews B* **78**, 195125 (2008), and references therein.

1　Electron Bands

When we start out learning quantum mechanics, we usually think in terms of single particles, such as electrons in atoms or molecules. This is historically how quantum mechanics was first developed as a rigorous theory.

In a sense, all of atomic, nuclear, and particle physics are similar, because they all involve interactions of just a few particles. Typically in these fields, one worries about scattering of one particle with one or two others, or bound states of just a few particles. In large nuclei, there may be around 100 particles.

Solid state physics requires a completely new way of thinking. In a typical solid, there are more than 10^{23} particles. It is hopeless to try to keep track of the interactions of all of these particles individually. The beauty of solid state physics, however, lies in the old physics definition of simplicity: "one, two, infinity." In many cases, infinity is simpler to study than three; we can often find exact solutions for an infinite number of particles, and 10^{23} is infinite to all intents and purposes.

Not only that, but *new phenomena* arise when we deal with many particles. Various effects arise that we would never guess just from studying the component particles such as electrons and nuclei. These "emergent" or "collective" effects are truly *fundamental* physical laws in the sense that they are universal paradigms. In earlier generations, many physicists took a reductionist view of nature, which says that we understand all things better when we break them into their constituent parts. In modern physics, however, we see some fundamental laws of nature arising only when many parts interact together.

1.1 Where Do Bands Come From? Why Solid State Physics Requires a New Way of Thinking

The concept of electron *bands* in solids, developed out of quantum mechanical theory in the early twentieth century, is an example of a universal idea that has wide application to a vast variety of materials, that fundamentally relies on the relationship of a large number of particles. As we will see, the theory of bands says that an electron can move freely, as though it was in vacuum, through a solid that is crowded with 10^{23} atoms per cubic centimeter. This goes against most people's intuition that an electron ought to scatter from all those atoms in its path. In fact, it does feel their presence, but their effect is taken into account in the calculation of the band energies, after which the presence of all those atoms can be largely ignored.

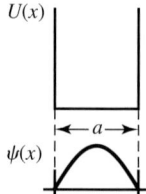

Fig. 1.1   A square well and its ground state wave function.

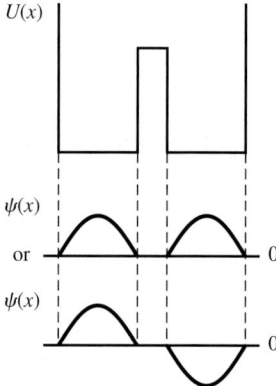

Fig. 1.2   Two independent square wells and their ground state wave functions.

1.1.1 Energy Splitting Due to Wave Function Overlap

To see how bands arise, we can use a very simple model. We start with the well-known example of a particle in a square potential, as shown in Figure 1.1. From introductory quantum mechanics, we know that the wave nature of the particle allows only discrete wavelengths. The time-independent Schrödinger equation for a particle with mass m is

$$-\frac{\hbar^2}{2m}\nabla^2\psi(x) + U(x)\psi(x) = E\psi(x), \tag{1.1.1}$$

which has the eigenstates

$$\psi(x) = A\sin(kx) \tag{1.1.2}$$

in the region where $U(x) = 0$, and is zero at the boundaries. The energies are

$$E = \frac{\hbar^2 k^2}{2m}, \tag{1.1.3}$$

where $k = N\pi/a$ and $N = 1, 2, 3, \ldots$.

Next consider the case of two square potentials separated by a barrier, as shown in Figure 1.2. If the barrier is high enough, or if the square-well potentials are far enough apart, then a particle cannot go from one well to the other, and we simply have two independent eigenstates of $\psi(x)$, one for each well, with the same energies. The eigenstate for

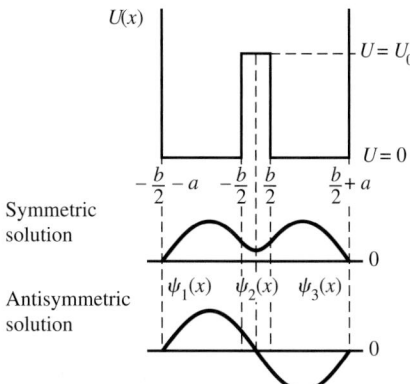

Fig. 1.3   Two coupled square wells and their ground state wave functions.

each well is the same if we multiply $\psi(x)$ by -1 or i or any other phase factor, since an overall phase factor does not change the energy or probabilities, which depend only on the magnitude of the wave function.

Suppose now that we bring the wells closer together, so that the barrier does not completely prevent a particle in one well from going to the other well, as shown in Figure 1.3. In this case, we say that the two regions are *coupled*. Then elementary quantum mechanics tells us that we cannot solve two separate Schrödinger equations for the two wells; we must solve one Schrödinger equation for the whole system. We write

$$U(x) = \begin{cases} \infty, & x < -\dfrac{b}{2} - a \\[2mm] 0, & -\dfrac{b}{2} - a < x < -\dfrac{b}{2} \\[2mm] U_0, & -\dfrac{b}{2} < x < \dfrac{b}{2} \\[2mm] 0, & \dfrac{b}{2} < x < \dfrac{b}{2} + a \\[2mm] \infty, & x > \dfrac{b}{2} + a \end{cases} \qquad (1.1.4)$$

and break the wave function into three parts,

$$\psi(x) = \psi_1(x), \quad -\dfrac{b}{2} - a < x < -\dfrac{b}{2}$$

$$\psi(x) = \psi_2(x), \quad -\dfrac{b}{2} < x < \dfrac{b}{2}$$

$$\psi(x) = \psi_3(x), \quad \dfrac{b}{2} < x < \dfrac{b}{2} + a \qquad (1.1.5)$$

with the boundary conditions

$$\psi_1\left(-b/2 - a\right) = 0, \quad \psi_3\left(b/2 + a\right) = 0,$$

$$\psi_1\left(-b/2\right) = \psi_2\left(-b/2\right), \quad \psi_2\left(b/2\right) = \psi_3\left(b/2\right)$$

$$\left.\frac{\partial\psi_1}{\partial x}\right|_{-\frac{b}{2}} = \left.\frac{\partial\psi_2}{\partial x}\right|_{-\frac{b}{2}}, \quad \left.\frac{\partial\psi_2}{\partial x}\right|_{\frac{b}{2}} = \left.\frac{\partial\psi_3}{\partial x}\right|_{\frac{b}{2}}.$$

Instead of two independent eigenstates, we now have **symmetric** and **antisymmetric** solutions for the eigenstates. Figure 1.3 shows the symmetric and antisymmetric states that arise from the ground states of the two wells. Of course, we could have also constructed symmetric and antisymmetric solutions in the case when the wells were far apart, as shown in Figure 1.2, but the energies of the symmetric and antisymmetric solutions would have been degenerate in that case. (Different wave functions with the same energy are called **degenerate** solutions.) We could write the solutions of two independent wells as any linear combination of the two solutions, and we would still obtain two solutions with the same energy.

When there is coupling of the wells, however, the energies of the symmetric and antisymmetric states are not the same. The antisymmetric solution has higher energy. This is a general rule: *the antisymmetric combination of two states almost always has higher energy than the symmetric combination.* To see why this is so, notice that the antisymmetric state must have a node at $x = 0$ since $\psi(-x) = -\psi(x)$. As shown in Figure 1.4, this means that the antisymmetric wave function must have slightly shorter wavelength components than the symmetric wave function. Recall that according to the Fourier theorem, any well-behaved function can be written as a sum of oscillating waves (see Appendix B). Fast changes of a function over short distance imply that the Fourier sum must include waves with shorter wavelength. The antisymmetric function must change faster within the barrier in order to go through the node. Since shorter wavelength corresponds to higher energy for all particles, this means that the antisymmetric wave function will have higher energy.

The more strongly coupled the wells are, the greater the energy splitting will be between the symmetric and antisymmetric states. This is because the barrier between the wells forces the symmetric wave function to fall toward zero inside the barrier, making it similar to the antisymmetric wave function. If the barrier is lower or thinner, it is more probable for the electron to be found in the barrier, and the symmetric wave function does not need

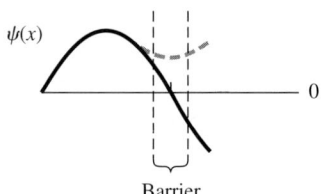

Fig. 1.4 Solid line: the antisymmetric ground state wave function of the coupled well system near $x = 0$. Dashed line: the symmetric ground state wave function of the coupled well near $x = 0$.

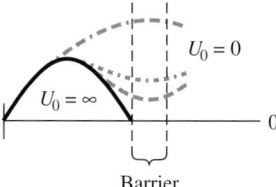

Fig. 1.5 The symmetric wave function of the coupled well system for various values of the barrier height. Solid line: $U_0 = \infty$. Dashed line: finite barrier height. Dashed-dotted line: lower barrier than in the case of the dashed line. Long dashed-dotted line: zero barrier.

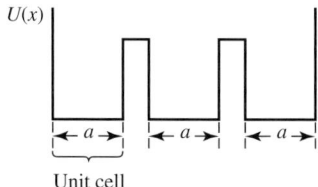

Fig. 1.6 Three coupled square wells.

to fall as far, as illustrated in Figure 1.5. If it bends less, it has longer-wavelength Fourier terms, which have lower energy.

Continuing on, imagine next that we have three square wells separated by small barriers, as shown in Figure 1.6. We define a **unit cell** as the repeated unit – in this case, the square well. Without even solving the Schrödinger equation for this system, we can see that there will be three different eigenstates that arise from linear combinations of the three ground states of the independent wells. For example, suppose we add together the single-well ground state wave functions with overall phase factors of either $+$ or $-$. There are 2^3 possibilities:

$$
\begin{array}{ccc|ccc}
+ & + & + & - & - & - \\
+ & + & - & - & - & + \\
+ & - & + & - & + & - \\
- & + & + & + & - & -
\end{array}
$$

but the second column is equivalent to the first column, since an overall phase factor of -1 does not change the wave function in any essential way. In addition, one of the four rows in one column can be written as a linear combination of the others; for example, $(+ + +)$ is the sum of $(- + +)$ and $(+ - +)$ and $(+ + -)$. This makes sense, because if the wells were independent we would have three independent states corresponding to the three independent wells, and when we allow coupling, we do not create new states out of nowhere.

Three wells is already too hard to bother to solve exactly. But following the same logic, we can jump directly to thinking about an infinite number of wells. Without solving this case exactly, we can see immediately that if we start with N degenerate eigenstates of the individual wells and allow coupling between neighboring wells, then we will still have

N eigenstates, but the states will have different energies depending on the exact linear combinations of the underlying single-cell states. States with more nodes will have higher energy, while states with fewer nodes will have lower energy. There will be one linear combination of the states with minimum energy, corresponding to all the single-cell state wave functions included with the same sign, and one linear combination with maximum energy, corresponding to the single-cell wave function in each cell having the opposite sign from its neighbor.

When there is a large number of cells, the difference in energy of a state by adding or subtracting one single node in the linear combination will be very small. Therefore, the states will be spaced closely together in energy. The large number of states will fall in an energy range which we call an energy **band**, as shown in Figure 1.7. For a large number of cells, the jumps in energy between the states will be so small that effectively the energy of an electron can change continuously in this range. In between the bands are energy **gaps**. These gaps arise from the gaps between the original single-cell states.

Exercise 1.1.1   Use Mathematica to solve the system of equations (1.1.1) and (1.1.4)–(1.1.6) for two coupled wells, for the case $2mU_0/\hbar^2 = 20$, $a = 1$, and $b = 0.1$. The calculation can be greatly simplified by assuming that the solution has the form $\psi_1(x) = A_1 \sin(Kx) + B_1 \cos(Kx)$, $\psi_2(x) = A_2(e^{\kappa x} \pm e^{-\kappa x})$, and $\psi_3(x) = \pm\psi_1(-x)$, where the $+$ and $-$ signs correspond to the symmetric and antisymmetric solutions, respectively. Since the overall amplitude of the wave doesn't matter, you can set $A_1 = 1$. In this case, there are four unknowns, B_1, A_2, K, and κ, and four equations, namely three independent boundary conditions and (1.1.1) in the barrier region, which gives κ in terms of K. B_1 and A_2 can be easily eliminated algebraically. You are left with a complicated equation for K; you can find the roots by first graphing both sides as functions of K to see approximately where the two sides are equal, and then using the Mathematica function FindRoot to get an exact value for K.

Plot the energy splitting of the two lowest energy states (the symmetric and antisymmetric combinations of the ground state) for various choices of the barrier

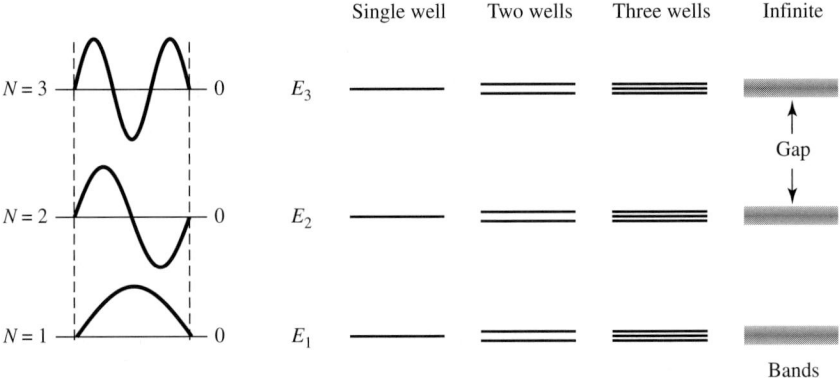

Energy states for coupled square wells.

thickness b. In the limit of infinite separation, they should have the same energy. Note that in the limit $b \to 0$, the symmetric and antisymmetric solutions of the two wells simply become the $N = 1$ and $N = 2$ solutions for a single square well. Plot the wave function for a typical value of b. Since the function has three parts, you will have to use the Mathematica function Show to combine the three function plots.

1.1.2 The LCAO Approximation

There is nothing special about the choice of square wells in the above examples. We could also have started with electron states in atoms or molecules, using atoms as our repeated cells. In this case, we would also find energy bands.

In the case of adjacent square wells, we could calculate the energy splitting exactly. In the case of atomic states, it is harder to do an exact calculation. A simple way of estimating the energy splitting of atomic states is to assume that the atomic orbitals are essentially unchanged. This is called the method of **linear combination of atomic orbitals** (LCAO). It is often an accurate approximation, because atoms in molecules and solids do not usually come too near to each other, so the atomic orbitals are not strongly distorted. In this case, we can write the overall wave function as

$$|\psi\rangle = c_1|\psi_1\rangle + c_2|\psi_2\rangle. \tag{1.1.6}$$

In general, $|\psi_1\rangle$ and $|\psi_2\rangle$ are not orthogonal, since the two wave functions overlap in the middle region between the atoms. We can expand the eigenvalue equation

$$H|\psi\rangle = E|\psi\rangle \tag{1.1.7}$$

as

$$\langle\psi_1|H|\psi\rangle = \langle\psi_1|E|\psi\rangle, \qquad \langle\psi_2|H|\psi\rangle = \langle\psi_2|E|\psi\rangle, \tag{1.1.8}$$

which is equivalent to

$$c_1\langle\psi_1|H|\psi_1\rangle + c_2\langle\psi_1|H|\psi_2\rangle = c_1E + c_2E\langle\psi_1|\psi_2\rangle$$
$$c_1\langle\psi_2|H|\psi_1\rangle + c_2\langle\psi_2|H|\psi_2\rangle = c_2E + c_1E\langle\psi_2|\psi_1\rangle. \tag{1.1.9}$$

The constants $E_1 = \langle\psi_1|H|\psi_1\rangle$ and $E_2 = \langle\psi_2|H|\psi_2\rangle$ are the unperturbed single-atom energies. There are two coupling terms, which we write as $U_{12} = \langle\psi_1|H|\psi_2\rangle = \langle\psi_2|H|\psi_1\rangle^*$, and the overlap integral, $I_{12} = \langle\psi_1|\psi_2\rangle$. These can be computed for the original orbital states. We then write

$$c_1E_1 + c_2(U_{12} - EI_{12}) = c_1E$$
$$c_1(U_{12}^* - EI_{12}^*) + c_2E_2 = c_2E. \tag{1.1.10}$$

Taking a perturbative approach, we assume that $E = \bar{E} + \Delta E$, where $\bar{E} = (E_1 + E_2)/2$ and ΔE is comparable to U_{12}. Assuming that the overlap integral I_{12} is small, because the orbitals do not overlap much in space, we drop the term $I_{12}\Delta E$ as negligible compared to U_{12}. We then have

$$\begin{pmatrix} E_1 & \tilde{U}_{12} \\ \tilde{U}_{12}^* & E_2 \end{pmatrix} \begin{pmatrix} c_1 \\ c_2 \end{pmatrix} = E \begin{pmatrix} c_1 \\ c_2 \end{pmatrix}, \tag{1.1.11}$$

where $\tilde{U}_{12} = U_{12} - \bar{E}I_{12}$. This is then an eigenvalue equation which we can solve. We find

$$E = \frac{E_1 + E_2}{2} \pm \sqrt{\left(\frac{E_1 - E_2}{2}\right)^2 + |\tilde{U}_{12}|^2}. \tag{1.1.12}$$

The lower-energy state is the **bonding** state, while the higher-energy state is the **antibonding** state, terms that may be familiar from chemistry.

The coupling term $\tilde{U}_{12}$ is almost always negative for identical orbitals. One can see this by writing

$$U_{12} = \int_{V_1} d^3r\, \psi_1^*(\vec{r})\left(U(\vec{r}) - \frac{\hbar^2\nabla^2}{2m}\right)\psi_2(\vec{r})$$
$$+ \int_{V_2} d^3r\, \psi_1^*(\vec{r})\left(U(\vec{r}) - \frac{\hbar^2\nabla^2}{2m}\right)\psi_2(\vec{r}), \tag{1.1.13}$$

where the first integral is over the volume primarily occupied by orbital ψ_1 and the second over the volume primarily occupied by ψ_2. In region V_2, the integral is nearly equal to

$$\bar{E}\int_{V_2} d^3r\, \psi_1^*(\vec{r})\psi_2(\vec{r}), \tag{1.1.14}$$

since the Hamiltonian acting on ψ_2 gives nearly the same state times its energy. Here we have set $\bar{E} = E_1 = E_2$ for identical orbitals. The integral for the region V_1 is generally much less than the integral over the region V_2, because the kinetic energy term $-\hbar^2\nabla^2\psi_2/2m$ will be negative for exponential decay of the wave function ψ_2 far from the center of the orbital. We then have

$$U_{12} < \bar{E}\int_{V_1+V_2} d^3r\, \psi_1^*(\vec{r})\psi_2(\vec{r}) = \bar{E}I_{12}, \tag{1.1.15}$$

and therefore, from the definition of $\tilde{U}_{12}$,

$$\tilde{U}_{12} < \bar{E}I_{12} - \bar{E}I_{12}, \tag{1.1.16}$$

which implies $\tilde{U}_{12} < 0$. When the eigenstates are computed for negative $\tilde{U}_{12}$, the lowest energy (bonding) state corresponds to the symmetric combination of the states, and the higher state corresponds to the antisymmetric combination, as we assumed in our discussion in Section 1.1.1.

Exercise 1.1.2   (a) Show that in the case of two identical atoms, the eigenstates of the LCAO model are the symmetric and antisymmetric linear combinations

$$|\psi\rangle = \frac{1}{\sqrt{2}}(|\psi_1\rangle \pm |\psi_2\rangle), \tag{1.1.17}$$

where the plus sign corresponds to the bonding state and the negative sign corresponds to the antibonding state. To simplify the problem, assume U_{12} is real. (Note that the wave function needs to be normalized.)

(b) The energies computed in (1.1.12) should actually be corrected slightly, because the change of the normalization of the wave function has not been taken into account. To account for the normalization, one should write

$$E = \frac{\langle \Psi | H | \Psi \rangle}{\langle \Psi | \Psi \rangle}, \tag{1.1.18}$$

where $|\Psi\rangle$ is the full eigenstate. Determine this corrected energy for the bonding and antibonding states of two identical atoms.

(c) (*Advanced*) If you did Exercise 1.1.1, then you can do a follow-up calculation to see how well the LCAO approximation works. Instead of a symmetric set of coupled wells, find the solution for the wave function of an electron in a single quantum well with an infinite barrier on one side and a barrier of infinite thickness, with height U_0, on the other side. Use this solution as the "atomic" state, and form symmetric and antisymmetric linear combinations of this. Determine the energies of the two states for $2mU_0/\hbar^2 = 100$ and $a = 1$, as the value of b is varied from 0 to 1. How well does the LCAO solution approximate the full solution? How important is the correction of part (b)?

1.1.3 General Remarks on Bands

The coupling term $\tilde{U}_{12}$ defined in the LCAO approximation, which determines the difference in energy between the bonding and antibonding orbitals, essentially gives the degree of band smearing in the solid. Just as the confined states in a single quantum well smear out into bands when many square wells are brought near each other, so also the confined states of an electron around an atom smear out when many atoms are brought near each other. This is why solids have electron energy bands.

Imagine starting with a large number of atoms very far apart. Each atom has distinct and independent electron orbitals. As shown in Figure 1.8, as the atoms get near each other, the interaction between the atoms leads to the appearance of bands, with gaps nearly equal to the gaps between the original atomic states. If we continue to push the atoms nearer to each other, these bands will widen, since the energy difference between the symmetric and antisymmetric combinations increases as the coupling increases, as discussed above, and the band gaps will shrink. Eventually, if we keep pushing the atoms closer to each other, the bands may cross. (It can be shown, however, that bands cannot cross in a one-dimensional system.)

In our square-well example, we used repeated, identical unit cells. It should not be hard to see, however, that if one or two cells were not the same as the others, it would not drastically change the overall argument. For example, if the atomic states are not the same in (1.1.12), we will not have symmetric and antisymmetric states, but we will still have two states as superpositions of the single-cell states with either the same or the opposite phase, which have increasing energy splitting with increasing coupling. Not only that, but if we had chosen the size of the cells randomly, within some range, we would still see bands and band gaps. Amorphous materials such as glasses can also have bands and band gaps.

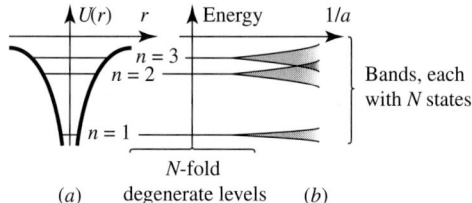

Fig. 1.8   (a) Schematic representation of nondegenerate electronic levels in an atomic potential. (b) The energy levels for N such atoms in a periodic array, plotted as a function of mean inverse interatomic spacing. When the atoms are far apart, the levels are nearly degenerate, but when the atoms are closer together, the levels broaden into bands.

Alloys are perhaps the best known examples of disordered materials that have well-defined bands and gaps.

One can therefore see that *the existence of electron bands and band gaps is not fundamentally related to periodicity*. Bands appear whenever a large number of cells are close enough together to have coupling between them. Band gaps exist whenever the coupling energy of the cells (which we can define as the difference in energy between the symmetric and antisymmetric states between two adjacent cells) is small compared to the energy jumps between states for an electron in a single cell. As we will see in Section 1.8.2, however, periodic structures have very sharply defined band gaps, while disordered materials have gaps with fuzzy boundaries.

The assumption of periodicity is an extremely powerful tool in solid state physics. Nevertheless, solid state physics does not begin and end with periodic structures. Many of the theories of solid state physics apply to amorphous and disordered systems.

The new physics which has arisen in the case of solids is that we cannot think in terms of interactions between single atoms. In introductory quantum mechanics, one typically considers scattering of single particles or atoms, but in solids, speaking of scattering between two atoms makes no sense, nor does it make sense to talk of an electron scattering with individual atoms. The electrons in a solid do not interact with single atoms; instead they are in a superposition of states belonging to all of the atoms in a macroscopic system. Each particle interacts with all the other particles.

Although solids are ubiquitous on Earth, it is actually somewhat surprising that they exist. For solids to exist, atoms must be packed at separations comparable to the electron orbital radius around a single atom, which is around 10^{-8} cm. This leads to densities of the order of 10^{24} atoms per cm^3. This is approximately 10^{28} times greater than the average density of the universe. Gravity must compact matter to an incredible degree, and the heat energy of compression must be lost, for solids to form.

1.2 The Kronig–Penney Model

Many of the properties of electron bands can be seen through a fairly simple, exactly solvable model, known as the Kronig–Penney model, which is an extension of the square-well

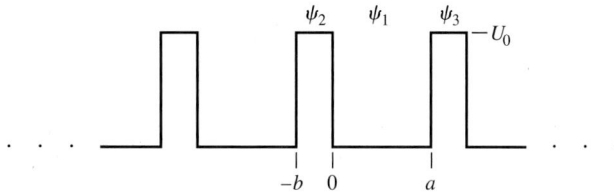

Fig. 1.9   The potential energy of the Kronig–Penney model.

structures we examined in Section 1.1.1. We imagine an electron in an infinite, perfectly periodic, one-dimensional structure, as shown in Figure 1.9.

As in the standard square-well model, we guess the form of the solution as follows:

$$\psi_1(x) = A_1 e^{iKx} + B_1 e^{-iKx}, \quad 0 < x < a$$

$$\psi_2(x) = A_2 e^{\kappa x} + B_2 e^{-\kappa x}, \quad -b < x < 0. \tag{1.2.1}$$

We only need to worry about these two regions, because the rest of the structure is identical to these.

Because every cell is identical, there is no reason for the wave function of an eigenstate in one cell to have greater magnitude than in any other cell. Therefore, it is reasonable to expect that the solution must be the same in every cell except possibly for an overall phase factor. The phase factor will in general be a function of the cell position, but constant within any given cell. Furthermore, the phase shift from one cell to the next should be the same, since there is no way to tell any two adjacent cells from another pair. We therefore set the phase factor equal to e^{ikX}, where X is the cell position and k is a constant. This implies

$$\psi_3(x) = \psi_2(x - a - b)e^{ik(a+b)}, \quad a < x < a + b \tag{1.2.2}$$

and the boundary conditions

$$\psi_1(0) = \psi_2(0), \qquad \left.\frac{\partial \psi_1}{\partial x}\right|_{x=0} = \left.\frac{\partial \psi_2}{\partial x}\right|_{x=0}$$

$$\psi_1(a) = \psi_3(a), \qquad \left.\frac{\partial \psi_1}{\partial x}\right|_{x=a} = \left.\frac{\partial \psi_3}{\partial x}\right|_{x=a}. \tag{1.2.3}$$

Plugging the definitions of ψ_1, ψ_2, and ψ_3 into these equations gives the matrix equation

$$\begin{pmatrix} 1 & 1 & -1 & -1 \\ iK & -iK & -\kappa & \kappa \\ e^{iKa} & e^{-iKa} & -e^{ik(a+b)}e^{-\kappa b} & -e^{ik(a+b)}e^{\kappa b} \\ iKe^{iKa} & -iKe^{-iKa} & -\kappa e^{ik(a+b)}e^{-\kappa b} & \kappa e^{ik(a+b)}e^{\kappa b} \end{pmatrix} \begin{pmatrix} A_1 \\ B_1 \\ A_2 \\ B_2 \end{pmatrix} = 0. \tag{1.2.4}$$

Setting the determinant of this matrix to zero gives the equation

$$\frac{(\kappa^2 - K^2)}{2\kappa K} \sinh(\kappa b) \sin(Ka) + \cosh(\kappa b) \cos(Ka) = \cos(k(a + b)). \tag{1.2.5}$$

The Schrödinger equation in the well region gives the energy $E = \hbar^2 K^2/2m$, and the Schrödinger equation in the barrier region gives $U_0 - E = \hbar^2 \kappa^2/2m$. Since κ and K both depend on E, Equation (1.2.5) can be solved for E for any given value of k.

We can greatly simplify the model by taking the limit $b \to 0$, $U_0 \to \infty$, in such a way that the product $U_0 b$ remains constant, which implies that $\kappa^2 b$ is a constant independent of E, and $\kappa b \to 0$. Then (1.2.5) reduces to

$$\frac{\kappa^2 b}{2K} \sin(Ka) + \cos(Ka) = \cos(ka). \qquad (1.2.6)$$

For any given k, we can solve this equation numerically for K, which gives us the energy $E = \hbar^2 K^2/2m$. Since (1.2.6) depends only on $\cos ka$, the solutions for $ka \pm 2\pi$ will be indistinguishable from the solution at ka. We therefore need only find the solutions of (1.2.6) for a range of k values such that ka varies by 2π. For each value of k, there are many solutions for K.

Figure 1.10 shows E vs k for a Kronig–Penney model, for k from $-\pi/a$ to π/a. Two general features stand out. The first is that there are gaps in the electron energy E, which occur when $|(\kappa^2 b/2K) \sin(Ka) + \cos(Ka)| > 1$; in other words, for some values of the total energy $E = \hbar^2 K^2/2m$ there is no corresponding value of k. These gaps appear at values of k that are multiples of π/a. A second feature is that the first derivative of E with respect to k vanishes at these points.

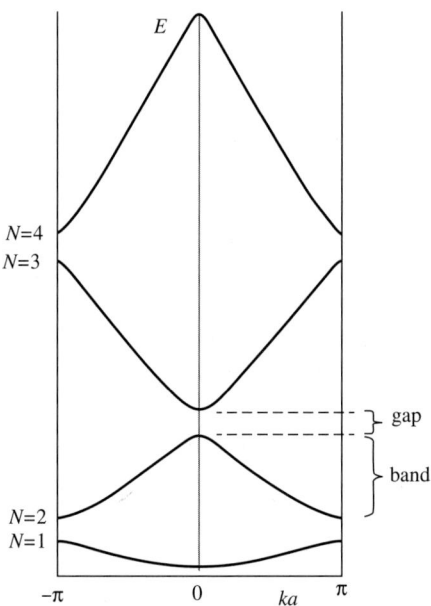

Fig. 1.10 Energy vs k for the Kronig–Penney model, for $\kappa^2 b = 4$. Different solutions of (1.2.6) for the same k are labeled by integers N.

The range $-\pi < ka < \pi$ is called the **Brillouin zone** of a periodic structure.[1] In the case of the Kronig–Penney model, it is fairly simple, since we are considering only a one-dimensional system. As we will see in Section 1.9, in three dimensions the Brillouin zone becomes more complicated.

Exercise 1.2.1   Verify the algebra leading to (1.2.5) and (1.2.6). Mathematica can be very helpful in simplifying the algebra.

Exercise 1.2.2   Find the zero-point energy, that is, $E = \hbar^2 K^2/2m$, at $k = 0$, using (1.2.6) in the limit $b \to 0$ and $U_0 \to \infty$ and $U_0 b$ finite but small. To do this, use the approximations for $\sin Ka \simeq Ka$ and $\cos Ka \simeq 1 - \frac{1}{2}(Ka)^2$, assuming Ka is very small at $k = 0$.

Exercise 1.2.3   Find the first gap energy at $ka = \pi$ using (1.2.6) in the limit $b \to 0$ and $U_0 b$ is small. You should write approximations for $\sin Ka$ and $\cos Ka$ near $Ka = \pi$, that is, $Ka \simeq \pi + (\Delta K)a$, where ΔK is small. You should find that you get an equation in terms of K that is factorizable into two terms that can equal zero. The difference between the energies $E = \hbar^2 K^2/2m$ for these two solutions for K is the energy gap.

Do your zero-point energy and gap energy vanish in the limit $U_0 b \to 0$?

In (1.2.2), we introduced the phase factor e^{ikX} for different cells, where X is the cell position $X = na$, in the limit $b \to 0$, and n is some integer. We can think of this phase factor as a plane wave that modulates the single-cell wave function; in other words,

$$\psi(x) = \psi_{\text{cell}}(x \bmod X)e^{ikX}, \tag{1.2.7}$$

where ψ_{cell} is the same for all cells and "$x \bmod X$" gives the position within each cell relative to the cell location $X = na$. We can therefore view Figure 1.10 as the dispersion relation for this overall plane wave.

We can get a feel for why the gaps appear at the points $ka = n\pi$ in the dispersion relation if we think of the physical effect of the repeated cells on the plane wave. The set of interfaces with spacing a make up a **Bragg reflector**. A Bragg reflector is a large number of equally spaced, partially reflecting, identical objects. As illustrated in Figure 1.11, if the distance between the objects is a, then the round-trip distance between two objects is $2a$. Therefore, the reflections of a traveling wave with wavelength $2a/n$ will all be in phase and add constructively. Even if only a small amount of the wave is reflected from any given object, a wave with wavelength $2a/n$ will be perfectly reflected in an infinite system. The incident wave plus the reflected wave traveling in the opposite direction make a standing wave. A standing wave has group velocity of zero; in other words, $v_g(k) = (1/\hbar)\partial\omega/\partial k = 0$. Since $E = \hbar\omega$ for electrons, this implies $\partial E/\partial k = 0$. (We will return to discuss group velocity in Section 3.3.)

Formally, we can see that the bands must have $\partial E/\partial k = 0$ at the symmetry points by implicit differentiation of (1.2.6). Setting $U_0 = \hbar^2\kappa^2/2m$ in the limit $U_0 \gg E$, we have

---

[1] It is sometimes called the "first" Brillouin zone, because Brillouin came up with a series of zones based on the symmetry of a system. (See Section 1.9.3.) It is typically called simply the Brillouin zone, however.

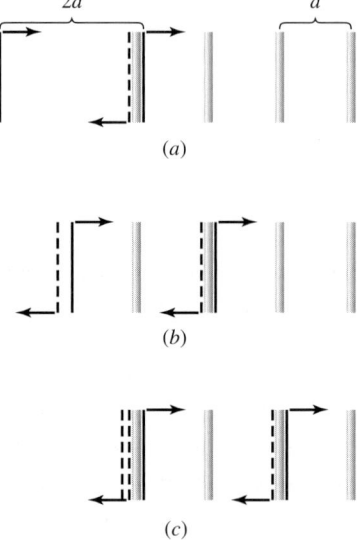

Fig. 1.11 Reflection of a wave from a Bragg reflector. (a) Two wavefronts (solid lines) of a wave with wavelength $2a$ approach a set of partially reflecting planes; the first wavefront emits a reflected wavefront (dashed line) moving in the opposite direction. (b) After the wavefronts have moved a distance a, the first wavefront emits another reflected wavefront. (c) After they have moved another distance a, the second wavefront emits a reflected wavefront also. The reflected waves from the first and second wavefronts are in phase. All of the partial reflections therefore add up constructively for waves with wavelength $2(a)$.

$$\frac{mU_0ba}{\hbar^2}\frac{\cos(Ka)}{Ka}dK - \frac{mU_0ba}{\hbar^2}\frac{\sin(Ka)}{(Ka)^2}dK - \sin(Ka)dK = -\sin(ka)dk, \quad (1.2.8)$$

or

$$F(K)dK = -\sin(ka)dk. \quad (1.2.9)$$

Using $dE = (\hbar^2 K/m)dK$, we then find that when $ka = n\pi$, then

$$\frac{\partial E}{\partial k} = \frac{\partial E}{\partial K}\frac{\partial K}{\partial k} = -\frac{\hbar^2 K}{m}\frac{\sin ka}{F(K(k))}. \quad (1.2.10)$$

Since $\sin ka = 0$ when $ka = n\pi$, $\partial E/\partial k$ goes to zero at the same points. (One might wonder if $F(K)$ can go to zero, but if $F(K) = 0$, then $\sin Ka$ and $\cos Ka$ must both have the same sign, in which case for the left side of (1.2.6), $|(\kappa^2 b/2K)\sin(Ka) + \cos(Ka)| > 1$, so that there is no real k which corresponds to this case.)

We can also see why the zone boundary is a boundary by realizing that the solution at $ka = \pm\pi$ corresponds to the single-well wave function multiplied by a phase factor of $e^{ika} = -1$ from one cell to the next. This is the maximum possible phase difference between cells. Increasing the phase angle ka beyond π is just the same as starting at phase angle $ka = -\pi$ and making the phase angle less negative.

In the case of weak coupling (large $U_0 b$), the energy bands of the Kronig–Penney model correspond to the single-well quantized states, with energy proportional to N^2. To see this, we can take the limit $U_0 b \to \infty$, in which case (1.2.6) becomes

$$\frac{U_0 b}{K} \sin Ka = 0, \qquad (1.2.11)$$

which implies $Ka = \pi N$ or $\lambda = 2a/N$. This is just what we expect from the discussion of Section 1.1 – in the limit of very weak coupling, that is, $U_0 b \to \infty$, we have the single-cell states, while when the coupling is increased, the gaps shrink, and these states are smeared out into bands. In the limit $U_0 b \to 0$, we are left with $K = k$ and $E = \hbar^2 k^2/2m$, which is, not surprisingly, just the energy of a free particle in vacuum, since there are no barriers left. Figure 1.12 shows the energy dispersion of the Kronig–Penney model with the higher bands plotted in adjacent zones. This is called an **extended zone** plot of the energy dispersion, while Figure 1.10 is called the **reduced zone** plot. As one may expect from Figure 1.12, when the barriers between the cells become small, the dispersion of the Kronig–Penney model approaches the free-electron dispersion.

In this Kronig–Penney model, the energy bands go up in energy forever, to $E = \infty$, because we assumed $U_0 = \infty$, which makes each cell a square well with single-state energies proportional to N^2. This is not true for bands arising from atomic orbitals, as in normal solids – the bound state energies of atoms are proportional to $-1/N^2$, not N^2. This means there is a maximum energy for the bands in a solid formed from atomic or molecular bound states. Above this energy, there will be a continuum of states with no

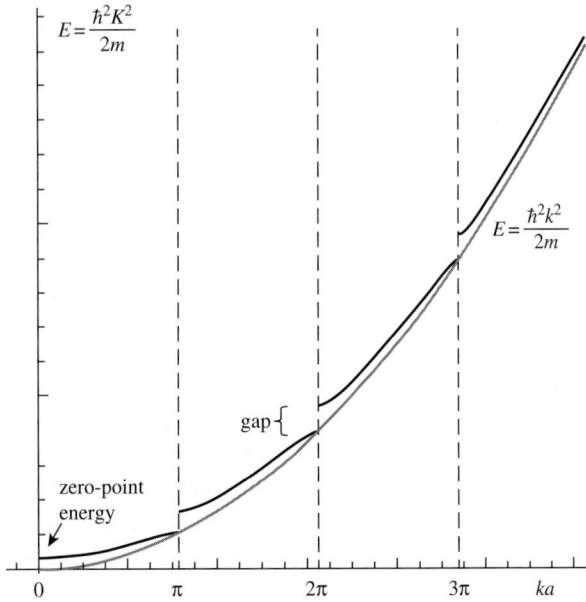

Fig. 1.12   Heavy lines: extended-zone plot for the Kronig–Penney model, for $\kappa^2 b = 4$. Gray line: the energy of a free electron.

energy gaps. This maximum energy is not necessarily the same as the energy of a free electron in vacuum, however. The energy of a motionless free electron, an infinite distance away from the solid, relative to the energy bands, depends on the properties of the surface of the material as well as the band energies.

Exercise 1.2.4   Use Mathematica to plot $\mathrm{Re}\,k$ as a function of $E = \hbar^2 K^2/2m$ using Equation 1.2.6. Assume that you have a set of units such that $\hbar^2/2m = 1$, set $a = 1$, and choose various values of $U_0 b$ from 0.1 to 3. This plot is just the Kronig–Penney reduced zone diagram turned on its side. Plot the first three bands. How do the gaps depend on your value of $U_0 b$? Then plot a close-up of the first band and band gap along with the free-electron dispersion, $k = \sqrt{E}$, on the same graph.

1.3 Bloch's Theorem

In Section 1.2, we used the periodicity of the system to guess a solution that was the same in every cell except for a phase factor that is constant within each cell. It turns out that this is a general property of *all* periodic systems, for any number of dimensions. This is known as **Bloch's theorem**. For any potential that is periodic such that $U(\vec{r} + \vec{R}) = U(\vec{r})$ for all $\vec{R} = N\vec{a}$, where $\vec{a}$ is some vector, the eigenstates of the Hamitonian have the property

$$\psi_{n\vec{k}}(\vec{r} + \vec{R}) = \psi_{n\vec{k}}(\vec{r})e^{i\vec{k}\cdot\vec{R}}, \qquad (1.3.1)$$

where n is a band index that we add because, as we have seen with the Kronig–Penney model, there can be more than one eigenstate with the same k.

This can be restated in another way. Multiplying through by a phase factor $e^{-i\vec{k}\cdot\vec{r}}$, we get

$$\psi_{n\vec{k}}(\vec{r} + \vec{R})e^{-i\vec{k}\cdot\vec{r}} = \psi_{n\vec{k}}(\vec{r})e^{i\vec{k}\cdot\vec{R}}e^{-i\vec{k}\cdot\vec{r}}$$

$$\psi_{n\vec{k}}(\vec{r} + \vec{R})e^{-i\vec{k}\cdot(\vec{R}+\vec{r})} = \psi_{n\vec{k}}(\vec{r})e^{-i\vec{k}\cdot\vec{r}}. \qquad (1.3.2)$$

This implies that $\psi_{n\vec{k}}(\vec{r})e^{-i\vec{k}\cdot\vec{r}}$ is a periodic function. We can therefore write the eigenstates as

$$\psi_{n\vec{k}}(\vec{r}) = \frac{1}{\sqrt{V}}u_{n\vec{k}}(\vec{r})e^{i\vec{k}\cdot\vec{r}}, \qquad (1.3.3)$$

where V is the volume of the crystal (introduced for normalization of the wave function) and $u_{n\vec{k}}(\vec{r})$ has the same periodicity as the potential. Note that here the phase factor depends on the continuous variable $\vec{r}$ instead of the discrete vector $\vec{R}$. The function $\psi_{n\vec{k}}(\vec{r})$ is called a **Bloch function**, and the function $u_{n\vec{k}}(\vec{r})$ can be called the **cell** function.

The power of this theorem is that in most cases, one never actually needs to compute the cell functions $u_{n\vec{k}}(\vec{r})$ for a given solid. Simply knowing that it has the same symmetry as the physical system is enough to compute selection rules using group theory, which we will study in Chapter 6. Often we do not even need to know the symmetry of the crystal. The fact that there is a plane wave part will allow us to use the quasiparticle picture of Chapter 2.

Proof. To prove Bloch's theorem, we could simply use the symmetry argument that led us to our guess for the solution of the Kronig–Penney model. In an infinite system in which all cells are identical, there is no reason why the wave function in any cell should be any different from the wave function in any other cell, except for a phase factor.

More rigorously, we can prove Bloch's theorem using the quantum mechanical rules of operators. Cohen-Tannoudji *et al.* (1977) give an elegant presentation of the basic properties of quantum mechanical operators, especially in section IID.

We define a translation operator $T_{\vec{R}}$ such that

$$T_{\vec{R}}\psi(\vec{r}) = \psi(\vec{r} + \vec{R}). \tag{1.3.4}$$

Using the periodicity of the potential, we can write

$$\begin{aligned} T_{\vec{R}}\left[U(\vec{r})\psi(\vec{r})\right] &= U(\vec{r} + \vec{R})\psi(\vec{r} + \vec{R}) \\ &= U(\vec{r})\psi(\vec{r} + \vec{R}) \\ &= U(\vec{r})T_{\vec{R}}\psi(\vec{r}). \end{aligned} \tag{1.3.5}$$

Also,

$$\begin{aligned} T_{\vec{R}}\nabla^2\psi(\vec{r}) &= \nabla^2\psi(\vec{r} + \vec{R}) \\ &= \nabla^2 T_{\vec{R}}\psi(\vec{r}). \end{aligned} \tag{1.3.6}$$

Therefore, $T_{\vec{R}}$ commutes with the Hamiltonian $H = -(\hbar^2/2m)\nabla^2 + U$ of the system. Commuting operators have common eigenstates, and therefore the eigenstates of the Hamiltonian are also eigenstates of the translation operator.

We define the eigenvalues of the translation operator as $C_{\vec{R}}$ such that

$$T_{\vec{R}}\psi(\vec{r}) = C_{\vec{R}}\psi(\vec{r}). \tag{1.3.7}$$

Two successive translations are the same as one translation, so we can write

$$\begin{aligned} T_{\vec{R}+\vec{R}'}\psi(\vec{r}) &= T_{\vec{R}}T_{\vec{R}'}\psi(\vec{r}) \\ C_{\vec{R}+\vec{R}'}\psi(\vec{r}) &= T_{\vec{R}}C_{\vec{R}'}\psi(\vec{r}) \\ &= C_{\vec{R}'}T_{\vec{R}}\psi(\vec{r}) \\ &= C_{\vec{R}}C_{\vec{R}'}\psi(\vec{r}) \end{aligned} \tag{1.3.8}$$

or

$$C_{\vec{R}+\vec{R}'} = C_{\vec{R}}C_{\vec{R}'}. \tag{1.3.9}$$

The eigenvalues of $T_{\vec{R}}$ must satisfy this relation.

The eigenvalues of $T_{\vec{R}}$ must also have unit magnitude. We can see this by writing

$$\int_{-\infty}^{\infty} d^3r \, \psi^*(\vec{r})\psi(\vec{r}) = \int_{-\infty}^{\infty} d^3r \, \psi^*(\vec{r}+\vec{R})\psi(\vec{r}+\vec{R})$$

$$= \int_{-\infty}^{\infty} d^3r \, T_{\vec{R}}\psi^*(\vec{r})T_{\vec{R}}\psi(\vec{r}), \qquad (1.3.10)$$

which is true because a change of the variable of integration from $\vec{r}$ to $\vec{r} - \vec{R}$ in the integral does not change its value, since the integral is over all space. We then have

$$\int_{-\infty}^{\infty} d^3r \, \psi^*(\vec{r})\psi(\vec{r}) = C_{\vec{R}}^* C_{\vec{R}} \int_{-\infty}^{\infty} d^3r \, \psi^*(\vec{r})\psi(\vec{r}), \qquad (1.3.11)$$

which implies

$$|C_{\vec{R}}|^2 = 1. \qquad (1.3.12)$$

For both (1.3.9) and (1.3.12) to hold true generally, we must have

$$C_{\vec{R}} = e^{i\vec{k}\cdot\vec{R}} \qquad (1.3.13)$$

for some real $\vec{k}$. This is the same as saying

$$T_{\vec{R}}\psi(\vec{r}) = \psi(\vec{r}+\vec{R}) = e^{i\vec{k}\cdot\vec{R}}\psi(\vec{r}), \qquad (1.3.14)$$

which is just the same as (1.3.1).

Exercise 1.3.1 Determine the cell function $u_{nk}(x)$ for the lowest band of the Kronig–Penney model in the limit $b \to 0$, with $a = 1$, $2mU_0b/\hbar^2 = 100$, and $\hbar^2/2m = 1$, for $k = \pi/2a$. Hint: What is the solution of the wave function in a flat potential?

1.4 Bravais Lattices and Reciprocal Space

As discussed in Section 1.1, solid state physics does not only deal with periodic structures. Nevertheless, the theory of periodic structures is extremely important because many solids do have periodicity. Solids that have periodic arrays of atoms are called **crystals**. Most metals and most semiconductors are crystals.

Crystals are common in nature because an ordered structure has lower entropy than a disordered structure, and lower entropy states are favored at low temperatures. Whether or not a system forms an ordered crystal at room temperature depends on the ratio of the thermal energy k_BT to the binding energy of two atoms. If k_BT is small compared to the binding energy, then the system is essentially in a zero-temperature state, even if it is quite hot compared to room temperature. We will return to discuss solid phase transitions in Section 5.4.

Bravais lattices. In order to fill all of space with a periodic structure, we take a finite volume of space, which we call the **primitive cell**, or **unit cell**, and make copies of it adjacent to each other by translating it without rotation through integer multiples of three vectors, $\vec{a}_1, \vec{a}_2$, and $\vec{a}_3$. These vectors, known as the **primitive vectors**, must be linearly independent, but need not be orthogonal. Some examples of lattices generated from primitive vectors are shown in Figure 1.13. The set of all locations of the unit cells is given by

$$\vec{R} = N_1 \vec{a}_1 + N_2 \vec{a}_2 + N_3 \vec{a}_3, \qquad (1.4.1)$$

where N_1, N_2, and N_3 are three integers. This set of all the vectors $\vec{R}$ makes up the **Bravais lattice** of the crystal. These vectors point to a set of *points* which define the origin of each primitive cell.

The primitive cell that is copied throughout space does not need to be cubic or rectangular; it can be any shape that will fill all space when copied periodically – it can be as complicated as the repeated elements of an Escher print. The most natural choice, however, is a parallelepiped with three edges equal to the primitive vectors.

A crystal can have more than one atom per primitive cell. Within each primitive cell, we can specify a **basis**, which is a set of vectors giving the location of the atoms relative to the origin of each cell. Figure 1.14 shows two examples of lattices with a basis. Table 1.1 gives the standard primitive vectors and basis vectors of some of the more common types of crystals.

The term "Bravais lattice" is typically used for just the set of points generated by translations of a single point through multiples of the primitive vectors. In this book, we will use the more general term **lattice** to refer to the set of all points generated by the Bravais lattice vectors plus the basis vectors within each unit cell.

Exercise 1.4.1   Use a program like Mathematica to create diagrams analogous to Figures 1.13 and 1.14 showing the location of the atoms for the last four crystal structures from Table 1.1. (In Mathematica, it is simple to create a set of spheres of radius r centered at points $\{x_1, y_1, z_1\}, \{x_2, y_2, z_2\}, \ldots$ using the command

Show[Graphics3D[Sphere[$\{\{x_1, y_1, z_1\}, \{x_2, y_2, z_2\}, \ldots\}], r$]]

Try using different viewpoint positions in the plotting. How many nearest neighbors does each atom have?

Exercise 1.4.2   Prove that in the wurtzite structure, each atom is equidistant from its four nearest neighbors.

The reciprocal lattice. As we saw in Section 1.2, in a one-dimensional periodic system like the Kronig–Penney model, the wavenumbers $k = \pm\pi/a$ have special properties because the set of cells with spacing a form a Bragg reflector, which perfectly reflects waves with wavelength $2a$. In a multi-dimensional system, there are many possible ways to form a set of periodic reflectors. As shown in Figure 1.15, every set of atoms that form a plane are part of a periodic reflector. Therefore, to determine the three-dimensional wave vectors that have the same role as the points $k = \pm\pi/a$ in a one-dimensional system, we need to find all the periodic, parallel sets of planes in the lattice.

Table 1.1 Common crystal structures	
Structure	**Standard primitive vectors and basis**
Simple cubic (sc)	$a\hat{x}, a\hat{y}, a\hat{z}$
Body-centered cubic (bcc)	$a\hat{x}, a\hat{y}, \frac{1}{2}a(\hat{x}+\hat{y}+\hat{z})$ *or* sc lattice with the basis $(0, \frac{1}{2}a(\hat{x}+\hat{y}+\hat{z}))$
Face-centered cubic (fcc)	$\frac{1}{2}a(\hat{y}+\hat{z}), \frac{1}{2}a(\hat{z}+\hat{x}), \frac{1}{2}a(\hat{x}+\hat{y})$ *or* sc lattice with the basis $(0, \frac{1}{2}a(\hat{x}+\hat{y}), \frac{1}{2}a(\hat{y}+\hat{z}), \frac{1}{2}a(\hat{z}+\hat{x}))$
Diamond	fcc lattice with the basis $(0, \frac{1}{4}a(\hat{x}+\hat{y}+\hat{z}))$
Simple hexagonal (sh)	$a\hat{x}, (\frac{1}{2}a\hat{x}+\frac{\sqrt{3}}{2}a\hat{y}), c\hat{z}$
Hexagonal close-packed (hcp)	sh lattice with the basis $\left(0, \frac{1}{2}a\hat{x}+\frac{1}{2\sqrt{3}}a\hat{y}+\frac{1}{2}c\hat{z}\right)$, where $c=\sqrt{\frac{8}{3}}a$
Graphite	$\frac{\sqrt{3}}{2}a\hat{x}+\frac{1}{2}a\hat{y}, -\frac{\sqrt{3}}{2}a\hat{x}+\frac{1}{2}a\hat{y}, c\hat{z}$ with the basis $\left(0, \frac{1}{2}c\hat{z}, \frac{1}{2\sqrt{3}}a\hat{x}+\frac{1}{2}a\hat{y}, -\frac{1}{2\sqrt{3}}a\hat{x}+\frac{1}{2}a\hat{y}+\frac{1}{2}c\hat{z}\right)$
Sodium chloride	fcc lattice with the basis $(0, \frac{1}{2}a(\hat{x}+\hat{y}+\hat{z}))$; the two basis sites have different atoms
Cesium chloride	sc lattice with the basis $(0, \frac{1}{2}a(\hat{x}+\hat{y}+\hat{z}))$; the two basis sites have different atoms
Zincblende	Diamond lattice but the two basis sites have different atoms
Wurtzite	sh lattice with the basis $\left(0, \frac{1}{2}a\hat{x}+\frac{1}{2\sqrt{3}}a\hat{y}+\frac{1}{2}c\hat{z},\right.$ $\left.\frac{3}{8}c\hat{z}, \frac{1}{2}a\hat{x}+\frac{1}{2\sqrt{3}}a\hat{y}+\frac{7}{8}c\hat{z}\right)$, where $c=\sqrt{\frac{8}{3}}a$
Perovskite	sc lattice with the basis $(0, \frac{a}{2}(\hat{x}+\hat{y}+\hat{z}),$ $\frac{a}{2}(\hat{x}+\hat{y}), \frac{a}{2}(\hat{y}+\hat{z}), \frac{a}{2}(\hat{x}+\hat{z}))$; the last three basis sites have identical atoms
Fluorite	$\frac{a}{2}(\hat{y}+\hat{z}), \frac{a}{2}(\hat{x}+\hat{z}), \frac{a}{2}(\hat{x}+\hat{y})$ with the basis $(0, \frac{a}{4}(\hat{x}+\hat{y}+\hat{z}), -\frac{a}{4}(\hat{x}+\hat{y}+\hat{z})$; the last two sites are identical
Cuprite	sc lattice with the basis $(0, \frac{a}{2}(\hat{x}+\hat{y}+\hat{z}),$ $\frac{a}{4}(\hat{x}+\hat{y}+\hat{z}), \frac{a}{4}(3\hat{x}+3\hat{y}+\hat{z}), \frac{a}{4}(3\hat{x}+\hat{y}+3\hat{z}),$ $\frac{a}{4}(\hat{x}+3\hat{y}+3\hat{z}))$; the first two sites are identical and the last four sites are identical

This problem is equivalent to finding all the plane waves that have the periodicity of the lattice. This takes us directly to the theory of Fourier transforms – the Fourier transform by definition gives us all the periodic waves that make up a given function (see Appendix B). We write the real-space lattice as a set of points given by Dirac δ-functions,

$$f(\vec{r}) = \sum_{\vec{R}} \sum_{i} \delta(\vec{r} - \vec{R} - \vec{b}_i), \qquad (1.4.2)$$

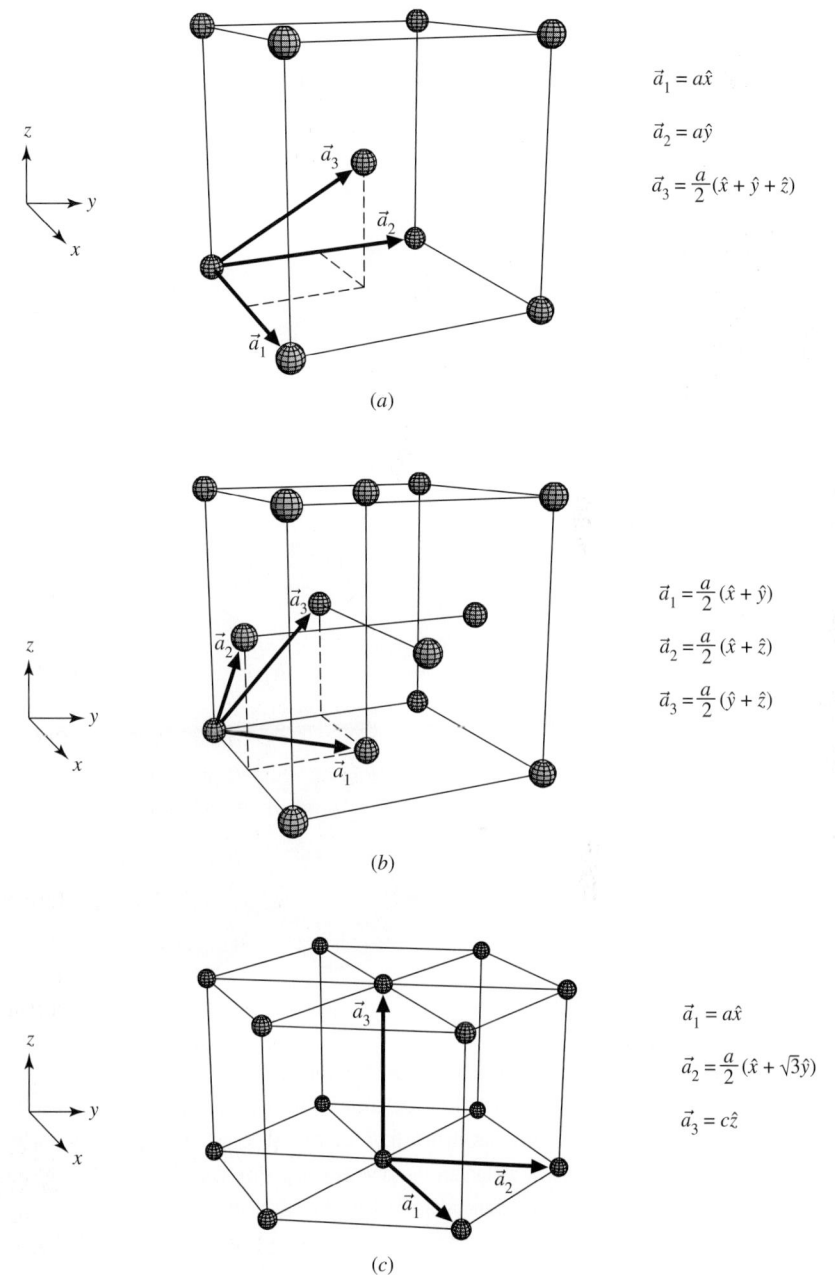

$$\vec{a}_1 = a\hat{x}$$

$$\vec{a}_2 = a\hat{y}$$

$$\vec{a}_3 = \frac{a}{2}(\hat{x} + \hat{y} + \hat{z})$$

(a)

$$\vec{a}_1 = \frac{a}{2}(\hat{x} + \hat{y})$$

$$\vec{a}_2 = \frac{a}{2}(\hat{x} + \hat{z})$$

$$\vec{a}_3 = \frac{a}{2}(\hat{y} + \hat{z})$$

(b)

$$\vec{a}_1 = a\hat{x}$$

$$\vec{a}_2 = \frac{a}{2}(\hat{x} + \sqrt{3}\hat{y})$$

$$\vec{a}_3 = c\hat{z}$$

(c)

Fig. 1.13 (a) Primitive vectors of a body-centered cubic (bcc) lattice. A parallelepiped that has these three vectors for sides will fill all of space with body-centered cubes when translated by integer multiples of the primitive vectors. (b) Primitive vectors for a face-centered cubic (fcc) lattice. (c) Primitive vectors for a simple hexagonal lattice.

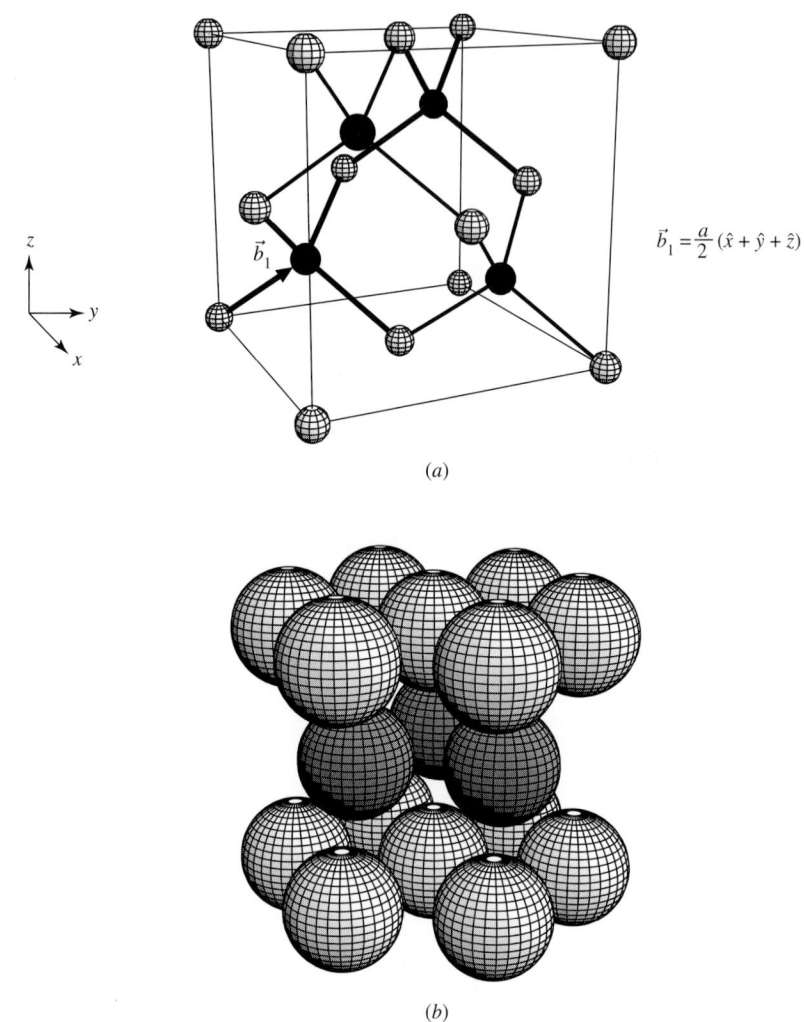

$$\vec{b}_1 = \frac{a}{2}(\hat{x} + \hat{y} + \hat{z})$$

(a)

(b)

Fig. 1.14 (a) Diamond lattice. The primitive vectors are the same as a face-centered cubic (fcc) lattice (light spheres), but in addition, next to each atom in the fcc lattice there is a second atom a short distance away (dark spheres), making a two-atom basis. This allows each atom to have four nearest neighbors at equal distances. (b) Close-packed hexagonal (hcp) lattice. The primitive vectors are the same as a simple hexagonal lattice, but there is an additional atom in the basis, giving two hexagonal planes shifted relative to each other.

where $\vec{R}$ are the Bravais lattice vectors that fill all of space and $\vec{b}_i$ are the basis vectors for the positions of the atoms within each unit cell. The Fourier transform is then given by

$$F(\vec{k}) = \int_{-\infty}^{\infty} d^3 r\, f(\vec{r}) e^{i\vec{k}\cdot\vec{r}}$$

$$= \sum_{\vec{R}} e^{i\vec{k}\cdot\vec{R}} \left(\sum_i e^{i\vec{k}\cdot\vec{b}_i} \right). \tag{1.4.3}$$

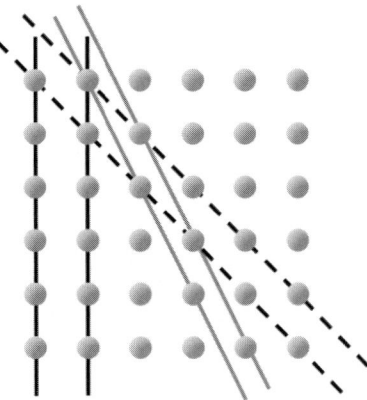

Fig. 1.15   Three of the many different sets of parallel planes that can form Bragg reflectors in a crystal.

The factor in the parentheses is known as the **structure factor**, and is the same for every primitive cell in the lattice.

The large number of different oscillating terms in the sum over $\vec{R}$ will cancel to zero unless $\vec{k}$ has a particular value $\vec{Q}$ such that

$$e^{i\vec{Q}\cdot\vec{R}} = 1, \tag{1.4.4}$$

or

$$\vec{Q} \cdot \vec{R} = 2\pi N, \tag{1.4.5}$$

for all $\vec{R}$, where N is some integer. This can be satisfied for

$$\vec{Q} = v_1\vec{q}_1 + v_2\vec{q}_2 + v_3\vec{q}_3, \tag{1.4.6}$$

where v_1, v_2, and v_3 are integers, and

$$\vec{q}_1 = \frac{2\pi(\vec{a}_2 \times \vec{a}_3)}{\vec{a}_1 \cdot (\vec{a}_2 \times \vec{a}_3)}$$

$$\vec{q}_2 = \frac{2\pi(\vec{a}_3 \times \vec{a}_1)}{\vec{a}_1 \cdot (\vec{a}_2 \times \vec{a}_3)}$$

$$\vec{q}_3 = \frac{2\pi(\vec{a}_1 \times \vec{a}_2)}{\vec{a}_1 \cdot (\vec{a}_2 \times \vec{a}_3)}. \tag{1.4.7}$$

The term in the numerator of these vectors, $\vec{a}_i \times \vec{a}_j$, gives a vector perpendicular to both $\vec{a}_i$ and $\vec{a}_j$, while the denominator is a normalization factor equal to the volume of the primitive cell parallelepiped. These choices of the $\vec{b}_i$ vectors ensure the condition

$$\vec{a}_i \cdot \vec{q}_j = 2\pi \delta_{ij}, \tag{1.4.8}$$

which implies

$$\vec{Q} \cdot \vec{R} = (v_1 N_1)\vec{a}_1 \cdot \vec{q}_1 + (v_2 N_2)\vec{a}_2 \cdot \vec{q}_2 + (v_3 N_3)\vec{a}_3 \cdot \vec{q}_3 = 2\pi N, \tag{1.4.9}$$

where N is an integer since the v_i and N_i are integers. This satisfies the condition (1.4.5).

The relation (1.4.6) implies that the Fourier transform of the lattice is nonzero for a specific set of $\vec{Q}$ vectors. At each of these values of $\vec{Q}$, the Fourier transform has a peak with height proportional to the number of Bravais lattice sites $\vec{R}$ in the crystal. These peaks correspond to plane waves of the form $e^{i\vec{Q}\cdot\vec{r}}$ with the same periodicity as the lattice in the direction of $\vec{Q}$. In other words, each of the vectors $\vec{Q}$ points in the direction normal to a set of parallel planes and has a magnitude equal to $n(2\pi/a')$, where a' is the distance between adjacent planes and n is an integer. This set of vectors $\vec{Q}$ defines a lattice just as the real-space $\vec{R}$ vectors do, and is called the **reciprocal lattice** of the crystal; it is the three-dimensional Fourier transform of the original lattice. The space that contains the reciprocal lattice, which has dimensions of inverse distance, is called **reciprocal space**, or "k-space."

The structure factor that appears in (1.4.3) is an overall multiplicative factor for the height of the reciprocal lattice peaks. When there is periodicity inside a unit cell, the structure factor can cause some peaks to have zero amplitude. This helps us to understand what would happen if we chose the "wrong" cell size. For example, consider the case of a lattice with spacing a between planes in the x-direction. In this case, the reciprocal lattice vectors have x-component $Q_x = 2\pi n/a$, for all integers n. What if we had treated this as a lattice with spacing $2a$ with a two-atom basis? The reciprocal lattice in this case would have points half as far apart, that is, $Q = 2\pi n/2a = \pi n/a$. But in this case, we have a structure factor given by

$$C(\vec{Q}) = \sum_i e^{i\vec{Q}\cdot\vec{b}_i}. \tag{1.4.10}$$

For the case of the two-atom basis $(0, a\hat{x})$, the sum is equal to

$$\sum_i e^{i\vec{Q}\cdot\vec{b}_i} = e^{iQ_x\cdot 0} + e^{iQ_x a}. \tag{1.4.11}$$

When $Q_x = \pi/a$, we then have

$$\sum_{\vec{r}_b} e^{i\vec{Q}\cdot\vec{r}_b} = e^{i(\pi/a)\cdot 0} + e^{(\pi/a)\cdot a} = 1 + (-1) = 0, \tag{1.4.12}$$

while for $Q_x = 2\pi/a$,

$$\sum_{\vec{r}_b} e^{i\vec{Q}\cdot\vec{r}_b} = e^{i(2\pi/a)\cdot 0} + e^{i(2\pi/a)\cdot a} = 1 + 1 = 2. \tag{1.4.13}$$

The structure factor removes the spurious new reciprocal lattice vectors that came from doubling the size of the unit cell, and increases the height of the remaining Fourier peaks to the same value we would have had if we had used a one-atom basis.

Exercise 1.4.3 Prove that you get the same reciprocal lattice peaks from a bcc crystal, whether you view it as a single Bravais lattice or as a simple cubic Bravais lattice with a two-site basis and the accompanying structure factor. (See Table 1.1.)

Hint: Notice that

$$\sum_n f\left(\frac{n}{2}\right) = \sum_n f(n) + \sum_n f\left(n + \frac{1}{2}\right).$$

Exercise 1.4.4   (a) Show that the volume of a Bravais lattice primitive cell is

$$V_{\text{cell}} = |\vec{a}_1 \cdot (\vec{a}_2 \times \vec{a}_3)|. \tag{1.4.14}$$

(b) Prove that the reciprocal lattice primitive vectors satisfy the relation

$$\vec{q}_1 \cdot (\vec{q}_2 \times \vec{q}_3) = \frac{(2\pi)^3}{\vec{a}_1 \cdot (\vec{a}_2 \times \vec{a}_3)}. \tag{1.4.15}$$

Hint: Write $\vec{q}_1$ in terms of the $\vec{a}_i$ and use the orthogonality relation (1.4.8).

With part (a) this proves that the volume of the reciprocal lattice cell is $(2\pi)^3/V_{\text{cell}}$.

(c) Show that the reciprocal lattice of the reciprocal lattice is the original real-space lattice, that is

$$2\pi \frac{\vec{q}_2 \times \vec{q}_3}{\vec{q}_1 \cdot (\vec{q}_2 \times \vec{q}_3)} = \vec{a}_1, \quad \text{etc.} \tag{1.4.16}$$

Useful vector relations:

$$\vec{A} \times (\vec{B} \times \vec{C}) = \vec{B}(\vec{A} \cdot \vec{C}) - \vec{C}(\vec{A} \cdot \vec{B})$$
$$(\vec{C} \times \vec{A}) \times (\vec{A} \times \vec{B}) = (\vec{C} \cdot (\vec{A} \times \vec{B})) \cdot \vec{A}$$
$$(\vec{A} \times \vec{B}) \cdot (\vec{C} \times \vec{D}) = (\vec{A} \cdot \vec{C})(\vec{B} \cdot \vec{D}) - (\vec{A} \cdot \vec{D})(\vec{B} \cdot \vec{C}). \tag{1.4.17}$$

(This exercise was originally suggested in Ashcroft and Mermin (1976).)

The Brillouin zone. In a one-dimensional system, the Brillouin zone was the range of ks from $-\pi/a$ to π/a. In a multi-dimensional system, the equivalent zone has boundaries defined by $\pm\pi/a'$, where a' is the distance between any set of parallel planes, given by $a' = 2\pi/|\vec{Q}|$. In other words, the wave vector of an electron wave in the solid cannot exceed half the magnitude of $\vec{Q}$ in any direction, because that wave vector corresponds to a wavelength of $2a$ that will be perfectly reflected by a Bragg reflector.

This leads to the following recipe for defining the Brillouin zone in a multi-dimensional periodic system:

1. Construct the reciprocal lattice using the formulas (1.4.7).
2. Starting from the origin, draw lines to the nearest neighbors, as shown in Figure 1.16(b).
3. **Bisect** these lines with **planes**, as shown in Figure 1.16(c).

These planes form the boundaries of the Brillouin zone. As we will see in Section 1.9.3, energy gaps open up at the boundaries of this Brillouin zone just as in the one-dimensional case.

The zone formed in this way is called the **Wigner–Seitz** cell. The Wigner–Seitz method can be used to form a primitive cell for any lattice, which can then be used to fill all space by translations of the lattice vectors. So, for example, it could be used instead of the parallelepiped we chose above for a Bravais lattice. The Wigner–Seitz cell has special importance for the reciprocal lattice, however, because the Wigner–Seitz cell corresponds to the Brillouin zone.

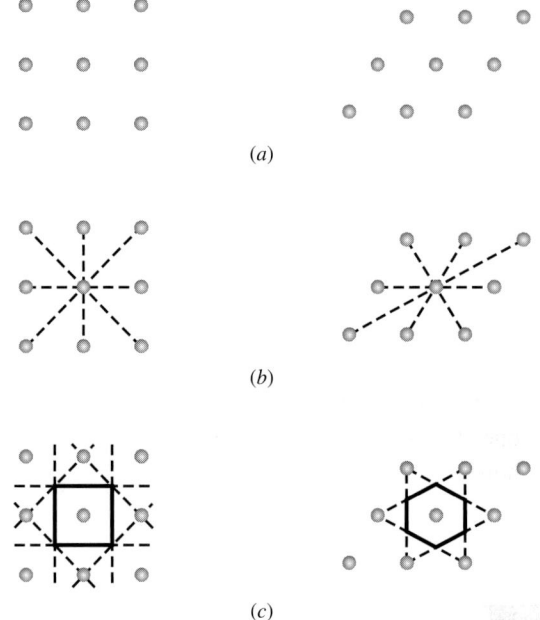

Fig. 1.16   (a) Two examples of reciprocal lattices in two dimensions. (b) Lines from the origin to the nearest neighbors in the reciprocal lattice. (c) These lines are bisected by planes to form the Wigner–Seitz cell.

Exercise 1.4.5   Show that the reciprocal lattice of a simple hexagonal lattice (see Table 1.1) is also a simple hexagonal, with lattice constants $2\pi/c$ and $4\pi/\sqrt{3}a$, rotated through $30°$ about the c-axis with respect to the real-space lattice.

Exercise 1.4.6   (a) A **graphene** lattice, or "honeycomb" lattice, is the same as the graphite lattice (see Table 1.1) but consists of only a two-dimensional sheet with lattice vectors $\vec{a}_1$ and $\vec{a}_2$ and a two-atom basis including only the graphite basis vectors in the $z = 0$ plane. Show that the reciprocal lattice vectors of this lattice are

$$\vec{q}_1 = \frac{4\pi}{\sqrt{3}a}\left(\frac{1}{2}\hat{x} + \frac{\sqrt{3}}{2}\hat{y}\right), \quad \vec{q}_2 = \frac{4\pi}{\sqrt{3}a}\left(-\frac{1}{2}\hat{x} + \frac{\sqrt{3}}{2}\hat{y}\right). \quad (1.4.18)$$

(Hint: Although this is a two-dimensional lattice, it is easiest to assume there is still a lattice vector $\vec{a}_3 = c\hat{z}$ and use this to calculate the reciprocal vectors in the plane.) Make a drawing of both the real-space and reciprocal-space lattices, and draw the Brillouin zone on the reciprocal space lattice.

(b) Show that the structure factor for the two-atom basis multiplies the peak in reciprocal space at $\vec{Q} = 0$ by 2, the peak at $\vec{Q} = \vec{q}_1$ by the factor $e^{-i\pi/3}$, and the peak at $\vec{Q} = 2\vec{q}_1$ by the factor $e^{i\pi/3}$. Label the reciprocal lattice vectors on your drawing by their peak height and show that the reciprocal lattice has the same symmetry as the real-space honeycomb lattice, and is not a simple hexagonal lattice.

(c) Show that the reciprocal lattice of graphene can be viewed as a simple hexagonal lattice with primitive vectors

$$\vec{q}_1' = \frac{4\pi}{a}\left(\frac{\sqrt{3}}{2}\hat{x} + \frac{1}{2}\hat{y}\right), \quad \vec{q}_2' = \frac{4\pi}{a}\left(\frac{\sqrt{3}}{2}\hat{x} - \frac{1}{2}\hat{y}\right), \tag{1.4.19}$$

and basis $(0, \vec{q}_1, 2\vec{q}_1)$.

(d) Show that the reciprocal lattice of this reciprocal lattice is a simple hexagonal lattice in real space with primitive vectors

$$\vec{a}_1' = \frac{a}{\sqrt{3}}\left(\frac{1}{2}\hat{x} + \frac{\sqrt{3}}{2}\hat{y}\right), \quad \vec{a}_2' = \frac{a}{\sqrt{3}}\left(-\frac{1}{2}\hat{x} + \frac{\sqrt{3}}{2}\hat{y}\right). \tag{1.4.20}$$

(e) Last, show that the structure factor of the reciprocal lattice in (c) eliminates one of every three real-space lattice points from the lattice of (d), leaving the original honeycomb lattice.

1.5 X-ray Scattering

The reciprocal lattice has a natural connection to x-ray scattering. Suppose a plane wave with wave vector $\vec{k}_0$ impinges on a crystal, as shown in Figure 1.17. We write this plane wave as

$$A_{\text{in}} = e^{i(\vec{k}_0 \cdot \vec{r} - \omega t)}. \tag{1.5.1}$$

Atoms in the crystal will lead to scattering of the incoming wave. In the Fraunhofer limit, a scattered wave far away can also be approximated by a plane wave with wave vector $\vec{k}$. We define the **scattering vector** as the difference between the incoming and outgoing (scattered) wave vectors:

$$\vec{s} = \vec{k} - \vec{k}_0. \tag{1.5.2}$$

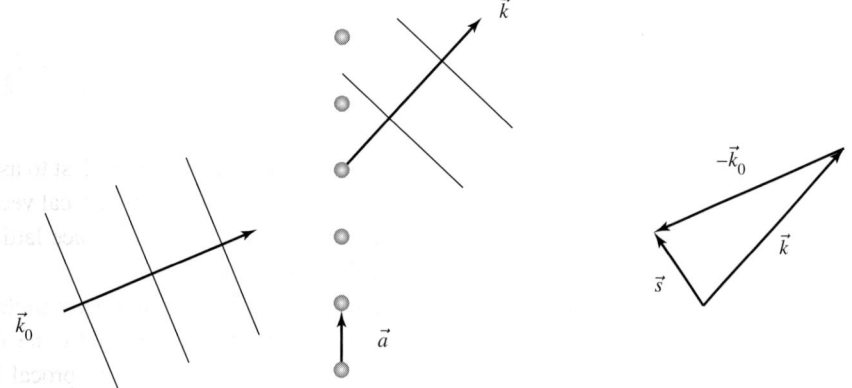

Fig. 1.17 Scattering of a wave from a row of reflectors.

If $\vec{a}$ is the vector from one atom to another, then the phase difference between the scattered waves from these two atoms will be

$$\delta = \vec{k} \cdot \vec{a} - \vec{k}_0 \cdot \vec{a} = (\vec{k} - \vec{k}_0) \cdot \vec{a} = \vec{s} \cdot \vec{a}. \tag{1.5.3}$$

The amplitude of the scattered wave from these two atoms will be proportional to

$$A_{\text{sum}} = \left(e^{i(0)} + e^{i\vec{s}\cdot\vec{a}} \right) e^{-i\omega t} \tag{1.5.4}$$

and therefore the intensity will be proportional to

$$I = A_{\text{sum}}^* A_{\text{sum}} = (1 + e^{-i\vec{s}\cdot\vec{a}})(1 + e^{i\vec{s}\cdot\vec{a}})$$
$$= 2(1 + \cos \vec{s} \cdot \vec{a}). \tag{1.5.5}$$

This is the standard interference pattern of two coherent sources.

In the case of many atoms, we sum over the phase factors of all the atoms, to get

$$A_{\text{sum}} = \sum_{\vec{R}} e^{i\vec{s}\cdot\vec{R}} e^{-i\omega t}, \tag{1.5.6}$$

which in the case of a simple Bravais lattice is

$$A_{\text{sum}} = \sum_{n_1,n_2,n_3} e^{i(n_1\vec{s}\cdot\vec{a}_1 + n_2\vec{s}\cdot\vec{a}_2 + n_3\vec{s}\cdot\vec{a}_3)} e^{-i\omega t}, \tag{1.5.7}$$

where $\vec{a}_1, \vec{a}_2, \vec{a}_3$ are the primitive vectors of the lattice.

Intensity peaks (that is, constructive interference of the x-rays) will occur when all three of the following conditions are satisfied:

$$\vec{s} \cdot \vec{a}_1 = 2\pi \nu_1, \quad \vec{s} \cdot \vec{a}_2 = 2\pi \nu_2,$$

and

$$\vec{s} \cdot \vec{a}_3 = 2\pi \nu_3, \tag{1.5.8}$$

where ν_i is an integer. This is the same as saying that

$$\vec{s} = \nu_1 \vec{q}_1 + \nu_2 \vec{q}_2 + \nu_3 \vec{q}_3, \tag{1.5.9}$$

where the $\vec{q}_i$ are the primitive vectors of the reciprocal lattice. In other words, comparing to the condition (1.4.6), $\vec{s}$ must be a reciprocal lattice vector $\vec{Q}$. The peaks in intensity of the x-ray scattering give the reciprocal lattice vectors directly.

The condition that *all three* of the conditions (1.5.8) must be satisfied is a stringent condition, because we have three equations and only two angles for the scattering x-rays that can be varied to define the output wave vector $\vec{k}$. The magnitude of $\vec{k}$ must be equal to the magnitude of $\vec{k}_0$ since scattering does not change the wavelength of the wave. In general, it is not possible to satisfy all three equations for every input wavelength and direction. Therefore, typical x-ray experiments use a *continuum* of inputs, either a broad range of wavelength (that is, a broad range of magnitudes of k), or a broad range of input angles.

In **Laue diffraction**, a "white light" x-ray source is used with a broad range of input wavelengths. Figure 1.18(a) shows an example of a Laue scattering pattern. The input

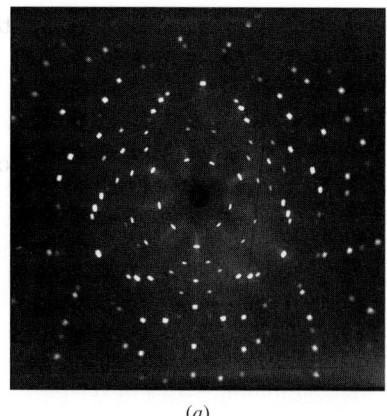 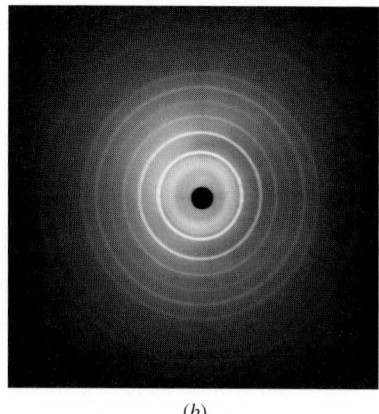

(a)   (b)

Fig. 1.18   (a) An example of Laue x-ray scattering, showing the threefold symmetry of the [111] axis of crystalline silicon. (b) An example of powder x-ray scattering, for table salt, NaCl. In each of these images, the center of the image corresponds to zero scattering (i.e., $\vec{k} = \vec{k}_0$). Both images used by permission of C.C. Jones, Union College, Schenectady, NY.

beam is held at a fixed direction relative to the crystal axes, and the angles of the diffracted spots are measured. Assuming that the input beam is along $\hat{z}$, we have

$$\vec{Q} = \vec{k} - \vec{k}_0$$
$$= k(\sin\theta\cos\phi, \sin\theta\sin\phi, \cos\theta - 1). \tag{1.5.10}$$

This method is useful for determining the orientation of crystals, because we can measure θ and ϕ of each spot in the diffraction pattern and match these to known reciprocal lattice vectors $\vec{Q}$. It does not give the absolute spacing of the atoms in a lattice, however, because one does not know the input wavelength that produced any given diffracted spot.

To get the absolute magnitudes of the lattice vectors, one can use the **powder diffraction** method. A monochromatic x-ray source is used, and a continuum of input angles over all possibilities. This is done by grinding the crystal into powder so that small crystallites randomly fall in all possible orientations. The scattered light in this case makes rings that are centered around the input wave vector $\vec{k}_0$, as shown in Figure 1.18(b). The diffraction angle corresponding to these rings is determined by the absolute magnitude of all the reciprocal lattice vectors, according to the following analysis.

For each reciprocal lattice vector $\vec{Q}$, we have

$$\vec{s} = \vec{k} - \vec{k}_0 = \vec{Q}, \tag{1.5.11}$$

which implies

$$|\vec{k} - \vec{k}_0|^2 = 2k^2 - 2\vec{k} \cdot \vec{k}_0 = Q^2$$
$$= 2k^2(1 - \cos\theta)$$
$$= 4k^2 \sin^2\frac{\theta}{2}, \tag{1.5.12}$$

where θ is the angle between $\vec{k}$ and $\vec{k}_0$. This then implies

$$2k \sin \frac{\theta}{2} = Q, \tag{1.5.13}$$

which can also be written as

$$\frac{4\pi}{\lambda} \sin \frac{\theta}{2} = |v_1 \vec{q}_1 + v_2 \vec{q}_2 + v_3 \vec{q}_3|. \tag{1.5.14}$$

The reciprocal lattice vectors can be grouped into sets that are multiples of the same vectors. We write the **Miller indices** as the lowest integers possible that give a reciprocal lattice vector in the same direction. So, for example, if a crystal has primitive reciprocal lattice vectors $\vec{q}_1, \vec{q}_2$, and $\vec{q}_3$, then the reciprocal lattice vector $\vec{Q}_M = 2\vec{q}_1 + \vec{q}_2 + \vec{q}_3$ would be written as [211]; this will generate an infinite series of reciprocal lattice vectors $n\vec{Q}_M$ for all integer n. Negative integers are written with an overline; for example, the vector $\vec{Q}_M = \vec{q}_1 - \vec{q}_2$ is written as [1$\bar{1}$0].

Each Miller vector corresponds to a set of parallel planes perpendicular to that vector, with spacing a' given by

$$\frac{2\pi}{a'} = |\vec{Q}_M|. \tag{1.5.15}$$

Thus, using (1.5.14), if the wavelength λ of the x-rays is known, the absolute spacing of each set of planes in a crystal can be read off for the scattering angles θ in the powder diffraction pattern.

Structure factor. As discussed in Section 1.4, a lattice can have a basis that consists of more than one atom at each lattice site. In this case, the amplitude of the scattered light is the sum over all lattice sites and over each atom in the basis,

$$A_{\text{sum}} = \sum_{\vec{R}} \sum_i e^{i\vec{s}\cdot\vec{R} + \vec{b}_i} e^{-i\omega t}$$

$$= \sum_{\vec{R}} e^{i\vec{s}\cdot\vec{R}} e^{-i\omega t} \left(\sum_i e^{i\vec{s}\cdot\vec{b}_i} \right). \tag{1.5.16}$$

The term in the parentheses is the structure factor. More generally, this structure factor can be written in terms of a continuous function over the entire primitive cell of the lattice,

$$C(\vec{Q}) = \int_{\text{cell}} d^3 r_b \, f(\vec{r}_b) e^{i\vec{Q}\cdot\vec{r}_b}, \tag{1.5.17}$$

where $f(\vec{r}_b)$ gives a relative weight of scattering of x-rays from different regions of the primitive cell. As with the structure factor introduced in Section 1.4, the x-ray structure can remove certain spurious peaks from the reciprocal lattice. Note that in Section 1.4 we used a structure factor that had only δ-functions, and because of this, the Fourier transform had uniform peak heights throughout reciprocal space. If the function $f(\vec{r}_b)$ is not made of δ-functions – for example, if it consists of one or more Gaussian functions – the Fourier transform will have peak heights that decrease with increasing magnitude of $\vec{Q}$. One of the main sources of structure factor in x-ray scattering is random motions of the atoms due to thermal vibrations, called the **dynamical structure factor**, or **Debye–Waller** factor. This effect is discussed in detail in Section 4.10.

Exercise 1.5.1   For the cubic crystal, there is a plane that contains all of the symmetry directions [100], [011], and [111]. Find the Miller indices of this plane. Sketch this plane in the cube and show the above symmetry directions.

Exercise 1.5.2   Determine the relative radius of the smallest five rings in a mono-chromatic powder diffraction measurement of a cubic crystal like that shown in Figure 1.18(b). Assume that the image plane is orthogonal to the input radiation.

1.6 General Properties of Bloch Functions

Even without knowing anything about the periodic potential in a particular crystal, Bloch's theorem allows us to make several general statements about the eigenstates of the system.

Bloch theorem as a Fourier series. We can use Fourier transform theory to express the cell function $u_{n\vec{k}}(\vec{r})$ in terms of the reciprocal lattice vectors. Since this function has the same periodicity as the lattice, we can use the Fourier transform formula

$$\begin{aligned} F_u(\vec{q}) &= \int_{-\infty}^{\infty} d^3r \, u_{n\vec{k}}(\vec{r}) e^{i\vec{q}\cdot\vec{r}} \\ &= \sum_{\vec{R}} e^{i\vec{q}\cdot\vec{R}} \left(\int_{\text{cell}} d^3r_b \, u_{n\vec{k}}(\vec{r}_b) e^{i\vec{q}\cdot\vec{r}_b} \right), \end{aligned} \qquad (1.6.1)$$

where in the second line we have written $\vec{r} = \vec{R} + \vec{r}_b$, and have broken the integral over all space into a sum over all Bravais lattice positions $\vec{R}$ and an integral over the relative coordinate $\vec{r}_b$ within each primitive cell. We have introduced a new reciprocal-space variable $\vec{q}$ because we are leaving $\vec{k}$ constant.

As discussed in Section 1.4, the Fourier transform (1.6.1) has nonzero values only when $\vec{q}$ is equal to a reciprocal lattice vector $\vec{Q}$. The cell function $u_{n\vec{k}}(\vec{r})$, which is the inverse Fourier transform of F_u, can therefore be written as a sum over the full set of reciprocal lattice vectors,

$$u_{n\vec{k}}(\vec{r}) = \sum_{\vec{Q}} C_{n\vec{k}}(\vec{Q}) e^{-i\vec{Q}\cdot\vec{r}}, \qquad (1.6.2)$$

where $C_{n\vec{k}}(\vec{Q})$ is a weight factor.[2] The dependence of $C_{n\vec{k}}(\vec{Q})$ on $\vec{k}$ gives an overall multiplier for the whole set of weight factors.

The full Bloch functions are then given by

$$\begin{aligned} \psi_{n\vec{k}}(\vec{r}) &= \frac{1}{\sqrt{V}} u_{n\vec{k}}(\vec{r}) e^{i\vec{k}\cdot\vec{r}} \\ &= \frac{1}{\sqrt{V}} \sum_{\vec{Q}} C_{n\vec{k}}(\vec{Q}) e^{i(\vec{k}-\vec{Q})\cdot\vec{r}}. \end{aligned} \qquad (1.6.3)$$

---

[2] See Appendix B for a discussion of how to convert the inverse Fourier transform of a periodic function into a sum, or Fourier series.

The Bloch functions have Fourier components only for $\vec{k} - \vec{Q}$. We can therefore write

$$\psi_{n\vec{k}}(\vec{r}) = \frac{1}{\sqrt{V}} \sum_{\vec{Q}} C_n(\vec{k} - \vec{Q})e^{i(\vec{k}-\vec{Q})\cdot\vec{r}}. \tag{1.6.4}$$

This is another way of expressing Bloch's theorem.

Orthogonality relation of Bloch functions. Using the above form of Bloch's theorem, we can write the inner product of two Bloch functions as

$$\langle \psi_{n\vec{k}} | \psi_{n\vec{k}'} \rangle = \int d^3r\, \psi^*_{n\vec{k}}(\vec{r})\psi_{n\vec{k}'}(\vec{r})$$

$$= \frac{1}{V} \int d^3r\, \sum_{\vec{Q}} C^*_n(\vec{k} - \vec{Q})e^{-i(\vec{k}-\vec{Q})\cdot\vec{r}} \sum_{\vec{Q}'} C_n(\vec{k}' - \vec{Q}')e^{i(\vec{k}'-\vec{Q}')\cdot\vec{r}}$$

$$= \sum_{\vec{Q},\vec{Q}'} \left(\frac{1}{V} \int d^3r\, e^{i(\vec{k}'-\vec{k}+\vec{Q}-\vec{Q}')\cdot\vec{r}} \right) C^*_n(\vec{k} - \vec{Q})C_n(\vec{k}' - \vec{Q}'). \tag{1.6.5}$$

The integral in the parentheses has an oscillating term that will average to zero when integrated over all space unless $\vec{k}' - \vec{k} + \vec{Q} - \vec{Q}' = 0$. This can only be the case if $\vec{k}' - \vec{k}$ is equal to a reciprocal lattice vector, which we will call $\vec{Q}''$. In that case, the sum over all $\vec{Q}'$ will always include one term that gives $\vec{Q} - \vec{Q}' = \vec{Q}''$. Then the exponential term will equal 1 for all $\vec{r}$, leaving a simple integral of the volume, which cancels the volume in the denominator. We then have

$$\langle \psi_{n\vec{k}} | \psi_{n\vec{k}'} \rangle = \left(\sum_{\vec{Q}} |C_n(\vec{k} - \vec{Q})|^2 \right) \delta_{\vec{k}',\vec{k}+\vec{Q}''}. \tag{1.6.6}$$

Assuming that the Bloch functions are normalized, the term in parentheses is just $\langle \psi_{n\vec{k}} | \psi_{n\vec{k}} \rangle = 1$. We therefore obtain

$$\langle \psi_{n\vec{k}} | \psi_{n\vec{k}'} \rangle = \delta_{\vec{k}',\vec{k}+\vec{Q}} \tag{1.6.7}$$

for any reciprocal lattice vector $\vec{Q}$. In other words, two Bloch functions are identical if $\vec{k}' = \vec{k} + \vec{Q}$ and are orthogonal otherwise. This implies that the Bloch functions are periodic in the wave vector $\vec{k}$, with the periodicity of each reciprocal lattice vector:

$$\psi_{n,\vec{k}+\vec{Q}}(\vec{r}) = \psi_{n,\vec{k}}(\vec{r}). \tag{1.6.8}$$

From basic quantum mechanics, we have the theorem that if any two eigenstates of a Hermitian operator have different eigenvalues, then they must be orthogonal. The Bloch states are eigenstates of the Hamiltonian, which is Hermitian, and therefore states in different energy bands are orthogonal. We can therefore generalize the relation (1.6.7) to

$$\langle \psi_{n\vec{k}} | \psi_{n'\vec{k}'} \rangle = \delta_{nn'}\delta_{\vec{k}',\vec{k}+\vec{Q}}. \tag{1.6.9}$$

Momentum and time reversal. As we have seen, the Bloch functions act in many ways like free plane waves. The momentum carried of a Bloch wave is not equal to just $\hbar\vec{k}$ as it is for a plane wave, however. The total momentum carried by an electron in a Bloch wave is

$$\langle \psi_{n\vec{k}} | \vec{p} | \psi_{n\vec{k}} \rangle = \int d^3 r \, \psi^*_{n\vec{k}}(\vec{r})(-i\hbar\nabla)\psi_{n\vec{k}}(\vec{r})$$

$$= \int d^3 r \, u^*_{n\vec{k}}(\vec{r})e^{-i\vec{k}\cdot\vec{r}}(-i\hbar\nabla)u_{n\vec{k}}(\vec{r})e^{i\vec{k}\cdot\vec{r}}$$

$$= -i\hbar \int d^3 r \, u^*_{n\vec{k}}(\vec{r})e^{-i\vec{k}\cdot\vec{r}} \left(\nabla u_{n\vec{k}}(\vec{r}) + i\vec{k} \right) e^{i\vec{k}\cdot\vec{r}}$$

$$= -i\hbar \int d^3 r \, u^*_{n\vec{k}}(\vec{r})\nabla u_{n\vec{k}}(\vec{r}) + \hbar\vec{k}. \tag{1.6.10}$$

If $u_{n\vec{k}}(\vec{r})$ is constant, we recover the plane-wave case.

In quantum mechanics, for a position-dependent wave function, we write the time-reversed state as the complex conjugate of the original state. This allows us to preserve the time-reversal property that the momentum of a wave changes sign on time reversal. Writing K for the time-reversal operator, we have

$$K\psi_{n\vec{k}}(\vec{r}) = \psi^*_{n\vec{k}}(\vec{r}), \tag{1.6.11}$$

and therefore

$$\int d^3 r \, K\psi^*_{n\vec{k}}(\vec{r})(-i\hbar\nabla)K\psi_{n\vec{k}}(\vec{r}) = \int d^3 r \, \psi_{n\vec{k}}(\vec{r})(-i\hbar\nabla)\psi^*_{n\vec{k}}(\vec{r})$$

$$= -\left(\int d^3 r \, \psi^*_{n\vec{k}}(\vec{r})(-i\hbar\nabla)\psi_{n\vec{k}}(\vec{r}) \right)^*$$

$$= -\langle \psi_{n\vec{k}} | \vec{p} | \psi_{n\vec{k}} \rangle, \tag{1.6.12}$$

where we have used the fact that the expectation value of $\vec{p}$ is real.

The time-reversed Bloch function is also a Bloch function, as we can see by applying the translation operator introduced in Section 1.3:

$$T_{\vec{R}}\psi^*_{n,\vec{k}}(\vec{r}) = \psi^*_{n,\vec{k}}(\vec{r} + \vec{R})$$

$$= \left(e^{i\vec{k}\cdot\vec{R}}\psi_{n\vec{k}}(\vec{r}) \right)^*$$

$$= e^{-i\vec{k}\cdot\vec{R}}\psi^*_{n\vec{k}}(\vec{r}). \tag{1.6.13}$$

The time-reversed state acted upon by the translation operator has the eigenvalue $e^{-i\vec{k}\cdot\vec{R}}$. If the time-reversed state lies in energy band n, and that band is nondegenerate, then this state must be equated to the Bloch state in band n with that eigenvalue, namely

$$\boxed{\psi^*_{n,\vec{k}}(\vec{r}) = \psi_{n,-\vec{k}}(\vec{r}).} \tag{1.6.14}$$

Hypothetically, the time-reversed state could lie in a different band n' and still have the same eigenvalue $e^{-i\vec{k}\cdot\vec{R}}$. However, that would violate the principle of microscopic time-reversibility in quantum mechanics. The energy of the time-reversed state is

$$\int d^3 r \, \psi_{n',\vec{k}}(\vec{r})H\psi^*_{n',\vec{k}}(\vec{r}) = \langle \psi_{n',\vec{k}} | H | \psi_{n',\vec{k}} \rangle^* \tag{1.6.15}$$

which, since H is Hermitian, is $E_{n'}(\vec{k})$. By time-reversibility, we require that the energy of the time-reversed state is the same as the original state $E_n(\vec{k})$, and therefore we must have $n = n'$. From this, it follows that

$$\boxed{E_n(-\vec{k}) = E_n(\vec{k}).} \qquad (1.6.16)$$

This is known as **Kramers' theorem** for nondegenerate bands.

The relation (1.6.14) implies that we can time-reverse the wave function by flipping the sign of $\vec{k}$. The case of degenerate bands, in particular spin degeneracy, is discussed in Section 1.13.

Critical points. Kramers' theorem, along with the periodicity property (1.6.8), implies that the zone boundary point $\vec{k} = \vec{Q}/2$ will have special properties. If we take the first derivative of $E_n(\vec{k})$ at $\vec{k} = \vec{Q}/2$ in the direction of $\vec{Q}$, we find

$$\left.\frac{\partial E_n(\vec{k})}{\partial k}\right|_{\vec{k}=\vec{Q}/2} = \lim_{dk \to 0} \frac{\langle \psi_{n,\vec{Q}/2+d\vec{k}}|H|\psi_{n,\vec{Q}/2+d\vec{k}}\rangle - \langle \psi_{n,\vec{Q}/2-d\vec{k}}|H|\psi_{n,\vec{Q}/2-d\vec{k}}\rangle}{2dk}. \qquad (1.6.17)$$

But

$$\begin{aligned}
\langle \psi_{n,\vec{Q}/2-d\vec{k}}|H|\psi_{n,\vec{Q}/2-d\vec{k}}\rangle &= \langle \psi_{n,\vec{Q}/2-d\vec{k}-\vec{Q}}|H|\psi_{n,\vec{Q}/2-d\vec{k}-\vec{Q}}\rangle \\
&= \langle \psi_{n,-\vec{Q}/2-d\vec{k}}|H|\psi_{n,-\vec{Q}/2-d\vec{k}}\rangle \\
&= \langle \psi_{n,\vec{Q}/2+d\vec{k}}|H|\psi_{n,\vec{Q}/2+d\vec{k}}\rangle \qquad (1.6.18)
\end{aligned}$$

and therefore, assuming that $E_n(\vec{k})$ is continuous,

$$\left.\frac{\partial E_n(\vec{k})}{\partial k}\right|_{\vec{k}=\vec{Q}/2} = 0. \qquad (1.6.19)$$

In other words, *the electron bands at the zone boundaries have either a maximum or a minimum* in the direction normal to the boundary, as illustrated in Figure 1.19.

The same is true at zone center, since

$$\left.\frac{\partial E_n(\vec{k})}{\partial k}\right|_{\vec{k}=0} = \lim_{dk \to 0} \frac{\langle \psi_{n,d\vec{k}}|H|\psi_{n,d\vec{k}}\rangle - \langle \psi_{n,-d\vec{k}}|H|\psi_{n,-d\vec{k}}\rangle}{2dk} = 0, \qquad (1.6.20)$$

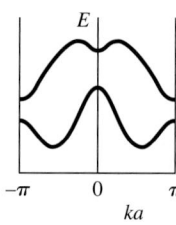

Fig. 1.19   An example of a possible band diagram. The slope of the bands must be zero at zone center and at the critical points on the zone boundaries.

by Kramers' theorem. Since this is true in every direction, we have

$$\nabla_{\vec{k}} E \big|_{\vec{k}=0} = 0. \tag{1.6.21}$$

On the zone boundaries, there will also be certain **critical points** at which the first derivative of the energy vanishes in all three directions, so that the gradient vanishes. These critical points correspond to a $\vec{k}$ at the midpoint between two reciprocal lattice vectors in each direction.

Taylor expansion of cell functions. In the same way, $\nabla_{\vec{k}} \psi_{n\vec{k}} = 0$ at zone center, according to theorem (1.6.14), which one can see easily by setting the imaginary part of ψ equal to zero at $k = 0$, which we are free to do at any one point, since ψ can have an arbitrary overall phase factor. Therefore,

$$\nabla_{\vec{k}} \psi_{n\vec{k}} \bigg|_{k=0} = \frac{1}{\sqrt{V}} (\nabla_{\vec{k}} u_{n\vec{k}} e^{i\vec{k}\cdot\vec{x}} + i\vec{x} u_{n\vec{k}} e^{i\vec{k}\cdot\vec{x}}) \bigg|_{\vec{k}=0} = 0, \tag{1.6.22}$$

which implies

$$\nabla_{\vec{k}} u_{n\vec{k}} \bigg|_{\vec{k}=0} = -i\vec{x} u_{n0}. \tag{1.6.23}$$

Writing a Taylor series for $u_{n\vec{k}}$, we have

$$u_{n\vec{k}} = u_{n0} + \nabla_{\vec{k}} u_{n\vec{k}} \bigg|_{\vec{k}=0} \cdot \vec{k} + \cdots = u_{n0} - i(\vec{x} \cdot \vec{k}) u_{n0} + \cdots, \tag{1.6.24}$$

which means that near zone center we can approximate

$$\boxed{u_{n\vec{k}} \approx u_{n0}(1 - i\vec{k} \cdot \vec{x}).} \tag{1.6.25}$$

We can generalize this result to other critical points as well.

Exercise 1.6.1   Construct the Brillouin zone for a three-dimensional simple cubic lattice, and use the theorems from this section to find the vector coordinates of the critical points in k-space.

1.7 Boundary Conditions in a Finite Crystal

So far, we have assumed that the periodic system under consideration is infinite. Real crystals, of course, have boundaries. We can take these boundaries into account by one of two common choices of boundary conditions. First, we can enforce $\psi(x) = \psi(L) = 0$ at each surface, which is realistic for electron energies comparable to the atomic bound state energies. This condition formally violates the assumption we used to prove Bloch's theorem, however, which is that the system is invariant under a translation $T_{\vec{R}}$, for any lattice vector $\vec{R}$. If the system is very large, however, this is not really a problem. Deep inside the crystal, the system is invariant under translations by a large number of lattice vectors, and therefore the Bloch wave functions will be very good approximations of the real eigenstates.

Alternatively, we can use *periodic* boundary conditions, $\psi(0) = \psi(L)$, $\partial\psi(0)/\partial x = \partial\psi(L)/\partial x$. This is known as the **Born–von Karman** boundary condition. While it is unphysical, it has the advantage that it formally satisfies the assumptions of Bloch's theorem, namely that the system is invariant under any translation $T_{\vec{R}}$. This boundary condition allows traveling wave solutions – a wave can exit one side and enter on the other side. This simulates the case of waves traveling in the same direction forever in an infinite medium.

The boundary conditions force a quantization condition on the possible values of the plane-wave vector $\vec{k}$, in addition to the constraint derived above which restricted the values of $\vec{k}$ to the Brillouin zone. In the case of the Born–von Karman boundary condition, the phase factor $e^{i\vec{k}\cdot\vec{x}}$ must be equal at $\vec{x} = 0$ and $\vec{x} = \vec{R}$, where $|\vec{R}| = L$. This condition is equivalent to

$$e^{i\vec{k}\cdot\vec{R}} = e^{i\vec{k}\cdot 0} = 1, \tag{1.7.1}$$

which implies that $\vec{k} \cdot \vec{R} = 2\pi N$, where N is an integer. The vector $\vec{R}$ must be a Bravais lattice vector, which implies

$$\vec{k} \cdot \vec{R} = \vec{k} \cdot (N_1\vec{a}_1 + N_2\vec{a}_2 + N_3\vec{a}_3)$$
$$= 2\pi N. \tag{1.7.2}$$

This condition is satisfied if $\vec{k}$ has the form

$$\vec{k} = \frac{\nu_1}{N_1}\vec{q}_1 + \frac{\nu_2}{N_2}\vec{q}_2 + \frac{\nu_2}{N_3}\vec{q}_3, \tag{1.7.3}$$

where the $\vec{q}_i$ are the reciprocal lattice primitive vectors defined in Section 1.4 and ν_1, ν_2, and ν_3 are integers, in which case

$$\vec{k} \cdot \vec{R} = (\nu_1 + \nu_2 + \nu_3)2\pi, \tag{1.7.4}$$

since $\vec{a}_i \cdot \vec{q}_j = 2\pi\delta_{ij}$.

The condition (1.7.3) implies that $\vec{k}$ can only have discrete values. This is just the same type of quantization that occurs due to the boundary conditions of any finite system. As illustrated in Figure 1.20, for periodic boundary conditions in a box with size L, a wave can only have discrete wavelengths $\lambda_n = L/\nu$, where ν is an integer, which implies $k_n = 2\pi/\lambda_n = 2\pi\nu/L$.

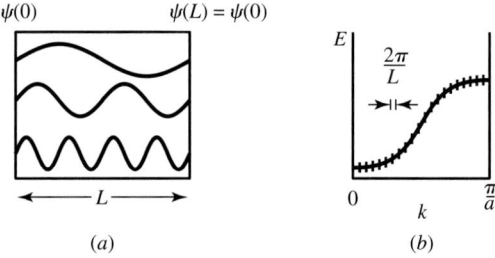

(a)   (b)

Fig. 1.20   (a) Boundary conditions of a finite system cause quantization of the possible wavelengths. (b) For this reason, an energy band actually consists of N discrete states, where N is the number of cells in the lattice.

Since $|v_1| \leq N_1/2$, $|v_2| \leq N_2/2$, and $|v_3| \leq N_3/2$ for $\vec{k}$-vectors in the Brillouin zone, there are only $N = N_1 N_2 N_3$ discrete states for $\vec{k}$ in the Brillouin zone. This means that if there are N primitive cells, there are N states in a band. This is the same statement we made in Section 1.1, namely N cell states lead to N band states. Along one dimension in the direction of primitive vector $\vec{a}_i$, these states are separated by $dk = 2\pi/(N_i|\vec{a}_i|)$. For large N, this spacing is so small that we can typically treat $\vec{k}$ as a continuous variable and ignore its discrete steps.

What if we had chosen the boundary condition of impenetrable boundaries, instead of the Born–von Karman periodic boundary conditions? In that case, we could have either an odd or even number of half-wavelengths within the size of the crystal, namely

$$e^{2i\vec{k}\cdot\vec{R}} = e^{i\vec{k}\cdot 0} = 1. \tag{1.7.5}$$

This would seem to lead to the conclusion that there are $2N$ states in a band in one dimension, or $8N$ states in a three-dimensional crystal. Which is it? Are there N or $8N$ states? We can resolve this by remembering that in the case of impenetrable boundaries, the allowed states are standing waves instead of traveling waves, in which case we count only positive k states. This means we restrict the counting of states to only the first octant of the three-dimensional space, which is one-eighth of the space that includes negative numbers. Therefore, we still have exactly N states for N cells.

As mentioned above, the spacing between the state vectors along one dimension is $\Delta k = 2\pi/(N_i|\vec{a}_i|)$. In three dimensions, instead of talking about the spacing between the states, we can talk about the volume per state in k-space. One way to calculate this is to notice that (1.7.3) defines a new lattice in reciprocal space that has lattice vectors $\vec{b}_1/N_1$, $\vec{b}_2/N_2$, and $\vec{b}_3/N_3$. According to the result of Exercise 1.4.4, the volume per cell of this lattice is

$$\begin{aligned}
\Delta^3 k &= \frac{(2\pi)^3}{|N_1\vec{a}_1 \cdot (N_2\vec{a}_2 \times N_3\vec{a}_3)|} \\
&= \frac{(2\pi)^3}{N_1 N_2 N_3 V_{\text{cell}}} \\
&= \frac{(2\pi)^3}{V},
\end{aligned} \tag{1.7.6}$$

where V is the total volume of the crystal.

Exercise 1.7.1   Suppose we have a ring with six identical atoms. This is a periodic sytem in one dimension, so the Bloch theorem applies. According to the LCAO approximation, discussed in Section 1.1.2, we write the wave function as a linear combination of the unperturbed atomic wave orbitals. This allows us to write a matrix for the eigenstates as follows:

$$\begin{pmatrix} E_0 & U & 0 & 0 & 0 & U \\ U & E_0 & U & 0 & 0 & 0 \\ 0 & U & E_0 & U & 0 & 0 \\ 0 & 0 & U & E_0 & U & 0 \\ 0 & 0 & 0 & U & E_0 & U \\ U & 0 & 0 & 0 & U & E_0 \end{pmatrix} \begin{pmatrix} c_1 \\ c_2 \\ c_3 \\ c_4 \\ c_5 \\ c_6 \end{pmatrix} = E \begin{pmatrix} c_1 \\ c_2 \\ c_3 \\ c_4 \\ c_5 \\ c_6 \end{pmatrix},$$

where $|\psi\rangle = c_1|\psi_1\rangle + c_2|\psi_2\rangle + c_3|\psi_3\rangle + c_4|\psi_4\rangle + c_5|\psi_5\rangle + c_6|\psi_6\rangle$. For convenience, we assume the coupling term U is real. In one dimension, the formulas for the boundary conditions take a simple form: We have $b = 2\pi/a$, and $k = vb/N = 2\pi v/Na$. Bloch's theorem tells us that the phase factor from one cell to the next will be e^{-ika}, which is $e^{i2\pi v/N}$. Show that the solution $c_{n+1} = e^{2\pi vi/6}c_n$ is an eigenvector of the above matrix, for $v = 1,\ldots,6$. (It is easy to do this in Mathematica.)

Note that if two eigenvalues are degenerate, then the eigenstates corresponding to these eigenvalues are not unique, and can be written as any linear superposition of the two states with the same eigenvalue. In the above solution, there are two sets of degenerate eigenstates corresponding to k traveling in opposite directions. (Since, in the reduced zone, we would take v from -3 to 3 instead of from 1 to 6.) One could therefore instead equally well write the coefficients as $2\cos ka = e^{ika} + e^{-ika}$, $2i\sin ka = e^{ika} - e^{-ika}$. Which eigenvectors does Mathematica give, if you use its eigenvector solver?

Suppose that instead of a ring, we have a chain of six atoms with the ends unconnected. How should you alter the above matrix to describe this system? For a specific choice of E_0 and U, namely $E_0 = 1$ and $U = 0.1$, solve for the eigenvalues of both the ring and the linear chain. How much do the eigenvalues differ in the two cases?

1.8 Density of States

In Section 1.7, we saw that for N atoms in a periodic structure, there are N states, which are equally spaced in k. How many states are in a given range of energy? This number is important to know in many calculations, as we will see later in this book.

To find the number of states in a given energy interval, we start with the general observation that for a very large number of states N, we can replace a sum over discrete states with a continuous integral as follows:

$$\sum_{\vec{k}} \to \frac{V}{(2\pi)^3} \int d^3k, \tag{1.8.1}$$

where we have normalized the integral by the volume in k-space per state, given by (1.7.6). In the limit that the volume of k-space is large compared to the volume per state, the number of states in a region of k-space is equal to the volume of the region times this normalization factor.

To find the number of k-states that fall within a given energy range $(E, E+dE)$, where dE is much less than E, we find the volume of the shell in k-space that has states with energies in the range $(E, E + dE)$, as illustrated in Figure 1.21. This is equal to

$$\mathcal{D}(E)dE = dE\frac{V}{(2\pi)^3} \int d^3k\, \delta(E(\vec{k}) - E). \tag{1.8.2}$$

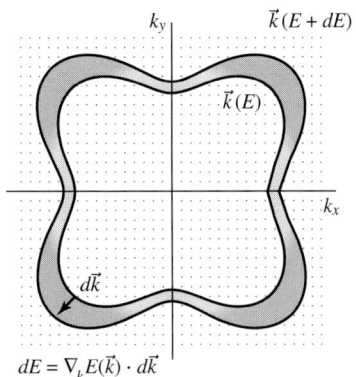

Fig. 1.21 The density-of-states factor $D(E)dE$ is the number of states (represented by the equally spaced dots) falling within the volume of k-space between two surfaces in k-space corresponding to $\vec{k} = \vec{k}(E)$ and $\vec{k} = \vec{k}(E + dE)$.

Here, $\mathcal{D}(E)$ is the **density of states**, equal to the number of k-states per unit energy at energy E. In other words, the density of states at a given energy E is given by the number of states per volume times the volume in k-space between two surfaces, the first surface consisting of all $\vec{k}$ such that $\vec{k} = \vec{k}(E)$, and the second surface corresponding to all $\vec{k}$ such that $\vec{k} = \vec{k}(E + dE)$.

The mathematical construct that relates dE to the distance between these two surfaces is the gradient. The change dE that corresponds to a change of the vector $\vec{k}$ by an amount dk in the direction normal to a constant-energy surface is given by the gradient $\nabla_{\vec{k}} E(\vec{k})$. This implies

$$\mathcal{D}(E)dE = dE \frac{V}{(2\pi)^3} \int d\sigma_{\vec{k}} \frac{1}{|\nabla_{\vec{k}} E(\vec{k})|}, \tag{1.8.3}$$

where $d\sigma_{\vec{k}}$ is a surface area element in k-space. In spherical coordinates, we can write this as

$$\mathcal{D}(E)dE = \frac{V}{(2\pi)^3} dE \int k^2(E, \phi, \theta) d\phi \, d(\cos\theta) \frac{1}{|\nabla_{\vec{k}} E(\vec{k})|}. \tag{1.8.4}$$

For each θ and ϕ, one must solve for the k at the angle that corresponds to energy E, and determine the gradient of E at that $\vec{k}$.

1.8.1 Density of States at Critical Points

As we saw in Section 1.6, $\nabla_{\vec{k}} E$ vanishes at zone center and at the boundaries of the Brillouin zone. This means that the density of states will have special properties at these points. It might seem that the density of states diverges at these points, but this is not always the case. For example, in the case of isotropic bands, the density of states formula (1.8.4) can be simplified to

$$\mathcal{D}(E)dE = \frac{V}{(2\pi)^3} 4\pi \, dE \frac{1}{|\nabla_{\vec{k}} E|} k^2(E). \tag{1.8.5}$$

Since the band at zone center must be at a minimum or maximum, we can expand the energy in powers of k as

$$E(k) = E_0 + \frac{1}{2} \left.\frac{\partial^2 E}{\partial k^2}\right|_{k=0} k^2 + \cdots . \tag{1.8.6}$$

The leading order of the gradient of E is therefore linear in k, which means that the density of states is proportional to k, which implies

$$\mathcal{D}(E)dE \propto \sqrt{(E - E_0)}dE. \tag{1.8.7}$$

The same thing occurs at the critical points on the zone boundaries where $\nabla_{\vec{k}} E$ vanishes, discussed in Section 1.6. In general, the band minimum or maximum at the critical point can be expanded in powers of k as

$$E(k) = E_0 + \frac{1}{2} \sum_{i,j} \frac{\partial^2 E}{\partial k_i \partial k_j}(k_i - k_i^{\text{crit}})(k_j - k_j^{\text{crit}}) + \cdots . \tag{1.8.8}$$

The quadratic dependence on k always leads to the same square root dependence of the density of states for a three-dimensional crystal near a critical point.

Below a minimum, or above a maximum, there are no states, so therefore there is a discontinuity in the first derivative of the density of states function. This is called a **van Hove singularity**. Figure 1.22 shows a typical density of states for electron bands in a three-dimensional crystal. A van Hove singularity can occur when there is a discontinuity in the first derivative in any direction; for example, at a saddle point.

At a band minimum or maximum in a three-dimensional system, a van Hove singularity leads only to a discontinuity in the first derivative of the density of states, but in a one-dimensional system, the density of states itself diverges at a van Hove singularity. In this case, the density of states is simply given by

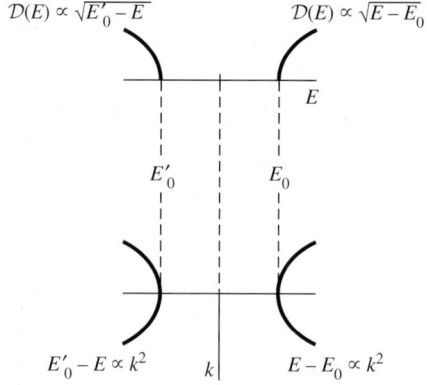

Lower plot: A typical band diagram with a minimum and a maximum at zone center, turned sideways. Upper plot: Density of states for these bands, in a three-dimensional crystal. At the band edge, there is a van Hove singularity in the first derivative.

$$\mathcal{D}(E)dE = dE\frac{L}{2\pi}\int dk\,\delta(E - E(k))$$

$$= dE\frac{L}{2\pi}\int dE'\frac{1}{|\partial E/\partial k|}\delta(E - E')$$

$$= dE\frac{L}{2\pi}\frac{1}{|\partial E/\partial k|}. \tag{1.8.9}$$

Near every band minimum or maximum, this diverges as $1/k$. This divergence does not lead to a physical infinity, however. To get a total number of states, one must integrate over some range of energy. Integrating this density of states over the singular point gives

$$\int_0^{\Delta E} dE\frac{1}{k} \sim \int_0^{\Delta E} dE\frac{1}{\sqrt{E}} \sim (\Delta E)^{1/2}, \tag{1.8.10}$$

which is finite.

Plotting the density of states as in Figure 1.22 is another way of picturing electron bands. In the gap regions, there are no states, while in the bands, there is a continuum of allowed states for the electrons.

Exercise 1.8.1   Show that if a band has a minimum but is not isotropic, that is,

$$E = E_0 + Ak_x^2 + Bk_y^2 + Ck_z^2,$$

that the density of states near $\vec{k} = 0$ is still proportional to $\sqrt{(E - E_0)}$. In this case, the Taylor expansion is

$$E = E_0 + \frac{1}{2}\sum_{i,j}\frac{\partial^2 E}{\partial k_i \partial k_j}k_i k_j + \cdots.$$

This gives a matrix for the quadratic term, which can always be diagonalized.

Exercise 1.8.2   Show that the density of states in an isotropic *two*-dimensional system near a band minimum or maximum does not depend on the energy of the electrons. The volume per state in k-space is $A/(2\pi)^2$, where A is the total area of the two-dimensional system.

Exercise 1.8.3   Plot the density of states for the lowest three bands of the one-dimensional Kronig–Penney model discussed in Section 1.2, for $\hbar^2/2m = 1$, $a = 1$, $b = 0$, and $U_0 b = 1$. You will need to solve for $\partial E/\partial k$ numerically, as we had to do for $E(k)$ in the Kronig–Penney model.

1.8.2 Disorder and Density of States

Density-of-states plots give us a natural way to look at the effect of disorder, that is, what happens to the electron bands when a crystal is not perfectly periodic. As discussed in Section 1.1, bands and band gaps appear whenever there is overlap of atomic orbitals, regardless of periodicity.

In the long wavelength limit (when the characteristic length of the disorder is much longer than the atomic lattice spacing), we can model disorder as regions with slightly larger or smaller spacing between atoms. We can then approximate the effect of the disorder by recalculating the band energy for a larger or smaller lattice spacing in each region. Larger spacing corresponds to less orbital overlap of adjacent atoms, which means less bonding–antibonding splitting. This corresponds to a smaller band gap; in other words, the upper, antibonding states will have lower energy and the lower, bonding states will have higher energy. This means that in a region of larger lattice spacing, there will be electron states inside the nominal energy gap.

In the absence of any other information, we can assume that the disorder is distributed randomly. In the long wavelength limit, we can view the disordered crystal as a set of perfectly ordered crystals with band gaps that are distributed according to a Gaussian distribution, according to the central limit theorem,

$$P(E_g) = \frac{1}{\sqrt{2\pi}(\Delta E)} e^{-(E_g(0)-E_g)^2/2(\Delta E)^2}, \tag{1.8.11}$$

where E_g is the band gap for a perfectly ordered crystal and ΔE_g is a characteristic range of energy fluctuations. The total density of states of the crystal will then be given by the convolution of this distribution with the density of states for a periodic structure,

$$\mathcal{D}(E) = \int dE_g \mathcal{D}(E - E_g) P(E_g). \tag{1.8.12}$$

The effect of the convolution is to smear out the band gaps of a solid. Disorder does not necessarily eliminate the existence of bands and band gaps, however. Figure 1.23(b) illustrates how a small degree of disorder smears the bands, while leaving them still much the same. In general, every real crystal has some degree of band smearing because there is always some degree of disorder.

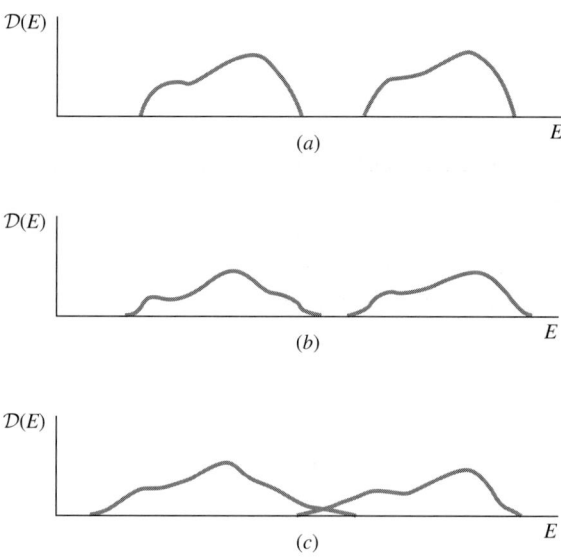

Fig. 1.23　Density of states of two electron bands in the presence of disorder. (a) No disorder. (b) Low disorder. (c) Larger disorder. In this last case, the band gap disappears, although the density of states is still lower in that region.

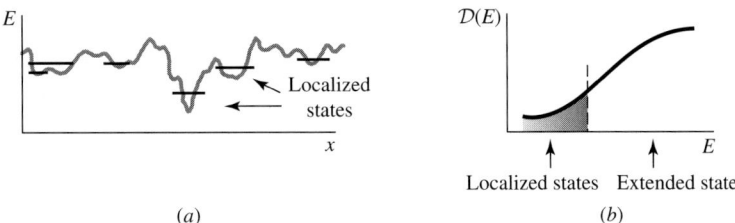

Fig. 1.24 (a) Random potential landscape, with bound states in the potential minima at low energy. (b) Mobility edge in the density of states.

A large degree of disorder can cause the tail of one band to overlap with the tail of another band, leading to a continuum of states filling the band gap, as illustrated in Figure 1.23(c). The density-of-states plot in this case gives us more information than a band diagram, because one often can see peaks in the density of states that are band-like, even when there are no well-defined band gaps.

Near a band edge, the density of states typically looks like a continuous function due to the smearing from disorder, but the nature of the states can be quite different. As the disorder gives a variation of the band gap, the effect of the disorder is to create local peaks and valleys in the potential energy felt by an electron, as illustrated in Fig. 1.24(a). The deepest valleys will in general be so far apart that the coupling between them is negligible. In this case, the electronic states in the valleys are **localized** states with discrete energies, similar to the states in a square well. We therefore expect that many of the lowest-energy states for electrons will correspond to localized electrons that cannot move. On large length scales, the average over many of these localized states will give a continuous density of states. Empirically, the density of localized states below the band gap in many solids is proportional to e^{-E/E_U}, where E_U is some characteristic energy. This is known as an **Urbach tail**.

As we saw in Section 1.1, the coupling between localized states increases continuously as they get nearer to each other, due to tunneling between the different energy minima. Therefore, bands of localized states can arise, just as they did for atomic states. These bands can allow electron motion across the crystal, but they will have low conductivity because of their low density of states.

Since trapped states below the band edge can conduct electrons, we might expect that the electrical conductivity decreases continuously toward zero as the energy of the electrons decreases. In fact, it can be proven that there is a sharp energy cutoff, called the **mobility edge**, below which the states do not contribute to conductivity at all, as illustrated in Figure 1.24(b). As discussed in Section 9.11, quantum interference can imply zero conductivity in the presence of disorder, an effect known as **Anderson localization**.

Exercise 1.8.4   If the distribution of energy gaps in a disordered, three-dimensional solid is given by (1.8.11), and the density of states is given by (1.8.7), use (1.8.12) to plot the density of states of a disordered three-dimensional material near a band minimum.

1.9 Electron Band Calculations in Three Dimensions

So far we have looked at the wave function of one electron in fixed external potential. In solids, however, the potential energy felt by an electron arises from the interaction of each electron with all the nuclei and all the other electrons. The potential energy felt by an electron is not only given by the positively charged nuclei, but also by the average negative charge of all the other electrons.

If a crystal has a certain periodicity, then the potential created from all these particles will have that periodicity, so all of the above theorems for periodic potentials still apply. Determining the exact nature of the electron bands, however, is a difficult task. We cannot simply solve for the eigenstate of a single electron in a fixed potential; we must solve for the eigenstates of the whole set of electrons, taking into account exchange between the identical electrons, which must be treated according to Fermi statistics. Methods of treating the Fermi statistics of electrons will be discussed in Chapter 8.

In general, the calculation of band structure is not an exact science. Typically, determining the band structure of a given solid involves interaction between experiment and theory – the crystal symmetry is determined by x-ray scattering, a band structure is calculated, this is corrected by other experiments such as optical absorption and reflectivity, etc. Many band structure calculations use experimental inputs from chemistry such as the electronegativity of ions, bond lengths, and so on. Calculating band structures from first principles, using nothing but the charge and masses of the nuclei, is still an area of frontier research, involving high-level math and supercomputers.

In this book, we will typically treat the electron bands as known functions for a given solid. Band diagrams have been published for many solids and tabulated in books, for example, Madelung (1996). At the same time, there are several useful approximation methods which allow us to write simple mathematical formulas for the bands without needing to go through all of the calculations to generate a band structure. The value of these approximations is not so much to predict the actual band structures quantitatively; rather, these models help to give us physical intuition about the nature of electron bands.

1.9.1 How to Read a Band Diagram

In a three-dimensional crystal, the full calculation of all the band energies involves finding the energy of each band at every point in the three-dimensional Brillouin zone. This is a large amount of information, which we need to present in a simple fashion for it to be useful.

As discussed in Section 1.6, there are certain critical points in the Brillouin zone which correspond to the points on the surfaces of the zone that are half way between the origin and another reciprocal lattice vector. Figure 1.25 gives the standard labeling of these critical points for common lattice structures. Typically, band structure calculations give the band energies along lines from the center of the Brillouin zone to one of these points, or from

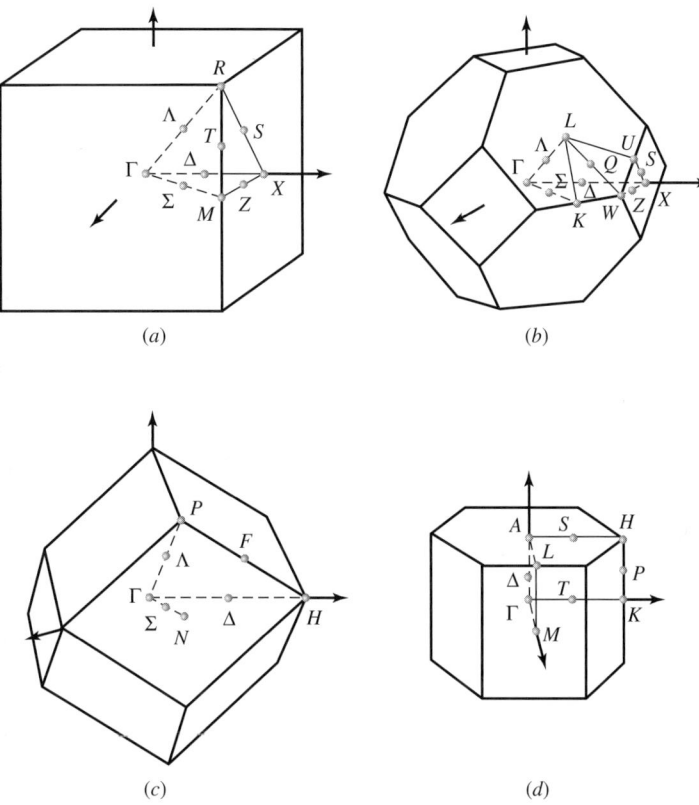

Fig. 1.25 Standard labeling of the high symmetry points in common Brillouin zones. (a) sc lattice. (b) fcc lattice. (c) bcc lattice. (d) sh lattice.

one of these points to another one. Figure 1.26(a) shows a typical band structure plot for silicon, a cubic crystal. The critical points are labeled according to the drawing shown in Figure 1.25(b). Note that the diagram is not symmetric about the Γ point because two different paths away from this point are plotted. Note also that the U/K point is not the midpoint between two reciprocal lattice points, and therefore the slope of the bands is not zero in every direction there; in particular, the slope is not zero along lines that are not normal to the zone boundary. The L and X points are critical points, and therefore if the bands are plotted with enough detail, one will see that they have zero slope at those points.

Figure 1.26(b) shows the density of states for the same crystal. As seen in this figure, van Hove singularities (discontinuities in the slope) correspond to critical points in the band structure. When bands overlap in energy, the density of states is just the sum of the density of states of the two bands. Notice that there is a gap in the density of states which corresponds to the energy gap in one band at the Γ point (zone center) and the minimum of the next higher band at the X point (the zone boundary in the [100] direction).

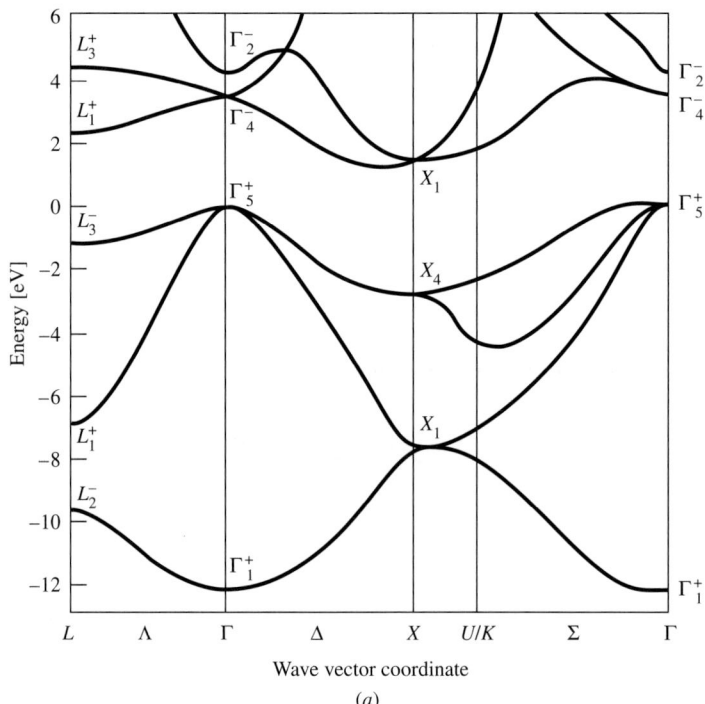

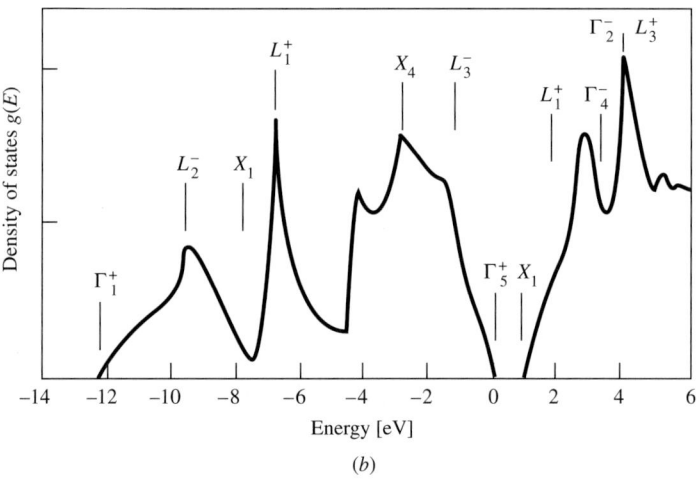

Fig. 1.26   (a) Calculated band structure of silicon, along the most important axes of symmetry in the Brillouin zone, using the fcc labeling of Figure 1.25(b). (From Chelokowski *et al.* (1976).) (b) Density of states for silicon from the same bands as in (a). Note how the points of symmetry in the Brillouin zone correspond to van Hove discontinuities of the first derivative of the density of states. The critical points of the bands are labeled in terms of group theory notation, which is discussed in Chapter 6.

1.9.2 The Tight-Binding Approximation and Wannier Functions

Although, as discussed above, in principle we must solve for the eigenstates of the full many-electron system, in many cases we can assume that some electrons are independent of others. One example is when the atoms in a lattice have tightly bound core electron orbitals that do not overlap much with the orbitals of adjacent atoms.

We begin by deducing another general theorem for the Bloch functions in a periodic system. According to (1.6.8), the Bloch functions are periodic in k-space. We can therefore follow a logic similar to that which we used in Section 1.6 for the cell functions $u_{n\vec{k}}(\vec{r})$ which are periodic in real space. Viewing $\psi_{n\vec{k}}(\vec{r})$ as a function of $\vec{k}$, keeping $\vec{r}$ constant, we write the inverse Fourier transform as

$$
\begin{aligned}
f_n(\vec{r}, \vec{r}') &= \frac{1}{(2\pi)^3} \int_{-\infty}^{\infty} d^3k \, \psi_{n\vec{k}}(\vec{r}) e^{-i\vec{k}\cdot\vec{r}'} \\
&= \frac{1}{(2\pi)^3} \sum_{\vec{Q}} e^{-i\vec{Q}\cdot\vec{r}'} \left(\int_{B.Z.} d^3k' \psi_{n\vec{k}'}(\vec{r}) e^{-i\vec{k}'\cdot\vec{r}'} \right),
\end{aligned}
\tag{1.9.1}
$$

where we have used the trick of writing $\vec{k} = \vec{Q} + \vec{k}'$ and breaking the integral over all reciprocal space into a sum over all reciprocal lattice positions $\vec{Q}$ times an integral only over $\vec{k}'$ in each unit cell, which in this case is the Brillouin zone. The sum over reciprocal lattice vectors here will give nonzero peaks only for $\vec{r}'$ equal to the Bravais lattice vectors $\vec{R}$. As in Section 1.4, the structure factor in this case will modulate the height of the peaks and remove spurious peaks.

Following the same logic as in Section 1.6, because the Bloch functions only have inverse Fourier components at the lattice vectors $\vec{R}$, we can write the Bloch functions as the Fourier series,

$$
\psi_{n\vec{k}}(\vec{r}) = \frac{1}{\sqrt{N}} \sum_{\vec{R}} \phi_n(\vec{r}, \vec{R}) e^{i\vec{k}\cdot\vec{R}},
\tag{1.9.2}
$$

where N is the number of lattice sites, and

$$
\phi_n(\vec{r}, \vec{R}) = \frac{\sqrt{N} V_{\text{cell}}}{(2\pi)^3} \int_{B.Z.} d^3k' \psi_{n\vec{k}'}(\vec{r}) e^{-i\vec{k}'\cdot\vec{R}},
\tag{1.9.3}
$$

where the integration is, again, over the Brillouin zone, and V_{cell} is the volume of the primitive cell in real space. The functions $\phi_n(\vec{r}, \vec{R})$ are called the **Wannier functions**. From the definition (1.3.3) of $\psi_{n\vec{k}}$ in terms of the cell function, we have

$$
\begin{aligned}
\phi_n(\vec{r}, \vec{R}) &= \frac{\sqrt{N} V_{\text{cell}}}{(2\pi)^3} \int_{-\infty}^{\infty} d^3k \frac{1}{\sqrt{V}} u_{n\vec{k}}(\vec{r}) e^{i\vec{k}\cdot\vec{r}} e^{-i\vec{k}\cdot\vec{R}} \\
&= \frac{\sqrt{N} V_{\text{cell}}}{(2\pi)^3} \int_{-\infty}^{\infty} d^3k \frac{1}{\sqrt{V}} u_{n\vec{k}}(\vec{r} - \vec{R}) e^{i\vec{k}\cdot(\vec{r}-\vec{R})}.
\end{aligned}
\tag{1.9.4}
$$

This implies $\phi_n(\vec{r}, \vec{R}) = \phi_n(\vec{r} - \vec{R})$, or

$$\psi_{n\vec{k}}(\vec{r}) = \frac{1}{\sqrt{N}} \sum_{\vec{R}} \phi_n(\vec{r} - \vec{R}) e^{i\vec{k}\cdot\vec{R}}. \tag{1.9.5}$$

In other words, the Bloch wave functions can be written as a sum of Wannier functions which are identical except that each Wannier function is centered around the midpoint of a different point in the real-space lattice.

This leads naturally to association of the Wannier functions with the atomic orbital functions. The LCAO approximation (linear combination of atomic orbitals), already discussed in Section 1.1.2, sets the Wannier functions equal to the unperturbed atomic orbitals. In the most general case, when there is more than one atomic orbital in the lattice basis, we approximate

$$\phi_n(\vec{r}) \simeq \sum_{i} c_i \Phi_{n,i}(\vec{r} - \vec{b}_i), \tag{1.9.6}$$

where the $\Phi_{n,i}(\vec{r})$ functions are the unperturbed atomic orbitals at each site $\vec{b}_i$ of the lattice basis, and the c_i are coefficients for the relative weights of the different orbitals.

For the case of a single orbital per unit cell, the Schrödinger equation gives us, using (1.9.5),

$$H\psi_{n\vec{k}} = \sum_{\vec{R}} e^{i\vec{k}\cdot\vec{R}} H\Phi_n(\vec{r} - \vec{R})$$

$$= E_n(\vec{k})\psi_{n\vec{k}} = E_n(\vec{k}) \sum_{\vec{R}} e^{i\vec{k}\cdot\vec{R}} \Phi_n(\vec{r} - \vec{R}). \tag{1.9.7}$$

The orbital functions Φ_n at different lattice sites are nearly orthogonal (the exact Wannier functions centered at different lattice sites are strictly orthogonal, per Exercise 1.9.1). Therefore, multiplying both sides by $\Phi_n^*(\vec{r})$ and integrating over space, we have

$$\langle \Phi_n | H | \psi_{n\vec{k}} \rangle = \sum_{\vec{R}} e^{i\vec{k}\cdot\vec{R}} \int d^3r \, \Phi_n^*(\vec{r}) H\Phi_n(\vec{r} - \vec{R})$$

$$= E_n(\vec{k}). \tag{1.9.8}$$

For tightly bound states, it is reasonable to include only the terms in the sum (1.9.8) for $\vec{R} = 0$ and the nearest neighbors. We then have

$$E_n(\vec{k}) = \int d^3r \, \Phi_n^*(\vec{r}) H\Phi_n(\vec{r})$$

$$+ \sum_{n.n.} e^{i\vec{k}\cdot\vec{R}} \left(\int d^3r \, \Phi_n^*(\vec{r}) H\Phi_n(\vec{r} - \vec{R}) \right), \tag{1.9.9}$$

where "$n.n.$" indicates only $\vec{R}$ values for nearest neighbors.

In a cubic lattice with lattice constant a, there are nearest neighbors at $\pm a\hat{x}$, $\pm a\hat{y}$, and $\pm a\hat{z}$. The energy of the band in this case can therefore be written as

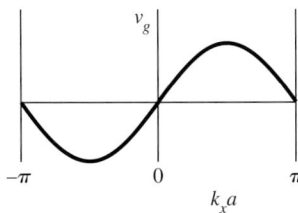

Fig. 1.27 The group velocity $(1/\hbar)\partial E/\partial k$ for a cubic lattice in the tight-binding approximation.

$$E_n(\vec{k}) = E_n + e^{ik_x a}U_{12} + e^{-ik_x a}U_{12} + e^{ik_y a}U_{12} + e^{-ik_y a}U_{12}$$

$$+ e^{ik_z a}U_{12} + e^{-ik_z a}U_{12}$$

$$= E_n(0) + 2U_{12}(\cos k_x a + \cos k_y a + \cos k_z a), \qquad (1.9.10)$$

where $E_n(0)$ is a constant approximately equal to the atomic orbital energy, and U_{12} is the coupling term, which is typically negative. This has the properties we deduced earlier for all Bloch functions, namely, $\partial E/\partial k = 0$ at a zone boundary, since $\sin(\pi/a) = 0$, which implies that the group velocity in the direction normal to a zone boundary vanishes, as discussed in Section 1.2. It also has the property that near $k = 0$, the velocity is proportional to k, as in the case of a free particle with $E = \hbar^2 k^2/2m$. Figure 1.27 shows the group velocity for this model of a cubic lattice as a function of k_x.

Exercise 1.9.1　Show that the Wannier functions centered at different lattice sites are orthogonal, that is,

$$\int d^3r\, \phi_n^*(\vec{r} - \vec{R}_m)\phi_{n'}(\vec{r} - \vec{R}_{m'}) = \delta_{nn'}\delta_{mm'}. \qquad (1.9.11)$$

Hint: This follows from the orthogonality relation (1.6.9) of the Bloch wave functions.

Exercise 1.9.2　Show that for the Hamiltonian $H = -(\hbar^2/2m)\nabla^2 + U(\vec{r})$, an equivalent way of writing (1.9.9) is

$$E_n(\vec{k}) = E_n(0) + \int d^3r\, \Phi_n^*(\vec{r})[U(\vec{r}) - U_0(\vec{r})]\Phi_n(\vec{r}) \qquad (1.9.12)$$

$$+ \sum_{n.n.} e^{i\vec{k}\cdot\vec{R}}\left(\int d^3r\, \Phi_n^*(\vec{r})[U(\vec{r}) - U_0(\vec{r})]\Phi_n(\vec{r} - \vec{R})\right),$$

where $E_n(0)$ is the unperturbed atomic orbital energy and $U_0(\vec{r})$ is the potential energy function of a single atom. In other words, the band energy depends on the difference of the periodic potential and the single-atom potential.

Hint: Note that $-(\hbar^2/2m)\nabla^2\Phi_n(\vec{r}) = (E_n - U_0(\vec{r}))\Phi_n(\vec{r})$.

Tight-binding model with a basis. We can generalize the above for when there is more than one atom per unit cell. For a two-atom basis $(0, \vec{b}_1)$, using (1.9.6), we have

$$H\psi_{n\vec{k}}(\vec{r}) = \sum_{\vec{R}} e^{i\vec{k}\cdot\vec{R}} H\left(c_0\Phi_{n0}(\vec{r}-\vec{R}) + c_1\Phi_{n1}(\vec{r}-\vec{b}_1-\vec{R})\right)$$

$$= E_n(\vec{k})\sum_{\vec{R}} e^{i\vec{k}\cdot\vec{R}}\left(c_0\Phi_{n0}(\vec{r}-\vec{R}) + c_1\Phi_{n1}(\vec{r}-\vec{b}_1-\vec{R})\right). \qquad (1.9.13)$$

As in Section 1.1.2, we want to form a matrix by projection of both sides of this equation onto each of the orbital functions in the unit cell. For the first orbital function, we have

$$\langle\Phi_{n0}|H|\psi_{n\vec{k}}\rangle = c_0\sum_{\vec{R}} e^{i\vec{k}\cdot\vec{R}}\int d^3r\, \Phi_{n0}^*(\vec{r})H\Phi_{n0}(\vec{r}-\vec{R})$$

$$+c_1\sum_{\vec{R}} e^{i\vec{k}\cdot\vec{R}}\int d^3r\, \Phi_{n0}^*(\vec{r})H\Phi_{n1}(\vec{r}-\vec{b}_1-\vec{R})$$

$$= c_0 E_n(\vec{k})\sum_{\vec{R}} e^{i\vec{k}\cdot\vec{R}}\int d^3r\, \Phi_{n0}^*(\vec{r})\Phi_{n0}(\vec{r}-\vec{R})$$

$$+c_1 E_n(\vec{k})\sum_{\vec{R}} e^{i\vec{k}\cdot\vec{R}}\int d^3r\, \Phi_{n0}^*(\vec{r})\Phi_{n1}(\vec{r}-\vec{b}_1-\vec{R}). \qquad (1.9.14)$$

As in the single-atom case discussed above, we assume that the orbitals on different atoms are nearly orthogonal, and that only the coupling terms from nearest neighbors are significant. Then we have

$$\langle\Phi_{n0}|H|\psi_{n\vec{k}}\rangle = c_0\int d^3r\, \Phi_{n0}^*(\vec{r})H\Phi_{n0}(\vec{r}) + c_0\sum_{n.n.} e^{i\vec{k}\cdot\vec{R}}\int d^3r\, \Phi_{n0}^*(\vec{r})H\Phi_{n0}(\vec{r}-\vec{R})$$

$$+c_1\sum_{n.n.} e^{i\vec{k}\cdot\vec{R}}\int d^3r\, \Phi_{n0}^*(\vec{r})H\Phi_{n1}(\vec{r}-\vec{b}_1-\vec{R})$$

$$= c_0 E_n(\vec{k}). \qquad (1.9.15)$$

Let us take the specific case of a honeycomb lattice, which is the structure of **graphene**, that is, a single two-dimensional sheet of graphite (see Table 1.1). The primitive vectors and basis of this lattice are shown in Figure 1.28. We make the simplifying assumption that there is only one relevant atomic orbital per lattice site, which we take as a radially symmetric s-orbital. As seen in Figure 1.28, each orbital Φ_{n0} (for example, at the origin)

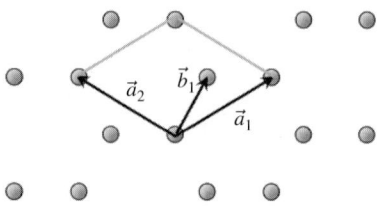

Fig. 1.28   Graphene ("honeycomb" lattice) primitive vectors $\pm(\sqrt{3}/2)a\hat{x} + (1/2)a\hat{y}$, and basis with one atom at the origin and the second at $\vec{b}_1 = (1/2\sqrt{3})a\hat{x} + (1/2)a\hat{y}$.

has nearest neighbors only of the orbital Φ_{n1}. For the atom at the origin, these are at $\vec{b}_1$, at $\vec{b}_1 - \vec{a}_1$, and at $\vec{b}_1 - \vec{a}_1 - \vec{a}_2$. We therefore have

$$c_0 E_{n0} + c_1 U_{12} \left(1 + e^{-i\vec{k}\cdot\vec{a}_1} + e^{-i\vec{k}\cdot(\vec{a}_1 + \vec{a}_2)}\right) = c_0 E_n(\vec{k}), \tag{1.9.16}$$

where U_{12} is the coupling integral between nearest neighbors. Similarly, we have for the projection $\langle \Phi_{n1} | H | \psi_{n\vec{k}} \rangle$,

$$c_0 U_{12} \left(1 + e^{i\vec{k}\cdot\vec{a}_1} + e^{i\vec{k}\cdot(\vec{a}_1 + \vec{a}_2)}\right) + c_1 E_{n1} = c_1 E_n(\vec{k}). \tag{1.9.17}$$

We write these as a matrix equation,

$$\begin{pmatrix} E_0 & U_{12} \left(1 + e^{-i\vec{k}\cdot\vec{a}_1} + e^{-i\vec{k}\cdot(\vec{a}_1 + \vec{a}_2)}\right) \\ U_{12} \left(1 + e^{i\vec{k}\cdot\vec{a}_1} + e^{i\vec{k}\cdot(\vec{a}_1 + \vec{a}_2)}\right) & E_0 \end{pmatrix} \begin{pmatrix} c_0 \\ c_1 \end{pmatrix}$$

$$= E_n(\vec{k}) \begin{pmatrix} c_0 \\ c_1 \end{pmatrix}. \tag{1.9.18}$$

Solving for the eigenvalues, we have

$$E_n(\vec{k}) = E_0 \pm U_{12} \sqrt{3 + 4 \cos \frac{\sqrt{3} k_x}{2} \cos \frac{k_y}{2} + 2 \cos k_y}. \tag{1.9.19}$$

The solution has the interesting property that at certain points in the Brillouin zone, the energy gap in the bands created by symmetric and antisymmetric combinations of the two orbitals goes to zero, and the electron energy is linear with k instead of quadratic, as shown in Figure 1.29.

General tight-binding model. Generalizing from the above example of a two-orbital basis, we can say that when there are N orbitals in the basis of the lattice, whether on the same atom or on different atoms in the unit cell, the energies of the bands arising from these orbitals are the eigenvalues of the $N \times N$ matrix, with elements given by

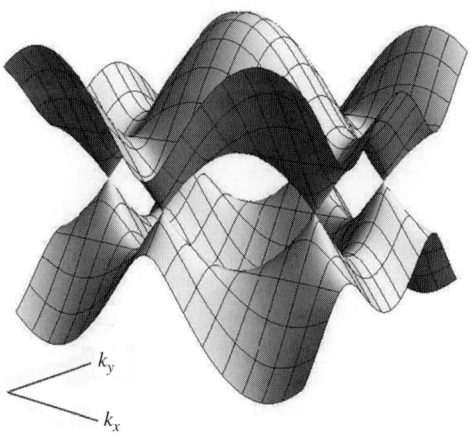

Fig. 1.29 Energy bands of the simple tight-binding model for graphene, from (1.9.19).

$$E_{ij} = \sum_{\vec{R}} e^{i\vec{k}\cdot\vec{R}} \int d^3r\, \Phi_i(\vec{r})H\Phi_j(\vec{r}-\vec{R}), \tag{1.9.20}$$

where i and j run from 1 to N. Here we have absorbed the basis vector location $\vec{b}_i$ into the definitions of the orbitals.

The lattice vectors $\vec{R}$ in the sum can be separated into $\vec{R} = 0$, which gives the overlap of different orbitals within the unit cell, the vectors $\vec{R}$ corresponding to the nearest neighbors of each atom in the unit cell (which may be different for different atoms in the unit cell), next-nearest neighbors, and so on. For most cases when the tight-binding approximation is valid, only the nearest neighbors and the next-nearest neighbors need to be included.

Exercise 1.9.3 As a follow-up to Exercise 1.4.6, show that the special point at which the gap energy goes to zero in (1.9.19) for the graphene lattice is one of the corners of the Wigner–Seitz cell for the Brillouin zone in reciprocal space.

Exercise 1.9.4 Calculate the energy band arising from a single orbital in a two-dimensional simple hexagonal lattice (see Table 1.1), using the tight-binding approximation, and plot the energy as a function of k_x and k_y. Assume that there is one coupling energy U_{12} for all nearest neighbors and set all other coupling energies to zero.

1.9.3 The Nearly Free Electron Approximation

At the opposite extreme from the tight-binding approximation, one can make the nearly free approximation for electrons in bands that arise from atomic orbitals with very large overlap. In this case, the electron states are almost the same as free plane waves. This approximation can work for the nearly free electrons in upper states even if there are tightly bound electrons in lower, core states. The effect of the core electrons is taken into account just as a change of the total classical charge of the nucleus.

In this approximation, we begin with the version of Bloch's theorem given in (1.6.4),

$$\psi_{n\vec{k}}(\vec{r}) = \frac{1}{\sqrt{V}} \sum_{\vec{Q}} C_n(\vec{k}-\vec{Q})e^{i(\vec{k}-\vec{Q})\cdot\vec{r}}, \tag{1.9.21}$$

where the $\vec{Q}$s are reciprocal lattice vectors. The nearly free electron approximation amounts to assuming that the wave function is nearly equal to a plane wave, $e^{i\vec{k}\cdot\vec{r}}$. This means that the leading term in the expansion is $C_n(\vec{k}) \approx 1$, which corresponds to $\vec{Q} = 0$, and higher-order terms of this expansion are small compared to this term.

We also write the periodic potential $U(\vec{r})$ as a Fourier series,

$$U(\vec{r}) = \sum_{\vec{Q}} U(\vec{Q})e^{-i\vec{Q}\cdot\vec{r}}. \tag{1.9.22}$$

We will assume that $U(\vec{Q} = 0)$ is zero, since this corresponds to a constant term which we can always remove by changing the definition of zero potential energy. Substituting these definitions into the Schrödinger equation, we have

$$\left[-\frac{\hbar^2}{2m}\nabla^2 + U(\vec{r}) - E_n(\vec{k})\right]\psi_{n\vec{k}}(\vec{r}) = 0$$

$$= \frac{1}{\sqrt{V}}\sum_{\vec{Q}}\left[\frac{\hbar^2}{2m}|\vec{k} - \vec{Q}|^2 + \sum_{\vec{Q}'}U(\vec{Q}')e^{-i\vec{Q}'\cdot\vec{r}} - E_n(\vec{k})\right]C_n(\vec{k} - \vec{Q})e^{i(\vec{k} - \vec{Q})\cdot\vec{r}}.$$

$$(1.9.23)$$

Multiplying by $(1/V)e^{-i(\vec{k} - \vec{Q}'')\cdot\vec{r}}$ and integrating over all space allows us to eliminate the exponential factors and one of the summations, because we know that $(1/V)\int d^3r\,e^{i\vec{k}\cdot\vec{r}} = \delta_{\vec{k},0}$ (see Appendix C). We obtain

$$\left[\frac{\hbar^2}{2m}|\vec{k} - \vec{Q}|^2 - E_n(\vec{k})\right]C_n(\vec{k} - \vec{Q}) + \sum_{\vec{Q}'}U(\vec{Q} - \vec{Q}')C_n(\vec{k} - \vec{Q}') = 0. \quad (1.9.24)$$

This is the main equation for the nearly free electron approximation. It is still exact, but as with the tight-binding model, we can solve it approximately by truncating the sum over $\vec{Q}'$ to only near neighbors of $\vec{Q} = 0$, on the assumption that $C_n(\vec{k} - \vec{Q}')$ falls off rapidly for increasing $\vec{Q}'$. In this case, we truncate in reciprocal space rather than real space, because a wave function spread out in real space, like a plane wave, is localized in k-space.

We can get a feel for this model by taking just one nearest neighbor, $\vec{Q}' = \vec{Q}_0$. If we set $\vec{Q} = 0$, we have

$$\left[\frac{\hbar^2 k^2}{2m} - E_n(\vec{k})\right]C_n(\vec{k}) + U(-\vec{Q}')C_n(\vec{k} - \vec{Q}') = 0. \quad (1.9.25)$$

We can also set $\vec{Q} = \vec{Q}_0$ in (1.9.24), which gives us, to first order,

$$\left[\frac{\hbar^2|\vec{k} - \vec{Q}'|^2}{2m} - E_n(\vec{k})\right]C_n(\vec{k} - \vec{Q}') + U(\vec{Q}')C_n(\vec{k}) = 0. \quad (1.9.26)$$

These equations can be written as

$$\begin{pmatrix} \dfrac{\hbar^2 k^2}{2m} & U^*(\vec{Q}_0) \\[2mm] U(\vec{Q}_0) & \dfrac{\hbar^2|\vec{k} - \vec{Q}_0|^2}{2m} \end{pmatrix}\begin{pmatrix} C_n(\vec{k}) \\[1mm] C_n(\vec{k} - \vec{Q}_0) \end{pmatrix} = E_n(\vec{k})\begin{pmatrix} C_n(\vec{k}) \\[1mm] C_n(\vec{k} - \vec{Q}_0) \end{pmatrix},$$

$$(1.9.27)$$

where we have used the relation $U(-\vec{Q}_0) = U^*(\vec{Q}_0)$, which follows from the properties of the inverse Fourier transform when $U(\vec{r})$ is real. We thus have a two-dimensional eigenvalue problem. Setting $\vec{k} = \vec{Q}_0/2 - \Delta\vec{k}$, we find the eigenvalues

$$E(\vec{k}) = \frac{\hbar^2|\Delta\vec{k}|^2}{2m} + \frac{\hbar^2(Q_0/2)^2}{2m} \pm \sqrt{\left(\frac{\hbar^2|\vec{Q}_0 \cdot \Delta\vec{k}|}{2m}\right)^2 + |U(\vec{Q}_0)|^2}. \quad (1.9.28)$$

In other words, in the region near a zone boundary, an energy gap opens, just as in the Kronig–Penney model.

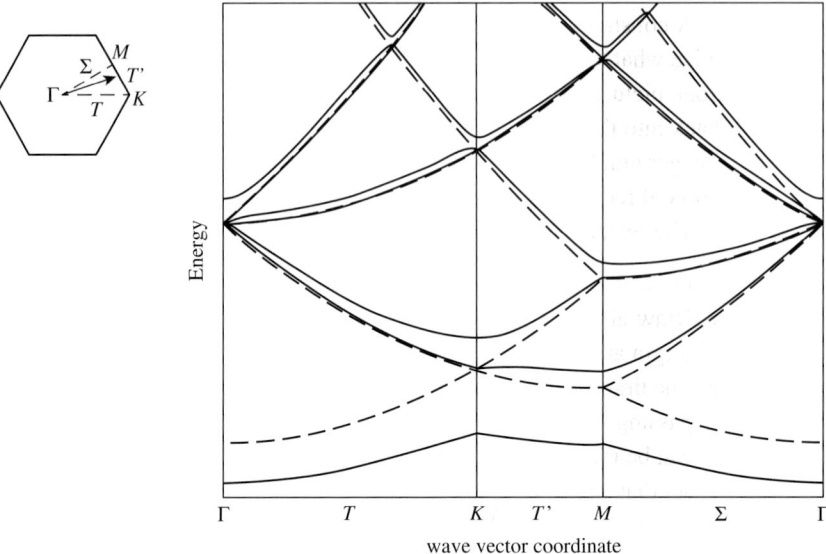

wave vector coordinate

Fig. 1.30 Solid lines: Energy bands of a hexagonal lattice in the nearly free electron approximation, found by diagonalizing a 19×19 matrix for $C_n(\vec{k})$ and the 18 nearest neighbors $C_n(\vec{k} - \vec{Q'})$ in the two-dimensional hexagonal reciprocal lattice. Dashed lines: free electron energy $E = \hbar^2 k^2 / 2m$ in the reduced zone of the same hexagonal lattice.

Figure 1.30 shows a numerical solution of the electron bands for a hexagonal crystal in the nearly free electron approximation. As seen in this figure, the energy of the bands tracks the free-particle energy $E(\vec{k}) = \hbar^2 k^2 / 2m$.

This model helps us to see what would happen to the bands if we used the wrong unit cell. In Section 1.4, we saw that if we decided to view a lattice with cell size a as a lattice with cell size $2a$ and a two-atom basis, the structure factor would ensure that no extra peaks would be predicted. It should also be the case that if we chose a double-size unit cell for computing the band structure, we should still get the same answer.

In (1.9.28), we see that the gap energy depends on the Fourier component of the potential that corresponds to the reciprocal lattice vector $\vec{Q}_0$. If we chose the wrong lattice spacing, $2a$, and a two-atom basis, then this would imply a zone boundary at $Q = \pi/2a$, which might lead us to expect a gap there. This would only occur if there is a Fourier component $U(\vec{Q}/2)$, however. But the periodic functions only have Fourier components corresponding to the reciprocal lattice vectors. Therefore, although we could draw the energy bands in a reduced zone with boundary $\vec{Q}/2$, there would be no gap in the bands at that zone boundary.

Higher-order Brillouin zones. As we have seen, in the nearly free electron approximation, the electron bands are nearly equal to the energy of a free electron in vacuum. When we plot this free electron energy in the reduced-zone scheme, we get a series of higher-energy bands. In three dimensions, the shape of these higher-energy bands can be very complicated.

A construction in terms of additional, higher-order Brillouin zones can help us to visualize what these higher-energy bands will look like. In essence, the higher-order Brillouin zone picture is just based on the fact that any part of the reciprocal lattice can be mapped back into the first Brillouin zone in the reduced-zone scheme, by subtracting or adding an integer number of reciprocal lattice vectors. The higher-order Brillouin zones tell us which parts of reciprocal space are mapped to which parts of the first Brillouin zone.

The recipe for drawing the higher-order Brillouin zones is as follows:

- Draw the reciprocal lattice.
- Draw all the Bragg planes, which consist of all planes which bisect a line between the origin and any other reciprocal lattice point.
- The first Brillouin zone is the set of all points that can be reached from the origin without passing through any Bragg planes. The second Brillouin zone is the set of all points that can be reached by passing through only one Bragg plane, the third Brillouin zone is the set that can be reached by passing through two Bragg planes, etc.
- The parts of a higher-order zone that lie outside the first zone can be mapped back into the first Brillouin zone in the reduced-zone scheme.

It can be proven that mapping each part of a higher-order Brillouin zone to the first Brillouin zone in the reduced-zone scheme will always completely fill up the first Brillouin zone; in other words, each higher-order zone is also a valid primitive cell for the reciprocal lattice.

In the case of nearly free electrons, the energy bands are just a perturbation of the free-electron energy surface. Therefore, the higher-energy bands will be a close approximation of the free-electron energy in the reduced-zone scheme. Figure 1.31 shows an example of how the high-order Brillouin zones generate the high-energy bands in the nearly free electron approximation, for a two-dimensional hexagonal lattice. As seen in this figure, the higher-energy bands have **folds** which arise from the shape of the original free-electron energy band.

Exercise 1.9.5   Construct the first four Brillouin zones for a two-dimensional square lattice, and show that mapping each part of the higher-order zone into the first Brillouin zone fills up the entire first zone.

1.9.4 $k \cdot p$ Theory

Another approximation method, which is very useful for understanding interactions between bands, uses a perturbation expansion of a different type. This method takes note of the fact that the critical points of the Brillouin zone have well-defined properties. If the energies at these critical points are known, then we can treat the band energy at a nearby point in the Brillouin zone as the sum of the energy at the critical point plus a small perturbation.

We begin by writing the Schrödinger equation in terms of the Bloch functions,

$$\left(\frac{p^2}{2m} + U(\vec{r}) \right) u_{n\vec{k}}(\vec{r})e^{i\vec{k}\cdot\vec{r}} = E_n(\vec{k}) u_{n\vec{k}}(\vec{r})e^{i\vec{k}\cdot\vec{r}}, \tag{1.9.29}$$

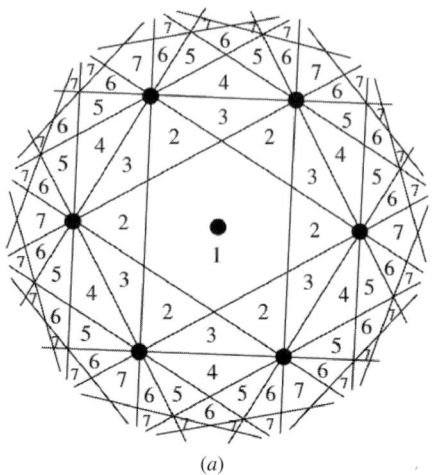

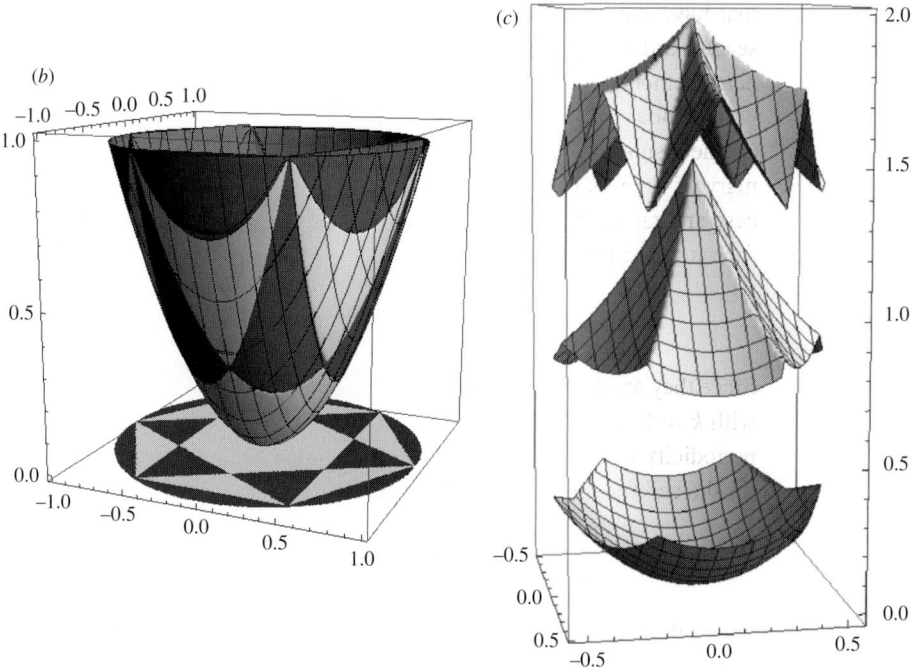

Fig. 1.31 (a) Brillouin zones for a two-dimensional hexagonal lattice. (b) Energy paraboloid for free electrons ($E = \hbar^2 k^2/2m$) above the k-plane for the two-dimensional hexagonal lattice. Regions of the free-electron energy surface falling in different Brillouin zones are shaded differently. (c) The first three zones of the energy paraboloid of (b) reduced to the first Brillouin zone. The bands have been separated in energy by an arbitrary constant.

where $\vec{p} = -i\hbar\nabla$. Since the derivative of $e^{i\vec{k}\cdot\vec{r}}$ is known, we can rewrite this as

$$\left(\frac{1}{2m}|\vec{p} + \hbar\vec{k}|^2 + U(\vec{r})\right) u_{n\vec{k}}(\vec{r}) = E_n(\vec{k})u_{n\vec{k}}(\vec{r}). \qquad (1.9.30)$$

We can then write this as the sum of three terms,

$$(H_0 + H_1 + H_2)u_{n\vec{k}}(\vec{r}) = E_n(\vec{k})u_{n\vec{k}}(\vec{r}), \qquad (1.9.31)$$

where

$$H_0 = \frac{p^2}{2m} + U(\vec{r})$$

$$H_1 = \frac{\hbar}{m}\vec{k}\cdot\vec{p}$$

$$H_2 = \frac{\hbar^2 k^2}{2m}. \qquad (1.9.32)$$

We have therefore reduced the problem to one which involves only the cell functions $u_{n\vec{k}}(\vec{r})$ that have the periodicity of the lattice. For small k, the terms H_1 and H_2 are first- and second-order perturbations on the Hamiltonian H_0, respectively. We can then use the standard methods of quantum mechanical perturbation theory to find the energies for small k.

In standard time-independent perturbation theory, the perturbed eigenstates are written as linear superpositions of the eigenstates of the unperturbed Hamiltonian H_0. (For a summary of time-independent perturbation theory, see Appendix E.) These eigenstates in our case are the $u_{n0}(\vec{r})$ functions for all n. This follows because we know that the Bloch wave functions are the eigenstates of the Hamiltonian, and if we set $k = 0$ in (1.9.29), we have

$$\left(\frac{p^2}{2m} + U(\vec{r})\right) u_{n0}(\vec{r}) = E_n(0)u_{n0}(\vec{r}). \qquad (1.9.33)$$

Since they are the eigenstates of the Hamiltonian in the restricted Hilbert space of functions with $k = 0$, the $u_{n0}(\vec{r})$ functions comprise an orthonormal basis for any function with the periodicity of the lattice, including any $u_{n\vec{k}}(\vec{r})$ function. To get the orthogonality relation for the cell functions, we use the definition (1.3.3) in (1.6.9) to obtain

$$\int d^3r\, \psi^*_{n0}(\vec{r})\psi_{m0}(\vec{r}) = \frac{1}{V}\sum_{\vec{R}}\int_{\text{cell}} d^3r\, u^*_{n0}(\vec{r})u_{m0}(\vec{r}) = \delta_{nm}, \qquad (1.9.34)$$

where the integration is over a single Bravais lattice unit cell. Since the sum over lattice vectors just gives us N integrations of the same thing, we therefore have

$$\frac{1}{V_{\text{cell}}}\int_{\text{cell}} d^3r\, u^*_{n0}(\vec{r})u_{m0}(\vec{r}) = \delta_{nm}, \qquad (1.9.35)$$

where $V = NV_{\text{cell}}$.

Perturbation theory for nondegenerate bands therefore gives us, to first order in k, the adjusted cell function

$$|u_{n\vec{k}}\rangle = |u_{n0}\rangle + \frac{\hbar}{m}\sum_{m\neq n}\frac{\vec{k}\cdot\langle u_{m0}|\vec{p}|u_{n0}\rangle}{E_n(0) - E_m(0)}|u_{m0}\rangle, \qquad (1.9.36)$$

and the energy, to first order in k,

$$E_n(\vec{k}) = E_n(0) + \frac{\hbar}{m}\vec{k} \cdot \langle u_{n0}|\vec{p}|u_{n0}\rangle. \tag{1.9.37}$$

Recall that $\vec{p}$ is an operator that acts on a quantum state, while $\vec{k}$ is simply a number. The matrix element is given by

$$\vec{p}_{nm} \equiv \langle u_{n0}|\vec{p}|u_{m0}\rangle = -i\hbar\frac{1}{V_{\text{cell}}}\int_{\text{cell}} d^3r\, u^*_{n\vec{k}}(\vec{r})\nabla u_{m\vec{k}}(\vec{r}). \tag{1.9.38}$$

The linear term in $E_n(\vec{k})$ in (1.9.37) will vanish because, as deduced in Section 1.6, the slope $\partial E/\partial k = 0$ at $\vec{k} = 0$; the bands have a maximum or minimum there. (In the case of degenerate bands, however, the first-order term may be important, as seen in Exercise 1.9.7.) We then go to second order in the perturbation expansion. To second order in k, we obtain

$$E_n(\vec{k}) = E_n(0) + \frac{\hbar^2 k^2}{2m} + \frac{\hbar^2}{m^2}\sum_{m\neq n}\frac{|\vec{k} \cdot \langle u_{m0}|\vec{p}|u_{n0}\rangle|^2}{E_n(0) - E_m(0)}. \tag{1.9.39}$$

The denominator $E_n(0) - E_m(0)$ in (1.9.39) gives the effect of **level repulsion**, which is very important in understanding band structure. As illustrated in Figure 1.32, if band m lies just above band n, then for k different from $\vec{k} = 0$, the second-order correction of the energy $E_n(\vec{k})$ will become large and negative; at the same time, the second-order correction to $E_m(\vec{k})$ at finite $\vec{k}$ will become large and positive. The energy denominator also implies that bands very far away in energy will have little effect on each other; in other words, if the atomic states from which the bands arise are far apart, they are unlikely to have strong effects on each other. This validates the assumption made in the nearly free electron approximation that core electron states should not have much effect on nearly free electron states at much higher energies.

The result (1.9.39) can also be written as

$$E_n(\vec{k}) = E_n(0) + \sum_{ij}\frac{\hbar^2}{2}M_{ij}^{-1}k_i k_j, \tag{1.9.40}$$

where

$$M_{ij}^{-1} = \frac{\delta_{ij}}{m} + \frac{2}{m^2}\sum_{m\neq n}\frac{\langle u_{n0}|p_i|u_{m0}\rangle\langle u_{m0}|p_j|u_{n0}\rangle}{E_n(0) - E_m(0)}. \tag{1.9.41}$$

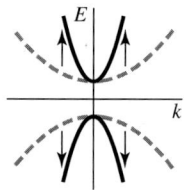

Fig. 1.32 Band repulsion in $k \cdot p$ theory. The two unperturbed bands (dashed lines) are repelled by each other away from $k = 0$ in the higher-order terms of the perturbation calculation, to give bands with lighter effective mass (solid lines).

In the case of isotropic bands, this becomes

$$E_n(\vec{k}) = E_n(0) + \frac{\hbar^2 k^2}{2m_{\text{eff}}} \tag{1.9.42}$$

where

$$\frac{1}{m_{\text{eff}}} = \frac{1}{m} + \frac{2}{m^2} \sum_m \frac{|p_{nm}|^2}{E_n(0) - E_m(0)}. \tag{1.9.43}$$

The k^2 dependence of the band depends on the **effective mass**, m_{eff}, which can be quite different from the free electron mass, m. In general, the more level repulsion occurs, the more the effective mass is changed from the bare electron mass. This is often also called the **renormalized** mass.

As discussed in Section 1.6, the full momentum of an electron moving in a crystal is given by $\langle \psi_{\vec{k}} | \vec{p} | \psi_{\vec{k}} \rangle$, where $\psi_{\vec{k}}$ is the full Bloch wave function. It is easy to show from (1.9.36) and (1.9.39) that the momentum carried by an electron in a Bloch state is

$$\langle \psi_{\vec{k}} | \vec{p} | \psi_{\vec{k}} \rangle = \frac{m}{\hbar} \nabla_{\vec{k}} E, \tag{1.9.44}$$

which is not equal to $\hbar \vec{k}$ unless the effective mass is the same as the free electron mass. The quantity $\hbar \vec{k}$ plays the role of an effective momentum, sometimes called the "crystal momentum."

The above analysis can also be used to expand the Bloch functions around other critical points in the Brillouin zone besides $\vec{k} = 0$. As we will see in Chapter 6, $k \cdot p$ theory has particular power when used along with group theory, because group theory allows us to generate selection rules for the critical points that have known symmetries. This can often be used to eliminate matrix elements from the calculation, showing which bands have an effect on each other and which do not.

The matrix element p_{nm} is intimately connected to the **optical transition rates.** As we will discuss in Section 5.2, the interaction Hamiltonian for radiative transitions is given by

$$H_{\text{int}} = -\frac{e}{m} \vec{A} \cdot \vec{p}. \tag{1.9.45}$$

The first-order optical transition rate between bands n and m is proportional to the matrix element

$$|\langle u_{m0} | \left(-\frac{e}{m} \vec{A} \cdot \vec{p} \right) | u_{n0} \rangle|^2 = \left(\frac{eA}{m} \right)^2 |\hat{\eta} \cdot \langle u_{m0} | \vec{p} | u_{n0} \rangle|^2, \tag{1.9.46}$$

where $\hat{\eta}$ is the polarization vector. In the isotropic case, we can write this simply as

$$\left(\frac{eA}{m} \right)^2 |p_{nm}|^2.$$

The unitless parameter

$$f_{nm} = \frac{2|p_{nm}|^2}{m(E_n - E_m)}, \tag{1.9.47}$$

which enters into both $k \cdot p$ theory and the optical transition rate, is called the **oscillator strength**, which is usually of the order of unity. As discussed in Section 7.4, the electron bands make up an oscillator which interacts with electromagnetic waves in the crystal. Measurement of the optical absorption spectrum of a material therefore tells us about its band structure.

Exercise 1.9.6   Prove the formula (1.9.44) for the electron momentum in the case of a band described by the $k \cdot p$ approximations (1.9.36) and (1.9.39).

Exercise 1.9.7   In the case of degenerate, isotropic bands in one dimension, Löwdin second-order perturbation theory (see Appendix E) says that the energies of the bands are given by the eigenvalues of the following matrix:

$$H_{ij}(k) = E_n(0)\delta_{ij} + \frac{\hbar k p_{ij}}{m} + \frac{\hbar^2 k^2}{2m}\delta_{ij} + \frac{\hbar^2 k^2}{m^2}\sum_l \frac{p_{il}p_{lj}}{E_n(0) - E_l(0)}, \qquad (1.9.48)$$

where i and j run over the range of degenerate states with energy E_n, and l is summed over all other states not degenerate with these. The matrix element $p_{ii} = \langle i|p|i\rangle = 0$, but $p_{ij} = \langle i|p|j\rangle$ can be nonzero for crystals without inversion symmetry.

Suppose that a band arises from two degenerate atomic states, and the matrix element $p_{12} = p_{21}$ is nonzero. Ignore the coupling to other states (e.g., assume the energies of other states are far from E_n). Find the 2×2 matrix H_{ij} for this system. Use Mathematica or a similar program to plot the eigenvalues of this matrix as functions of k. Set $E_1 = E_2 = 0$, $p_{12} = 0.5$, and $\hbar = m = 1$.

Does this result violate the theorem deduced in Section 1.6, that the first derivative of the energy bands vanishes at the Brillouin zone center? To explore this, suppose that there is a small interaction between the bands; in particular, set $E_{11}(0) = \Delta E = -E_{22}(0)$, where $\Delta E = 0.01$ or some other small number. Plot the eigenvalues of this matrix. What happens where the bands crossed? What is the slope of the bands there? What will happen in the limit $\Delta E \to 0$?

This type of band structure is called a **camelback** structure and is common in crystals without inversion symmetry.

1.9.5 Other Methods of Calculating Band Structure

The accurate calculation of band structures is still a frontier topic of research, and beyond the scope of this book. Several methods have been developed over the years, with increasing levels of accuracy.

As mentioned at the beginning of this section, the energy of the electrons depends not only on the charge of nuclei, but on the Coulomb interactions of all the electrons with each other. The wave function of the electrons determines the location of the electron charge, but the location of the electron charge determines the wave function of the electrons. For a correct solution, the electron wave functions and the potential energy due to the electron charge must be self-consistent. This is fundamentally a many-body physics problem; we will discuss many-body calculations in Chapter 8.

We have already seen in Section 1.6 that the Bloch states of different bands are orthogonal. Since the core electrons are nearly the same as the atomic states, which have slow variation near the atomic nucleus, this means that the electron wave functions for higher levels will tend to have strong spatial oscillations near a nucleus, so that the overlap integral $\int \psi_n^* \psi_m d^3r$ will vanish. This leads to problems for numerical calculations.

One way to solve for the higher band states without using rapidly oscillating wave functions is the **pseudopotential** method. In this method, instead of using just the potential $U(\vec{r})$ of the bare nucleus, a new $U(\vec{r})$ is used which includes the effects of the Coulomb repulsion and Pauli exclusion of the core electrons, to repel the electrons in higher states from the core region.

Using this new $U(\vec{r})$, the upper electron states can be calculated using the nearly free electron approximation; the inner, core electron states are assumed to remain nearly the same as the atomic core states. This strong distinction between the two types of states is one of the major assumptions of this method.

There is no exact way of calculating the potential $U(\vec{r})$; in this method one simply starts with a guess and then improves $U(\vec{r})$ by iteration. This can be done either by comparing the calculated band structure to experimental data or by adjusting $U(\vec{r})$ to give self-consistency. Once the valence electron states are calculated, the local charge density due to these electrons can be calculated, which is proportional to $\rho(\vec{r}) = \psi^*(\vec{r})\psi(\vec{r})$. The Coulomb repulsion from this charge density then gives an adjustment to $U(\vec{r})$. Eventually, the adjusted $U(\vec{r})$ will not change upon iteration, when it is consistent with the charge density of the valence states.

The band structure of silicon in Figure 1.26(a) was calculated using a pseudopotential method. Notice how the bands have the character of nearly free electrons – for example, the lowest energy band is nearly parabolic and the next energy band has a maximum at zone center, as in Figure 1.30. In general, pseudopotential methods give reasonable predictions of many band structure parameters, but still require some experimental input for realistic calculations.

To be even more accurate, not only the Coulomb energy but also the **exchange** and **correlation** energy of the electrons should be accounted for self-consistently. These terms come from many-body interactions of the electrons, which we will study in Chapter 8. It can be shown that these effects also depend only on the local density $\rho(\vec{r})$. **Density-functional theory** (DFT), discussed in Section 8.12, uses a variational method like pseudopotential theory to guess a solution and then iterate for self-consistency. Modern computational power has made very accurate calculations of band structure possible by means of DFT. In general, methods that aim to predict the band structure using only the atomic number and mass of the atoms are known as **ab initio** calculations. For a review of band-structure calculation methods, see Bernhol (1999).

1.10 Angle-Resolved Photoemission Spectroscopy

One of the most powerful tools for determining the band structure of a material is the **photoemission** process, by which an incoming photon kicks an electron out of the solid.

In vacuum, the electron will travel ballistically with the momentum and energy it had when it left the material. A current of electrons ejected in this way from the material can then be analyzed for their direction of motion and kinetic energy. This measurement is known as **angle-resolved photoemission spectroscopy** (ARPES).

Typically, the momentum of the photon is negligible compared to the momentum of the electron. The absorption of the photon can therefore be viewed as a "vertical" process, in which the electron moves to higher energy while staying at nearly the same k-vector. The high-energy electron can then have enough energy to overcome the work function of the material and leave the crystal.

In thinking of the process by which the electron leaves the solid, the question immediately arises of what conservation rules to apply. We have already seen that $\hbar k$ is not the true momentum of an electron; this is given by (1.6.10),

$$\langle \psi_{\vec{k}} | \vec{p} | \psi_{\vec{k}} \rangle = \hbar \vec{k} - i\hbar \int d^3r \, u_{n\vec{k}}^* \nabla u_{n\vec{k}}. \tag{1.10.1}$$

When the electron crosses the boundary of the solid, do we conserve momentum, or do we conserve $\hbar k$? The answer is that we conserve $\hbar k$ in the direction parallel to the surface, not the total electron momentum. This can be understood as a consequence of the wave nature of the electrons, in analogy with Snell's law, which is discussed in detail in Chapter 3. We write $\vec{k} = \vec{k}_\parallel + k_\perp \hat{z}$, where $\vec{k}_\parallel$ is the wave vector component parallel to the surface and $k_\perp$ is the component perpendicular to the surface. The spacing of the wave fronts along a direction $\vec{x}$ on the surface is given by the condition $\vec{k} \cdot \vec{x} = k_\parallel x = 2\pi n$, where n is an integer. The distance between points of phase 2π is therefore $\Delta x = 2\pi / k_\parallel$. This spacing must be the same for the wave both inside and outside of the solid, a condition generally known as **phase matching**. Although $\vec{k}_\parallel$ is conserved, the total momentum of the electron is in general not conserved. Therefore, the crystal must recoil slightly, taking up the difference of the real momentum when the electron leaves.

We therefore have two rules, conservation of energy and conservation of the k-vector parallel to the surface. The direction of an ejected electron can be measured, yielding two angles, namely the angle θ relative to the normal to the surface, and an azimuthal angle ϕ which gives the direction parallel to the plane of the surface. The two components of $\vec{k}_\parallel$ are then given by

$$\vec{k}_\parallel = (k \sin\theta \cos\phi, k \sin\theta \sin\phi). \tag{1.10.2}$$

The magnitude of k of the electron in the vacuum outside the material can be known by measuring the kinetic energy E_{kin} of the ejected electrons, which yields

$$k = \sqrt{2mE_{kin}/\hbar^2}. \tag{1.10.3}$$

We therefore can know $\vec{k}_\parallel$ fully. For a two-dimensional material, the band energy E_n is only a function of $\vec{k}_\parallel$. We can then obtain the band structure of a two-dimensional material directly, knowing the photon energy $\hbar\omega$:

$$E_n(\vec{k}_\parallel) = E_{kin} - \hbar\omega. \tag{1.10.4}$$

ARPES is capable of absolute band mapping of two-dimensional electronic states for this reason.

Equation (1.10.4) gives the band energy relative to the vacuum energy, where $E_{kin} = 0$. In practice, it is often easier to determine the energy relative to the Fermi level of the electrons in the material, which can be defined by accurately measuring the electric potential difference between the sample and the analyzer. We will discuss the Fermi statistics of electrons in detail in Chapter 2.

For three-dimensional, bulk materials, we must make some additional assumptions in order to use ARPES to study band energies. In general, one must assume some model for the band structure and fit the data to this model. For example, we can make a simple model that a band near zone center has energy given by

$$E_n(\vec{k}) = \frac{\hbar^2 |\vec{k}|^2}{2m_{\text{eff}}} - E_0, \tag{1.10.5}$$

where m_{eff} is an effective mass (see Section 1.9.4) and E_0 gives the energy at zone center, which is negative relative to the energy of an electron at rest in vacuum. For a fixed value of $k_\parallel$, there will be a range of kinetic energies corresponding to different values of $k_\perp$. The number of electrons emitted with a given energy E_{kin} is proportional to

$$N(E_{kin}) \propto \int_{-\infty}^{\infty} dk_\perp \, \delta(E_{kin} - E_n - \hbar\omega). \tag{1.10.6}$$

To resolve the δ-function that enforces energy conservation, we must change the variable of integration to an energy. We write

$$E_n = \frac{\hbar^2 (k_\parallel^2 + k_\perp^2)}{2m_{\text{eff}}} - E_0, \tag{1.10.7}$$

which has the Jacobian

$$\frac{\partial E_n}{\partial k_\perp} = \frac{\hbar^2 k_\perp}{m_{\text{eff}}} = \frac{\hbar^2}{m_{\text{eff}}} \sqrt{2m_{\text{eff}}(E_n + E_0)/\hbar^2 - k_\parallel^2}. \tag{1.10.8}$$

We thus have

$$N(E_{kin}) \propto \int_{\hbar^2 k_\parallel^2 / 2m_{\text{eff}} - E_0}^{\infty} dE_n \, \frac{m_{\text{eff}}}{\hbar^2 \sqrt{2m_{\text{eff}}(E_n + E_0)/\hbar^2 - k_\parallel^2}} \delta(E_{kin} - E_n - \hbar\omega)$$

$$= \frac{m_{\text{eff}}}{\hbar^2 \sqrt{2m_{\text{eff}}(E_{kin} - \hbar\omega + E_0)/\hbar^2 - k_\parallel^2}} \Theta(E_{kin} - \hbar\omega + E_0 - \hbar^2 k_\parallel^2 / 2m_{\text{eff}}), \tag{1.10.9}$$

where $\Theta(E)$ is the Heaviside function (see Appendix C).

Recalling that $k_\parallel^2 = k^2 \sin^2 \theta$, we can write

$$N(E_{kin}) \propto \frac{1}{\sqrt{E_{kin}(1 - (m/m_{\text{eff}}) \sin^2 \theta) - \hbar\omega + E_0}}$$
$$\times \Theta(E_{kin}(1 - (m/m_{\text{eff}}) \sin^2 \theta) - \hbar\omega + E_0).$$

This is a peaked function with a maximum at $E_{kin} = (\hbar\omega - E_0)/(1 - (m/m_{\text{eff}}) \sin^2 \theta)$. The effective mass m_{eff} can be found by fitting this function to the angle-resolved ARPES data,

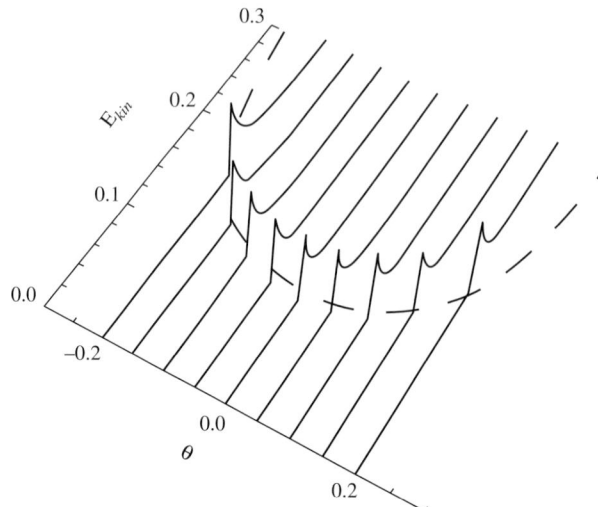

Fig. 1.33 ARPES emission spectra as a function of angle, for the simple effective-mass model given in the text, for $m/m_{\mathrm{eff}} = 10$ and $\hbar\omega - E_0 = 0.1$. The dashed line follows the curve $E_{kin} = (\hbar\omega - E_0)/(1 - (m/m_{\mathrm{eff}})\sin^2\theta)$.

as shown in Figure 1.33. Similar to the van Hove singularities discussed in Section 1.8, the infinity in this equation does not cause a problem because it is integrable.

Complicating effects. The above discussion gives the general way that ARPES can be used to deduce band structure. However, there are several additional effects that must be taken into account in analyzing real data.

- *Internal scattering and surface reflection.* Photoemission can be approximately described as a three-step process. In the first step, electrons in bands are excited by photon absorption into nearly free electron states with the same mass as electrons outside the material. In the second step, these electrons propagate to the surface. In the third step, the electrons are transmitted through the surface to the vacuum outside.

   Breaking photoemission down into these steps allows us to account for ways in which electrons can be lost before they reach the outside. First, electrons propagating toward the surface may scatter with imperfections in the lattice, changing their energy and momentum. The scattering rate for these processes may depend on the energy of the electrons (as discussed in Chapter 5.) The scattering rates for electrons as a function of energy have been measured and tabulated for many materials, and are often presented as a "universal curve." Second, the electrons may be reflected from the surface, and the probability of reflection may also depend on the electron energy.

- *Line broadening.* The energy spectra of the optical transitions can be broadened by many-body effects, as discussed in Section 8.4. A simple way to see why this broadening occurs is to recall that the uncertainty principle of quantum mechanics does not allow the energy of a state to be defined more accurately than $\hbar/\tau$, where τ is the time spent by an electron in the state. Interactions with other electrons (or with phonons, discussed in Chapter 4) can reduce the time spent in a state.

- *Space charge effects.* If the current of photoemitted electrons is large, they will repel each other, leading to blurring of the angle-resolved and energy-resolved data. This effect can be mitigated experimentally by reducing the fluence of photons, so that the photoemission current is sufficiently low that the Coulomb interaction among photoelectrons is negligible.
- *Fermi level.* Clearly, only states in bands that are occupied by electrons can contribute to the photoemission. ARPES measurements therefore only work for states below the Fermi level of the electrons. As discussed in Chapter 2, when the temperature increases, the Fermi level is smeared out over a range of order $E_F \pm k_B T$.

Exercise 1.10.1   Suppose that a band energy is given by (see Section 1.9.2)

$$E_n(\vec{k}) = -E_0 + U_{12}(\cos k_x a + \cos k_y a + \cos k_z a). \qquad (1.10.10)$$

Determine the spectral line shape $N(E_{kin})$ for angle-resolved photoemission from this band, following the approach for the parabolic band discussed in this section.

1.11 Why Are Bands Often Completely Full or Empty? Bands and Molecular Bonds

One often finds that electron bands in solids are either *full* or *empty*. Why?

As discussed in Section 1.1, in a solid with N atoms, there will be N band states for each atomic state. If a single atomic orbital gives rise to a single band, then there will be $2N$ states in that band, if we count both spin states of the orbital. If that atomic orbital is occupied by two electrons, then the N atoms will contribute $2N$ electrons which will exactly fill the $2N$ states in the band. If the atomic orbital is not occupied, then the band will be completely empty. This tells us already why it is generally no accident that a band can be completely full or empty. The number of states and the number of electrons are directly related.

We can gain insight into the nature of electron bands by looking closely at the nature of molecular bonds in the LCAO approximation, which, as discussed in Section 1.9.2, is used in the tight-binding approximation. When the overlap of orbitals is very strong, the LCAO approximation will break down, but in many cases it is quite good.

1.11.1 Molecular Bonds

In general, electronic bands and molecular bonds are intimately related. Harrison (1980) gives an excellent discussion of the relation of chemical bonds and bands in detail.

As discussed in Section 1.1.2, when two electron orbitals overlap, two new states are created that can be approximated as symmetric and antisymmetric linear combinations of the original orbitals. As mentioned in Section 1.1.2, these are called the bonding and antibonding states, respectively.

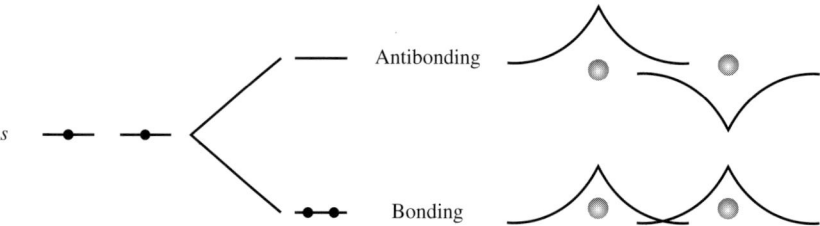

Fig. 1.34 Bonding of two atoms with partially filled s-orbitals; for example, two hydrogen atoms or two atoms from Group I of the periodic table.

Figure 1.34 shows how bonding occurs in the case of two atoms with overlapping s-orbitals, in the LCAO approximation. If there is one electron per atom, then the electrons from both atoms can fall into the lowest state, thus reducing the total energy of the pair of atoms. This is why we say they are **bonded**. Separating the two atoms would increase the total energy of the system, that is, would require work.

If the two s-orbitals were filled, in other words, if each original orbital had two electrons, then the energy splitting due to the wave function overlap would not lead to bonding. Two of the electrons would fall into the lower, bonding state, while the remaining two would have to go into the higher, antibonding state, because of the Pauli exclusion principle. Since, according to (1.1.12), the average energy of the two states remains the same, there is no decrease of the total energy of the atoms. This is why, for example, helium atoms do not form homo atomic molecules.

Figure 1.35 shows the case of two atoms with partially filled, overlapping s- and p-orbitals. Without knowing the exact location of the atoms, we cannot say how the orbitals will split when the atomic states overlap, but we can say that, in general, there will be an equal number of states shifting upward and downward by the same amounts. This follows from the general mathematical theorem that the sum of the eigenvalues of a matrix is equal to the trace of the matrix, no matter how large the off-diagonal elements are. In the LCAO approximation, we construct a square matrix as in (1.1.11), in which the diagonal elements are the unperturbed atomic state energies, and the off-diagonal elements are the coupling integrals. If these are nonzero, the energy eigenvalues will shift, but the sum of all the shifted energies will remain the same. This means that if some states are shifted to lower energy, other states must shift upward by the same amount.

If the total number of electrons in the atomic states is less than half of the total number of states, then all of the electrons can lower their energy when the atoms get near to each other. This leads to stable bonding. As illustrated in Figure 1.35, if the total number of electrons in the two types of orbitals is equal to eight, then the lower, bonding orbitals will be completely full and the upper, antibonding orbitals will be completely empty. This is known as a "full shell" in chemistry terminology.

If the total number of electrons in these states is greater than eight, the atoms can still bond, but the energy of bonding will be much less, because one or more of the electrons must occupy an upper, antibonding orbital; this extra energy used will partially cancel out the energy gained by the electrons going into a lower, bonding orbital. If there are fewer

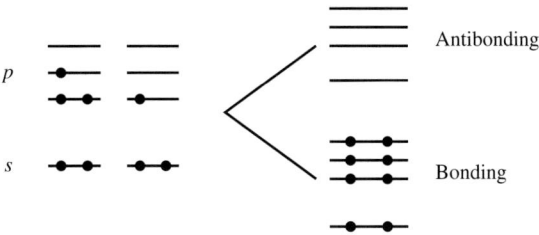

Fig. 1.35 Bonding of two atoms with partially filled s- and p-orbitals.

than eight electrons, the bonding energy will be less, because there are fewer electrons that can fall into the lower-energy bonding orbitals.

Exercise 1.11.1 Suppose that two identical, neighboring atoms have substantial overlap of one s-orbital and one p-orbital (we can assume, for example, that the p-orbitals in the x-direction overlap, and that the other two p-orbitals of the two atoms, pointing in the y- and z-directions, have very little overlap).

(a) In a mathematical program like Mathematica, construct a 4×4 matrix of the LCAO matrix elements:

$$E_s = \langle \psi_{1s} | H | \psi_{1s} \rangle = \langle \psi_{2s} | H | \psi_{2s} \rangle, \; U_{ss} = \langle \psi_{1s} | H | \psi_{2s} \rangle$$
$$E_p = \langle \psi_{1p} | H | \psi_{1p} \rangle = \langle \psi_{2p} | H | \psi_{2p} \rangle, \; U_{sp} = \langle \psi_{1s} | H | \psi_{2p} \rangle,$$
$$U_{pp} = \langle \psi_{1p} | H | \psi_{2p} \rangle, \quad \text{etc.} \tag{1.11.1}$$

Note that the atomic s- and p-orbitals of a single atom are orthogonal. Assume that the coupling integrals are all real.

(b) Let $E_s = 1$, $E_p = 1.3$, $U_{pp} = 3U_{ss}$, and $U_{sp} = U_{ss}$. Plot the eigenvalues of the matrix as a function of U_{ss} for U_{ss} in the range 0–1. Will this system allow bound states?

(c) Solve for the eigenvectors at $U_{ss} = 0.6$. What is the character of the states – which are mostly p-like? Which are mostly s-like? Which are bonding? Which are antibonding?

(d) What happens to the character of the bonds if you change the relative weights of the coupling terms U_{ss}, U_{sp}, and U_{pp}? Try plotting the eigenvalues vs U_{ss} for some different values of the ratios.

Exercise 1.11.2 Using the arguments of this section, can you give a reasonable explanation why the elements F and Cl are gases, that is, why a molecule of two F or Cl atoms would be very weakly bound? In particular, without knowing details of the molecular states, why would you expect F_2 to have about half the binding energy of O_2? (The actual ratio is about one-third).

It would seem that in the LCAO approximation there is no reason for noble gases like Ne and Ar to ever form molecules or solids. Solids of these materials are possible, however. They are bound by the much weaker **van der Waals** force.

1.11.2 Classes of Electronic Structure

We now move from single molecular bonds to bands in solids. As discussed above, if there are N pairs of atoms, and each pair has eight electrons in bonding states and no electrons in the antibonding states, then the energy bands arising from these orbitals will have $8N$ states and $8N$ electrons in the lower energy band (in other words, a completely full band), and $8N$ the states in the upper band with no electrons (a completely empty band).

This situation is so common that a standard terminology has been defined. The lower, full band is typically called the **valence** band, while the upper, empty band is called the **conduction** band. If this is the case, then the material cannot conduct electricity at low temperature. The state of the system can only change if some electrons can change their states. According to the Pauli exclusion principle, if an electron changes state, it must go into an unoccupied state. If the energy gap between the valence and conduction bands is small enough, electrons could change state by jumping up to the empty conduction band, but at low temperature the electrons in the full band have no nearby empty states into which they can move. The valence band will remain unchanged, with no acceleration of the electrons and no current flow.

This leads to the following definitions: An **insulator** is a material with a full valence band and empty conduction band, with an energy gap of the order of 3 eV or greater. A **semiconductor** is a material that has a band gap from around 0.3 eV to 3 eV. In both cases, the gap energy is large compared to $k_B T$ at room temperature (about 0.026 eV), so that no current flows at room temperature in pure materials. One can also talk of wide band-gap semiconductors, with gaps that lie between those of regular semiconductors and insulators, and narrow band-gap semiconductors, with gaps comparable to or smaller than $k_B T$ at room temperature. There are no sharp distinctions between these definitions. The distinction between a wide band-gap semiconductor and an insulator, in particular, is somewhat arbitrary.

The situation of having eight electrons shared by eight bonding orbitals is favored by combinations of elements from Group I of the periodic table binding with elements from Group VII, elements from Group II binding with elements from Group VI, elements from Group III binding with elements from Group V, and elements from Group IV binding with themselves. This is why one often hears of II–VI or III–V semiconductors, and why silicon, germanium, and silicon carbide (in Group IV) are also semiconductors. Bonds between elements from Groups I and VII are typically called **ionic** bonds, while II–VI, III–V, and IV–IV bonds are called **covalent**. In general, as with the distinction between insulators and semiconductors, the distinction between these different types of bonds is somewhat vague. Chemists speak of the **degree of ionicity** in a bond.

The typical energies of bands and band gaps in solids have the same order of magnitude as the typical energies of atomic orbitals. The Rydberg energy of a hydrogen atom is Ry = 13.6 eV; this leads to the following energies of the higher levels:

$$n = 2: \quad \frac{-13.6 \text{ eV}}{2^2} = -3.4 \text{ eV}$$

$$n = 3: \quad \frac{-13.6 \text{ eV}}{3^2} = -1.5 \text{ eV}. \tag{1.11.2}$$

This indicates that a typical band gap in a solid is a few eV. For nuclei with larger atomic number Z, the $n = 1$ shell has lower energy, since the energy for a single electron bound to a nucleus with Z protons is $E_1 = -Z^2 \text{Ry}$. For higher electron states, however, most of the extra charge of the nucleus is canceled out by the charge of the core electrons in low-energy states, so that the effective charge of the nucleus plus core electrons is just a few electronic charges. This will leave the transitions between the atomic-level states in the range of a few eV. Typical values of band gaps of semiconductors are 1–2 eV, although it is quite common to find wide-band gap-semiconductors and insulators with gaps of the order of 3–5 eV. Alloys with light elements such as C and N, in the second row of the periodic table, will tend to have larger gaps, since they involve transitions between $n = 2$ and $n = 3$ states. Spreading of the bands due to the overlap of orbitals, as we discussed in Section 1.1.3, will tend to reduce energy gaps.

In contrast to insulators and semiconductors, a true **metal**, or **conductor**, will always conduct electricity, because it has bands that are only partially filled with electrons even in the ground state. In this case, there is always an empty state arbitrarily close in energy to the occupied states. This can occur, for example, when there are seven or fewer electrons in the eight bonding states derived from s- and p-orbitals. The element Be, for example, is a metal even though it has a filled s-orbital, because the nearby p-states are empty. Materials with partially filled d-orbitals have the same property.

Partially filled bands which give metallic behavior can also arise even when there are eight electrons in the bands generated from s- and p-orbitals, if the bands are so close in energy that they cross at some points in the Brillouin zone. For example, tin is a metal even though it belongs to Group IV like silicon and germanium. In this case, the bands are close together because they are built from atomic orbitals of high n that lie near each other. **Semimetals** have a very small overlap of their bands, so that they have a very small density of electron states in the overlap region. Therefore, although they always conduct, they can have very poor conduction properties.

1.11.3 sp^3 Bonding

We have used the LCAO method several times in this chapter. The LCAO method assumes that we already know the location of the atoms; the coupling integrals are functions of the relative positions of the atoms. Predicting the exact location at which atoms will arrange themselves, however, is a difficult task; in general, there may be more than one stable crystal structure for the same set of atoms; these different structures, which may have different symmetries, are known as **polytypes**.[3] We can make some general statements about crystal bonds, however.

Bonds tend to repel each other for the simple reason that clouds of negatively charged electrons have a repulsive Coulomb force. Therefore, if one atom has two bonds to adjacent atoms, the bonds will likely be in a straight line. If the atom has three bonds, they will most

---

[3] This is the basis of Kurt Vonnegut's famous "Ice-9" in the novel *Cat's Cradle*.

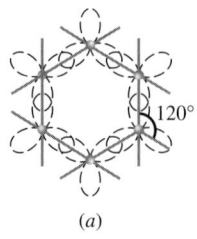 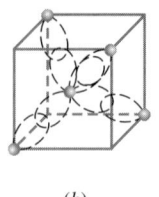

(a)   (b)

Fig. 1.36 (a) Threefold bonding leading to planes with hexagonal symmetry. (b) Fourfold bonding leading to tetragonal symmetry.

likely arrange themselves in a plane at $120°$ from each other. If the atom is bonded to four other atoms, the bonds will most likely arrange themselves to point to the four corners of a tetrahedron.

Threefold bonding naturally leads to crystals with hexagonal symmetry, as illustrated in Figure 1.36(a). Fourfold bonding naturally leads to crystals with tetragonal symmetry, as shown in Figure 1.36(b). The tetragonal bonds naturally fit into a cubic symmetry, because the four non-adjacent corners of a cube correspond to the corners of a tetragon.

Because it is the basis of bonding for many crystals and a good example of the relation of bands and bonds, we now look closely at example of tetrahedral bonding known as sp^3 bonding. We write four states, called sp^3 orbitals, as linear superpositions of s and p atomic orbitals. These are linear combinations of these orbitals as follows:

$$\psi_1 = \tfrac{1}{2}\left(\Phi_s + \Phi_x + \Phi_y + \Phi_z\right) \quad [111]$$

$$\psi_2 = \tfrac{1}{2}\left(\Phi_s + \Phi_x - \Phi_y - \Phi_z\right) \quad [1\bar{1}\bar{1}]$$

$$\psi_3 = \tfrac{1}{2}\left(\Phi_s - \Phi_x + \Phi_y - \Phi_z\right) \quad [\bar{1}1\bar{1}]$$

$$\psi_4 = \tfrac{1}{2}\left(\Phi_s - \Phi_x - \Phi_y + \Phi_z\right). \quad [\bar{1}\bar{1}1].$$

(1.11.3)

These point along the directions shown in the tetrahedrally bonded crystal of Figure 1.36(b). Similarly, we can write superpositions of an s-orbital and two p-orbitals, known as sp^2 bonds. In this case, these will form three bonds $120°$ apart in a plane, as shown in Figure 1.36(a), with a leftover p-orbital sticking out perpendicular to the plane, in the z-direction (also commonly known as the c-direction).

The unit cell in the sp^3-bonded tetragonal structure consists of two atoms connected by four sp^3 bonds. By our previous LCAO analysis, if the two atoms in the unit cell have a total of eight electrons in the s- and p-orbitals, then there will be eight electrons per unit cell to go in the eight bonding states per unit cell, and therefore the band that arises from these bonding states will also be full. For this reason, sp^3-bonded materials make up a broad class of semiconductors and insulators, including diamond (carbon), silicon, and all kinds of III–V and II–VI compounds.

Exercise 1.11.3   The unperturbed p-orbitals point along the following vectors:

$$\vec{v}_x = \{1, 0, 0\}$$

$$\vec{v}_y = \{0, 1, 0\}$$

$$\vec{v}_z = \{0, 0, 1\}. \tag{1.11.4}$$

We could imagine forming an orbital that points in the [111] direction by the linear combination $\Phi_x + \Phi_y + \Phi_z$, and [1$\bar{1}\bar{1}$] as $\Phi_x - \Phi_y - \Phi_z$, etc. Show that if we form four orbitals in the sp^3 directions given in (1.11.3), namely [111], [1$\bar{1}\bar{1}$], [$\bar{1}1\bar{1}$], and [$\bar{1}\bar{1}1$], from just superpositions of p-orbitals, the resulting orbitals are not linearly independent.

Exercise 1.11.4   Show that if the original atomic orbitals are orthonormal, then the linear combinations of (1.11.3) are also orthonormal.

Tight-binding model of sp^3 bands. We can model the bands of a crystal with sp^3 bonds using the tight-binding model introduced in Section 1.9.2. Using the general formula (1.9.20), we now account for eight orbitals, namely an s-orbital and three p-orbitals on each atom of the two-atom basis of the crystal. This will give us an 8×8 matrix which we must diagonalize.

It is easiest to write the four orbitals on each atom in terms of the four sp^3 states given above. This makes the coupling terms easy, as each sp^3 orbital can be assumed to have significant coupling only to the orbital pointing along the same axis on the neighboring atom. We have the complication, however, that although the sp^3 states are orthogonal, they are not eigenstates of the atomic Hamiltonian. Therefore, we must account for coupling between these states on a single atom. Assuming that the p-orbitals are degenerate with energy E_p, and that the s-orbitals have a different energy E_s, we have

$$\langle \psi_1 | H | \psi_1 \rangle = \frac{1}{4}\big((\langle\Phi_s| + \langle\Phi_x| + \langle\Phi_y| + \langle\Phi_z|)H(|\Phi_s\rangle + |\Phi_x\rangle + |\Phi_y\rangle + |\Phi_z\rangle)\big)$$

$$= \frac{1}{4}(E_s + 3E_p) \equiv \bar{E} \tag{1.11.5}$$

$$\langle \psi_1 | H | \psi_2 \rangle = \frac{1}{4}\big((\langle\Phi_s| + \langle\Phi_x| + \langle\Phi_y| + \langle\Phi_z|)H(|\Phi_s\rangle + |\Phi_x\rangle - |\Phi_y\rangle - |\Phi_z\rangle)\big)$$

$$= \frac{1}{4}(E_s - E_p) \equiv \delta. \tag{1.11.6}$$

The p-orbitals typically have higher energy than the s-orbitals, so δ is negative. All four of the sp^3 orbitals are degenerate with energy $\bar{E}$, and the coupling between any two sp^3 orbitals on the same atom is δ.

Ignoring the k-dependence (that is, considering only the Brillouin zone center), we therefore have the matrix equation

$$
\begin{pmatrix}
\bar{E} & \delta & \delta & \delta & U_{12} & 0 & 0 & 0 \\
\delta & \bar{E} & \delta & \delta & 0 & U_{12} & 0 & 0 \\
\delta & \delta & \bar{E} & \delta & 0 & 0 & U_{12} & 0 \\
\delta & \delta & \delta & \bar{E} & 0 & 0 & 0 & U_{12} \\
U_{12} & 0 & 0 & 0 & \bar{E} & \delta & \delta & \delta \\
0 & U_{12} & 0 & 0 & \delta & \bar{E} & \delta & \delta \\
0 & 0 & U_{12} & 0 & \delta & \delta & \bar{E} & \delta \\
0 & 0 & 0 & U_{12} & \delta & \delta & \delta & \bar{E}
\end{pmatrix}
\begin{pmatrix}
c_{11} \\
c_{21} \\
c_{31} \\
c_{41} \\
c_{12} \\
c_{22} \\
c_{32} \\
c_{42}
\end{pmatrix}
= E
\begin{pmatrix}
c_{11} \\
c_{21} \\
c_{31} \\
c_{41} \\
c_{12} \\
c_{22} \\
c_{32} \\
c_{42}
\end{pmatrix},
$$

$$(1.11.7)$$

where U_{12} is the coupling term between two overlapping sp^3 orbitals (which is typically negative, as discussed in Section 1.1.2), and the coefficients c_{ij} give the relative weight of the four sp^3 states for each of the two atoms in the unit cell.

Diagonalization gives the following eigenstates, written in terms of the original atomic orbitals:

eigenvalues	eigenstates
$\bar{E} - |U_{12}| - 3|\delta|$	$\Phi_s^{(1)} + \Phi_s^{(2)}$
$\bar{E} - |U_{12}| + |\delta|$	$\Phi_x^{(1)} + \Phi_x^{(2)}, \ \Phi_y^{(1)} + \Phi_y^{(2)}, \ \Phi_z^{(1)} + \Phi_z^{(2)}$
$\bar{E} + |U_{12}| - 3|\delta|$	$\Phi_s^{(1)} - \Phi_s^{(2)}$
$\bar{E} + |U_{12}| + |\delta|$	$\Phi_x^{(1)} - \Phi_x^{(2)}, \ \Phi_y^{(1)} - \Phi_y^{(2)}, \ \Phi_z^{(1)} - \Phi_z^{(2)}.$

$$(1.11.8)$$

We thus see that although the bonds are not made purely of either s-states and p-states, there will nevertheless be four electron bands corresponding to pure s- and p-states in bonding or antibonding combinations.

When $|U_{12}|$ is large compared to $|\delta|$, as is often the case in III–V semiconductors, there will be a gap between a full valence band arising from p-states and an empty conduction band arising from s-orbitals.

Exercise 1.11.5 Show that the eigenstates of (1.11.7) are those given in (1.11.8). To do this, you will need to first find the eigenvectors of (1.11.7) in terms of the eight sp^3 states, then use (1.11.3) to write these in terms of the original atomic states. Finally, show that the degenerate eigenstates involving p-orbitals can be written as linear combinations of the states given in (1.11.8).

1.11.4 Dangling Bonds and Defect States

In a perfectly periodic, infinite crystal, we can imagine that every atomic orbital is involved in a bond of the type discussed above. However, in any real crystal there will be some orbitals that are not. One reason is disorder, which always occurs at some level. We have already discussed the case of long-range disorder in Section 1.8.2. In the short-range limit, we can speak of **point defects** at single lattice sites in a crystal. These can can consist either of missing atoms in the lattice ("**vacancies**"), extra atoms where they should not

Table 1.2 Types of local defects	
impurity	atom that does not belong to the stoichiometry of the main crystal
vacancy	missing atom in a lattice
interstitial	an extra atom in between other atoms in a unit cell
dislocation	an atom that is moved away from its place in the unit cell

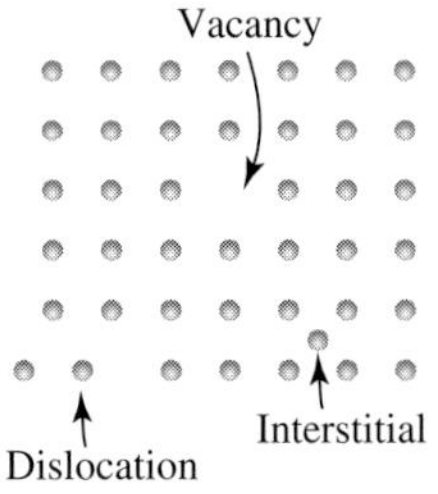

Fig. 1.37 Examples of defects in a solid.

be ("**interstitials**"), atoms of a different type, giving the wrong stoichiometry for a crystal ("**impurities**"), and shifts of one part of the lattice relative to another part ("**dislocations**"). Figure 1.37 illustrates some of these point defects, which are listed in Table 1.2. Defects and dislocations play a major role in many aspects of solid state physics, as we will see in the coming chapters.

Defect states tend to have "dangling bonds," that is, orbitals that do not substantially overlap with other atomic orbitals in the crystal. Because of this, there will be defect states with energies that fall inside the band gaps of the crystal. We can understand this by realizing that for an orbital with no overlap with a neighboring orbital, there will be no symmetric–antisymmetric energy splitting. Since the appearance of bands and band gaps is deeply connected to the overlap integrals that give the symmetric–antisymmetric splitting, orbitals with little or no overlap will look very much like the original atomic orbitals.

When there are just a few of these defects compared to the number of atoms in the whole crystal, these defect states will be mostly isolated from each other. Since they are localized to small regions, the defect states will have discrete energies, like the confined states in a square well. Thus, in addition to the Urbach tail discussed in Section 1.8.2 which describes long-range disorder, isolated defects can give sharp lines in the density of states corresponding to particular sets of defect states.

Defect states are closely related to **surface states**, which we will examine in Section 1.12. Like defects, atoms on the surface of a crystal have orbitals that stick out into space and do not overlap substantially with other atomic orbitals. This leads to surface states that fall within the energy gaps of the bulk crystal.

A further complication of surfaces is **surface reconstructions**. The atoms at a surface typically do not stay in the same relative positions as in the bulk; they can lower their energy by moving to new positions, giving new bonds and states that do not correspond to either the bulk states or the original atomic states. The physics of surfaces is an ongoing field of research; Oura *et al.* (2003) give a good overview of this topic.

1.12 Surface States

As discussed in Section 1.3, Bloch's theorem is based on the assumption of invariance under a given set of translations; that is, it assumes that the properties of the system are the same if we observe a location that is moved from the present point by any translation that belongs to the lattice. But every real crystal is finite; there are boundaries on the outsides. In Section 1.7, we looked at two ways to treat a finite crystal: either to assume fictional, periodic (Born–von Karman) boundary conditions, so that our imaginary crystal effectively has no surfaces, or to create standing waves with nodes at the surfaces of the crystal, as the sum of two Bloch waves with k and $-k$ in opposite directions.

Kronig–Penney model of surface states. There is another way to satisfy the boundary conditions at the surfaces. Let us return to the Kronig–Penney model we looked at in Section 1.2. The solutions were found to satisfy (1.2.5), that is,

$$\frac{(\kappa^2 - K^2)}{2\kappa K} \sinh(\kappa b) \sin(Ka) + \cosh(\kappa b) \cos(Ka) = \cos(k(a+b)), \qquad (1.12.1)$$

where both K and κ depend on the energy E. In Section 1.2, we treated k as a free parameter which we picked, and then we solved for E to get the electron bands.

Suppose, instead, that we pick the energy E, and solve for k. Clearly, if the left side of (1.12.1) is greater than 1, then k cannot be real. This condition corresponds to energies inside the band gap. In that case, the inverse cosine function will give us a value of k that is complex. Figure 1.38 shows the real and imaginary parts of k as a function of E

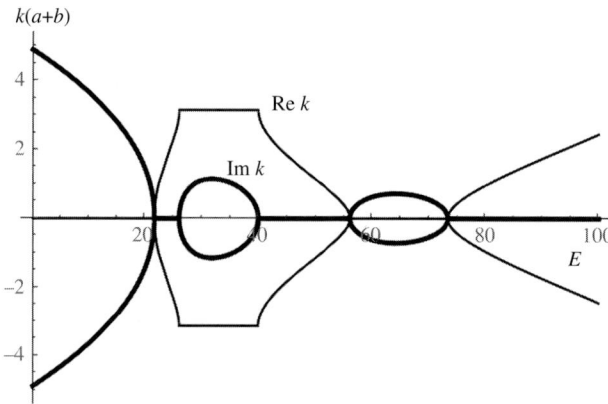

Fig. 1.38 The real (thin curve) and imaginary (heavy curve) parts of the wave vector k as a function of E, for the Kronig–Penney model given by (1.12.1), for $a = 0.9$, $b = 0.1$, $U_0 = 100$, and $\hbar^2/2m = 1$.

found using (1.12.1). When k is complex, the wave will have the form $\psi(x) \sim e^{ik_R x}e^{-k_I x}$, where $k_R = \text{Re } k$ and $k_I = \text{Im } k$. This means that the wave has a decaying part. It therefore cannot be a solution for an infinite periodic system, but it can be a solution if there is a boundary. In this case, the solution will be nonzero near the boundary and decay exponentially into the bulk. Positive k_I corresponds to decay from boundary on the left, while negative k_I corresponds to a state decaying from a boundary on the right. This is another way of deriving the existence of surface states, which we have already encountered in Section 1.11.4.

We cannot pick E to be any value, however. For surface states, we have the additional constraint of the boundary condition that the wave function must satisfy. Suppose that there is an infinite barrier at $x = x_0$. Then we have, for the Kronig–Penney wave function of Section 1.2,

$$\psi_1(x_0) = A_1 e^{iKx_0} + B_1 e^{-iKx_0} = 0, \tag{1.12.2}$$

where A_1 and B_1 depend on E and k through the matrix equation (1.2.4). We thus have two equations which we can solve for the two unknowns, E and k. For the Kronig–Penney model, there is just one solution within each band gap. Figure 1.39 gives an example of a surface state for the Kronig–Penney model that satisfies this boundary condition.

When we move beyond a one-dimensional model, we obtain bands of surface states. To see how, recall that for the one-dimensional model of Section 1.2 we used $E = \hbar^2 K^2/2m$ and $E = U_0 - \hbar^2 \kappa^2/2m$, which were obtained from the Schrödinger equation. Suppose that the system is translationally invariant in a direction y orthogonal to x. Then we can write the wave function as $\psi(x, y) = \psi_{KP}(x)e^{ik_y y}$, where $\psi_{KP}(x)$ is the Kronig–Penney wave function used above. Then Schrödinger's equation gives

$$E = \frac{\hbar^2 K^2}{2m} + \frac{\hbar^2 k_y^2}{2m} \quad \Rightarrow \quad K = \sqrt{2mE/\hbar^2 - k_y^2} \tag{1.12.3}$$

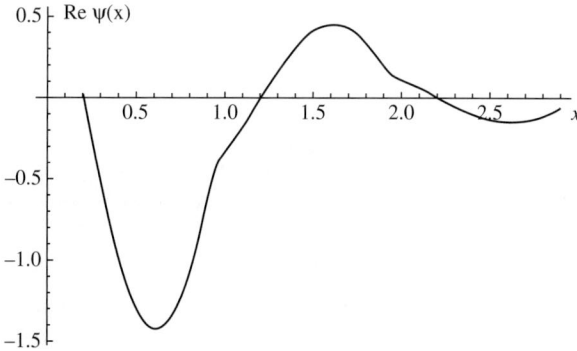

Fig. 1.39 The real part of the wave function for a surface state with $\psi(x_0) = 0$ for $x_0 = 0.2$, for the Kronig–Penney model with $a = 0.9$, $b = 0.1$, $U_0 = 100$, and $\hbar^2/2m = 1$. The energy of the state is $E = 14.9$, falling in the lowest energy gap of Figure 1.38.

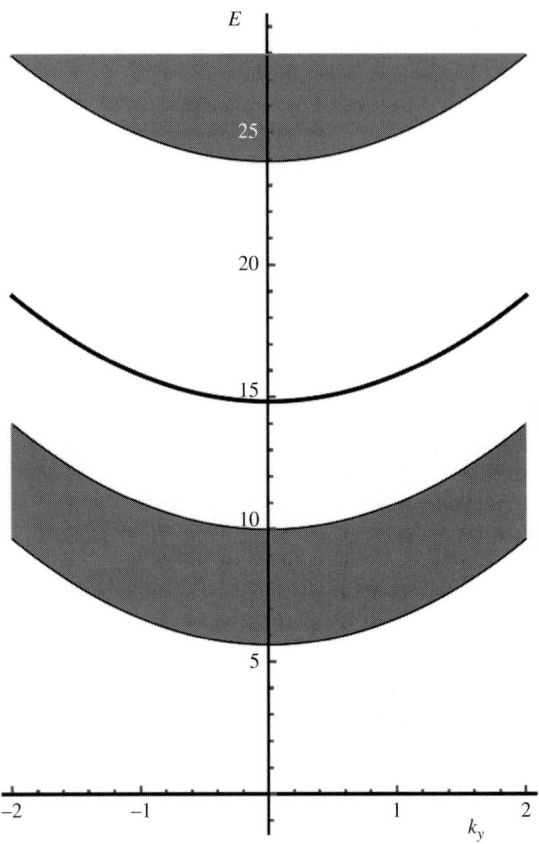

Fig. 1.40 The energies of the bands (gray regions) and surface states (solid line) for the Kronig–Penney model with the same parameters as Figures 1.38 and 1.39.

and

$$E = U_0 - \frac{\hbar^2 \kappa^2}{2m} + \frac{\hbar^2 k_y^2}{2m} \quad \Rightarrow \quad \kappa = \sqrt{2m(U-E)/\hbar^2 + k_y^2}. \qquad (1.12.4)$$

Picking any value for k_y, we can then solve for the energy subject to the boundary conditions as above. We will then have a surface-state band $E(k_y)$ that lies inside the band gap of the Kronig–Penney model, as shown in Figure 1.40.

LCAO model of surface states. Another model for the overlap of neighboring atomic states is the LCAO model, which is used in the tight-binding model for periodic crystals, discussed in Section 1.9.2. The LCAO model will give the same behavior as the Kronig–Penney model in the limit of weak coupling. In the limit of strong coupling, the LCAO model no longer approximates the Kronig–Penney model, as the Kronig–Penney model approaches the nearly free electron model of Section 1.9.3.

We can gain a lot of intuition about surface states and their relation to bulk states by solving a simple LCAO model. In the tight-binding model, only the unit cell is treated, and the wave number k is used to characterize different traveling Bloch states in an infinite crystal. In the present case, we will solve the LCAO model for the entire system, with a small number of unit cells, so that we can treat the bulk states and surface states on an equal footing.

We consider a one-dimensional chain of identical atoms, each with a symmetric s-orbital and an antisymmetric p-orbital. For a chain with periodic boundary conditions, the LCAO Hamiltonian is

$$
H = \begin{pmatrix}
E_s & 0 & U_{ss} & U_{sp} & 0 & 0 & \cdots & U_{ss} & -U_{sp} \\
0 & E_p & -U_{sp} & -U_{pp} & 0 & 0 & & U_{sp} & -U_{pp} \\
U_{ss} & -U_{sp} & E_s & 0 & U_{ss} & U_{sp} & & & \\
U_{sp} & -U_{pp} & 0 & E_p & -U_{sp} & -U_{pp} & & & \\
0 & 0 & U_{ss} & -U_{sp} & E_s & 0 & & & \\
0 & 0 & U_{sp} & -U_{pp} & 0 & E_p & & & \\
\vdots & & & & & & \ddots & & \\
U_{ss} & U_{sp} & & & & & & E_s & 0 \\
-U_{sp} & -U_{pp} & & & & & & 0 & E_p
\end{pmatrix} .
$$

$$(1.12.5)$$

This matrix has blocks with diagonal terms for E_s and E_p, which correspond to the original atomic eigenstates, and coupling terms in blocks, with terms for the various couplings of the s- and p-orbitals. The coupling term for the s- and p-orbitals have opposite signs for neighboring atoms on the left and right of an atom, because the p-orbital is antisymmetric. For the same reason, we expect U_{pp} to have the opposite sign of U_{ss}. (We take all terms as real, for simplicity.) The blocks in the upper right-hand corner and the lower left-hand corner give the periodic boundary conditions, corresponding to the Born–von Karman conditions discussed in Section 1.7.

The eigenvalues of this Hamiltonian are shown in Figure 1.41(a), for a chain of 30 atoms, as the coupling is increased. As seen in this figure, as the coupling increases, the atomic states first broaden into bands and then overlap, as discussed in Section 1.1.3 in connection with Figure 1.8. When the bonding–antibonding energy exceeds the splitting between the original orbitals, a new gap opens up, similar to the gap seen in sp^3-bonded crystals, discussed in Section 1.11.3.

We can now ask what happens when we drop the periodic boundary conditions, which in this model corresponds to dropping the upper right and lower left blocks coupling atom 1 to atom N at the other end of the chain. Figure 1.41(b) shows the eigenvalues of the matrix in this case. As seen in this figure, the bands look overall the same, but there are two degenerate states that lie in the middle of the band gap at high coupling. These correspond to localized surface states on the right end and left end of the atomic chain. Their existence is robust against the details of the values of the coupling parameters, although the existence of the band gap at high coupling energy depends crucially on the asymmetry of the coupling of the s and p states which gives the minus signs in the matrix elements of (1.12.5). This asymmetry also occurs in covalent sp^3-bonded crystals.

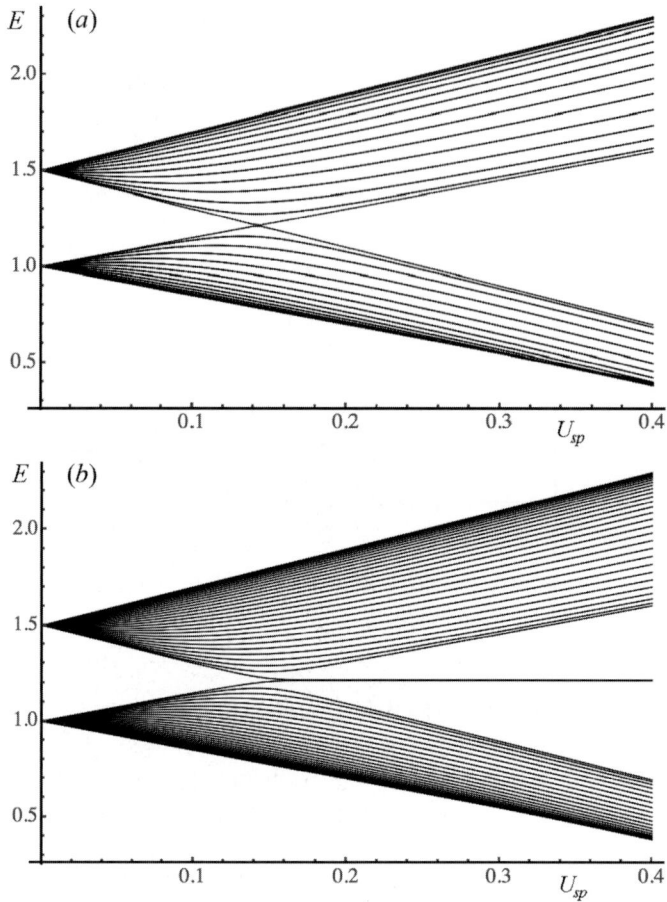

Fig. 1.41 (a) The energies of the LCAO model of bands arising from a single-atom chain of 30 atoms each with two relevant atomic orbitals, one of which is symmetric (s-like) and one of which is antisymmetric (p-like), for the Hamiltonian (1.12.5) in the text, with the parameters $E_s = 1.0$, $E_p = 1.5$, $U_{ss} = 0.75U_{sp}$, and $U_{pp} = U_{sp}$.
(b) The same model but with periodic boundary conditions removed, leading to the appearance of surface states.

The exact states corresponding to the surface states and bulk bands can be found from the eigenvectors of the Hamiltonian. Without looking at these in detail, we can say generally that the surface states include terms mostly from the ends of the chain, while the bulk states look similar to extended Bloch waves.

Surface states play a major role in the new field of **topological insulators**, and topological effects of electron bands in general. In a topological insulator, the surface states can be conducting while the bulk of the material is an insulator. For a review of topological considerations of bands, see Hasan and Kane (2010). The spin-orbit effects discussed in Section 1.13 can also play an important role in the topology of the bands.

Exercise 1.12.1   The Su–Schrieffer–Heeger (SSH) linear chain model also has two orbitals per unit cell, but envisions these as two different atoms with only one relevant atomic orbital on each. Each atom couples differently to its neighbors on the right and on the left, with coupling terms v and w, respectively, giving the Hamiltonian

$$H = \text{const.} + \begin{pmatrix} 0 & v & 0 & 0 & . & . & . & 0 & w \\ v & 0 & w & 0 & & & & 0 & 0 \\ 0 & w & 0 & v & & & & & \\ 0 & 0 & v & 0 & & & & & \\ . & & & & . & & & & \\ . & & & & & . & & & \\ . & & & & & & . & & \\ 0 & 0 & & & & & & 0 & v \\ w & 0 & & & & & & v & 0 \end{pmatrix}$$

(1.12.6)

for periodic boundary conditions. For free ends, the same matrix is used but with the w terms in the upper right and lower left corners dropped.

(a) Show that the SSH Hamiltonian is equivalent to (1.12.5) for the special case $U_{ss} = U_{pp} = U_{sp}$. To do this, show that a simple transformation of the basis converts the SSH matrix into (1.12.5). Hint: Consider what basis transformation will make one 2×2 block on the diagonal of (1.12.6) into the 2×2 diagonalized block at the same location in the matrix (1.12.5).

(b) Show, for as many atoms as your computer resources can handle, that the SSH model gives qualitatively the same behavior as seen in Figure 1.41.

1.13 Spin in Electron Bands

So far in this chapter we have not discussed electron spin at all. In many cases, the spin of the electrons can be treated simply by doubling the number of electron states, assuming that each k-state has two degenerate spin states. Even when the states are not strictly degenerate, the energy splitting between the different spin states in crystals is typically very small in the absence of a magnetic field. We will return to discuss spin and magnetic effects at length in Chapter 10, but sometimes, even in the absence of a magnetic field, spin effects play an important role.

The contribution of spin to the electron band energies comes from the **spin–orbit** interaction energy. The spin–orbit coupling is a result of relativistic quantum theory, as discussed in Appendix F. For an electron in the presence of a potential $U(\vec{r})$, it is written

$$H_{SO} = \frac{\hbar}{4m^2 c^2} (\nabla U \times \vec{p}) \cdot \vec{\sigma}, \tag{1.13.1}$$

where $\vec{p}$ is the momentum operator acting on the electron orbital wave function, and $\vec{\sigma}$ is the Pauli spin operator, defined in terms of the Pauli matrices,

$$\sigma_x = \begin{pmatrix} 0 & 1 \\ 1 & 0 \end{pmatrix}, \quad \sigma_y = \begin{pmatrix} 0 & -i \\ i & 0 \end{pmatrix}, \quad \sigma_z = \begin{pmatrix} 1 & 0 \\ 0 & -1 \end{pmatrix}. \tag{1.13.2}$$

For a central potential, $\nabla U(\vec{r}) = (1/r)(dU/dr)\vec{r}$, and therefore the spin–orbit term is often written as

$$H_{SO} = \xi(\vec{r})\vec{L} \cdot \vec{S}, \tag{1.13.3}$$

where $\vec{L} = \vec{r} \times \vec{p}$ is the angular momentum operator, $\vec{S} = \hbar\vec{\sigma}/2$ is the spin angular momentum, and the factor $\xi(\vec{r})$ is given by

$$\xi(\vec{r}) = \frac{1}{2m^2c^2r}\frac{dU}{dr}. \tag{1.13.4}$$

In a typical crystal, however, the potential $U(\vec{r})$ is not centered around a single nucleus as it is in atomic calculations, so it is more general to use the form (1.13.1). Recall that $\vec{p} = -i\hbar\nabla$ when acting on the band wave functions.

We can estimate the order of magnitude of the spin–orbit effect by taking the natural unit of energy U as the effective Rydberg of the atomic orbitals and the natural unit of distance as the effective Bohr radius. The effective Rydberg is given by $Z_{\text{eff}}^2\text{Ry}$, where Ry is the hydrogen Rydberg and Z_{eff} is the effective atomic number, assuming that the core electrons cancel out some of the charge of the nucleus. For example, an element in Group IV of the periodic table will have $Z_{\text{eff}} = 4$. The effective Bohr radius is a_0/Z_{eff}, where a_0 is the hydrogen Bohr radius. Assuming that the angular momentum and spin quantum numbers are of the order of unity, the spin–orbit interaction energy is then

$$\langle H_{SO} \rangle \sim \frac{\hbar^2}{2m^2c^2}\frac{Z_{\text{eff}}^4\text{Ry}}{a_0^2} \sim 0.1 \text{ eV}. \tag{1.13.5}$$

This can give significant splittings of band energies.

1.13.1 Split-off Bands

We can see how the spin–orbit term enters the band energy calculations by using the tight-binding model of Section 1.9.2. The full Hamiltonian H used in the tight-binding model must include the spin–orbit interaction (1.13.1) introduced here.

As discussed in Section 1.9.2 (and also in Section 1.11.3), when there are several orbitals in the basis of a lattice, one must construct a matrix for the couplings between the different states, and then find the band energies as the eigenstates of this matrix. We start by including only the couplings between orbitals on the same atom. There are nine possible coupling terms between the three p-orbitals, Φ_x, Φ_y, and Φ_z, and for each of these there are three spinor terms, for σ_x, σ_y, and σ_z. Thus, for example, two of the terms of (1.13.1) are

$$\int d^3r\, \Phi_x^*(\vec{r})H_{SO}\Phi_x(\vec{r}) = \frac{-i\hbar^2}{m^2c^2}\sigma_z\int d^3r\, \Phi_x^*(\vec{r})[(\nabla_xU)\nabla_y - (\nabla_yU)\nabla_x]\Phi_x(\vec{r}),$$

$$\int d^3r\, \Phi_x^*(\vec{r})H_{SO}\Phi_y(\vec{r}) = \frac{-i\hbar^2}{m^2c^2}\sigma_z\int d^3r\, \Phi_x^*(\vec{r})[(\nabla_xU)\nabla_y - (\nabla_yU)\nabla_x]\Phi_y(\vec{r}). \tag{1.13.6}$$

To resolve all these integrals, it is useful to think in terms of the **parity** of the functions relative to $\vec{r} = 0$. An integrand with overall negative parity, that is, which is antisymmetric relative to 0, will give an integral that vanishes when integrated from $-\infty$ to $+\infty$. Since the integrals in the coupling terms are three-dimensional, integrands with net negative parity in any of the three spatial directions will give vanishing integrals.

Because $U(\vec{r})$ is symmetric relative to the center of any lattice site, its derivative in any direction is antisymmetric, and therefore $\nabla_x U$ has negative parity relative to x. A p-orbital has negative parity along its axis and positive parity in the two perpendicular directions; the derivative of a p-orbital will flip the parity. When all the parities are taken into account, the only nonzero terms remaining give the following form of the spin–orbit Hamiltonian, acting on the three states Φ_x, Φ_y, and Φ_z:

$$H_{SO} = U_0 \begin{pmatrix} 0 & -i\sigma_z & i\sigma_y \\ i\sigma_z & 0 & -i\sigma_x \\ -i\sigma_y & i\sigma_x & 0 \end{pmatrix}, \tag{1.13.7}$$

where U_0 is a constant with units of energy, and the σ_i are the Pauli matrices acting on the electron spin. Here, we have used the fact that H_{SO} must be Hermitian, so that the eigenvalues are real.

We can write this out explicitly in terms of the spins, using six states, which we write as $\Psi_1 = \Phi_x\alpha_\uparrow$, $\Psi_2 = \Phi_x\alpha_\downarrow$, $\Psi_3 = \Phi_y\alpha_\uparrow$, etc., where $\alpha_\uparrow$ and $\alpha_\downarrow$ are the two pure electron spin states. Then

$$H_{SO} = U_0 \begin{pmatrix} 0 & 0 & -i & 0 & 0 & 1 \\ 0 & 0 & 0 & i & -1 & 0 \\ i & 0 & 0 & 0 & 0 & -i \\ 0 & -i & 0 & 0 & -i & 0 \\ 0 & -1 & 0 & i & 0 & 0 \\ 1 & 0 & i & 0 & 0 & 0 \end{pmatrix}. \tag{1.13.8}$$

Diagonalization of this matrix gives four degenerate states with eigenvalue U_0,

$$\begin{aligned} \Psi_{\frac{3}{2},+\frac{3}{2}} &= \frac{1}{\sqrt{2}}(\Phi_x + i\Phi_y)\alpha_\uparrow \\ \Psi_{\frac{3}{2},+\frac{1}{2}} &= \frac{1}{\sqrt{6}}[(\Phi_x + i\Phi_y)\alpha_\downarrow - 2\Phi_z\alpha_\uparrow] \\ \Psi_{\frac{3}{2},-\frac{1}{2}} &= \frac{1}{\sqrt{6}}[(\Phi_x - i\Phi_y)\alpha_\uparrow + 2\Phi_z\alpha_\downarrow] \\ \Psi_{\frac{3}{2},-\frac{3}{2}} &= \frac{1}{\sqrt{2}}(\Phi_x - i\Phi_y)\alpha_\downarrow, \end{aligned} \tag{1.13.9}$$

and two degenerate states with eigenvalue $-2U_0$,

$$\begin{aligned} \Psi_{\frac{1}{2},+\frac{1}{2}} &= \frac{1}{\sqrt{3}}[(\Phi_x + i\Phi_y)\alpha_\downarrow + \Phi_z\alpha_\uparrow] \\ \Psi_{\frac{1}{2},-\frac{1}{2}} &= \frac{1}{\sqrt{3}}[(\Phi_x - i\Phi_y)\alpha_\uparrow - \Phi_z\alpha_\downarrow]. \end{aligned} \tag{1.13.10}$$

The labels of these states have been written in terms of the total angular momentum $\vec{J} = \vec{L} + \vec{S}$, where $\vec{L}$ is the orbital angular momentum and $\vec{S}$ is the spin. One can easily see how this assignment arises by recalling that the $L = 1$ spherical harmonics are

$$Y_{1,1} = -\sqrt{\frac{3}{8\pi}}(x+iy)$$

$$Y_{1,0} = \sqrt{\frac{3}{4\pi}}z$$

$$Y_{1,-1} = \sqrt{\frac{1}{8\pi}}(x-iy). \tag{1.13.11}$$

The p-orbitals have the same parity properties as the $x, y,$ and z functions, and thus have the same mapping to $L = 1$ angular momentum states. When combined with the $\frac{1}{2}$-spin, the total angular momentum is either $\frac{3}{2}$ or $\frac{1}{2}$.

The spin–orbit interaction therefore implies that the degeneracy of the six p-states is split in this crystal. There will be one fourfold-degenerate band with total angular momentum $J = \frac{3}{2}$, and a **split-off** band with two states with angular momentum $J = \frac{1}{2}$. A split-off band due to spin–orbit interaction is a common feature in crystals with bonds formed from p-orbitals. For example, as seen in Section 1.9.2, in tetrahedral crystals with sp^3 bonds, there are bands made of p-orbitals, which will be split by the spin–orbit term.

The accounting of the parities which we used above is much more easily done using the group theory formalism of Chapter 6; specific cases involving the spin–orbit interaction are addressed in Section 6.6.

Exercise 1.13.1   Prove explicitly using parity that H_{SO} has the form given in (1.13.7), for three degenerate p-orbitals and a central potential $U(r)$.

1.13.2 Spin–Orbit Effects on the k-Dependence of Bands

We now move on to the next order of the tight-binding approximation (1.9.20), namely terms for nearest neighbors. For simplicity, we will use the example of a simple cubic lattice, in which each atom has six nearest neighbors, as illustrated in Figure 1.42. For nearest neighbors in the x-direction, we must compute terms of the following type:

$$U_{xx} = \frac{-i\hbar^2}{m^2c^2}\int d^3r\,\Phi_x^*(\vec{r})[(\nabla_x U)\nabla_y - (\nabla_y U)\nabla_x]\Phi_x(\vec{r}-a\hat{x}), \tag{1.13.12}$$

$$U_{xy} = \frac{-i\hbar^2}{m^2c^2}\int d^3r\,\Phi_x^*(\vec{r})[(\nabla_x U)\nabla_y - (\nabla_y U)\nabla_x]\Phi_y(\vec{r}-a\hat{x}). \tag{1.13.13}$$

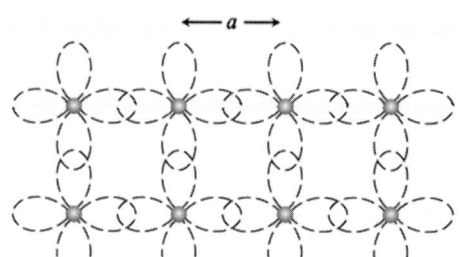

Fig. 1.42   Adjacent p-orbitals in one plane of a simple cubic lattice.

Because the two orbitals $\Phi_n(\vec{r})$ and $\Phi(\vec{r} - a\hat{x})$ are not centered on the same location, parity in the x-direction is no longer a concern. However, parity in the y- and z-directions still matters, and U_{xx} vanishes due to negative parity in the y-direction. However, U_{xy} is nonzero.

If we change the variable x in (1.13.13) to $-x$, then the x-integral becomes

$$I(a) = \int_{-\infty}^{\infty} dx \, \Phi_x^*(x, y, z)[(\nabla_x U)\nabla_y - (\nabla_y U)\nabla_x]\Phi_y(x - a, y, z).$$

$$= \int_{-\infty}^{\infty} dx \, \Phi_x^*(-x, y, z)[(-\nabla_x U)\nabla_y + (\nabla_y U)\nabla_x]\Phi_y(-x - a, y, z). \quad (1.13.14)$$

Since $\Phi^*(-x, y, z) = -\Phi(x, y, z)$ and $\Phi_y(-x, y, z) = \Phi_y(x, y, z)$, we therefore have $I(a) = I(-a)$. Therefore, the sum of the two nearest-neighbor terms in the x-direction in the tight-binding formula (1.9.20) is

$$i\sigma_z U_{xy}(e^{ik_x a} + e^{-ik_x a}) = 2i\sigma_z U_{xy} \cos k_x a. \quad (1.13.15)$$

We take into account only nearest-neighbor terms with orthogonal p-orbitals, assuming that terms with parallel orbitals have negligible overlap in space. Accounting for these nearest-neighbor terms, the spin–orbit interaction (1.13.7) is then

$$H_{SO} = \begin{pmatrix} 0 & -i\sigma_z U(k_x, k_y) & i\sigma_y U(k_x, k_z) \\ i\sigma_z U(k_x, k_y) & 0 & -i\sigma_x U(k_y, k_z) \\ -i\sigma_y U(k_x, k_z) & i\sigma_x U(k_y, k_z) & 0 \end{pmatrix}, \quad (1.13.16)$$

where $U(k_i, k_j) = U_0 + 2U_{xy}(\cos k_i a + \cos k_j a)$.

If $\vec{k}$ is small, we can assume that the eigenstates are still well approximated by the states given in (1.13.9) and (1.13.10). For $\vec{k}$ in the x-direction, the energies of these states (when properly normalized) are

$$\langle \Psi_{\frac{3}{2}, \pm \frac{3}{2}} | H_{SO} | \Psi_{\frac{3}{2}, \pm \frac{3}{2}} \rangle = U_0 + 2U_{xy}(1 + \cos k_x a)$$

$$\langle \Psi_{\frac{3}{2}, \pm \frac{1}{2}} | H_{SO} | \Psi_{\frac{3}{2}, \pm \frac{1}{2}} \rangle = U_0 + \tfrac{2}{3}U_{xy}(5 + \cos k_x a)$$

$$\langle \Psi_{\frac{1}{2}, \pm \frac{1}{2}} | H_{SO} | \Psi_{\frac{1}{2}, \pm \frac{1}{2}} \rangle = -2U_0 - \tfrac{8}{3}U_{xy}(2 + \cos k_x a). \quad (1.13.17)$$

Note that states that have the same magnitude of angular momentum but opposite k-direction have the same energy. This is a consequence of the fact that the cubic system we have chosen is **centrosymmetric**; that is, $U(-\vec{r}) = U(\vec{r})$. (This is not the same as having a central potential, which requires $U(\vec{r}) = U(|\vec{r}|)$.) Centrosymmetry led to the form of the coupling terms (1.13.15), which gave only terms with $\cos k_x a$.

If $U(\vec{r})$ has an antisymmetric term, the diagonal terms for the p-states in (1.13.16) will no longer necessarily vanish. Then $\nabla_x U$ will not change sign on the change of variable $x \to -x$, in which case we have a term with $I(-a) = -I(a)$ in (1.13.14), which gives a contribution proportional to $\sin k_x a$. We take the leading-order contribution as only those terms that couple nearest-neighbor orbitals aligned along the same direction as the vector separating them, because these correspond to the maximal spatial overlap of the orbital wave functions. Let us assume that $U(\vec{r})$ has antisymmetric terms along the x- and y-axes. Then, accounting again for parity in all directions, the contribution due to the antisymmetric terms is

$$H_{SO}^{(a)} = \begin{pmatrix} 2U_{xx}\sigma_z \sin k_x a & 0 & 0 \\ 0 & 2U_{xx}\sigma_z \sin k_y a & 0 \\ 0 & 0 & 0 \end{pmatrix}. \tag{1.13.18}$$

When this is applied to the states (1.13.9) and (1.13.10) for $\vec{k}$ in the x-direction, we obtain

$$\langle \Psi_{\frac{3}{2},\pm\frac{3}{2}} | H_{SO} | \Psi_{\frac{3}{2},\pm\frac{3}{2}} \rangle = \pm U_{xx}(\sin k_x a + \sin k_y a)$$

$$\langle \Psi_{\frac{3}{2},\pm\frac{1}{2}} | H_{SO} | \Psi_{\frac{3}{2},\pm\frac{1}{2}} \rangle = \pm\frac{1}{3} U_{xx}(\sin k_x a + \sin k_y a)$$

$$\langle \Psi_{\frac{1}{2},\pm\frac{1}{2}} | H_{SO} | \Psi_{\frac{1}{2},\pm\frac{1}{2}} \rangle = \pm\frac{2}{3} U_{xx}(\sin k_x a + \sin k_y a). \tag{1.13.19}$$

Note that in this case, states with opposite angular momentum have opposite energy shift. In other words, the bands act as though there is an effective magnetic field proportional to the magnitude of $\vec{k}$. This is generally a possibility in non-centrosymmetric crystals. Figure 1.43 illustrates the effect of the terms (1.13.17) and (1.13.19) on the p-states in our example.

If we look at only one of the states in Figure 1.43, it violates the version of Kramers' rule (1.6.16) we deduced in Section 1.6, because the $\sin k_x a$ term gives opposite energy shift for $\vec{k} \to -\vec{k}$. That version of Kramers' rule did not take into account spin. The more general version of Kramers' rule, which will be proven in Section 6.9, is

$$\boxed{E_{n,-m_J}(-\vec{k}) = E_{n,m_J}(\vec{k}),} \tag{1.13.20}$$

where m_J is the projection of the total angular momentum $\vec{J} = \vec{L} + \vec{S}$. It is easy to see that this rule is satisfied for the bands in Figure 1.43.

This general form of Kramers' theorem can be viewed as a simple consequence of time-reversal symmetry. Time reversal of an electron in a band state corresponds to flipping its propagation direction from $\vec{k}$ to $-\vec{k}$ and also flipping the sign of its total angular

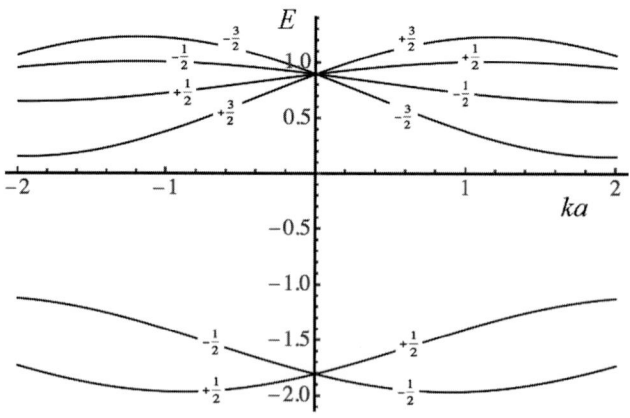

Fig. 1.43   Energy of the six states (1.13.9) and (1.13.10) for the spin–orbit terms (1.13.16) and (1.13.18), for $U_0 = 0.5$, $U_{xy} = 0.1$, and $U_{xx} = 0.5$.

momentum. In the absence of any terms in the Hamiltonian that break time-reversal symmetry (such as a real magnetic field), the eigenstates of the Hamiltonian must also satisfy time-reversal symmetry.

Exercise 1.13.2   (a) Show explicitly that parity rules give the spin–orbit term (1.13.18) for $U(\vec{r})$ antisymmetric in the x- and y-directions and symmetric in the z-directions.

(b) What would this term be if $U(\vec{r})$ had an antisymmetric term only in the x-direction?

References

N.W. Ashcroft and N.D. Mermin, *Solid State Physics* (Holt, Rinehart, and Winston, 1976).

J. Bernhol, "Computational materials science: the era of applied quantum mechanics," *Physics Today* **52**, 30 (1999).

J.R. Chelokowski, D.J. Chadi, and M.L. Cohen, "Calculated valence-band densities of states and photoemission spectra of diamond and zinc-blende semiconductors," *Physical Review* B **8**, 2786 (1973).

C. Cohen-Tannoudji, B. Diu, and F. Laloë, *Quantum Mechanics* (Wiley, 1977).

M.Z. Hasan and C.L. Kane, "Topological Insulators," Reviews of Modern Physics **82**, 3045 (2010).

W. Harrison, *Electronic Structure* (W.H. Freeman and Co., 1980).

O. Madelung, *Introduction to Solid State Theory* (Springer, 1978).

K. Oura, V.G. Lifshits, A.A. Saranin, and M. Katayama, *Surface Science: An Introduction* (Springer, 2003).

2 Electronic Quasiparticles

In this chapter, we will eventually discuss a fair amount of semiconductor and transistor technology. Unfortunately, many physicist students don't study these in detail, thinking that they are "applied" science rather than "fundamental" science. But as we will see, to properly understand that technology, one must engage with many very fundamental concepts. One such concept is the idea of the "renormalized vacuum," in which the entire crystal in its ground state is viewed as a new vacuum, and only excitations out of the ground state count as particles of interest.

The physics of semiconductors led to some of the most fascinating fundamental physics results of the twentieth century, including the fractional quantum Hall effect, in which electrons act as though they have charge of some fraction of e. We will review this at the end of this chapter, but quite a bit of foundation must be laid to get there.

2.1 Quasiparticles

We saw in Section 1.11 that it is very common to have a solid with one or more entirely full bands and some entirely empty bands. Suppose that we have two bands with minima and maxima at the center of the Brillouin zone, as shown in Figure 2.1, which is a fairly typical band structure for a semiconductor. In the ground state of the crystal, the lower band, known as the **valence** band, is completely full, and the upper band, known as the **conduction** band, is completely empty. None of the electrons in the lower band can change its state because all the nearby states are filled. If we put energy into the system, however, we can promote an electron from a state in the lower band to a state in the upper band. In this case, the electron in the upper band can move freely into other states in the same band with very little energy change. We call this a **free electron**.

At the same time, an empty state is left in the valence band. If another electron from the same band moves into this state, it will leave an empty state in a new place, as illustrated in Figure 2.1.

Rather than keeping track of all the electrons in the band that move to fill the empty state, we can simply keep track of where the empty state goes. We call this empty state a **hole**. One can think of it in the same way as a bubble in a glass of water. When you look at a bubble rising in a glass of water, you don't think "the water fell, to fill the empty spot, leaving an empty spot further up," although this of course is what happens. It is much easier simply to think, "the bubble moved up."

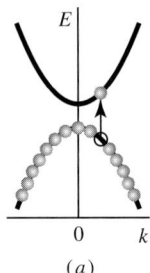

Fig. 2.1 A typical semiconductor band structure near zone center. (a) The excitation process. An electron can be promoted from the full, lower band to the empty, upper band, leaving behind an empty state. (b) Motion of a hole, which is equivalent to motion of electrons in the valence band with the opposite k-vector.

This leads to a new way of thinking about the electrons in a material. Instead of worrying about all the electrons frozen by Pauli exclusion in the lower band, we can define a new "vacuum" state which is equal to the ground state of the crystal. If we put energy in, we "create" two new particles, which are really excitations of the original ground state. These new particles can be called **quasiparticles**. The created quasiparticles can also annihilate each other if an electron in the upper band falls back down into a hole in the lower band. This process is called **recombination**.

This picture of the ground state of a system being a vacuum (that is, a **renormalized** vacuum) and the excitations being new particles that are created and destroyed is a very general and important concept in condensed matter physics. It is used not only for semiconductors but for many other many-body systems, as we will see in later chapters.

Calling them quasiparticles makes them sound as if they are not real. They are real, however, in the sense that they transport energy, charge, and mass. Actually, one can argue that every normal particle is also a quasiparticle. In relativistic quantum mechanics, the field equation for the electron is the Dirac equation, which implies the existence of negative-energy states of electrons as well as positive-energy states (see Appendix F). In order to avoid a collapse of all particles to infinite negative energy, Dirac hypothesized that the negative-energy states are all filled with electrons in the same way as the lower band in Figure 2.1, and this prevents positive-energy particles from falling into the negative-energy states. This is called the **Dirac sea**.[1] In this case, just as in the bands of Figure 2.1, an electron can be promoted from the lower band to the upper band with the input of sufficient energy. This leads to the existence of a free, positive-energy electron, as well as an empty state in the lower band. This empty state, or hole, is called a positron. Not only the electrons but every fermion has an antiparticle; these antiparticles are the **antimatter** of high-energy physics. The infinite number of electrons at negative energy doesn't matter, because these are just part of the ground state of the system, which we define as our vacuum.

Since all the normal, positive-energy electrons we know are actually excitations out of the Dirac vacuum state, does that mean that normal electrons in a vacuum are not real?

---

[1] As the terminology suggests, it is a lot easier to keep track of a "bubble" in the "sea" than to keep track of all the water.

This gets into philosophical questions. What is it that makes something real? In the minds of many physicists, *any* system can be redefined as a vacuum when it is in its ground state, and the excitations out of this state are the real particles of interest. The energy of the ground state contributes only a constant, even if it is a large or infinite constant, which can be removed by a simple shift of the definition of zero energy.

The philosophy of quasiparticles forms the basis of an important debate in modern physics, the question of whether there exist fundamental particles that are the "building blocks of the universe." Could *all* the particles we know be quasiparticle excitations of a deeper-lying system? If so, the search for "fundamental" particles becomes much less interesting. For a summary of this debate, see Pines and Laughlin (2000).

It has become fashionable among high-energy physicists to adopt only the quasiparticle picture of electrons and positrons and to ignore the negative-energy Dirac sea. This is not because the concept of the Dirac sea has been dispensed with; it is just because the Dirac sea can be ignored, as the ground state of the system. As in solid state physics, once one switches to the quasiparticle picture, the underlying system does not enter into the mathematics.

Some students have the misconception that holes do not exist in any particular place – that they are only a k-space mathematical construction. This error arises from thinking only in terms of the energy band picture in which the eigenstates are k-states, as we have done in Figure 2.1. Holes are not restricted to remain in eigenstates any more than electrons are. For both electrons and holes, we can construct a wave packet that represents a particle restricted to a certain location, by putting the particle in a superposition of many eigenstates. In a vacuum, a wave packet in one place can be created from a superposition of plane waves, and in the same way, a wave packet in a crystal can be created from a superposition of Bloch waves. Electrons and holes in crystals can certainly move from one place to another just like any electron or positron in vacuum.

The free electron is also not more real than the hole. Both are quasiparticles with the same footing in the equations. The free electron can disappear by falling back into the valence band, just as the hole can disappear by being filled with an electron. To treat both of these quasiparticles on an equal footing, they are often grouped together under the names **charge carriers**, **free carriers**, or simply **carriers**.

The quasiparticle picture does not apply only to electrons and holes in semiconductors and insulators. As will be discussed in Section 2.4, we can also think of metals in terms of quasiparticles. We will also see, in later chapters, that other excitations of the solid, such as sound vibrations and electromagnetic field waves, can be taken into account in the quasiparticle picture. In general, all the excitations of a system above the ground state can be viewed as quasiparticles.

2.2 Effective Mass

As discussed in Section 1.8, at zone center and at every critical point on a zone boundary, there is a maximum or minimum of the electron energy bands, and away from these points, the energy varies as $(\vec{k} - \vec{k}_{\text{crit}})^2$. This is the same form of dependence as expected for free particles in vacuum, for speeds much less than the speed of light,

$$E = \frac{\hbar^2 k^2}{2m}, \tag{2.2.1}$$

where m is the mass of the electron in vacuum. As we saw in the discussion of $k \cdot p$ theory in Section 1.9.4, in the case of isotropic bands, the curvature of the bands in solids near a band minimum takes the form

$$E = E_0 + \frac{\hbar^2 k^2}{2m_{\text{eff}}}, \tag{2.2.2}$$

where m_{eff} is an **effective mass** which can be quite different from the vacuum electron mass. Once we have taken into account this effective mass due to the band structure, a free electron near the conduction band minimum will behave exactly like a free particle in vacuum.

Because the Bloch states of the crystal are eigenstates, a free electron in a perfect crystal moves **without scattering** in a straight line through the crystal, behaving just like a particle in a vacuum with mass m_{eff}, despite the presence of the 10^{23} or so closely packed atoms of the crystal. It is important to remember that we are talking about a **quasiparticle** that does this. Of course, the underlying electrons interact with each other and the atoms constantly, but all of these interactions are taken into account in the band energies that give rise to the effective mass. The quasiparticle itself does not scatter unless it interacts with other quasiparticles or with an imperfection in the crystal. In the latter case, we treat the imperfection (which can be a single atom defect, or a large number of atoms out of place, known as a dislocation, discussed in Section 1.11.4) as an independent object sitting in the "vacuum" of a perfect crystal.

Near a valence band **maximum**, the energy of the electrons can often be approximated as

$$E = E_0 - \frac{\hbar^2 k^2}{2m_{\text{eff}}}, \tag{2.2.3}$$

where m_{eff} is another effective mass which depends on the curvature of the valence band. In this case, the electrons have **negative** mass. A hole, however, is the absence of an electron. When an electron moves to the right in the valence band, this corresponds to a hole moving to the left. In the same way, an electron losing energy corresponds to a hole gaining energy. Therefore, for the quasiparticle holes in this case, we can write

$$E = \frac{\hbar^2 k^2}{2m_{\text{eff}}}, \tag{2.2.4}$$

where we have defined $E_0 = 0$ as the energy at the top of the valence band. In other words, near a valence band maximum, a hole acts like a particle of *positive* mass. It, too, will move in a straight line through the crystal, just like a positively charged particle in a vacuum.

Note that there is no constraint on how heavy or light the effective mass of the free electrons and holes can be. For example, electrons in two-dimensional sheets of graphene have an effective mass of zero. As seen in Figure 1.29 of Section 1.9.2, which gives the bands of graphene calculated using a tight-binding model, there are several points $\vec{k}_i$ in the Brillouin zone where the bands cross, with a linear dispersion $E = \hbar v(|\vec{k} - \vec{k}_i|)$. This type of linear dispersion can be seen as the zero-mass limit of the relativistic energy equation,

$$E = \sqrt{(mc^2)^2 + (cp)^2}, \tag{2.2.5}$$

using $p = \hbar k$ and the velocity v from the band dispersion instead of the speed of light, c.

Exercise 2.2.1 Calculate the effective mass of an electron at zone center in the lowest energy band of the Kronig–Penney model given in Section 1.2, in the limit $b \to 0$ and $U_0 \to \infty$, and $U_0 b$ finite but small.

Exercise 2.2.2 If the free electrons in a semiconductor have effective mass $m_c = 0.1 m_0$, where m_0 is the mass of the electron in vacuum, what is their average speed at $T = 300$K? If the free holes in the same material have effective mass $m_v = 1.5 m_0$, what is their average speed at the same temperature? Assume a Maxwell–Boltzmann distribution of the electron energies.

Exercise 2.2.3 The effective mass of free carriers can be measured using cyclotron resonance. As discussed in Section 2.9, the resonance frequency (also known as the cyclotron frequency) of a charged particle in a magnetic field is, in MKS units,

$$\omega_c = \frac{|q|B}{m}, \tag{2.2.6}$$

where q is the charge and m is the effective mass of the particle. If a material is placed in a DC magnetic field and then a small oscillating field is applied, the free carriers will resonate at the cyclotron frequency. Knowing the magnitude of B then allows one to deduce the value of m from the resonance frequency.

Calculate the cyclotron frequency of an electron with effective mass $0.1\ m_0$ in the presence of a magnetic field B $= 1$ T, where m_0 is the mass of the free electron in vacuum. Some convenient unit conversions to remember are 1 T $= 1$ V-s/m^2 and $m_0 = 0.5 \times 10^6$ eV/c^2.

In the case of an isotropic effective mass, the density of states of the quasiparticles becomes very easy to calculate. From (1.8.2), we have

$$\mathcal{D}(E)dE = dE\frac{V}{(2\pi)^3}\int d^3k\ \delta(E_{\vec{k}} - E)$$

$$= dE\frac{V}{(2\pi)^3}\int 4\pi k^2 dk\ \delta(E_{\vec{k}} - E). \tag{2.2.7}$$

Assuming $E_{\vec{k}} = E_0 + \hbar^2 k^2/2m$, and therefore $dE_{\vec{k}} = \hbar^2 k dk/m$, this implies

$$\mathcal{D}(E)dE = dE\frac{4\pi V}{(2\pi)^3}\int k\frac{m}{\hbar^2}\ \delta(E_{\vec{k}} - E)dE_{\vec{k}}$$

$$= dE\frac{4\pi V}{(2\pi)^3}\sqrt{\frac{2m(E - E_0)}{\hbar^2}}\frac{m}{\hbar^2}, \tag{2.2.8}$$

or

$$\mathcal{D}(E)dE = dE \frac{V}{2\pi^2} \frac{m\sqrt{2m(E - E_0)}}{\hbar^3}, \tag{2.2.9}$$

which is proportional to $\sqrt{E - E_0}$, as we already deduced in Section 1.8. The total density of states must also be multiplied by the level degeneracy of the states. For example, to take into account the possibility of two spin states for each k-state, we must multiply the above density of states by a factor of 2.

The effective mass picture will break down at high energy and high momentum. Near a zone boundary, an electron band will have $\partial E/\partial k = 0$, which means that E cannot keep increasing as k^2 indefinitely. Recall also that according to the Bloch theorem, all momenta in reciprocal space are indistinguishable from the same momentum plus or minus any reciprocal lattice vector. Therefore, the momentum of quasiparticles in a periodic crystal is not strictly conserved; it is only conserved *modulo* a reciprocal lattice vector. In other words, if a quasiparticle with momentum $\hbar\vec{k}$ picks up an additional momentum $\hbar\Delta\vec{k}$ due to some scattering process, then the new momentum $\hbar\vec{k}'$ will be

$$\hbar\vec{k}' = \hbar(\vec{k} + \Delta\vec{k}) \bmod \vec{Q}, \tag{2.2.10}$$

where $\vec{Q}$ is a reciprocal lattice vector. In particular, if a scattering process would lead to a momentum outside the Brillouin zone, the momentum "wraps around" to a momentum inside the zone. This is called an **umklapp process** ("umklapp" is the German word for "fold over.")

Because the quasiparticle momentum is not strictly conserved, it is sometimes called **quasimomentum** or **crystal momentum** instead of momentum. Again, this does not mean the momentum is not real. It just means that it is the momentum of a quasiparticle, which has definite quantum states that are not equivalent to free particle states in a vacuum.

2.3 Excitons

In Section 2.2, we said that putting energy into a solid led to the creation of two new quasiparticles, a free electron and a hole. The free electron acts just like an electron in vacuum, with a different mass but with the same charge. The hole, on the other hand, is the absence of an electron. This means that adding a hole is the same as subtracting one negative charge; in other words, a hole has positive charge equal and opposite to that of the electron. The two quasiparticles move in opposite directions in the presence of the same electric field.

In the new vacuum of the crystal in the quasiparticle picture, we therefore create two particles with equal and opposite charge whenever we promote an electron from a valence band to a conduction band. These two quasiparticles will therefore feel a Coulomb force that attracts them toward each other. If the energies of the electron and hole are low enough, they can form a bound state. The pair will move jointly. This complex is called an **exciton**. Excitons are quite common in all kinds of materials, especially in biological and organic

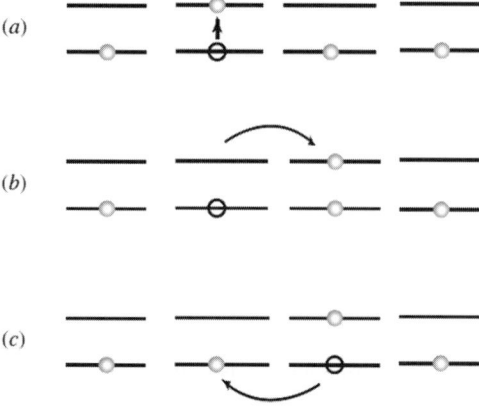

Fig. 2.2   Exciton motion in the Frenkel picture. The atoms in a crystal are represented by a set of two-level systems. (a) An electron is excited into a higher state. (b) The electron can now move into a states of the same energy in a nearby atom or molecule. (c) The extra negative charge on the nearby atom tends to push electrons in the valence states into the orginal atom.

materials; excitons are also important for many optical effects. For a general review of exciton properties, see, for example, Knox (1963).

Excitons are classified into two limits: Frenkel excitons and Wannier excitons. In the **Frenkel** exciton limit, the electron and hole are so tightly bound to each other that they both sit at the same lattice site. Figure 2.2 shows how we can think of the Coulomb attraction in this case. If an electron is promoted to an excited state, it can move into adjacent, empty sites. This leads to a extra negative charge on the adjacent atom. By Coulomb repulsion, an electron in the lower state will be pushed into the empty state on the originally excited atom. This corresponds to the hole moving to follow the excited electron.[2]

In **Wannier** excitons, the electron and hole are much less strongly coupled. In this case, the electron and hole orbit each other over many lattice sites, just like an electron and positron orbit each other to form a positronium atom in vacuum. Again, because the electron and hole are quasiparticles which exist in the new vacuum state, there is no scattering from the atoms in the crystal; the effect of the underlying crystal appears only in the effective masses of the quasiparticles and the dielectric constant of the medium that renormalizes the Coulomb interaction.

The Wannier exciton is therefore essentially an atom completely analogous to hydrogen or positronium. Following the standard formalism for the hydrogen atom, (e.g. Cohen-Tannoudji *et al.*, 1977; ch. 7), the energy spectrum for the bound electron and proton is the Rydberg energy plus the kinetic energy associated with the center of mass,

$$E_{\mathrm{H}} = \frac{-\mathrm{Ry}}{N^2} + \frac{\hbar^2 k^2}{2(m_0 + m_P)}, \tag{2.3.1}$$

---

[2]   A variant type of Frenkel exciton is the **charge transfer exciton**. In some materials, the hole and the electron have the lowest energy when they sit in adjacent lattice sites instead of the same site.

where N is the principal quantum number, m_0 is the electron mass, and m_P is the proton mass. The Rydberg is defined as (in MKS units)

$$Ry = \frac{e^2}{8\pi \epsilon_0 a_0},$$ (2.3.2)

where ϵ_0 is the vacuum dielectric constant, and the Bohr radius is defined as

$$a_0 = \frac{4\pi \hbar^2 \epsilon_0}{e^2 m_r},$$ (2.3.3)

for the reduced mass $m_r = m_0 m_P/(m_0 + m_P)$.

For an exciton, we write down the same solution, but use the effective masses of the free electron and hole and replace the factor e^2/ϵ_0 by e^2/ϵ, where ϵ is the low-frequency permittivity of the solid. This implies

$$E_{ex} = \frac{-Ry_{ex}}{N^2} + \frac{\hbar^2 k^2}{2(m_c + m_v)},$$ (2.3.4)

where

$$Ry_{ex} = \frac{e^2}{8\pi \epsilon a_{ex}}$$ (2.3.5)

and

$$a_{ex} = \frac{4\pi \hbar^2 \epsilon}{e^2 m_r}$$ (2.3.6)

with $m_r = m_c m_v/(m_c + m_v)$, and m_c and m_v are the conduction-band and valence band effective masses, respectively. The rescaling of the Coulomb interaction $e^2/\epsilon_0 \rightarrow e^2/\epsilon$ in these equations implies that the binding energy of the exciton will be several orders of magnitude less than that of hydrogen or positronium. In the semiconductor Cu_2O, for example, which has relatively isolated conduction and valence bands, so that the electron and hole masses very nearly equal the free electron mass m_0, these equations imply a binding energy of 13.6 eV/$2(\epsilon/\epsilon_0)^2$. The permittivity of this material is $\epsilon = 7\epsilon_0$, which implies $Ry_{ex} = 13.6$ eV/$2 \cdot (49) = 0.138$ eV, which is very close to the experimental value of 0.150 eV. The difference between the expected and actual value comes from corrections to the orbital wave functions due to the fact that the excitonic Bohr radius is comparable to the lattice constant.

The most important material property controlling the binding energy of excitons is therefore the dielectric constant. Table 2.1 gives a list of typical exciton binding energies and excitonic Bohr radii. Biological materials and molecular solids typically have very low dielectric constants, because they have a lot of empty space, and therefore excitons in these materials tend to have large binding energy and small radius; in other words, excitons in these materials tend to be Frenkel excitons.

Exciton bound states exist *below* the energy gap of a semiconductor or insulator. The binding energy of an exciton is subtracted from the energy needed to create a free electron and hole.

Table 2.1 Exciton parameters in various materials

	ϵ/ϵ_0	Binding energy (meV)	Approximate radius (Å)
KCl	4.6	580	3
CuCl	5.6	190	7
Cu_2O	7.1	150	7
Si	11.4	12	50
GaAs	13.1	4	150

An exciton is the quantum of electronic excitation in a solid (thus the name **exciton**). This is true even if the excitation energy is above the band gap. In this case, the electron and hole will be a correlated pair. If there were no scattering of the electron and hole into other states, the unbound electron and hole created from a single excitation would stay correlated with each other forever. These correlated, unbound electron–hole pair states are sometimes important for understanding optical effects; they can be treated with hypergeometric confluent functions as discussed by Landau and Lifshitz (1977: s. 36). Eventually, scattering effects will cause the electron and hole to become uncorrelated if they are not bound to each other.

In principle, we can also work in the exciton picture in the case of metals. In this case, however, the presence of free carriers will lead to screening of the Coulomb interaction between an electron and hole, as is discussed in Section 5.5. Strong screening will effectively wash out the Coulomb attraction which makes the excitons stable.

Excitons cease to exist when the electron and the hole meet at the same point in space and the electron fills the valence state of the hole. The rate of exciton recombination is therefore proportional to the probability of this happening, which is given by the square of the electron–hole orbital wave function at zero separation. The full quantum mechanical method for calculating the exciton wave function is given in Section 11.6. For a Wannier exciton, the probability of the electron and hole being at zero separation is given by the hydrogenic wave function value

$$|\phi(0)|^2 = \frac{1}{\pi a_{ex}^3}. \tag{2.3.7}$$

This implies that smaller excitons tend to have faster decay rates. Typical radiative lifetimes for excitons range from hundreds of picoseconds to milliseconds. In some cases, excitons can be metastable. The presence of an exciton can create such a strong electric field at a lattice site that the crystal structure is distorted, which in turn leads to trapping of the electron and hole at different lattice sites so that they cannot recombine. This is called a **self-trapped exciton** and is important in understanding radiation damage of many materials.

Exercise 2.3.1　Calculate the excitonic Bohr radius expected in the semiconductor ZnO, which has an average dielectric constant of 8.3, conduction-band effective mass of $0.275m_0$, and effective hole mass of $0.59m_0$, where m_0 is the vacuum electron mass.

2.4 Metals and the Fermi Gas

Suppose that in some solid, there is a partially filled band as illustrated in Figure 2.3. At low temperature, the electrons will fall into the lowest energy states, but because of the Pauli exclusion principle, only one electron can occupy each state. At $T = 0$, the electrons will fill up all the states below some energy E_F, which is called the **Fermi level**, and all the states above this level will be empty. This is called the **Fermi sea**. As discussed in Section 1.11.2, this system will conduct electricity at low temperatures, because electrons at the top of the Fermi sea can be accelerated by an electric field into nearby, empty states with slightly higher energy, as illustrated in Figure 2.3.

At first glance, there seems to be an inconsistency. We have written down the energy of the electrons as simply the free-particle energy $E = \hbar^2 k^2/2m$, where m is the effective mass of the band, but what about the energy due to all the repulsive Coulomb interactions of the negatively charged electrons? In a gas of electrons of substantial density, we would expect a strong effect due to the electron charge.

The answer is that there is indeed a contribution of the electron–electron Coulomb interaction to the energy of the electrons, as well as the Coulomb interaction of the electrons with the positively charged nuclei in the solid, but this energy is already taken into account in the shape of the band. As discussed in Section 1.9.5, a proper calculation of the band structure of a material must include the effects of the electrons on each other self-consistently. Once we have a given band, almost the entire effect of the Coulomb interactions is accounted for in the value of the effective mass and the band gaps. (There will be a small, additional effect of electron–electron correlations, as will be discussed in Section 8.15.) Therefore, in our model of the electrons, the effect of Coulomb repulsion of the electrons in their ground state is ignored. We don't ignore the Coulomb interaction of electrons and holes in an exciton, as discussed in Section 2.3, because that is an excited state. Similarly, when a metal is at finite temperature, there can be effects of the Coulomb interactions of the electrons, which will be discussed in Chapter 8. But as we will see in Section 2.4.2, in many metals the zero-temperature approximation works well even at room temperature.

It is also possible that the states near the maximum of an electron band can be empty, for example, as illustrated in Figure 2.4 for the case of two bands that cross in energy.

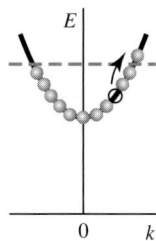

Fig. 2.3   A Fermi sea in a metal. Electrons can increase in energy by moving from states below the Fermi level to unoccupied states above the Fermi level.

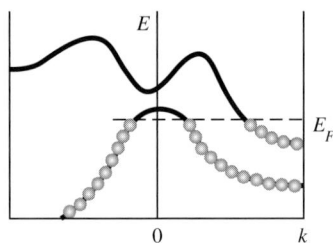

Fig. 2.4   A hole Fermi sea in a metal, in the case of band crossing.

The bottom band in this case can arise from a full molecular state and the upper band from an empty molecular state, but if the bands cross, electrons in the lower band can lower their energy by falling into the minimum of the upper band that occurs at the zone boundary.

In this case, we can think of the empty states at the top of the valence band as a Fermi sea of *holes*. This population of holes will have all the same conduction properties as a Fermi sea of electrons. In each case, the particles will have an effective mass given by the curvature of the bands.

In thinking about these metallic systems, we have a choice about the quasiparticle picture we want to use. We could define the quasiparticles as excitations out of the ground state, which for a metal is a partially filled band. In this case, the vacuum is the $T = 0$ state of all the electrons. We can create a quasiparticle either as a free electron above the Fermi level, or as a hole in the Fermi sea below the Fermi level, as illustrated in Figure 2.3. As in the case of excitations in a semiconductor, the addition of energy creates an electron–hole *pair*, when an electron below the Fermi level is promoted to a state above the Fermi level, leaving a hole behind.

Alternatively, we can define the vacuum as a state that includes only completely full or empty bands, and define the quasiparticles as all those particles that contribute to the conduction of the material, that is, all the quasiparticles in a partially filled band. These can be either electrons or holes, depending on which picture is more convenient. This freedom of defining the vacuum energy is just the same as the freedom we have in defining the $E = 0$ point of the potential energy of a system. For example, in the case of gravitational potential energy, we are free to define any height above the surface of the earth as $E = 0$. So also in the electron bands, we are free to define $E = 0$ at any point, whichever is most convenient. With semiconductors, the ground state (filled valence band, empty conduction band) is the most obvious choice. In metals, it is often convenient to choose the same vacuum state as in semiconductors (filled core states, empty conduction states), which is not the ground state of the system, and treat all the quasiparticles in the partially filled bands as free particles.

It is therefore often a valid approximation to think of metals as having carriers that act just the same as noninteracting fermions in a vacuum, but with a renormalized mass. The properties of this system are reviewed in the following subsections. For the general theory of Fermi–Dirac statistics, see, for example, Huang (1963).

2.4.1 Isotropic Fermi Gas at $T = 0$

The Fermi energy of an electron gas depends on the total number of particles. For an isotropic system, we can calculate this dependence simply in terms of the effective mass of the particles.

Since each electron fills one state, if we find the total number of states below the Fermi level, we have found the total number of particles. At $T = 0$, all the electrons are below the Fermi level and none is above the Fermi level. Using the density of states for an isotropic mass calculated in Section 2.2, the total number of electrons is therefore

$$
\begin{aligned}
N &= \int_0^{E_F} \mathcal{D}(E)dE \\
&= \int_0^{E_F} 2\frac{V}{2\pi^2}\frac{m}{\hbar^3}\sqrt{2mE}\,dE \\
&= 2V\frac{\sqrt{2}}{2\pi^2}\frac{m^{3/2}}{\hbar^3}\left(\frac{2}{3}E_F^{3/2}\right),
\end{aligned}
\tag{2.4.1}
$$

where we have assumed that there are two spin states for every electron k-state. The density of the electrons is therefore

$$
n = \frac{N}{V} = \frac{2\sqrt{2}}{3\pi^2}\frac{m^{3/2}}{\hbar^3}E_F^{3/2}.
\tag{2.4.2}
$$

We can then solve for E_F as a function of the density,

$$
E_F = \left[n\frac{(3\pi^2)}{2\sqrt{2}}\frac{\hbar^3}{m^{3/2}}\right]^{2/3}.
\tag{2.4.3}
$$

For a density of $n \sim 10^{23}$ cm^{-3} and electron mass on the order of the free electron mass in vacuum, this implies $E_F \sim 10$ eV.

To understand what this means for a macroscopic solid, let us ask what would happen if we compressed the crystal volume by 1%. This implies that the electron density will increase by 1%, which means that the Fermi energy will increase by two-thirds of 1%, or around 60 meV. This is well above the typical thermal energy k_BT at room temperature of 26 meV.

Another way of looking at it is to consider at the compressibility (that is, the amount of pressure it would take to change the volume of the gas). The pressure for a closed system is defined as

$$
P = -\frac{\partial U}{\partial V},
\tag{2.4.4}
$$

where U is the total energy. This is simply the weighted integral

$$
\begin{aligned}
U &= \int_0^{E_F} \mathcal{D}(E)E\,dE \\
&= \int_0^{E_F} 2\frac{V}{2\pi^2}\frac{\sqrt{2}m^{3/2}}{\hbar^3}E^{3/2}\,dE
\end{aligned}
$$

$$= 2V \frac{\sqrt{2}}{2\pi^2} \frac{m^{3/2}}{\hbar^3} \left(\frac{2}{5} E_F^{5/2} \right)$$

$$= 2V \frac{\sqrt{2}}{5\pi^2} \frac{m^{3/2}}{\hbar^3} \left[\frac{N}{V} \frac{(3\pi^2)}{2\sqrt{2}} \frac{\hbar^3}{m^{3/2}} \right]^{5/3}. \tag{2.4.5}$$

From this, it follows that the pressure is simply

$$P = \frac{2}{3} \frac{U}{V},$$

and the bulk compressibility is

$$B = -V \frac{\partial P}{\partial V}$$

$$= \frac{5}{3} P = \frac{2}{3} n E_F. \tag{2.4.6}$$

The pressure needed to change the volume by the fraction $\Delta V/V = 1\%$ is then

$$\Delta P = B(\Delta V/V)$$

$$= \frac{2}{3} n E_F (0.01), \tag{2.4.7}$$

which for a typical solid density is 8 eV$\times 10^{23}$ cm$^{-3} \times 0.01 \simeq 10^4$ atm! This means that to compress a Fermi gas of electrons at a typical solid density by 1%, we would need a pressure of over 10,000 atm. This is correct – typical bulk compressibilities of solids lie in the range of a few times 10^{10} N/m^2. To significantly alter the structure of a crystal, **diamond-anvil** cells are used which can reach megabar pressures in tiny volumes.

This shows that the Fermi pressure of the electrons is extremely important in the stability of solids. We have already seen in Section 1.1.2 that bringing atoms close together increases the bonding–antibonding splitting. This would seem to imply that bringing the atoms closer and closer together would continue to decrease the energy of the crystal. In the same way, if we ignored the bonding energy, the Coulomb potential between the charged particles would also give a net decrease of energy as the atoms get closer together. This can be seen by a simple model, shown in Figure 2.5, in which we have an infinite series of alternating charges, all spaced equally a distance l apart. The Coulomb potential felt by any one charge is

$$U = 2 \frac{-q^2}{l} + 2 \frac{q^2}{2l} + 2 \frac{-q^2}{3l} + \cdots$$

$$= -2 \frac{q^2}{l} \left(1 - \frac{1}{2} + \frac{1}{3} \cdots \right) = -2 \frac{q^2}{l} \ln 2, \tag{2.4.8}$$

| +q | −q | +q | −q | +q | −q |

$\longleftarrow$ l $\longrightarrow$

Fig. 2.5 A linear Madelung model for the Coulomb energy of a solid.

which is increasingly negative as l decreases. This is known as the **Madelung potential**. A three-dimensional lattice will also have a net Coulomb energy that is increasingly negative as l decreases. The Madelung potential is a reasonable approximation for electrons that are delocalized and not bound to any one nucleus; for example, nearly free electrons in the conduction band of a metal, or for strongly ionic crystals. In the covalent bonding picture, the Coulomb energy is taken into account in the Rydberg binding energy of the individual atoms.

These considerations would make it seem that there is nothing to prevent a solid from collapsing to infinite density. As the atoms get closer together, however, at some point the Fermi energy becomes significant, overcoming the bonding energy and the Coulomb potential. As seen above, at around $n \sim 10^{23}$ cm^{-3}, the Fermi energy of a metal is comparable to the Rydberg of the atomic states. The picture for a semiconductor or insulator is not much different. One still has to put the same number of electrons in the same volume of space, and so the density n above can be approximately replaced by $1/a^3$, where a is the unit cell size. This is why typical densities of crystals are not much greater than 10^{23} cm^{-3}. The Fermi pressure is what makes solids stable.

Exercise 2.4.1 (a) Compute the root-mean-squared velocity of a Fermi gas at $T = 0$ as a function of the density.

(b) What is the root-mean-squared magnitude of the wave vector k in cm^{-1} for a $T = 0$ Fermi gas of free electrons with density 10^{22} cm^{-3} and effective mass $m_c = 0.1m_0$, where m_0 is the free electron mass?

2.4.2 Fermi Gas at Finite Temperature

When $T \neq 0$, the electrons no longer all sit below the Fermi energy. Electrons can be thermally excited by an energy on the order of $k_B T$. This means that electrons within $k_B T$ below the Fermi edge can jump to states about $k_B T$ above the Fermi edge. From statistical mechanics, the exact formula for the occupation number of a state with energy E is

$$f(E) = \frac{1}{e^{(E-\mu)/k_B T} + 1}, \tag{2.4.9}$$

where μ is the chemical potential of the gas, which is equal to the Fermi energy E_F when $T = 0$. Figure 2.6 shows this distribution. As seen in this figure, the distribution differs from the $T = 0$ distribution only for energies within $k_B T$ of the Fermi level. As $k_B T \to 0$,

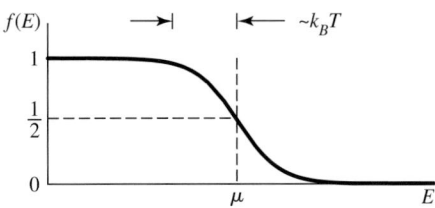

The Fermi–Dirac distribution.

the thermally excited region narrows until the distribution function is equal to 1 for all energies below E_F and 0 for all energies above E_F.

As we found in the previous subsection, typical Fermi energies in solids are of the order of several eV, compared to $k_B T = 0.026$ eV at room temperature. This means that we can often use the $T = 0$ approximation even at room temperature, because the depletion region is a small perturbation of the $T = 0$ distribution.

For $T \neq 0$, the total number of particles in the gas is now given by

$$
\begin{aligned}
N &= \int_0^\infty f(E)\mathcal{D}(E)dE \\
&= \int_0^\infty \frac{1}{e^{(E-\mu)/k_B T} + 1} \, 2\frac{V}{2\pi^2} \frac{m}{\hbar^3} \sqrt{2mE} \, dE.
\end{aligned}
\tag{2.4.10}
$$

As illustrated in Figure 2.7, if the number of particles is constant, then as the temperature rises, the chemical potential μ must fall below the original value of E_F. This is because the Fermi–Dirac distribution is antisymmetric relative to $E = \mu$; that is, as T increases, the decrease of $f(\mu - E)$ below 1 is exactly equal to the increase of $f(\mu + E)$ above zero. This follows from the simple mathematical fact that

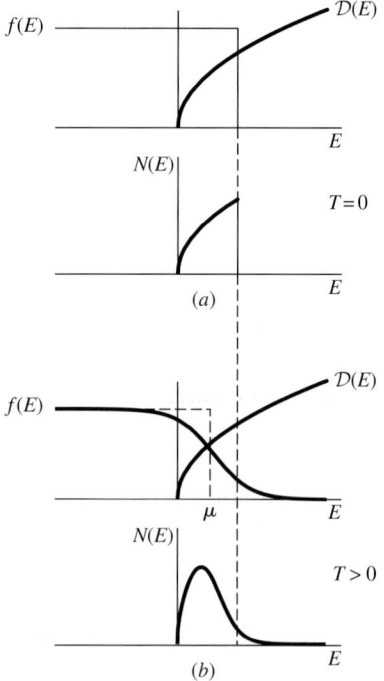

(a)

(b)

Fig. 2.7   Noninteracting electron energy distribution. The total number of particles at a given energy is the product of the density of states and the Fermi–Dirac occupation number. In (a), the Fermi–Dirac distribution $f(E)$ and the resulting $N(E)$ is plotted for $T = 0$. In (b), $f(E)$ is plotted for finite T. For a fixed number of particles, the chemical potential μ must decrease relative to its $T = 0$ value in order to keep the integral $N = \int N(E)dE = \int f(E)\mathcal{D}(E)dE$ constant.

$$1 - \frac{1}{e^{-x} + 1} = \frac{1}{e^x + 1}. \tag{2.4.11}$$

Since there are more states above μ than below, if the number of particles per state above μ increases by the same amount that the number of particles per state below μ decreases, there will be a net increase of the total number of particles unless the value of μ decreases.

As the temperature rises, the chemical potential can fall to well below the single-particle ground state energy. In this case, the factor $e^{(E-\mu)/k_BT}$ is much larger than 1 for all E, in which case the Fermi–Dirac distribution can be approximated as simply

$$f(E) \sim e^{-E/k_BT} e^{\mu/k_BT}. \tag{2.4.12}$$

This is just the Maxwell–Boltzmann distribution. At high temperature, a Fermi gas acts like a classical gas with Maxwell–Boltzmann statistics.

2.5 Basic Behavior of Semiconductors

As discussed in Section 2.1, a semiconductor is a material with an energy gap between a full band and an empty band. Since the electrons in a full valence band cannot move into different states because of Pauli exclusion, none of them can respond to a force, and therefore a semiconductor does not conduct electricity at $T = 0$. A semiconductor will conduct electricity only when some electrons have been excited into the upper, conduction band. In this case, not only can these free electrons in the upper band move into nearby, empty states, but also, electrons in the valence band can move into the empty states left behind in the lower band. In other words, holes in the valence band can also conduct electricity. The electron and hole quasiparticles carry the entire current, which is why they are often called charge carriers.

The conductivity of a semiconductor is therefore proportional to the number of free carriers, including both free electrons and holes. If there are no other influences, the number of electrons in the conduction band just depends on the statistical probability of an electron in the valence band being thermally excited to the conduction band. As we will see, this probability depends strongly on the temperature and the gap energy. The sensitivity of the conductivity to these factors is what makes semiconductors so interesting.

In Section 2.1, we gave the simple picture of a semiconductor with a valence band maximum and conduction band minimum at the center of the Brillouin zone. This is called a **direct gap** semiconductor. There is no reason why both the valence band maximum and conduction band minimum must be at the center of the zone, however. The minimum of the upper band could be at one of the critical points on the zone boundaries, or it could lie at some other point inside the Brillouin zone, as shown in Figure 2.8. A semiconductor with valence band maximum and conduction band minimum at different points in the Brillouin zone is called an **indirect gap** semiconductor. In this case, an electron in the valence band must change both energy *and* momentum in order to go from the valence band maximum to the conduction band minimum. Of course, an electron in the valence band can also be promoted directly upward to a higher energy state at the same momentum. In this case, the

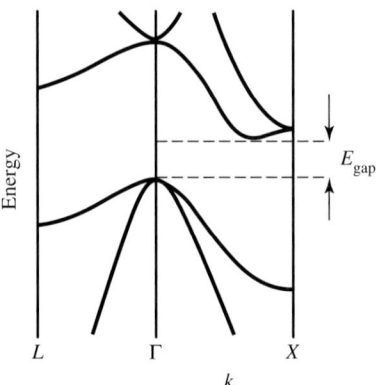

Fig. 2.8 Band structure of the indirect gap semiconductor silicon in the region of the band gap.

excited electron will tend to fall down into the conduction band minimum, gaining momentum but losing energy. Whether the gap is direct or indirect will not affect the probability of occupation of the bands in equilibrium.

2.5.1 Equilibrium Populations of Electrons and Holes

At $T = 0$, there will obviously be no electrons in the conduction band of a semiconductor. To determine the number at finite temperature requires a little work, though, because a thermal excitation that lifts an electron from the valence band to the conduction band creates *two* free particles; in other words, in the quasiparticle picture, the number of particles is not conserved.

The problem essentially reduces to finding the chemical potential of the electrons and holes. In the case of a metal, the chemical potential was equal to the highest energy of the occupied electron states at $T = 0$. Where do we put the chemical potential for a semiconductor at $T = 0$? Do we put it at the top of the valence band? Or do we put it at the bottom of the conduction band?

As in the case of a metal, we write the total number of electrons in the conduction band of a semiconductor as

$$N_e = \int_{E_c}^{\infty} f(E)\mathcal{D}_c(E)dE, \tag{2.5.1}$$

where $\mathcal{D}_c(E)$ is the density of states in the conduction band at energy E, the function $f_e(E)$ is the occupation number, given by the Fermi–Dirac distribution (2.4.9), and μ_e is the chemical potential of the electrons. The upper bound of the integral is not, of course, really infinity, but simply high enough to extend past any states occupied thermally by electrons. Similarly, the total number of holes in the valence band is

$$N_h = \int_{-\infty}^{E_v} (1 - f(E))\mathcal{D}_v(E)dE, \tag{2.5.2}$$

where $\mathcal{D}_v(E)$ is the density of states of the valence band. As discussed in Section 2.4.2, the Fermi–Dirac distribution has the property that the depletion at a given energy below the chemical potential is equal to the occupation at the same energy above the chemical potential; in other words,

$$1 - f(E) = 1 - \frac{1}{e^{(E-\mu)/k_B T} + 1}$$

$$= \frac{1}{e^{(\mu-E)/k_B T} + 1}. \tag{2.5.3}$$

As also mentioned in Section 2.4.2, we can make the approximation

$$f(E) = \frac{1}{e^{(E-\mu)/k_B T} + 1} \simeq e^{-E/k_B T} e^{\mu/k_B T} \tag{2.5.4}$$

for the Fermi–Dirac distribution, which is valid when $E - \mu$ is large compared to $k_B T$. This will be true, for example, when μ is somewhere in the band gap, and the band gap is large compared to $k_B T$. We then have, by changing the variables of integration,

$$N_e = \int_0^\infty e^{-(E+E_c)/k_B T} e^{\mu/k_B T} \mathcal{D}_c(E + E_c) dE$$

$$N_h = \int_0^\infty e^{-(E-E_v)/k_B T} e^{-\mu/k_B T} \mathcal{D}_v(E_v - E) dE. \tag{2.5.5}$$

Multiplying these two together, we have

$$N_e N_h = \left(\int_0^\infty e^{-E/k_B T} \mathcal{D}_c(E + E_c) dE \right) \left(\int_0^\infty e^{-E/k_B T} \mathcal{D}_v(E_v - E) dE \right) e^{-E_g/k_B T}, \tag{2.5.6}$$

where $E_g = E_c - E_v$, or, if we define

$$N_Q^{(e)} \equiv \int_0^\infty e^{-E/k_B T} \mathcal{D}_c(E + E_c) dE$$

$$N_Q^{(h)} \equiv \int_0^\infty e^{-E/k_B T} \mathcal{D}_v(E_v - E) dE, \tag{2.5.7}$$

we can write

$$\boxed{N_e N_h = N_Q^{(e)} N_Q^{(h)} e^{-E_g/k_B T}.} \tag{2.5.8}$$

This is known in chemistry as a **mass action equation** or in plasma physics as a **Saha equation**; this type of equation is a simple consequence of chemical equilibrium between different species that can convert into each other. The integrals in (2.5.7) can be called the density-of-states factors, or the **effective density of states**, because they count the number of states available within energy $k_B T$ of the band edge. For a typical three-dimensional, parabolic band, we have the density of states (1.8.7), that is, $\mathcal{D}_c(E) \propto \sqrt{E - E_c}$, which implies $\mathcal{D}_c(E + E_c) \propto \sqrt{E}$ and similarly, $\mathcal{D}_v(E_v - E) \propto \sqrt{E}$.

Since each free electron comes from creating one hole, we know that $N_h = N_e$, which means we can take the square root to obtain

$$\boxed{N_e = \sqrt{N_Q^{(e)} N_Q^{(h)}}\, e^{-E_g/2k_B T}.} \tag{2.5.9}$$

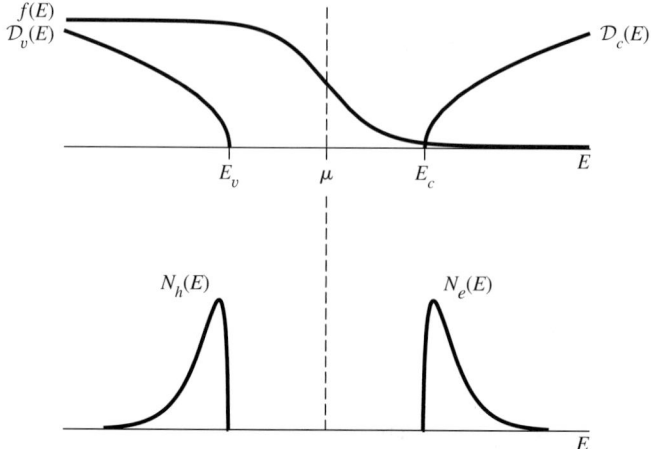

Fig. 2.9 Electron and hole energy distributions for a semiconductor at finite T. For $\mu \approx (E_c + E_v)/2$, the number of holes in the valence band equals the number of excited electrons in the conduction band.

If the density-of-states factors for the electrons and holes are equal (e.g., the effective masses of the valence and conduction bands are equal), then we have simply

$$N_e = N_Q^{(e)} e^{-E_g/2k_B T}.$$ (2.5.10)

Comparing this to the equation for N_e in (2.5.5), we have

$$-E_c + \mu = -\frac{E_g}{2}$$ (2.5.11)

which implies

$$\mu = E_c - \frac{E_g}{2}.$$ (2.5.12)

In other words, for equal effective masses, the chemical potential lies exactly in the middle of the band gap, as illustrated in Figure 2.9. If the density of states of the valence band is greater than that of the conduction band, the chemical potential will be pushed higher toward the conduction band. If E_g is large compared to $k_B T$, this will be a small correction, since the shift depends only logarithmically on the ratio of the density of states factors.

Exercise 2.5.1 Determine the value of the chemical potential in the case when both the conduction and valence band are simple isotropic bands, and the hole effective mass is four times the conduction band effective mass, in the limit of low temperature, $k_B T \ll E_g$.

2.5.2 Semiconductor Doping

In a typical semiconductor with a band gap of 1–2 eV, very few electrons will be thermally excited enough to jump from the valence band to the conduction band. It is possible to get free carriers in another way, however, by selective **doping** of the material.

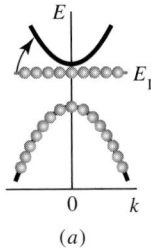 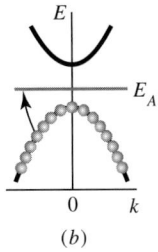

(a)   (b)

Fig. 2.10   (a) Donor and (b) acceptor states in a doped semiconductor near zone center. Excitations to empty states can occur with much lower energies than in an undoped semiconductor.

As discussed in Section 1.11.2, typical semiconductors are made from II–VI, III–V or Group IV elements, which have eight electrons per unit cell filling eight molecular bonding states. Suppose that in a III–V semiconductor, we replace a single atom with one from Group VI. This atom will have an extra electron and an extra positive charge on its nucleus. It makes sense that the extra electron will tend to stay near the nucleus with the extra charge, but if the energy of the extra electron is high enough, it may leave the dopant atom and migrate through the crystal as a free quasiparticle.

As in the case of excitons, if the dielectric constant of the material is high, then the bound states of carriers at an impurity can be viewed as hydrogen-like states of a carrier orbiting a charged particle. If a dopant nucleus has one extra positive charge relative to the atom it replaces in the lattice, then a negative electron will see a single positive charge sitting in the background "vacuum" of the normal band structure of the material. It can then orbit this positive charge in a hydrogenic state with a binding energy given by (2.3.5). The reduced mass in this case will just be the effective mass of the electron, because the impurity cannot move, and therefore it effectively has infinite mass.

In the same way, a dopant nucleus with one *less* charge than the one it replaces in the lattice will look like a single negative charge to the quasiparticles. Therefore, a hole can orbit this atom in a hydrogenic bound state. Again, the negative impurity has effectively infinite mass because it cannot move, and so the binding energy of the hole will simply be proportional to the mass of the hole, according to (2.3.5) and (2.3.6).

At high temperature, that is, when $k_B T$ is much greater than the binding energy of a carrier orbiting an impurity, the carriers will no longer be bound to the impurities and will move freely through the crystal. Therefore, an atom with more positive charge on its nucleus than the one it replaces contributes an electron to the conduction band. This type of impurity is therefore called a charge **donor**, and a semiconductor doped with this type of atom is called *n*-**type** (for "negative"). In the same way, an atom with a nucleus with more negative charge than the one it replaces contributes a hole to the valence band, which is the same as accepting an electron from the valence band. This type of impurity is therefore called an electron **acceptor**, and a semiconductor doped with this type of atom is called "*p*-**type**" (for "positive").

Donor states lie just below the conduction band, while acceptor states lie just above the valence band, as shown in Figure 2.10. At low temperature, the donor states are filled with electrons, while the acceptor states are all empty (filled with holes.) Because bound states around the impurities are localized states which cannot move, they effectively have infinite mass. Therefore, impurity states are drawn as horizontal lines; recall that light mass corresponds to large band curvature, and therefore infinite mass corresponds to zero curvature.

Sometimes people want to draw exciton states on the band diagrams just like donor and impurity states. This is not so simple, however. If an exciton becomes unbound, it contributes *both* an electron and a hole to the conduction. One could draw an exciton as a donor and an acceptor state simultaneously, but it is better to simply remember that an exciton is a *two*-particle state and therefore it cannot be properly drawn on the energy band diagram for *single* particles. Instead, one should start from scratch and draw the proper excitation energy dispersion curve.

If there are both types of impurities in the same material, then extra electrons from donors will fall down to fill the acceptor states. If there are more donors than acceptors, then there will be free electrons left over which can contribute to conduction, while if there are more acceptors than donors, there will be extra holes. The conductivity σ of the system is therefore proportional to the difference between the two doping concentrations. A semiconductor with nearly equal concentrations of donors and acceptors is called **compensated**. To make a semiconductor highly insulating, it is easier to compensate the impurities than to remove the impurities and make a pure semiconductor, because it is nearly impossible to have a perfectly impurity-free crystal.

2.5.3 Equilibrium Populations in Doped Semiconductors

At $T = 0$, each additional electron and hole will be in its ground state, orbiting the impurity atom from which it came. At higher temperature, however, these quasiparticles can leave their bound states and move freely through the material. The probability of this happening depends on the binding energy of the carrier on the impurity. As we did in Section 2.5.1, we can write a mass action equation for the equilibrium concentration of the quasiparticles. For example, for electrons from donor atoms, at low temperature we write

$$N_e N_h = N_Q^{(e)} N_D e^{-(E_c - E_D)/k_B T}, \tag{2.5.13}$$

where N_D is the number of donor states available to the holes. At high temperature, however, this equation will break down, because the assumption that $k_B T \ll (E_c - E_D)$ made in the derivation of (2.5.8) will no longer be true. As discussed at the end of Section 2.5.1, if the number of donor states is much less than the number of states in the conduction band, as is typically the case in doped semiconductors, the chemical potential will be pushed down toward the donor state energy. In this case, we must use the Fermi–Dirac occupation number for the holes in the donor states,

$$N_h = N_D \frac{1}{e^{(\mu - E_D)/k_B T} + 1}. \tag{2.5.14}$$

This can be rewritten as

$$
\begin{aligned}
N_h &= N_D \left(1 - \frac{1}{e^{(\mu - E_D)/k_B T} + 1} \right) e^{-(\mu - E_D)/k_B T} \\
&= N_D \left(1 - f_h(E_D) \right) e^{-(\mu - E_D)/k_B T} \\
&= (N_D - N_h) e^{-(\mu - E_D)/k_B T}.
\end{aligned}
\tag{2.5.15}
$$

Therefore, the above mass-action law will be revised to

$$N_e N_h = N_Q^{(e)}(N_D - N_h)e^{-(E_c - E_D)/k_B T}. \tag{2.5.16}$$

The number of neutral donors, that is, the number of donors with electrons orbiting them, is just equal to $N_D^0 = (N_D - N_h)$. Therefore, we could equally well write

$$\boxed{\frac{N_e N_h}{N_D^0} = N_Q^{(e)} e^{-\text{Ry}_D/k_B T},} \tag{2.5.17}$$

where $\text{Ry}_D = E_c - E_D$ is the binding energy of an electron in orbit around a donor. This is the same mass action law we would have found if we had started with electrons, holes, and neutral donors as independent quasiparticles, in the limit that the hole and donor mass are much greater than the free electron mass, just like the case of dissociation of hydrogen atoms into protons and electrons.

We can solve for the number of electrons in the conduction band in equilibrium by noting that $N_h = N_e$ and $N_D^0 = N_D - N_e$. Then we have

$$N_e^2 - (N_D - N_e)N_Q^{(e)} e^{-\text{Ry}_D/k_B T} = 0 \tag{2.5.18}$$

or

$$N_e = \frac{1}{2} e^{-\text{Ry}_D/k_B T} N_Q^{(e)} \left(\sqrt{1 + 4e^{\text{Ry}_D/k_B T} N_D/N_Q^{(e)}} - 1 \right). \tag{2.5.19}$$

In the limit of low temperature, this becomes

$$N_e = \sqrt{N_D N_Q^{(e)}} e^{-\text{Ry}_D/2k_B T}, \tag{2.5.20}$$

just as in the case of carriers excited across the band gap, while in the limit of $k_B T \gg \text{Ry}_D$, we have $N_e \to N_D$.

Since the binding energy Ry_D is of the same order of magnitude as the exciton binding energy in the system (i.e., 5–50 meV), instead of the band gap energy, 1–2 eV, this means that the system will be a conductor at relatively low temperatures. In particular, typical doped semiconductors are conductors at room temperature, since $k_B T \simeq 26$ meV is comparable to the binding energy of carriers at impurities. This shows why doping is so important – by selective control of the doping levels, the same semiconductor material can be turned into either an insulator, a conductor with negative free charges, or a conductor with positive free charges. This is true even for relatively low levels of doping; for example, doping levels of 10^{16} cm^{-3} compared to typical atomic densities around 10^{22} cm^{-3}, that is, one in one million atoms different from the rest!

The same type of mass-action equation can be applied to excitons. Excitons will dissociate into free electrons and holes when the temperature is high enough, just as a hydrogen atom will dissociate into electrons and protons. In the case of excitons, however, we cannot always assume that the electron effective mass is much lighter than the other effective masses. In the general case, if we assume that the exciton and free carrier lifetimes are long enough so that we can treat the total number of electrons and holes as conserved, then we write

$$\frac{N_e N_h}{N_{ex}} = \frac{N_Q^{(e)} N_Q^{(h)}}{N_Q^{(ex)}} e^{-\mathrm{Ry}_{ex}/k_B T}, \tag{2.5.21}$$

where the density of states factors are calculated for the respective effective masses of the free carrier and exciton states as in (2.5.7).

As shown in many thermodynamics textbooks (e.g. Reif 1965), the mass-action equation can be derived simply from the equilibrium condition $\mu_e + \mu_h = \mu_{ex}$, and from the Maxwellian (low-density) formula for the total density,

$$N_i = e^{-(E_i-\mu)/k_B T} N_Q^{(i)}, \tag{2.5.22}$$

where E_i is the ground state energy of species i. A mass-action equation of this form breaks down in the limit when the binding energy is very small – it can easily be shown by solving (2.5.21) for N_{ex}, subject to the constraints $N_e = N_h$ and $N_{ex} = N - N_e$, that N_{ex} goes to a constant, and not zero, when $\mathrm{Ry}_{ex} \to 0$. The reason is that exciton states with extremely shallow binding are still counted as exciton states, when in reality such states would be unstable, with very large orbits, and electrons and holes in such orbits would be indistinguishable in practice from free electrons and holes. As a first-order correction, one can subtract from N_Q^{ex} the states with energy greater than Ry_{ex}, that is, a fraction approximately equal to $e^{-\mathrm{Ry}_{ex}/k_B T}$, in which case the mass-action equation becomes

$$\frac{N_e N_h}{N_{ex}} = \frac{N_Q^{(e)} N_Q^{(h)} e^{-\mathrm{Ry}_{ex}/k_B T}}{N_Q^{(ex)}(1 - e^{-\mathrm{Ry}_{ex}/k_B T})} = \frac{N_Q^{(e)} N_Q^{(h)}}{N_Q^{(ex)}} \frac{1}{e^{\mathrm{Ry}_{ex}/k_B T} - 1}, \tag{2.5.23}$$

which has the expected behavior that $N_{ex} \to 0$ when $\mathrm{Ry}_{ex} \to 0$. Much theoretical effort has been given, especially in the astrophysics community, to justifying this approach more rigorously, for example using Planck–Larkin partition functions, which give higher-order corrections (see, e.g., Ebeling *et al.* 1976; Kraft *et al.* 1986).

Exercise 2.5.2   Use a plotting package like Mathematica to plot the ratio of free electrons to electrons bound to impurities (neutral donors) as a function of T, using (2.5.19), for the realistic case of doping density $N_D = 10^{16}$ cm^{-3} and $\mathrm{Ry}_D = 5$ meV, and effective mass $m = 0.1m_0$. At what temperature is N_e 99% of N_D?

2.5.4 The Mott Transition

As discussed above, a single impurity in a semiconductor can be modeled as an atom with hydrogen-like states sitting in the background vacuum of a perfect crystal. If we start with these states as the eigenstates of the quasiparticles, then we can again go through exactly the same argument as we did in Section 1.1: If the impurity states overlap each other substantially, then *bands* of impurity states will arise.

If we keep increasing the doping level of a semiconductor, we therefore expect that, at some point, we will see dramatic changes in the conductivity of the material. At low temperature and low doping level, the semiconductor should not conduct electricity, because all of the carriers are confined in the lowest bound states of the impurities. At high dopant concentration, however, the system will become a metal when the wave functions overlap. This is called a **Mott transition** from insulator to conductor. Doped semiconductors in this limit are labeled as $n+$ or $p+$ to indicate that they have high doping leading to metallic behavior. (The plus signs mean lots of dopants, and have nothing to do with the charge.)

The basic condition for a Mott transition is that the spacing between the impurities is comparable to the Bohr radius a of the electron bound state. Since the spacing between the impurities is proportional to $n^{-1/3}$, where n is the impurity concentration, this condition is

$$na^3 \sim 1. \tag{2.5.24}$$

For a hydrogen-like bound state in a dielectric medium, we have from (2.3.6)

$$a = \frac{4\pi\epsilon\hbar^2}{e^2 m}, \tag{2.5.25}$$

where we have set the reduced mass m_r equal to the effective mass m of the free carrier, since the impurity nucleus effectively has infinite mass. This implies an orbital radius a of around 100 Å for typical semiconductors; in other words, a typical Mott transition density is 10^{18} cm^{-3}. Above this concentration, the impurity states will form a band of their own which will make the material metallic.

Another way of looking at the Mott transition is that it occurs when the Rydberg of the bound states is small compared to the Fermi energy of free electrons at the same density. In this case, most of the electrons are frozen by Pauli exclusion and cannot freely orbit a single atom. From (2.4.3), we have

$$E_F = \left[n \frac{(3\pi^2)}{\sqrt{2}} \frac{\hbar^3}{m^{3/2}} \right]^{2/3} = \left(n \frac{(3\pi^2)}{\sqrt{2}} \right)^{2/3} \frac{\hbar^2}{m}, \tag{2.5.26}$$

while from (2.3.5) we have

$$\text{Ry}_{\text{imp}} = \frac{e^2}{8\pi\epsilon a}. \tag{2.5.27}$$

If the Fermi energy is much larger than the Rydberg energy (e.g., 10 times greater), we have

$$\left(n \frac{(3\pi^2)}{\sqrt{2}} \right)^{2/3} \frac{\hbar^2}{m} = 10 \frac{e^2}{8\pi\epsilon a}$$

$$2a \left(n \frac{(3\pi^2)}{\sqrt{2}} \right)^{2/3} \frac{4\pi\epsilon\hbar^2}{e^2 m} = 10, \tag{2.5.28}$$

or

$$na^3 = \frac{10}{6\pi^2} \sim 1. \tag{2.5.29}$$

The Mott transition also occurs for excitons. At high density, the exciton wave functions will overlap, and the excitons will cease to exist as individual quasiparticles, converting instead to a conducting plasma. This is also true of atoms in vacuum, which can undergo a transition to plasma at high density. However, the transition from an exciton gas to a plasma of free electrons and holes is generally much more complicated than this simple formula, because of collisional processes, as we will see in Section 8.11.2.

Exercise 2.5.3   Show that the exact relation between density and average interparticle spacing is given by

$$\bar{r} = 0.554 n^{-1/3} \tag{2.5.30}$$

and not simply $\bar{r} = n^{-1/3}$, as is often assumed. To do this, first define $Q(r)dr$ as the probability that the nearest neighbor of a particle lies between r and $r + dr$, and $P(r)$ as the probability of there being no neighbor closer than r,

$$P(r) = 1 - \int_0^r Q(r')dr'. \tag{2.5.31}$$

If n is the average density of the particles, it follows that

$$Q(r)dr = P(r)(n4\pi r^2 dr), \tag{2.5.32}$$

that is, the product of the probability of no particle up to r, times the probability of a particle being between r and $r + dr$. Show that the mean distance between particles is

$$\bar{r} = \int_0^\infty Q(r)r\,dr = \left(\frac{3}{4\pi n}\right)^{1/3} \Gamma\left(\frac{4}{3}\right). \tag{2.5.33}$$

This exercise was orginally suggested by Ridley (1988).

2.6 Band Bending at Interfaces

So far, we have almost entirely dealt with the properties of spatially uniform systems. The world is not uniform, however, and so in numerous situations we must deal with contacts between different types of materials. Without interfaces, nothing would ever happen – the world would be one single thing.

2.6.1 Metal-to-Metal Interfaces

The simplest type of interface is a metal-to-metal contact. As discussed in Section 2.4.2, the $T = 0$ approximation is valid for most metals. We assume that all the carriers are in the lowest possible states, filling the states up to a Fermi energy, E_F.

What will happen if two metals in contact have different intrinsic Fermi levels relative to the vacuum energy, that is, different work functions, as shown in Figure 2.11(a)? Free electrons from the metal with the higher Fermi level can fall down into empty states in the other metal. In so doing, however, they increase the negative charge on that side and leave

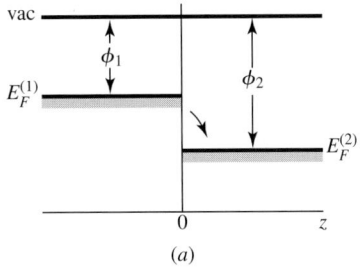

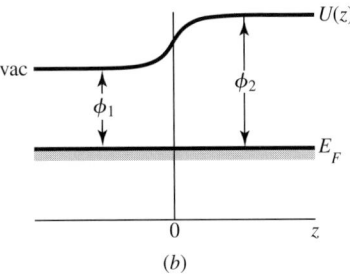

Fig. 2.11   Energy levels at the interface of two different metals. The work function ϕ of a metal gives the difference between a free electron in vacuum and the Fermi level E_F of the metal. a) When the metals are brought together, if the intrinsic Fermi levels are different, then electrons will diffuse from one side of the junction to the other. b) The same metal-metal interface in equilibrium. Charge transfer across the junction stops when the Fermi levels are equal. The work functions do not change, which implies that the vacuum potential $V(z)$ must vary across the interface.

behind a net positive charge on the other side. Eventually, when enough electrons have changed sides, the additional negative charge on the one side will be enough to repel any new electrons from coming in. At this point, the system will be in equilibrium.

Without doing any math, we can already predict that equilibrium will be reached when the Fermi levels of the two metals are equal. For this to happen, the charge transfer must cause the electric potential to vary across the boundary, as shown in Figure 2.11(b). The potential will be constant far from the junction, and vary rapidly near the junction region. This is known as **band bending**.

To determine the thickness of the band bending region, we can use the standard Poisson equation for electric potential as a function of the local charge,

$$\nabla^2 V(\vec{r}) = -\frac{\rho(\vec{r})}{\epsilon}. \tag{2.6.1}$$

In one dimension, this is simply

$$\frac{\partial^2 V}{\partial z^2} = -\frac{\rho(z)}{\epsilon}. \tag{2.6.2}$$

We can solve this approximately by assuming that all the mobile charge is transferred from one side to the other over some distance d from the interface. Then the charge distribution is

$$\rho(z) = \begin{cases} 0, & z < -d \\ en, & -d < z < 0 \\ -en, & 0 < z < d \\ 0, & z > d \end{cases}, \tag{2.6.3}$$

where n is the density of free carriers. This implies

$$V(z) = \begin{cases} V_0, & z < -d \\ V_0 - (en/2\epsilon)(z+d)^2, & -d < z < 0 \\ V_0 - \Delta + (en/2\epsilon)(z-d)^2, & 0 < z < d \\ V_0 - \Delta, & z > d \end{cases}, \tag{2.6.4}$$

where Δ is the voltage difference in the intrinsic Fermi levels of the metals. Continuity at the boundary implies

$$-(en/2\epsilon)d^2 = -\Delta + (en/2\epsilon)d^2 \tag{2.6.5}$$

or

$$d = \sqrt{\frac{\Delta\epsilon}{en}}. \tag{2.6.6}$$

For typical initial Fermi level differences of a few eV and carrier concentrations of around 10^{21} cm^{-3} (note that only electrons in the partially filled bands contribute to the charge transfer), this implies a distance d of a few angstroms. The potential is almost perfectly constant everywhere else inside the metal, as we would expect for a good conductor.

This implies that putting two dissimilar metals in contact generates a voltage difference, called a **contact potential**. This potential difference does not drive a current, however, as the carriers are in equilibrium.

2.6.2 Doped Semiconductor Junctions

Suppose now that we join two doped semiconductors together, one which is n-type and one which is p-type, as shown in Figure 2.12(a). Electrons in the donor states on one side can fall into empty acceptor states on the other side. The same thing will happen as in the case of a metal-to-metal contact: The transfer of charge from one side to the other will lead to band bending that brings the two Fermi levels to be equal. More generally, we can say that the two chemical potentials will be equal in equilibrium; the Fermi level is defined as the chemical potential in the $T = 0$ limit.

As discussed in Section 2.5.3, the chemical potential of the n-doped semiconductor will lie in the gap between the bottom of the conduction band and the donor states, very near to the donor state energy; similarly, the chemical potential of a p-doped material will be slightly below the empty acceptor states. The bands on the two sides of the junction will therefore bend until the chemical potentials of conduction and valence bands of the two materials far from the interface come to the same value, as shown in Figure 2.12(b). Note that this means that the n-doped side will have high voltage (positive charge), and the p-doped side will have low voltage, since the negative sign of the electron charge implies

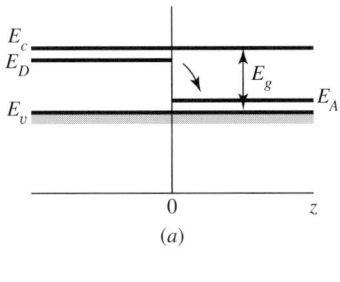

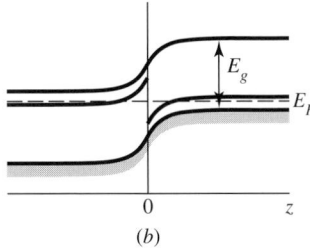

Fig. 2.12 Doped semiconductor junction. (a) As in the case of a metal, electrons in donor states can transfer across the junction into empty hole states. (b) After transfer, the bands are bent so that the Fermi levels are equal.

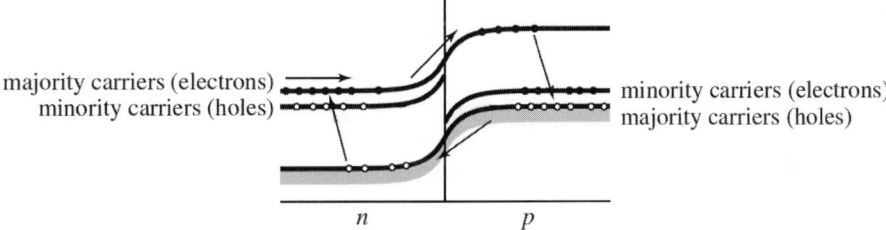

Fig. 2.13 Current flow from right to left in a p–n junction.

low electron potential energy $U(z)$ corresponds to high electric potential. In other words, the n-doped side loses some electrons, becoming positively charged, and the p-doped side gains some electrons.

This system, known as a **p–n junction**, is the basis of all kinds of semiconductor devices. Figure 2.13 shows the current flow in a p–n junction, corresponding to positive current going from right to left. Consider first the flow of the free electrons from left to right. Electrons in the conduction band will have two choices. They can tunnel through the barrier at the junction (created by the band bending in the junction region) and move into the empty acceptor states on the right side, or they can jump up to the conduction band on the other side of the junction. The tunneling rate will typically be very low for two reasons: the barrier exponentially suppresses the electron wave function from passing through, and the number of available states to tunnel into on the other side is low. Typical concentrations of dopants in semiconductors are 10^{18} cm^{-3} or less, compared to total electron densities of

10^{22} cm^{-3}. Electrons coming from the left must either tunnel into an empty acceptor state or a free hole state, and the total number of both of these is equal to the number of acceptors.

If the electrons jump up to the conduction band on the other side of the junction, they must climb a barrier approximately equal to the band gap of the semiconductor. From Maxwellian statistics, the probability of a particle gaining an energy E by random processes is proportional to $e^{-E/k_B T}$, where T is the temperature and k_B is Boltzmann's constant, equal to 1.38×10^{-23} J/K. The current that jumps up the barrier will therefore be proportional to

$$I \sim e^{-(E_g - qV)/k_B T}, \tag{2.6.7}$$

assuming Maxwellian statistics, where V is the externally applied voltage between the two sides, if there is any. Although this path is deterred by the energy barrier, there are many states in the conduction band available to the electrons during their entire path from left to right. Once they have climbed the barrier, electrons on the right side will eventually drop down to the lower band and recombine with the holes coming in from the right. Therefore, this current is known as the **recombination current**.

Now consider electrons moving in the opposite direction, from right to left. On one hand, there are many electrons in the valence band on the right. They cannot move easily into the valence band on the left, because that band is full of electrons. They can move to the conduction band on the left via tunneling through the barrier, but as for the electrons moving from left to right, the need for quantum tunneling will give a very small rate of crossing the barrier. The electrons in the valence band on the right side of the interface are lower in energy than the conduction band on the left, so they must jump up in energy to tunnel through the barrier. For both reasons, the flow of electrons in this direction will tend to be very small. The current due to these elections is called the **generation current**, because it generates free carriers on the left side.

We can apply all the same arguments to free holes moving the opposite way. On one hand, they can tunnel through a barrier, with suppressed rate. On the other hand, they can climb a barrier, where there are many more available states in the valence band on the other side. (Recall that down on this diagram corresponds to higher energy for holes, since holes are like bubbles that rise to higher electron energy.)

The current on the n-side is predominantly carried over long distances by free electrons. These are called the **majority carriers** for the n-type region; the holes in the donor states are called the **minority carriers**. Of course, there are just as many electrons as holes in the semiconductor (far away from the depletion region), because the overall material is charge-neutral, and each free electron is created when one breaks away from orbiting a donor, leaving behind a positively charged impurity (trapped hole). The holes in the donor states are called the "minority" carriers because they carry less current; they must move by hopping from one donor to the next, which is much slower than the motion of free carriers. On the opposite side, in the p-doped region, the roles are reversed: The free holes in the valence band are the majority carriers, and the electrons in the acceptor states are the minority carriers.

If the voltage drop across the junction is large and negative (that is, the right-hand side of Figure 2.13 has lower voltage, corresponding to shifting that side up, since this is a diagram for electron energy), then the barrier will be higher, and the current that jumps the

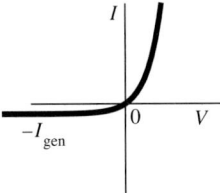

Fig. 2.14   Current–voltage characteristic of a p–n junction.

barrier will be negligible. Only the generation current will occur, and this will be nearly independent of the applied voltage, depending only on the details of the depletion region (which can, in fact, depend weakly on the applied voltage). This limiting value of the current is called the **saturation** current.

The total current through the junction will be the sum of the two types of current, which we write as

$$I = I_0 e^{(qV - E_g)/k_B T} - I_s, \qquad (2.6.8)$$

where I_s is the saturation current, E_g is the band gap, and I_0 is an overall factor for the recombination current. The total current must be zero at $V = 0$, which means we can set $I_0 e^{-E_g/k_B T} = I_s$, to obtain

$$\boxed{I = I_s \left(e^{qV/k_B T} - 1 \right).} \qquad (2.6.9)$$

This current–voltage relation is plotted in Figure 2.14. This is the electrical characteristic that defines an ideal **diode**. Note that it is **nonlinear** – the current is not proportional to the voltage. Current can easily pass in one direction, but not in the other. The p–n junction is also the basis of the operation of bipolar transistors, which we will examine in Section 2.7.1.

This junction can also act as a **photodiode**, because light energy that promotes electrons from the valence band to the conduction band will greatly increase the number of free carriers, and therefore the total current. Alternatively, if a current is forced through the junction, it will emit light. Often an insulating layer is added between the p and n regions so make the region with built-in electric field bigger. This is then called a p-i-n photodiode.

Exercise 2.6.1   (a) Calculate the band bending depth d for a p–n junction in the semiconductor GaAs, with doping concentration $n = 10^{16}$ cm^{-3} and dielectric constant $\epsilon = 14\epsilon_0$, using (2.6.6).

(b) Suppose that instead of having equal doping concentrations, the p-type side is doped at concentration $n = 5 \times 10^{16}$ cm^{-3} and the n-type is doped at $n = 10^{16}$ cm^{-3}. How does this change the junction?

2.6.3 Metal–Semiconductor Junctions

In general, to make circuits with semiconductor elements, one must connect metal wires to semiconductor devices, which means we must worry about the interfaces of semiconductors and metals. Figure 2.15 shows the case when the intrinsic Fermi level of the

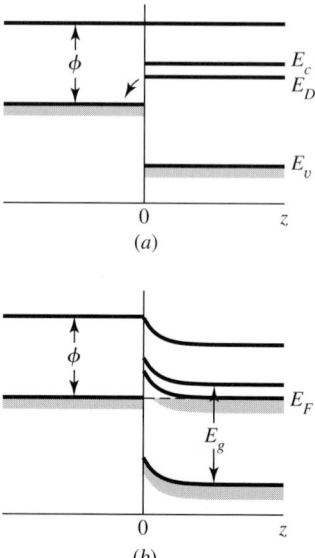

Fig. 2.15   Schottky contact, (a) at initial contact, and (b) in equilibrium.

semiconductor (near the donor states) is higher than that of the metal. Our prescription is to bend the bands to match the Fermi levels, which leads to the band structure shown in Figure 2.15(b). Note that although the bands bend in the metal, the Fermi level is flat, because the electrons can equilibrate to a common Fermi level in a partially filled band. As seen in this figure, after equilibration, electrons in the metal must jump over or tunnel through a barrier in order to reach the conduction band of the semiconductor. Moreover, there is a depletion region in the semiconductor with no carriers in the conduction states. This is called a **Schottky barrier**.

On the other hand, if the Fermi level of the metal is higher than that of the semiconductor, as shown in Figure 2.16, then the band bending will not lead to a barrier. This is an example of an **ohmic contact**, because the conduction through the junction will obey Ohm's law. The difference between these two types of contacts depends on the difference of the work functions of the semiconductor and metal.

The picture of an ohmic contact in Figure 2.16 is actually rarely the case in real metal–semiconductor contacts. This is because of the existence of surface states, or interface states, at the semiconductor surface. As discussed in Section 1.12, these typically have energies inside the band gap of the semiconductor. Some of these may be occupied, giving a Fermi level in the midst of the surface states. Figure 2.17 shows the case of an n-type semiconductor surface facing a vacuum (or air). The equilibration of the Fermi levels in the bulk and the surface states will lead to band bending just as in the cases discussed above. Figure 2.17(b) shows a Schottky barrier created by band bending due to surface states.

In a metal–semiconductor junction, these dangling bonds are not automatically elimi-nated, because there may be no natural bonding between the dissimilar materials. In this

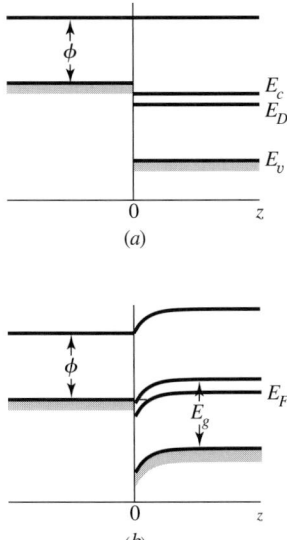

Fig. 2.16 Ohmic contact, (a) at initial contact, and (b) in equilibrium.

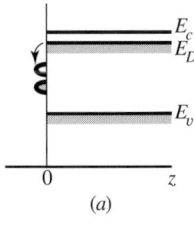

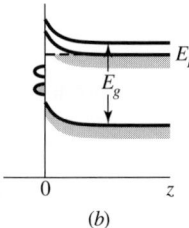

Fig. 2.17 (a) The energy bands of electrons at the surface of a semiconductor. (b) Band bending after equilibration of the electrons.

case, the Fermi level of the metal will equal the Fermi level of the surface states of the semiconductor, just as in the case of two metals discussed in Section 2.6.1. This is known as **Fermi level pinning**. Since the surface state energies lie within the gap, this means that the scenario of Figure 2.15 is almost always the case when there are surface states, leading to a Schottky barrier. Furthermore, since the energy of the surface states is determined by the type of dangling bond of the semiconductor, the height of the Schottky barrier will be independent of the type of metal used. The height of the Schottky barrier will be determined simply by the difference in energy of the Fermi level of the surface states and the Fermi level of the bulk semiconductor.

To create an ohmic contact in the presence of interface states, different approaches are used. A method that works well for silicon is to create a heavily doped region on the surface which has metallic properties, as discussed in Section 2.5.4. This layer will have the same Fermi level as a lightly doped layer in the same semiconductor; for example, an $n+$ heavily doped silicon layer will have no significant band bending relative to an n-doped region of silicon. The only band bending will occur at the interface of the heavily doped region and the metal contact. The thickness of this band-bending region decreases with increasing charge density, as seen in (2.6.6) of Section 2.6.1. For heavy doping, this thickness can be so small that electrons easily tunnel through any Schottky barrier that forms.

Another way to create an ohmic contact is to use a graded-alloy interface. In this method, a narrow-band-gap semiconductor is used at the surface, and then the alloy fraction is varied continuously, so that the band gap increases continuously from the narrow band gap at the surface to the final, larger band gap of the bulk semiconductor. This gradient of band gap can often be created by annealing – a metal is placed on the surface, and then the surface is heated, so that the metal diffuses into the semiconductor. When the distance over which the band gap increases is much greater than the intrinsic depletion depth d (which depends on the doping density, as discussed in Section 2.6.1), the height of the Schottky barrier is greatly reduced, and can be less than the thermal energy of the electrons, $k_B T$.

Exercise 2.6.2   Draw a schematic of the bands for a metal–semiconductor junction in which the metal has a Fermi level that lies (a) below, and (b) above the energy of the acceptors in a p-type semiconductor.

2.6.4 Junctions of Undoped Semiconductors

As discussed in Section 2.5.1, the chemical potential of an undoped semiconductor lies in the middle of the gap between the highest occupied state and the lowest unoccupied state. If the bands have the same effective mass, the chemical potential will lie exactly in the middle; if not, the chemical potential will be pushed slightly up or down toward the band with lower density of states.

Suppose that we make a junction between two semiconductors with different band gaps. A first guess for the band alignment is to line up the chemical potentials at the centers of the two band gaps, as shown in Figure 2.18. This turns out to be a good approximation, but not strictly correct. The reason is that the interfaces can have interface states and trapped surface charge, because the atoms of the two materials do not line up to bond perfectly with each other. Stress due to mismatch of the lattice constants in the different materials may also lead to shifts of the band energies. Because of this, the **band offsets** between two different semiconductor materials typically need to be measured or looked up in the

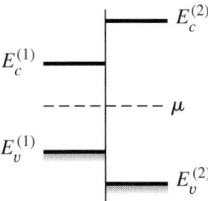

Fig. 2.18   Heterojunction of two semiconductors with different band gaps.

literature. These band offsets are typically recorded as the energy difference between the top of the valence bands of the two materials.

Exercise 2.6.3   Draw a schematic of the bands for a p–i–n structure, which consists of three layers in the sequence: p-doped semiconductor, insulator (i.e., undoped semi-conductor) with higher band gap, and n-doped semiconductor with the same band gap as the first semiconductor, before and after equilibration of the carriers, for two cases: (a) the insulator width much greater than the depletion distance in the doped semiconductors, and (b) the insulator width much less than the depletion distance.

Remember that the band offsets of the bands at the junction must remain the same no matter how the bands bend.

2.7 Transistors

It is not an overstatement to say that the physics of band bending has been used to revolutionize the world through the invention of the transistor. A transistor is generically a three-terminal device in which the electrical input at one terminal controls the current flow between the other two. This allows electrical logic, in which the output of one device becomes the controller of the next.

2.7.1 Bipolar Transistors

Figure 2.19(a) shows the electron band diagram of a set of three semiconductor regions doped n–p–n. The band bending has already been taken into account as we did for a p–n junction in Section 2.6.2. Focusing on the electron conduction band, we see a hill for the electrons between two conducting regions. Electrons will jump over this barrier with probability proportional to $e^{-(E_g-qV)/k_BT}$.

Suppose we now make electrical contact separately to all three regions, as shown in Figure 2.20. This is known as a **bipolar transistor**. With this wiring, the voltage of the middle region, which we call the **base**, can be controlled separately. If the voltage in this middle region is increased (making it more positive, which corresponds to a lower barrier for the electrons), the current over the barrier will increase exponentially with

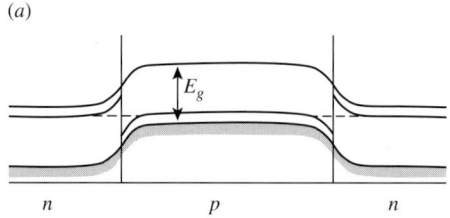

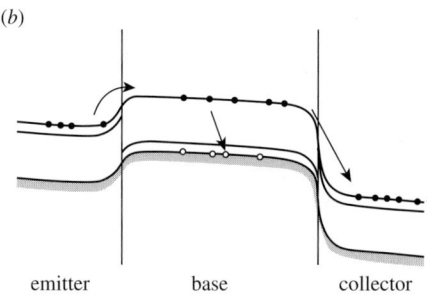

Fig. 2.19 (a) Electron band diagram of an n–p–n transistor in equilibrium. (b) Bands of the same transistor when $V_C > V_B > V_E$. The main path for electrons is from the conduction band of the emitter to the conduction band of the collector; a small fraction of the free electrons recombines with holes in the p-type region coming from the base connection. There is no common Fermi level because the system is not in equilibrium; current flows.

increasing gate voltage, just as the current in a diode increases as its internal barrier is reduced.

Figure 2.19(b) shows the bands of an n–p–n transistor when there is a voltage drop between the **collector** and the **emitter**. Electrons leave the emitter and are received by the collector, which means that positive current flows from the collector to the emitter. The voltage between the base and emitter controls the current between the collector and emitter because, as seen in Figure 2.19(b), electrons coming from the emitter must jump the barrier on that side first. Afterward they can fall down on the other side into the n-type region on the collector side. Since they fall down in energy, the potential energy drop between the base and collector doesn't affect the current. The device therefore has the property that the voltage on the base controls the current between the outer two terminals. This allows one electrical circuit to control another.

The arrows in Figure 2.20 show the direction of the currents in normal operation. Current conservation for the assigned directions of current implies $I_E = I_C + I_B$. The current into the p-type base enters as holes. As shown in Figure 2.19(b), some of the electrons coming from the emitter will recombine with these holes, but most will flow over the barrier and end up at the collector. The ratio of I_C to I_B is therefore some large factor β, which depends on the design of the transistor. This is a crucial difference of a transistor from simply having two p–n diodes connected back to back. In a bipolar transistor, the central region is thin enough that most electrons from the emitter make it through to the collector without recombining with holes from the base input.

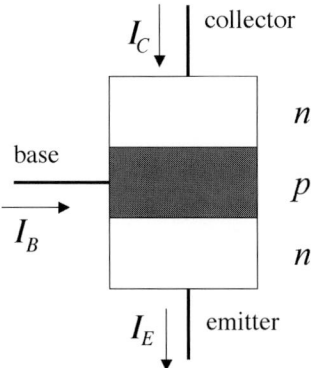

Electrical contacts for an *n–p–n* transistor.

The **Ebers–Moll model** for an ideal, symmetric *n–p–n* transistor approximates each junction in the transistor as obeying the relation (2.6.9). For the base current, there are two terms which add, corresponding to holes recombining with electrons coming from both sides:

$$I_B = \frac{I_s}{\beta}\left(e^{qV_{BE}/k_BT} - 1\right) + \frac{I_s}{\beta}\left(e^{qV_{BC}/k_BT} - 1\right)$$

$$= \frac{I_s}{\beta}\left(e^{qV_{BE}/k_BT} + e^{qV_{BC}/k_BT}\right) - \frac{2I_s}{\beta}. \tag{2.7.1}$$

For the collector current, there are two terms for electrons jumping the two sides of the central barrier, giving contributions with opposite signs. Some of the electrons coming from the emitter recombine with holes from the base, so we must subtract this small amount that never makes it to the collector. The total collector current is therefore

$$I_C = I_s\left(e^{qV_{BE}/k_BT} - 1\right) - I_s\left(e^{qV_{BC}/k_BT} - 1\right) - \frac{I_s}{\beta}\left(e^{qV_{BE}/k_BT} - 1\right)$$

$$= I_s\left(e^{qV_{BE}/k_BT} - e^{qV_{BC}/k_BT}\right) - \frac{I_s}{\beta}\left(e^{qV_{BE}/k_BT} - 1\right). \tag{2.7.2}$$

The emitter current has a similar form, such that $I_E = I_C + I_B$. When $V_{BC} < 0$, that is, when the collector voltage V_C is higher than V_B, the terms proportional to e^{V_{BC}/k_BT} in each of these equations become much less than the other terms and can be neglected, consistent with the statement above that V_{BE} controls the current under normal conditions.

A *p–n–p* transistor has the dopings reversed: the outer regions are *p*-type and the middle region is *n*-type. In this case, the majority carrier in the outer *p*-regions is holes in the valence band, and the base current is carried by electrons in the conduction band. The barrier for the holes is reduced when the center voltage is made more negative. Therefore, the base voltage must be *lower* than that of the source of the holes in order for current to flow between the *p*-type regions.

Bipolar transistors are typically characterized by their *I–V* curve as a function of the current flowing into the base. Figure 2.21 shows the relationship of the collector current I_c to the collector–emitter voltage drop V_{CE}, for several choices of the current into the base,

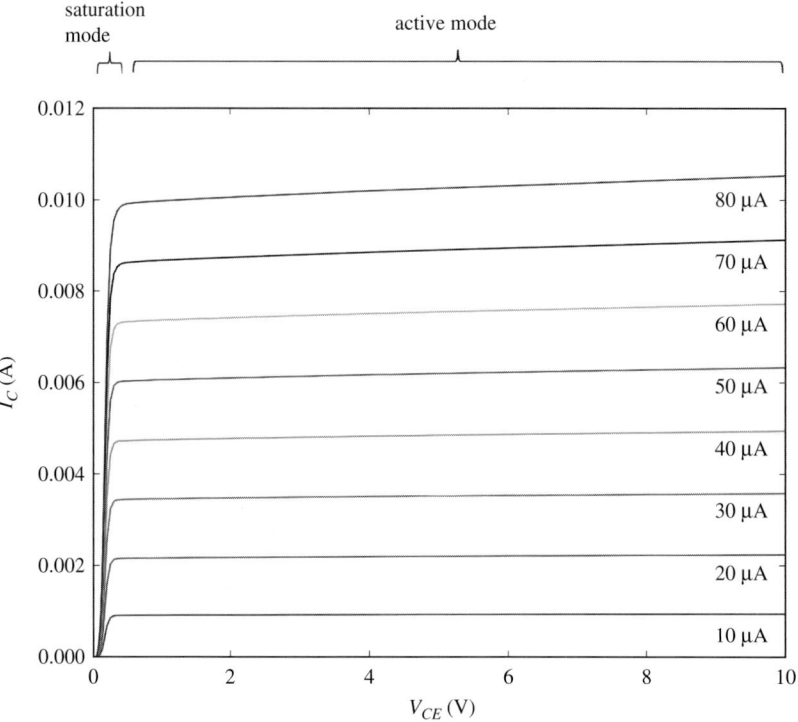

Fig. 2.21 Thin lines: Collector current I_C for an n–p–n bipolar transistor as a function of collector–emitter voltage drop, for the choices of base current I_B as labeled. Data for a Motorola 2N3904 transistor, courtesy of the Istvan Danko, University of Pittsburgh.

I_B, which in turn is a function of the voltage drop V_{BE}, according to the Ebers–Moll relation (2.7.1). As seen in Figure 2.21, when V_{CE} is large, the current I_C is nearly independent of V_{CE}, because V_C merely controls the amount the electrons drop down in energy after they have jumped the barrier from the emitter to the base, as discussed above. This is known as the **active mode** of the transistor – the current I_C is directly controlled by the base. When V_{CE} is low enough that $V_{BE} > V_{CE}$, the collector current I_C drops rapidly, as the reverse current of electrons jumping from the collector to the base becomes comparable to the current in the other direction. This is known as the **saturation mode**, or "fully on" mode – the current path from the collector to the emitter looks like a simple current path with low resistance. The total current must be zero at $V_{CE} = 0$, when the forward and reverse currents are equal.

Exercise 2.7.1 Use a program like Mathematica to plot curves like those shown in Figure 2.21 for an n–p–n bipolar transistor using the Ebers–Moll model, with $\beta = 150$. Assuming that the emitter is grounded ($V_E = 0$), you will need to first generate a function that finds the value of V_B needed to give a specific value of I_B for a given input value of V_C. In what ways do the data for the real transistor, shown in Figure 2.21, deviate from the Ebers–Moll model?

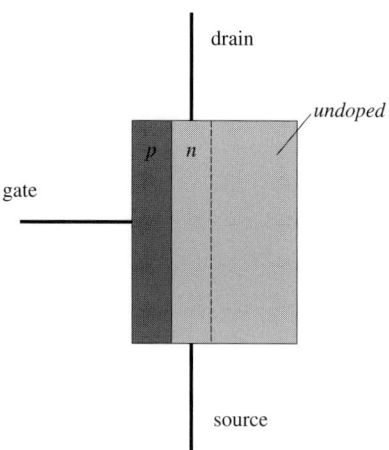

drain

undoped

gate

source

Fig. 2.22   Schematic of an *n*-channel JFET.

2.7.2 Field Effect Transistors

The standard type of transistor on a chip is a **field-effect transistor** (FET). In a bipolar transistor, a barrier is created which the charge carriers must jump over, and this barrier can be raised or lowered by a control voltage. In a FET, a conducting channel is created, and the number of carriers in this channel is modulated by the control voltage.

Instead of the names "base," "emitter," and "collector" used for the contacts of *n–p–n* and *p–n–p* transistors, it is standard to use the terms **gate**, **source**, and **drain** for FETs. There is nothing particularly deep about this change of nomenclature, just a recognition that FETs operate differently from bipolar transistors.

JFETs. A common type of FET is a **junction field-effect transistor** (JFET). Figure 2.22 shows a typical construction. Note that although this structure resembles a *p–n* diode, the current in a JFET stays entirely in the *n*-doped region, in a channel between the source and drain, rather than flowing across the *p–n* junction.

Figure 2.23(a) shows the band structure for the *n*-channel JFET shown in Figure 2.22, and Figure 2.23(b) shows the same bands after the band bending due to equilibration. The same procedure is followed as in Section 2.6, namely, we bend the bands to line up the chemical potentials. As discussed in Section 2.6.3, the Fermi level of an undoped semiconductor lies in the middle of the energy gap between the valence and conduction band. To do the band bending properly, we must line this level up with the Fermi level of the *p* and *n* regions.

As seen in Figure 2.23(b), the *n*-doped region in the middle forms a channel for free electrons to flow. If the voltage on the gate contact is made more negative, raising the potential energy of electrons in the *p*-type region, it will lift the channel relative to the conduction band of the undoped material, making the channel shallower and wider, until at some point the channel is no longer a minimum which holds electrons. At this point, the conduction will be switched off.

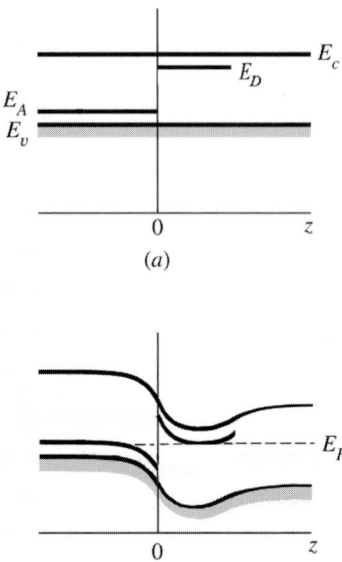

Fig. 2.23 (a) Band structure of the *n*-channel JFET shown in Figure 2.22 before equilibration, with zero applied gate voltage. (b) The same JFET bands after equilibration.

The *n*-channel JFET therefore conducts through the main channel when the gate voltage is zero or positive, and stops conducting when the gate voltage goes negative. This is similar to an *n–p–n* bipolar transistor, except that the switching threshold is offset, so that the normal state of the transistor is "on" at $V = 0$, and it only stops conducting when the gate voltage goes below a negative threshold. A *p*-channel JFET works just the opposite of an *n*-channel JFET, conducting at zero voltages and turning off when the gate goes positive.

For low voltage difference between the source and drain of a JFET, the conduction between these two contacts is ohmic, that is, the current increases linearly with the voltage gradient, since the channel acts simply as a resistor. At higher voltage difference between the source and the drain, there must be significant positive charge at one end of the conducting channel and negative charge at the other end, as shown in Figure 2.24. The negative charge at one end raises the potential energy for electrons at that end, which has the effect of raising the channel. This ultimately will have the same effect as using the gate to increase the potential energy of the electrons – the channel will be raised and flattened, carrying less charge. Thus, increasing the source–drain voltage beyond some point will no longer increase the current. Figure 2.25 shows a typical set of current–voltage relations for a JFET, for various values of the gate voltage. At low source–drain voltage, all the curves have the same slope, which corresponds to the intrinsic channel resistance $R = V_{DS}/I_D$, while at high voltage they all saturate at a current level that depends on the gate voltage.

MOSFETs. Another common type of FET is a **metal–oxide–semiconductor field-effect** transistor (MOSFET). An example of a MOSFET structure is shown in Figure 2.26. Although the oxide in a MOSFET acts as an insulator, over long periods of time the electrons can tunnel through this barrier and come to equilibrium. Figure 2.27(a) shows

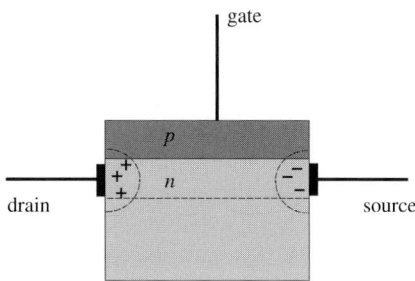

Fig. 2.24   JFET with voltage gradient between the source and drain.

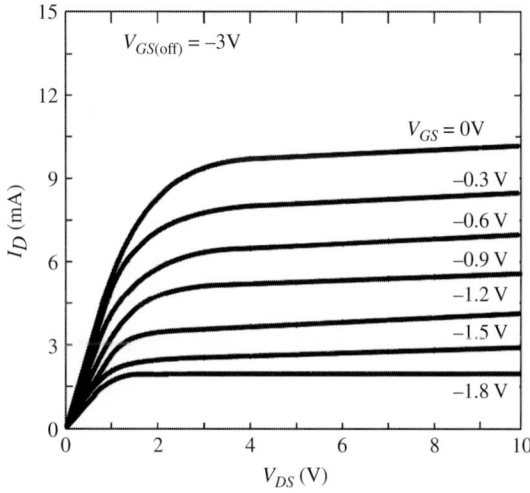

Fig. 2.25   Typical current–voltage map for an *n*-channel JFET. V_{DS} is the voltage drop between the drain and source, and V_{GS} is the voltage drop between the gate and source. From the data sheet for a Vishay 2N3819.

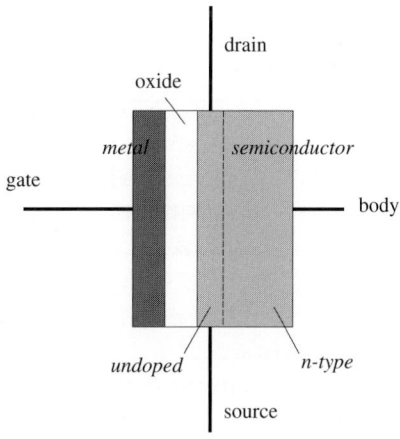

Fig. 2.26   Schematic of a simple *n*-channel MOSFET (which uses *p*-type substrate).

the band structure of the structure Figure 2.26, in the direction perpendicular to the layers, before any electron tunneling has occurred, and Figure 2.27(b) shows the conduction band after the system has come to equilibrium, following the band bending rules of Section 2.6. The bands in the semiconductor are bent to match the Fermi level of the metal with the Fermi level in the bulk of the doped semiconductor, as electrons from the metal tunnel through the insulator into the semiconductor, leading to a charge displacement. The oxide layer is typically thin enough that the chemical potential of the oxide layer doesn't matter; instead the constraint is that the slope of $U(z)$ remains constant across an interface.

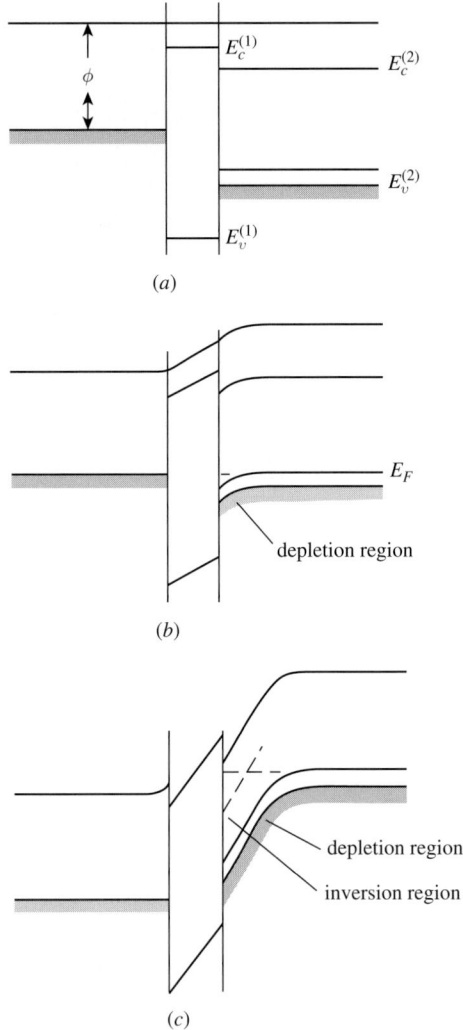

Fig. 2.27 (a) Electron band structure of an n-channel MOSFET without electron equilibration. (b) The same structure after equilibration. (c) The same structure with an applied positive voltage on the metallic gate. The tilted dashed line indicates the middle of the band gap in the depletion region, which corresponds to the spatially varying Fermi level in this region when there are no free carriers, while the horizontal dashed line indicates the Fermi level of the carriers in the body, far from the interface.

As seen in Figure 2.27(b), in equilibrium there is a depletion region where there are no free holes in the valence band near the interface of the oxide and semiconductor, due to the band bending. Since the source and drain contacts are at the surface, contacting only this depletion region, there is no charge to carry current between the source and drain. This is known as the **depletion mode** of the MOSFET.

If the gate voltage is raised, that is, the potential energy of the electrons at the metal is decreased, as illustrated in Figure 2.27(c), the conduction band near the interface can be pulled down so far that electrons will accumulate in the minimum of the conduction band at the interface. This is known as **inversion mode**, also called **enhancement mode** (the opposite of depletion) – the region at the interface acts like an n-doped region, with current carried by free electrons in the conduction band. Since the source and drain contacts are n-type, current can flow freely between the source and drain. This device is therefore known as an n-channel MOSFET. High gate voltage switches on the conducting channel when the gate voltage is above the inversion threshold, and switches off the conducting channel otherwise.

The threshold voltage V_T at which a MOSFET will go into inversion mode is determined by the point at which the bulk Fermi level crosses the Fermi level of the depletion region. For an n-channel MOSFET, the Fermi level in the bulk is pinned close to the valence band by the presence of the acceptors. The depletion region acts essentially like an undoped semiconductor, since the acceptors there are all filled. The Fermi level of the depletion region is therefore in the middle of the energy gap, as discussed in Section 2.6.3. When this level crosses the acceptor level in the bulk, the channel at the interface will have carriers that conduct current.

As seen in Figure 2.27(c), when this occurs, the conduction band in the interface region still has energy above the valence band in the bulk of the substrate. It may seem strange that electrons can accumulate in the conduction band at the interface when it has higher energy than the valence band in the bulk. One way to think of this is that both electrons and holes can diffuse from the bulk into the inversion region, but when the Fermi level of the bulk is above the middle of the gap of the inversion region, the probability of an electron doing this is much greater than the probability for a hole. Therefore, even if the probability of a jump upward is low, the probability of an electron jumping up to the conduction band is higher than a hole jumping up in the same region (recall that holes gain energy by moving *down* on this type of diagram), and therefore over time a negative charge will build up in the inversion region.

The current in a MOSFET is subject to saturation due to pinch-off just as in a JFET. This is often a useful feature. Overall, the I–V characteristic of an n-channel MOSFET looks very much like that of an n-channel JFET, except it is offset by the threshold voltage V_T, so that it does not conduct at $V = 0$. For gate voltages just a little above the threshold, the channel current increases linearly with an ohmic region like that of a JFET, with current proportional to $V_{GS} - V_T$.

Because of the oxide barrier between the gate and conducting channel, there is essentially zero current through the insulator from the base to the channel between the source and drain. This is one of the most useful properties of a MOSFET. The fact that it does not pass any current through its conducting channel until the gate voltage is above a threshold value also makes it very useful for digital logic applications, in which we want to draw very little current.

Exercise 2.7.2   A MESFET (metal–semiconductor FET) is made using a Schottky barrier at a metal-doped semiconductor junction (see Section 2.6.2) instead of an oxide barrier, as in a MOSFET, or p–n junction as in a JFET. For an n-doped channel, the drain and source are connected to the conducting channel using $n+$ doped regions, and an undoped substrate is used. Sketch the band bending for the junctions of a metal (with Fermi level inside the gap of the semiconductor), n-doped semiconductor, and undoped semiconductor in equilibrium. From this sketch, what do you expect the I–V characteristic of this device to look like; in particular, will the channel conduct at $V_{GS} = 0$? Will this device operate more like a MOSFET or a JFET?

2.8 Quantum Confinement

So far, most of the behavior we have been studying has applied to three-dimensional materials, since, after all, we live in a three-dimensional world. Modern solid state physics, however, allows the possibility of studying physics in lower dimensions. These are systems that are not just approximately one- or two-dimensional, such as a string or a piece of paper, but systems which are truly one- or two-dimensional, in the sense that particles have no degrees of freedom in one or two spatial dimensions. Such systems are called **quantum confined** systems. Bastard (1988) gives a comprehensive overview of the physics of quantum confined structures.

The technology now exists to deposit layers of materials that are controlled in thickness down to a single layer of atoms. A typical method of depositing such layers is molecular-beam epitaxy (MBE), in which atoms or molecules are evaporated into a high vacuum and allowed to shoot in a beam toward a solid substrate. If the substrate temperature is hot enough, the atoms will tend to wet the surface smoothly.

Suppose that two types of semiconductor have been deposited in layers as shown in Figure 2.28. The two semiconductors have different band gaps. In the plane of the layers, the electrons and holes are free to move, but in the direction perpendicular to the planes, the

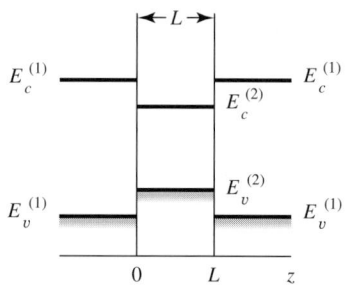

Fig. 2.28   A quantum well structure produced by three semiconductor layers.

electrons and holes feel a confining square potential just like the one-dimensional square well studied in Section 1.1. We can use the same simple Schrödinger equation to solve for the electron and hole states.

If we ignore the leakage of the carrier wave functions into the barrier layers and assume they are impenetrable, we can solve for the confined states easily. For a well of width L, the allowed wavelengths are $\lambda = 2L, L, 2L/3$, etc. The energies of the confined states are therefore

$$E_N = \frac{\hbar^2 k_N^2}{2m} = \frac{\pi^2 \hbar^2 N^2}{2mL^2}, \qquad (2.8.1)$$

where m is the effective mass of the particle and N is an integer. The energy difference between the $N = 1$ and $N = 2$ level is therefore

$$\Delta E = \frac{\pi^2 \hbar^2 (4 - 1)}{2mL^2} = \frac{3\pi^2 \hbar^2}{2mL^2}. \qquad (2.8.2)$$

If this energy is much larger than $k_B T$, then the carriers will all occupy the lowest confined state. If we arbitrarily pick $\Delta E = 10 k_B T$, we can solve for the thickness L that will cause this to happen at a given temperature. We obtain

$$L = \sqrt{\frac{3\pi^2 \hbar^2}{20 m k_B T}}. \qquad (2.8.3)$$

For $T = 300$ K and a typical carrier mass on the order of the free electron mass, this implies L is of the order of a few nanometers. Thicknesses this thin are easily obtainable by modern methods. In this case, carriers in the wells have no degree of freedom of motion in the direction perpendicular to the place of the layer, because they must all be in the same quantum state. We say they are **quantum confined**, and a structure that confines the particles to a two-dimensional plane in this way is called a **quantum well**.

We could go further and create barriers of higher-band-gap materials in the other dimensions, as well. By creating barriers with high-gap material in other directions, we can remove as many spatial dimensions of freedom as we want. A system with two confined dimensions (i.e., one spatial degree of freedom) is called a **quantum wire**, and a structure that confines the carriers in all three dimensions, with no spatial degrees of freedom, is called a **quantum dot**. A quantum dot, in a sense, is just a large molecule. The study of these types of structures is typically called **nanoscience**, and it has become a major topic of research of modern solid state physics.

One reason for the interest in these types of systems is that instead of simply taking whatever nature gives us, we can use quantum confinement to give the exact band gaps and band structure that we want. This is sometimes called "band structure engineering." Another reason for pursuing the technology of quantum confined systems is that they have altered densities of states, which give them unique optical properties. The density of states of lower-dimensional systems are discussed in Section 2.8.1.

As discussed in Section 2.6.3, one of the important parameters in a semiconductor heterostructure is the **band offset**, that is, the difference in energy between the top of the valence bands of the two materials. Naively, one would expect that the chemical potentials of the two materials would be the same in equilibrium, which would imply that the middle of the band gaps of the materials line up. This does not always happen, however.

Disorder in quantum-confined systems. One problem of quantum-confined systems is their sensitivity to disorder. One source of disorder arises from variations in the thickness of the quantum-confined material.

Suppose, for example, that we ask how much the energy of an electron will change in a quantum well if the width of the quantum well varies by one atom. From (2.8.1), we have

$$E_N = \frac{\hbar^2 k_N^2}{2m} = \frac{\pi^2 \hbar^2 N^2}{mL^2}. \tag{2.8.4}$$

The change in E_N for a width change ΔL is therefore

$$\Delta E_N = \frac{2\pi^2 \hbar^2 N^2}{mL^2} \frac{\Delta L}{L}. \tag{2.8.5}$$

A typical lattice constant of a semiconductor is around 5 Å. For a well width of 5 nm, the fractional change $\Delta L/L$ corresponding to one atom variation is therefore about 10%, which means that for a typical confinement energy of 100 meV, the variation in energy due to one lattice constant width variation is 20%, or 20 meV. Therefore, unless the layers are perfectly smooth, there will be fluctuations of the energy comparable to the kinetic energy at room temperature. This is not out of the question, however. Modern deposition methods can make layers that are smooth to within a single lattice constant of a material.

Another source of disorder is impurities, including the dopants used to make n-type or p-type semiconductors. If a quantum well is only a few tens of atoms in width, then the presence of an impurity atom can substantially change the effective band gap at that point, because the fraction of the material that is different is significant compared to the total number of atoms across the structure.

This type of disorder can be minimized by **modulation doping**. In this scheme, the dopant atoms are put in the barrier instead of in the well. The extra carriers from the dopants then fall into the quantum confined area, which has very few impurities.

Another, related source of disorder comes from the ternary alloys used to make the semiconductors of different band gaps. For example, GaAs has a band gap of 1.52 eV, while AlAs has a band gap of 2.2 eV. By mixing together Ga and Al with As, to make $Ga_x Al_{1-x} As$, the band gap can be tuned in this range. This type of alloy has intrinsic disorder, however, because the atoms of different types are randomly distributed in the material. Because of the randomness, regions with a higher concentration of one type of atom than another will occur. The band gap will therefore have intrinsic fluctuations, as discussed in Section 1.8.2. In a quantum-confined system, this effect is magnified, because the size of a fluctuation of the alloy concentration can be comparable to the confinement width of some structures.

2.8.1 Density of States in Quantum-Confined Systems

For the case of isotropic mass, we can deduce some simple rules for the density of states, using the definitions of Section 1.8. We have already deduced the density of states for a

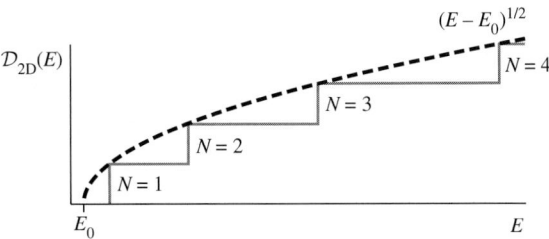

Fig. 2.29   Density of states of a two-dimensional quantum confined system.

three-dimensional system in Section 2.2. For a two-dimensional system with effective mass m, we have

$$D(E)dE = dE\frac{A}{(2\pi)^2}\int_0^\infty 2\pi k dk\, \delta(E_{\vec{k}} - E),\tag{2.8.6}$$

where A is the area of the system. Since $dE_{\vec{k}} = \hbar^2 k dk/m$, this implies

$$D(E)dE = dE\frac{A}{2\pi}\int_0^\infty \frac{m}{\hbar^2}\delta(E_{\vec{k}} - E)dE_{\vec{k}}$$

$$= dE\frac{A}{2\pi}\frac{m}{\hbar^2}\Theta(E - E_0^{(2D)}),\tag{2.8.7}$$

where $\Theta(E)$ is the Heaviside function, equal to 1 for $E > 0$ and 0 otherwise, and $E_0^{(2D)}$ is the energy of the $k = 0$ state (which includes the band gap energy and the quantum confinement energy discussed above). In other words, the density of states of a two-dimensional system is constant above the quasiparticle minimum energy. This assumes that all the particles are in the lowest confined energy state. For each confined state, there will be another density of states factor, so that the total density of states of a quantum well system will look like Figure 2.29. The total density of states is proportional to the number of levels occupied, N, while the confinement energy of each level is proportional to N^2, according to (2.8.1), which implies that the density of states is approximately proportional to $(E - E_0^{(3D)})^{1/2}$, the result for a three-dimensional system. In the limit $L \to \infty$, the quantum confined density of states converges to the 3D density of states.

In the same way, one can show that the density of states for a one-dimensional isotropic system is proportional to $1/k \sim (E - E_0^{(1D)})^{-1/2}$. Although the density of states goes to infinity at $E = E_0^{(1D)}$, this does not cause the total *number* of states to become infinite, because the integral $\int D(E)dE$ is integrable at $E = E_0^{(1D)}$.

The density of states for a zero-dimensional system (a quantum dot) is just a δ-function, since in this case there is simply an atomic-like series of discrete states. Table 2.2 gives a summary of the different dependences (spin degeneracy is not included).

Exercise 2.8.1   Calculate the density of states for the one-dimensional case and show that your result agrees with that given in Table 2.2.

Table 2.2 Density of states for particles with isotropic mass

Dimensions	Density of states $\mathcal{D}(E)dE$
3	$\dfrac{V}{2\pi^2}\dfrac{\sqrt{2}m^{3/2}\sqrt{E - E_0^{(3D)}}}{\hbar^3}dE$
2	$\dfrac{A}{2\pi}\dfrac{m}{\hbar^2}\Theta(E - E_0^{(2D)})dE$
1	$\dfrac{L}{2\pi}\dfrac{m^{1/2}}{\sqrt{2}\hbar\sqrt{E - E_0^{(1D)}}}dE$
0	$\delta(E - E_0^{(0D)})dE$

Exercise 2.8.2 (a) In the $T = 0$ approximation, the total number of particles that corresponds to a Fermi energy E_F is given by

$$N = \int_0^{E_F} \mathcal{D}(E)dE. \tag{2.8.8}$$

Calulate the area density of the particles needed to have a Fermi energy of $E_F = 100$ meV for a two-dimensional isotropic electron gas with effective mass $m = m_0$, where m_0 is the free electron mass.

(b) Suppose that we produce the two-dimensional electron gas of part (a) in a quantum well with thickness 10 nm and infinitely high barriers. Then the total volume of a quantum well with area 1 mm^2 is 10^{-5} mm^3. Compare the total number of electrons needed in this system to get a Fermi energy of 100 meV, and the total number of electrons needed for a three-dimensional isotropic electron gas with the same total volume.

Exercise 2.8.3 In the case of modulation doping, donor electrons can fall down from states in the barrier into a quantum-confined state in the quantum well. The Fermi level in the barrier material is nearly the same as the barrier conduction band level. Draw a schematic of the band bending in the case of a wide quantum well sandwiched between two n-doped barriers.

2.8.2 Superlattices and Bloch Oscillations

The ability to make various semiconductor heterostructures means that we can make all types of band structures analogous to the one-dimensional Kronig–Penney model we studied in Section 1.2.

As we discussed above, a semiconductor layer sandwiched between two other semiconductor layers with larger band gaps makes a quantum well which has a potential energy profile in one dimension that is exactly the same as the one-dimensional square well studied in Section 1.1. In the same way, if we have a series of layers sandwiched between other layers with higher band gaps, then the potential energy profile in the direction perpendicular to the planes is a series of coupled quantum wells just like that of the Kronig–Penney model studied in Section 1.2. This is known as a **superlattice**. In this case, just as we have seen in numerous examples, the overlap of the wave functions can lead to the formation of energy bands. We now have the possibility of bands within bands, or **mini-bands**, as shown in Figure 2.30.

If the layers are made in a periodic pattern with the same thicknesses, then Bloch's theorem applies again. The periodic layer structure acts as a set of Bragg reflectors for the electron waves just as atomic layers do. We can therefore treat the electrons or holes as free particles and redraw the electron bands in a smaller zone, as shown in Figure 2.30. Since the electrons and holes in the effective mass approximation act as free particles, we can use the nearly free electron approximation of Section 1.9.3 to determine the band gaps.

There are numerous possibilities, therefore, of altering the band structure of materials by the appropriate choice of material thickness. As discussed above, we have the possibility of band structure engineering, designing whatever band structure is needed for a particular device.

Exercise 2.8.4   In many semiconductors such as GaAs, the conduction band has a conduction band minimum at zone center, and an indirect gap with higher energy at another minimum, at a critical point on the zone boundary. The zone-center minimum is typically called the Γ-valley, while the other, indirect minima are called the X-valley and the L-valley.

Suppose that we make a superlattice with well material GaAs, which has a Γ-valley 1.5 eV above the valence band maximum, and barrier material AlAs, which has a Γ-valley 2.6 eV higher than the GaAs valence band maximum, but an [X-valley] only 1.7 eV higher than the GaAs valence band maximum. If the GaAs quantum well width is small enough, the confined state of the Γ-valley can be pushed higher than the AlAs X-valley, as shown in Figure 2.31. In this case, electrons will fall from the GaAs Γ-valley into the X-valley of the barriers, so that the AlAs forms a quantum well for the electrons, with the GaAs layers acting as barriers. The valence-band holes will remain confined in the GaAs layers, however. This is called a **Type II** superlattice. A structure in which both the electrons and holes are in the same layer is called a **Type I** superlattice.

Assuming that there is negligible penetration of the wave function of the Γ-valley electrons into the barriers, estimate the GaAs quantum well thickness at which the above structure will convert from a Type I to a Type II structure. The conduction electron effective mass is approximately $0.06m_0$ for GaAs.

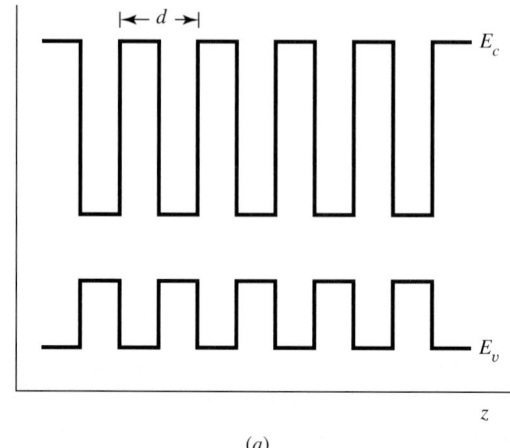

Fig. 2.30   (a) Spatial band profile of a superlattice structure. (b) The original conduction and valence bands, for a quantum well material with lattice constant a. (c) Mini-bands arising from the quantum well states, in the reduced zone. The additional periodicity of the system leads to new Bloch states.

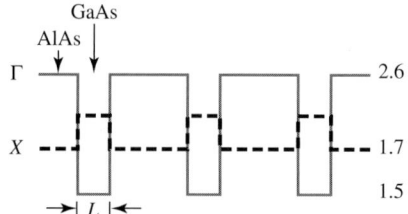

Fig. 2.31   Conduction band energies for a Type II superlattice, discussed in Exercise 2.8.3.

Bloch oscillations. Superlattices have been used to demonstrate (Waschke *et al.* 1993) the basic effect of **Bloch oscillations**, in which a DC electric field generates AC electromagnetic radiation due to the reflections of the electron waves in the periodic potential. This effect is a generic property of all periodic systems (see Krieger and Iafrate 1986), but is not observed in standard bulk solids because the oscillation frequency in bulk materials is too low, for reasons we will see below.

The basic effect can be understood as follows. As an electric field accelerates the electrons in a periodic solid, their quasimomentum $\hbar k$ increases. If k increases past the boundary of the first Brillouin zone, it will be wrapped around to the opposite side of the Brillouin zone. Assuming that the electron remains in the same band, it will therefore undergo periodic motion. (Transitions to other bands, which can occur in high fields, are discussed by Krieger and Iafrate 1986.) To put it another way, as the electrons accelerate, their wavelength becomes shorter, until they have wavelength so short that they undergo Bragg reflection (discussed in Section 1.2) from the periodic structure and thus are reflected to the opposite direction. They then accelerate back the other way again due to the static electric field, and so on.

To prove this, we assume a one-dimensional system, and write the time-dependent Schrödinger equation,

$$-\frac{\hbar^2}{2m}\frac{\partial^2}{\partial x^2}\psi + U(x)\psi = i\hbar\frac{\partial}{\partial t}\psi. \tag{2.8.9}$$

We assume that the potential $U(x)$ in (2.8.9) has the same periodicity as the medium. Now we guess the solution of the form of a Bloch state $\psi_k = u_k(x)e^{i(kx-\omega t)}$, where $u_k(x)$ is a Bloch cell function with the periodicity of the lattice. Substituting this into (2.8.9), we have

$$\frac{\hbar^2}{2m}\left(-\frac{\partial^2 u_k}{\partial x^2} - 2ik\frac{\partial u_k}{\partial x} + k^2 u_k\right) + U(x)u_k = (\hbar\omega)\,u_k. \tag{2.8.10}$$

This is a differential equation that we can, in principle, solve for the periodic function $u_k(x)$ for any given value of k.

Now let us add a term to the potential energy for a constant electric field, namely $-qEx$. We guess now that the solution of the time-dependent Schrödinger equation will have the form of a Bloch state $\psi_k = u_{k(t)}(x)e^{i(k(t)x-\omega(t)t)}$, with an explicit time dependence for k and ω. Substituting this into the Schrödinger equation (2.8.9), we have

$$\frac{\hbar^2}{2m}\left(-\frac{\partial^2 u_k}{\partial x^2} - 2ik\frac{\partial u_k}{\partial x} + k^2 u_k\right) + U(x)u_k - qExu_k$$

$$= i\hbar\frac{\partial u_k}{\partial k}\frac{\partial k}{\partial t} + i\hbar\left(ix\frac{\partial k}{\partial t} - i\omega - it\frac{\partial\omega}{\partial t}\right)u_k \tag{2.8.11}$$

$$= -x\frac{\partial(\hbar k)}{\partial t}u_k + i\frac{\partial u_k}{\partial k}\frac{\partial(\hbar k)}{\partial t} + \left(\hbar\omega + t\frac{\partial(\hbar\omega)}{\partial t}\right)u_k.$$

We can solve this equation by first setting

$$\frac{\partial(\hbar k)}{\partial t} = qE = F, \tag{2.8.12}$$

which is just what we would expect for the change of momentum with a force. We then have

$$i\frac{\partial u_k}{\partial k}\frac{\partial(\hbar k)}{\partial t} + (\hbar\omega)u_k + t\frac{\partial\omega}{\partial k}\frac{\partial(\hbar k)}{\partial t}u_k$$

$$= \frac{\hbar^2}{2m}\left(-\frac{\partial^2 u_k}{\partial x^2} - 2ik\frac{\partial u_k}{\partial x} + k^2 u_k\right) + U(x)u_k. \quad (2.8.13)$$

We assume that $k = 0$ at $t = 0$, which gives $\hbar k = qEt$, or $t = \hbar k/qE$. Using this in (2.8.13), and rearranging, we have

$$(\hbar\omega)u_k = \frac{\hbar^2}{2m}\left(-\frac{\partial^2 u_k}{\partial x^2} - 2ik\frac{\partial u_k}{\partial x} + k^2 u_k\right)$$

$$+ \left(U(x) + \frac{i}{u_k}\frac{\partial u_k}{\partial k}qE - k\frac{\partial(\hbar\omega)}{\partial k}\right)u_k. \quad (2.8.14)$$

This has exactly the same form as (2.8.10), but instead of just $U(x)$, we have the last term in parentheses on the right-hand side, with two extra terms. However, each of these terms is also periodic in x, since $u_k(x)$ has the periodicity of $U(x)$, and the third term in the parenthesis has no x-dependence at all. Thus, we have an equation for a Bloch function with an altered periodic potential, which can be solved self-consistently for $u_k(x)$. The overall wave function is the product of this periodic function times a plane wave factor e^{ikx}, with $k = qEt/\hbar$. When k equals π/a, Bragg reflection will occur, taking k from π/a to $-\pi/a$. Then k will continue to increase at the same rate.

The total time to go through the whole allowed range of k values is found by taking the total range of k, equal to $2\pi/a$, and dividing by the rate of advance of k, which from the above is $dk/dt = qE/\hbar$. We then have the period $T = (2\pi/a)/(qE/\hbar)$, or, using $\omega = 2\pi/T$, the oscillation frequency

$$\boxed{\omega_B = \frac{a|q|E}{\hbar},} \quad (2.8.15)$$

which is proportional to the electric field E and the lattice constant a.

The larger the lattice constant, the less time it takes for the electrons to speed up to have wavelength comparable to it; therefore larger lattice constant corresponds to shorter oscillation period. Lattice constants of tens of nanometers, which are typical for semiconductor superlattices, give oscillations in the THz frequency range for DC electric fields of kV/cm, that is, a few volts across tens of microns; this has technological relevance as a source of THz radiation. In principle, Bloch oscillations exist in any periodic system, including bulk crystals with periodic atomic lattices; the atomic spacing of angstroms in typical solids implies frequencies in the GHz range for comparable electric fields. For these longer periods, however, scattering of the electrons breaks up the coherence of the oscillations. As we will discuss in Chapter 5, typical scattering times for electrons in solids are of the order of tens of picoseconds.

According to the above analysis, if there were no scattering of the electrons, the electrons in a crystal would never carry DC current, and instead simply oscillate back and forth in response to a DC field. Scattering therefore plays an essential role in electrical conduction.

Exercise 2.8.5   Show that in the case when $U(x) = 0$ and $u_k = 1$, that is, the states are plane waves $\psi = e^{i(kx - \omega t)}$ in a vacuum, and both k and ω are time-dependent, the solution of (2.8.14) for $k = 0$ at $t = 0$ implies $k = qEt/\hbar$ and $\omega = (qEt)^2/6\hbar m$. Show that this implies that the average value of the energy, defined by

$$\int_{-\infty}^{\infty} dx\, \psi^* H \psi = \int_{-\infty}^{\infty} dx\, \psi^* i\hbar \frac{\partial \psi}{\partial t}, \tag{2.8.16}$$

is equal to $\hbar^2 k^2/2m$. In other words, the kinetic energy grows in time in this case, and there is no Bragg reflection.

2.8.3 The Two-Dimensional Electron Gas

In Section 2.6, we discussed the effect of band bending at an interface. This band bending can also be used to create quantum confined states.

Consider a metal–oxide–semiconductor field-effect transistor (MOSFET), discussed in Section 2.7.2. Figure 2.32 shows just the conduction band of the FET in inversion mode, from Figure 2.27(c). It is possible, by controlling the doping density and other design parameters, to make the inversion region so narrow that it acts as a quantum well that confines the electrons.

If the region of band bending is narrow enough, the quantized states in the well will be separated by energy greater than $k_B T$, and the electrons will all lie in the lowest quantized level. As discussed above, the electrons in this case have truly two-dimensional motion. Therefore, we have a two-dimensional electron gas (2DEG) in the well, and the conductivity of the system depends only on this two-dimensional system. A nice feature of this system is that the Fermi level of the electrons in the 2DEG, and therefore the electron density, can be tuned simply by changing the voltage of the metal gate.

The 2DEG is a playpen for all kinds of interesting physical effects, including the fractional quantum Hall effect. We will return to the 2DEG in Section 2.9.3 to discuss some of the fascinating effects which occur in this system.

2.8.4 One-Dimensional Electron Transport

As discussed at the beginning of Section 2.8, **quantum wires** (QWRs) are one-dimensional conductors. There are many ways to make QWRs. One way is to start with a quantum well and etch away two sides using nanolithographic techniques, until a narrow line only a few

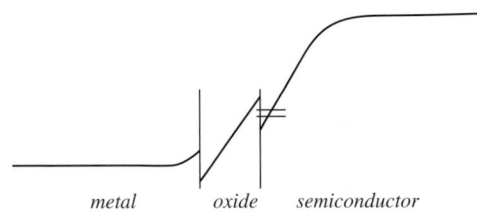

metal   *oxide*   *semiconductor*

Fig. 2.32   Conduction band of a MOSFET in inversion mode. Quantized levels of the inversion layer are shown as horizontal lines.

nanometers wide is left. After this etching, new barrier material can be deposited over the entire structure. (It is also possible to leave the etched surfaces alone, but as discussed in Section 1.11.4, free surfaces often have dangling bonds which change the band structure.)

In recent years, various methods of chemistry have been used to create molecular chains that act as QWRs. These include carbon nanotubes (which are essentially a single sheet of graphite, rolled up into a tube only a few nanometers across) and conducting organic polymer chains. Common metals such as gold can also be drawn into very thin whiskers that are only a few nanometers across. In all of these cases, the electronic states will be quantized in two dimensions while continuous in the other dimension, along the wire.

A surprising property of QWRs is quantization of current; that is, the current through a QWR depends only on universal constants of nature, independent of the details of the geometry of the wire. We can prove this in the following calculation.

From Table 2.2, accounting for the electron spin degeneracy, the number of quantum states in a one-dimensional system in an energy range dE is

$$D(E)dE = 2\frac{L}{2\pi}\frac{m^{1/2}}{\sqrt{2}\hbar\sqrt{E}}dE = \frac{mL}{\pi\hbar^2 k}dE. \tag{2.8.17}$$

Each of these quantum states can carry current as an electron moves in that state down the wire. Assuming that the electrons move without scattering, the current for a given state is given simply by the charge of the electron divided by the transit time across the wire. The transit time is just the length of the wire L divided by the velocity v of the electron, and therefore for a given value of k, the current is

$$I_k = \frac{e}{(L/v)} = \frac{e}{L}\left(\frac{p}{m}\right) = \frac{e\hbar k}{Lm}. \tag{2.8.18}$$

The total current carried in an energy range ΔE is therefore equal to the number of quantum states in that range times the current per state:

$$I = \sum_k I_k = \left(\Delta E\frac{mL}{\pi\hbar^2 k}\right)\frac{e\hbar k}{Lm} = \frac{e}{\pi\hbar}\Delta E. \tag{2.8.19}$$

The dependence on k has dropped out. Essentially, although states with higher k carry more current, since the electron moves faster, this is canceled out by the fact that there are more states with low k for a given energy range.

To convert this result to a formula for conductivity, we suppose that the wire is connected between two conductors with Fermi levels, as shown in Figure 2.33. For there to be a steady current, the two gases must be at different potential energies. The difference in energy is given by $\Delta U = e|\Delta V|$. The range of energy of electrons flowing in the wire is then just the range $\Delta E = \Delta U$ shown in Figure 2.33. Electrons with energies below this range will flow equally in both directions, giving no net contribution to the current.

The formula (2.8.19) can then be written

$$I = 2\frac{e^2}{h}\Delta V, \tag{2.8.20}$$

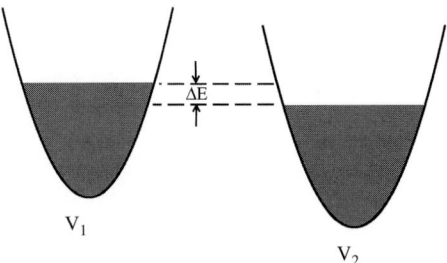

Fig. 2.33 Two electron gases with different potential energies, connected by a quantum wire.

or more generally,

$$I = 2 \sum_i t_i \frac{e^2}{h} \Delta V, \tag{2.8.21}$$

where the sum is over quantized states that contribute to the current, and t_i is the transmission coefficient for each state, which depends on the reflection of the electron waves from the contacts to the wire. These depend on the wave properties of the electrons just like the reflection of coefficient of light at a boundary.

This is one version of the **Landauer** formula. As advertised, the intrinsic conductivity of each quantum channel involves only universal constants of nature. As we will see in Section 2.9.3, this result has a deep connection to the quantized conductivity seen in two-dimensional electron gases in a magnetic field.

2.8.5 Quantum Dots and Coulomb Blockade

A quantum dot (QD) confines electrons in all three dimensions. In this way, QDs are just like atoms or molecules with quantum states; some people call quantum dots "artificial atoms." QDs can be fabricated in numerous ways. One can start with a quantum well and etch it to leave small pillars that are a few nanometers across. Another method that works surprisingly well is to simply deposit onto the surface of a solid a number of atoms less than enough to cover the whole surface with one atom thickness. In this case, the atoms will tend to form small beads, just like water beading up on a surface. These beads can then be covered by a barrier layer with a different band gap. One advantage of this method is that it does not require etching; in general it is difficult to grow new layers with techniques like molecular beam epitaxy after etching has been done, since the etching alters the surface and typically introduces oxygen. Quantum dots can also be fabricated by wet chemistry. The QDs start out as sediment particles a few nanometers across, which are allowed to drop from a liquid solution onto a surface. The liquid is then removed, leaving the dots on the surface. In some cases, a barrier material may be added at this point to cover all the dots; in other cases the sediment particles may be coated with a barrier material via chemical reaction while still in solution. Finally, organic molecules a few nanometers across may be treated as QDs.

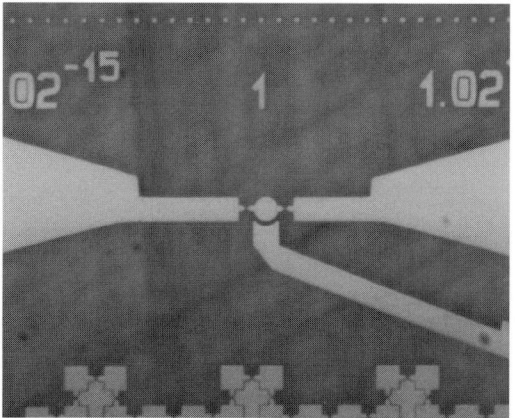

Fig. 2.34   A quantum dot with tunneling barriers into and out of the dot, and a gate contact to change its potential. From A.G. Sheer, University of Konstanz.

Since QDs are zero-dimensional, there is no current conduction to speak of within a dot. There are two ways to make QDs active in devices, however. One way is to allow tunneling of charge into and out of a QD via thin barriers. Another way is to send photons into the dot, which excite electrons from the valence band to the conduction band.

Exercise 2.8.6   Quantum dots are rarely exactly symmetric. Suppose that a QD is rectangular with a length of 7.5 nm, width of 8 nm, and height of 6 nm. Compute the lowest five confined state energies of an electron in this dot, for an electron effective mass of 0.1 times the vacuum electron mass, assuming infinite barrier height around the dot.

Coulomb blockade. Figure 2.34 shows an example of QD structure fabricated by etching away material to leave behind a very small metal dot (the central circle in the image) with a radius of a few nanometers, and three metal contacts separated from the dot by thin tunneling barriers.

Figure 2.35(a) shows an example of the intrinsic conduction band structure for the dot with the two tunneling barriers and two surrounding metal contacts. An electron in the dot has discrete energy levels because of the quantum confinement in all three dimensions. The bands are tilted due to the voltage difference of the contact on the left and the contact on the right. In this configuration of the bands, an electron can tunnel from the left-hand contact into the upper confined state of the quantum dot. We assume that the lower state is filled, being below both contact Fermi energies, and therefore it does not participate in conduction.

In Figure 2.35(b), the quantized energy levels of the dot are shifted upward due to one additional electron being in the dot. The energy levels are shifted up because the Coulomb repulsion of the electrons gives them higher potential energy. In this case, no new electron can tunnel from the left contact into the dot, since it would have to gain energy to go up

(*a*)

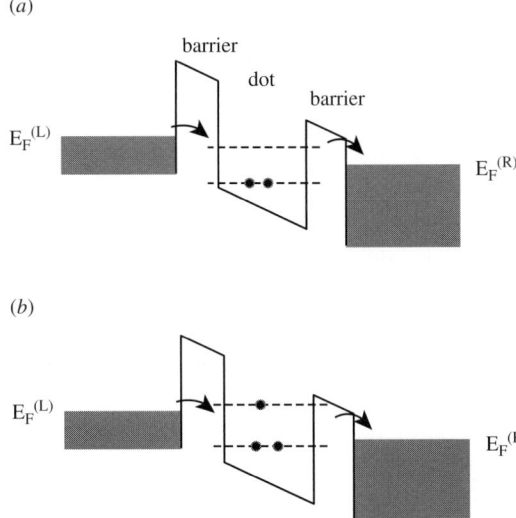

(*b*)

Fig. 2.35 (a) Band structure for a quantum dot with tunneling barriers between two electron reservoirs. (b) The altered band structure when an electron in the dot raises the quantized energy levels in the dot.

to the upper quantized state. (We assume that the system is at low enough temperature that the electrons have small probability of jumping to higher levels.) The electron in the upper level can still tunnel out to the right in this case. However, if the tunneling barrier is high enough, this may take some time, so the dot can be left in a metastable state with a single electron in the upper quantized level. The conduction through the dot therefore corresponds to a series of single electrons, tunneling one at a time into the dot and then out. As long as one electron is in the dot, no more can come in.

This is known as a **Coulomb blockade**. A single electron changes the effective resistance of the current path through the dot by orders of magnitude, since the tunneling current depends very sensitively on the relative energies of the states. This can also be used for a single-electron transistor, since one electron in the dot controls whether the dot conducts or not. There are many different versions of single-electron transistors, all of which use the fact that a single electron can significantly shift the quantized energy levels in a dot.

The sensitivity of the states of a dot to single electron charge can also lead to an unwanted effect. Sometimes impurities in a solid can lead to trapped single electrons or holes. If one of these is near a quantum dot, it can substantially shift the states of the dot.

Exercise 2.8.7 Calculate the Coulomb potential energy of two electrons separated by 10 nm, in a solid with dielectric constant of 10. How does this energy compare to the typical energy level spacing of 10–100 meV in nanostructures?

2.9 Landau Levels and Quasiparticles in Magnetic Field

As discussed in many quantum mechanics textbooks (e.g., Cohen-Tannoudji *et al.* 1977: s. E$_{IV}$), a charged particle in the presence of a magnetic field will undergo quantized circular motion in the plane perpendicular to the magnetic field. We start by looking at the basic physics using a semiclassical picture.

In a magnetic field, an electron will feel a force perpendicular to its velocity and perpendicular to the magnetic field. Classically, the force is equal to

$$\vec{F} = q\vec{v} \times \vec{B},\tag{2.9.1}$$

where q is the electron charge, $\vec{v}$ is the velocity of the electron, and $\vec{B}$ is the magnetic field. This implies that in the absence of any other forces, an electron will undergo circular motion. This is known as a **cyclotron orbit**. Taking the force $\vec{F}$ as a centripetal force, the condition for circular motion is

$$F = \frac{mv^2}{r} = |q|vB,\tag{2.9.2}$$

where F, v, and B are all magnitudes. We solve for the radius r to obtain

$$r = \frac{mv}{|q|B}.\tag{2.9.3}$$

This is known as the **cyclotron radius**, after the cyclotron devices used in the twentieth century to accelerate electrons to high speeds. The cyclotron radius of the orbit will get smaller with increasing magnetic field.

The time for an electron to complete an orbit is just given by the distance traveled divided by the speed,

$$T = \frac{2\pi r}{v} = \frac{2\pi m}{|q|B},\tag{2.9.4}$$

or

$$\omega_c = \frac{2\pi}{T} = \frac{|q|B}{m},\tag{2.9.5}$$

which is known as the **cyclotron frequency** (used previously in Section 2.2).

Let us now bring in the wave nature of the electrons. From the time-dependent quantum mechanics for a particle in magnetic field (e.g., Baym 1969: 76), we have the general formula

$$\psi(t) = e^{i\theta(t)}\psi(0),\tag{2.9.6}$$

where

$$\theta(t) = \frac{q}{\hbar} \int_{\vec{x}(0)}^{\vec{x}(t)} \vec{A}(\vec{x}) \cdot d\vec{x}.\tag{2.9.7}$$

This means that a particle traveling on a path $\vec{x}(t)$ will aquire a phase factor given by the path integral of the $\vec{A}$-field over the trajectory of the particle.

In a Landau orbit, the particle moves in a circular path. Therefore, the phase change is given by a path integral over a closed path, which is equal to an area integral according to Stokes' theorem,

$$\oint \vec{A} \cdot d\vec{x} = \int_S \vec{B} \cdot d\sigma \equiv \Phi. \tag{2.9.8}$$

The integral Φ is the flux of the magnetic field that passes through the orbit of the particle.

Since the wave function must be single-valued, the total phase change of the wave function going around the circular path must be equal to $2\pi\nu$, where ν is an integer. The total phase change cannot be equal to zero, because this would mean that the gradient $\nabla\psi$ of the wave function would be zero, which in turn would mean that the momentum of the particle is zero, contradicting the assumption that we have an orbiting particle. This implies

$$\frac{|q\Phi|}{\hbar} = 2\pi\nu, \tag{2.9.9}$$

with ν a non-negative integer. We can rewrite this as

$$|\Phi| = \frac{2\pi\nu\hbar}{|q|} = \nu\Phi_0, \tag{2.9.10}$$

where

$$\Phi_0 = \frac{h}{|q|} \tag{2.9.11}$$

is called a **flux quantum**. For an electron with $|q| = e$, it is equal to 4.14×10^{-15} T-m^2, or 4.14×10^{-7} gauss-cm^2. This number, which depends only on universal constants of nature, comes up in numerous calculations with magnetic field.

The flux quantization also implies energy quantization. For a homogeneous magnetic field, the relation of the cyclotron radius to the velocity we deduced above implies

$$\begin{aligned}
E = \frac{1}{2}mv^2 &= \frac{1}{2}m\left(\frac{qBr}{m}\right)^2 = \frac{q^2}{2\pi m}(B\pi r^2)B \\
&= \frac{q^2}{2\pi m}\Phi B \\
&= \frac{q^2}{2\pi m}\frac{2\pi\nu\hbar}{|q|}B \\
&= \hbar\omega_c\nu. \tag{2.9.12}
\end{aligned}$$

A fully quantum mechanical calculation gives a zero-point offset (see Section 2.9.1), but we can see from this simple semiclassical calculation that the quantization of the energies arises from the constraint that the electron wave function must come back to its original value around any closed path.

As we will see in Chapter 5, there are many different scattering processes for electrons in a solid. The Landau orbit picture described above is only accurate when the cyclotron radius is much smaller than the mean free path of the electrons for scattering processes. At very weak magnetic field, the electron will scatter many times before it completes an orbit.

Exercise 2.9.1   Calculate the magnetic field needed to have a Landau orbit with radius less than an electron coherence length of 100 nm. What is the energy of the lowest Landau level for this magnetic field?

2.9.1 Quantum Mechanical Calculation of Landau Levels

In this section, we will use a full quantum mechanical model for the energy states of an electron in a magnetic field. As we will see, we get nearly the same results as the semiclassical calculation. We also will be able to derive the number of allowed quantum states.

We begin with the Hamiltonian for a particle in an electromagnetic field,

$$H = \frac{1}{2m}(\vec{p} - q\vec{A})^2, \tag{2.9.13}$$

with q the charge of the particle and m the effective mass. This gives the time-independent Schrödinger equation,

$$\frac{1}{2m}(-i\hbar\nabla - q\vec{A})^2\psi = E\psi. \tag{2.9.14}$$

There are various choices for the gauge of the A-field. For a magnetic field in the z-direction, for convenience we pick the Coulomb gauge with $\vec{A} = Bx\hat{y}$. We then guess the form of the solution

$$\psi = e^{i(k_y y + k_z z)}\phi(x), \tag{2.9.15}$$

which, when substituted into (2.9.14), gives

$$-\frac{\hbar^2}{2m}\frac{\partial^2\phi}{\partial x^2} + \frac{1}{2}m\omega_c^2(x - x_0)^2\phi = \left(E - \frac{\hbar^2 k_z^2}{2m}\right)\phi, \tag{2.9.16}$$

where we have defined

$$x_0 = \frac{\hbar k_y}{m\omega_c} \tag{2.9.17}$$

and

$$\omega_c = \frac{|q|B}{m}, \tag{2.9.18}$$

which is the cyclotron frequency, used previously in Section 2.2. Equation (2.9.16) is just the equation for a one-dimensional harmonic oscillator, which has the quantized eigenvalues

$$E = \frac{\hbar^2 k_z^2}{2m} + \hbar\omega_c\left(\nu - \frac{1}{2}\right), \tag{2.9.19}$$

where ν is an integer that runs from 1 to infinity (the mathematics of the quantized harmonic oscillator are reviewed in Appendix D). Figure 8.25 shows these energy levels as a function of magnetic field. These are called **Landau levels**. If the B-field is large enough, the energy splitting between the Landau levels, $\hbar\omega_c$, will be large compared to $k_B T$, and the electrons will all lie in the lowest possible Landau level. Note that although we have treated the y-axis and x-axis differently in our choice of $\vec{A}$, the energy of the eigenstates

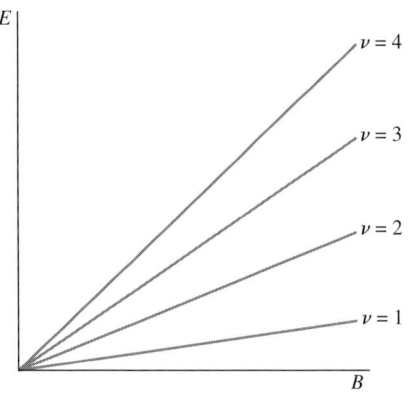

Fig. 2.36 Landau levels of charged particles in a magnetic field.

does not depend on this choice, as is proper since the value of $\vec{B}$ does not depend on the choice of gauge for $\vec{A}$.

The number of electrons that can fit in a given Landau level depends on the total number of states of a Landau level. To calculate this, we can no longer use $\vec{k}$ as the appropriate quantum number, as we did in Section 1.8, to calculate the density of states. The proper quantum numbers are k_z and ν, the Landau level number. Suppose the dimensions of the system are $-L_x/2 < x < L_x/2$ in the x-direction and $-L_y/2 < y < L_y/2$ in the y-direction. Then for the quantized motion in the plane perpendicular to the magnetic field, x_0 runs from $-L_x/2$ to $L_x/2$, and therefore, from definition (2.9.17), k_y runs from $-m\omega_c L_x/2\hbar$ to $+m\omega_c L_x/2\hbar$.

Since the form of the solution (2.9.15) is a plane wave along y, and this plane wave is subject to the boundary condition $-L_y/2 < y < L_y/2$, the same logic used in Section 1.8 implies that k_y can only have values

$$k_y = \frac{2\pi N_y}{L_y}, \tag{2.9.20}$$

where N_y is an integer. The total number of states is therefore given by the total range of k_y divided by the distance in k-space per k_y state, that is,

$$\begin{aligned} N &= \frac{m\omega_c L_x/\hbar}{2\pi/L_y} \\ &= \frac{m\omega_c L_x L_y}{2\pi\hbar}. \end{aligned} \tag{2.9.21}$$

This implies that the total number of states in a Landau level in the dimensions perpendicular to the magnetic field is

$$\boxed{N = \frac{|q|BA}{h},} \tag{2.9.22}$$

where $A = L_x L_y$ is the area of the plane. (This does not take into account spin; see the discussion of spin below.) The total density of states is therefore proportional to the magnetic field, and therefore also proportional to the cyclotron energy of the Landau level. In the z-dimension perpendicular to the plane, the density of states is still given by the one-dimensional formula from Section 2.8.1, $\mathcal{D}(E)dE = (L/2\pi)(\sqrt{m/2})dE/(\hbar\sqrt{E - E_0})$.

The number of states (2.9.22) has a natural interpretation in terms of the semiclassical cyclotron orbits discussed above. Because electrons are fermions, two electrons with the same spin in the same Landau level cannot orbit the same flux quantum. We must have one electron of a given spin per flux quantum. The number of electron states in a two-dimensional plane is therefore given by the total flux divided by the flux per state,

$$N = \frac{BA}{\Phi_0}$$
$$= \frac{|q|BA}{h}, \tag{2.9.23}$$

which is just the same as the result deduced above.

The number of states per Landau level in a plane perpendicular to the magnetic field is a constant that depends only on the strength of the B-field. Notice also that the number of states does not depend on the effective mass of the particles, just as the value of the flux quantum does not depend on the effective mass.

Exercise 2.9.2   Calculate the total number of free electrons that can occupy a single Landau level for magnetic field 10 T in a solid cube with dimension 1 cm, if the conducting electrons have effective mass $0.1m_0$, at temperature $T = 1$ K.

Spin splitting in Landau levels. The above calculation for the Landau levels of an electron relied only on the mass and the charge of the electron; we did not take into account the spin of the electrons. When we take into account the spin of the electron, the Zeeman effect (derived in Appendix F) leads to a shift of the electron energy given by

$$E_s = \pm\frac{g}{2}\mu_B B = \pm g \left(\frac{\hbar e}{4m_0}\right) B, \tag{2.9.24}$$

where B is the magnetic field, $\mu_B = \hbar e/2m_0$ is the **Bohr magneton** with e the electric charge and m_0 the vacuum electron mass, and g is the **Landé g-factor**. The energy shifts to the positive for electrons with spin in the opposite direction to the magnetic field and to the negative for spin in the same direction as the magnetic field. This is known as the **Zeeman** splitting.

In vacuum, $g \simeq 2.002$. (The Dirac equation, discussed in Appendix F, gives exactly 2; the slight deviation from this value is due to higher-order field theory corrections.) Therefore, in vacuum, the Zeeman shift downward of the lowest spin state almost perfectly cancels the Landau level shift upward. In solids, however, both the effective mass of the electron used in (2.9.3) and the g-factor used in (2.9.24) depend on the properties of the electron bands. A method of calculating the g-factor of electrons in a solid will be presented in Section 10.2. In a typical semiconductor, the Zeeman splitting is much less than the Landau level energy. For example, in GaAs, the effective mass of an electron is

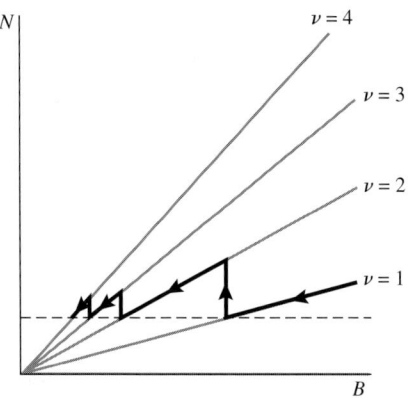

Fig. 2.37 Thin lines: the number of states in the Landau levels of a conductor as a function of magnetic field. Dashed line: the critical number, equal to the number of fermions in the system. Heavy line: the number of available states in the lowest occupied Landau level of the system as a function of magnetic field.

about $0.07m_0$ and the g-factor is about 0.4, leading to spin splitting about 1/80 of the lowest Landau level energy. Under most conditions for electrons in solids one can therefore assume that each Landau level has two spin states with nearly equal energy.

2.9.2 De Haas–Van Alphen and Shubnikov–De Haas Oscillations

Suppose that we have a two-dimensional conductor, for example a FET, as described in Section 2.8.3, and we keep the number of conducting electrons fixed and vary the B-field perpendicular to the plane while the system is kept at low temperature. In this case, the Fermi level of the free electrons will oscillate up and down as B varies. This effect can be understood in terms of the density of states of the Landau levels of the system.

We begin by imagining that the system is at high magnetic field, with all the electrons in the lowest Landau level. According to (2.9.22), the total number of states in a Landau level is proportional to B. Therefore, if the B-field is high enough, the number of states in the lowest Landau level can always be made large enough to hold all the electrons.

Figure 2.37 illustrates what happens as the magnetic field is reduced. As the number of available states in the lowest Landau level is reduced, according to (2.9.22), at some point the number of states in that Landau level exactly equals the number of electrons. At this point, if the magnetic field is reduced any further, extra electrons will have to go into the next highest Landau level, and therefore the Fermi level must jump up by an amount $\hbar\omega_c$. Most of the electrons will be in a full Landau level, while just a few of the electrons will be in the next-higher, mostly empty Landau level. If the B-field continues to decrease, the number in the upper level will increase as the number of states in the lower level decreases, until eventually both levels are completely full. At this point, the Fermi level must once again jump up as the number of states in the two levels falls below the total number of electrons.

The period of the oscillation is found by setting the total number of electrons in the system equal to the number of states at the critical points. From (2.9.22), we have

$$N = \nu \frac{2eB_{\text{crit}}A}{h},$$

(2.9.25)

or

$$\frac{1}{B_{\text{crit}}} = \nu \frac{2eA}{Nh},$$

(2.9.26)

where we have assumed each level has spin degeneracy of 2. The period of the oscillations in $1/B$ is therefore equal to $2e/nh$, where n is the number of electrons per area.

Since the electrons in a full Landau level cannot move into any other states in the same level due to Pauli exclusion, the conductivity of the system is proportional to the number of electrons in the topmost, partially filled Landau level. The conductivity of the system will therefore oscillate as the B-field is varied. This effect, known as **Shubnikov–de Haas** oscillations, can be used to measure the density of free carriers in a semiconductor. At the critical magnetic field for the jump, we know that the Landau level is full, and therefore the total number of carriers in the system is equal to the number of states at that magnetic field.

In the same way, only the free carriers in the topmost Landau level contribute to the magnetic moment of the system. Therefore, the magnetic moment of the system,

$$\mu = \frac{\partial U}{\partial B},$$

(2.9.27)

will vary strongly as the magnetic field varies. This is known as the **De Haas–van Alphen** effect.

The Shubnikov–De Haas and De Haas–van Alphen effects also occur for three-dimensional systems. In this case, the electrons can move in the z-direction with unlimited speed. The temperature of the system effectively limits the number of available states in the z-direction, however. The density of states of the Landau level is multiplied by the one-dimensional density of states factor (2.5.7),

$$N_Q^{(1d)} = \frac{L}{2\pi} \int \frac{\sqrt{m/2}}{\hbar\sqrt{E - E_0}} e^{-E/k_B T} dE,$$

(2.9.28)

which depends on T, but not on B. This then just becomes a constant multiplicative factor for the total number of states available in a Landau level at a given B.

Exercise 2.9.3   Calculate explicitly the magnetic moment of a two-dimensional system with a half-filled lowest Landau level.

2.9.3 The Integer Quantum Hall Effect

As we saw in Section 2.8.3, a MOSFET can be used to create a two-dimensional electron gas. Suppose that we apply a magnetic field perpendicular to the plane of the gas. If we apply a voltage in the plane of the carriers, we expect to observe the Hall effect, as illustrated in Figure 2.38. Charged particles moving perpendicular to a magnetic field

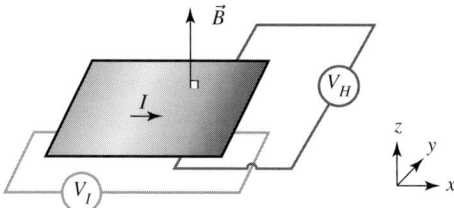

Fig. 2.38 The geometry of a Hall effect measurement.

are accelerated in the direction perpendicular to their motion, which means that negative charge will tend to pile up along one side of the plane, leading to a voltage in the direction perpendicular to the current. This is known as the **Hall voltage**. Note that the sign of the Hall voltage depends on the sign of the majority carriers in the system. Positive carriers (holes) will be accelerated to the right, while negative carriers (free electrons) moving in the opposite direction (corresponding to current in the same direction) will accelerate to their left, thereby giving the opposite polarity of the Hall voltage.

In the previous sections, we looked at the effect of varying the magnetic field while keeping the number of carriers in the 2DEG constant. Alternatively, we could keep the magnetic field constant and vary the number of carriers in the 2DEG by varying the gate voltage of the FET, which controls the amount of band bending. In this case, we also expect to see oscillations in the properties of the system as the number of carriers passes through the critical numbers defined by filling all the states in an integer number of Landau levels. In particular, we expect the conductivity properties of the system to have special properties at these points. When all the electrons exactly fill all the states in an integer number of Landau levels, then by Pauli exclusion, none of the electrons can change state, just like a full valence band.

A surprising result of the experiments is that the properties of the system do not merely oscillate. As shown in Figure 2.39, there are plateaus in the voltages measured in both the parallel and perpendicular directions at the points where the number of electrons equals the number of states in an integer number of Landau levels. Even more surprising, when the system is in one of these plateaus, the Hall resistance of the system is equal to a constant, $R_H = h/e^2\nu$, where ν is the number of Landau levels filled and h and e are Planck's constant and the electron charge, universal constants of nature, independent of the exact size and shape or material properties of the system. This is known as the **integer quantum Hall effect**.

We can understand how this quantized resistance arises by examining the current that flows in the plane. The Hall effect occurs when there is a balance of the magnetic force and the electric force on the electrons,

$$E_y = v_x B_z, \tag{2.9.29}$$

where v_x is the velocity of the electrons. The velocity is related to the current by the relation,

$$J_x = nqv_x, \tag{2.9.30}$$

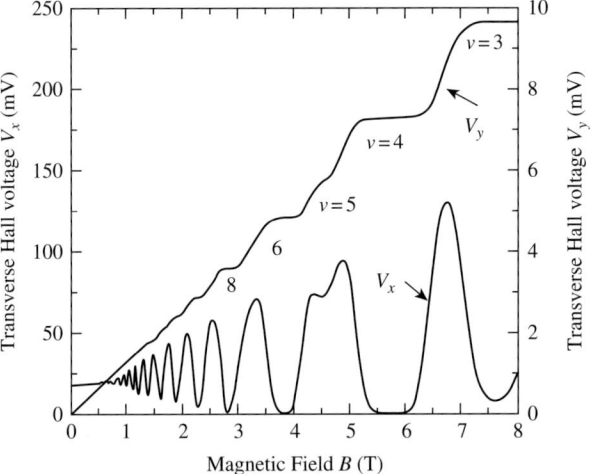

Fig. 2.39 Experimental voltages measured for the integer quantum Hall effect, as a function of B, for constant current. From Cage *et al.* 1985.

where $n = \nu N/A$ is the number of carriers per area in the two-dimensional plane, with ν the number of occupied Landau levels. We therefore have

$$J_x = \nu \frac{N}{A} q \frac{E_y}{B_z}. \tag{2.9.31}$$

The total magnetic flux $\Phi = B_z A$ is equal to $N\Phi_0$, where Φ_0 is the flux quantum contained in the orbit of a single electron, as discussed in Section 2.9.1. Therefore

$$J_x = \nu \frac{Nq}{N\Phi_0} E_y$$
$$= \nu \left(\frac{q|q|}{h} \right) E_y, \tag{2.9.32}$$

where we have used the definition of the flux quantum (2.9.11). Since J_x has units of amperes/cm in a two-dimensional system, we can convert this to an equation of current and voltage. Multiplying both sides by the length in the y-direction, we have

$$|I_x| = \nu \left(\frac{q^2}{h} \right) |\Delta V_y| \equiv \frac{|\Delta V_y|}{R_H}. \tag{2.9.33}$$

When the Landau levels have two spin states (see the discussion at the end of Section 2.9.1), the right side will be multiplied by 2.

In principle, as magnetic field is tuned, there is just one point at which the number of electrons in the 2DEG exactly matches the number of states in a Landau level. Naively one would expect that if the number of electrons in the system no longer exactly matches the number of states in an integer number of Landau levels, the conductivity properties will be different. Experimentally, however, the properties of the system are those of exactly filled Landau levels even as the magnetic field or electron density is varied over a wide range.

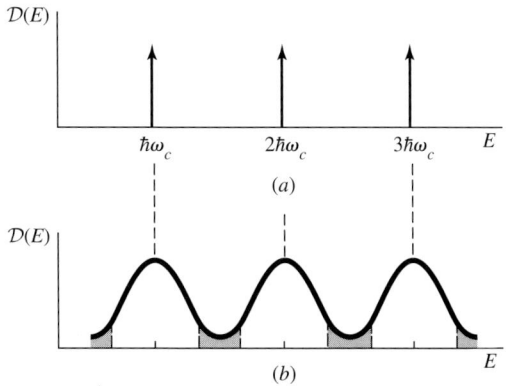

Fig. 2.40   (a) Density of states of Landau levels in a perfect two-dimensional system. (b) Density of states in a two-dimensional system with disorder. The shaded regions indicate localized states.

To understand the system properly, we must remember that the system is not a perfect two-dimensional plane. As discussed in Sections 1.8.2 and 2.8, all real systems have disorder. One might expect that this disorder would make the Hall resistance even less likely to have the exact value h/e^2, but the effect is the opposite.

Figure 2.40(a) shows the density of states of two-dimensional Landau levels in a perfect system, while Figure 2.40(b) shows the density of states in a real system. As discussed in Section 1.8.2, a reasonable hypothesis allows us to categorize the states in a disordered system into two classes. In a random potential landscape, low-energy states will be **localized** states, confined in energy minima, while above some energy cutoff known as the **mobility edge**, the electronic states will be **extended** states. When there is a maximum to the potential energy, states above some threshold will not be confined.

The states in a magnetic field will have the same properties. For electrons to have Landau orbits, they must be free to move in the plane. Therefore, localized states will not contribute to the Landau levels. As illustrated in Figure 2.40(b), between the Landau levels there is some range of energy in which the states are all localized, known as a **mobility gap**.

As discussed in Section 1.1.2, adding disorder to a system does not change the total number of states. If there are N states in an ideal Landau level, there will be N states in the same Landau level in the presence of disorder. This means that as the Fermi level varies through this range, there will still be one exactly filled Landau level, until the Fermi level hits the range of the free states in the next higher Landau level.

The disorder in the system is therefore essential for the observation of the integer quantum Hall effect. The existence of a range of localized states allows the Fermi level to vary over a wide range while keeping an exactly filled Landau level. If there were no localized states, an integer number of Landau levels would be filled only at certain exact voltages.

Because of this conspiracy of nature to force the Hall resistance to exactly h/e^2, these measurements can be used as a measure of this fundamental ratio of constants of nature. This ratio is now measured routinely to accuracy of parts per billion using quantum Hall measurements.

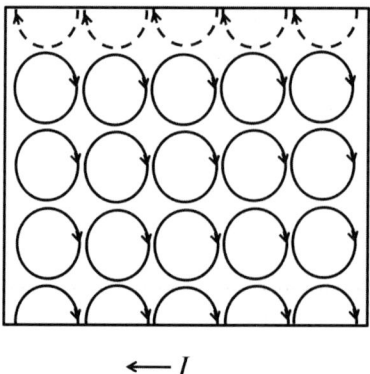

$\longleftarrow I$

Fig. 2.41   The electron motion in two-dimensional electron gas in a full Landau level, which occurs in the integer quantum Hall effect. Only electrons bouncing along the lower edge lead to net current. Edge states along the upper edge (indicated by dashed lines) are depleted of electrons.

Topological considerations. There is an alternative way of looking at the quantized Hall resistance. Let us return to the semiclassical picture of electrons in cyclotron orbits, illustrated in Figure 2.41. Electrons moving in circular orbits in the bulk of the electron gas will not contribute to a net current. The only orbits that give a net flow of current from one side to the other are those that reflect off the boundaries of the system, known as **edge states**. Electrons in these states will keep bouncing along one edge.

The current is then constrained to move in a one-dimensional channel. This then looks just like a one-dimensional quantum wire (QWR), discussed in Section 2.8.4. Recall from that discussion that we derived the Landauer formula for the conduction of a single quantum channel, accounting for two degenerate spin states,

$$I = 2\frac{e^2}{h}\Delta V. \tag{2.9.34}$$

This is the same as the quantum Hall formula (2.9.33) for a single Landau level when two spin states are included.

The fact that the quantum Hall resistance and the Landauer resistance are the same is not an accident. Both can be viewed as elementary examples of a **topological** effect. The one-dimensional channel in both the QWR and the quantum Hall effect is stable against changes of its geometry, since a bend in the path will still leave it one-dimensional. Therefore, its properties cannot depend on geometric factors. In general terms, when the geometry does not matter, one can say that h/e^2 is the natural unit for one-dimensional resistance, since the current $I \sim e/\Delta t$, where Δt is the transit time of an electron across a channel, and $\Delta t \sim h/\Delta E$ in quantum mechanics. This gives $I \sim e(\Delta E)/h \sim e(e\Delta V)/h \sim (e^2/h)\Delta V$.

Topological effects in the conduction properties of materials have become a major topic of study; in more complicated materials, conducting surface states can play the same role relative to non-conducting bulk, three-dimensional states that the edge states play in the quantum Hall effect in relation to the non-conducting bulk of the two-dimensional electron gas.

Exercise 2.9.4 What kind of current and voltage sensitivity is required to observe the integer quantum Hall effect? To answer this, suppose that a typical structure is 1 micron in width, and a Hall voltage of 10 μV is observed. What current does this correspond to, in amperes, for the first Landau level?

2.9.4 The Fractional Quantum Hall Effect and Higher-Order Quasiparticles

The integer quantum Hall effect is perhaps not too surprising. Experimentalists working with 2DEG systems were surprised to observe, however, that as they increased the magnetic field even further, at very low temperature they saw plateaus corresponding to only one-third of the states in a Landau level being filled, or alternatively, to a full Landau level of particles with charge of $e/3$. Just as the orginal Landau level series corresponded to the number of states $N/A = \nu eB/h$, where $\nu = 1, 2, 3, \ldots$, if the experiments are done with high magnetic field and high resolution, as shown in Figure 2.42, a new Landau series corresponding to $\nu = \frac{1}{3}, \frac{2}{3}, \frac{3}{3}(= 1), \frac{4}{3}, \ldots$ is observed. Plateaus corresponding to series of other odd-integer fractional charges, such as $\frac{1}{5}$ and $\frac{1}{7}$, are also observed. How can we understand this?

A full treatment of the fractional quantum Hall effect requires understanding the many-body wave function of the electrons; for a general review see Laughlin (1999). We can get a basic understanding of this effect, though, by thinking in terms of the interactions of the electrons.

Suppose that we want to write down the correct wave function for two electrons, taking into account the fact that the state must satisfy the Fermi–Dirac statistics of the electrons

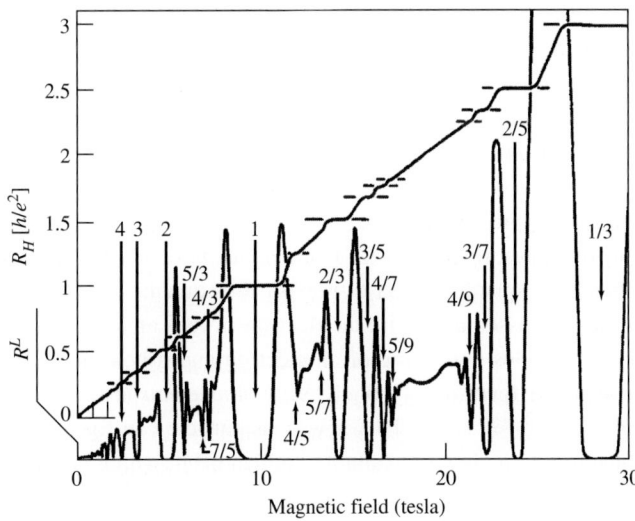

Fig. 2.42   Hall resistance and longitudinal resistance measured for the fractional quantum Hall effect, as a function of B, for constant current. From Stormer and Tsui (1997).

(that is, must change sign on interchange of the electrons) and must be an eigenstate of the total angular momentum operator,

$$L_z = \sum_i -i\hbar \left(x_i \frac{\partial}{\partial y_i} - y_i \frac{\partial}{\partial x_i} \right). \tag{2.9.35}$$

The only form of wave function that satisfies both conditions is

$$\psi(z_1, z_2) \propto (z_1 - z_2)^l, \tag{2.9.36}$$

where $z_n = x_n + iy_n = r_n e^{i\theta_n}$ is a complex number corresponding to the position of electron n in the plane, and l is an odd number. If we set $z_1 = 0$, then we have

$$\psi(0, z) \propto r^l e^{il\theta}. \tag{2.9.37}$$

This means that the wave function of an electron, relative to another electron, must have l zeros in the azimuthal direction. As discussed in Section 2.9.1, the phase change of the electron corresponds to the number of flux quanta contained in its orbit. Therefore, an electron in a state with $l = 3$ corresponds to an electron orbit confining three flux quanta.

In a full many-body calculation of the wave function of N electrons, we must minimize the energy for a wave function with the form

$$\psi(z_1, \ldots, z_N) \propto \prod_{n<m}^{N} (z_n - z_m)^l. \tag{2.9.38}$$

Without delving into this calculation, one can already see that the ground state will be one in which every electron is bound to l flux quanta, where l is an odd integer. Since we must put energy into the system to tear away a flux quantum from an electron, we can view the flux quanta as effectively positively charged particles which are attracted to the negative charge of the electron. Since the stable condition is l flux quanta per electron, each flux quantum will have an effective charge of $+e/l$. For example, three flux quanta bound to one electron corresponds to each flux quantum having $+e/3$ charge. In other words, we can account for the interaction of the electrons by writing a Hamiltonian with an effective charge for the flux quanta.

We therefore can go to a new quasiparticle picture. When the electron is bound to three flux quanta, it has two extra flux quanta compared to the one flux quantum that is always associated with an electron in its Landau orbit. Just as excitons are composite quasiparticles consisting of an electron and a hole bound together, we can define a new **composite fermion** as a bound state of an electron and two flux quanta, with a total charge of $-e + \frac{2}{3}e = -\frac{1}{3}e$. This new quasiparticle can then be seen as confining a single flux quantum in its lowest available Landau orbit, which is determined using the new effective charge.[3]

---

[3] It is also possible to think of the quasiparticle as having the same charge e but having "absorbed" two flux quanta, so that the effective magnetic field it feels is one-third of the original. This picture gives the same results for the Hall plateaus. See Jain (2003).

In this new picture, we can now use the same argumentation for the integer quantum Hall effect in Section 2.9.3 to understand the Hall effect of the new quasiparticles. A quasiparticle with charge $-e/3$ will have a set of Landau levels with three times the energy spacing and one-third the number of states. These new Landau levels will also be separated by mobility gaps, just as the integer Landau levels were. We no longer worry about the underlying electrons (or the lattice of atoms, for that matter) and only worry about the new $\frac{1}{3}$-charge quasiparticles. The same approach applies for electrons with $l = 5$ or higher, which correspond to composite fermions composed of an electron bound to four or some larger even number of flux quanta. A great number of experiments have confirmed this picture. Furthermore, new quasiparticles consisting of bound states of composite fermions can also be formed.

Again, it is tempting to think that these new quasiparticles are not "real." But they are real in the sense that they carry charge and have well-defined Landau levels. The fractional charge of these quasiparticles has been dramatically demonstrated by the observation of shot noise of the composite fermions (De Picciotto *et al.* 1997; Samindayar *et al.* 1997). (As discussed in Section 9.5, the amplitude of shot noise depends on the charge of the particles.) In other words, the composite fermions carry charge in "lumps" that give "clicks" just as electrons do. This affects our understanding of all particles. Even if we observe clicks in detection apparatus for the arrival of particles, it does not mean that they are indivisible, fundamental entities. As discussed in Section 2.1, all particles may ultimately be built out of other particles in some underlying field.

This discussion highlights the two different types of quasiparticle picture that we have been working with. In one picture, we define the ground state of a system as the vacuum, and define the excitations out of the ground state as the quasiparticles. This was the case for free electrons and holes in a semiconductor. In a second picture, we define the vacuum as the state with all bands either empty or full. The ground state of the system then consists of a number of additional quasiparticles created in the empty band. This was the case for a metal – in Section 2.4 we treated the state of the system in which there is an empty conduction band, and any number of full bands at lower energy, as the vacuum, and treated the free electrons in the conduction band as quasiparticles created in that vacuum. We could, alternatively, adopt the first picture, and define the ground state of the metal, with the electrons already in the conduction band, as the vacuum, and consider holes created below the Fermi level and electrons excited out of the ground state as the quasiparticles. We will return to these two pictures of a metal in Section 5.5.1.

The quasiparticles in the fractional quantum Hall effect are defined in the second picture. We start with an empty band and create the composite quasiparticles in that band; they are not excitations out of the ground state, but instead form the ground state. The unifying concept in all cases is that we can take an enormous amount of information about the underlying system and bury it in the definition of the vacuum and the quasiparticles, and then have a simple system in which only the quasiparticles are relevant.

Exercise 2.9.5 Prove that the wave function (2.9.38) is an eigenstate of the total angular momentum operator for N particles.

References

G. Bastard, *Wave Mechanics Applied to Semiconductor Heterostructures* (Halsted Press, 1988).

G. Baym, *Lectures on Quantum Mechanics* (Benjamin-Cummins, 1969).

M.E. Cage, R.F. Dziuba, and B.F. Field, "A test of the quantum Hall effect as a resistance standard," *IEEE Transactions on Instrumentation and Measurement* **IM-34**, 301 (1985).

C. Cohen-Tannoudji, B. Diu, and F. Laloë, *Quantum Mechanics* (Wiley, 1977).

R. De Picciotto, M. Reznikov, M. Heiblum et al., "Direct observation of a fractional charge," *Nature* **389**, 162 (1997).

W. Ebeling, W.-D. Kraft, and D. Kremp, *Theory of Bound States and Ionization Equilibria in Plasmas and Solids* (Akademie-Verlag, 1976).

K. Huang, *Statistical Mechanics* (Wiley, 1963).

J.K. Jain, "The role of analogy in unraveling the fractional quantum Hall effect mystery," *Physica E* **20**, 79 (2003).

R.S. Knox, *Theory of Excitons* (Academic Press, 1963).

W.-D. Kraft, D. Kremp, W. Ebeling, and G. Röpke, *Quantum Statistics of Charged Particle Systems* (Plenum, 1986).

J.B. Krieger and G.J. Iafrate, "Time evolution of Bloch electrons in a homogeneous electric field," *Physical Review* B **33**, 5494 (1986).

L.D. Landau and E.M. Lifshitz, *Quantum Mechanics*, 3rd edition, J.B. Sykes and J.S. Bell, trans. (Pergamon Press, 1977).

R. B. Laughlin, "Nobel lecture: fractional quantization," *Reviews of Modern Physics* **71**, 863 (1999).

D. Pines and R. Laughlin, "The theory of everything," *Proceedings of the National Academy of Sciences* **97**, 28 (2000).

B.K. Ridley, *Quantum Processes in Semiconductors*, 2nd edition, (Oxford, 1988).

F. Reif *Fundamentals of Statistical and Thermal Physics* (McGraw-Hill, 1969).

L. Samindayar, D.C. Glattli, Y. Jin, and B. Etienne, "Observation of the $e/3$ fractionally charged Laughlin quasiparticle," *Physical Review Letters* **79**, 2526 (1997).

H.L. Stormer and D.C. Tsui, "Composite fermions in the fractional quantum Hall effect," in *Perspectives in Quantum Hall Effects*, S. Das Sarma and A. Pinczuk, eds. (Wiley, 1997).

C. Waschke, H. G. Roskos, R. Schwedlem. et al., "Coherent submillimeter-wave emission from Bloch oscillations in a semiconductor superlattice," *Physical Review Letters* **70**, 3319 (1993).

3 Classical Waves in Anisotropic Media

So far, we have studied the properties of electrons in solids. In computing these properties, we have effectively assumed that the solid is a static, unchanging background. In reality, however, the atoms in a solid constantly move in response to sound waves and electromagnetic waves. Treating the atoms in the solid as static is a reasonable approximation if they do not move far from their equilibrium positions. Eventually, however, we must concern ourselves with the interactions of electrons with these waves. Before we do this, we must first write down the proper description of the waves themselves.

In elementary physics, we learn simple wave theory, in which all media are isotropic, and all waves obey Snell's law. Most solids are **anisotropic**, however; that is, some of their properties are not the same in all directions. This should not be surprising, because crystals have symmetry axes that point in certain directions, and therefore the overlap of the atomic orbitals will be different in different directions. In general, there is no particular reason to expect that material properties will be the same in all directions. The isotropic model works well only for gases, fluids, and strongly disordered solids.

In anisotropic media, Snell's law breaks down for all types of waves. For example, a ray entering a medium perpendicular to the surface may be refracted to move off at an angle, in complete violation of Snell's law, as shown in Figure 3.1.

Both sound and light have unusual properties in solids. We will examine sound waves first, and then light waves. In each case, the symmetry properties of crystals greatly simplify the analysis of waves in these media.

3.1 The Coupled Harmonic Oscillator Model

There are several approximations often made in the theory of vibrations in solids. As discussed above, often the assumption is made that materials are isotropic, with the same properties in all directions. Relaxing this assumption gives us the theory of anisotropic materials, the subject of this chapter. Another assumption is that the vibrations are *linear*, that is, that the frequency of a vibration does not change as its amplitude gets larger. When this breaks down, we move to the field of **nonlinear** dynamics. Nonlinear effects will be discussed in Chapters 5 and 7. In this chapter, we stick entirely to the linear approach, which amounts to assuming that the amplitude of vibrations is not too large.

Last, we can treat a system as a *continuum*, or we can treat it as made of tiny, discrete cells. In this section, we explicitly treat the case of discrete unit cells. In Section 3.4, we

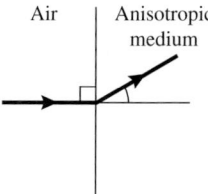

Air   Anisotropic
medium

Fig. 3.1   An example of a light ray that does not obey Snell's law in an anisotropic crystal.

will treat the continuum limit, which amounts to assuming that the vibrations have low frequency and long wavelength. As we will discuss in Chapter 4, there is debate about whether free space itself is a continuum or discrete; because the continuum model is the low-frequency limit of a discrete system, every discrete system will look continuous at low enough energies.

3.1.1 Harmonic Approximation of the Interatomic Potential

When we talked about crystal bonds in Chapter 1, we showed how the overlap of molecular orbitals can lead to a reduction of the total energy. The reason why atoms cannot keep getting closer is Pauli exclusion. As atoms get closer, the electrons are forced in closer proximity. As shown in Section 2.4.1, when the available volume per electron decreases, the Fermi energy increases rapidly.

The total interaction potential between two atoms is often approximated by the Lennard–Jones, or "6–12" potential, illustrated in Figure 3.2. At long range, atoms attract each other due to covalent bonding, as discussed in Section 1.11.2, or due to van der Waals attraction. At short distance, the repulsion due to the interaction of the electrons in core states quickly becomes dominant.

As in many cases in physics, the minimum in energy can be approximated as a harmonic potential using the Taylor approximation,

$$U(x) = U_0 + \frac{1}{2} \frac{\partial^2 U}{\partial x^2}\bigg|_{x_0} (x - x_0)^2$$

$$\equiv U_0 + \frac{1}{2} K(x - x_0)^2, \tag{3.1.1}$$

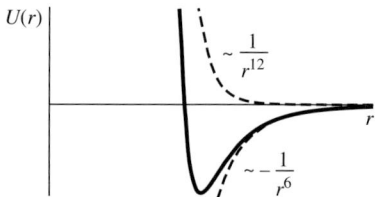

$U(r)$

$\sim \dfrac{1}{r^{12}}$

r

$\sim -\dfrac{1}{r^{6}}$

Fig. 3.2   A typical atomic interaction potential, known as the 6–12, or Lennard–Jones potential.

where x_0 is the equilbrium atomic spacing. When the amplitude of the motion is small enough, the parabolic approximation is always a good approximation of any minimum. This means that the force between two atoms is well approximated by Hooke's law for springs for low-amplitude oscillations, $F = -K(x - x_0)$. In fact, Hooke's law for springs comes from this fact about interatomic forces. The metal of springs obeys Hooke's law because the atoms in the metal have a harmonic potential.

These same oscillations are responsible for the sound waves in solids, which we will eventually quantize into **phonons**. We will model the interactions between atoms as springs, which, as we see here, is a very reasonable approximation for low-amplitude oscillations.

Exercise 3.1.1   Assuming that we can model the potential energy of the bond between two hydrogen atoms with the Lennard–Jones potential,

$$U(R) = Ar^{-12} - Br^{-6}, \tag{3.1.2}$$

what values do you find for A and B, given the experimentally determined values of the depth of the potential minimum $\Delta E = 4.7$ eV, and the mean interatomic distance $x_0 = 0.74$ Å? Approximating the potential energy as a parabola centered at the displacement x_0, what is the value of the force constant? What natural frequency $\omega_0 = \sqrt{K/m}$ does this imply for the vibration of two hydrogen atoms?

Since the oscillation frequency of atoms depends on the exact interatomic potential between them, and this depends in turn on the character of the bond between the atoms, it should come as no surprise that the vibrational characteristics of solid media are in general anisotropic, that is, different in different directions. The most common materials that are effectively isotropic on the macroscopic scale are strongly disordered materials.

3.1.2 Linear-Chain Model

To begin our study of vibrational waves in solids, we start with a simple, one-dimensional model, just as we started our study of electronic states with the simple, one-dimensional Kronig–Penney model. Since the interatomic potential can always be approximated as a harmonic potential, as discussed in Section 3.1.1, we consider a set of atoms connected by springs, as shown in Figure 3.3. We allow for the possibility that there are two different types of atoms with different masses, as we might find in a lattice with a two-atom basis, and two types of springs, with spring constants K_1 and K_2.

Hooke's law tells us that the force exerted by a spring is proportional to the net change of its length. Therefore, the forces on any two atoms in the unit cell are

$$M_1 \ddot{x}_n = K_2(y_n - x_n) - K_1(x_n - y_{n-1})$$
$$M_2 \ddot{y}_n = K_1(x_{n+1} - y_n) - K_2(y_n - x_n). \tag{3.1.3}$$

We guess a solution of the form

$$x_n(t) = x_0 e^{i(kan - \omega t)} \tag{3.1.4}$$
$$y_n(t) = y_0 e^{i(kan - \omega t)}.$$

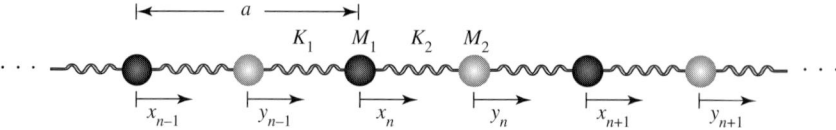

Fig. 3.3 The linear chain model with a two-atom basis.

Substituting these into (3.1.3) yields

$$-\omega^2 M_1 x_0 = K_2(y_0 - x_0) - K_1(x_0 - y_0 e^{-ika})$$
$$-\omega^2 M_2 y_0 = K_1(e^{ika} x_0 - y_0) - K_2(y_0 - x_0). \tag{3.1.5}$$

For $k = 0$, we therefore have the matrix equation

$$\begin{pmatrix} -\dfrac{K_1 + K_2}{M_1} & \dfrac{K_1 + K_2}{M_1} \\[2ex] \dfrac{K_1 + K_2}{M_2} & -\dfrac{K_1 + K_2}{M_2} \end{pmatrix} \begin{pmatrix} x_0 \\ y_0 \end{pmatrix} = -\omega^2 \begin{pmatrix} x_0 \\ y_0 \end{pmatrix}, \tag{3.1.6}$$

which has the determinant equation

$$-(K_1 + K_2)(M_1 + M_2)\omega^2 + M_1 M_2 \omega^4 = 0, \tag{3.1.7}$$

with the eigenvalues

$$\omega_1 = 0$$
$$\omega_2 = \sqrt{(K_1 + K_2)(M_1 + M_2)/(M_1 M_2)} \tag{3.1.8}$$

and the eigenvectors

$$\vec{x}_1 = \begin{pmatrix} 1 \\ 1 \end{pmatrix} \quad \text{and} \quad \vec{x}_2 = \begin{pmatrix} -M_2/M_1 \\ 1 \end{pmatrix}. \tag{3.1.9}$$

The first solution corresponds to a simple translation of all the atoms in the same direction, while the second corresponds to beating of the two atoms of the unit cell in opposite directions.

If k is not equal to zero, then we must solve the more general matrix equation,

$$\begin{pmatrix} -\dfrac{K_1 + K_2}{M_1} & \dfrac{K_1 e^{-ika} + K_2}{M_1} \\[2ex] \dfrac{K_1 e^{ika} + K_2}{M_2} & -\dfrac{K_1 + K_2}{M_2} \end{pmatrix} \begin{pmatrix} x_0 \\ y_0 \end{pmatrix} = -\omega^2 \begin{pmatrix} x_0 \\ y_0 \end{pmatrix}, \tag{3.1.10}$$

which has the determinant equation

$$-(K_1 + K_2)(M_1 + M_2)\omega^2 + M_1 M_2 \omega^4 + 2K_1 K_2 - 2K_1 K_2 \cos ka = 0. \tag{3.1.11}$$

The solutions of this equation are shown in Figure 3.4 for various choices of the ratio of K_1 and K_2. Note that we have plotted the solutions in the same Brillouin zone used for the

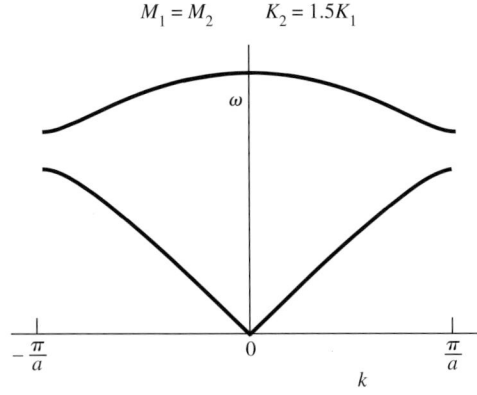

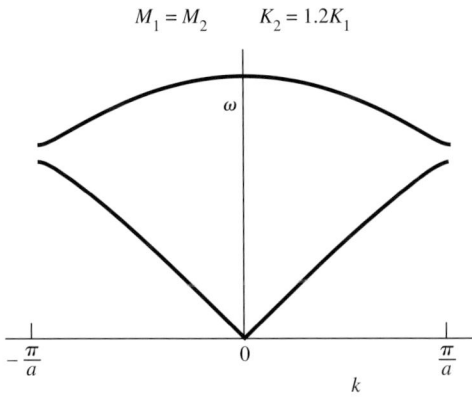

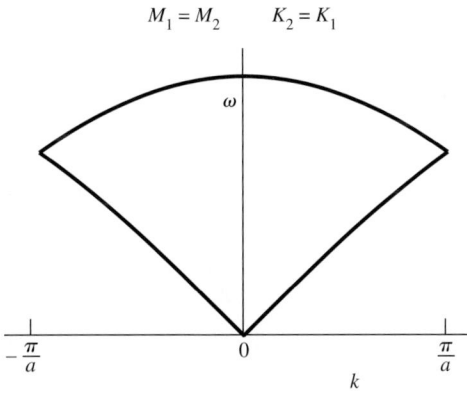

Fig. 3.4   Solutions of the classical two-atom linear-chain model, for various choices of the spring constants K_1 and K_2.

electronic states. The reason is the same: A sound wave with k outside the Brillouin zone is indistinguishable from one within the Brillouin zone, as can be seen by noticing that the only k-dependence of the frequencies in (3.1.11) is in the term $\cos ka$, which is periodic with period $2\pi/a$.

At the zone boundaries, when $K_1 \neq K_2$, the group velocity is $\partial\omega/\partial k = 0$, just as in the case of electron waves. Physically, as in the case of electron waves, a sound wave will experience Bragg diffraction from a periodic array when its wavelength is equal to twice the lattice spacing, so that the sound wave becomes a standing wave composed of two waves going in opposite directions, with group velocity of zero. Mathematically, we can see that $\partial\omega/\partial k = 0$ if we take the derivative of (3.1.11), to find

$$- 2(K_1 + K_2)(M_1 + M_2)\omega d\omega + 4M_1 M_2 \omega^3 d\omega = -2K_1 K_2 a \sin ka\, dk \qquad (3.1.12)$$

or

$$\frac{\partial\omega}{\partial k} = \frac{2K_1 K_2 a \sin ka}{2(K_1 + K_2)(M_1 + M_2)\omega - 4M_1 M_2 \omega^3}, \qquad (3.1.13)$$

which vanishes when $ka = N\pi$.

The group velocity does not vanish when $k = 0$ and $\omega = 0$, however, because when both ω and k are small, $\cos ka \simeq 1 + \frac{1}{2}(ka)^2$ and (3.1.11) becomes

$$- (K_1 + K_2)(M_1 + M_2)\omega^2 = 2K_1 K_2 (ka)^2, \qquad (3.1.14)$$

which implies $\omega = vk$, where $v = \sqrt{(2K_1 K_2 a^2)/(K_1 + K_2)(M_1 + M_2)}$, and therefore $\partial\omega/\partial k$ is a constant. This dependence of ω on k is just the same as normal sound with velocity v, and therefore this lower branch of the eigenmodes is called the **acoustic** branch. The upper branch, which has $\omega \neq 0$ when $k = 0$ and which corresponds to vibrations within a unit cell, is typically called the **optical** branch, because, as we will see, for example, in Section 5.1.3, this branch often interacts strongly with electromagnetic waves.

The plots of Figure 3.5 show what happens if we really only have one type of atom in the chain (i.e., $M_1 = M_2$ and $K_1 = K_2$) but we pick the wrong unit cell, with two atoms, instead of just one. In this case, the gap at $k = \pi/a$ closes up, and we are left with a dispersion curve that does not really have $\partial\omega/\partial k = 0$ until $k = 2\pi/a$, which is the proper

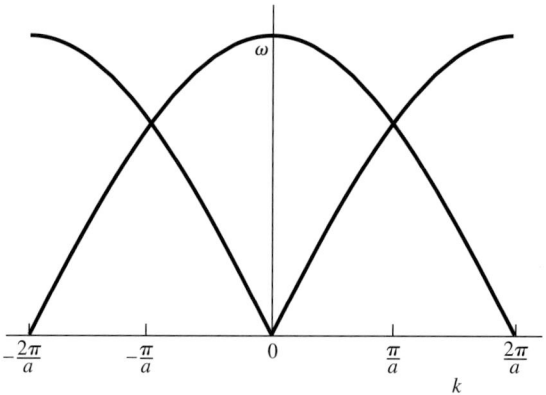

Fig. 3.5   Solutions of the linear-chain model with a two-atom basis when both atoms are identical.

zone boundary for a unit cell size of $a/2$. The upper branch is then seen to be just the continuation of the lower branch, displaced by $k = 2\pi/a$.

Exercise 3.1.2 The optical mode vibrational frequency of GaAs is 8 THz. In the linear chain model, knowing the masses of the Ga and As atoms from the periodic table, what does this imply for the force constant of the springs between the atoms? (Assume the two force constants are equal.)

What does this imply for the sound speed of this material? The unit cell size of GaAs is $a = 5.65$ Å.

Exercise 3.1.3 Show analytically that if the two masses are equal and the two spring constants are equal, then the dispersion relation becomes the simple relation for a linear monatomic chain,

$$\omega = 2\sqrt{K/M}\sin(ka/2), \tag{3.1.15}$$

where a is the spacing between the atoms.

3.1.3 Vibrational Modes in Higher Dimensions

The same approach can be used for higher dimensions. Let us consider a two-dimensional lattice of atoms connected by springs, as illustrated in Figure 3.6. To make things easier, we consider only a simple basis with one atom per unit cell, and assume that all of the spring constants are the same. The system is made more complex by the possibility of interactions between atoms in several different directions, however.

The direction of the force exerted by a spring is always along the direction between the two atoms it connects. In general, if both atoms have moved, the direction of the force on one atom will change. If we assume that the displacement of an atom is small compared to the distance between atoms, though, then we can treat the direction of the springs as

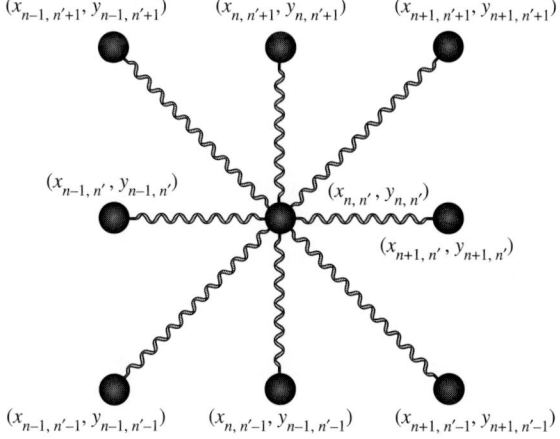

Fig. 3.6 Nearest-neighbor and next-nearest-neighbor interactions in a two-dimensional lattice with a one-atom basis.

unchanged by the motion of the atoms. For example, the force for the diagonal spring along positive x and y is

$$\vec{F} = -K\Delta l \left(\frac{1}{\sqrt{2}} \hat{x} + \frac{1}{\sqrt{2}} \hat{y} \right), \tag{3.1.16}$$

where l is the unstretched length of the spring. The change in the length is approximately

$$\Delta l = l - \sqrt{(l/\sqrt{2} + \Delta x)^2 + (l/\sqrt{2} + \Delta y)^2}$$

$$\simeq l - l\sqrt{1 + \sqrt{2}(\Delta x + \Delta y)/l}$$

$$\simeq \frac{1}{\sqrt{2}} \frac{\Delta x + \Delta y}{l}. \tag{3.1.17}$$

Thus, writing $x_{n,n'}$ for the displacement of the atom at position (n, n') in the $\hat{x}$-direction and $y_{n,n'}$ for the displacement in the $\hat{y}$-direction, Hooke's law for the eight closest neighbors of gives us

$$M\ddot{x}_{n,n'} = K(x_{n+1,n'} - x_{n,n'}) - K(x_{n,n'} - x_{n-1,n'})$$

$$+ \frac{1}{2}K(x_{n+1,n'+1} - x_{n,n'} + y_{n+1,n'+1} - y_{n,n'})$$

$$- \frac{1}{2}K(x_{n,n'} - x_{n-1,n'-1} + y_{n,n'} - y_{n-1,n'-1})$$

$$+ \frac{1}{2}K(x_{n+1,n'-1} - x_{n,n'} - y_{n+1,n'-1} + y_{n,n'})$$

$$- \frac{1}{2}K(x_{n,n'} - x_{n-1,n'+1} - y_{n,n'} - y_{n-1,n'+1}), \tag{3.1.18}$$

and a similar equation for $\ddot{y}_{n,n'}$.

As in the previous section, we guess a solution of the form

$$x(t) = x_0 e^{i(\vec{k}\cdot\vec{R} - \omega t)}$$

$$y(t) = y_0 e^{i(\vec{k}\cdot\vec{R} - \omega t)}, \tag{3.1.19}$$

where we must now keep account of a two-dimensional vector $\vec{k} = (k_x, k_y)$, and the lattice vector $\vec{R} = (an, an')$. Substituting these into the equations of motion, we obtain

$$-\omega^2 M x_0 = K x_0 (e^{ik_x a} + e^{-ik_x a} - 2) + \frac{K}{2}(x_0 + y_0)(e^{i(k_x a + k_y a)} + e^{-i(k_x a + k_y a)} - 2)$$

$$+ \frac{K}{2}(x_0 - y_0)(e^{i(k_x a - k_y a)} + e^{-i(k_x a - k_y a)} - 2)$$

$$-\omega^2 M y_0 = K y_0 (e^{ik_y a} + e^{-ik_y a} - 2) + \frac{K}{2}(x_0 + y_0)(e^{i(k_x a + k_y a)} + e^{-i(k_x a + k_y a)} - 2)$$

$$- \frac{K}{2}(x_0 - y_0)(e^{i(k_x a - k_y a)} + e^{-i(k_x a - k_y a)} - 2). \tag{3.1.20}$$

Let us pick $\vec{k}$ along the $\hat{x}$-direction. The equations of motion then correspond to the matrix equation

$$\frac{K}{M} \begin{pmatrix} 4\cos ka - 4 & 0 \\ 0 & 2\cos ka - 2 \end{pmatrix} \begin{pmatrix} x_0 \\ y_0 \end{pmatrix} = -\omega^2 \begin{pmatrix} x_0 \\ y_0 \end{pmatrix}. \tag{3.1.21}$$

This is already diagonal, with eigenvectors $(1, 0)$ and $(0, 1)$. When k is low, so that we can approximate $\cos ka \simeq 1 - \frac{1}{2}(ka)^2$, the eigenvalues are

$$\omega_L = \sqrt{2}\omega_0 ka$$
$$\omega_T = \omega_0 ka, \tag{3.1.22}$$

where $\omega_0 = \sqrt{K/m}$. Here we have labeled the two modes L and T for **longitudinal** and **transverse**. In the one-dimensional linear-chain model with two atoms per unit cell, we had two eigenmodes, which corresponded to an acoustic mode and an optical mode, and the optical mode corresponded to motion of the two atoms relative to each other, in the low-frequency limit. In the present case, the two eigenvectors have different meanings. Both are acoustic modes, with dispersion in the low-frequency limit given by $\omega = vk$, but the two modes have two different **polarizations**. The lower-frequency mode has $x_0 = 0$ and $y_0 \neq 0$, which means that the motion of the atoms is perpendicular to $\vec{k}$, or transverse, as illustrated in Figure 3.7(a). The higher-frequency mode has $x_0 \neq 0$ and $y_0 = 0$; in other words, the motion of the atoms is in the same direction as $\vec{k}$, or longitudinal, as illustrated in Figure 3.7(b).

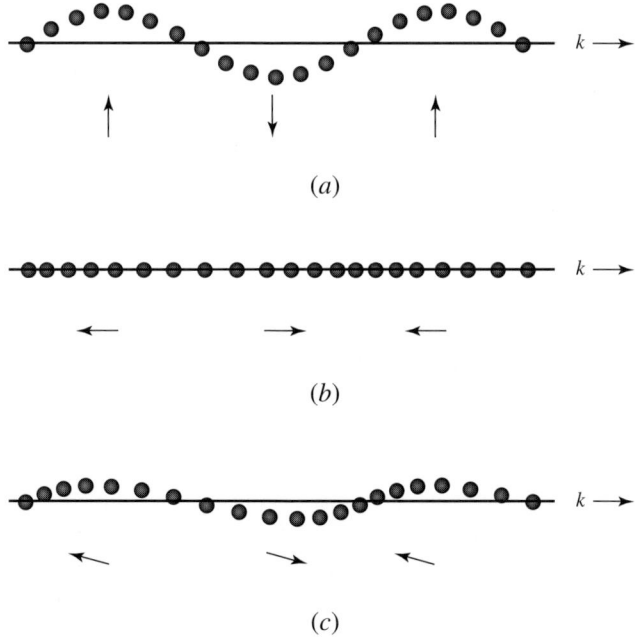

(a)

(b)

(c)

Fig. 3.7   (a) Atomic motion for a purely transverse acoustic wave. (b) Atomic motion for a purely longitudinal wave. (c) Atomic motion for a mixed-polarization wave.

In three dimensions, there are three possible polarizations for acoustic waves, unlike electromagnetic waves which can only have transverse modes. In general, the three polarization modes do not have to be purely longitudinal or transverse; the polarization vector can be in some other direction, as illustrated in Figure 3.7(c). Along a symmetry direction of the Brillouin zone, however, it is common to have two degenerate, purely transverse modes, and one longitudinal mode with faster velocity. Away from such a symmetry direction, the two lower modes continue to be mostly transverse, and the high mode mostly longitudinal, and so we still call them L and T modes.

Notice that the transverse mode in our model has lower speed even though all the spring constants were the same. This is because in the longitudinal mode, an atom pushes directly against the spring connecting it to its nearest neighbor, while the transverse mode relies on restoring force from atoms further away. It is generally the case that longitudinal modes have higher frequency in solids, for this reason.

Exercise 3.1.4   Use the matrix solver in a program like Mathematica to solve (3.1.20) for a general $\vec{k}$, and plot ω versus k, for $\vec{k}$ in two different directions in the plane. Determine the polarization vector of the two modes in each case.

The vibrational spectra of real, three-dimensional solids can be accurately calculated using a three-dimensional model that is essentially the same as those above. If the crystal structure is known, then the positions of all the atoms in the unit cell at zero displacement can be written down, and the interactions between the atoms can be modeled as springs, just as in the linear-chain model. The matrix equation in this case will have a dimension equal to the number of atoms in the unit cell times three, since each atom has three spatial degrees of freedom. Since a molecule in a solid is not normally free to spin,[1] rotational degrees of freedom are usually not taken into account.

In the linear-chain model of Section 3.1.2, we had a unit cell with two degrees of freedom, and we found two eigenmodes for each k. In the two-dimensional model, the two degrees of freedom corresponded to different directions of motion of the single atom in the unit cell. In general, since the dimension of the matrix equation for the vibration is equal to the number of degrees of freedom, the number of eigenmodes at a given $\vec{k}$ equals the number of degrees of freedom in the unit cell. Furthermore, because a zero-frequency acoustic mode corresponds to a simple translation, a crystal will always have a number of acoustic modes equal to the number of simple translations it can undergo. In particular, in three dimensions, there will always be exactly three acoustic modes, which can have different polarizations, that is, directions of the oscillation of the atoms relative to the k-vector, and different speeds.

For example, a unit cell of two atoms in three dimensions will have six degrees of freedom, and six vibrational modes. Three of these will be acoustic modes, and three will be optical. A crystal with a larger unit cell can have many more modes. For example, the crystal SrF_2, which has a fluorite lattice with a unit cell of three atoms (see Table 1.1), which gives nine degrees of freedom. Three of these are acoustic modes, and the six remaining

---

[1] "Buckyball" solids, that is, solids composed of 60 atom carbon spheres, are one notable exception – at high temperatures buckyballs can spin freely while keeping their position in a lattice.

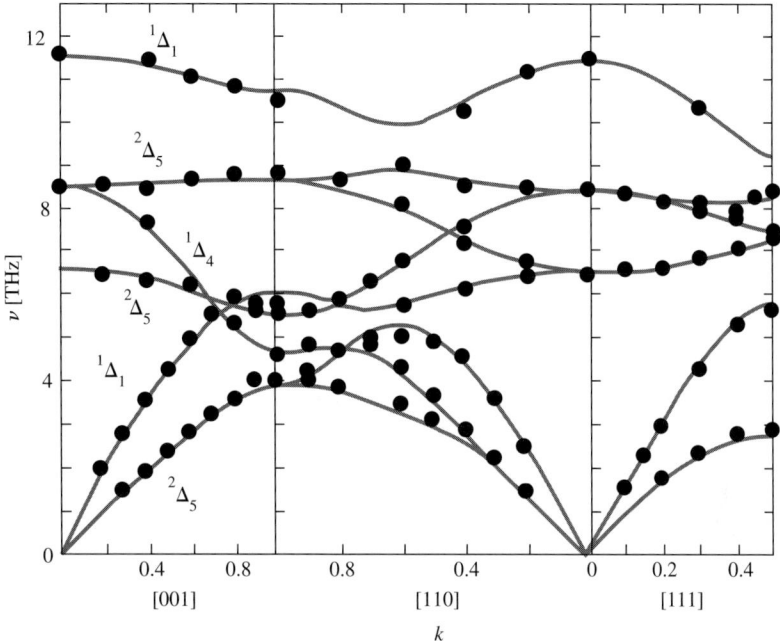

Fig. 3.8 Vibrational modes of SrF_2, which has a fluorite lattice. The data points are the experimental results of neutron scattering experiments, while the solid lines are the result of a multi-parameter spring-constant model. The curves along the [001] direction are labeled using the group theory notation discussed in Chapter 6; the superscripts in front of the symbols give the degeneracies of the modes. From Elcombe (1972).

modes are optical modes. Figure 3.8 shows the vibrational modes of SrF_2, measured by neutron scattering (for details on this method, see Section 3.2).

The linear-chain model of Section 3.1.2 included only nearest-neighbor interactions – each atom was connected by springs only to its nearest neighbors, while in the two-dimensional model we also included next-nearest-neighbor interactions. Models of the vibrational modes can be given greater sophistication by including high-order interactions to even further atoms. Some models also split an atom into the nucleus and a shell of electrons around it, which can move separately. This allows polarization of the atom as an additional degree of freedom. The electron shell can then be coupled to neighboring shells by additional springs.

Each additional spring constant gives another parameter by which the theoretical model can be fit to the data. Figure 3.8 shows a fit of a spring model to the vibrational mode data of SrF_2. These models are useful in predicting the frequencies of vibrational modes that cannot be directly measured. By fitting to the frequencies which are known, the model gives predictions of the entire vibrational spectrum.

An *ab initio* model of the ground state of a crystal (see Section 1.9.5) can also give a prediction of the vibrational spectrum by calculating the second derivative of the total energy with respect to perturbations of the atomic positions. The spring constants of the atomic interactions are not, after all, free parameters, but come from the same Hamiltonian as used for the electronic band structure.

3.2 Neutron Scattering

In Section 1.5, we examined the method of x-ray scattering for determining the positions of the atoms in a crystal. Neutron scattering is a similar method that is very useful for determining the vibrational spectrum of a crystal, because neutrons carry much more momentum for the particle energy – actually, neutron scattering is useful for determining the excitation spectrum of all kinds of condensed matter systems. Vibrational spectra determined from neutron scattering have been tabulated for many materials; for example, Bilz and Kress (1979) give a compilation for numerous insulators.

As with x-rays, we write the amplitude of a scattered wave as the sum of the phase factors for the scattered waves from all the sites in the crystal:

$$A_{\text{sum}} = \sum_l e^{i\vec{s}\cdot\vec{r}_l} e^{-i\omega t}, \tag{3.2.1}$$

where $\vec{s} = \vec{k} - \vec{k}_0$ is the scattering vector and $\vec{r}_l$ is the position in the lattice.

If there is a vibration in the lattice, the positions of the atoms will change in time. We write the displacement of a given atom in three dimensions as

$$\vec{u}_l = \vec{u}_0 e^{i(\vec{q}\cdot\vec{r}_l - \omega_{\vec{q}} t)}, \tag{3.2.2}$$

where $\vec{u}_0$ is a vector with magnitude equal to the amplitude of the vibration, which points in the direction of the motion of the atom, and $\omega_{\vec{q}}$ is the vibrational frequency. Then the amplitude of the scattered wave is

$$A_{\text{sum}} = \sum_l e^{i\vec{s}\cdot(\vec{r}_l + \vec{u}_l)} e^{-i\omega t}. \tag{3.2.3}$$

If the amplitude $\vec{u}_0$ is small compared to the wavelength (i.e., $\vec{s}\cdot\vec{u}_l \ll 1$), then we can approximate

$$e^{i\vec{s}\cdot\vec{u}_l} \simeq 1 + i\vec{s}\cdot\vec{u}_l, \tag{3.2.4}$$

and therefore

$$A_{\text{sum}} \simeq \sum_l e^{i\vec{s}\cdot\vec{r}_l} e^{-i\omega t} + i\vec{s}\cdot\sum_l \vec{u}_l e^{i\vec{s}\cdot\vec{r}_l} e^{-i\omega t}. \tag{3.2.5}$$

The first term is just the original elastic scattering term for a static lattice. The second term represents inelastic scattering from the vibrational waves. Substituting in the definition (3.2.2), we write the second term as

$$A_{\text{inel}} = i(\vec{s}\cdot\vec{u}_0) \sum_l e^{i(\vec{s}+\vec{q})\cdot\vec{r}_l} e^{-i(\omega+\omega_{\vec{q}})t}. \tag{3.2.6}$$

By the same logic as followed in Section 1.5, constructive interference will occur when

$$\vec{s} + \vec{q} = v_1 \vec{b}_1 + v_2 \vec{b}_2 + v_3 \vec{b}_3, \tag{3.2.7}$$

that is, when the scattering vector is equal to a reciprocal lattice vector minus the vibrational wave vector. If we know the reciprocal lattice vectors already, then we can deduce the vibrational wave vector.

At the same time, if we measure the energy of the scattered neutrons, we can deduce the frequency of the vibration, which then allows us to determine the entire dispersion relation for the frequency of the vibrations as a function of wave vector. The change in energy of the neutrons is given by

$$\Delta E = \hbar \omega_{\vec{q}} = \frac{\hbar^2 \vec{k}^2}{2m} - \frac{\hbar^2 \vec{k}_0^2}{2m}. \tag{3.2.8}$$

Note that the $(\vec{s} \cdot \vec{u}_0)$ factor in (3.2.6) also means that the scattering amplitude depends on the direction of $\vec{u}_0$; that is, it is polarization-sensitive.

We can see why neutrons are much more useful than x-rays for determining a vibrational spectrum by comparing the frequency shifts. X-rays have frequency of 10^{18} Hz while sound waves have frequency around 10^{13} Hz or less. To detect the frequency of the vibrational waves requires frequency resolution of the scattered x-rays of around one part in a million. On the other hand, a neutron with a wavelength comparable to a lattice spacing of a typical crystal (i.e., a few angstroms) has frequency around 10^{14} Hz, which is much more comparable to the vibrational frequencies of a solid.

Exercise 3.2.1   Suppose that instead of the form (3.2.2), we use the form $\vec{u}_l = \vec{u}_0 \cos(\vec{q} \cdot \vec{r}_l - \omega_{\vec{q}} t)$, which has no imaginary part, and therefore corresponds to a real wave in the medium. Show that this implies that the scattered wave will be shifted in frequency either up or down by the frequency $\omega_{\vec{q}}$, that is, energy can be either absorbed from or emitted into the sound waves.

3.3 Phase Velocity and Group Velocity in Anisotropic Media

Just as we found the frequency spectrum of electron waves in the Brillouin zone of a three-dimensional lattice, we now have the frequency spectrum of vibrational waves in the same lattice. Many of the same theorems apply; for example, Kramers' theorem and the property that the gradient of ω vanishes at the critical points, as deduced in Section 1.6 (except at the point $\omega = 0$ at zone center – we will return to discuss this special point in Section 3.9).

In Section 1.2, this result was discussed in terms of the group velocity of a wave hitting a Bragg reflector. The same applies here – a sound wave hitting a Bragg reflector will undergo complete reflection when its wavelength is twice the Bragg reflector spacing, leading to a standing wave. We have not yet looked at the general properties of the group velocity, however. Often these concepts are discussed only in the context of an isotropic or one-dimensional medium. Here we look a little closer at these concepts in the case when a medium can be anisotropic.

To start, suppose that we have a wave pulse that consists of a plane wave modulated by a slowly varying envelope function A,

$$y(\vec{r}, t) = e^{-i(\vec{k}\cdot\vec{r}-\omega t)} A(\vec{r}, t). \tag{3.3.1}$$

The wave function y can be anything, such as sound, light, or electron waves. In standard, linear wave theory, ω and k are related by the simple relation $\omega = vk$. As we have seen, however, in anisotropic solids ω can be a complicated function $\omega(\vec{k})$.

At time $t = 0$, we assume that the wave pulse is centered at $\vec{r} = 0$ and has a Gaussian envelope function

$$A(\vec{r}, 0) = e^{-\alpha r^2}. \tag{3.3.2}$$

We want to know what happens to this wave pulse at later times, if $\omega(\vec{k})$ is not a simple, linear function of k.

By the Fourier theorem, any wave can be viewed as a sum of plane waves. We can therefore solve this problem by taking the Fourier transform of the modulated wave and seeing what happens to each component plane wave. The Fourier transform of the wave at $t = 0$ is given by

$$
\begin{aligned}
F(\vec{k}') &= \int_{-\infty}^{\infty} d^3r \left(e^{-i\vec{k}\cdot\vec{r}} e^{-\alpha r^2} \right) e^{i\vec{k}'\cdot\vec{r}} \\
&= \int_{-\infty}^{\infty} d^3r \, e^{-\alpha r^2} e^{i(\vec{k}'-\vec{k})\cdot\vec{r}} \\
&= \left(\frac{\pi}{\alpha}\right)^{3/2} e^{-|\vec{k}-\vec{k}'|^2/4\alpha},
\end{aligned} \tag{3.3.3}
$$

where we have used the standard result from Appendix B, that the Fourier transform of a Gaussian is a Gaussian.

The Fourier transform gives the relative weight of each plane wave component in the pulse. Each component plane wave has a single k, and therefore travels without dispersion. To find the wave form at later t, we can therefore just multiply each component plane wave at $t = 0$ with the phase factor $e^{i\omega t}$, remembering that ω can depend on $\vec{k}$:

$$
\begin{aligned}
y(\vec{r}, t) &= \left[\frac{1}{2\pi} \int_{-\infty}^{\infty} d^3k' F(\vec{k}') e^{-i\vec{k}'\cdot\vec{r}} \right] e^{i\omega(\vec{k}')t} \\
&= \frac{1}{2\pi} \int_{-\infty}^{\infty} d^3k' \left(\frac{\pi}{\alpha}\right)^{3/2} e^{-|\vec{k}-\vec{k}'|^2/4\alpha} e^{-i(\vec{k}'\cdot\vec{r}-\omega(\vec{k}')t)}.
\end{aligned} \tag{3.3.4}
$$

Since ω is a function of $\vec{k}$, we expand

$$
\begin{aligned}
\omega(\vec{k}') &= \omega(\vec{k}) + \nabla_{\vec{k}}\omega\big|_{\vec{k}} \cdot (\vec{k}' - \vec{k}) + \cdots \\
&\simeq \omega + \vec{v}_g \cdot \vec{\kappa},
\end{aligned} \tag{3.3.5}
$$

where we have defined $\vec{k}' = \vec{k} + \vec{\kappa}$. Then we have

$$
\begin{aligned}
y(\vec{r}, t) &\simeq \frac{1}{2\pi} \left(\frac{\pi}{\alpha}\right)^{3/2} \int_{-\infty}^{\infty} d^3\kappa \, e^{-\kappa^2/4\alpha} e^{-i(\vec{\kappa}\cdot\vec{r}-\vec{\kappa}\cdot\vec{v}_g t)} e^{-i(\vec{k}\cdot\vec{r}-\omega t)} \\
&= e^{-i(\vec{k}\cdot\vec{r}-\omega t)} \frac{1}{2\pi} \left(\frac{\pi}{\alpha}\right)^{3/2} \int_{-\infty}^{\infty} d^3\kappa \, e^{-\kappa^2/4\alpha} e^{-i\vec{\kappa}\cdot(\vec{r}-\vec{v}_g t)} \\
&= e^{-i(\vec{k}\cdot\vec{r}-\omega t)} e^{-\alpha|\vec{r}-\vec{v}_g t|^2},
\end{aligned} \tag{3.3.6}
$$

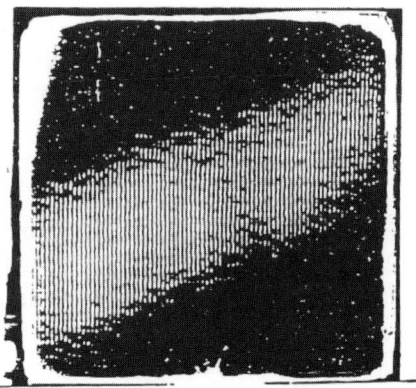

Fig. 3.9   Motion of a wave in quartz with a group velocity different from the phase velocity. The wave fronts move in the direction of the phase velocity, while the energy flows in the direction of the group velocity. Reproduced from Staudte and Cook (1967), with the permission of the Acoustical Society of America.

where the final integral has been resolved as the inverse Fourier transform of a Gaussian which is a function of $(\vec{r} - \vec{v}_g t)$.

The solution (3.3.6) is the original plane wave modulated by an envelope function centered at a point $\vec{r}$ that moves with velocity $\vec{v}_g$. From this, we see that the wave fronts move with the velocity $v_\phi = \omega/k$ in the direction of $\vec{k}$, while the energy of the pulse moves with velocity $\vec{v}_g = \nabla_{\vec{k}}\omega$. Of particular significance, the group velocity $\vec{v}_g$ is not necessarily in the same direction as the wave vector $\vec{k}$. The wave fronts always move in the direction of $\vec{k}$, but the energy of the wave can move in a different direction, as seen, for example, in Figure 3.9. In the rest of this chapter, we will see two examples in which this happens, namely, sound and light waves in anisotropic crystals.

Exercise 3.3.1   Determine the envelope function $A(x, t)$ of a Gaussian wavepacket with frequency ω and wavenumber k propagating in a one-dimensional medium (e.g., an optical pulse in a fiber optic) when $\partial^2\omega/\partial k^2$ is significant (i.e., approximate $\omega(k)$ to second order, instead of only to first order as in (3.3.5)). What happens to the width of the wavepacket over time?

3.4 Acoustic Waves in Anisotropic Crystals

In the previous sections, we formulated the vibrational spectrum of acoustic waves in terms of the microscopic forces between atoms. In the low-frequency limit, we can instead use a model of the solid as continuous; that is, not take into account the underlying atomic structure except in determining the overall crystal symmetry. This will be valid as long as the wavelength of the waves is long compared to the unit cell size a.

In this model, we cannot take into account the higher-frequency optical vibrational modes discussed in Section 3.1.2, because in treating the crystal as a continuum, there are no degrees of freedom that could allow vibrations internal to the unit cell, which give rise to the optical modes. The continuum model is very useful for acoustic waves, however. In particular, it allows us to deduce the acoustic wave spectrum from macroscopic quantities measured in static loading experiments, known as the **elastic constants**. Alternatively, measurements of the sound velocity can be used to deduce the static elastic properties of the material.

In three dimensions, Hooke's law is generalized in terms of matrix algrebra. Although the math is difficult at first, crystal symmetry will allow us to greatly simplify the equations. For a comprehesive discussion of vibrational modes in anisotropic crystals, see Auld (1973) or Dove (1993).

3.4.1 Stress and Strain Definitions: Elastic Constants

To begin, we write down a generalization of Hooke's law, $F = -Kx$, for a three-dimensional, continuous medium, as follows:

$$\tilde{\sigma} = \tilde{C}\tilde{\varepsilon}, \tag{3.4.1}$$

that is,

$$\sigma_{ij} = \sum_{lm} C_{ijlm}\varepsilon_{lm}. \tag{3.4.2}$$

Here, the term that corresponds to the force is $\tilde{\sigma}$, which is a 3×3 matrix called the **stress tensor**, which plays the role of the force in Hooke's law. It is related to the force on a surface by the relation

$$\vec{F} = \tilde{\sigma} \cdot \hat{n}A, \tag{3.4.3}$$

where $\hat{n}$ is a unit vector normal to the surface and A is the area of the surface. Since we are dealing with a continuous medium, the stress has units of pressure, that is, force per unit area.

Figure 3.10 shows how the stresses apply to a small volume element. The volume element here is not the same as the unit cell of a lattice; here we are assuming continuum mechanics, in which we assume that a volume element is much larger than a unit cell of the underlying lattice, but still small compared to the wavelength of an acoustic wave.

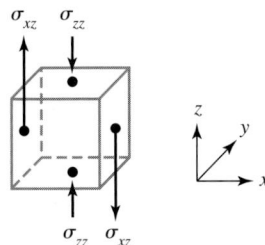

Fig. 3.10 Stresses on a volume element.

The first index of the stress refers to the face of the volume element to which the force is applied, while the second index refers to the direction of the force. Each force is assumed to be accompanied by an equal and opposite force, as shown in Figure 3.10. Thus, an element σ_{zz} of the stress tensor corresponds to two equal and opposite forces along the z-axis, which tends to squeeze the crystal, while a stress σ_{xz} corresponds to two forces along the z-axis but displaced along the x-axis, leading to a twisting of the volume element. If these forces were unbalanced, the volume element would have a net torque and would have to spin; therefore it is normally assumed that $\sigma_{ij} = \sigma_{ji}$; for example, if there is a σ_{xy} stress, there must be an equal and opposite σ_{yx} that cancels the torque. For short periods of time, however, the torque can be unbalanced, leading to rotational motion.

The displacement of the medium in response to the stress is another 3×3 matrix, $\tilde{\varepsilon}$, called the **strain tensor**. The strain matrix gives the fractional change of the dimensions of a volume element, and is therefore unitless. Relating these two in (3.4.2) is a $3 \times 3 \times 3 \times 3$ **double tensor** $\widetilde{C}$, which consists of the crystal **elastic constants**, which depend on the spring constants of the medium. To match the units of stress, the elastic constants have units of pressure.

Stresses and strains can be categorized in two types. The first is a **hydrostatic** stress or strain, which has equal terms along the diagonal:

$$\tilde{\sigma} = \sigma \begin{pmatrix} 1 & 0 & 0 \\ 0 & 1 & 0 \\ 0 & 0 & 1 \end{pmatrix}. \tag{3.4.4}$$

This corresponds to an equal pressure in all directions, as would be experienced by an object immersed in pressurized water. By contrast, a **shear** stress or strain matrix has a trace of zero.

The strain matrix is defined in terms of the displacement $\vec{u}$ of the local medium, which we used in the spring models of Sections 3.1.2 and 3.1.3. In principle, one could have an unbalanced strain, of the form

$$\tilde{\varepsilon} = \tau \begin{pmatrix} 0 & 1 & 0 \\ 0 & 0 & 0 \\ 0 & 0 & 0 \end{pmatrix}, \tag{3.4.5}$$

with

$$\varepsilon_{lm} = \frac{\partial u_l}{\partial x_m}. \tag{3.4.6}$$

This is known as a **simple shear**. It can be decomposed into the sum of a symmetric and an antisymmetric strain,

$$\begin{pmatrix} 0 & 1 & 0 \\ 0 & 0 & 0 \\ 0 & 0 & 0 \end{pmatrix} = \begin{pmatrix} 0 & \frac{1}{2} & 0 \\ \frac{1}{2} & 0 & 0 \\ 0 & 0 & 0 \end{pmatrix} + \begin{pmatrix} 0 & \frac{1}{2} & 0 \\ -\frac{1}{2} & 0 & 0 \\ 0 & 0 & 0 \end{pmatrix}. \tag{3.4.7}$$

A shear strain corresponding to a symmetric matrix is called a **pure shear**. An antisymmetric pure shear implies rotation of the medium, as illustrated in Figure 3.11. Since we

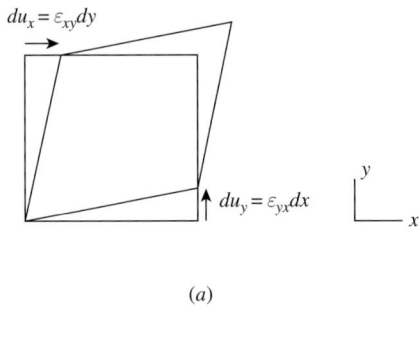

(a)

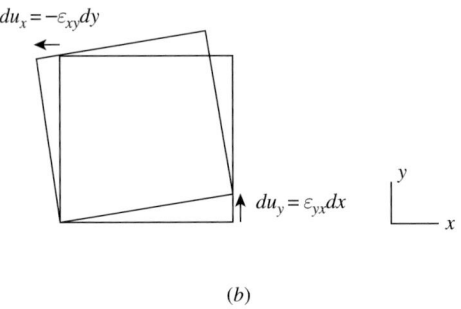

(b)

Fig. 3.11   Change of a cubic volume element by (a) a pure shear, and (b) the antisymmetric part of a simple shear.

normally assume, as discussed above, that the medium is irrotational, we enforce symmetry by using the definition

$$\varepsilon_{lm} = \frac{1}{2}\left(\frac{\partial u_l}{\partial x_m} + \frac{\partial u_m}{\partial x_l}\right). \tag{3.4.8}$$

This definition implies $\varepsilon_{ij} = \varepsilon_{ji}$, as for the stresses.

A matrix can have a trace of zero either by having terms on the diagonal that sum to zero, for example,

$$\tilde{\varepsilon} = \tau \begin{pmatrix} 2 & 0 & 0 \\ 0 & -1 & 0 \\ 0 & 0 & -1 \end{pmatrix}, \tag{3.4.9}$$

or by having only off-diagonal terms, for example,

$$\tilde{\varepsilon} = \tau \begin{pmatrix} 0 & 1 & 0 \\ 1 & 0 & 0 \\ 0 & 0 & 0 \end{pmatrix}. \tag{3.4.10}$$

When the axes of a system are rotated to a different direction, a pure shear matrix always continues to have a trace of zero. The general matrix for a rotation around the unit vector $\hat{u}$ by an angle θ is

$$R(\theta) = \begin{pmatrix} c + (1-c)u_x^2 & (1-c)u_xu_y - su_z & (1-c)u_xu_z + su_y \\ (1-c)u_yu_x + su_z & c + (1-c)u_y^2 & (1-c)u_yu_z - su_x \\ (1-c)u_zu_x - su_y & (1-c)u_zu_y + su_x & c + (1-c)u_z^2 \end{pmatrix},$$

$$(3.4.11)$$

where $c = \cos\theta$ and $s = \sin\theta$. Suppose that we have a pure shear along the x-axis, given by (3.4.9). We can rotate this into the [111] axis by performing a 35.26° rotation around the y-axis and then a 45° rotation around the z-axis. The product of these two rotations is the rotation matrix

$$R = \begin{pmatrix} \frac{1}{\sqrt{3}} & -\frac{1}{\sqrt{2}} & -\frac{1}{\sqrt{6}} \\ \frac{1}{\sqrt{3}} & \frac{1}{\sqrt{2}} & -\frac{1}{\sqrt{6}} \\ \frac{1}{\sqrt{3}} & 0 & \sqrt{\frac{2}{3}} \end{pmatrix}. \qquad (3.4.12)$$

Transforming the shear (3.4.9) by this, we obtain

$$R\tilde{\varepsilon}R^{-1} = \begin{pmatrix} 0 & 1 & 1 \\ 1 & 0 & 1 \\ 1 & 1 & 0 \end{pmatrix}. \qquad (3.4.13)$$

Any stress or strain can always be written as a sum of a hydrostatic plus a shear term. For example, a **uniaxial** strain, which consists of a compression along only one axis, can be decomposed as follows:

$$\begin{pmatrix} 1 & 0 & 0 \\ 0 & 0 & 0 \\ 0 & 0 & 0 \end{pmatrix} = \frac{1}{3}\begin{pmatrix} 1 & 0 & 0 \\ 0 & 1 & 0 \\ 0 & 0 & 1 \end{pmatrix} + \frac{1}{3}\begin{pmatrix} 2 & 0 & 0 \\ 0 & -1 & 0 \\ 0 & 0 & -1 \end{pmatrix}. \qquad (3.4.14)$$

Therefore, in practice we often do not need to refer to the full strain matrices; we can simply talk of the magnitude of the hydrostatic and the shear strains. For a purely harmonic medium, that is, one which obeys the generalized Hooke's law (3.4.2), a hydrostatic stress leads to a hydrostatic strain, which corresponds to a volume change of the unit cell, but no change of the crystal symmetry.[2] On the other hand, a pure shear strain does not change the unit cell volume, but it does change the crystal symmetry.

Exercise 3.4.1   Show explicitly that the rotation matrix (3.4.12) corresponds to the two successive rotations given in the text, that is, that it is the rotation that transforms the [100]-axis into the [111]-axis.

---

[2] As we will discuss in Section 5.4, solid state phase transitions that correspond to a change of crystal symmetry can occur under extremely high hydrostatic stress, due to anharmonic terms in the Hamiltonian.

Table 3.1 Elastic constant matrix in reduced notation for a medium with cubic symmetry

$$
\begin{pmatrix}
C_{11} & C_{12} & C_{12} & 0 & 0 & 0 \\
C_{12} & C_{11} & C_{12} & 0 & 0 & 0 \\
C_{12} & C_{12} & C_{11} & 0 & 0 & 0 \\
0 & 0 & 0 & C_{44} & 0 & 0 \\
0 & 0 & 0 & 0 & C_{44} & 0 \\
0 & 0 & 0 & 0 & 0 & C_{44}
\end{pmatrix}
$$

Because the $3 \times 3 \times 3 \times 3$ double tensor $\widetilde{C}$ is an unwieldy four-dimensional matrix, it is standard to use **reduced notation** for the stress and strain as follows:

$$
\begin{aligned}
\varepsilon_1 &= \varepsilon_{xx} & \sigma_1 &= \sigma_{xx} \\
\varepsilon_2 &= \varepsilon_{yy} & \sigma_2 &= \sigma_{yy} \\
\varepsilon_3 &= \varepsilon_{zz} & \sigma_3 &= \sigma_{zz} \\
\varepsilon_4 &= 2\varepsilon_{yz} & \sigma_4 &= \sigma_{yz} \\
\varepsilon_5 &= 2\varepsilon_{xz} & \sigma_5 &= \sigma_{xz} \\
\varepsilon_6 &= 2\varepsilon_{xy} & \sigma_6 &= \sigma_{xy}.
\end{aligned}
\tag{3.4.15}
$$

As discussed above, we do not need nine terms, because $\varepsilon_{ij} = \varepsilon_{ji}$ and $\sigma_{ij} = \sigma_{ji}$. Note that ε_4, ε_5, and ε_6 are multiplied by 2, while the stresses are not, to take into account that there are two terms ε_{ij} and ε_{ji} that contribute in the matrix multiplication to each term σ_{ij}.

The 81 possible terms in $\widetilde{C}$ are therefore cut down to 36. We can then write the generalized Hooke's law in terms of a 6×6 C-matrix operating on six-component vectors, as follows:

$$
\sigma_I = \sum_J C_{IJ} \varepsilon_J.
\tag{3.4.16}
$$

Further simplification is possible if we assume that the potential energy is quadratic in all position variables, which implies that the C-matrix must be symmetric, that is, $C_{IJ} = C_{JI}$. (See Exercise 3.4.2.) As discussed in Section 3.1.1, for low-amplitude displacements from an energy minimum, this approximation is always valid. In this case, instead of 36 constants, we have just 21 $(= 6 + 5 + 4 + 3 + 2 + 1)$.

Further reduction of the number of elastic constants can be done by knowing the symmetry of the crystal. In general, many of the elastic constants will be zero or equal to other elastic constants. The website that accompanies this book (www.cambridge.org/snoke) gives the form of the elastic constant C-matrix for all of the possible crystal symmetries. For example, Table 3.1 gives the form of the elastic constant matrix in reduced matrix for a cubic crystal.

As seen in this table, a cubic crystal has only *three* independent elastic constants. An isotropic medium has the same form of the elastic constant matrix as a cubic crystal, but in addition the elastic constants are subject to the constraint $C_{44} = (C_{11} - C_{12})/2$. A crystal with no rotational symmetries, which has 21 independent elastic constants, is known as a **triclinic** system.

Often, we know the stress and want to find the strain. In this case, it is convenient to define the **compliance tensor**, $S = C^{-1}$, that is,

$$\varepsilon_I = \sum_J S_{IJ}\sigma_J. \tag{3.4.17}$$

The number of independent compliance constants is equal to the number of independent elastic constants.

Exercise 3.4.2   Prove the statement above, that if the potential energy of the system is quadratic, of the form

$$\frac{U}{V} = \frac{1}{2}\sum_{ij}\sigma_{ij}\varepsilon_{ij}, \tag{3.4.18}$$

then the elastic constant matrix C_{ijlm} must be symmetric with respect to interchanging the first two indices with the last two indices.

Exercise 3.4.3   (a) Determine the compliance tensor S in terms of the elements of the cubic, or isotropic, elastic constant matrix C given in Table 3.1.

(b) If a uniaxial stress is applied along the [110] axis, what strain is created?

(c) Show that for an isotropic medium, if the direction of a uniaxial stress is originally along the x-axis, creating a strain, and then the direction of the stress is rotated about the y-axis by any angle θ, the strain created, relative to the new uniaxial stress axis, is still the same.

Exercise 3.4.4   Verify the statement above, that a hydrostatic stress leads to a hydrostatic strain, using the cubic (but not necessarily isotropic) compliance tensor deduced in part (a) of the prior exercise. Is it also the case that a pure shear stress leads to a pure shear strain?

The elastic constants in the C-matrix are related to two simple "engineering" constants defined for an isotropic medium. **Young's modulus** gives the strain produced by a given stress in the same direction, that is, the fractional increase in the length of a beam due to a tensile stress, as illustrated in Figure 3.12(a). For an isotropic medium, this is equal to

$$Y = \frac{\sigma_{xx}}{\varepsilon_{xx}} = C_{11}. \tag{3.4.19}$$

For typical solids, this is on the order of 10^{11} dynes/cm^2 = 100 kbar. In other words, a pressure of 1 kbar is typically needed to create a strain of 1%, that is, a fractional change of the volume of 1%.

When a solid is stressed, it does not experience strain only in the same direction as the stress. As illustrated in Figure 3.12(b), a compressive stress will also cause expansion of the solid in the perpendicular directions as the solid is squeezed outward. **Poisson's ratio** is defined as the ratio of the transverse strain (fractional expansion) to the fractional compression, which for an isotropic medium is equal to

$$\nu = -\frac{\varepsilon}{\varepsilon'} = -\frac{S_{12}}{S_{11}} = \frac{C_{12}}{C_{11} + C_{12}}. \tag{3.4.20}$$

This is a unitless ratio, of the order of 0.3 for typical solids.

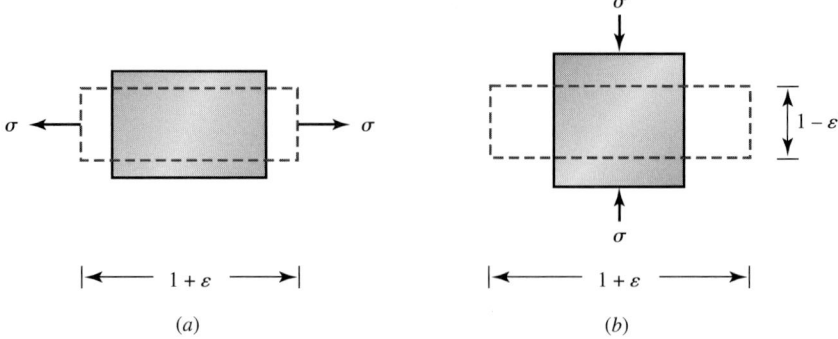

Fig. 3.12   (a) Young's modulus definition. (b) Poisson's ratio definition.

Since an isotropic medium has only two independent elastic constants, these values tell us all we need to know about the elastic deformations of an isotropic medium. One can also define **Lamé coefficients** λ and μ, which are related to the elastic constants by the following:

$$C_{11} = \lambda + 2\mu$$
$$C_{12} = \lambda$$
$$C_{44} = \mu. \tag{3.4.21}$$

Last, the **bulk modulus** is often defined, which gives pressure needed per fractional change in volume:

$$B = \lambda + \frac{2}{3}\mu = \frac{1}{3}(C_{11} + 2C_{12}). \tag{3.4.22}$$

The Lamé coefficient μ is sometimes called the **shear modulus**. In media that will not support shear, such as water, $\mu = 0$. All of these different terms are just different notations to reflect the basic fact that an isotropic medium has only two independent parameters for its linear elastic response.

3.4.2 The Christoffel Wave Equation

Starting with the generalized Hooke's law of (3.4.2), we can deduce a wave equation for an anisotropic medium just as we did in Section 3.1.2 for a linear chain. In addition to Hooke's law, we also need an equivalent for the force law $F = ma$. We write this as

$$\sum_j \frac{\partial \sigma_{ij}}{\partial x_j} = \rho \ddot{u}_i, \tag{3.4.23}$$

where ρ is the mass density of the medium and u_i is the displacement of a volume element in the i-direction. Note that we are concerned with the *gradient* of the stress, $\partial \sigma_{ij}/\partial x_j$, since if there is no gradient of the stress, there will be no net force on a volume element, because, as discussed in Section 3.4.1, at each location we assume the static loading condition of equal forces in opposite directions. A constant stress leads simply to a compression, not motion.

Combining (3.4.2) and (3.4.23), we therefore have

$$\rho \ddot{u}_i = \sum_{jlm} C_{ijlm} \frac{\partial \varepsilon_{lm}}{\partial x_j}$$

$$= \frac{1}{2} \sum_{jlm} C_{ijlm} \left(\frac{\partial^2 u_l}{\partial x_j \partial x_m} + \frac{\partial^2 u_m}{\partial x_j \partial x_l} \right)$$

$$= \sum_{jlm} C_{ijlm} \frac{\partial^2 u_l}{\partial x_j \partial x_m}, \tag{3.4.24}$$

where we have used the definition (3.4.8)

$$\varepsilon_{lm} = \frac{1}{2} \left(\frac{\partial u_l}{\partial x_m} + \frac{\partial u_m}{\partial x_l} \right) \tag{3.4.25}$$

and we have made use of the fact that l and m are dummy variables, so that we can switch them in the second summation.

As before, we guess a plane-wave solution of the form

$$\vec{u} = \vec{u}_0 e^{i(\vec{k} \cdot \vec{r} - \omega t)}, \tag{3.4.26}$$

where $\vec{u}_0$ is the polarization vector. Unlike an electromagnetic wave, the polarization does not have to be purely transverse; it can have a longitudinal component of the oscillation in the direction of the propagation. Substituting this form of the solution into (3.4.24), we find

$$\rho \omega^2 u_{0i} = \sum_{jlm} C_{ijlm} k_j k_m u_{0l} \tag{3.4.27}$$

or

$$\boxed{\sum_{jlm} \left(C_{ijlm} k_j k_m - \rho \omega^2 \delta_{il} \right) u_{0l} = 0.} \tag{3.4.28}$$

This is called the **Christoffel equation**; it is the generalized classical wave equation in an anisotropic, continuous medium. There will be three polarization vectors $\vec{u}^0$ which are the eigenvectors of the matrix constructed from $\tilde{\tilde{C}}$ and $\vec{k}$, with eigenvalues $\rho \omega^2$.

For a cubic crystal, the form of the elastic constant matrix shown in Table 3.1 implies that the Christoffel wave equation has the form, for a wave propagating in the z-direction,

$$\begin{pmatrix} C_{44}k^2 - \rho \omega^2 & & \\ & C_{44}k^2 - \rho \omega^2 & \\ & & C_{11}k^2 - \rho \omega^2 \end{pmatrix} \begin{pmatrix} u_{0x} \\ u_{0y} \\ u_{0z} \end{pmatrix} = 0 \tag{3.4.29}$$

since $C_{44} = C_{(xz)(xz)} = C_{(yz)(yz)}$ and $C_{11} = C_{(zz)(zz)}$. This implies that waves with different polarizations will have two different speeds: waves with **transverse** polarization will have speed $v = \omega/k = \sqrt{C_{44}/\rho}$, while waves with **longitudinal** polarization will have $v = \sqrt{C_{11}/\rho}$. In general, C_{11} is larger than C_{44}, so that longitudinal acoustic waves are faster than transverse acoustic waves. This makes sense because the spring constant of two atoms pressing closer to each other, as occurs in longitudinal waves, is likely to be stiffer than the spring constant for atoms to slip past each other, as occurs in transverse waves. Note that both polarizations are possible as long as the shear elastic constant C_{44} is nonzero, which means they are also allowed in an isotropic medium.

In general, acoustic waves in anisotropic media do not have to be purely transverse or longitudinal; they can have a component of each, depending on the direction the wave is traveling. Also, the two transverse polarizations do not in general have to have the same velocity, although one expects that when a wave is propagating along a symmetry axis of a crystal such that two axes perpendicular to the k-vector are identical, then the transverse waves should be identical.

Exercise 3.4.5 Write down the Christoffel wave equation for a wave in a cubic crystal propagating along the [111] direction. For the choice $C_{11} = 1$, $C_{12} = C_{44} = 0.5$, and $\rho = 1$, use a program like Mathematica to find the eigenvectors of this matrix, which correspond to the polarizations of the propagating waves, and the velocity of each polarization.

3.4.3 Acoustic Wave Focusing

For a given choice of wave vector, we can solve the Christoffel equation for the eigenvectors and eigenvalues, which give the polarizations and the natural frequency ω of each polarization. This gives a function $\omega(\vec{k})$ defined for all $\vec{k}$.

Figure 3.13 shows a possible surface generated from the function $\omega(\vec{k})$ by fixing the value of ω and plotting all $\vec{k}$ that correspond to this frequency. This is known as the **slowness surface** of the wave, because in any direction $\vec{k}$, the distance of the surface from the origin is proportional to the inverse of the phase velocity.

In general, a warped slowness surface like this can cause all manner of strange propagation effects. The group velocity vector $\vec{v}_g = \nabla_{\vec{k}}\omega$ is always perpendicular to this surface. As seen in this figure, the group velocity is often not parallel to $\vec{k}$. As discussed in Section 3.3, this means that the energy flow of the wave will not be in the same direction as the motion of the wave fronts.

At certain points of the slowness surface, namely the inflection points, the group velocity for many $\vec{k}$ vectors will be nearly parallel, as illustrated in Figure 3.13. We therefore expect that a large fraction of the acoustic energy will flow in these directions. This effect has been observed, as shown dramatically in Figure 3.14, and is known as **phonon focusing**, that is, acoustic wave focusing. This is a bit of a misnomer, since the sound is not focused to a single point as with a lens; a better name would be **acoustic caustics**. A **caustic** is the concentration of energy flow seen whenever there is a change of direction of the group velocity vector. Usually a curved surface produces a line of focused energy instead of a single point. In optics, light passing through any curved interface between two media

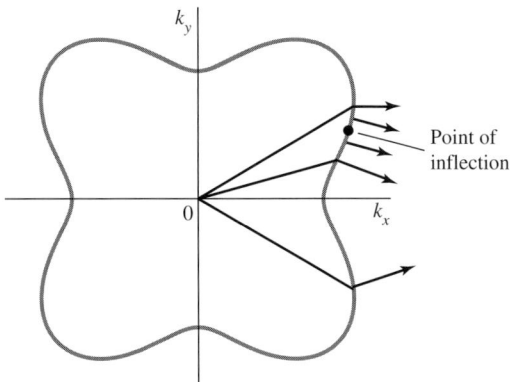

Fig. 3.13   Group velocity direction determined as the normal to a slowness surface.

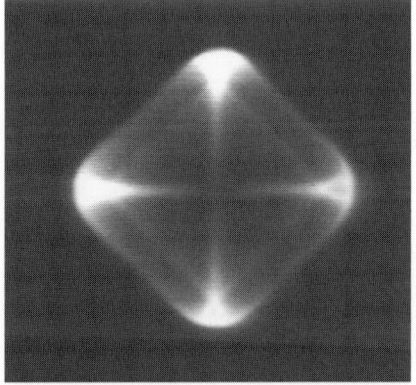

Fig. 3.14   Experimental examples of phonon focusing, that is, acoustic caustics. The images give the energy pattern detected on one face of a cubic crystal (germanium) from a heat source in the center on the opposite side. The lower image is a higher-resolution close-up of the central square pattern in the upper image, rotated by 45 degrees. From Northrup and Wolfe (1980).

forms caustics; for example, the lines of light seen on the bottom of a swimming pool due to the refraction of the Sun's light through the curved water surface. Gravitational lensing due to the curvature of space caused by massive objects also creates caustics in the light propagation from distant stars (e.g., Knudson *et al.* 2001).

As discussed in Section 5.4, scattering of sound waves becomes very strong at high frequency. Impurities and defects in crystals also scatter sound waves. Therefore, patterns like this are seen only at low temperature or at low frequency, in very pure crystals. For a review of acoustic caustic effects, see Wolfe (2005).

3.5 Electromagnetic Waves in Anisotropic Crystals

So far in this chapter, we have discussed acoustic waves in solids. All of the same mathematics can be applied to electromagnetic waves, as well. This is the basis of understanding **birefringent** optical materials, which have odd properties such as double refraction, in which light hitting a surface splits into two rays going in two directions, one of which does not obey Snell's law. For example, a ray at zero angle of incidence may be refracted away from the normal, as shown in Figure 3.1. This effect is important in numerous technological applications in optics (see, e.g., Iizuka 2002 or Yariv 1989).

3.5.1 Maxwell's Equations in an Anisotropic Crystal

To begin, we must go back to Maxwell's equations to ask how the properties of the medium enter into the wave speed. In differential form, Maxwell's equations (in the MKS system) are

$$\nabla \cdot \vec{E} = \frac{\rho}{\epsilon_0}$$

$$\nabla \cdot \vec{B} = 0$$

$$\nabla \times \vec{E} = -\frac{\partial \vec{B}}{\partial t}$$

$$\nabla \times \vec{B} = \mu_0 \epsilon_0 \frac{\partial \vec{E}}{\partial t} + \mu_0 \vec{J}, \tag{3.5.1}$$

where ρ is the charge density, $\vec{J}$ is the current density, and ϵ_0 and μ_0 are the vacuum permittivity and permeability, respectively. In a typical solid, there is no net charge, so $\rho = 0$. The entire effect of the medium on the light waves therefore comes from the current density term $\vec{J}$.

We can derive Maxwell's wave equation by taking the time derivative of the fourth equation,

$$\nabla \times \frac{\partial \vec{B}}{\partial t} = \mu_0 \epsilon_0 \frac{\partial^2 \vec{E}}{\partial t^2} + \mu_0 \frac{\partial \vec{J}}{\partial t}, \tag{3.5.2}$$

substituting for the time derivative of $\vec{B}$ from the third of Maxwell's equations to obtain

$$-\nabla \times \nabla \times \vec{E} = \mu_0\epsilon_0 \frac{\partial^2 \vec{E}}{\partial t^2} + \mu_0 \frac{\partial \vec{J}}{\partial t}. \tag{3.5.3}$$

From a vector identity, we have $\nabla \times \nabla \times \vec{E} = \nabla \cdot (\nabla \cdot \vec{E}) - \nabla^2 \vec{E}$, and from the first Maxwell's equation we have $\nabla \cdot \vec{E} = 0$ when $\rho = 0$, which therefore implies

$$\boxed{\nabla^2 \vec{E} = \mu_0\epsilon_0 \frac{\partial^2 \vec{E}}{\partial t^2} + \mu_0 \frac{\partial \vec{J}}{\partial t}.} \tag{3.5.4}$$

This is Maxwell's wave equation in general form. In a vacuum, $\vec{J} = 0$ and we have simply

$$\nabla^2 \vec{E} = \frac{1}{c^2} \frac{\partial^2 \vec{E}}{\partial t^2}, \tag{3.5.5}$$

where $c = 1/\sqrt{\mu_0\epsilon_0}$.

In general, however, in a medium, the charge will move in response to the electric field. This motion will then lead to a current term $\vec{J}$, which will have an effect back on the electric field. To take into account the current, we must solve Maxwell's wave equation self-consistently. If the medium is not a conductor, then the effect of the electric field will be to **polarize** the medium, that is, to cause negative and positive charges to be pulled away from each other. When the charges are moving relative to each other, there will also be a current.

The simplest way to take into account the effect of the electric field on the medium is to make the **linear response** assumption, which is that the induced polarization is proportional to the electric field:

$$\boxed{\vec{P} = \chi\epsilon_0\vec{E}.} \tag{3.5.6}$$

Here, χ is a unitless parameter called the **electrical susceptibility**, which gives the constant of proportionality. A great deal of experimental solid state physics is concerned with the electrical susceptibility, because it controls the optical properties of the medium. There are also other susceptibilities such as magnetic susceptibility (see Section 10.4); in general, one can define a susceptibility as the constant of proportionality between the response of a system and the driving force in any system where one assumes a linear response.

The polarization $\vec{P}$ has units of dipole moment per unit volume, or

$$\frac{\text{charge} \times \text{distance}}{\text{distance}^3} = \frac{\text{charge}}{\text{distance}^2}, \tag{3.5.7}$$

since the dipole moment is equal to the charge moved times the distance over which it is moved. If the polarization oscillates, then there will be a current produced. The current density is equal to the amount of charge moved through a given area per unit time. In terms of the polarization, it is simply given by

$$\boxed{\vec{J} = \frac{\partial \vec{P}}{\partial t},} \tag{3.5.8}$$

with units of charge per second per distance squared. We can therefore write

$$\frac{\partial \vec{J}}{\partial t} = \chi \epsilon_0 \frac{\partial^2 \vec{E}}{\partial t^2} \tag{3.5.9}$$

and therefore

$$\nabla^2 \vec{E} = \mu_0 \epsilon_0 (1 + \chi) \frac{\partial^2 \vec{E}}{\partial t^2}. \tag{3.5.10}$$

This is the adjusted wave equation for a polarizable medium. We define the **permittivity** of the medium as

$$\boxed{\epsilon = \epsilon_0 (1 + \chi).} \tag{3.5.11}$$

which implies

$$\nabla^2 \vec{E} = \mu_0 \epsilon \frac{\partial^2 \vec{E}}{\partial t^2}. \tag{3.5.12}$$

We can also define the **dielectric constant**, $\kappa = \epsilon / \epsilon_0$, and the **index of refraction**,

$$\boxed{n = \sqrt{\epsilon / \epsilon_0} = \sqrt{1 + \chi},} \tag{3.5.13}$$

which then gives

$$\nabla^2 \vec{E} = \frac{n^2}{c^2} \frac{\partial^2 \vec{E}}{\partial t^2}, \tag{3.5.14}$$

which is the same as the vacuum wave equation (3.5.5) but with a speed c/n.

In (3.5.6), we took χ as a simple constant of proportionality, a single number. This allowed us to write ϵ as a simple number. In general, however, in an anisotropic medium, the constant of proportionality χ will depend on which direction the electric field points. Along some bond directions, for example, the electrons may move quite easily, while between planes of a crystal, electrons may not be able to move far at all. To allow for this, we write the dielectric constant as a 3×3 tensor, that is,

$$\vec{P} = \tilde{\chi} \epsilon_0 \vec{E} \tag{3.5.15}$$

or

$$P_i = \sum_j \chi_{ij} \epsilon_0 E_j, \tag{3.5.16}$$

which implies

$$\epsilon_{ij} = \epsilon_0 (\delta_{ij} + \chi_{ij}). \tag{3.5.17}$$

(Unlike the case of elastic constants, we do not need to keep track of *four* indices – at least, not yet.) The wave equation (3.5.12) then becomes

$$\boxed{\nabla^2 \vec{E} = \mu_0 \tilde{\epsilon} \frac{\partial^2 \vec{E}}{\partial t^2}.} \tag{3.5.18}$$

As we did in deriving the Christoffel equation, we assume a plane-wave solution,

$$\vec{E} = \vec{E}_0 e^{i(\vec{k} \cdot \vec{r} - \omega t)}. \tag{3.5.19}$$

The curl and divergence for a plane wave become simply the dot and cross products

$$\nabla \cdot \vec{E} = i\vec{k} \cdot \vec{E}$$
$$\nabla \times \vec{E} = i\vec{k} \times \vec{E} \tag{3.5.20}$$

and the time derivative is simply $\partial \vec{E}/\partial t = -i\omega \vec{E}$. The wave equation (3.5.18) then becomes

$$\vec{k} \times (\vec{k} \times \vec{E}) + \omega^2 \mu_0 \tilde{\epsilon} \vec{E} = 0 \tag{3.5.21}$$

or

$$\sum_j \left((k_i k_j - k^2 \delta_{ij}) + \omega^2 \mu_0 \epsilon_{ij} \right) E_j = 0. \tag{3.5.22}$$

This is the electromagnetic equivalent of the Christoffel equation (3.4.28).

By a proper choice of axes, the dielectric tensor can always be diagonalized. These are the **principal axes** of the medium. In this basis, the dielectric tensor is

$$\tilde{\epsilon} = \begin{pmatrix} \epsilon_1 & 0 & 0 \\ 0 & \epsilon_2 & 0 \\ 0 & 0 & \epsilon_3 \end{pmatrix}. \tag{3.5.23}$$

Suppose that a wave is traveling in the z-direction, i.e., $k_1 = k_2 = 0$, $k_3 = k$. Then we have

$$\begin{pmatrix} n_1^2 \dfrac{\omega^2}{c^2} - k^2 & 0 & 0 \\ 0 & n_2^2 \dfrac{\omega^2}{c^2} - k^2 & 0 \\ 0 & 0 & n_3^2 \dfrac{\omega^2}{c^2} \end{pmatrix} \begin{pmatrix} E_1 \\ E_2 \\ E_3 \end{pmatrix} = 0. \tag{3.5.24}$$

This has the solutions

$$\left(n_1^2 \frac{\omega^2}{c^2} - k^2 \right) E_1 = 0$$
$$\left(n_2^2 \frac{\omega^2}{c^2} - k^2 \right) E_2 = 0$$
$$E_3 = 0. \tag{3.5.25}$$

Since electromagnetic waves are transverse, there are only two propagating modes, which correspond to the two different polarizations. The two polarizations propagate with different speeds.

Exercise 3.5.1   Write down the Christoffel electromagnetic equation for a wave traveling in the direction $(\hat{x} + \hat{y})/\sqrt{2}$ for a crystal with dielectric tensor (3.5.23).

3.5.2 Uniaxial Crystals

A special case of anisotropic crystals is the **uniaxial** crystal. In this case, one principal axis is different from the other two, that is, $n_1 = n_2 = n_o$ and $n_3 = n_e$. The special axis is called the **extraordinary axis**, or sometimes the **optic axis** or the c-axis (labeling the other two

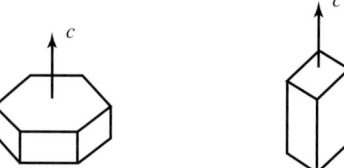

Fig. 3.15 Definition of the c-axis for the unit cell of two examples of uniaxial crystals.

axes "a" and "b"). As illustrated in Figure 3.15, one axis is selected out as special in both hexagonal and tetragonal symmetry.

In this case, the dielectric tensor becomes even simpler. The wave equation is

$$
\begin{pmatrix}
n_o^2 \dfrac{\omega^2}{c^2} - k_2^2 - k_3^2 & k_1 k_2 & k_1 k_3 \\[2mm]
k_1 k_2 & n_o^2 \dfrac{\omega^2}{c^2} - k_1^2 - k_3^2 & k_2 k_3 \\[2mm]
k_1 k_3 & k_2 k_3 & n_e^2 \dfrac{\omega^2}{c^2} - k_1^2 - k_2^2
\end{pmatrix}
\begin{pmatrix}
E_1 \\ E_2 \\ E_3
\end{pmatrix}
= 0.
$$

(3.5.26)

The condition for a solution is that the determinant of the matrix is zero. After some algebra, we find this is

$$
\left(k^2 - n_o^2 \frac{\omega^2}{c^2} \right) \left((k_1^2 + k_2^2) n_o^2 \frac{\omega^2}{c^2} + k_3^2 n_e^2 \frac{\omega^2}{c^2} - n_o^2 n_e^2 \frac{\omega^4}{c^4} \right) = 0.
$$

(3.5.27)

There are two solutions:

$$
\boxed{k^2 = n_o^2 \frac{\omega^2}{c^2},}
$$

(3.5.28)

which is the equation for a sphere in k-space; and

$$
\boxed{\frac{k_1^2}{n_e^2} + \frac{k_2^2}{n_e^2} + \frac{k_3^2}{n_o^2} = \frac{\omega^2}{c^2},}
$$

(3.5.29)

which is the equation for an ellipsoid in k-space. Both of these are the slowness surfaces that give all the $\vec{k}$-vectors corresponding to a given frequency ω.

Exercise 3.5.2   Prove that the determinant of the wave equation matrix for a uniaxial crystal is that given in (3.5.27). This can be easily done with a program like Mathematica.

As in the case of the slowness surfaces for sound waves, the group velocity $\vec{v}_g = \nabla_k \omega$, which corresponds to the direction normal to the slowness surface, gives the direction of the propagation of the energy, that is, the direction of the *rays* used in geometrical optics. The wave fronts continue to move in the direction of $\vec{k}$, but the energy (ray) moves in the direction of $\vec{v}_g$, which often is not parallel to $\vec{k}$.

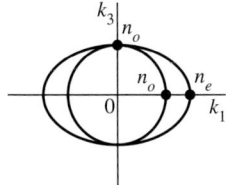

Fig. 3.16   Slowness surface for a uniaxial crystal.

Figure 3.16 gives an example of a slowness surface of a uniaxial crystal in the $k_1 - k_3$ plane, with both the spherical and ellipsoidal solutions. For the spherical surface, the group velocity is always parallel to $\vec{k}$. For the ellipsoid, if $k_2 = 0$, we have

$$\omega = c\sqrt{\frac{k_1^2}{n_e^2} + \frac{k_3^2}{n_o^2}}.$$ (3.5.30)

The gradient $\nabla_{\vec{k}}\omega$ in the $k_1 - k_3$ plane is therefore given by

$$\nabla_{\vec{k}}\omega = \left(\frac{\partial\omega}{\partial k_1}, \frac{\partial\omega}{\partial k_3}\right) = \frac{c^2}{\omega}\left(\frac{k_1}{n_e^2}, \frac{k_3}{n_o^2}\right).$$ (3.5.31)

If we write

$$k_1 = k\sin\theta$$
$$k_3 = k\cos\theta,$$ (3.5.32)

where θ is the angle relative to the optic axis, then we have

$$\vec{v}_g = \nabla_k\omega = \frac{c^2 k}{\omega}\left(\frac{\sin\theta}{n_e^2}, \frac{\cos\theta}{n_o^2}\right).$$ (3.5.33)

Therefore, if $\vec{k}$ has an angle θ relative to the optic axis, the group velocity vector will have the angle θ' given by

$$\boxed{\tan\theta' = \frac{n_o^2}{n_e^2}\tan\theta.}$$ (3.5.34)

Example. Suppose that light is incident on a uniaxial crystal with $n_e = 1.6, n_o = 1.5$, cut with the c-axis at $\theta = 30°$ from the normal of the surface, as shown in Figure 3.17. What happens to light incident normal on the surface?

Using (3.5.34), we have

$$\tan\theta = \frac{1}{\sqrt{3}}$$

$$\tan\theta' = \frac{1}{\sqrt{3}}\frac{(1.5)^2}{1.6^2} = 0.51$$

or $\theta' = 26.9°$. Relative to the normal, the ray therefore deviates by $30 - 26.9 = 3.1$ degrees. In other words, the light hitting the surface splits into two – one ray (called the **ordinary**

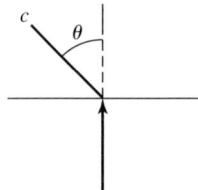

Example of light incident on a uniaxial crystal.

ray, corresponding to the spherical slowness surface) continues on in the same direction, and the other ray (called the **extraordinary** ray, corresponding to the ellipsoidal slowness surface) moves off at an angle of $3.1°$. This is the effect of **double refraction**.

Exercise 3.5.3   Determine the angle of the extraordinary ray, relative to the normal, for light incident normal to the surface of a uniaxial crystal with $n_e = 1.45, n_o = 1.65$, with the crystal cut with the c-axis at $45°$ relative to the normal.

Generalized Snell's law. In general, if the incident ray is not normal to the surface, we must come up with a replacement for Snell's law to determine the angle of refraction. Figure 3.18 shows this complicated situation. Inside the crystal we have drawn the double slowness surface of the anisotropic medium, while outside we have drawn the slowness surface of vacuum, which consists simply of a sphere with radius $k = \omega/c$.

The general rule for refraction at an interface is that the in-plane component of $\vec{k}$ must be conserved. This corresponds to the condition that the wave fronts must line up on both sides of the interface. The perpendicular component of the wave vector need not be conserved. (This means, incidentally, that the momentum of a photon refracted at a surface is not conserved – the surface must recoil in response to the refraction.)

For the ordinary ray, this condition means that we equate $k \sin \theta$ on the outside slowness surface with $k' \sin \theta'$ on the inside spherical surface. This implies

$$\frac{\omega}{c} \sin \theta = \frac{\omega}{c} n_o \sin \theta'$$

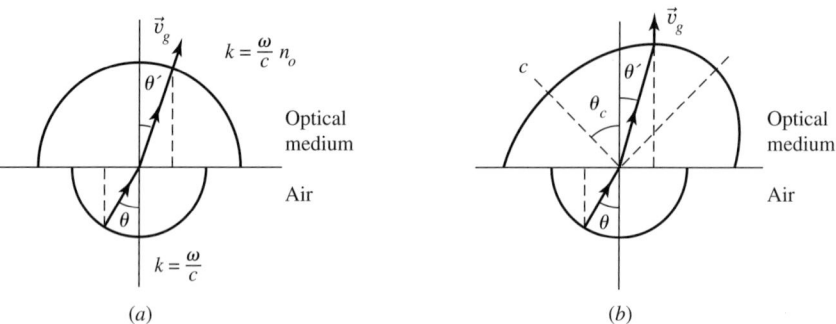

Generalized Snell's law construction. (a) An ordinary ray. (b) An extraordinary ray, when the c-axis is not normal to the surface.

or

$$\sin \theta = n_o \sin \theta'. \tag{3.5.35}$$

This is just the standard Snell's law. Figure 3.18(a) shows the matching of the in-plane $\vec{k}$-components in this case.

For the extraordinary ray, we must work a little harder. Figure 3.18(b) shows this case. From (3.5.29), the equation for the ellipsoid inside the crystal is

$$\frac{(k' \sin \tilde{\theta})^2}{n_e^2} + \frac{(k' \cos \tilde{\theta})^2}{n_o^2} = \frac{\omega^2}{c^2}, \tag{3.5.36}$$

where $\tilde{\theta}$ is the angle relative to the optic axis. For the angle relative to the normal, we must subtract off the cut angle (the angle of the optic axis relative to the normal) θ_c, that is, $\theta' = \tilde{\theta} - \theta_c$.

Matching the in-plane components inside and outside gives

$$k \sin \theta = k' \sin \theta'. \tag{3.5.37}$$

We therefore have two equations and two unknowns, k' and θ'. We first solve (3.5.37) for k' in terms of θ'

$$k' = \frac{k \sin \theta}{\sin \theta'}. \tag{3.5.38}$$

Substituting this into (3.5.36), we have

$$\frac{k^2 \sin^2 \theta}{\sin^2 \theta'} \frac{\sin^2 \tilde{\theta}}{n_e^2} + \frac{k^2 \sin^2 \theta}{\sin^2 \theta'} \frac{\cos^2 \tilde{\theta}}{n_o^2} = k^2 \tag{3.5.39}$$

that is,

$$\frac{\sin^2 \theta}{\sin^2 \theta'} \left(\frac{\sin^2 \tilde{\theta}}{n_e^2} + \frac{\cos^2 \tilde{\theta}}{n_o^2} \right) = 1 \tag{3.5.40}$$

or

$$\sin \theta = n(\tilde{\theta}) \sin \theta', \tag{3.5.41}$$

where $n(\tilde{\theta})$ is defined by

$$\frac{1}{n^2(\tilde{\theta})} = \frac{\sin^2 \tilde{\theta}}{n_e^2} + \frac{\cos^2 \tilde{\theta}}{n_o^2}. \tag{3.5.42}$$

This is the generalized Snell's law for the case of a uniaxial crystal.

Exercise 3.5.4   A uniaxial crystal with $n_e = 1.45, n_o = 1.65$, is cut with the c-axis at $45°$ relative to the normal. If light is incident on the surface at $45°$ ($90°$ relative to the c-axis), determine the angles, relative to the normal, of the ordinary and extraordinary refracted rays.

3.5.3 The Index Ellipsoid

In the example of the previous section, what is the difference between the ordinary and extraordinary rays? What determines whether light goes one way or the other?

The answer is that the polarization of the rays determines whether they are ordinary or extraordinary. As discussed in Section 3.5.1, the different eigenmodes of the wave equation correspond to different polarizations. To find the normal modes that correspond to the allowed polarizations, we can use the construct of the **index ellipsoid**. Note that this ellipsoid is *not* the same as the slowness surface ellipsoid discussed in the previous section. In particular, for a uniaxial crystal, it does not have the same shape.

To begin, we note that any symmetric second-rank tensor $\tilde{A}$ (i.e., any real, symmetric square matrix) can be represented by a *surface* that consists of all $\vec{x}$ such that

$$\sum_{i,j} A_{i,j} x_i x_j = 1. \tag{3.5.43}$$

If the matrix is diagonalized, then this becomes simply

$$A_{11} x_1^2 + A_{22} x_2^2 + A_{33} x_3^2 = 1. \tag{3.5.44}$$

In other words, if we look at the surface, the principal axes correspond to the eigenvectors of the matrix. A principal axis is defined as a direction in which the normal to the surface is parallel to $\vec{x}$.

We now apply this construction to the dielectric tensor. It is more convenient to work with the inverse matrix,

$$\tilde{\eta} = \epsilon_0 \tilde{\epsilon}^{-1}, \tag{3.5.45}$$

where $\tilde{\epsilon}^{-1}$ is the inverse matrix of the dielectric tensor. In its principal coordinates, this is just

$$\tilde{\epsilon}^{-1} = \begin{pmatrix} \dfrac{1}{\epsilon_1} & 0 & 0 \\ 0 & \dfrac{1}{\epsilon_2} & 0 \\ 0 & 0 & \dfrac{1}{\epsilon_3} \end{pmatrix}, \tag{3.5.46}$$

and therefore

$$\tilde{\eta} = \begin{pmatrix} \dfrac{1}{n_1^2} & 0 & 0 \\ 0 & \dfrac{1}{n_2^2} & 0 \\ 0 & 0 & \dfrac{1}{n_3^2} \end{pmatrix}. \tag{3.5.47}$$

This matrix is therefore represented by the surface

$$\boxed{\dfrac{x_1^2}{n_1^2} + \dfrac{x_2^2}{n_2^2} + \dfrac{x_3^2}{n_3^2} = 1.} \tag{3.5.48}$$

For a uniaxial crystal, this is

$$\frac{x_1^2}{n_o^2} + \frac{x_2^2}{n_o^2} + \frac{x_3^2}{n_e^2} = 1. \tag{3.5.49}$$

This is the index ellipsoid. Note that, in contrast with the slowness surface (3.5.29), the length of the special axis is proportional to n_e, as opposed to n_o. Figure 3.19 illustrates the difference.

The $\tilde{\eta}$ matrix makes it easy to rewrite the wave equation (3.5.21) in terms of the $\vec{D}$ field. As in isotropic electromagnetism, we write $\vec{D} = \epsilon \vec{E}$, so here we write

$$\vec{D} = \tilde{\epsilon}\vec{E} \tag{3.5.50}$$

and therefore

$$\vec{E} = \tilde{\epsilon}^{-1}\vec{D} = \frac{\tilde{\eta}}{\epsilon_0}\vec{D}. \tag{3.5.51}$$

We define $\hat{u}$ as the unit vector in the direction of $\vec{k}$, that is, $\vec{k} = k\hat{u}$. The wave equation (3.5.21) therefore becomes

$$k^2\hat{u} \times \left(\hat{u} \times \frac{\tilde{\eta}}{\epsilon_0}\vec{D}\right) + \omega^2\mu_0\vec{D} = 0, \tag{3.5.52}$$

or

$$(\tilde{u}\tilde{u}\tilde{\eta})\vec{D} = -\frac{\omega^2}{c^2k^2}\vec{D} \equiv -\frac{1}{n^2}\vec{D}, \tag{3.5.53}$$

with

$$\tilde{u} = \begin{pmatrix} 0 & -u_z & u_y \\ uz & 0 & -u_x \\ -u_y & u_x & 0 \end{pmatrix}. \tag{3.5.54}$$

This is now a standard eigenvalue problem. The eigenvectors correspond to the polarizations of the propagating wave, and the eigenvalues correspond to the index of refraction felt by each mode.

To find the eigenvectors, we could diagonalize the matrix numerically using standard methods, or we can use a geometric method based on the index ellipsoid. Note that

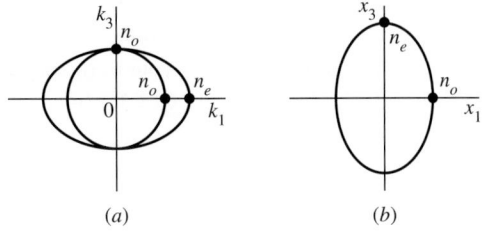

(a)   (b)

Fig. 3.19   (a) Slowness surface, and (b) index ellipsoid for the same uniaxial crystal, for the case $n_e > n_o$.

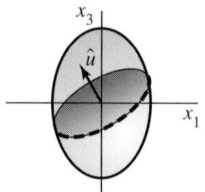

Fig. 3.20   Geometric construction for finding the polarizations that are eigenmodes of the wave equation, using the index ellipsoid of a uniaxial crystal.

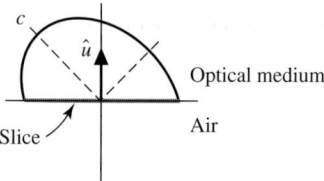

Fig. 3.21   Index ellipsoid geometric construction for finding the principal polarizations for a wave incident on a uniaxial crystal with the *c*-axis not normal to the surface, as in Figure 3.17.

$\hat{u} \times (\hat{u} \times \tilde{\eta}\vec{D})$ is the projection of the vector $\tilde{\eta}\vec{D}$ into a plane normal to $\hat{u}$. In other words, it is a slice through the index ellipsoid perpendicular to $\hat{u}$, as illustrated in Figure 3.20. The cross-section of an ellipsoid is an ellipse.

Since, as discussed above, the eigenvectors of a matrix correspond to the principal axes of the surface contruction, the eigenvectors of the propagating wave are therefore the principal axes of this ellipse.

Example. In Section 3.5.2, we have already determined the directions of the rays that are double refracted from light incident normal to a uniaxial crystal with $n_e = 1.6, n_o = 1.5$, cut with the *c*-axis at $30°$ from the normal of the surface (Figure 3.17). What are the polarizations of the rays?

To determine this, we draw the index ellipsoid in the plane of incidence, as shown in Figure 3.21. Since $\hat{u}$ is normal to the surface, the slice of the ellipsoid is parallel to the surface.

This slice is an ellipse, with one principal axis perpendicular to the plane of incidence (vertical, or out of the page), and the other in the plane of incidence, perpendicular to the first axis. Since the ellipsoid is a surface of rotation around its extraordinary axis, it has dimension n_o in the vertical direction (out of the page). The dimension in the perpendicular direction is given by $n(\theta_c)$, defined by (3.5.42).

The vertical polarization is therefore the ordinary ray, which obeys the normal Snell's law, while the horizontal polarization (in the plane of incidence) is the extraordinary ray, which is refracted according to the generalized Snell's law at angle $3.1°$ from the normal, according to the calculation of Section 3.5.2.

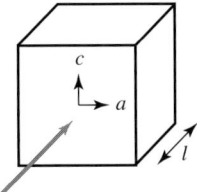

Fig. 3.22 Phase retarder geometry for a uniaxial crystal.

Exercise 3.5.5   Determine the principal polarizations for a wave traveling in the direction $(\hat{x} + \hat{y})/\sqrt{2}$ in a crystal with dielectric tensor (3.5.23). A program like Mathematica can easily diagonalize a 3×3 matrix. Hint: Convert the Christoffel equation (3.5.22) to an equation for $\vec{D} = \tilde{\epsilon}\vec{E}$.

Exercise 3.5.6   A uniaxial crystal is cut with the c-axis parallel to the surface, as shown in Figure 3.22, with $n_e = 1.7$ and $n_o = 1.77$. For light incident normal to the surface, determine the polarizations of the two propagating waves and the index of refraction felt by each. At what angle does each of the outgoing rays propagate?

   This geometry is the standard configuration for a **phase retarder** plate. By selecting the proper thickness l, one polarization can be delayed relative to the other by a fixed phase.

3.6 Electro-optics

Recall that in starting our discussion of electromagnetic waves in anisotropic media, we made the assumption (3.5.6) that the polarization induced in the medium is linearly proportional to the electric field. This is the assumption of linear optics. In actuality, however, this assumption is almost never perfectly true. The polarization can in general be a complicated function of the electric field. To approximate this dependence, we can use a perturbation series,[3]

$$P = \chi\epsilon_0 E + 2\chi^{(2)}E^2 + 4\chi^{(3)}E^3 + \cdots . \tag{3.6.1}$$

There are two cases when these higher-order terms become important. One case is when the electric field strength is very high, as in ultrafast laser pulses. This case will be studied in Chapter 7. The second case is when a strong DC electric field is applied to an optical medium, as for example in an electro-optic modulator used for communications. In this case, one can view the effect of the DC electric field as changing the index of refraction felt by the propagating wave. Then it is more convenient to expand the index of refraction in a power series.

---

[3] The prefactors of the coefficients here are the standard notation, defined for convenience in later calculations.

In general, since we must use a matrix for the index of refraction in an anisotropic medium, we must use matrices for this expansion. The standard expansion is

$$\eta_{ij} = \eta_{ij}^0 + \sum_k r_{ijk} E_k + \sum_{kl} s_{ijkl} E_k E_l + \cdots,$$ (3.6.2)

where $\tilde{\eta}$ is the index ellipsoid matrix introduced in the previous section. The r_{ijk} coefficients are known as **Pockels** coefficients, while the s_{ijkl} coefficients are known as **Kerr** coefficients. In the isotropic case, this reduces to

$$\frac{1}{n^2(E)} = \frac{1}{n^2(0)} + rE + sE^2 + \cdots$$ (3.6.3)

or

$$n(E) = n(0) - \frac{1}{2} r n^3 E - \frac{1}{2} s n^3 E^2 + \cdots,$$ (3.6.4)

where we have used the approximation

$$\frac{1}{\sqrt{1 - \alpha}} \simeq 1 - \frac{1}{2}\alpha$$ (3.6.5)

for small α, assuming that nrE and nsE^2 are small. Given the field $\vec{E}$, then, we can calculate a new index ellipsoid and determine the change of the normal modes of the wave.

This brings us back to tensors with three and four indices, as in the case of acoustic waves. We again adopt reduced notation,

$$xx \leftrightarrow 1$$
$$yy \leftrightarrow 2$$
$$zz \leftrightarrow 3$$ (3.6.6)
$$yz \leftrightarrow 4$$
$$xz \leftrightarrow 5$$
$$xy \leftrightarrow 6,$$

and write $\tilde{r}$ and $\tilde{s}$ as two-index tensors by grouping pairs of indices together:

$$r_{(xy),z} \equiv r_{63}$$ (3.6.7)

and

$$s_{(xy),(xz)} \equiv s_{65}.$$ (3.6.8)

This means that the $\tilde{r}$ tensor will be a 6×3 matrix in reduced notation, and the $\tilde{s}$ tensor will be a 6×6 matrix in reduced notation.

As in the case of acoustic waves, the symmetry of crystals generally reduces these matrices down to just a few independent terms. Tables for the electro-optic tensors which show the nonzero terms have been compiled. For example, for a cubic crystal without inversion symmetry, the electro-optic tensor $\tilde{r}$ in reduced notation has the form shown in Table 3.2,

Table 3.2 Electro-optic matrix for T_d symmetry

$$\begin{pmatrix} 0 & 0 & 0 \\ 0 & 0 & 0 \\ 0 & 0 & 0 \\ r_{41} & 0 & 0 \\ 0 & r_{41} & 0 \\ 0 & 0 & r_{41} \end{pmatrix}$$

with only one independent constant r_{41} in the positions shown, and all other terms zero. For a DC electric field in the z-direction, the equation for the index ellipsoid then becomes

$$\frac{x_1^2}{n^2} + \frac{x_2^2}{n^2} + \frac{x_3^2}{n^2} + 2r_{41}x_1x_2E_3 = 1, \tag{3.6.9}$$

that is,

$$\tilde{\eta} = \begin{pmatrix} \dfrac{1}{n^2} & r_{41}E_3 & 0 \\ r_{41}E_3 & \dfrac{1}{n^2} & 0 \\ 0 & 0 & \dfrac{1}{n^2} \end{pmatrix}. \tag{3.6.10}$$

Diagonalizing this matrix gives the eigenvectors $\hat{z}$, $\hat{x} + \hat{y}$, and $\hat{x} - \hat{y}$, and the eigenvalues, respectively,

$$\frac{1}{n_z^2} = \frac{1}{n^2}$$

$$\frac{1}{n_{x+y}^2} = \frac{1 - n^2 r_{41}E_3}{n^2}$$

$$\frac{1}{n_{x-y}^2} = \frac{1 + n^2 r_{41}E_3}{n^2}. \tag{3.6.11}$$

To lowest order, the change in index felt by the last two modes is $\Delta n = \frac{1}{2}n^3 r_{41}E$. The electric field therefore turns the isotropic medium into a birefringent medium. This is the basis of many electro-optic devices. Typical values of r are very small, however, on the order of a few picometers per volt, which means that very large electric fields must be used.

Exercise 3.6.1 The electro-optic matrix for lithium niobate, LiNbO$_3$, has the form given in Table 3.3, with four independent constants. Write down the index ellipsoid matrix $\tilde{\eta}$ for an electric field along the z-direction (i.e., the x_3 direction, which is the same as the c-axis).

(a) For a wave propagating in the x-direction, what are the natural polarizations and what are the indices of refraction felt by each?

Table 3.3	Electro-optic matrix for C_{3v} symmetry	
r_{11}	0	r_{13}
$-r_{11}$	0	r_{13}
0	0	r_{33}
0	r_{42}	0
r_{42}	0	0
0	$-r_{11}$	0

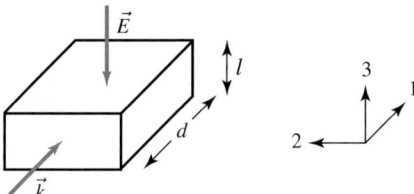

Fig. 3.23   Electro-optic modulator.

(b) Figure 3.23 shows a typical electro-optic modulator geometry. Electro-optic devices are often characterized by their **half-wave voltage**, that is, the voltage needed to produce a phase shift of π between the two different polarizations. Show that for a crystal with length d in the x-direction, and thickness l in the z-direction across which the voltage is applied, the half-wave voltage of a LiNbO$_3$ crystal is given by

$$V_\pi = \frac{l\lambda_0}{drn^3},$$

where λ_0 is the vacuum wavelength of the light and r is the appropriate electro-optic constant. If r $= 30 \times 10^{-12}$ m/V, and l and d are both 2 mm, what is the half-wave voltage?

3.7 Piezoelectric Materials

So far we have treated electromagnetic waves and acoustic waves as completely separate. In general, however, it should be no surprise that these two can interact. When an atom moves, the charge contained in it moves, which can in principle create a polarization. In the same way, an electric field that acts on the charged electrons and nuclei will cause motion, which can lead to strain. A material in which an electric field can produce a strain, and in which a strain can produce an electric field, is known as a **piezoelectric** material.

Let us return to our model of dielectric polarization introduced in Section 3.5.1. We consider a linear chain of atoms as shown in Figure 3.24. In a normal dielectric medium,

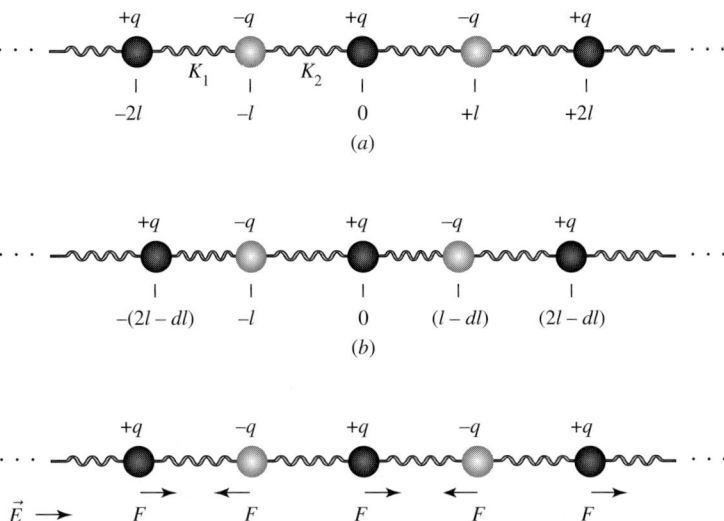

Linear chain model of a piezoelectric material, (a) equilibrium, (b) under compressive strain, with $K_2 \gg K_1$, (c) under an applied electric field.

we expect the positive charge to be pulled in the direction of an electric field in the medium, while the negative charge of the electrons is pulled in the opposite direction. Microscopically, the total polarization of the medium can be defined as

$$P = \frac{1}{V} \sum_i x_i q_i, \tag{3.7.1}$$

where q_i is the net charge on each atom i and x_i is the position of each atom, and V is the volume. For the one-dimensional chain in equilibrium, shown in Figure 3.24(a), the total polarization adds up to

$$P = \frac{1}{V}\Big[0 + (-l)(-q) + (l)(-q) + (-2l)(q) + (2l)(q) + \cdots\Big] = 0. \tag{3.7.2}$$

Suppose that all the masses of the atoms are equal, and so are the spring constants $K_1 = K_2 = K$, which measure the stiffness of the springs. In this case, the only difference between the atoms is the opposite charge on each. Then when an electric field is applied, each positive charge will move a small distance dl in one direction; each negative charge will move the same distance in the opposite direction. According to Hooke's law for springs, the magnitude of the distance dl is determined by $F = qE = 2K(dl)$, since each electron is connected to two springs. The total polarization will then be

$$P = \frac{1}{V}\Big[(dl)(q) + (-l - dl)(-q) + (l - dl)(-q) + (-2l + dl)(q) + (2l + dl)(q) + \cdots\Big]$$
$$= \frac{N}{V}q(dl) = \frac{N}{V}\frac{q^2}{2K}E, \tag{3.7.3}$$

where N is the number of atoms. When an external electric field is applied, there is a net total polarization proportional to the electric field, as assumed in Section 3.5.1.

Let us now consider the asymmetric case. Suppose that in Figure 3.24(a) the spring constant K_1 is much less than the spring constant K_2, so that we can ignore compression of the K_2 spring. If a compressive force is applied to the chain, as shown in Figure 3.24(b), the weak spring will compress to length $l - dl$. The total polarization in this case is then

$$
\begin{aligned}
P &= \frac{1}{V}\Big[0 + (-l)(-q) + (l - dl)(-q) + (-2l + dl)(q) + (2l - dl)(q) + \\
&\quad + (-3l + dl)(-q) + (3l - 2dl)(-q) + \cdots\Big] \\
&= \frac{N}{4V}(dl)q,
\end{aligned}
\tag{3.7.4}
$$

where N is the number of atoms. The compressive strain has led to a change of the polarization, and has generated an internal electric field even in the absence of an externally applied electric field.

Conversely, suppose an electric field is applied to the chain, as shown in Figure 3.24(c). In this case, it is easy to see that the electric field will lead to a compression of the chain, that is, a strain, when $K_1 \neq K_2$. In the limit when the K_2 springs are so stiff that their compression is negligible, there will be a compression of all the K_1 springs.

We thus have a material in which a compression leads to an electric field, and an electric field leads to a compression. This is known as the **piezoelectric** effect. It is not uncommon, since all it requires is an asymmetry of the bond strengths between different atoms in a solid, and that occurs in many materials. The strength of the piezoelectric effect can vary greatly, however. A typical piezoelectric constant (which is given the symbol e) is of the order of 10^{-5} C/cm^2. One way to think of this odd unit of C/cm^2 is to think of it as the pressure per unit electric field, with pressure in units of energy per volume (noting that 1 joule is equal to 1 coulomb-volt), and electric field in units of volts/cm:

$$
\frac{\dfrac{\text{C V}}{\text{cm}^3}}{\dfrac{\text{V}}{\text{cm}}} = \frac{\text{C}}{\text{cm}^2}.
\tag{3.7.5}
$$

In general, crystals have different piezoelectric constants for electric field applied in different directions in the crystal.

The interaction of the electric and elastic fields can be summarized in the following equations:

$$
\boxed{D_i = \sum_j \epsilon_{ij} E_j + \sum_{jl} e_{ijl}\varepsilon_{jl},}
\tag{3.7.6}
$$

$$
\boxed{\sigma_{ij} = -\sum_l e_{ijl} E_l + \sum_{lm} C_{ijlm}\varepsilon_{lm},}
\tag{3.7.7}
$$

where e_{ijl} is a new matrix of material constants known as the piezoelectric tensor. Just as with the elastic constants and the electro-optic constants, the symmetry of the material determines the number of independent piezoelectric constants. Table 3.4 gives an example of the piezoelectric tensor for a cubic crystal. The matrices for different crystal symmetries are given on the website that accompanies this book (www.cambridge.org/snoke). Some crystal symmetries have no nonzero piezoelectric constants. As one can see from Figure 3.24, if a crystal has inversion symmetry, there will be no piezoelectric effect.

Exercise 3.7.1　(a) GaAs is a piezoelectric material with piezoelectric constant $e = 1.6 \times 10^{-5}$ C/cm^2. What pressure is generated on the crystal by a potential drop of 5 V applied across a slab of GaAs with thickness 1 mm?

　　(b) The relevant compliance constant of GaAs is $S = 1.7 \times 10^{-12}$ cm^2/dyne. How much does the surface of a 1 mm GaAs crystal actually move under the electric field calculated in part (a)?

The linear chain shown in Figure 3.24 is also intrinsically unstable to a spontaneous distortion. In the case when all the springs are undistorted, the potential energy of a single mass is given by the infinite sum for the Coulomb energy of all the two-body interactions, which we have already discussed in Section 2.4.1, as the **Madelung potential**,

$$U = 2\frac{-q^2}{l} + 2\frac{q^2}{2l} + 2\frac{-q^2}{3l} + \cdots = -2\frac{q^2}{l}\ln 2. \qquad (3.7.8)$$

If the medium is distorted so that the weak springs have length $l - dl$, then the total Coulomb energy is

$$U = \left(\frac{-q^2}{l} + \frac{-q^2}{l - dl}\right) + \left(\frac{q^2}{2l - dl} + \frac{q^2}{2l - dl}\right) - \left(\frac{-q^2}{3l - dl} + \frac{-q^2}{3l - 2dl}\right) + \cdots$$

$$\approx 2\frac{-q^2}{l} + \frac{-q^2}{l}\left(\frac{dl}{l}\right) + 2\frac{q^2}{2l} + 2\frac{q^2}{2l}\left(\frac{dl}{2l}\right) + \frac{-q^2}{3l} + 3\frac{-q^2}{3l}\left(\frac{dl}{3l}\right) + \cdots$$

$$= -2\frac{q^2}{l}\ln 2\left(1 + \frac{dl}{l}\right). \qquad (3.7.9)$$

The total energy of N masses is therefore lowered by $2N(q^2/l^2)(dl)$, while the energy of compression of N springs is $\frac{1}{2}NK_1(dl)^2$. Depending on the spring constant, then, the chain can lower its energy significantly by distorting into unequal spacings. This is known as a **Peierls transition**.

A Peierls transition can occur even if the spring constants are equal. The fact that the Coulomb energy decreases linearly while the spring distortion energy grows as the square of the displacement means that there will always be some range of distortion over which the total energy is lower for a distorted lattice than for one with equal spacing.

Exercise 3.7.2　GaAs has cubic symmetry (T_d in the group theory notation of Chapter 6), leading to the piezoelectric tensor shown in Table 3.4, with $e_{41} = 1.6 \times 10^{-5}$ C/cm^2.

Table 3.4 Piezoeletric tensor for a cubic crystal (T or T_d symmetry)

$$\begin{pmatrix} 0 & 0 & 0 \\ 0 & 0 & 0 \\ 0 & 0 & 0 \\ e_{41} & 0 & 0 \\ 0 & e_{41} & 0 \\ 0 & 0 & e_{41} \end{pmatrix}$$

The elastic constants of GaAs are $C_{11} = 11.8 \times 10^{11}$ dyne/cm^2, $C_{12} = 5.3 \times 10^{11}$ dyne/cm^2, and $C_{44} = 5.9 \times 10^{11}$ dyne/cm^2. For a voltage of 5 V applied in the [100] direction across a slab of GaAs with thickness 1 mm, what strain will be created, assuming there is no applied stress?

Suppose that you want to generate an electric field by applying a stress to a GaAs crystal. What is a possible direction for a stress that would generate an electric field? Can you do this with a stress in the [100] direction?

3.8 Reflection and Transmission at Interfaces

Some scientists have said that everything interesting in physics involves an interface. The real world is not infinite and homogeneous. We must therefore understand the interaction of waves with surfaces and interfaces.

As we already know intuitively, when a wave hits an interface, part of it can be reflected and part of it transmitted. In Section 3.5.2, we determined the angle of the transmitted wave in the case of an electromagnetic wave hitting an anisotropic medium, a method that can also be applied to acoustic waves. We did not determine the amplitudes of the reflected and transmitted waves, however, for a given input wave. How much energy is reflected compared to how much is transmitted? We can work this out using the boundary conditions at the interface.

3.8.1 Optical Fresnel Equations

We first work this out for an electromagnetic wave. For simplicity, we assume that the medium is isotropic. The boundary conditions are therefore simple, as deduced, for example, by Jackson (1999: s. I.5). The tangential component of the E-field is continuous across a boundary, while the tangential component of the H-field ($H = B/\mu$, where μ is the magnetic permeability) is continuous. For a nonmagnetic medium, $\vec{H} = \vec{B}/\mu_0$ in a nonmagnetic medium, which means that the tangential component of the B-field is continuous across a boundary. (See Section 7.3 for some of the strange optical effects that can occur in a magnetic medium.)

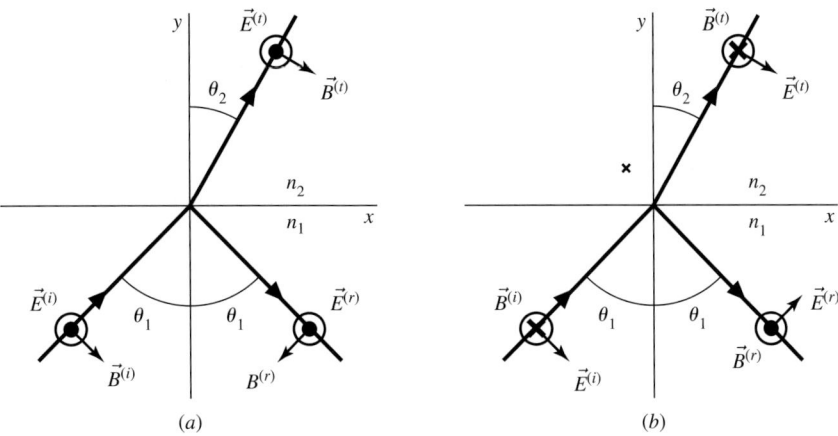

Fig. 3.25   (a) *s*-polarization for an electromagnetic wave incident on a surface. (b) *p*-polarization.

We must consider the cases of two different polarizations of the incident wave separately. Figure 3.25 shows the two polarizations. The wave with electric field polarized perpendicular to the plane of incidence (and therefore parallel to the interface, if the interface is flat) is sometimes called *s*-polarization (from the German word for perpendicular, "senkrecht"). The wave with electric field parallel to the plane of incidence is called, in the same nomenclature, *p*-polarization (for "parallel," which is the same word in German as in English). Note that *s*- and *p*-polarizations have nothing to do with *s* and *p* symmetry in orbitals. This notational similarity has confused more than one graduate student.

We write the incident wave in the form

$$E^{(i)}(\vec{r}, t) = E_0^{(i)} e^{i(\vec{k}\cdot\vec{r} - \omega t)}, \tag{3.8.1}$$

and similar forms for the reflected and transmitted waves. For *s*-polarization, the boundary conditions of continuity imply

$$E_0^{(i)} + E_0^{(r)} = E_0^{(t)}, \tag{3.8.2}$$

and for the tangential component of the *B*-field, which is perpendicular to the *E*-field in a transverse plane wave,

$$B_0^{(i)} \cos\theta_1 - B_0^{(r)} \cos\theta_1 = B_0^{(t)} \cos\theta_2. \tag{3.8.3}$$

Since from Maxwell's equations we have $i\vec{k} \times \vec{E}_0 = i\omega\vec{B}_0$ and therefore $B_0 = (n/c)E_0$, in an isotropic medium (3.8.3) becomes

$$n_1 E_0^{(i)} \cos\theta_1 - n_1 E_0^{(r)} \cos\theta_1 = n_2 E_0^{(t)} \cos\theta_2. \tag{3.8.4}$$

We define the **reflection coefficient** as the ratio of the amplitudes of the incident and reflected wave, $r = E_0^{(r)}/E_0^{(i)}$, and the **transmission coefficient**, $t = E_0^{(t)}/E_0^{(i)}$. We can then solve the two equations (3.8.2) and (3.8.4) for $E_0^{(r)}$ and $E_0^{(t)}$ in terms of $E_0^{(i)}$ to obtain

$$r_\perp = \frac{n_1 \cos \theta_1 - n_2 \cos \theta_2}{n_1 \cos \theta_1 + n_2 \cos \theta_2}$$

$$t_\perp = 1 + r_\perp.$$

$$(3.8.5)$$

Similarly, we can obtain for the parallel polarization,

$$r_\parallel = \frac{n_1 \cos \theta_2 - n_2 \cos \theta_1}{n_1 \cos \theta_2 + n_2 \cos \theta_1}$$

$$t_\parallel = \frac{n_1}{n_2}(1 - r_\parallel).$$

$$(3.8.6)$$

Equations (3.8.5) and (3.8.6) are known as the **Fresnel** equations. At normal incidence, $\theta_1 = 0$, the Fresnel equations reduce to the simple form

$$r_\perp = r_\parallel = \frac{n_1 - n_2}{n_1 + n_2}$$

$$t_\perp = t_\parallel = \frac{2n_1}{n_1 + n_2}.$$

$$(3.8.7)$$

Note that since these are equations relating *amplitudes*, the transmission coefficient can be larger than 1. This does not violate energy conservation since the energy flow is given by the Poynting vector,

$$\vec{S} = \vec{E}^* \times \vec{H}.$$

$$(3.8.8)$$

Using the relation $E_0 = (c/n)B_0$, the time-averaged energy flow per area is therefore

$$\langle |S| \rangle = \frac{1}{2}n\sqrt{\epsilon_0/\mu_0}E_0^2 = \frac{1}{2}\frac{E_0^2}{Z} = \frac{1}{2}ZH_0^2.$$

$$(3.8.9)$$

The value $Z = (\sqrt{\mu_0/\epsilon_0})/n$ is known as the **impedance** of the medium. (For $n = 1$, the intrinsic impedance of vacuum $\sqrt{\mu_0/\epsilon_0} = 377 \ \Omega$.)

To show that there is energy conservation for the incoming and outgoing waves at the boundary, we must take into account the fact that waves at different angles will have different cross-sectional areas into which their power flows. For a constant area A on the surface shared in common by the incoming and outgoing waves, the cross-sectional areas of the waves will be $A \cos \theta_1$ for the incoming and reflected waves, and $A \cos \theta_2$ for the refracted wave. The energy conservation condition is therefore

$$\langle S_i \rangle A_i = \langle S_r \rangle A_r + \langle S_t \rangle A_t,$$

$$\Rightarrow n_1 |E_0^{(i)}|^2 \cos \theta_1 = n_2 |E_0^{(t)}|^2 \cos \theta_2 + n_1 |E_0^{(r)}|^2 \cos \theta_1,$$

$$(3.8.10)$$

or

$$\frac{n_2 \cos \theta_2}{n_1 \cos \theta_1}t^2 + r^2 = 1,$$

$$(3.8.11)$$

which is satisfied by the coefficients given by the Fresnel equations.

In terms of the impedance, the Fresnel relations for normal incidence can be written as

$$r = \frac{Z_2 - Z_1}{Z_1 + Z_2}, \qquad t = 1 + r, \tag{3.8.12}$$

which has the same form as (3.8.7), except that the sign of the reflection coefficient is reversed. Note that *high* index of refraction corresponds to *low* impedance, and vice versa.

Exercise 3.8.1   Using the Fresnel equations, determine **Brewster's angle**, the angle of incidence at which no *p*-polarized light can be reflected ($r_\parallel \to 0$). Before polaroid plastic polarizers were invented, polarized light was produced in the lab by using a stack of plates at Brewster's angle. The partial polarization of light reflected from smooth surfaces is also why polaroid sunglasses work. The sunglasses, which are polarizers, selectively eliminate light reflected from surfaces.

3.8.2 Acoustic Fresnel Equations

The same approach can be taken with acoustic waves to determine the amplitude coefficients of reflection and transmission. Instead of the boundary conditions of Maxwell's equations, we have two analogous requirements. First, the force on the surface must be continuous. In terms of the stress discussed in Section 3.4.1, this condition is

$$\tilde{\sigma}^{(1)} \cdot \hat{n} = \tilde{\sigma}^{(2)} \cdot \hat{n}. \tag{3.8.13}$$

It is easy to understand why this must be true, since the stress inside the crystal must balance the external load. In the case of a free surface, this means that the force normal to the surface must be zero.

In addition, we have a constraint on the displacement of the medium. In the case when there is no slip, such as when two solids are bonded together, the displacement of the medium must be continuous across the boundary:

$$\vec{u}^{(1)} = \vec{u}^{(2)}. \tag{3.8.14}$$

In the case of a non-viscous liquid or gas next to a solid, only the normal component must be continuous.

Figure 3.26 shows the geometry of reflection and refraction for transverse acoustic waves at a boundary. In general, the reflection and refraction rules are more complicated than for electromagnetic waves, because there are three polarizations which can couple to each other. The simplest case is a transverse acoustic wave, shown in Figure 3.26(a). A transverse acoustic plane wave in *s*-polarization can be treated the same as the *s*-polarized electromagnetic wave shown in Figure 3.25. We set the *z*-direction as the transverse polarization direction, so that the transverse plane wave is written as $\vec{u} = u_0 \hat{z} e^{i(\vec{k}\cdot\vec{r} - \omega t)}$. For a cubic crystal (or isotropic medium), we use the elastic constants of Table 3.1. Then from the definitions (3.4.2) and (3.4.8) given in Section 3.4.1, the stresses generated by the transverse wave are

$$\sigma_{xz} = C_{44}\epsilon_{xz} = C_{44}\left(\frac{\partial u_z}{\partial x}\right) = C_{44}ik_x u_z = C_{44}iku_0 \sin\theta \tag{3.8.15}$$

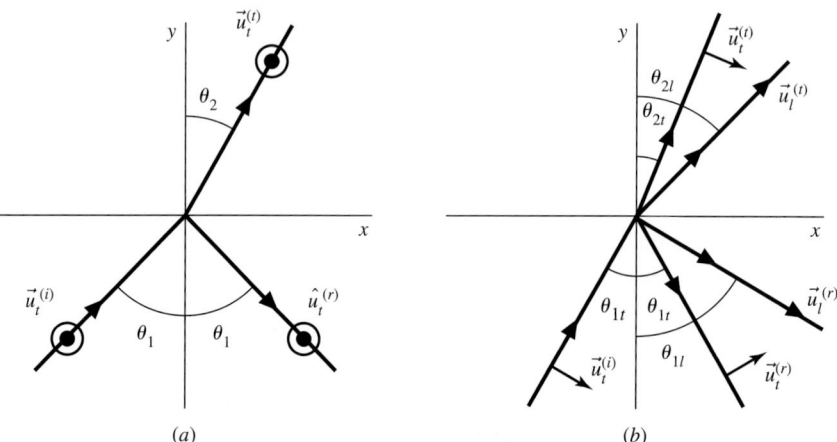

(a) Geometry of reflection and refraction for an incident s-polarized transverse acoustic wave (also known as an SH wave). (b) Geometry for an incident p-polarized transverse acoustic wave (also known as an SV wave).

and

$$\sigma_{yz} = C_{44}\epsilon_{yz} = C_{44}\left(\frac{\partial u_z}{\partial y}\right) = C_{44}ik_y u_z = C_{44}iku_0 \cos\theta. \tag{3.8.16}$$

Only σ_{yz} is normal to the surface. The boundary conditions at $\vec{r} = 0$ therefore become

$$C_{44}^{(1)}k_1 \cos\theta_1 u_0^{(i)} - C_{44}^{(1)}k_1 \cos\theta_1 u_0^{(r)} = C_{44}^{(2)}k_2 \cos\theta_2 u_0^{(t)} \tag{3.8.17}$$

or, since $k = \omega/v = \omega\sqrt{\rho/C_{44}}$,

$$Z_1 \cos\theta_1 u_0^{(i)} - Z_1 \cos\theta_1 u_0^{(r)} = Z_2 \cos\theta_2 u_0^{(t)}, \tag{3.8.18}$$

where $Z = \sqrt{\rho C_{44}}$ is the acoustic impedance.

In addition, the continuity of the displacement requires

$$u_0^{(i)} + u_0^{(r)} = u_0^{(t)}. \tag{3.8.19}$$

Solving these equations for the reflection and transmission coefficients, we have, analogous to (3.8.5),

$$r_\perp = \frac{Z_1 \cos\theta_1 - Z_2 \cos\theta_2}{Z_1 \cos\theta_1 + Z_2 \cos\theta_2} \tag{3.8.20}$$

$$t_\perp = 1 + r_\perp.$$

As with electromagnetic waves, the acoustic energy flow in a medium depends on its impedance. The acoustic Poynting vector is (see, e.g., Auld 1973: V.1, 155)

$$\vec{S} = -\vec{v}^* \cdot \tilde{\sigma}, \tag{3.8.21}$$

where $\vec{v} = \partial \vec{u}/\partial t$. Using (3.8.15) and (3.8.16) for a transverse wave, we obtain

$$\langle \vec{S} \rangle = -\tfrac{1}{2} i \omega u_0 (C_{44} i k u_0 \sin \theta \hat{x} + C_{44} i k u_0 \cos \theta \hat{y})$$

$$= \tfrac{1}{2} \omega u_0^2 C_{44} k (\sin \theta \hat{x} + \cos \theta \hat{y})$$

$$= \tfrac{1}{2} \sqrt{\rho C_{44}} \, u_0^2 \omega^2 \hat{k}, \tag{3.8.22}$$

or

$$\langle \vec{S} \rangle = \frac{1}{2} Z v_0^2 \hat{k}, \tag{3.8.23}$$

where $v_0 = \omega u_0$ and $\hat{k}$ is a unit vector in the direction of $\vec{k}$. As seen here, the stress field $\tilde{\sigma}$ and the velocity field $\vec{v}$ play complementary roles analogous to the electric field $\vec{E}$ and magnetic field $\vec{H}$, respectively, of electromagnetic waves.

In the case of a p-polarized transverse acoustic wave, life is more complicated. As seen in Figure 3.26(b), the displacement in this direction has a component in both the transverse and the longitudinal polarizations of the transmitted and reflected waves. An incoming wave with this polarization can therefore generate two different polarizations of outgoing waves. This is a major difference from electromagnetic waves.

To determine the relative amplitude of all of these four outgoing waves, we again must start with the same boundary conditions (3.8.13) and (3.8.14) and match the components at the boundary. We will not go through this lengthy algebraic calculation here (for the general case, see Auld 1973: V. 2, ch. 9). One relatively simple case is an acoustic wave in an isotropic medium reflecting off a free (unloaded) surface. In this case, the transmitted amplitudes are zero, and the reflection coefficients are found to be

$$r_{lt} = \frac{2(v_l/v_t) \sin 2\theta_{1t} \cos 2\theta_{1t}}{\sin 2\theta_{1t} \sin 2\theta_{1l} + (v_l/v_s)^2 \cos^2 2\theta_{1t}} \tag{3.8.24}$$

$$r_{tt} = -\frac{\sin 2\theta_{1t} \cos 2\theta_{1l} - (v_l/v_s)^2 \cos^2 2\theta_{1t}}{\sin 2\theta_{1t} \sin 2\theta_{1l} + (v_l/v_s)^2 \cos^2 2\theta_{1t}} \tag{3.8.25}$$

for a transverse incident wave, with $\sin \theta_{1l} = (v_l/v_t) \sin \theta_{1t}$.

In the same way, a longitudinally polarized acoustic wave will have a component parallel to a p-polarized transverse acoustic wave. For reflection of an incident longitudinal wave from a free surface, the coefficients are found to be

$$r_{ll} = \frac{\sin 2\theta_{1t} \sin 2\theta_{1l} - (v_l/v_s)^2 \cos^2 2\theta_{1t}}{\sin 2\theta_{1t} \sin 2\theta_{1l} + (v_l/v_s)^2 \cos^2 2\theta_{1t}} \tag{3.8.26}$$

$$r_{tl} = \frac{2(v_l/v_t) \sin 2\theta_{1l} \cos 2\theta_{1t}}{\sin 2\theta_{1t} \sin 2\theta_{1l} + (v_l/v_s)^2 \cos^2 2\theta_{1t}}, \tag{3.8.27}$$

where, again, $\sin \theta_{1l} = (v_l/v_t) \sin \theta_{1t}$.

Frustratingly, the common nomenclature for acoustic waves uses the letters s and p, but in a different way from the standard optical notation. A transverse s-polarized wave is commonly called an SH wave, a transverse p-polarized wave is called an SV wave, and a longitudinal wave is called a P wave. Note that a P acoustic wave has no analogy with a p-polarized transverse wave! The S here stands for a **shear** wave (horizontal and vertical) while the P stands for a **pressure** wave.

Exercise 3.8.2   Another simple case of acoustic waves is the case of longitudinal waves incident normal to a surface, in which case there is no coupling to transverse waves. Show that in this case the equations (3.8.20) also apply, with $Z = \sqrt{\rho C_{11}}$.

Exercise 3.8.3   Verify the equations (3.8.24) through (3.8.27) for a plane wave reflecting from a free surface in an isotropic medium, using the boundary conditions (3.8.13) and (3.8.14). A program like Mathematica can help to solve the four equations with five unknowns.

3.8.3 Surface Acoustic Waves

The reflection Fresnel equations (3.8.24) through (3.8.27) present an interesting puzzle. Suppose that the denominator is zero, which implies an infinite reflection coefficient. This means that even for infinitesimal incident amplitude, there will be a finite traveling wave. What does this mean physically?

The only way for the denominator to vanish is for one of the terms to be negative. This is possible if k_y is imaginary for both the transverse and longitudinal waves, that is, if both waves decay exponentially away from the surface. Then, for the geometry of Figure 3.26(b), we can write, for the transverse reflected wave,

$$k_y^{(t)} \equiv i\alpha_t = k^{(t)} \cos \theta_{1t} = \frac{\omega}{v_t} \cos \theta_t \tag{3.8.28}$$

and

$$k_z^{(t)} \equiv k_R = k^{(t)} \sin \theta_{1t} = \frac{\omega}{v_t} \sin \theta_t, \tag{3.8.29}$$

and similarly for the longitudinal reflected wave,

$$k_y^{(l)} \equiv i\alpha_l = \frac{\omega}{v_l} \cos \theta_{1l} \tag{3.8.30}$$

and

$$k_z^{(t)} \equiv k_R = \frac{\omega}{v_l} \sin \theta_{1l}. \tag{3.8.31}$$

The in-plane k-component k_R is assumed to be the same for both polarizations. Then the term $\sin 2\theta_t \sin 2\theta_l = 4 \sin \theta_t \sin \theta_l \cos \theta_t \cos \theta_l$ will be negative because each of the cosine terms is imaginary. Substituting these definitions into the resonance condition

$$\sin 2\theta_t \sin 2\theta_l + (v_l/v_s)^2 \cos^2 2\theta_t = 0, \tag{3.8.32}$$

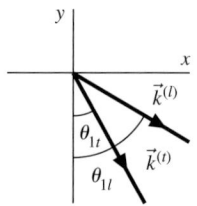

Fig. 3.27   Geometry for a surface acoustic wave, which can be seen as a resonance of reflection from a free surface.

we find after some algebra that it can be expressed as follows:

$$\left(\frac{v_R}{v_t}\right)^6 - 8\left(\frac{v_R}{v_t}\right)^4 + 8\left[3 - 2\left(\frac{v_t}{v_l}\right)^2\right]\left(\frac{v_R}{v_t}\right)^2 - 16\left[1 - \left(\frac{v_t}{v_l}\right)^2\right] = 0,$$

(3.8.33)

where $v_R = \omega/k_R$.

The picture we have is a wave traveling with speed v_R along the surface, with amplitude that decays exponentially away from the surface. The speed v_R is not equal to either v_l or v_t for waves in the bulk of the crystal. For typical materials, v_R lies in the range $0.85v_t$ to $0.95v_t$.

This resonance is known as a **Rayleigh wave**, or a **surface acoustic wave** (SAW). Essentially, a wave becomes trapped at the surface because it has the wrong frequency and wavelength to travel in the bulk.

One reason why Rayleigh waves are very important technologically is because they are extremely sensitive to surface pressure. Note that equations (3.8.24) through (3.8.27) were deduced assuming a free boundary (i.e., $\tilde{\sigma} \cdot \hat{n} = 0$). If something exerts a pressure on the surface, then we must go back to the condition (3.8.13) for a given pressure. Since the Rayleigh wave solution is a sharp resonance, small changes in the bounary condition will give large changes in the wave behavior. This effect can therefore be used as a means of sensing slight pressures; for example, SAWs are used in some schemes for detecting finger touches on interactive computer monitors.

Exercise 3.8.4   Earthquakes can also generate surface acoustic waves. Compute the Rayleigh wave speed for a SAW in rock that has $v_l = 10$ km/s and $v_t = 5$ km/s.

3.9 Photonic Crystals and Periodic Structures

So far in this chapter, we have discussed anisotropy in solids due to their crystal structure. Another type of anisotropy is possible – engineered anisotropy. It is possible to create layered optical materials using methods similar to those used for semiconductor heterostructures, discussed in Section 2.8. In particular, it is possible to create structures with periodic variations in one, two, or all three dimensions. Figure 3.28 shows an example.

In Sections 1.3–1.6, we proved a number of theorems for periodic systems, starting with Bloch's theorem. The proof of this theorem used the fact that if the Hamiltonian commutes with the translation operator $T_{\vec{R}}$, then the eigenstates of the Hamiltonian are also eigenstates of the translation operator. This allowed us to deduce that $\psi(\vec{r} + \vec{R}) = e^{i\vec{k}\cdot\vec{R}}\psi(\vec{r})$, which is one way of stating Bloch's theorem.

This proof relied only on the wave nature of electrons. We can therefore generalize it for any wave. Given a standard wave equation (ignoring anisotropy),

$$\nabla^2\psi = \frac{1}{v^2}\frac{\partial^2}{\partial t^2}\psi,$$

(3.9.1)

Fig. 3.28 A photonic crystal. Courtesy of Sandia National Labs.

we allow for the possibility that the wave speed v depends on the position. This will be the case, for example, for electromagnetic waves if the susceptibility depends on position:

$$
\nabla^2 \vec{E} = \mu_0 \epsilon_0 (1 + \chi(\vec{r})) \frac{\partial^2 \vec{E}}{\partial t^2}
$$
$$
= \frac{n(\vec{r})^2}{c^2} \frac{\partial^2 \vec{E}}{\partial t^2}. \tag{3.9.2}
$$

Using the general wave equation (3.9.1), assuming $\psi(\vec{r}, t) = \psi(\vec{r})e^{i\omega t}$, we therefore write

$$
v^2(\vec{r}) \nabla^2 \psi = -\omega^2 \psi. \tag{3.9.3}
$$

If the system is periodic, the translation operator $T_{\vec{R}}$ commutes both with the scalar $v(\vec{r})$ and with ∇^2. As we deduced in Section 1.3, the eigenvalues $C_{\vec{R}}$ of the translation operator have unit magnitude and the property $C_{\vec{R}+\vec{R}'} = C_{\vec{R}} C_{\vec{R}'}$, which implies $C_{\vec{R}} = e^{i\vec{k}\cdot\vec{R}}$. This implies

$$
T_{\vec{R}} \psi(\vec{r}) = \psi(\vec{r} + \vec{R}) = e^{i\vec{k}\cdot\vec{R}} \psi(\vec{r}), \tag{3.9.4}
$$

which is Bloch's theorem. As discussed in Section 1.6, there are several corollaries to Bloch's theorem, including Kramers' rule, which implies that for the eigenvalues,

$$
\omega^2(-\vec{k}) = \omega^2(\vec{k}) \tag{3.9.5}
$$

and

$$
\nabla_{\vec{k}} \omega^2(\vec{k}) \Big|_{\vec{k} = \vec{Q}/2} = 0, \tag{3.9.6}
$$

where $\vec{Q}$ is a reciprocal lattice vector. Applying the chain rule of differentiation, we have

$$
2\omega(\vec{k}) \, \nabla_{\vec{k}} \omega(\vec{k}) \Big|_{\vec{k} = \vec{Q}/2} = 0. \tag{3.9.7}
$$

This implies that if $\omega \neq 0$, then the gradient of $\omega(\vec{k})$ must vanish at the critical points, just as we deduced for electron waves in Section 1.6.

This analysis applies to both optical and acoustic waves that obey (3.9.1). One implication of this analysis is that the lattice vibrations in a crystal have the same critical points in the Brillouin zone as the electron wave functions. This behavior is seen for a one-dimensional system in Section 3.1.2. Another implication is that an engineered periodicity in the optical properties of a system will produce band gaps and critical points with zero group velocity just as in the band structure of electrons. Materials with this kind of engineered periodicity are called **photonic crystals**. Essentially, such a material consists of a crystal in which the lattice sites are occupied by materials with different dielectric constants. Then, in any direction there exists a set of Bragg mirrors that will perfectly reflect light over some range of wavelengths.

A difference between electron waves and acoustic or optical waves is seen in the behavior at $\omega = 0$. As seen in (3.9.7), if the frequency ω vanishes, then the gradient of ω need not. This is why the acoustic sound velocity does not equal zero at $\omega = 0$, as seen, for example, in the one-dimensional lattice vibration model of Section 3.1.2.

The case of $\omega = 0$ can only occur if ψ is translationally invariant. This can be seen easily if we write the one-dimensional wave equation as

$$\left(v^2(r) \frac{\partial^2}{\partial r^2} + \omega^2 \right) \psi = 0. \tag{3.9.8}$$

If $\omega = 0$ and $v(r) \neq 0$, then we have

$$\frac{\partial^2 \psi}{\partial r^2} = 0. \tag{3.9.9}$$

If we impose the boundary condition $|\psi(x = \pm\infty)| \neq \infty$, then we must also have $\partial \psi / \partial r = 0$. This is the case for acoustic waves when $k = 0$ and $\omega = 0$. The eigenstate of the vibration is a simple translation of the entire crystal. An electromagnetic wave with $\omega = 0$ is just a DC field.

Light with frequency inside a photonic band gap cannot propagate. It is possible, however, to have defect states with frequency that lies inside a photonic band gap, just as we discussed electronic defect states that lie inside a band gap. These states, which cannot propagate in the medium, are known as localized states.

Transfer matrix method. The reflectivity and transmission of a one-dimensional structure consisting of multiple layers of dielectric can be calculated using a **transfer matrix** method, which is easily adaptable to a computer.

Consider the single interface shown in Figure 3.29. For any element, we can view the electric field as the sum of left-going and right-going waves. For an interface between two media with different indices of refraction, we write the outgoing waves in terms of the reflection and transmission coefficients of the incoming waves:

$$E_{2+} = t_{12}E_{1+} + r_{12}E_{2-}$$
$$E_{1-} = t_{12}E_{2-} + r_{12}E_{1+}, \tag{3.9.10}$$

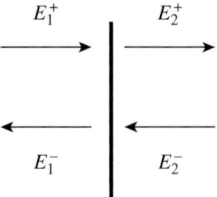

Fig. 3.29   Incoming and outgoing waves for an interface.

where the reflection and transmission coefficients are given by the Fresnel equations of Section 3.8.1. This set of equations can be solved for E_{2-} and E_{2+} in terms of E_{1-} and E_{1+}:

$$\begin{pmatrix} E_{2+} \\ E_{2-} \end{pmatrix} = \frac{1}{t_{21}} \begin{pmatrix} t_{12}t_{21} - r_{12}r_{21} & r_{21} \\ -r_{12} & 1 \end{pmatrix} \begin{pmatrix} E_{1+} \\ E_{1-} \end{pmatrix}. \tag{3.9.11}$$

The matrix here is called the **transfer matrix**, relating the field on one side to the field on the other side.

For the homogeneous medium of thickness between two interfaces, the wave on one side is related to the wave on the other side simply by the transfer matrix equation

$$\begin{pmatrix} E_{2+} \\ E_{2-} \end{pmatrix} = \begin{pmatrix} e^{ikd} & 0 \\ 0 & e^{-ikd} \end{pmatrix} \begin{pmatrix} E_{1+} \\ E_{1-} \end{pmatrix}, \tag{3.9.12}$$

where k is the wave number in the medium, related to the vacuum wave number k_0 by $k = k_0 n$, where n is the index of refraction of the medium.

To determine the total reflectivity of a stack of layers, one can simply write the waves in the final region as the wave in the first region times the product of the transfer matrices:

$$\begin{pmatrix} rE_0 \\ E_0 \end{pmatrix} = \prod_i M_i \begin{pmatrix} 0 \\ tE_0 \end{pmatrix}, \tag{3.9.13}$$

where the M_i matrices have appropriate constants for the various interfaces and homogeneous regions. To find the net effect of a system, one can take the first region to have only an **outgoing** wave E_t, and solve for the amplitudes of the incident and reflected waves E_i and E_r on the other side of the system. The transmission coefficient is then simply $t = E_t/E_i$, and the reflection coefficient is $r = E_r/E_i$.

Exercise 3.9.1   Using the transfer matrix method, determine the reflectivity spectrum, $|E^{(r)}|^2/|E^{(i)}|^2$ vs k, for a periodic stack of 20 layers, alternating with $d_1 = 500$ nm, $n_1 = 3$, and $d_2 = 750$ nm, $n_2 = 2$, surrounded by air on both sides. Do you see a photonic band gap, that is, a range of frequencies with no propagating waves?

References

B.A. Auld, *Acoustic Fields and Waves in Solids*, Vols. I and II (Wiley, 1973).

H. Bilz and W. Kress, *Phonon Dispersion Relations in Insulators* (Springer, 1979).

M.T. Dove, *Introduction to Lattice Dynamics* (Cambridge University Press, 1993).

M.M. Elcombe, "The lattice dynamics of strontium fluoride," *Journal of Physics* C **5**, 2702 (1972).

K. Iizuka, *Elements of Photonics*, Vol. I (Wiley, 2002).

J.D. Jackson, *Classical Electrodynamics*, 3rd ed. (Wiley, 1999).

A. Knudson, K.U. Ratnatunga, and R.E. Griffiths, "Investigation of gravitational lens mass models," *Astronomical Journal* **122** 103 (2001).

G. Northrup and J.P. Wolfe, "Ballistic phonon imaging in germanium," *Physical Review* B **22**, 6196 (1980).

J.H. Staudte and W.D. Cook, "Visualization of quasilongitudinal and quasitransverse elastic waves," *Journal of the Acoustical Society* **41**, 1547 (1967).

J.P. Wolfe, *Imaging Phonons: Acoustic Wave Propagation in Solids* (Cambridge University Press, 2005).

A. Yariv, *Quantum Electronics* (Wiley, 1989).

4 Quantized Waves

In the standard treatment of quantum mechanics, the student is introduced to the wave function, which is then used to determine the probability of detecting a particle. The connection between the wave and the particle is not at all clear, however, leading to many of the conundrums debated under the heading of wave–particle duality.

The Schrödinger equation and the standard discussion of waves and particles are actually approximations of the deeper-lying theory of quantum mechanics known as **field theory**. In field theory, the existence of particles is seen as a natural consequence of the mathematics of waves. Field theory starts with the field, which is the medium of waves, and **quantizes** these waves into particles. Particles in this view are simply the eigenstates of the excitations of the field.

Field theory is indispensable in solid state theory, because we must deal with the statistics and the interactions of many particles. By starting with the underlying fields, the proper treatment of many-particle systems appears naturally.

4.1 The Quantized Harmonic Oscillator

All of the basic properties of field theory can be seen starting with the physical model of the harmonic oscillator. Because this system is treated in many introductory textbooks, we present here only the basic results; for an extended treatment, see, for example, Cohen-Tannoudji *et al.* (1977: ch. 5).

As discussed in Section 3.1.1, the interaction between two atoms in a solid is well modeled as a simple harmonic oscillator. Near an energy minimum, the dependence of the energy on position can be approximated as $U = \frac{1}{2}K(x - x_0)^2$. Since we are free to set $x_0 = 0$, we can write simply

$$U = \frac{1}{2}Kx^2. \tag{4.1.1}$$

The kinetic energy of an atom with mass M is just $p^2/2M$, and therefore we write the Hamiltonian

$$H = \frac{p^2}{2M} + \frac{1}{2}Kx^2, \tag{4.1.2}$$

which can be rewritten as

$$H = \frac{p^2}{2M} + \frac{1}{2}M\omega_0^2 x^2, \tag{4.1.3}$$

where $\omega_0 = \sqrt{K/M}$ is the **natural frequency** of the oscillator.

In quantum mechanics, the x and p observables are represented by operators subject to the commutation relation,

$$[x, p] = i\hbar. \tag{4.1.4}$$

This relation arises simply from the Fourier transform properties of a wave, and from equating the momentum p with $\hbar k$, where k is the spatial frequency of the wave, since the k operator has basis functions of the form e^{ikx} in the x basis, and therefore the k operator is equivalent to $-i\partial/\partial x$, which implies

$$[x, -i\frac{\partial}{\partial x}]f(x) = -ix\frac{\partial}{\partial x}f(x) + if(x) + ix\frac{\partial}{\partial x}f(x)$$
$$= if(x). \tag{4.1.5}$$

As in all cases of non-commuting observables, this commutation relation can be used to directly deduce an uncertainty relation, in this case $\Delta x \Delta p \geq \hbar/2$ (see, e.g., Cohen-Tannoudji *et al.* 1977: s. CIII).

Appendix D reviews the mathematics of the quantum single harmonic oscillator. In summary, we define the operators

$$\hat{x} = \sqrt{\frac{M\omega_0}{\hbar}}x$$

$$\hat{p} = \sqrt{\frac{1}{M\hbar\omega_0}}p, \tag{4.1.6}$$

so that the Hamiltonian takes on the simple form

$$H = \frac{1}{2}(\hat{x}^2 + \hat{p}^2)\hbar\omega_0 \tag{4.1.7}$$

and the commutation relation between $\hat{x}$ and $\hat{p}$ is

$$[\hat{x}, \hat{p}] = i. \tag{4.1.8}$$

The quantum mechanics of the harmonic oscillator revolve around the definition of the new operators,

$$a = \frac{1}{\sqrt{2}}(\hat{x} + i\hat{p})$$

$$a^\dagger = \frac{1}{\sqrt{2}}(\hat{x} - i\hat{p}), \tag{4.1.9}$$

with the commutation relation

$$[a, a^\dagger] = 1. \tag{4.1.10}$$

As we will see, these also become fundamental in quantum field theories. Using these, the Hamiltonian becomes

$$H = \left(a^\dagger a + \tfrac{1}{2}\right)\hbar\omega_0. \tag{4.1.11}$$

It can be proven (see Appendix D) that the eigenstates of this Hamiltonian have energies

$$E = \left(N + \tfrac{1}{2}\right)\hbar\omega_0, \qquad (4.1.12)$$

where N is a non-negative integer. We will write these eigenstates here as $|N\rangle$. They are called **Fock states** or **number states**.

The operators a and $a^\dagger$ are known as the **destruction** and **creation** operators, respectively, because they have the following action on the eigenstates of the Hamiltonian:

$$a|N\rangle = \sqrt{N}|N-1\rangle$$
$$a^\dagger|N\rangle = \sqrt{1+N}|N+1\rangle. \qquad (4.1.13)$$

In other words, $a^\dagger$ "creates" a quantum of excitation energy in the system, while a "destroys" one quantum. The wave function of the particle in the harmonic oscillator has been *quantized* into states which correspond to definite numbers of excitations.

We can get a feel for why these quantized states exist by remembering that a wave in any confined geometry has definite eigenstates due to the boundary conditions. As illustrated in Figure 4.1, in a system with constant potential energy and fixed walls a distance L apart, the allowed wavelengths are $\lambda = 2L/N$, which implies momentum $p = \hbar k = 2\pi\hbar/\lambda = \pi\hbar N/L$. For a particle with mass M, this gives the familiar result

$$E = \frac{p^2}{2M} = \frac{\pi^2\hbar^2 N^2}{2ML^2}. \qquad (4.1.14)$$

As illustrated in Figure 4.2, a harmonic potential does not have walls with a well-defined separation L. The size of the confinement region depends on the energy of the particle, because a particle with higher energy moves higher up in potential energy and therefore farther from the center. We can approximate the size of the confinement region felt by the wave by equating the potential energy at maximum distance with the total energy

$$E = \frac{1}{2}KL^2, \qquad (4.1.15)$$

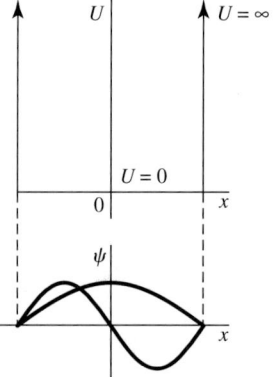

Fig. 4.1   The allowed eigenstates of a wave in a confining potential of fixed width.

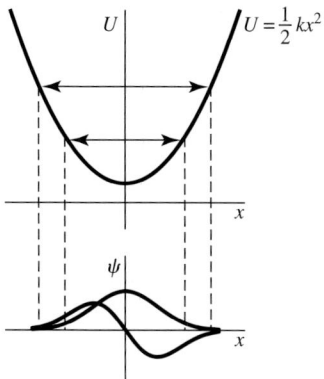

Fig. 4.2 The eigenstates of a harmonic potential, analogous to the eigenstates of a confining potential with fixed width.

which implies

$$L = \sqrt{2E/K}. \tag{4.1.16}$$

If we require the same wave resonance condition $\lambda = 2L/N$, we have

$$E = \frac{p^2}{2M} = \frac{\pi^2 \hbar^2 N^2}{2ML^2} = \frac{\pi^2 \hbar^2 N^2}{2M(2E/K)}. \tag{4.1.17}$$

Solving this equation for E gives

$$E = \frac{\pi}{2} N\hbar \sqrt{\frac{K}{M}} = \frac{\pi}{2} N\hbar\omega_0, \tag{4.1.18}$$

which is very close to the correct solution for the eigenenergies of the harmonic oscillator, given in (4.1.12). The eigenstates of the harmonic oscillator have definite energies for the same reason that the states of a box do – only waves with definite wavelength can fit in the confined geometry.

Exercise 4.1.1 Prove that the commutation relation (4.1.10) follows from the definitions (4.1.9) and the commutation relation (4.1.4).

4.2 Phonons

Instead of a single harmonic oscillator, consider a linear chain of identical atoms connected by springs, as illustrated in Figure 4.3. For simplicity, we assume that all the atoms are identical, instead of the two-atom model discussed in Section 3.1.2.

The Hamiltonian for this system is given by

$$H = \sum_n \left(\frac{1}{2}\frac{p_n^2}{M} + \frac{1}{2}K(x_n - x_{n-1})^2 \right), \tag{4.2.1}$$

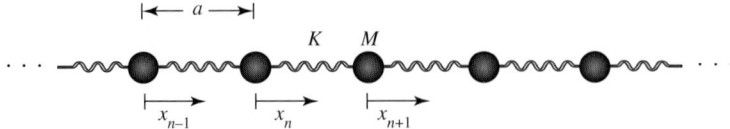

Fig. 4.3 Linear chain of identical masses and springs.

where the index n labels the individual atoms. The classical solution to this problem, following the methodology of Section 3.1.2, is

$$x_n(t) = \text{Re } x_0 e^{i(kan - \omega_k t)}, \qquad (4.2.2)$$

with

$$\omega_k = \omega_0 \sin(ka/2) \qquad (4.2.3)$$

and

$$\omega_0 = 2\sqrt{K/M}. \qquad (4.2.4)$$

We define the following **normal variables** which are sums of the quantum mechanical operators:

$$x_k = \frac{1}{\sqrt{N}} \sum_n x_n e^{-ikan}$$

$$p_k = \frac{1}{\sqrt{N}} \sum_n p_n e^{-ikan}, \qquad (4.2.5)$$

where N is the total number of atoms.

We can rewrite the Hamiltonian (4.2.1) in terms of these new operators by using a few mathematical tricks. First, we expand the sum over the x_n operators in (4.2.1) as

$$\frac{1}{2}K \sum_n (x_n - x_{n-1})^2 = \frac{1}{2}K \sum_n (x_n^2 - 2x_n x_{n-1} + x_{n-1}^2). \qquad (4.2.6)$$

Since the sum is over all n, the index $(n-1)$ is just a dummy variable which can be replaced by n or $(n+1)$, if the chain is infinite, or finite with periodic boundary conditions. We therefore can write

$$\frac{1}{2}K \sum_n (x_n - x_{n-1})^2 = \frac{1}{2}K \sum_n (2x_n^2 - 2x_n x_{n-1})$$

$$= K \sum_n \left(x_n^2 - \frac{1}{2}(x_n x_{n-1} + x_n x_{n+1}) \right). \qquad (4.2.7)$$

Next, we can write

$$\sum_n x_n^2 = \sum_{n,n'} x_n x_{n'} \delta_{n,n'}$$

$$\sum_n x_n x_{n\pm1} = \sum_{n,n'} x_n x_{n'} \delta_{n\pm1,n'}, \qquad (4.2.8)$$

where $\delta_{n,n'}$ is the Kronecker delta, equal to 1 when $n = n'$ and 0 otherwise. In the limit $N \to \infty$, the Kronecker delta can be written in terms of the identity

$$\delta_{n,n'} = \frac{1}{N} \sum_k e^{ika(n-n')}, \tag{4.2.9}$$

where $k = (v/N)(2\pi/a)$, with v an integer ranging from $-N/2$ to $N/2$. Substituting this for the Kronecker delta in the terms (4.2.8), we have

$$\sum_n x_n^2 = \sum_k \left(\frac{1}{\sqrt{N}} \sum_n x_n e^{-ikan} \right)^{\dagger} \left(\frac{1}{\sqrt{N}} \sum_{n'} x'_n e^{-ikan'} \right)$$

$$\sum_n x_n x_{n\pm 1} = \sum_k \left(\frac{1}{\sqrt{N}} \sum_n x_n e^{-ika(n\pm 1)} \right)^{\dagger} \left(\frac{1}{\sqrt{N}} \sum_{n'} x'_n e^{-ikan'} \right), \tag{4.2.10}$$

or

$$\sum_n x_n^2 = \sum_k x_{-k} x_k$$

$$\sum_n x_n x_{n\pm 1} = \sum_k |x_k|^2 e^{\pm ika}, \tag{4.2.11}$$

and therefore

$$\frac{1}{2} K \sum_n (x_n - x_{n-1})^2 = K \sum_k |x_k|^2 (1 - \cos ka)$$

$$= 2K \sum_k |x_k|^2 \sin^2(ka/2)$$

$$= 2K \sum_k |x_k|^2 \left(\frac{\omega_k}{\omega_0} \right)^2$$

$$= \frac{1}{2} M \sum_k |x_k|^2 \omega_k^2, \tag{4.2.12}$$

where we have used the definition (4.2.3) for ω_k. The same identity (4.2.11) can be applied to the sum over p_n operators, to finally give

$$\boxed{H = \sum_k \left(\frac{|p_k|^2}{2M} + \frac{1}{2} M \omega_k^2 |x_k|^2 \right).} \tag{4.2.13}$$

Comparing this to the Hamiltonian (D.6), we see that the Hamiltonian for the linear chain is a sum of independent Hamiltonians for *single* harmonic oscillators. Each wave vector k corresponds to a different harmonic oscillator, with a different natural frequency given by the classical eigenfrequency ω_k.

It therefore makes sense to quantize each of these modes just as we did the single harmonic oscillator. We write

$$H = \sum_k \left(a_k^\dagger a_k + \tfrac{1}{2} \right) \hbar\omega_k, \tag{4.2.14}$$

with

$$a_k = \frac{1}{\sqrt{2}}(\hat{x}_k + i\hat{p}_k)$$

$$a_k^\dagger = \frac{1}{\sqrt{2}}(\hat{x}_k^\dagger - i\hat{p}_k^\dagger) \tag{4.2.15}$$

and

$$\hat{x}_k = \sqrt{\frac{M\omega_k}{\hbar}} x_k$$

$$\hat{p}_k = \sqrt{\frac{1}{M\hbar\omega_k}} p_k. \tag{4.2.16}$$

The commutation relation which expresses uncertainty is therefore

$$[a_k, a_{k'}^\dagger] = \delta_{k,k'}, \tag{4.2.17}$$

which implies that the new operators have the action

$$a_k|N_k\rangle = \sqrt{N_k}|N_k - 1\rangle,$$
$$a_k^\dagger|N_k\rangle = \sqrt{1 + N_k}|N_k + 1\rangle. \tag{4.2.18}$$

In this formulation, we call the energy quanta of the sound field **phonons**. These particles appear simply as the eigenstates of the Hamiltonian of the field.

Note that each mode has the zero-point energy $\hbar\omega_k/2$, so that even when all N_k are zero, the sum over all k is infinite. This is not really a problem, because as discussed in Section 2.1, we are always free to define the ground state of a system as the vacuum and define the particles as excitations out of it.

To fully describe the sound field, we now need **Fock states** with a separate value of N_k defined for each possible k, that is, $|\ldots, N_{k_1}, N_{k_2}, N_{k_3}, \ldots\rangle$, equivalent to

$$|\psi_i\rangle = \prod_k \frac{\left(a_k^\dagger\right)^{N_k}}{\sqrt{N_k!}}|0\rangle. \tag{4.2.19}$$

The function N_k is the **distribution function**, giving the occupation number of each k-state. The vacuum state $|0\rangle$ is the zero-particle state, which in the case of a solid means the ground state of the system, as discussed in Section 2.1. The creation and destruction operators for one k do not affect the number of particles in a state with different k. The same holds true if we generalize the formalism to three dimensions. In this case, we write $N_{\vec{k}}$ for each possible plane wave in the system.

Exercise 4.2.1   Prove the commutation relation (4.2.17) using the definitions of a_k and $a_k^\dagger$ and the commutation relation $[x_n, p_{n'}] = i\hbar\delta_{n,n'}$. The identity (4.2.9) is useful.

Exercise 4.2.2   Show that (4.2.13) is equal to (4.2.14) by substituting into (4.2.14) the definitions of a_k and $a_k^\dagger$. You will need to use the definitions of x_k and p_k in terms of x_n and p_n, and the relation $[x_n, p_{n'}] = i\hbar\delta_{n,n'}$.

Exercise 4.2.3   Prove the identity (4.2.9) by computing the sum for finite N and then taking the limit $N \to \infty$.

Continuum limit. The fact that our model of the system consisted of discrete atoms does not affect whether or not there are phonons. We can equally well quantize the sound field in the continuum limit, in which we let the distance $a \to 0$. The mass M can be rewritten in terms of the mass density ρ,

$$M = \rho V_{\text{cell}} = \rho(ahw), \tag{4.2.20}$$

where h and w are the dimensions of the unit cell in the directions perpendicular to x, and the spring constant can be rewritten in terms of the effective elastic constant C introduced in Section 3.4.1, which gives the force per area F/A, and the strain ε, which gives the fractional change of the unit cell, that is,

$$\frac{F}{A} = C\varepsilon = -C\frac{x}{a} \quad \leftrightarrow \quad F = -Kx,$$

$$\to K = C\frac{A}{a} = C\frac{hw}{a}. \tag{4.2.21}$$

Substituting these definitions into (4.2.3), we have

$$\omega_k = \lim_{a \to 0} 2\sqrt{\frac{K}{M}} \sin(ka/2) = 2\sqrt{\frac{C/a}{\rho a}} \frac{ka}{2} = \sqrt{\frac{C}{\rho}} k, \tag{4.2.22}$$

or $\omega_k = vk$ as in standard continuum theory.

Local amplitude operators. The fact that we can write down the state of the system in terms of Fock states with an integer number of phonons does not mean that the system must be in such a state. As we will see in Section 4.4, it often makes sense to talk of the classical wave amplitude, even though a measurement of the energy will always give quantized values. We therefore want to be able to write the real-space displacement of the medium in terms of the phonon operators.

Since the displacement x_n is a real observable, it must be Hermitian. Therefore, according to definitions (4.2.5) and (4.2.16), we must have

$$\hat{x}_k^\dagger = \hat{x}_k^* = \hat{x}_{-k},$$
$$\hat{p}_k^\dagger = \hat{p}_k^* = \hat{p}_{-k}, \tag{4.2.23}$$

and consequently, from (4.2.15),

$$\hat{x}_k = \frac{1}{\sqrt{2}}\left(a_k + a_{-k}^\dagger\right),$$

$$\hat{p}_k = \frac{-i}{\sqrt{2}}\left(a_k - a_{-k}^\dagger\right). \tag{4.2.24}$$

The definition (4.2.5) implies

$$x_n = \frac{1}{\sqrt{N}} \sum_k x_k e^{ikan},$$ (4.2.25)

and therefore, by substitution,

$$x_n = \frac{1}{\sqrt{N}} \sum_k \sqrt{\frac{\hbar}{2M\omega_k}} \left(a_k + a^\dagger_{-k}\right) e^{ikan},$$ (4.2.26)

or, in the continuum limit,

$$x(r) = \sum_k \sqrt{\frac{\hbar}{2\rho V \omega_k}} \left(a_k e^{ikr} + a^\dagger_k e^{-ikr}\right),$$ (4.2.27)

where we have defined the continuous variable $r = na$, and $V = NV_{\text{cell}}$ is the total volume. Since the summation is over the dummy variable k, we have switched $k \to -k$ in the second term (ω_k depends only on the magnitude of k). In the same way, we can write for the velocity of the medium in the continuum limit,

$$\dot{x}(r) = \frac{p(r)}{M} = -i \sum_k \sqrt{\frac{\hbar \omega_k}{2\rho V}} \left(a_k e^{ikr} - a^\dagger_k e^{-ikr}\right).$$ (4.2.28)

So far, we have considered an acoustic wave only in one dimension. In many cases, we will want to keep track of the full three-dimensionality of the systems. In this case, we write

$$\vec{x}(\vec{r}) = \sum_{\vec{k},\lambda} \hat{\eta}_{\vec{k}\lambda} \sqrt{\frac{\hbar}{2\rho V \omega_{\vec{k}\lambda}}} \left(a_{\vec{k}\lambda} e^{i\vec{k}\cdot\vec{r}} + a^\dagger_{\vec{k}\lambda} e^{-i\vec{k}\cdot\vec{r}}\right)$$

$$\dot{\vec{x}}(\vec{r}) = -i \sum_{\vec{k},\lambda} \hat{\eta}_{\vec{k}\lambda} \sqrt{\frac{\hbar \omega_{\vec{k}\lambda}}{2\rho V}} \left(a_{\vec{k}\lambda} e^{i\vec{k}\cdot\vec{r}} - a^\dagger_{\vec{k}\lambda} e^{-i\vec{k}\cdot\vec{r}}\right),$$ (4.2.29)

where $\hat{\eta}_{\vec{k}\lambda}$ is a unit polarization vector, and λ is an index to label the polarization mode. For each $\vec{k}$, there are three possible phonon polarizations, as discussed in Section 3.4.2, so that the index λ runs from 1 to 3. For each polarization, we must account for a different possible complex amplitude, and therefore we must have separate creation and destruction operators for each polarization.

4.3 Photons

Because we have generated phonons from quantized sound waves, some people view them as not really real, that is, an "epi-phenomenon" in philosophical terms, while they view photons as truly fundamental particles. The approach of field theory in generating photons is exactly the same as that for generating phonons, however.

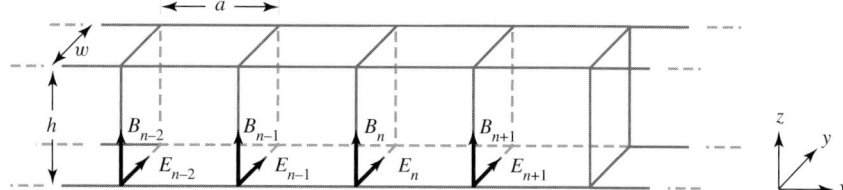

A linear chain of vacuum cells.

To start, we consider a model of space consisting of discrete elements of length a and height h and width w, as shown in Figure 4.4. We assume that the electric and magnetic fields are constant within each element, and that the electric field points in the $\hat{y}$ direction and the magnetic field points in the $\hat{z}$ direction, as shown in the figure.

Maxwell's equations in vacuum, in integral form, are

$$\int_S \vec{E} \cdot d\vec{\sigma} = 0$$

$$\int_S \vec{B} \cdot d\vec{\sigma} = 0$$

$$\oint \vec{E} \cdot d\vec{l} = -\frac{\partial}{\partial t} \int \vec{B} \cdot d\vec{\sigma}$$

$$\oint \vec{B} \cdot d\vec{l} = \frac{1}{c^2} \frac{\partial}{\partial t} \int \vec{E} \cdot d\vec{\sigma}. \tag{4.3.1}$$

The first two equations imply that $\vec{E}$ does not vary in the $\hat{y}$ direction and $\vec{B}$ does not vary in the $\hat{z}$ direction. Integrating over a loop around the front side of one of the volume elements, the fourth equation gives

$$(B_n - B_{n-1})h = -\frac{1}{c^2} \frac{\partial}{\partial t}(E_n h a). \tag{4.3.2}$$

Using the third equation, we can obtain equations for B_n and B_{n-1} by integrating around loops on the bottoms of the elements:

$$(E_n - E_{n-1})w = -\frac{\partial}{\partial t}(B_n w a)$$

$$(E_{n-1} - E_{n-2})w = -\frac{\partial}{\partial t}(B_{n-1} w a). \tag{4.3.3}$$

Taking the time derivative of (4.3.2), and substituting this into (4.3.3), we have

$$\frac{1}{a}(E_n - E_{n-1}) - \frac{1}{a}(E_{n-1} - E_{n-2}) = \frac{1}{c^2} \frac{\partial^2 E_n}{\partial t^2} a \tag{4.3.4}$$

or

$$\frac{1}{a^2}(E_n - 2E_{n-1} + E_{n-2}) = \frac{1}{c^2} \frac{\partial^2 E_n}{\partial t^2}. \tag{4.3.5}$$

In the limit $a \to 0$, this becomes

$$\frac{\partial^2 E}{\partial x^2} = \frac{1}{c^2} \frac{\partial^2 E}{\partial t^2}. \tag{4.3.6}$$

This is the well-known Maxwell's wave equation for classical electromagnetic waves in one dimension, which has the solutions

$$E_n = \text{Re } E_0 e^{i(kan - \omega_k t)}, \tag{4.3.7}$$

with $\omega = ck$.

To see how quantum mechanics affects this solution, we can write down the classical Hamiltonian for electromagnetic field,

$$H = V_{\text{cell}} \sum_n \left(\frac{1}{2} \epsilon_0 E_n^2 + \frac{1}{2} \frac{1}{\mu_0} B_n^2 \right), \tag{4.3.8}$$

where $V_{\text{cell}} = ahw$ is the volume of each element. We can write $\vec{E}$ and $\vec{B}$ in terms of the vector potential $\vec{A}$ according to the definitions

$$\vec{E} = -\frac{\partial \vec{A}}{\partial t}$$

$$\int_S \vec{B} \cdot d\vec{\sigma} = \oint \vec{A} \cdot d\vec{l}. \tag{4.3.9}$$

The first equation implies $\vec{A} \| \vec{E}$, which means for the volume element considered above, the latter equation can be written

$$B_n wa = -(A_n - A_{n-1})w. \tag{4.3.10}$$

Substituting into the Hamiltonian (4.3.8), we have

$$H = V_{\text{cell}} \sum_n \left(\frac{1}{2} \epsilon_0 (\dot{A}_n)^2 + \frac{1}{2} \frac{1}{\mu_0 a^2} (A_n - A_{n-1})^2 \right). \tag{4.3.11}$$

Recall that for the sound wave, we had the Hamiltonian (4.2.1),

$$H = \sum_n \left(\frac{1}{2} M \dot{x}_n^2 + \frac{1}{2} K (x_n - x_{n-1})^2 \right). \tag{4.3.12}$$

The one-to-one correspondence of these two suggests that we should treat them the same way. Not only that, we know that the vector potential A acts as a momentum. In classical electrodynamics, the momentum of a charged particle moving in a magnetic field is equal to (see, e.g., Goldstein *et al.* 2002: 57)

$$p = M\dot{x} + qA. \tag{4.3.13}$$

If there is an uncertainty relation of the classical momentum and x, then there must also be an uncertainty relation between A and its conjugate. In a harmonic oscillator, p and x are 90° out of phase, just as A and E are 90° out of phase in an electromagnetic wave.[1] We can

---

[1] This analysis assumes transverse electromagnetic waves in the Coulomb gauge. In general, the commutation relation is $[A_i(\vec{r}), \dot{A}_j(\vec{r}')] = (i\hbar/\epsilon_0)\delta_{ij}\delta^T(\vec{r} - \vec{r}')$, where $\delta^T(\vec{r})$ is the "transverse" δ-function, which picks out only the transverse part of a wave. See Loudon 1973, pages 145–147. For the finite cell method used here, $\delta^T(0) = 1/V_{\text{cell}}$.

therefore apply the same quantization procedure as in Section 4.2, replacing $M \leftrightarrow V_{\text{cell}}\epsilon_0$, $K \leftrightarrow V_{\text{cell}}/\mu_0 a^2$, $A \leftrightarrow x$, and $E \leftrightarrow -\dot{x}$. We write

$$H = \sum_k \left(a_k^\dagger a_k + \frac{1}{2} \right) \hbar \omega_k,$$ (4.3.14)

with

$$a_k = \frac{1}{\sqrt{2}} \left(\sqrt{\frac{\epsilon_0 V_{\text{cell}} \omega_k}{\hbar}} A_k - i \sqrt{\frac{\epsilon_0 V_{\text{cell}}}{\hbar \omega_k}} E_k \right)$$

$$a_k^\dagger = \frac{1}{\sqrt{2}} \left(\sqrt{\frac{\epsilon_0 V_{\text{cell}} \omega_k}{\hbar}} A_k^* + i \sqrt{\frac{\epsilon_0 V_{\text{cell}}}{\hbar \omega_k}} E_k^* \right)$$ (4.3.15)

and the normal variables

$$A_k = \frac{1}{\sqrt{N}} \sum_n A_n e^{-ikan}$$

$$E_k = \frac{1}{\sqrt{N}} \sum_n E_n e^{-ikan}$$ (4.3.16)

and the mode frequency

$$\omega = 2\frac{c}{a} \sin(ka/2).$$ (4.3.17)

In the continuum limit $a \to 0$, this mode frequency becomes, naturally, $\omega = ck$, and the vector potential is

$$A(r) = \sum_k \sqrt{\frac{\hbar}{2\epsilon_0 V \omega}} \left(a_k e^{ikr} + a_k^\dagger e^{-ikr} \right),$$ (4.3.18)

and the electric field is

$$E(r) = i \sum_k \sqrt{\frac{\hbar \omega}{2\epsilon_0 V}} \left(a_k e^{ikr} - a_k^\dagger e^{-ikr} \right).$$ (4.3.19)

Although we used an imaginary cell volume V_{cell} for the photon Hamiltonian, when we compute real amplitudes in the continuum limit, we always encounter $V = V_{\text{cell}} N$, so the cell size drops out. As with phonons, we can write down fully three-dimensional electromagnetic fields, as follows:

$$\vec{A}(\vec{r}) = \sum_{\vec{k},\lambda} \hat{\eta}_{\vec{k}\lambda} \sqrt{\frac{\hbar}{2\epsilon_0 V \omega}} \left(a_{\vec{k}\lambda} e^{i\vec{k}\cdot\vec{r}} + a_{\vec{k}\lambda}^\dagger e^{-i\vec{k}\cdot\vec{r}} \right)$$

$$\vec{E}(\vec{r}) = i \sum_{\vec{k},\lambda} \hat{\eta}_{\vec{k}\lambda} \sqrt{\frac{\hbar \omega}{2\epsilon_0 V}} \left(a_{\vec{k}\lambda} e^{i\vec{k}\cdot\vec{r}} - a_{\vec{k}\lambda}^\dagger e^{-i\vec{k}\cdot\vec{r}} \right),$$ (4.3.20)

where $\hat{\eta}_{\vec{k}\lambda}$ is a unit polarization vector. In this case, there are only two allowed polarizations, so the polarization index λ runs from 1 to 2. As discussed in Section 3.5.1, in an anisotropic crystal, ω and ϵ_0 must be replaced by $\omega_{\vec{k}\lambda}$ and $\tilde{\epsilon}$, which depend on the direction of $\vec{k}$ and the polarization and not only on the magnitude of $\vec{k}$.

The field theory of photons has been presented in detail by Louisell (1973) and Mandel and Wolf (1995). Most of this formalism also applies to phonons. In the following sections, we present some of the most useful results of quantum statistics.

4.4 Coherent States

The operators a_k and $a_k^\dagger$ are called the creation and destruction operators because of their roles in the algebra. What do these operators actually measure, though?

The answer is that they correspond to a measurement of the *complex amplitude* of a wave. The eigenstates of a_k correspond to states with definite phase and amplitude, that is,

$$a_k|\alpha_k\rangle = \alpha_k|\alpha_k\rangle, \tag{4.4.1}$$

where $\alpha_k = A_k e^{i\theta_k}$ is a complex number which gives the complex amplitude of the wave.

By itself, the operator a_k is not Hermitian, which means it cannot be an experimental observable. The sum $(a_k + a_k^\dagger)$ is Hermitian, however, and corresponds to

$$\langle\alpha_k|(a_k + a_k^\dagger)|\alpha_k\rangle = 2A_k \cos\theta_k\langle\alpha_k|. \tag{4.4.2}$$

This is the real amplitude of a field, which can be measured. The state $|\alpha_k\rangle$, which is an eigenstate of a_k, is called a **coherent state**, because it has a definite phase and amplitude.

By the definition (4.4.1), the product $a_k^\dagger a_k$ therefore gives

$$\langle\alpha_k|a_k^\dagger a_k|\alpha_k\rangle = A_k e^{-i\theta_k} A_k e^{i\theta_k} \langle\alpha_k|\alpha_k\rangle = A_k^2. \tag{4.4.3}$$

Recall from (4.2.18) that product $a_k^\dagger a_k$ acting on a Fock number state gives the value

$$\langle N_k|a_k^\dagger a_k|N_k\rangle = \sqrt{N_k}\sqrt{N_k}\langle N_k|N_k\rangle = N_k. \tag{4.4.4}$$

In other words, the product $a_k^\dagger a_k$ gives the number of particles in a state. We write this as the **number operator** $\hat{N}_k \equiv a_k^\dagger a_k$ (the hat distinguishes this from the simple occupation number, N_k). As we see in (4.4.3), the same measurement gives the amplitude squared of a coherent state. In other words, a measurement of the square of the amplitude is a measurement of the total number of particles.

Clearly, a coherent state is not also an eigenstate of the number operator. In terms of the Fock number states, a coherent state is equal to the superposition

$$|\alpha_k\rangle = e^{-|\alpha_k|^2/2} \sum_{N_k=0}^{\infty} \frac{\alpha_k^{N_k}}{\sqrt{N_k!}}|N_k\rangle. \tag{4.4.5}$$

The set of all coherent states is a complete set. In other words, any state can be written as a superposition of coherent states, but the states are not orthogonal, since

$$\langle \alpha_k | \alpha_k' \rangle \neq 0 \quad \text{if} \quad \alpha_k \neq \alpha_k', \tag{4.4.6}$$

unlike Fock states, which are all orthonormal.

Displacement operator. The parallel of the creation operator, for a coherent state, is the **displacement operator**, $D_k(z)$, where z is a complex number. This unitary operator is defined as

$$D_k(z) = e^{za_k^\dagger - z^* a_k}, \tag{4.4.7}$$

and has the properties

$$D_k^\dagger(z) = D_k^{-1}(z) = D_k(-z). \tag{4.4.8}$$

It can be represented in factorized form using the Campbell–Baker–Hausdorf formula of quantum mechanics (see, e.g., Mandel and Wolf 1995: s. 10.11.5), which applies to any two operators A and B which do not necessarily commute, but whose commutator commutes with both A and B:

$$e^{A+B} = e^A e^B e^{-[A,B]/2} = e^B e^A e^{[A,B]/2}. \tag{4.4.9}$$

The displacement operator can therefore be written as

$$D_k(z) = e^{-|z|^2/2} e^{za_k^\dagger} e^{-z^* a_k} = e^{|z|^2/2} e^{-z^* a_k} e^{za_k^\dagger}. \tag{4.4.10}$$

The displacement operator has its name because transformation of the operator a_k by the displacement operator leads to a simple shift by value z:

$$D_k^\dagger(z) a_k D_k(z) = a_k + z. \tag{4.4.11}$$

Using this relation, it is easy to see that the displacement operator has the action on the vacuum state $|0\rangle$ of creating a coherent state. We write

$$\begin{aligned}
a_k \left(D_k(\alpha_k) |0\rangle \right) &= \left(D_k(\alpha_k) D_k^\dagger(\alpha_k) \right) a_k D_k(\alpha_k) |0\rangle \\
&= D_k(\alpha_k)(a_k + \alpha_k) |0\rangle \\
&= \alpha_k \left(D_k(\alpha_k) |0\rangle \right).
\end{aligned} \tag{4.4.12}$$

The state $D_k(\alpha_k)|0\rangle$ is therefore an eigenstate of the operator a_k with eigenvalue α_k, and thus

$$D_k(\alpha_k) |0\rangle = |\alpha_k\rangle. \tag{4.4.13}$$

This shows that the coherent state can be thought of as the ground state of a displaced oscillator, in which the position is shifted by value α_k.

Exercise 4.4.1   Verify that the superposition of Fock states given in (4.4.5) for the coherent state has the eigenvalue given in (4.4.1).

Exercise 4.4.2   The state (4.4.5) is a Poisson distribution of photons. Assuming that only one k-state is occupied, calculate the uncertainty defined by

$$\Delta N_k = \sqrt{\langle (\hat{N}_k - \bar{N}_k)^2 \rangle}. \tag{4.4.14}$$

Show it is equal to the expected value for a Poisson distribution, $\Delta N_k = \sqrt{N_k}$.

Exercise 4.4.3 Prove that the relation (4.4.11) is true. You will need the quantum mechanical relation

$$[A, f(B)] = [A, B] f'(B). \tag{4.4.15}$$

Exercise 4.4.4 Show explicitly using formula (4.4.10) that $D_k(\alpha_k)|0\rangle = |\alpha_k\rangle$.

Time dependence of a coherent state. The time evolution of the complex amplitude can be found using Schrödinger's equation:

$$
\begin{aligned}
\frac{\partial}{\partial t}\alpha_k = \frac{\partial}{\partial t}\langle\alpha_k|a_k|\alpha_k\rangle &= \left(\frac{\partial}{\partial t}\langle\alpha_k|\right)a_k|\alpha_k\rangle + \langle\alpha_k|a_k\left(\frac{\partial}{\partial t}|\alpha_k\rangle\right) \\
&= \left(-\frac{1}{i\hbar}\langle\alpha_k|H\right)a_k|\alpha_k\rangle + \langle\alpha_k|a_k\left(\frac{1}{i\hbar}H|\alpha_k\rangle\right) \\
&= \frac{1}{i\hbar}\langle\alpha_k|[a_k, H]|\alpha_k\rangle.
\end{aligned} \tag{4.4.16}
$$

The Hamiltonian has the form (4.2.14). Only the term in H with $\hat{N}_k = a_k^\dagger a_k$ has a nonzero commutator with a_k, so that we have

$$
\begin{aligned}
\frac{\partial}{\partial t}\alpha_k &= -i\omega_k\langle\alpha_k|[a_k, \hat{N}_k]|\alpha_k\rangle \\
&= -i\omega_k\langle\alpha_k|a_k|\alpha_k\rangle \\
&= -i\omega_k\alpha_k,
\end{aligned} \tag{4.4.17}
$$

which implies

$$\alpha_k(t) = A_k e^{-i\omega_k t}. \tag{4.4.18}$$

Therefore, in a coherent state, the real and imaginary components of the amplitude are

$$
\begin{aligned}
A_R &= \frac{1}{2}A_k\left(e^{-i\omega_k t} + e^{i\omega_k t}\right) = A_k\cos\omega_k t \\
A_I &= \frac{1}{2i}A_k\left(e^{-i\omega_k t} - e^{i\omega_k t}\right) = -A_k\sin\omega_k t,
\end{aligned} \tag{4.4.19}
$$

as illustrated in Figure 4.5. We can represent the state of the system as a vector (A_R, A_I) of the complex plane, known as a **phasor**, as shown in Figure 4.5. This vector rotates clockwise with angular frequency ω_k.

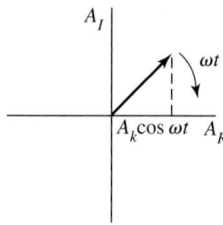

Fig. 4.5 The field oscillation represented by a phasor.

More generally, we can write operators for the real and imaginary components of the phasor, given by

$$\hat{A}_R = \text{Re } a_k = \frac{a_k + a_k^\dagger}{2}$$

$$\hat{A}_I = \text{Im } a_k = \frac{a_k - a_k^\dagger}{2i}. \qquad (4.4.20)$$

Number-phase uncertainty. There is no rule that a system must always be in a Fock state, that is, that it must have a definite number of particles. A coherent state is just as physically realizable. Experiments that measure the number of photons will force the system into a definite Fock state, but experiments which measure phase will not. In fact, a measurement of phase causes maximal uncertainty in the number of particles. This **number-phase uncertainty** is expressed by the commutation relation, which is easy to prove using the definition of $\hat{N}_k$,

$$[\hat{N}_k, a_k] = -a_k, \qquad (4.4.21)$$

that is, the number operator and the complex amplitude operator do not commute, and therefore cannot be simultaneously observed. Since the phase θ_k is found by a measurement of complex amplitude, we cannot determine the phase exactly if we know the number of particles exactly.

In a coherent state, the uncertainty in the real component is given by

$$
\begin{aligned}
\langle (\Delta A_R)^2 \rangle &= \langle \alpha_k | (\hat{A}_R - \bar{A}_R)^2 | \alpha_k \rangle \\
&= \langle \alpha_k | (\hat{A}_R^2 - 2\hat{A}_R\bar{A}_R + \bar{A}_R^2) | \alpha_k \rangle \\
&= \langle \alpha_k | \left(\tfrac{1}{4}(a_k^\dagger a_k^\dagger + a_k a_k + 2a_k^\dagger a_k + 1) \right. \\
&\qquad \left. -(a_k^\dagger + a_k)A_k \cos \omega_k t + A_k^2 \cos^2 \omega_k t \right) | \alpha_k \rangle \\
&= \frac{A_k^2}{4} \left(e^{2i\omega_k t} + e^{-2i\omega_k t} + 2 \right) + \tfrac{1}{4} \\
&\qquad -A_k^2 \left(e^{i\omega_k t} + e^{-i\omega_k t} \right) \cos \omega_k t + A_k^2 \cos^2 \omega_k t \\
&= \tfrac{1}{4}, \qquad (4.4.22)
\end{aligned}
$$

that is, $\Delta A_R = \tfrac{1}{2}$. The same applies to ΔA_I. Thus, although a coherent state is not an eigenstate of the Hamiltonian, the uncertainty in the complex amplitude does not increase in time.

We can represent the total uncertainty as an area in the complex plane equal to $\Delta A_R \Delta A_I = \tfrac{1}{4}$, as shown in Figure 4.6(a). This is the minimum possible total uncertainty. All states other than the coherent state have larger total uncertainty. The uncertainty principle along one axis can be reduced, however, if the uncertainty in the perpendicular direction is increased. For example, as shown in Figure 4.6(b), the uncertainty in the phase can be reduced by increasing the uncertainty in the number, keeping the total area the same. Figure 4.6(c) shows a Fock state, which has minimal amplitude uncertainty and maximal phase uncertainty.

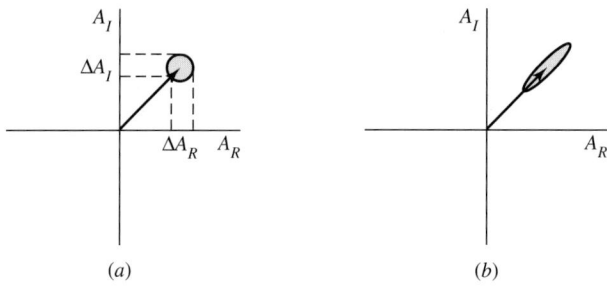

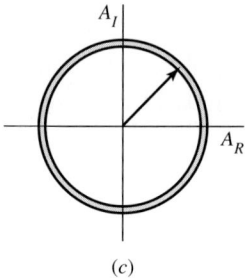

Fig. 4.6 (a) Uncertainty of a coherent state in the phasor picture. (b) A squeezed wave in the phasor picture. (c) A Fock state in the phasor picture.

States which have one axis of uncertainty traded off against another are called **squeezed** states. Many experiments have demonstrated the effect of reducing the uncertainty of one component below the value of $\frac{1}{2}$ by increasing the other.

Classical correspondence. Coherent states allow us to relate the amplitude of a wave to the number of particles. Suppose we have a coherent phonon state with $\alpha_k = A_k e^{-i\omega_k t}$. From (4.2.27), we have

$$
\begin{aligned}
\langle \hat{x}(r) \rangle &= \sum_{k'} \sqrt{\frac{\hbar}{2\rho V \omega_{k'}}} \langle \alpha_k | \left(a_{k'} e^{ik'r} + a_{k'}^\dagger e^{-ik'r} \right) | \alpha_k \rangle \\
&= \sum_{k'} \sqrt{\frac{\hbar}{2\rho V \omega_{k'}}} \left(A_k e^{i(k'r - \omega_k t)} + A_k e^{-i(k'r - \omega_k t)} \right) \delta_{k',k} \\
&= \sqrt{\frac{\hbar}{2\rho V \omega_k}} \, 2A_k \cos(kr - \omega_k t).
\end{aligned}
\tag{4.4.23}
$$

This corresponds to a classical wave with $x(r,t) = x_0 \cos(kr - \omega_k t)$. Equating these two, we can solve for A_k, which yields

$$
\boxed{A_k = \sqrt{\frac{\rho V \omega_k}{2\hbar}} x_0.}
\qquad \text{(phonons)}
\tag{4.4.24}
$$

In other words, the number of phonons is related to the classical wave amplitude by the relation

$$\langle \hat{N}_k \rangle = \langle a_k^\dagger a_k \rangle = A_k^2 = \frac{\rho V \omega_k}{2\hbar} x_0^2. \qquad \text{(phonons)} \qquad (4.4.25)$$

We can apply the same type of analysis to photons in the continuum limit, to obtain

$$A_k = \sqrt{\frac{\epsilon_0 V}{2\hbar \omega}} E_0, \qquad \text{(photons)} \qquad (4.4.26)$$

where E_0 is the electric field amplitude and

$$\langle \hat{N}_k \rangle = \langle a_k^\dagger a_k \rangle = \frac{\epsilon_0 V}{2\hbar \omega} E_0^2. \qquad \text{(photons)} \qquad (4.4.27)$$

Exercise 4.4.5   In a solid with density 1 g/cm^3, a sound wave has frequency 1 MHz, velocity 5×10^5 cm/s, and intensity 1 W/cm^2. What is the root-mean-squared displacement of the atoms?

Exercise 4.4.6   What is the amplitude in V/cm of a laser beam with diameter 2 mm, with 10^{20} photons in a cavity of length 1 m? Assume each photon is in the visible range with energy $\hbar\omega \simeq 2$ eV. What is its intensity in W/cm^2?

4.5 Spatial Field Operators

As discussed above, a state with an integer number of particles is not the only physically realizable state. A coherent state is also possible, and in general, all manner of super-positions of states are possible. Quantum mechanics says only that if energy is being measured, then an integer number of quanta must be detected. If another quantity is being measured, or if nothing is being measured, the system does not need to be in an energy eigenstate.

Since the Fourier transform theorem says that any spatial function can be viewed as a sum of single-frequency states, we can write any spatial wave function as a sum of momentum eigenstates. Therefore, we can make a creation operator for a wave in any spatial state by writing a sum of momentum-state creation and destruction operators. One particularly useful operator is the Fourier transform of a δ-function, $\delta(r)$, which corresponds to a particle at a single point. These spatial creation and destruction operators are sometimes known simply as **field operators**. We write, for a one-dimensional system,

$$\psi^\dagger(r) = \sum_k \frac{e^{-ikr}}{\sqrt{L}} a_k^\dagger$$

$$\psi(r) = \sum_k \frac{e^{ikr}}{\sqrt{L}} a_k. \qquad (4.5.1)$$

The spatial operator $\psi^\dagger(r)$ creates a particle at exactly the point r. Alternatively, as with the momentum creation and destruction operators, the operator $\psi(r)$ can be viewed as measuring the *complex amplitude* of the wave at point r. For a three-dimensional system, the spatial field operators become

$$\psi^\dagger(\vec{r}) = \sum_{\vec{k}} \frac{e^{-i\vec{k}\cdot\vec{r}}}{\sqrt{V}} a_{\vec{k}}^\dagger$$

$$\psi(\vec{r}) = \sum_{\vec{k}} \frac{e^{i\vec{k}\cdot\vec{r}}}{\sqrt{V}} a_{\vec{k}}. \tag{4.5.2}$$

In general, a particle can be created with any wave function $\phi(\vec{r})$ by creating the superposition

$$|\phi\rangle = \int d^3r\, \phi(\vec{r})\psi^\dagger(\vec{r})|0\rangle, \tag{4.5.3}$$

where $|0\rangle$ is the vacuum state. The plane wave $\phi(\vec{r}) = (1/\sqrt{V})e^{i\vec{k}\cdot\vec{r}}$ can also be created this way. Using the definition of $\psi^\dagger(\vec{r})$, this resolves to

$$|\phi\rangle = \int d^3r\, \frac{1}{\sqrt{V}}e^{i\vec{k}\cdot\vec{r}} \left(\frac{1}{\sqrt{V}} \sum_{k'} e^{-i\vec{k}'\cdot\vec{r}} a_{\vec{k}}^\dagger \right) |0\rangle$$

$$= \frac{1}{V} \sum_{k'} \left(\int d^3r\, e^{i(\vec{k}-\vec{k}')\cdot\vec{r}} \right) a_{\vec{k}}^\dagger |0\rangle. \tag{4.5.4}$$

Using the identity

$$\int d^3r\, e^{i(\vec{k}-\vec{k}')\cdot\vec{r}} = (2\pi)^3 \delta(\vec{k}-\vec{k}') \tag{4.5.5}$$

and converting the sum over $\vec{k}$ to an integral, this becomes simply

$$|\psi\rangle = a_{\vec{k}}^\dagger|0\rangle, \tag{4.5.6}$$

as one would expect.

Note the difference between a plane wave state created by the creation operator $a_{\vec{k}}^\dagger$ and a coherent state with amplitude $A_{\vec{k}}^2 = 1$, created by the displacement operator. The former has exactly one particle, and indefinite phase (though it has a definite wavelength). The latter has definite phase, and is a superposition of different number states, with average number equal to 1.

Exercise 4.5.1   Show that the spatial field operators $\psi(r)$ and $\psi^\dagger(r)$ defined in (4.5.1) have commutation relations similar to those of the momentum state operators, that is,

$$[\psi(r), \psi^\dagger(r')] = \delta(r - r'). \tag{4.5.7}$$

You will need the identity

$$\delta(r) = \frac{1}{L} \sum_{k} e^{ikr}. \tag{4.5.8}$$

Many-particle states. A state with many particles can be generated by operating successively on the vacuum state with creation operators:

$$|\psi\rangle = \prod_i \frac{1}{\sqrt{N_i!}} \left(\int d^3 r_i \, \phi_i(\vec{r}) \psi^\dagger(\vec{r}_i) \right)^{N_i} |0\rangle, \tag{4.5.9}$$

where the $\phi_i(\vec{r})$ are the various single-particle eigenstates. The normalization of this state requires that we compute

$$\langle\psi|\psi\rangle = \langle 0| \prod_j \frac{1}{\sqrt{N_j!}} \left(\frac{1}{\sqrt{V}} \int d^3 r_j \, \phi_j^*(\vec{r}_j) \psi(\vec{r}_j) \right)^{N_j} \tag{4.5.10}$$

$$\times \prod_i \frac{1}{\sqrt{N_i!}} \left(\frac{1}{\sqrt{V}} \int d^3 r_i \, \phi_i(\vec{r}_i) \psi^\dagger(\vec{r}_i) \right)^{N_i} |0\rangle.$$

This will generate a product of destruction operators followed by a product of creation operators. This can be turned into a normal-ordered product by commuting each destruction operator successively through all of the creation operators to its right, generating a commutator $\delta(\vec{r}_i - \vec{r}_j)$ each time. This commutator will eliminate one integral over space, and generate a term,

$$\int d^3 r \, \phi_i^*(\vec{r})\phi_j(\vec{r}) = \delta_{ij}, \tag{4.5.11}$$

because we assume orthonormality of the single-particle eigenstates. Therefore, although many possible terms arise from the many commutations, only the ones in which each wave function $\phi_i(\vec{r})$ is matched with its complex conjugate will be nonzero. For $N_i > 1$, that is, for multiple occupation of a single state, there will be $N_i!$ such terms, giving a factor which cancels the $N_i!$ in the denominator. Therefore, $\langle\psi|\psi\rangle = 1$ as expected.

Coherence. Just as the combination $a_k^\dagger a_k$ gave us the square of the amplitude in state k, the combination $\psi^\dagger(r)\psi(r)$ gives the total density of particles at point r, and is known as the **density operator**. The function $\psi^\dagger(r)\psi(r)$ does not tell us all the information about a wave, however. In general, the wave can be in a superposition of many possible states. It is therefore common to define another entity, known as the **density matrix**, as follows:

$$\tilde{\rho} = \begin{pmatrix} \psi^\dagger(r_1)\psi(r_1) & \psi^\dagger(r_2)\psi(r_1) & \psi^\dagger(r_3)\psi(r_1) & \cdots \\ \psi^\dagger(r_1)\psi(r_2) & \psi^\dagger(r_2)\psi(r_2) & \psi^\dagger(r_3)\psi(r_2) & \cdots \\ \psi^\dagger(r_1)\psi(r_3) & \psi^\dagger(r_2)\psi(r_3) & \psi^\dagger(r_3)\psi(r_3) & \cdots \\ & \vdots & & \\ & \vdots & & \end{pmatrix}. \tag{4.5.12}$$

Density matrices will be discussed in detail in Section 9.1. In addition to the density operators on the diagonal, the density matrix also has **off-diagonal** elements such as $\psi^\dagger(r_1)\psi(r_2)$, which correspond to the probability of taking one particle from r_1 and returning it to r_2. This is known as a **single-particle correlation function**. If the wave is in a coherent state, this function will be nonzero for all r_1 and r_2:

$$
\begin{aligned}
\langle \alpha_k | \psi^\dagger(r_1) \psi(r_2) | \alpha_k \rangle &= \langle \alpha_k | \left(a_k^\dagger \frac{e^{-ikr_1}}{\sqrt{L}} \right) \left(\frac{e^{ikr_2}}{\sqrt{L}} a_k \right) | \alpha_k \rangle \\
&= \langle \alpha_k | A_k e^{-i\theta_k} \frac{e^{-ikr_1}}{\sqrt{L}} \frac{e^{ikr_2}}{\sqrt{L}} A_k e^{i\theta_k} | \alpha_k \rangle \\
&= \frac{\langle \hat{N}_k \rangle}{L} e^{-ik(r_1 - r_2)}.
\end{aligned} \tag{4.5.13}
$$

Therefore, coherence is sometimes called **off-diagonal, long-range order** (ODLRO), to distinguish it from diagonal long-range order, which is the same as spatial periodicity.

4.6 Electron Fermi Field Operators

We have seen that phonons and photons arise as the energy eigenstates of waves in vibrational and electromagnetic fields. It is therefore natural to suppose that *all* particles have a similar origin, from the quantization of an underlying field. In particular, it is natural to view electrons as having the same status.

The uncertainty relation of the creation and destruction operators for fermions is expressed in terms of an **anticommutation** relation instead of a commutation relation. We use curly brackets instead of square brackets to express this:

$$
\begin{aligned}
\{b_k, b_{k'}^\dagger\} &= b_k b_{k'}^\dagger + b_{k'}^\dagger b_k = \delta_{k,k'}, \\
\{b_k, b_{k'}\} &= \{b_k^\dagger, b_{k'}^\dagger\} = 0.
\end{aligned} \tag{4.6.1}
$$

Here we have used the letter b instead of a for the creation and destruction operators, to avoid confusion about which type of operator we are using. The fermion operators have the same role as boson operators for phonons and photons, however. They create and destroy electrons with the following rules:

$$
\begin{aligned}
b_k | N_k \rangle &= \sqrt{N_k} | N_k - 1 \rangle \\
b_k^\dagger | N_k \rangle &= \sqrt{1 - N_k} | N_k + 1 \rangle.
\end{aligned} \tag{4.6.2}
$$

The latter relation ensures the law of Pauli exclusion, that only one electron can occupy a single quantum state. All of the standard algebra of field operators applies to electrons. Note, however, that we cannot construct a coherent state (4.4.5) from fermion operators, because we cannot have a product of two fermion operators on the same k-state. This is the primary reason why phonons and photons are associated with classical waves while electrons are not.

Since there are many possible k-states, we can define Fock states and field operators for fermions just as for boson particles like phonons and photons. The many-body Fock number states are written $| \ldots, N_{k_1}, N_{k_2}, N_{k_3}, \ldots \rangle$ just as for phonons and photons, but with the constraint that all N_k must have values only of either 0 or 1.

In Section 2.1, we introduced the concept of holes in a valence band as missing electrons. In some cases, it can be convenient to define hole creation and destruction operators distinct from the electron operators. A hole is the removal of an electron, and therefore the hole creation operator corresponds to an electron destruction operator. Including spin, we write

$$b^\dagger_{h,\vec{k}\uparrow} = b_{e,-\vec{k}\downarrow}, \qquad b_{h,\vec{k}\uparrow} = b^\dagger_{e,-\vec{k}\downarrow}. \tag{4.6.3}$$

The removal of a negative-charge electron with momentum $\vec{k}$ and positive spin is effectively the same as adding positive charge, negative momentum, and down spin.

For electrons, to construct the spatial field operators we want to explicitly include the fact that the electron states are Bloch waves. In general, as discussed in Section 4.5, we can write down a spatial field operator as a sum over any complete set of eigenstates of a system. In this case, instead of using plane waves as the basis states, we use Bloch functions, as follows:

$$\Psi^\dagger_n(\vec{r}) = \sum_{\vec{k}} \frac{e^{-i\vec{k}\cdot\vec{r}}}{\sqrt{V}} u^*_{n\vec{k}}(\vec{r}) b^\dagger_{n\vec{k}}$$

$$\Psi_n(\vec{r}) = \sum_{\vec{k}} \frac{e^{i\vec{k}\cdot\vec{r}}}{\sqrt{V}} u_{n\vec{k}}(\vec{r}) b_{n\vec{k}}, \tag{4.6.4}$$

where n is the electron band index. Here we have used Ψ instead of ψ to distinguish electron fermion operators from boson spatial field operators. Similar to the boson case, we can generate a many-particle state with

$$|\psi\rangle = \prod_i \left(\int d^3 r_i \, \phi_i(\vec{r}_i) \Psi^\dagger(\vec{r}_i) \right) |0\rangle. \tag{4.6.5}$$

The normalization of this many-body wave function requires that each state ϕ_i be unique, that is, no more than one fermion can be put into a single-particle state.

Viewing electrons as the eigenstates of an underlying field, while just as straightforward as that quantization of phonons and photons, leads to several philosophical questions. In the case of phonons and photons, the field of interest could be clearly understood in classical terms – the vibrational field in the case of phonons, and the electromagnetic field in the case of photons. What is the field of interest in the case of electrons? Is it real?

It is quite natural to say that in addition to the electromagnetic field, there is also a **matter field** which exists throughout space. This field is just as real as the electromagnetic field, in terms of its role in the underlying mathematics. The primary difference between the two is not in their realness, but in the fact that the creation and destruction operators obey a commutation relation in the case of phonons and photons (bosons), and an anticommutation relation in the case of electrons (fermions). Therefore, in the case of electrons a coherent state as discussed in Section 4.4 cannot be created, and therefore one can never observe a macroscopic coherent electron wave as one can with light, unless the electrons pair into bosons, as discussed in Chapter 11.

There seems to be a chicken and egg problem, however. Which is more fundamental, the photon or electron field? In the case of phonons and photons, we deduced the commutation relations of the particle creation and destruction operators from the commutation relation $[x, p] = i\hbar$ for the electrons or atoms which made up the harmonic oscillator. Is the electronic field somehow more fundamental than the phonon and photon fields? Some physicists have argued so (e.g., Mead 2001). On the other hand, the fact that the boson and fermion fields have the same mathematical structure except for a positive or negative sign seems to indicate that they have the same underlying basis in reality.

Exercise 4.6.1 Show that the commutation relation for the electron spatial field operators is

$$\{\Psi_n(\vec{r}), \Psi_{n'}^\dagger(\vec{r}')\} = \delta(\vec{r} - \vec{r}')\delta_{n,n'}, \qquad (4.6.6)$$

using the normalization relation for the Bloch cell functions

$$\sum_k u_{n\vec{k}}^*(\vec{r})u_{n\vec{k}}(\vec{r}') = V\delta(\vec{r} - \vec{r}'). \qquad (4.6.7)$$

Exercise 4.6.2 Prove, using the result of the previous exercise, that the fermion many-particle state (4.6.5) is normalized if

$$\int d^3r \, \phi_j^*(\vec{r})\phi_i(\vec{r}) = \delta_{ij}.$$

4.7 First-Order Time-Dependent Perturbation Theory: Fermi's Golden Rule

Time-dependent perturbation theory in quantum mechanics is presented in several standard textbooks (e.g., Sakurai 1995; Baym 1969). Here we review this method and apply it specifically to the many-body Fock states of quantized waves.

We begin by assuming that the Hamiltonian of a quantum mechanical system is given by

$$H = H_0 + V_{\text{int}}, \qquad (4.7.1)$$

where H_0 is the single-particle Hamiltonian and V_{int} is a term which is small compared to H_0. This allows us to assume that the eigenstates of the system are approximately still the same as the eigenstates of H_0. We then talk of transitions between these eigenstates due to V_{int}, which we call the **interaction term** in the Hamiltonian.

We assume that the system has been prepared at time t_0 in a quantum mechanical state $|\psi_{t_0}\rangle$. At a later time t, the state is written as $|\psi_t\rangle$.

The Schrödinger equation gives the time evolution of the system as

$$i\hbar\frac{\partial}{\partial t}|\psi_t\rangle = (H_0 + V_{\text{int}})|\psi_t\rangle. \qquad (4.7.2)$$

In the **interaction representation**, we define a new state $|\psi(t)\rangle$ (written with the t in parentheses rather than subscript) given by

$$|\psi(t)\rangle = e^{iH_0t/\hbar}|\psi_t\rangle, \qquad (4.7.3)$$

and a new operator

$$V_{\text{int}}(t) = e^{iH_0 t/\hbar} V_{\text{int}} e^{-iH_0 t/\hbar}. \tag{4.7.4}$$

In this representation, the Schrödinger equation is rewritten as

$$i\hbar \frac{\partial}{\partial t} |\psi(t)\rangle = V_{\text{int}}(t) |\psi(t)\rangle, \tag{4.7.5}$$

which has the advantage of not depending on H_0. As discussed in Chapter 2, in many cases we treat the ground state of a system as a new vacuum, and deal only with excitations out of this vacuum. The interaction picture allows us to do this naturally.

Integrating both sides of this equation, we have

$$|\psi(t)\rangle = |\psi(t_0)\rangle + \frac{1}{i\hbar} \int_{t_0}^{t} V_{\text{int}}(t') |\psi(t')\rangle dt'. \tag{4.7.6}$$

To first order in t, we can approximate $|\psi(t')\rangle = |\psi(t_0)\rangle$, which implies

$$|\psi(t)\rangle = \left(1 + \frac{1}{i\hbar} \int_{t_0}^{t} V_{\text{int}}(t') dt'\right) |\psi(t_0)\rangle. \tag{4.7.7}$$

To find $|\psi(t)\rangle$ to second order in t, one can substitute this approximation for $|\psi(t)\rangle$ into (4.7.6). By repeated substitution, one can show that

$$\begin{aligned}
|\psi(t)\rangle \;=\; & \left(1 + (1/i\hbar) \int_{t_0}^{t} dt' \, V_{\text{int}}(t') \right. \\
& \left. + (1/i\hbar)^2 \int_{t_0}^{t} dt' \int_{t_0}^{t'} dt'' \, V_{\text{int}}(t') V_{\text{int}}(t'') + \cdots \right) |\psi(t_0)\rangle,
\end{aligned} \tag{4.7.8}$$

which we can write in shorthand as

$$|\psi(t)\rangle = e^{-(i/\hbar) \int_{t_0}^{t} V_{\text{int}}(t') dt'} |\psi(t_0)\rangle. \tag{4.7.9}$$

The operator $S(t, t_0) \equiv e^{-(i/\hbar) \int_{t_0}^{t} V_{\text{int}}(t') dt'}$ is unitary, since the wave function must be normalized at all times.

This formula can be used to deduce the rate of transitions between states. Suppose that the system at time $t_0 = 0$ is in a state $|i\rangle$, which is an eigenstate of H_0. Then the probability amplitude for it being in another eigenstate $|n\rangle$ of H_0 at time t is, to lowest order in t,

$$\begin{aligned}
\langle n | \psi_t \rangle &= e^{-(i/\hbar) E_n t} \langle n | \psi(t) \rangle \\
&= e^{-(i/\hbar) E_n t} \frac{1}{i\hbar} \int_0^t \langle n | V_{\text{int}}(t') | i \rangle \, dt' \\
&= e^{-(i/\hbar) E_n t} \frac{1}{i\hbar} \langle n | V_{\text{int}} | i \rangle \int_0^t e^{(i/\hbar)(E_n - E_i) t'} \, dt',
\end{aligned} \tag{4.7.10}$$

where we have kept only the first-order term of the expansion (4.7.8), and we have used the fact that at time t_0 the initial state is the same in both representations, $|\psi(t_0)\rangle = |\psi_0\rangle = |i\rangle$.

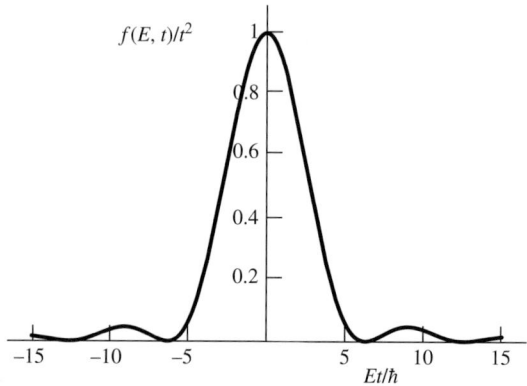

Fig. 4.7   The peak function $f(E, t) = \sin^2(Et/2\hbar)/(E/2\hbar)^2$ which appears in the calculation of Fermi's golden rule.

The eigenstates of H_0 are assumed to be orthonormal. The total probability of being in state $|n\rangle$ at time t is therefore

$$|\langle n|\psi_t\rangle|^2 = \frac{1}{\hbar^2}|\langle n|V_{\text{int}}|i\rangle|^2 \left|\int_0^t e^{(i/\hbar)(E_n - E_i)t'} dt'\right|^2$$

$$= \frac{1}{\hbar^2}|\langle n|V_{\text{int}}|i\rangle|^2 \left|\frac{e^{(i/\hbar)(E_n - E_i)t} - 1}{(i/\hbar)(E_n - E_i)}\right|^2$$

$$= |\langle n|V_{\text{int}}|i\rangle|^2 \frac{\sin^2[(E_n - E_i)t/2\hbar]}{[(E_n - E_i)/2]^2}. \tag{4.7.11}$$

The function $\sin^2(Et/2\hbar)/(E/2)^2$ is a peak with oscillating wings, as shown in Figure 4.7. As time increases, the zeroes of the oscillations get closer together in energy, so that the width of the central peak decreases, and the height of the peak increases, so that it looks more and more like a δ-function. Using the form (C.3) of the δ-function given in Appendix C, we have

$$\lim_{t \to \infty} \frac{\left|e^{ixt} - 1\right|^2}{x^2} = \lim_{t \to \infty} \frac{\sin^2(xt/2)}{(x/2)^2}$$

$$= \delta(x)2\pi t. \tag{4.7.12}$$

We can therefore write

$$\boxed{\frac{\partial}{\partial t}|\langle n|\psi_t\rangle|^2 = \frac{2\pi}{\hbar}|\langle n|V_{\text{int}}|i\rangle|^2\delta(E_n - E_i).} \tag{4.7.13}$$

This is one form of **Fermi's golden rule**.

To use this formula, we must integrate over the final states to get a finite value. We find the total rate of depletion of the initial state by integrating over all possible final states other than the initial state:

$$\frac{\partial}{\partial t}\sum_{n\neq i}|\langle n|\psi_t\rangle|^2 = \frac{2\pi}{\hbar}\sum_{n\neq i}|\langle n|V_{\text{int}}|i\rangle|^2\delta(E_n - E_i). \tag{4.7.14}$$

If there is a continuum of possible final states, we can switch the sum to the integral

$$\frac{\partial}{\partial t}\sum_{n\neq i}|\langle n|\psi_t\rangle|^2 = \int \mathcal{D}(E_n)dE_n \, |\langle n|V_{\text{int}}|i\rangle|^2\delta(E_n - E_i), \tag{4.7.15}$$

where $\mathcal{D}(E_n)$ is the density of states, defined in Section 1.8. The δ-function then removes the integral to give us

$$\frac{\partial}{\partial t}\sum_{n\neq i}|\langle n|\psi_t\rangle|^2 = \frac{2\pi}{\hbar}\left[|\langle n|V_{\text{int}}|i\rangle|^2\mathcal{D}(E_n)\right]_{E_n=E_i}. \tag{4.7.16}$$

This is another useful form of Fermi's golden rule. The total rate of transitions from initial state $|i\rangle$ to states with energy E_n is proportional to the square of the matrix element of the Hamiltonian between the initial and final states, times the density of final states, for energy-conserving transitions.

Suppose now that the eigenstates of H_0 are many-body Fock states. If there are interactions between these states, then the Hamiltonian will have terms which do not only contain the number operator $N = a_{\vec{k}}^\dagger a_{\vec{k}}$, but also have unbalanced creation and destruction operators which do not conserve the number in a k-state; for example,

$$V_{\text{int}} = \sum_{\vec{k},\vec{k}'}A_{\vec{k},\vec{k}'}\, a_{\vec{k}'}^\dagger a_{\vec{k}}, \tag{4.7.17}$$

where $A_{\vec{k},\vec{k}'}$ gives the strength of the interaction. (The exact forms of various interaction Hamiltonians will be studied in Chapter 5.) Each term in the sum corresponds to destroying a particle in state k and recreating it in state k'. If the initial state is $|i\rangle = |\ldots, N_i, \ldots, N_f, \ldots\rangle$ and the final state is $|f\rangle = |\ldots, N_i - 1, \ldots, N_f + 1, \ldots\rangle$, Fermi's golden rule gives us

$$\frac{\partial}{\partial t}|\langle f|\psi_t\rangle|^2 = \frac{2\pi}{\hbar}\left|\langle f|\sum_{\vec{k},\vec{k}'}A_{\vec{k},\vec{k}'}\, a_{\vec{k}'}^\dagger a_{\vec{k}}\, |i\rangle\right|^2\delta(E_f - E_i)$$

$$= \frac{2\pi}{\hbar}\left|A_{i,f}\sqrt{1+N_f}\sqrt{N_i}\right|^2\delta(E_f - E_i)$$

$$= \frac{2\pi}{\hbar}|A_{i,f}|^2(1+N_f)N_i\,\delta(E_f - E_i). \tag{4.7.18}$$

Since the eigenstates are orthogonal, only the term $A_{i,f}$ survives out of the interaction energy sum. The factor N_i expresses the simple fact that the rate of scattering out of a state is proportional to the number of particles in it in the first place. The $(1+N_f)$ term expresses the well-known law of **stimulated scattering** of bosons, that the probability of a particle transition to a final state f depends on the number of particles in the final state. If the number in the final state N_f is large, then the amplification of the transition rate can be quite large. Of course, if we had used fermion states instead of bosons, then we would have a factor $(1 - N_f)$ instead, which expresses the law of Pauli exclusion, that two fermions cannot occupy the same state. We can therefore summarize that the time scale for the transition per particle, $\tau_{i \to f}$, is given by

$$\frac{1}{\tau_{i \to f}} = \frac{1}{N_i} \frac{\partial}{\partial t} |\langle \psi_f | \psi_t \rangle|^2 \propto \begin{cases} (1 + N_f) & \text{bosons} \\ (1 - N_f) & \text{fermions.} \end{cases} \tag{4.7.19}$$

A tremendous amount of physics is included in these two statements. All of chemistry is dependent on the principle of Pauli exclusion, while all of the physics of Planck radiation, lasers, and superfluids are implied by the principle of stimulated scattering.

In deriving Fermi's golden rule, we made two assumptions about the time scale. In (4.7.10), we assumed that t was small, so that we could ignore higher-order terms. On the other hand, in using (4.7.12), we assumed that t was long enough to treat the oscillating function as a δ-function. Both of these conditions can be true if t falls in the proper intermediate time range. The first condition means that t must be short compared to the time for the initial state to be significantly depleted. As time goes on, the transitions out of this state will decrease its amplitude, and the initial state must be recomputed. The second condition means that the time must be long compared to $\hbar/\Delta E_n$, where ΔE_n is the energy range for significant changes in the matrix element and density of states. The more rapidly these vary with E_n, the longer t must be for it to be valid to treat the oscillating function as a δ-function in the integral (4.7.15). As seen above, the matrix element of a Fock state depends on the number of particles at a given energy, and therefore rapid changes in the energy distribution function of the particles will also affect this condition.

If the initial state is very rapidly depleted, and if the matrix element varies rapidly with energy, it is possible that no range of t will satisfy both conditions. In this case, Fermi's golden rule will break down. Fermi's golden rule is therefore valid only under the condition

$$\tau_{i \to f} \gg \hbar/\Delta E_{\text{char}}, \tag{4.7.20}$$

where $\tau_{i \to f}$ is the transition time as calculated using Fermi's golden rule and ΔE_{char} is a characteristic energy range over which the properties of the system (distribution function, density of states, scattering matrix element) have significant variation.

The field of study is known as **quantum kinetics** deals with the situations in which Fermi's golden rule breaks down. For example, when an ultrafast laser pulse excites the electrons in a solid, their population can change very rapidly on short time scales; in this case the evolution of the electrons must be computed taking into account the phase and not only the amplitudes of the particles. In another example, at very low temperatures the characteristic energy range is very small because $k_B T$ is low. This relates to Anderson localization, which we will discuss in Chapter 9.

Exercise 4.7.1 Prove the statement given in this section that the time evolution operator $S(t, t_0) = e^{-(i/\hbar) \int_{t_0}^{t} V_{\text{int}}(t')dt'}$ is unitary, under the assumption that the norm of the wave function $\langle \psi_t | \psi_t \rangle = 1$ at all times.

Exercise 4.7.2 In theoretical work that eventually led to the laser, Einstein considered a set of N two-level atoms, as shown in Figure 4.8, in a box with N_k photons, where

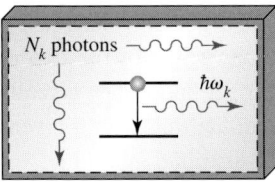

Fig. 4.8 A two-level system in a sealed box with N_k photons.

the energy $\hbar\omega_k$ corresponds to the energy difference between the two electron levels. Based on the rules (4.7.19) above, one can straightforwardly write down **rate equations** for the populations of electrons in the different states and the photons:

$$\frac{\partial \bar{N}_{hi}}{\partial t} = \frac{1}{N}\sum_i \left[-AN_{i,hi}(1 - N_{i,lo})(1 + N_k) + BN_{i,lo}N_k(1 - N_{i,hi}) \right]$$

$$\frac{\partial \bar{N}_{lo}}{\partial t} = \frac{\partial N_k}{\partial t} = \frac{1}{N}\sum_i \left[AN_{i,hi}(1 - N_{i,lo})(1 + N_k) - BN_{i,lo}N_k(1 - N_{i,hi}) \right],$$

$$(4.7.21)$$

where $N_{lo,i}$ and $N_{hi,i}$ are the electron occupation numbers for atom i and the bars indicate their average occupation numbers.

(a) Show that the rules for fermions imply that the $(1 - N)$ factors can be dropped, under the assumption that $N_{i,lo} + N_{i,hi} = 1$.

(b) Show that in equilibrium these equations are satisfied if $A = B$, $\bar{N}_{up}/\bar{N}_{lo} = e^{-\hbar\omega_k/k_BT}$, and N_k is a Planck distribution. (See Section 4.8.1 for a derivation of this distribution from the quantum kinetics.) This was one of the original arguments for stimulated emission.

Exercise 4.7.3   Find the rate of transitions from initial state $|i\rangle$ with N_p phonons in state $\vec{p}$, and N_q in state $\vec{q}$, and none in any other states, to the final many-body state $|f\rangle$ with $N_p - 1$ phonons in state $\vec{p}$, and $N_q - 1$ in state $\vec{q}$, and one phonon in state $\vec{p}'$, for the interaction term

$$V_{\text{int}} = \sum_{\vec{k},\vec{k}',\vec{k}''} \left(A_{\vec{k},\vec{k}'} a^\dagger_{\vec{k}''} a_{\vec{k}'} a_{\vec{k}} + A^*_{\vec{k},\vec{k}'} a^\dagger_{\vec{k}} a^\dagger_{\vec{k}'} a_{\vec{k}''} \right), \qquad (4.7.22)$$

where $\vec{k} + \vec{k}' = \vec{k}''$. This type of term for inelastic phonon scattering will be derived in Section 5.4. Note that the total number of phonons is not conserved.

4.8 The Quantum Boltzmann Equation

In Section 4.7, we deduced the rate for transitions from a single state to one other state, assuming a continuous density of states, namely, Fermi's golden rule. What if there are many particles, all undergoing transitions simultaneously? In this case, it is not obvious how to apply Fermi's golden rule. For example, if two particles collide and scatter via an elastic interaction, which of the two particles should we count as the particle in the initial state? The proper initial state is really the many-body state for all the particles. What we want is a way of starting with a given distribution of particles in different states and evolving the whole distribution in time.

We suppose that the initial state is a superposition of Fock states:

$$|\psi_i\rangle = \sum_n \alpha_n |\psi_n\rangle = \alpha_1 |N_1 N_2, \cdots\rangle + \alpha_2 |N_1', N_2', \cdots\rangle + \cdots . \tag{4.8.1}$$

Note that we do *not* assume an equilibrium distribution of particles; we can use any instantaneous nonequilibrium state. For convenience, in the following we drop the vector notation in the subscripts for the k-states.

We want to know the evolution of the number of particles in state $\vec{k}$ as a function of time. We therefore use the time-dependent perturbation theory of Section 4.7. If the initial state of the system is $|\psi_i\rangle$, and the state of the system at some later time t is $|\psi_t\rangle$, the change in the number of particles is given by

$$
\begin{aligned}
\Delta\langle \hat{N}_k\rangle &= \langle \psi_t | \hat{N}_k | \psi_t\rangle - \langle \psi_i | \hat{N}_k | \psi_i\rangle \\
&= \langle \psi(t) | e^{iH_0 t/\hbar} \hat{N}_k e^{-iH_0 t/\hbar} | \psi(t)\rangle - \langle \psi_i | \hat{N}_k | \psi_i\rangle \\
&= \langle \psi_i | e^{(i/\hbar)\int_0^t V_{\text{int}}(t')dt'} \hat{N}_k e^{-(i/\hbar)\int_0^t V_{\text{int}}(t')dt'} | \psi_i\rangle - \langle \psi_i | \hat{N}_k | \psi_i\rangle \\
&= \langle \psi_i | e^{(i/\hbar)\int_0^t V_{\text{int}}(t')dt'} [\hat{N}_k, e^{-(i/\hbar)\int_0^t V_{\text{int}}(t')dt'}] | \psi_i\rangle .
\end{aligned}
\tag{4.8.2}
$$

The operator $\hat{N}_k$ commutes with H_0 since we assume H_0 is a sum of number operators, of the form (4.2.14). If $\hat{N}_k$ commutes with V_{int}, then there is no change in $\langle \hat{N}_k\rangle$ over time. In the last line of (4.8.2), we have also used the fact that the operator $S(t,0) = e^{(i/\hbar)\int_0^t V_{\text{int}}(t')dt'}$ is unitary, so that its adjoint is its inverse.

Using (4.7.8) we write out the series expansion,

$$
\begin{aligned}
\Delta\langle \hat{N}_k\rangle = \langle \psi_i| \left(1 - (1/i\hbar)\int_0^t V_{\text{int}}(t')dt' + \cdots \right) & \left((1/i\hbar)\int_0^t dt' [\hat{N}_k, V_{\text{int}}(t')] \right. \\
& \left. + (1/i\hbar)^2 \int_0^t dt' \int_0^{t'} dt'' [\hat{N}_k, V_{\text{int}}(t')V_{\text{int}}(t'')] + \cdots \right) |\psi_i\rangle .
\end{aligned}
\tag{4.8.3}
$$

Let us now examine the terms of this series for a specific interaction term V_{int}. In Chapter 5, we will derive many possible interaction terms. Here, we pick a number-conserving interaction term that corresponds to the collision of two particles. This is written as

$$V_{\text{int}} = \frac{1}{2V} \sum_{k_1, k_2, k_3} U_{k_1, k_2, k_3, k_4} a_{k_4}^\dagger a_{k_3}^\dagger a_{k_2} a_{k_1}, \tag{4.8.4}$$

where the summation is not over $\vec{k}_4$ because it is implicitly assumed that momentum is conserved, so that $\vec{k}_4 = \vec{k}_1 + \vec{k}_2 - \vec{k}_3$. We assume the interaction energy U is symmetric on exchange of $\vec{k}_1$ with $\vec{k}_3$ or $\vec{k}_2$ with $\vec{k}_4$.

The lowest-order term of (4.8.3) is found by multiplying the leading term in the right-hand series multiplied by the leading "1" in the left-hand series. This is

$$\frac{1}{i\hbar} \int_0^t dt' \langle \psi_i | [\hat{N}_k, V_{\text{int}}(t')] | \psi_i \rangle \tag{4.8.5}$$

$$= \frac{1}{i\hbar} \int_0^t dt' \langle \psi_i | \left(\hat{N}_k e^{iH_0 t'/\hbar} V_{\text{int}} e^{-iH_0 t'/\hbar} - e^{iH_0 t'/\hbar} V_{\text{int}} e^{-iH_0 t'/\hbar} \hat{N}_k \right) | \psi_i \rangle.$$

The exponential terms with H_0 in (4.8.5) just give factors which depend on the total energy of the states, without changing the number of particles in any of the k-states. The $\hat{N}_k$ operator also does not change the numbers in the states. For the most general form of the initial state (4.8.1), we will therefore have terms of the form

$$\langle \psi_i | a_{k_1}^\dagger a_{k_2}^\dagger a_{k_3} a_{k_4} | \psi_i \rangle = \alpha_1^* \alpha_2 \langle N_1, N_2, \cdots | a_{k_1}^\dagger a_{k_2}^\dagger a_{k_3} a_{k_4} | N_1', N_2', \cdots \rangle + \cdots \tag{4.8.6}$$

This is an example of a **correlation function**, which we will discuss at length in Chapter 9. In a large system with many particles and many different possible configurations, we expect this correlation function to average to nearly zero, because the coefficients α_n are in general complex and can have negative and positive real and imaginary parts. The exception is when the same Fock state is used for both the bra and ket, giving

$$\langle a_{k_1}^\dagger a_{k_2}^\dagger a_{k_3} a_{k_4} \rangle = \sum_n |\alpha_n|^2 \langle \psi_n | a_{k_1}^\dagger a_{k_2}^\dagger a_{k_3} a_{k_4} | \psi_n \rangle, \tag{4.8.7}$$

which is just the weighted average for the probabilities $|\alpha_n|^2$ for each possible Fock state; each of these terms is positive. In this case, however, the commutator $[\hat{N}_k, V_{\text{int}}]$ gives two terms which cancel out for every state n, because $\hat{N}_k$ acts the same to the left or to the right.

We therefore move on to the next order in the expansion (4.8.3). There are two terms which are second order in V_{int}. One of them comes from multiplying the second-order term in the right-hand series by the leading order "1" in the left-hand series. This gives a term proportional to $\langle \psi_i | [\hat{N}_k, V_{\text{int}}(t') V_{\text{int}}(t'')] | \psi_i \rangle$. We expect this term will be negligible for the same reason that the first order was: When two different terms in the superposition of Fock states in (4.8.1) are taken, there will be a high-order correlation function (this time with eight operators) which will average to zero; if the same term is used in both the bra and ket, giving a factor $|\alpha_n|^2$ which is nonzero, the two terms in the commutator will cancel.

The remaining second-order term is

$$\Delta \langle \hat{N}_k \rangle = \langle \psi_i | \frac{1}{\hbar^2} \left(\int_0^t dt' V_{\text{int}}(t') \right) \left(\int_0^t dt'' [\hat{N}_k, V_{\text{int}}(t'')] \right) | \psi_i \rangle. \tag{4.8.8}$$

By the same argument as above, we expect that only terms which couple a Fock state to itself are non-negligible. Inserting a sum over the complete set of all possible Fock states, and using the definition of the interaction terms (4.7.4), we then have

$$\Delta \langle \hat{N}_k \rangle = \sum_n |\alpha_n|^2 \sum_m \frac{1}{\hbar^2} \left(\int_0^t dt' e^{(i/\hbar)(E_n - E_m)t'} \right) \left(\int_0^t dt'' e^{(i/\hbar)(E_m - E_n)t''} \right)$$

$$\times \langle \psi_n | V_{\text{int}} | \psi_m \rangle \langle \psi_m | [\hat{N}_k, V_{\text{int}}] | \psi_n \rangle$$

$$= \sum_n |\alpha_n|^2 \sum_m \left(\frac{e^{(i/\hbar)(E_n - E_m)t} - 1}{E_n - E_m} \right) \left(\frac{e^{-(i/\hbar)(E_n - E_m)t} - 1}{E_n - E_m} \right)$$

$$\times \langle \psi_n | V_{\text{int}} | \psi_m \rangle \langle \psi_m | [\hat{N}_k, V_{\text{int}}] | \psi_n \rangle, \tag{4.8.9}$$

where $|\psi_n\rangle$ and $|\psi_m\rangle$ are Fock states. The time-dependent factors are resolved using the identity (4.7.12). Although this identity assumes the limit $t \to \infty$, we can still consider a time interval dt which is short enough that the change in the population $\Delta \langle \hat{N}_k \rangle$ is small. We therefore have

$$\boxed{\frac{d \langle \hat{N}_k \rangle}{dt} = \sum_n |\alpha_n|^2 \frac{2\pi}{\hbar} \sum_m \langle \psi_n | V_{\text{int}} | \psi_m \rangle \langle \psi_m | [\hat{N}_k, V_{\text{int}}] | \psi_n \rangle \delta(E_n - E_m).} \tag{4.8.10}$$

This is the most general equation for the change of the expectation value of the occupation number N_k in time, for the general initial state (4.8.1).

Let us now calculate the commutator for the interaction (4.8.4). We can resolve the commutator in (4.8.10) by using the relations

$$[\hat{N}_k, a_{k'}] = -a_k \delta_{k,k'}$$

$$[\hat{N}_k, a_{k'}^\dagger] = a_k^\dagger \delta_{k,k'}, \tag{4.8.11}$$

which are valid, surprisingly, for both boson and fermion creation and destruction operators. For the four-operator term in the interaction (4.8.4), we have

$$\hat{N}_k a_{k_4}^\dagger a_{k_3}^\dagger a_{k_2} a_{k_1} = a_{k_4}^\dagger a_{k_3}^\dagger a_{k_2} a_{k_1} \delta_{k,k_4} + a_{k_4}^\dagger a_{k_3}^\dagger a_{k_2} a_{k_1} \delta_{k,k_3}$$

$$- a_{k_4}^\dagger a_{k_3}^\dagger a_{k_2} a_{k_1} \delta_{k,k_2} - a_{k_4}^\dagger a_{k_3}^\dagger a_{k_2} a_{k_1} \delta_{k,k_1}$$

$$+ a_{k_4}^\dagger a_{k_3}^\dagger a_{k_2} a_{k_1} \hat{N}_k. \tag{4.8.12}$$

Thus

$$[\hat{N}_k, V_{\text{int}}] = \frac{1}{2V} \sum_{k_1, k_2} \left(U_{k_1, k_2, k', k} \, a_k^\dagger a_{k'}^\dagger a_{k_2} a_{k_1} + U_{k_1, k_2, k, k'} \, a_{k'}^\dagger a_k^\dagger a_{k_2} a_{k_1} \right.$$

$$\left. - U_{k', k, k_2, k_1} \, a_{k_1}^\dagger a_{k_2}^\dagger a_k a_{k'} - U_{k, k', k_2, k_1} \, a_{k_1}^\dagger a_{k_2}^\dagger a_{k'} a_k \right)$$

$$= \frac{1}{2V} \sum_{k_1, k_2} (U_D \pm U_E)(a_k^\dagger a_{k'}^\dagger a_{k_2} a_{k_1} - a_{k_1}^\dagger a_{k_2}^\dagger a_k a_{k'}), \tag{4.8.13}$$

where $\vec{k}' = \vec{k}_1 + \vec{k}_2 - \vec{k}$, and $U_D = U_{k_1, k_2, k', k}$ is the direct interaction term and $U_E = U_{k_1, k_2, k, k'}$ is the exchange term, and the $+$ sign is for bosons and the $-$ sign is for fermions. (We assume $U_{k_1, k_2, k', k} = U_{k, k', k_2, k_1}$, which is to say, we assume time-reversal symmetry.) For hard-sphere scattering, also known as s-wave scattering, U is a constant, which means s-wave scattering is enhanced by a factor of 4 for bosons and is forbidden for fermions.

Since the Fock states are orthonormal, the matrix element $\langle\psi_m|[\hat{N}_k, V_{\text{int}}]|\psi_n\rangle$ for a given term in the sum (4.8.13) determines the state $|\psi_m\rangle$, so that we no longer sum over all final states $|\psi_m\rangle$ in (4.8.10). The summation over k-states in the definition (4.8.4) of V_{int} is eliminated in the term $\langle\psi_n|V_{\text{int}}|\psi_m\rangle$ in (4.8.10) because only four terms which couple $|\psi_n\rangle$ to $|\psi_m\rangle$ survive. Since a destruction operator a_k acting to the right on a state with N_k particles in the single-particle state $\vec{k}$ gives a factor $\sqrt{N_k}$, and a creation operator $a_k^\dagger$ gives a factor $\sqrt{1 \pm N_k}$ (where the + sign is for bosons and the − sign is for fermions), we have

$$\frac{d\langle N_k\rangle}{dt} = \frac{2\pi}{\hbar} \frac{1}{2V^2} \sum_{k_1,k_2} (U_D \pm U_E)^2 \delta(E_{k_1} + E_{k_2} - E_k - E_{k'}) \tag{4.8.14}$$
$$\times \langle N_{k_1} N_{k_2}(1 \pm N_k)(1 \pm N_{k'}) - N_k N_{k'}(1 \pm N_{k_1})(1 \pm N_{k_2})\rangle,$$

where the average $\langle\ldots\rangle$ is the weighted average over the Fock states $|\psi_n\rangle$ which are included in the initial state $|\psi_i\rangle$. We have dropped the hat on N_k on the left side because $\langle\psi_i|\hat{N}_k|\psi_i\rangle$ is equal to $\langle N_k\rangle$ for the same weighted average. Equation (4.8.15) has the same condition of validity as Fermi's golden rule, namely the time scale for depletion of any given state must be long compared to $\hbar/\Delta E_f$, where ΔE_f is the range of final states which can be considered smooth.

We can make a very powerful approximation by treating the average of the product which occurs in (4.8.15) as a product of averages:

$$\langle N_{k_1} N_{k_2}(1 \pm N_k)(1 \pm N_{k'}) - N_k N_{k'}(1 \pm N_{k_1})(1 \pm N_{k_2})\rangle$$
$$\simeq \langle N_{k_1}\rangle\langle N_{k_2}\rangle(1 \pm \langle N_k\rangle(1 \pm \langle N_{k'}\rangle)) - \langle N_k\rangle\langle N_{k'}\rangle(1 \pm \langle N_{k_1}\rangle)(1 \pm \langle N_{k_2}\rangle). \tag{4.8.15}$$

This relies on the assumption that there are no special correlations between the occupation numbers of different states; this in turn follows from the assumption that the correlation functions $\langle a_{k_1}^\dagger a_{k_2}^\dagger a_{k_3} a_{k_4}\rangle$ are negligible. (This is shown in Snoke *et al.* 2012.) We therefore can write the equation entirely in terms of the distribution function $\langle N_k\rangle$ for the k-states:

$$\boxed{\begin{aligned}\frac{d\langle N_k\rangle}{dt} = \frac{2\pi}{\hbar}\frac{1}{2V^2} \sum_{\vec{k}_1,\vec{k}_2} (U_D \pm U_E)^2 \big(\langle N_{k_1}\rangle\langle N_{k_2}\rangle(1 \pm \langle N_k\rangle)(1 \pm \langle N_{k'}\rangle) \\ - \langle N_k\rangle\langle N_{k'}\rangle(1 \pm \langle N_{k_1}\rangle)(1 \pm \langle N_{k_2}\rangle)\big)\,\delta(E_{k_1} + E_{k_2} - E_k - E_{k'}).\end{aligned}}$$

$$\tag{4.8.16}$$

This is the **quantum Boltzmann equation** for two-body scattering, which gives the total rate of change of the probability distribution.[2] We see once again that we have the same final states factors $(1 \pm N_k)$ as we did in computing the transition rates for single particles using Fermi's golden rule.

Exercise 4.8.1   Prove the relations (4.8.11) using the known commutation relation $[a, a^\dagger] = 1$ for bosons and the anticommutation relation $\{a, a^\dagger\} = 1$ for fermions.

---

[2] Technically, this is a **Fokker–Planck** equation, and a Boltzmann equation includes spatial evolution (see Section 5.11). However, because of the role this integral plays in the Boltzmann transport equation, this is commonly called the quantum Boltzmann equation, and we defer to this nomenclature.

Exercise 4.8.2   Write down the quantum Boltzmann equation for the evolution of a population of fermionic particles interacting with a population of phonons through an interaction term of the form

$$V_{\text{int}} = \sum_{k,k_1} U_k (a_k b^\dagger_{k_1+k} b_{k_1} + a^\dagger_k b^\dagger_{k_1-k} b_{k_1}), \tag{4.8.17}$$

where $a^\dagger_{k_3}$ is the (bosonic) phonon creation operator, and k_3 is determined by momentum conservation. (The form of this term will be derived in Chapter 5.) This interaction term corresponds to a change of the momentum of an electron due to phonon emission; the number of phonons is not conserved.

4.8.1 Equilibrium Distributions of Quantum Particles

An important implication of the quantum Boltzmann equation (4.8.10) is that it is *deterministic*. Given any initial distribution $\langle N_k \rangle$ for all $\vec{k}$, we can, in principle, find the distribution at all later times t by solution of this equation. This provides an important connection with the field of thermodynamics. Historically, at the time of Maxwell and Boltzmann there was considerable debate over the so-called H-theorem, which says that a system far from equilibrium must approach equilibrium. Boltzmann justified this assumption using a classical rate equation, which he justified by a statistical argument that was never fully accepted (contributing eventually to Boltzmann's suicide). The quantum Boltzmann equation gives a deterministic evolution of the quantum wave function, however. One may interpret the wave function statistically, but the evolution of the expectation values of the particle occupation numbers is not random – they move deterministically toward equilibrium.

Figure 4.9 shows the numerical solution of the quantum Boltzmann equation for an isotropic, nonequilibrium distribution of bosons at low density, scattering with each other via a two-body interaction of the form used in the previous section. Since the system is isotropic, that is, the distribution function $\langle N_k \rangle$ depends only on the magnitude of $\vec{k}$ and not on the direction, we can write the distribution function simply as $f(E_k) = \langle N_k \rangle$. In this case, the angles in the integral (4.8.16) can be eliminated by analytical integration, to reduce this integral to just a double integral over two energies, which can then be computed numerically. Figure 4.9 has been generated by computing the rate $df(E)/dt$ for all E at each point in time, and then updating each $f(E)$ according to

$$f(E) \rightarrow f(E) + \frac{df(E)}{dt} dt, \tag{4.8.18}$$

where dt is some small time interval, chosen such that the change of $f(E)$ is small during any one update. According to the calculation shown in Figure 4.9, after the particles have scattered an average of five times each, the distribution does not change significantly. The system has come to equilibrium.

We can easily determine the equilibrium distribution by looking at the quantum Boltzmann equation, which is to say, using Fermi's golden rule. In equilibrium, we must have

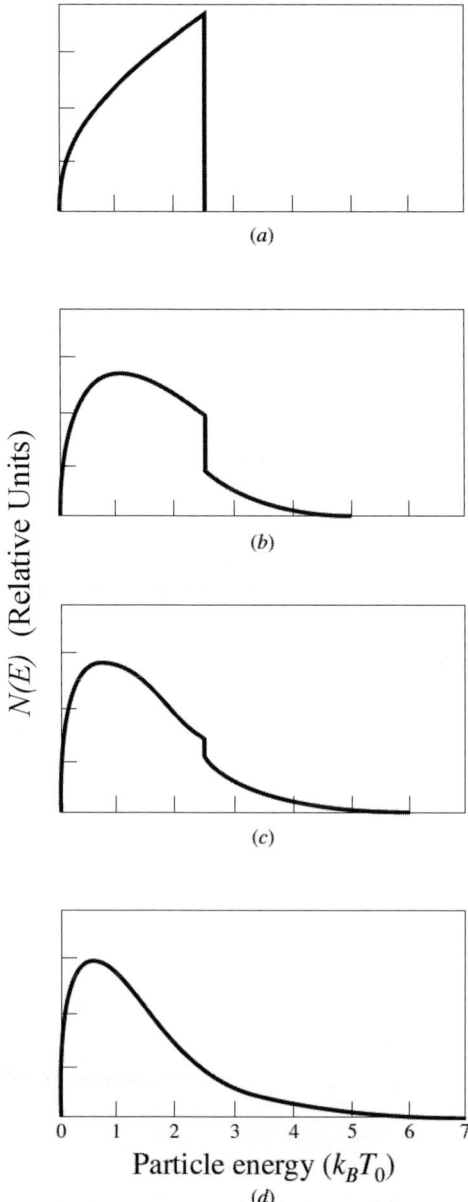

Fig. 4.9 Nonequilibrium evolution of an isotropic gas of bosons at low density interacting with the potential (4.8.4), found by numerically solving (4.8.16). (a) The initial state corresponds to equal occupation of all states up to a cutoff, in other words, the number of particles $N(E)$ is proportional to the density of states $D(E) \propto \sqrt{E}$. The horizontal energy axis is given in units of $k_B T_0$, where T_0 is the final temperature. (b) After an average of one scattering event per particle. (c) After an average of two scattering events per particle. (d) After four scattering events per particle. From Snoke and Wolfe (1989).

$$\frac{df(E_k)}{dt} = 0 \tag{4.8.19}$$

for all $\vec{k}$. This will occur if for all the different scattering processes,

$$f(E_{k_3})f(E_{k_2})(1 \pm f(E_{k_1}))(1 \pm f(E_k)) - f(E_k)f(E_{k_1})(1 \pm f(E_{k_2}))(1 \pm f(E_{k_3})) = 0. \tag{4.8.20}$$

This is the principle of **detailed balance**. It can easily be verified that a distribution of the form

$$f(E) = \frac{1}{e^{\alpha + \beta E} \mp 1} \tag{4.8.21}$$

satisfies this condition, where the $-$ sign is for bosons and the $+$ sign for fermions. The constants α and β are determined by the conditions

$$N = \sum_k f(E_k) = \int f(E)\mathcal{D}(E)dE \tag{4.8.22}$$

and

$$U = \sum_k E_k f(E_k) = \int E f(E)\mathcal{D}(E)dE. \tag{4.8.23}$$

We can equate α and β to standard thermodynamics quantities by noting that if $\alpha \gg 1$, the equilibrium distribution is equal to $f(E) = e^{-\alpha}e^{-\beta E}$. Equating this to the standard equilibrium distribution $f(E) = e^{\mu/k_B T}e^{-E/k_B T}$, we then have $\beta = 1/k_B T$ and $\alpha = -\mu/k_B T$, or

$$f(E) = \frac{1}{e^{(E-\mu)/k_B T} \mp 1}, \tag{4.8.24}$$

which is the well-known equilibrium distribution for quantum particles; when the sign is negative it is called the Bose–Einstein distribution and when the sign is positive it is called the Fermi–Dirac distribution. This derivation using the quantum Boltzmann equation shows in a natural way why the boson and fermion equilibrium formulas are the same except for the sign in the denominator. It is easy to see that the equilibrium solution of the quantum Boltzmann equation will be the same even if we invoke some other process besides the two-body elastic scattering mechanism assumed here.

In the limit $\mu \ll E_0$, which corresponds to low particle density, both the Fermi–Dirac and the Bose–Einstein distribution equal $f(E) = e^{-(E-\mu)/k_B T}$, which is known as the Maxwell–Boltzmann distribution. The equilibrium solution shown in Figure 4.9 for the distribution of the particles at late times converges to this Maxwell–Boltzmann distribution times the density of states of the particles, which for the three-dimensional gas in the model is proportional to $\sqrt{E}$, as discussed in Section 1.8.1. Note that Boltzmann's constant k_B does not appear in (4.8.15), however. The energy distribution at late times is determined entirely by the constraints of energy and number conservation. The constant k_B just gives the constant of proportionality between the energy per particle and the temperature scale we have chosen.

In the above discussion, we have used the example of a two-body scattering mechanism with the interaction energy (4.8.4). The general conclusions we have made about equilibrium do not depend on the form of the interaction, however. We could just as easily have used a different interaction, such as electron–phonon scattering, or one of the other scattering mechanisms which we will discuss in Chapter 5. Figure 4.10 shows an example of the evolution of a population of excitons in a semiconductor, evolving by exciton–phonon interaction. Modern ultrafast optics experiments can resolve the energy distribution of carriers on time scales that are short compared to the time to equilibrate. Therefore, we can observe the approach to equilibrium according to the quantum Boltzmann equation. As seen in Figure 4.10, the *path* to equilibrium is quite different from the two-body scattering, but the end result is still a Maxwell–Boltzmann distribution.

Note that the deterministic nature of the quantum Boltzmann equation does not mean that there are no fluctuations. The average value $\langle N_k \rangle$ evolves deterministically, but even if the initial state is a Fock state with a definite number of particles in each state, the state of the system at a later time may be a superposition of different Fock states, and the value $\langle (\hat{N}_k - \langle N_k \rangle)^2 \rangle$ may be nonzero.

Exercise 4.8.3   Show that (4.8.21) satisfies the detailed balance condition (4.8.20). A critical step in the proof is to invoke energy conservation, $E_k + E_{k_1} = E_{k_2} + E_{k_3}$.

4.8.2 The H-Theorem and the Second Law

We know that quantum mechanics is time-reversible on the microscopic scale. Yet we have derived an equation with irreversibility. How did the irreversibility enter in? Clearly it must lie in an approximation we have made. It does *not* stem from using second-order perturbation theory, however. Nor does it stem from using the $t \to \infty$ limit for small changes in time.

In deriving the result (4.8.16), we dropped correlation functions of the form $\rho_{k_1,k_2,k_4,k_4} \equiv \langle a_{k_1}^\dagger a_{k_2}^\dagger a_{k_3} a_{k_4} \rangle$, arguing that they are negligible. This amounted to keeping just the information about the occupation numbers $\langle \hat{N}_k \rangle$. The correlation functions cannot be strictly zero, however. We can write an evolution equation similar to (4.8.2),

$$\Delta \langle \rho_{k_1,k_2,k_4,k_4} \rangle = \langle \psi_t | \rho_{k_1,k_2,k_4,k_4} | \psi_t \rangle - \langle \psi_i | \rho_{k_1,k_2,k_4,k_4} | \psi_i \rangle \tag{4.8.25}$$
$$= \langle \psi(t) | \rho_{k_1,k_2,k_4,k_4}(t) | \psi(t) \rangle - \langle \psi_i | \rho_{k_1,k_2,k_4,k_4} | \psi_i \rangle.$$

It can be shown that this evolution equation leads to nonzero values for $\langle \rho_{k_1,k_2,k_4,k_4} \rangle$ even if the initial state is a pure Fock state. However, under normal circumstances, the magnitude of $\langle \rho_{k_1,k_2,k_4,k_4} \rangle$ will remain very small.[3] The quantum Boltzmann equation approach amounts to setting all these correlation terms strictly to zero after each time step. This amounts to an erasure of information, which ultimately leads to the irreversibility in the system.

The quantum Boltzmann equation implies Boltzmann's "H-theorem," which is the basis of the Second Law of Thermodynamics, namely, that entropy never decreases in a closed

---

[3]   The exception is bosonic systems at high occupation number, such as a Bose–Einstein condensate or a coherent sound wave. In such cases, macroscopic coherence can occur. See Snoke *et al.* (2012).

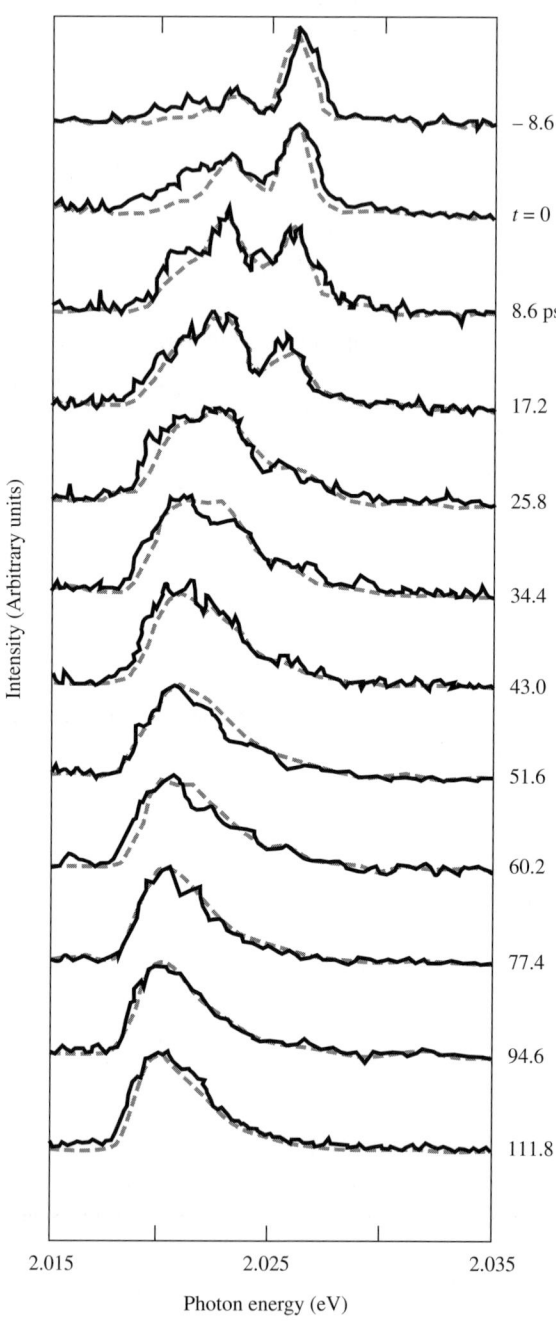

Fig. 4.10 Solid lines: energy distribution of excitons in the semiconductor Cu_2O measured at various times following a short (2 ps) laser pulse. Dashed lines: solution of the quantum Boltzmann equation for the time evolution of the population via phonon scattering (see Section 5.1). From Snoke *et al.* (1991).

system. We can define a semiclassical entropy in terms of the distribution function of the particles. For classical particles, this is

$$S = -k_B \sum_k \langle \hat{N}_k \rangle \ln\langle N_k \rangle. \tag{4.8.26}$$

For fermions and bosons, this is modified to (see, e.g., Band 1955)

$$S = -k_B \sum_k \left(\langle \hat{N}_k \rangle \ln\langle N_k \rangle \mp (1 \pm \langle N_k \rangle) \ln(1 \pm \langle N_k \rangle) \right), \tag{4.8.27}$$

where the upper sign is for bosons and the lower sign is for fermions. Assuming conservation of the total number of particles, the time derivative of this is

$$\frac{\partial S}{\partial t} = -k_B \sum_k \frac{\partial \langle N_k \rangle}{\partial t} \ln\left(\frac{\langle N_k \rangle}{1 \pm \langle N_k \rangle} \right). \tag{4.8.28}$$

The time derivative of $\langle N_k \rangle$ is given by the quantum Boltzmann equation (4.8.16), so that we have

$$\frac{\partial S}{\partial t} = -k_B \sum_{k,k_1,k_2} C \ln\left(\frac{\langle N_k \rangle}{1 \pm \langle N_k \rangle} \right) \Big[\langle N_{k_1} \rangle \langle N_{k_2} \rangle (1 \pm \langle N_k \rangle)(1 \pm \langle N_{k'} \rangle)$$
$$- \langle N_k \rangle \langle N_{k'} \rangle (1 \pm \langle N_{k_1} \rangle)(1 \pm \langle N_{k_2} \rangle) \Big], \tag{4.8.29}$$

where C is a positive, real factor that contains the matrix element and conservation of energy δ-function.

If we pick four particular states $\vec{k}, \vec{k}', \vec{k}_1$, and $\vec{k}_2$, then the total of all terms in the sum involving the same four states is

$$\left(\ln\left(\frac{\langle N_k \rangle}{1 \pm \langle N_k \rangle} \right) + \ln\left(\frac{\langle N_{k'} \rangle}{1 \pm \langle N_{k'} \rangle} \right) - \ln\left(\frac{\langle N_{k_1} \rangle}{1 \pm \langle N_{k_1} \rangle} \right) - \ln\left(\frac{\langle N_{k_2} \rangle}{1 \pm \langle N_{k_2} \rangle} \right) \right)$$
$$\times \Big[\langle N_{k_1} \rangle \langle N_{k_2} \rangle (1 \pm \langle N_k \rangle)(1 \pm \langle N_{k'} \rangle) - \langle N_k \rangle \langle N_{k'} \rangle (1 \pm \langle N_{k_1} \rangle)(1 \pm \langle N_{k_2} \rangle) \Big]$$
$$= \ln\left(\frac{\langle N_k \rangle \langle N_{k'} \rangle (1 \pm \langle N_{k_1} \rangle)(1 \pm \langle N_{k_2} \rangle)}{\langle N_{k_1} \rangle \langle N_{k_2} \rangle (1 \pm \langle N_k \rangle)(1 \pm \langle N_{k'} \rangle)} \right) \tag{4.8.30}$$
$$\times \Big[\langle N_{k_1} \rangle \langle N_{k_2} \rangle (1 \pm \langle N_k \rangle)(1 \pm \langle N_{k'} \rangle) - \langle N_k \rangle \langle N_{k'} \rangle (1 \pm \langle N_{k_1} \rangle)(1 \pm \langle N_{k_2} \rangle) \Big].$$

If the in-scattering term in the square brackets is larger than the out-scattering term, then the denominator of the logarithm is larger than the numerator, making the logarithm negative. Conversely, if the in-scattering is less than the out-scattering term, the term in the square brackets is negative. Since the whole sum in (4.8.29) consists of terms like this, the total sum is less than or equal to zero, and therefore $\partial S/\partial t > 0$. This is the standard form of the H-theorem.

Time reversal. It is instructive to ask what we would find for a quantum Boltzmann equation going backward in time. Analogous to the derivation starting with (4.8.2), we would have

$$\Delta\langle \hat{N}_k \rangle = \langle \psi_{-t} | \hat{N}_k | \psi_{-t} \rangle - \langle \psi_i | \hat{N}_k | \psi_i \rangle. \tag{4.8.31}$$

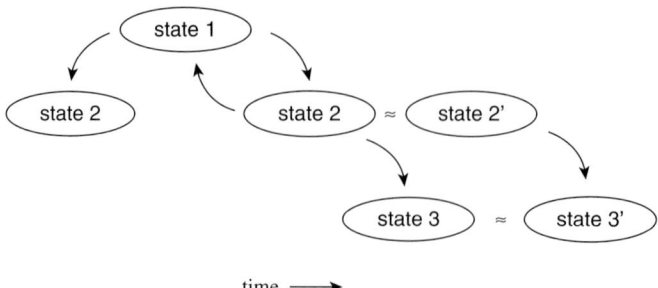

Fig. 4.11 Approximation steps in the iterative method of evolving the many-body wave state using the quantum Boltzmann equation.

The time dependence comes entirely in the formula (4.7.12),

$$\lim_{t\to\infty} \frac{\left(e^{-ixt} - 1\right)\left(e^{ixt} - 1\right)}{x^2} = \lim_{t\to\infty} \frac{\sin^2(xt/2)}{x^2}$$
$$= \delta(x)2\pi |t|, \qquad (4.8.32)$$

which is manifestly symmetric in time. Suppose that we have an initial state $|\psi_i\rangle$, which evolves to a higher-entropy state forward in time. If we had no other knowledge, we would also have to assume that it evolved *from* a higher-entropy state at an earlier time. Yet we know that this never happens. How can we resolve this apparent contradiction?

The loss of reversibility in the quantum Boltzmann equation is illustrated in Figure 4.11. The system starts in state 1, in which all of the correlation functions such as ρ_{k_1,k_2,k_3,k_4}, defined above, are strictly zero. The true evolution of the full quantum mechanical solution gives state 2, which has nonzero values for these correlation functions. If this state 2 were run backwards in time, the true evolution would take this state back to state 1. However, in the iterated quantum Boltzmann equation approach, we approximate state 2 by state 2′, which has all correlation functions strictly equal to zero. State 2′ then evolves to state 3′, which is a good approximation of state 3, the real state reached by state 2. Thus, if we continue forward in time, we have a series of successive approximations which are very good approximations of the evolution of the full quantum mechanical solution. If we evolve backward in time, however, our dropping of the correlation information means that the backward evolution will not be a good approximation of the real backwards evolution. The tiny values of the correlation functions, which are dropped in the iterated Boltzmann approach, carry information that is crucial for recovering the true time-reversed behavior.

4.9 Energy Density of Solids

From the equilibrium distributions derived in Section 4.8.1, we can calculate the equilibrium distribution of phonons and photons. Since these are bosons, they have the

Bose–Einstein distribution in equilibrium. These particles are not subject to the number conservation condition (4.8.22), however. In this case, entropy is maximized for the largest number of particles, which implies $\alpha = 0$, and we have simply

$$f(E) = \frac{1}{e^{E/k_B T} - 1},\tag{4.9.1}$$

which is known as the Planck distribution.

The total energy of a system is given simply by the formula

$$U = \int_0^\infty E f(E) \mathcal{D}(E) dE,\tag{4.9.2}$$

and heat capacity per volume, at constant volume, is given by the relation

$$\frac{C_V}{V} = \frac{1}{V}\frac{\partial U}{\partial T}.\tag{4.9.3}$$

The specific heat is the energy change per unit mass instead of per unit volume. By using the Planck distribution for phonons and photons, knowing their density of states, we can make definite predictions for the total energy and heat capacity of solids.

4.9.1 Density of States of Phonons and Photons

In Sections 2.2 and 2.8.1, we deduced the density of states for various electron systems. We can do the same for phonons and photons.

For an isotropic system, such as a vacuum, we have from (1.8.5),

$$\mathcal{D}(E) dE = \frac{V}{(2\pi)^3} 4\pi \, dE \, \frac{1}{|\nabla_{\vec{k}} E|} k^2(E).\tag{4.9.4}$$

For acoustic phonons at low frequency, and for photons in an isotropic system, we have the simple relation

$$E = \hbar\omega = \hbar v k,\tag{4.9.5}$$

where v is the phase velocity of the wave. The gradient is therefore a constant, $|\nabla_{\vec{k}} E| = \hbar v$, which yields

$$\begin{aligned} \mathcal{D}(E) dE &= \frac{V}{(2\pi)^3} 4\pi \, dE \, \frac{1}{\hbar v} \frac{E^2}{(\hbar v)^2} \\ &= \frac{V}{2\pi^2 \hbar^3 v^3} E^2 dE.\end{aligned}\tag{4.9.6}$$

In the case of photons, we multiply this number by a factor of 2, to take into account the fact that there are two allowed polarizations, and use $v = c/n$, where n is the index of refraction. This yields

$$\mathcal{D}(E) dE = \frac{V n^3}{\pi^2 \hbar^3 c^3} E^2 dE.\tag{4.9.7}$$

Similarly, for acoustic phonons at low temperature, if we assume that all three phonon branches have approximately the same sound speed, then we have

$$\mathcal{D}(E)dE = \frac{3V}{2\pi^2\hbar^3 v^3}E^2 dE. \tag{4.9.8}$$

In the case of phonons, the gradient goes to zero at the Brillouin zone boundary, as discussed in Section 3.9. This means that, like the electronic density of states, there will be a van Hove singularity in the phonon density of states.

4.9.2 Planck Energy Density

Setting $E = \hbar\omega$, the total radiant energy at a given photon frequency is equal to

$$U(\omega)d\omega = (\hbar\omega)f(\omega)\mathcal{D}(\omega)d\omega, \tag{4.9.9}$$

or

$$U(\omega)d\omega = \frac{1}{e^{\hbar\omega/k_B T} - 1}\left(\frac{\hbar V n^3}{\pi^2 c^3}\right)\omega^3 d\omega. \tag{4.9.10}$$

Integrating over all frequencies, the total energy is therefore

$$U = \int_0^\infty U(\omega)d\omega = \frac{1}{15}\left(\frac{\pi^2 V n^3}{\hbar^3 c^3}\right)(k_B T)^4. \tag{4.9.11}$$

This is the **Stefan–Boltzmann** law. It implies that any object has electromagnetic radiation with a total energy which depends on the temperature of the object. The T^4 dependence is quite strong.

Moreover, the Planck distribution implies that an object at a fixed temperature will radiate photons into empty space. This is known as **blackbody** radiation. For a surface element of area dA, the average velocity of photons perpendicular to the element is $\langle v\cos\theta\rangle_\theta = \frac{1}{2}v$, where $v = c/n$ is the speed of light in the medium. Since at any point in time half of the radiant energy is moving toward the surface, on average, the total radiant emission per area will be

$$I = \frac{dE}{dAdt} = \left(\tfrac{1}{4}c/n\right)\frac{U}{V} = \frac{1}{60}\left(\frac{\pi^2 n^2}{\hbar^3 c^2}\right)(k_B T)^4. \tag{4.9.12}$$

The Planck radiation spectrum was taken early in the history of quantum mechanics as definite evidence of the particle nature of light. In the modern picture, the Planck radiation spectrum comes from the Bose–Einstein energy distribution, which as we have seen, comes from the commutation properties of the creation and destruction operators $a_{\vec{k}}$ and $a_{\vec{k}}^\dagger$. These commutation properties in turn come from the commutation relation $[x, p] = i\hbar$ used for the oscillator, as discussed in Section 4.1. As discussed in that section, this relation comes simply from the properties of Fourier transforms, when p is identified with the spatial frequency k. The Planck spectrum can therefore be seen not so much as arising from the particle nature of light as from the wave nature of electrons and other oscillators.

Exercise 4.9.1   Convert (4.9.10) into an equation for the radiant energy per unit wavelength for a blackbody radiation source. From this equation, determine the peak emission wavelength, that is, the wavelength at which the radiant energy per wavelength is maximum, for $T = 300$ K.

4.9.3 Heat Capacity of Phonons

Just as we did for photons, we can write down the total energy for acoustic phonons. Using the energy $E = \hbar v k$ and the density of states (4.9.8), we have

$$U = \int_0^\infty Ef(E)\mathcal{D}(E)dE = \frac{1}{10}\left(\frac{\pi^2 V}{\hbar^3 v^3}\right)(k_B T)^4. \tag{4.9.13}$$

The heat capacity is given by

$$\frac{C_V}{V} = \frac{1}{V}\frac{\partial U}{\partial T} = \frac{2}{5}\left(\frac{\pi^2}{\hbar^3 v^3}\right)k_B^4 T^3. \tag{4.9.14}$$

This formula works well at low temperature, but it breaks down at high temperature because the upper bound of the integration cannot really be infinity; the phonons have an upper maximum energy at the Brillouin zone boundary.

In general, to calculate the phonon energy properly, we must use the proper dispersion relation of the phonons in the Brillouin zone. We can make a simple approximation, however, by assuming that the acoustic phonon branch remains linear up to the Brillouin zone boundary. This is known as the **Debye** approximation. Instead of worrying about the complexities of the shape of the Brillouin zone boundary, we can approximate that the Brillouin zone is a sphere in reciprocal lattice space with radius k_D, which is chosen such that the number of k-states in the sphere equals the number of k-states in the actual Brillouin zone. As discussed in Section 1.7, the volume of k-space per state is $(2\pi)^3/V$, and therefore the total number of states in a sphere of radius k_D is equal to the total volume of the sphere divided by this number,

$$N = \frac{\frac{4}{3}\pi k_D^3}{(2\pi)^3/V} = V\frac{k_D^3}{6\pi^2}. \tag{4.9.15}$$

Switching variables in the integral from ω to k, we can write the total energy as

$$U = \int_0^{k_D} Ef(E)\mathcal{D}(E)dE = \int_0^{k_D}(\hbar v k)\frac{1}{e^{\hbar v k/k_B T} - 1}\left(\frac{3V}{2\pi^2}\right)k^2 dk. \tag{4.9.16}$$

At high T, when $k_B T \gg \hbar v k$, that is, when the $k_B T$ is much larger than the energy of a phonon at the Brillouin zone boundary, the occupation number of the phonons is just

$$\frac{1}{e^{\hbar v k/k_B T} - 1} = \frac{1}{1 + \hbar v k/k_B T + \cdots - 1} \simeq \frac{k_B T}{\hbar v k}, \tag{4.9.17}$$

which implies

$$U = \int_0^{k_D} (\hbar v k) \frac{k_B T}{\hbar v k} \left(\frac{3V}{2\pi^2}\right) k^2 dk$$

$$= k_B T \left(\frac{V}{2\pi^2}\right) k_D^3. \tag{4.9.18}$$

Inserting the definition of k_D from (4.9.15), we have

$$U = k_B T \left(\frac{V}{2\pi^2}\right) \frac{6\pi^2 N}{V}$$

$$= 3k_B T N \tag{4.9.19}$$

and therefore

$$\frac{C_V}{V} = \frac{3k_B N}{V} = \text{const.} \tag{4.9.20}$$

This is the same heat capacity we would get if we simply assumed that we had N isolated oscillators. By the equipartition theorem of thermodynamics, each degree of freedom in a system has energy $k_B T/2$. For a harmonic oscillator with three degrees of freedom in x and three degrees of freedom in p, the total energy per oscillator in equilibrium is therefore $3k_B T$. The total energy of N oscillators is therefore $3N k_B T$. This is known as the **Dulong–Petit** approximation. It implies that the heat capacity of a solid at high temperature is essentially constant.

We can do one better by including the optical phonon branches, if there are any, which have approximately constant energy across the Brillouin zone. The **Einstein** approximation takes the optical phonon as having constant energy. For the simple case in which there is one optical phonon branch with the same energy as the acoustic branch at the Brillouin zone boundary, we have

$$U = \int_0^{k_D} \left[(\hbar v k) \frac{1}{e^{\hbar v k/k_B T} - 1} + (\hbar v k_D) \frac{1}{e^{\hbar v k_D/k_B T} - 1} \right] \left(\frac{3V}{2\pi^2}\right) k^2 dk. \tag{4.9.21}$$

In the high-temperature limit, this formula just gives $C_V/V = 6k_B N/V$, because there are now twice as many degrees of freedom, namely two atoms per unit cell. Recall from Section 3.1.2 that optical phonon branches arise when there is more than one atom per unit cell.

The parameter k_D is sometimes given in terms of a **Debye temperature**, Θ_D defined by

$$k_B \Theta_D = \hbar v k_D. \tag{4.9.22}$$

From (4.9.14), we then have in the low-temperature limit,

$$\frac{C_V}{V} = \frac{2}{5} \left(\frac{\pi^2}{(k_B \Theta_D/k_D)^3} \right) k_B^4 T^3$$

$$= \frac{2}{5} \pi^2 k_D^3 k_B \left(\frac{T}{\Theta_D}\right)^3$$

$$= \frac{12}{5} \pi^4 \frac{k_B N}{V} \left(\frac{T}{\Theta_D}\right)^3. \tag{4.9.23}$$

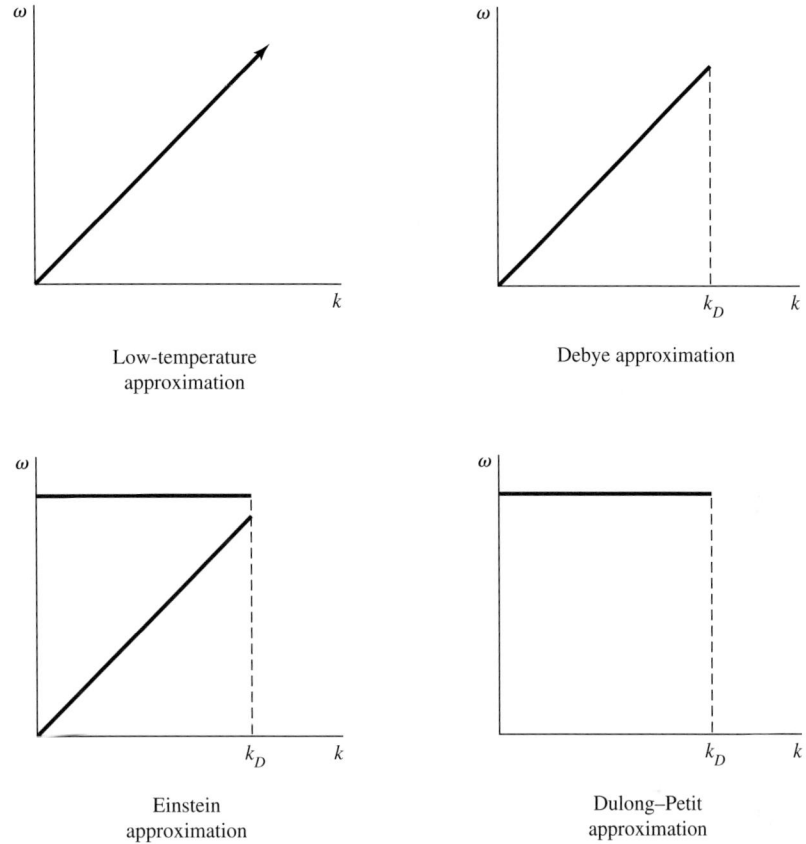

Fig. 4.12   The phonon dispersion used in the various approximations of the heat capacity of the phonons in a solid.

The Debye temperature effectively gives the crossover temperature between the low-temperature and the high-temperature approximations.

The temperature dependences calculated here are borne out by experiments. Figure 4.13 shows a typical heat capacity curve as a function of temperature. At low temperature, the heat capacity is proportional to T^3, while at high temperature it flattens.

A heat capacity for photons can be calculated from (4.9.11) in the same way. It should be obvious that at low temperature the phonon heat capacity is much larger than the photon heat capacity, because the photon speed which goes into the total energy is so much larger, and it is raised to the third power in the denominator. The photon frequency does not have an upper bound, however, while the phonon frequency does, and so at very high temperatures the photon energy can dominate over the sound energy (though the material will probably no longer be a solid).

Exercise 4.9.2   Calculate the phonon specific heat, in units of J/K-kg, of a solid with speed of sound $v = 4 \times 10^5$ cm/s, and density $\rho = 5$ g/cm^3, at a temperature of 77 K (much less than the Debye temperature).

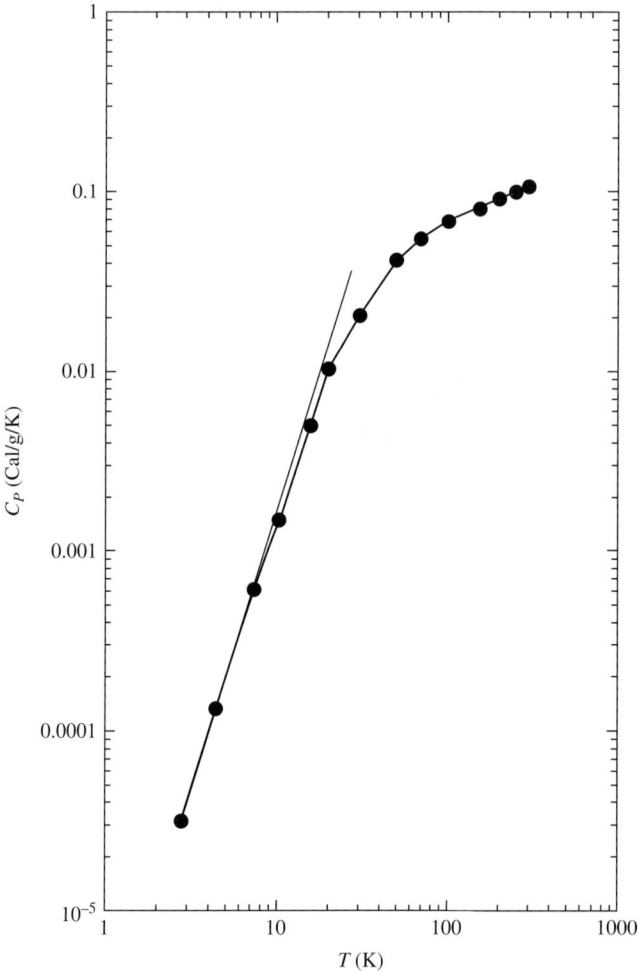

Fig. 4.13 The specific heat of Cu_2O as a function of temperature. The thin straight line is the T^3 power law. Data from Hu and Johnston (1951), and Gregor (1962).

Exercise 4.9.3 Calculate the temperature at which the photon energy density in a solid will exceed the phonon energy density in the same solid, if the solid has a Debye temperature of 300 K, phonon velocity of 4×10^5 cm/s, and an index of refraction of $n = 2$, assuming that the solid does not melt.

4.9.4 Electron Heat Capacity: Sommerfeld Expansion

In a semiconductor or insulator, the electrons in the valence band cannot change state. Therefore, to first order, they contribute nothing to the heat capacity. In metals, however, the electrons can contribute to the heat capacity. Using the density of states of Section 2.2, we have for a metal at finite temperature

$$U = \int_0^\infty Ef(E)\mathcal{D}(E)dE$$

$$= \int_0^\infty E\frac{1}{e^{(E-\mu)/k_BT} + 1} \, 2\frac{V}{2\pi^2}\frac{m}{\hbar^3}\sqrt{2mE} \, dE. \tag{4.9.24}$$

When k_BT is much smaller than the Fermi energy, we can approximate this formula by an expansion in k_BT. We can write this in the form

$$U = \int_{-\infty}^\infty H(E)f(E)dE, \tag{4.9.25}$$

where $f(E)$ is the Fermi–Dirac function, and $H(E)$ vanishes below $E = 0$. In general, any formula of this form can be expanded by the following procedure. We integrate by parts to obtain

$$U = -\int_{-\infty}^\infty K(E)\frac{\partial f(E)}{\partial E}dE, \tag{4.9.26}$$

where

$$K(E) = \int_{-\infty}^E H(E')dE'. \tag{4.9.27}$$

Since the Fermi–Dirac function $f(E)$ varies rapidly only near $E = \mu$, we expand in this region,

$$K(E) = K(\mu) + (E - \mu)\frac{\partial K}{\partial E}\Big|_{E=\mu} + \frac{1}{2}(E - \mu)^2\frac{\partial^2 K}{\partial E^2}\Big|_{E=\mu} + \cdots. \tag{4.9.28}$$

Since $\partial f/\partial E$ is even with respect to $E = \mu$, the linear term in this expansion will vanish when substituted into (4.9.26). The two leading terms are therefore

$$U = -K(\mu)\int_{-\infty}^\infty \frac{\partial f}{\partial\varepsilon}d\varepsilon - \frac{1}{2}\frac{\partial^2 K}{\partial E^2}\Big|_{E=\mu}(k_BT)^2\int_{-\infty}^\infty \varepsilon^2\frac{\partial f(\varepsilon)}{\partial\varepsilon}d\varepsilon + \cdots,$$

$$= -K(\mu)(-1) - \frac{1}{2}\frac{\partial^2 K}{\partial E^2}\Big|_{E=\mu}(k_BT)^2\left(-\frac{\pi^2}{3}\right) + \cdots, \tag{4.9.29}$$

where $\varepsilon = (E - \mu)/k_BT$. We obtain, replacing $K(E)$ with its definition in terms of the original function $H(E)$,

$$U = \int_0^\mu H(E)dE + \frac{\pi^2}{6}(k_BT)^2\frac{\partial H}{\partial E}\Big|_{E=\mu} + \cdots. \tag{4.9.30}$$

This is known as a **Sommerfeld** expansion. For the particular case of the electron heat capacity under consideration here, we substitute $H(E) = E\mathcal{D}(E)$. Taking into account the dependence of μ on T (see Exercise 4.9.4), we obtain

$$\frac{C_V}{V} = \frac{1}{V}\frac{\partial U}{\partial T} = \frac{\pi^2}{3}k_B^2 T\frac{\mathcal{D}(E_F)}{V} = \frac{\pi^2}{2}\left(\frac{k_BT}{E_F}\right)\frac{k_BN}{V}, \tag{4.9.31}$$

where we have used (2.4.2) for $n = N/V$, the total electron density.

Comparing (4.9.20) and (4.9.31), we see that if each unit cell contributes one or two free electrons in a metal, then the ratio of the heat capacity of the electrons to that of the phonons at room temperature is approximately a factor $k_B T / E_F$, which is small for typical metals. Therefore, the heat capacity of the free electrons in a metal is small compared to the contribution of the phonons to the total heat capacity.

Exercise 4.9.4 Fill in the missing steps from (4.9.30) to (4.9.31). First, write a Sommerfeld expansion to second order in T for the total electron number N. The lowest-order terms of both U and N can then be expanded as

$$\int_0^\mu H(E)dE \simeq \int_0^{E_F} H(E)dE + H(E_F)(\mu - E_F). \tag{4.9.32}$$

Using the constraint that N is constant for all temperatures, you can obtain a relation for μ in terms of T and substitute this into the equation for U, and then take its derivative, keeping terms to lowest order in T.

4.10 Thermal Motion of Atoms

Phonons in equilibrium have the Planck distribution function (4.9.1). We can find the rms displacement $\sqrt{\langle x^2 \rangle}$ of the atoms by using (4.4.25) to compare the occupation number to an amplitude. We start by recalling that the displacement of an atom can be written in terms of the normal variable x_k, defined in (4.2.5), with the inverse relation (4.2.25). This then implies the square of the displacement is

$$|x_n|^2 = \frac{1}{N} \sum_{k,k'} x_k x_{k'}^* e^{ikan} e^{-ik'an}. \tag{4.10.1}$$

The average excursion is then given by

$$\langle x^2 \rangle = \frac{1}{N} \sum_n |x_n|^2$$

$$= \frac{1}{N} \sum_{k,k'} \left(\frac{1}{N} \sum_n e^{i(k-k')an} \right) x_k x_{k'}^*$$

$$= \frac{1}{N} \sum_k |x_k|^2, \tag{4.10.2}$$

where we have used a δ-function identity for the term in the parentheses to eliminate the sum over k'.

As we did in Section 4.4, we can relate the mode amplitude x_k to the wave amplitude x_0 to the single-atom amplitude by equating a coherent state to a classical oscillation. We write $x_n = x_0 e^{i(kan - \omega_k t)}$, so that

$$x_k = \frac{1}{\sqrt{N}} \sum_n x_n e^{-ikan} = \frac{1}{\sqrt{N}} \sum_n x_0 e^{i(kan - \omega_k t)} e^{-ikan}$$

$$= \frac{1}{\sqrt{N}} \sum_n x_0 e^{-i\omega_k t} = \sqrt{N} x_0 e^{-i\omega_k t}, \tag{4.10.3}$$

and therefore

$$|x_k|^2 = N x_0^2. \tag{4.10.4}$$

Using (4.4.25), we then have

$$\langle x^2 \rangle = \sum_k \langle N_k \rangle \frac{2\hbar}{\rho V \omega_k}$$

$$= \frac{2\hbar^2}{\rho V} \int_0^\infty \frac{f(E)}{E} \mathcal{D}(E) dE. \tag{4.10.5}$$

Exercise 4.10.1   Use (4.10.5) to find the rms displacement of the atoms in a crystal of silicon at $T = 300$ K, for a phonon speed of $v = 5 \times 10^5$ cm/s and density of 2.3 g/cm³. For the unit cell size of silicon 5.43 Å, what fraction of the unit cell size is the rms displacement?

Debye–Waller effect. In Sections 1.5 and 3.2, we calculated the diffraction patterns for x-ray and neutron scattering from crystals assuming a purely classical model of the atoms and their vibrational motion. If we want to understand the effect of temperature on the diffraction patterns, we must turn to a quantum mechanical description, since the thermal vibrations should be described as a Planck distribution of phonons. We expect that thermal fluctuations will add disorder to the system, and therefore the diffraction pattern will be degraded. This calculation provides a good example of using all the tools of the quantum formalism developed earlier in this chapter.

In Section 1.5, we introduced the concept of the structure factor, which determines the intensity of the diffracted spots in the case of a lattice with a basis of more than one atom, equal to the sum $\sum e^{i\vec{s}\cdot\vec{r}_b}$, where $\vec{r}_b$ are the positions of the atoms in the unit cell. We can generalize this concept to the case when a single atom has some probability of being in more than one place in the unit cell. In this case, we want the thermal average of the quantum mechanical expectation value of this factor. We sum over all possible quantum states, with each state weighted by its probability at a given temperature. This gives us the **dynamical structure factor**, or **Debye–Waller factor**,

$$\langle e^{i\vec{s}\cdot\vec{r}} \rangle_T \equiv \sum_i \frac{e^{-\beta E_i}}{Z} \langle i | e^{i\vec{s}\cdot\vec{r}} | i \rangle, \tag{4.10.6}$$

where the partition function is

$$Z = \sum_i e^{-\beta E_i}, \tag{4.10.7}$$

with $\beta = 1/k_B T$. The position $\vec{r}$ is the same as the position vector $\vec{x}$ given in terms of phonon creation and destruction operators in (4.2.29).

Since we are only concerned about the motion of one atom, we can pick the equilibrium position of this atom at the origin, and write

$$\vec{x} = \sum_{\vec{k},\lambda} \hat{\eta}_{\vec{k}\lambda} \sqrt{\frac{\hbar}{2\rho V \omega_{\vec{k}\lambda}}} \left(a_{\vec{k}\lambda} + a^{\dagger}_{\vec{k}\lambda}\right). \tag{4.10.8}$$

Combining all the prefactors of the creation and destruction operators together into a constant $C_{\vec{k}}$, we can write generically

$$i\vec{s} \cdot \vec{r} = i \sum_{\vec{k}} C_{\vec{k}}(a^{\dagger}_{\vec{k}} + a_{\vec{k}}). \tag{4.10.9}$$

In the thermal average, we must allow for all possible numbers of phonons in every $\vec{k}$-state; in other words, a given state $|i\rangle$ is equal to a many-particle Fock state $|N_1, N_2, \ldots\rangle$, which we can write as the product state $|N_1\rangle|N_2\rangle \ldots$ (since the terms in the Hamiltonian which correspond to different phonon modes commute, they belong to different Hilbert spaces). Relative to the zero-point energy, the total energy E_i of a state i is equal to the sum over all $\vec{k}$-states of the energy per phonon times the number of phonons in that state,

$$E_i = \sum_{\vec{k}} \hbar \omega_{\vec{k}} \hat{N}_{\vec{k}}. \tag{4.10.10}$$

Because both this sum and the sum (4.10.9) appear in an exponent, and because the creation and destruction operators for a given $\vec{k}$ commute with all the creation and destruction operators for other $\vec{k}$-vectors, the total thermal average can be factored into a product of the thermal averages for each state $\vec{k}$,

$$\langle e^{i\vec{s}\cdot\vec{r}}\rangle_T = \prod_{\vec{k}} \sum_{N_{\vec{k}}=0}^{\infty} \frac{e^{-\beta\hbar\omega_{\vec{k}}N_{\vec{k}}}}{Z_{\vec{k}}} \langle N_{\vec{k}}|e^{C_{\vec{k}}(a^{\dagger}_{\vec{k}}+a_{\vec{k}})}|N_{\vec{k}}\rangle, \tag{4.10.11}$$

with

$$Z_{\vec{k}} = \sum_{N_{\vec{k}}=0}^{\infty} e^{-\beta\hbar\omega_{\vec{k}}N_{\vec{k}}} = \frac{1}{1 - e^{-\beta\hbar\omega_{\vec{k}}}}, \tag{4.10.12}$$

where we have summed the infinite geometrical series in the standard way.

The expectation value of the exponential can be resolved using the Campbell–Baker–Hausdorf identity (4.4.9),

$$e^{A+B} = e^A e^B e^{-[A,B]/2}, \tag{4.10.13}$$

which allows us to write

$$\langle N_{\vec{k}}|e^{iC_{\vec{k}}(a^{\dagger}_{\vec{k}}+a_{\vec{k}})}|N_{\vec{k}}\rangle = \langle N_{\vec{k}}|e^{iC_{\vec{k}}a^{\dagger}_{\vec{k}}}e^{iC_{\vec{k}}a_{\vec{k}}}|N_{\vec{k}}\rangle e^{-C_{\vec{k}}^2/2}$$

$$= \langle N_{\vec{k}}| \left(1 + iC_{\vec{k}}a^{\dagger}_{\vec{k}} - \frac{1}{2}C_{\vec{k}}^2(a^{\dagger}_{\vec{k}})^2 + \cdots\right) \left(1 + iC_{\vec{k}}a_{\vec{k}} - \frac{1}{2}C_{\vec{k}}^2(a_{\vec{k}})^2 + \cdots\right) |N_{\vec{k}}\rangle$$

$$\times \left(1 - \frac{1}{2}C_{\vec{k}}^2 + \frac{1}{8}C_{\vec{k}}^4 + \cdots\right). \tag{4.10.14}$$

Only terms with equal numbers of creation and destruction operators survive in the expansion. Thus, grouping terms by power of the constant $C_{\vec{k}}$, we obtain

$$\langle N_{\vec{k}}|e^{iC_{\vec{k}}(a_{\vec{k}}^{\dagger}+a_{\vec{k}})}|N_{\vec{k}}\rangle = 1 - C_{\vec{k}}^{2}\left(N_{\vec{k}}+\frac{1}{2}\right) + C_{\vec{k}}^{4}\left(\frac{1}{4}(N_{\vec{k}}^{2}-N_{\vec{k}})+\frac{1}{2}N_{\vec{k}}+\frac{1}{8}\right) + \cdots . \tag{4.10.15}$$

From this expansion, we see that we can compute (4.10.11) by performing the thermal average of powers of $N_{\vec{k}}$. We will do this here for just the first few powers, and extrapolate our results to higher powers.

In general, for any boson mode, the average number per mode is

$$\langle N\rangle = \frac{1}{Z}\sum_{N=0}^{\infty}Ne^{-\varepsilon N}$$

$$= -\frac{1}{Z}\frac{\partial}{\partial\varepsilon}Z$$

$$= \frac{1}{e^{\varepsilon}-1} \equiv \bar{N}, \tag{4.10.16}$$

where we have used the abbreviation $\varepsilon = \beta\hbar\omega_{\vec{k}}$, and the partition function (4.10.12). (This is another way to deduce the result (4.8.24) for bosons.) The thermal average of N^{2} for a boson mode is

$$\langle N^{2}\rangle = \frac{1}{Z}\sum_{N=0}^{\infty}N^{2}e^{-\varepsilon N}$$

$$= \frac{1}{Z}\frac{\partial^{2}}{\partial\varepsilon^{2}}Z$$

$$= \frac{1}{e^{\varepsilon}-1}\left(1+\frac{2}{e^{\varepsilon}-1}\right)$$

$$= \bar{N}(1+2\bar{N}). \tag{4.10.17}$$

The thermal average (4.10.11) therefore becomes, using the expansion (4.10.15),

$$\langle e^{i\vec{s}\cdot\vec{r}}\rangle_{T} = \prod_{\vec{k}}\left(1 - \frac{C_{\vec{k}}^{2}}{2}(2\bar{N}_{\vec{k}}+1) + \frac{1}{2}\left(\frac{c_{\vec{k}}^{2}}{2}(2\bar{N}_{\vec{k}}+1)\right)^{2} + \cdots\right). \tag{4.10.18}$$

This suggests that to all orders the expansion is equal to that of an exponential, and this can be proven (Maradudin 1963: s. VII.2). We thus have the simple result

$$\langle e^{i\vec{s}\cdot\vec{r}}\rangle_{T} = \exp\left(-\sum_{\vec{k}}\frac{C_{\vec{k}}^{2}}{2}(2\bar{N}_{\vec{k}}+1)\right) = e^{\frac{1}{2}\langle(i\vec{s}\cdot\vec{r})^{2}\rangle_{T}}. \tag{4.10.19}$$

This is typically written as e^{-W}. Substituting in the definition of $C_{\vec{k}}$ and the Planck distribution for $\bar{N}_{\vec{k}}$, we obtain

$$\boxed{W = \frac{1}{2}\sum_{\vec{k}}(\vec{s}\cdot\hat{\eta}_{\vec{k}\lambda})^{2}\,\frac{\hbar}{2\rho V\omega_{\vec{k}\lambda}}\coth\hbar\omega_{\vec{k}}/2k_{B}T.} \tag{4.10.20}$$

The effect of the thermal fluctuations is to reduce the intensity of the diffracted spots by the factor e^{-2W}.

Exercise 4.10.2   Show that the Debye–Waller factor e^{-W} vanishes in the case of a one- or two-dimensional system, but is finite in three dimensions. This reflects the fact, discussed in Sections 10.5, 10.7, and 11.5, that thermal fluctuations prevent long-range order in one and two dimensions. Although there may be locally ordered regions, the fluctuations wash out any overall crystal order.

References

W. Band, *An Introduction to Quantum Statistics* (D. van Nostrand, 1955).

G. Baym, *Lectures on Quantum Mechanics* (Benjamin Cummings, 1969).

C. Cohen-Tannoudji, B. Diu, and F. Laloë, *Quantum Mechanics* (Wiley, 1977).

H. Goldstein, C. Poole, and J. Safko, *Classical Mechanics*, 3rd ed. (Addison-Wesley, 2002).

L.V. Gregor, "The heat capacity of cuprous oxide from 2.8 to 21° K," *Journal of Physical Chemistry* **66**, 1645 (1962).

J.H. Hu and H.L. Johnston, "Low temperature heat capacities of inorganic solids. IX. Heat capacity and thermodynamic properties of cuprous oxide from 14 to 300° K," *Journal of the American Chemical Society* **73**, 4550 (1961).

R. Loudon, *The Quantum Theory of Light* (Oxford University Press, 1973).

W.H. Louisell, *Quantum Statistical Properties of Radiation* (John Wiley and Sons 1973).

L. Mandel and E. Wolf, *Optical Coherence an Quantum Optics* (Cambridge University Press, 1995).

A.A. Maradudin, *Theory of Lattice Dynamics in the Harmonic Approximation* (Academic Press, 1963).

C. Mead, *Collective Electrodynamics* (MIT Press, 2001).

J.J. Sakurai, *Modern Quantum Mechanics*, revised edition (Addison-Wesley, 1995).

D.W. Snoke and J.P. Wolfe, "Population dynamics of a Bose gas near saturation," *Physical Review B* **39**, 4030 (1989).

D.W. Snoke, D. Braun, and M. Cardona, "Carrier thermalization in Cu_2O: phonon emission by excitons," *Physical Review* B **44**, 2991 (1991).

D.W. Snoke, G.Q. Liu, and S. Girvin, "The basis of the Second Law of Thermodynamics in quantum field theory," *Annals of Physics* **327**, 1825 (2012).

5 Interactions of Quasiparticles

We now have a panoply of quasiparticles that can exist in solids: electrons, holes, phonons, and photons. In the quasiparticle picture, we do not worry about the underlying solid; we only worry about the interactions of these particles with each other.

To properly understand solids, then, we must understand the interactions of each of these with all the others: electron–electron interaction, electron–phonon interaction, electron–photon interaction, phonon–phonon interaction, phonon–photon interaction, and photon–photon interaction. Figure 5.1 illustrates the various interactions. We will put off phonon–photon, photon–photon, and higher-order electron–photon interactions until Chapter 7, when we discuss nonlinear optics.

We must add one more entity to the list of things which can interact in a solid. This is the category of **defects** and **dislocations** in a crystal. No crystal is perfect; as discussed in Section 1.8.2, one or two atoms may not sit in the right location, or an extra atom of a different type from the rest of the lattice may be inserted. Defects may also occur as continuous lines of mislocated atoms, or as planes. As we have continued to do in the quasiparticle picture, we view the underlying lattice of atoms as perfect, identify the ground state of this lattice as the vacuum, and think of defects as new objects which have been created in that vacuum.

Defects, like electrons and phonons, can also move through the crystal. The total energy of the crystal with a single defect is essentially the same whether the defect is in one place or another. Their effective mass is generally much higher than that of other quasiparticles, however, and therefore one is often justified in treating them as immobile.

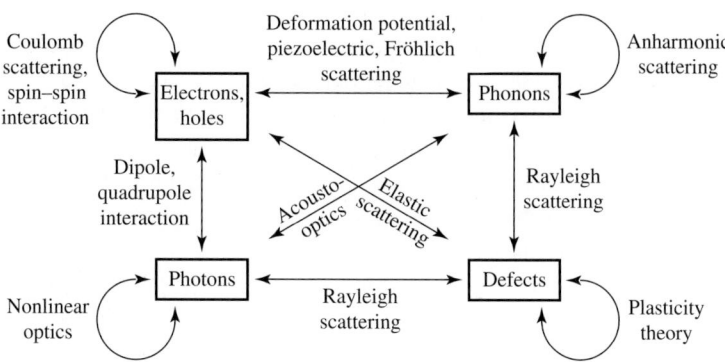

Fig. 5.1 Interactions among the quasiparticles in a solid.

The interactions between the quasiparticles control all of the **temporal dynamics** of the system. As we have seen in Chapter 4, Fermi's golden rule, and more generally, the quantum Boltzmann equation, give us the time evolution of a system using the energy of interaction between states.

5.1 Electron–Phonon Interactions

Up to this point, we have written down the quasiparticles as eigenstates of the Hamiltonian of the system. The Hamiltonian for which we found the eigenstates was not the full Hamiltonian, however. In each case, we approximated the Hamiltonian by ignoring higher-order terms. For example, in determining the electron states of a crystal, we assumed that it was perfectly periodic. If there are vibrations of the crystal, however, the perfect periodicity will be lost. We can write the full Hamiltonian as the sum of the energy for a perfectly periodic crystal plus the difference in energy due to vibrations. Since vibrations in a solid correspond to phonons, this extra term will lead to electron–phonon coupling.

It is easy to see how such a term can come about. As discussed in Section 2.4.1, the electron energy increases dramatically when the volume per electron is reduced. This goes both ways. As discussed in Section 3.1.1, the electronic charge determines the effective spring constant between the atoms. A change of the electron distribution can therefore change the energy of vibration of the atoms.

5.1.1 Deformation Potential Scattering

Distortions of the lattice of a solid are decribed in terms of strain. As discussed in Chapter 3, all strains can be viewed as a sum of a hydrostatic strain plus a shear (traceless) strain. A hydrostatic strain does not change the crystal symmetry; it only changes the overall volume of the unit cell. Therefore, hydrostatic strain can change the energy of the electrons in each band, but it does not change the shape of the overall band structure of the crystal.[1]

On the other hand, shear deformations do not change the overall volume of the unit cell, but they do change the symmetry of the crystal. Therefore, shear deformations do not change the total energy, but they can change the band structure. In particular, shear can have two effects: **mixing** of bands, that is, the electron eigenstates in the presence of shear strain may be superpositions of the eigenstates of the unstrained crystal; and **splitting** of bands, that is, eigenstates which are degenerate in the unstrained crystal may become separated in energy in the presence of shear strain. Splitting of degenerate states on reduction of symmetry is discussed in Chapter 6.

Let us start with the simple case of a single electron band in a crystal subjected to a purely hydrostatic deformation. We can assume that to lowest order, the change of the electron energy is linear with the strain. Therefore, we can write the interaction energy

[1] An exception is phase transitions at very high pressure, as discussed in Section 5.4.

at every point $\vec{x}$ in terms of a **deformation potential** D which multiplies the hydrostatic strain,

$$H_{\text{int}} = D \operatorname{Tr} \tilde{\varepsilon} = D(\varepsilon_{xx} + \varepsilon_{yy} + \varepsilon_{zz}). \tag{5.1.1}$$

The hydrostatic deformation potential D can be measured in static stress experiments by measuring the shift of the electron bands under hydrostatic pressure. Typical magnitudes of the deformation potentials are a few eV. In other words, as deduced in Section 2.4.1, if a strain distorts the lattice by 1%, the electron energies will vary by tens of meV.

We can write the strain in terms of the quantized operators using the formalism of Chapter 4. From Section 3.4.2, the strain is related to the displacement of the medium by the relation

$$\varepsilon_{ij} = \frac{1}{2} \left(\frac{\partial u_i}{\partial x_j} + \frac{\partial u_j}{\partial x_i} \right). \tag{5.1.2}$$

The displacement u can be related to the quantized phonon amplitude by formula (4.2.29) from Section 4.2. We write (using u for displacement instead of x)

$$\vec{u}(\vec{x}) = \sum_{\vec{k},\lambda} \hat{\eta}_{\vec{k}\lambda} \sqrt{\frac{\hbar}{2\rho V \omega_{\vec{k}\lambda}}} \left(a_{\vec{k}\lambda} e^{i\vec{k}\cdot\vec{x}} + a^{\dagger}_{\vec{k}\lambda} e^{-i\vec{k}\cdot\vec{x}} \right), \tag{5.1.3}$$

where $\hat{\eta}_{\vec{k}\lambda}$ is a unit polarization vector, and λ is the polarization index. For a hydrostatic strain, we then have

$$H_{\text{int}} = D \left(\frac{\partial u_x}{\partial x} + \frac{\partial u_y}{\partial y} + \frac{\partial u_z}{\partial z} \right). \tag{5.1.4}$$

Substituting in the displacement (5.1.3), we have

$$H_{\text{int}} = \sum_{\vec{k},\lambda} D \left(\hat{\eta}_{\vec{k}\lambda} \cdot \vec{k} \right) \sqrt{\frac{\hbar}{2\rho V \omega_{\vec{k}\lambda}}} \, i \left(a_{\vec{k}\lambda} e^{i\vec{k}\cdot\vec{x}} - a^{\dagger}_{\vec{k}\lambda} e^{-i\vec{k}\cdot\vec{x}} \right). \tag{5.1.5}$$

Note that the dot product $(\hat{\eta}_{\vec{k}\lambda} \cdot \vec{k})$ is equal to k for a purely longitudinal phonon, and 0 for a purely transverse phonon. In other words, a pure hydrostatic deformation will only couple electrons to phonons with at least some longitudinal polarization component. In the following, we will treat only purely longitudinal phonons and drop the summation over λ.

The energy (5.1.5) is the energy per electron at one point $\vec{x}$ in space. To account for the quantum mechanical nature of the electrons, we must integrate this energy times the probability of finding an electron at all points in space. We write the total energy of deformation as

$$H_D = \int d^3x \, \Psi_n^{\dagger}(\vec{x}) H_{\text{int}}(\vec{x}) \, \Psi_n(\vec{x}). \tag{5.1.6}$$

Here we have used the spatial field operators $\Psi_n^{\dagger}(\vec{x})$ and $\Psi_n(\vec{x})$ to give us the probability amplitude of the electrons, where n is the electron band index. The spatial field operators are defined using the Bloch functions for a periodic solid, discussed in Section 4.6,

$$\Psi_n(\vec{x}) = \sum_{\vec{k}} \frac{e^{i\vec{k}\cdot\vec{x}}}{\sqrt{V}} u_{n\vec{k}}(\vec{x})\, b_{n\vec{k}},$$

$$\Psi_n^\dagger(\vec{x}) = \sum_{\vec{k}} \frac{e^{-i\vec{k}\cdot\vec{x}}}{\sqrt{V}} u_{n\vec{k}}^*(\vec{x})\, b_{n\vec{k}}^\dagger, \tag{5.1.7}$$

where $b_{n\vec{k}}^\dagger$ and $b_{n\vec{k}}$ are the fermion creation and destruction operators.

Inserting (5.1.5) into (5.1.6), we have, for longitudinal phonons,

$$H_D = \sum_{\vec{k}_1,\vec{k}_2} \int d^3x \left(\frac{e^{-i\vec{k}_2\cdot\vec{x}}}{\sqrt{V}} u_{n\vec{k}_2}^*(\vec{x})\, b_{n\vec{k}_2}^\dagger\right) H_{\text{int}}(\vec{x}) \left(\frac{e^{i\vec{k}_1\cdot\vec{x}}}{\sqrt{V}} u_{n\vec{k}_1}(\vec{x})\, b_{n\vec{k}_1}\right)$$

$$= \sum_{\vec{k},\vec{k}_1,\vec{k}_2} Dk \sqrt{\frac{\hbar}{2\rho V \omega_k}} \left[i\, a_{\vec{k}} b_{n\vec{k}_2}^\dagger b_{n\vec{k}_1} \left(\frac{1}{V}\int d^3x\, e^{i(\vec{k}_1-\vec{k}_2+\vec{k})\cdot\vec{x}} u_{n\vec{k}_2}^*(\vec{x}) u_{n\vec{k}_1}(\vec{x})\right) \right.$$

$$\left. - i\, a_{\vec{k}}^\dagger b_{n\vec{k}_2}^\dagger b_{n\vec{k}_1} \left(\frac{1}{V}\int d^3x\, e^{i(\vec{k}_1-\vec{k}_2-\vec{k})\cdot\vec{x}} u_{n\vec{k}_2}^*(\vec{x}) u_{n\vec{k}_1}(\vec{x})\right) \right]. \tag{5.1.8}$$

The integral over $\vec{x}$ can be simplified by making the assumption that the wavelengths of the waves involved are much longer than the size of the unit cell. In this case, we can treat the plane-wave factor as essentially constant over any given unit cell, and convert the integral over all space into a sum over all the lattice sites, in which each lattice site contributes the integral over a unit cell times the plane-wave factor for that cell. Taking $k \ll \pi/a$ allows us to write $u_{n\vec{k}}(\vec{x}) \approx u_{n0}(\vec{x})$,

$$\frac{1}{V}\int d^3x\, e^{i(\vec{k}_1-\vec{k}_2+\vec{k})\cdot\vec{x}} u_{n\vec{k}_2}^*(\vec{x}) u_{n\vec{k}_1}(\vec{x})$$

$$\simeq \frac{1}{N} \sum_{\vec{R}} e^{i(\vec{k}_1-\vec{k}_2+\vec{k})\cdot\vec{R}} \frac{1}{V_{\text{cell}}} \int_{\text{cell}} d^3x\, u_{n0}^*(\vec{x}) u_{n0}(\vec{x})$$

$$= \delta_{\vec{k}_2,\vec{k}_1+\vec{k}}, \tag{5.1.9}$$

where we have used the normalization relation (1.9.35) for the Bloch cell functions of the electrons, and (4.2.9) to convert the sum over plane waves to a Kronecker-δ. We therefore have, finally,

$$\boxed{H_D = \sum_{\vec{k},\vec{k}_1} Dk \sqrt{\frac{\hbar}{2\rho V \omega_k}}\, i \left(a_{\vec{k}} b_{n,\vec{k}_1+\vec{k}}^\dagger b_{n,\vec{k}_1} - a_{\vec{k}}^\dagger b_{n,\vec{k}_1-\vec{k}}^\dagger b_{n,\vec{k}_1}\right).} \tag{5.1.10}$$

The creation and destruction operators give us a simple way of looking at the interaction. The Hamiltonian can be viewed as the sum of all possible momentum-conserving scattering processes, in which an electron starts with momentum $\vec{k}_1$ and either emits or absorbs a phonon with momentum $\vec{k}$. Note that the electron–phonon interaction does not conserve the total number of phonons – phonons appear when they are emitted and disappear when they are absorbed by the electrons. Since the electron operators occur only in pairs $b^\dagger b$,

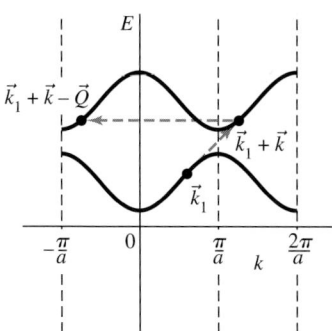

Fig. 5.2 An Umklapp process in electron scattering, as illustrated in the repeated zone picture. The electron begins in state $\vec{k}_1$ and scatters by absorbing a phonon with wave vector $\vec{k}$ into a state at $\vec{k}_1 + \vec{k}$, which lies outside the first Brillouin zone. This is equivalent to a state with wave vector $\vec{k}_1 + \vec{k} - \vec{Q}$ in the first Brillouin zone, where $\vec{Q}$ is a reciprocal lattice vector.

the number of electrons is conserved. Recall that for Bloch waves, the states $\vec{k}$ and $\vec{k} + \vec{Q}$ are the same, where $\vec{Q}$ is a reciprocal lattice vector. Therefore, a scattering process which takes an electron to some $\vec{k}$ outside the Brillouin zone will wrap around to a final electron state with crystal momentum $\vec{k} - \vec{Q}$, as shown in Figure 5.2. This is known as an **umklapp** process, already introduced in Section 2.2.

Exercise 5.1.1   Show that the interaction Hamiltonian (5.1.10) is Hermitian.

To account for shear deformations, we must generalize the above discussion to incorporate a deformation potential *operator*, $\tilde{D}$, which operates on the electron band states, that is,

$$H = \tilde{D}\tau, \tag{5.1.11}$$

where τ is the appropriate shear strain, which can also be given in terms of the quantized phonon operators. In this case, we must use the total probability amplitude,

$$\Psi(\vec{x}) = \sum_n \Psi_n(\vec{x}), \tag{5.1.12}$$

which is a sum over all electron bands. Then the total energy of the deformation is

$$H_{\text{int}} = \int d^3x \, \Psi^\dagger(\vec{x}) \, H(\vec{x}) \, \Psi(\vec{x}) = \sum_{n,n'} H_{n,n'} \tag{5.1.13}$$

where $H_{n,n'}$ is the Hamiltonian matrix defined by

$$H_{n,n'} = \int d^3x \, \Psi^\dagger_{n'}(\vec{x}) \, H(\vec{x}) \, \Psi_n(\vec{x}). \tag{5.1.14}$$

Since the operator $\tilde{D}$ acts on the electron band states, the integral analogous to (5.1.9) over the Bloch wave functions will have the term

$$\frac{1}{V_{\text{cell}}} \int_{\text{cell}} d^3x \, u^*_{n'0}(\vec{x}) \, \tilde{D} \, u_{n0}(\vec{x}) = \langle n' | \tilde{D} | n \rangle, \tag{5.1.15}$$

which is equal to $D\delta_{n,n'}$ for a hydrostatic strain, but in general has off-diagonal elements for a shear strain. The generalized deformation Hamiltonian matrix is therefore

$$H_{n,n'} = \sum_{\vec{k},\lambda,\vec{k}_1} \langle n'|\tilde{D}|n\rangle k \sqrt{\frac{\hbar}{2\rho V\omega_k}}\, i\left(a_{\vec{k}\lambda}\, b^\dagger_{n',\vec{k}_1+\vec{k}}\, b_{n\vec{k}_1} - a^\dagger_{\vec{k}\lambda}\, b^\dagger_{n',\vec{k}_1-\vec{k}}\, b_{n\vec{k}_1}\right). \qquad (5.1.16)$$

Shear deformation can therefore lead to coupling between bands, or between states with mixtures of different symmetries (as in $k \cdot p$ theory, discussed in Section 1.9.4). Like the hydrostatic deformation potential, the shear deformation potentials can be measured in static stress experiments.

Exercise 5.1.2   In a cubic crystal, a triply degenerate band with p-symmetry has the following deformation Hamiltonian, known as the **Pikus–Bir** Hamiltonian (derived in Section 6.11.2):

$$H_{PB} = a\mathrm{Tr}\,\tilde{\varepsilon} + $$
$$\begin{pmatrix} b(2\varepsilon_{xx} - \varepsilon_{yy} - \varepsilon_{zz}) & d\varepsilon_{xy} & d\varepsilon_{xz} \\ d\varepsilon_{xy} & b(2\varepsilon_{yy} - \varepsilon_{xx} - \varepsilon_{zz}) & d\varepsilon_{yz} \\ d\varepsilon_{xz} & d\varepsilon_{xz} & b(2\varepsilon_{zz} - \varepsilon_{xx} - \varepsilon_{yy}) \end{pmatrix},$$
$$(5.1.17)$$

operating on the basis of the three degenerate states $|p_x\rangle$, $|p_y\rangle$, and $|p_z\rangle$. The constant a is a hydrostatic deformation potential, while b and d are shear deformation potentials.

(a) Show that for a uniaxial strain along the [001] direction, the hydrostatic shift is $3a\varepsilon$, where $\varepsilon = \frac{1}{3}(\varepsilon_{xx} + \varepsilon_{yy} + \varepsilon_{zz})$ is the hydrostatic strain, and the splitting of the states is equal to $|3b\tau|$, where $\tau = (\frac{1}{2}\varepsilon_{xx} + \frac{1}{2}\varepsilon_{yy} - \varepsilon_{zz})$. For a uniaxial strain along the [111] direction, show that the splitting of the states is equal to $|3d\tau|$, where τ is the shear strain in this direction. (See Section 3.4.1 for definitions.)

(b) Construct the shear deformation operator $\tilde{D}$ for the Pikus–Bir Hamiltonian above explicitly for the transverse phonon case $\vec{k} = k\hat{x}, \hat{\eta}_{k,y} = \hat{y}$, and show that the Hamiltonian in this case gives a nonzero electron–phonon interaction energy. Do the same for the transverse phonon along [110], $\vec{k} = k(\hat{x} + \hat{y})/\sqrt{2}, \hat{\eta}_{k,y} = (\hat{x} - \hat{y})/\sqrt{2}$. Show that this shear term also gives a nonzero contribution for longitudinal phonons.

5.1.2 Piezoelectric Scattering

As discussed in Section 3.7, in some crystals a strain creates an electric polarization, which leads to an internal electric field. If this occurs, it will obviously have a strong interaction with the electrons in the medium.

Assuming that the dielectric constant is isotropic, we have from Section 3.7,

$$E_i = \frac{D_i}{\epsilon} = \frac{1}{\epsilon} \sum_{jl} e_{ijk}\varepsilon_{jl}, \qquad (5.1.18)$$

where e_{ijk} is the piezoelectric tensor.

As with the deformation potential interaction, we can determine the strain for a plane wave using (5.1.2) and (5.1.3):

$$\varepsilon_{ij} = \frac{1}{2}\left(\frac{\partial u_i}{\partial x_j} + \frac{\partial u_j}{\partial x_i}\right) \tag{5.1.19}$$

and

$$\vec{u}(\vec{x}) = \sum_{\vec{k},\lambda} \hat{\eta}_{\vec{k}\lambda} \sqrt{\frac{\hbar}{2\rho V \omega_{\vec{k}\lambda}}} \left(a_{\vec{k}\lambda} e^{i\vec{k}\cdot\vec{x}} + a_{\vec{k}\lambda}^{\dagger} e^{-i\vec{k}\cdot\vec{x}}\right), \tag{5.1.20}$$

where $\hat{\eta}_{\vec{k}\lambda}$ is a unit polarization vector and λ is the polarization index.

For a symmetric piezoelectric tensor, with $e_{ijl} = e_{ilj}$, the electric field (5.1.18) becomes

$$E_i = \frac{1}{\epsilon}\sum_{jl} e_{ijl}\frac{\partial u_j}{\partial x_l},$$

$$= \frac{1}{\epsilon}\sum_{\vec{k},\lambda}\sum_{jl} e_{ijl}\left(k_j\hat{\eta}_{(\vec{k}\lambda)l}\right)\sqrt{\frac{\hbar}{2\rho V \omega_{\vec{k}\lambda}}}\, i\left(a_{\vec{k}\lambda} e^{i\vec{k}\cdot\vec{x}} - a_{\vec{k}\lambda}^{\dagger} e^{-i\vec{k}\cdot\vec{x}}\right), \tag{5.1.21}$$

where $\hat{\eta}_{(\vec{k}\lambda)l}$ is the lth component of the polarization unit vector $\hat{\eta}_{\vec{k}\lambda}$.

From classical electrodynamics, the potential energy of the electric field of a plane wave is given by

$$\phi = \frac{-\vec{k}\cdot\vec{E}}{ik^2}, \tag{5.1.22}$$

since $\nabla\vec{E} = i\vec{k}\cdot\vec{E}$, $\nabla^2\phi = -k^2\phi$, and $\vec{E} = -\nabla\phi$. This implies

$$H = -e\phi$$

$$= \frac{e}{\epsilon}\sum_{\vec{k},\lambda}\sum_{ijl} e_{ijl}\frac{k_i k_j}{k^2}\hat{\eta}_{(\vec{k}\lambda)l}\sqrt{\frac{\hbar}{2\rho V \omega_{\vec{k}\lambda}}}\left(a_{\vec{k}\lambda} e^{i\vec{k}\cdot\vec{x}} + a_{\vec{k}\lambda}^{\dagger} e^{-i\vec{k}\cdot\vec{x}}\right). \tag{5.1.23}$$

The deformation-potential interaction Hamiltonian was proportional to $k/\sqrt{\omega}$, or, in the case when $\omega = vk$, proportional to $\sqrt{k}$. The piezoelectric interaction Hamiltonian, by contrast, is proportional to $1/\sqrt{\omega}$, or $1/\sqrt{k}$. The piezoelectric interaction will therefore be more important at low k, that is, at low temperature.

As in the case of the hydrostatic deformation-potential interaction, to get the total energy we multiply by the probability of finding an electron at each point $\vec{x}$, and integrate over all $\vec{x}$. Following the same procedures, we obtain

$$\boxed{\begin{aligned}H_{\text{piezo}} &= \frac{e}{\epsilon}\sum_{\vec{k},\lambda}\sum_{\vec{k}_1}\sum_{ijl} e_{ijl}\frac{k_i k_j}{k^2}\hat{\eta}_{(\vec{k}\lambda)l}\sqrt{\frac{\hbar}{2\rho V \omega_{\vec{k}\lambda}}}\\ &\quad\times\left(a_{\vec{k}}b_{n,\vec{k}_1+\vec{k}}^{\dagger}b_{n\vec{k}_1} + a_{\vec{k}}^{\dagger}b_{n,\vec{k}_1-\vec{k}}^{\dagger}b_{n\vec{k}_1}\right).\end{aligned}} \tag{5.1.24}$$

Exercise 5.1.3 For the piezoelectric tensor for a cubic crystal given in Table 3.4, show that the piezoelectric interaction Hamiltonian couples electrons to transverse phonons in the [110] direction but not longitudinal phonons in the same direction.

5.1.3 Fröhlich Scattering

Recall from Section 3.1.2 that in crystals with a two-atom basis, optical phonons exist which correspond to oscillation of the two atoms in the unit cell relative to each other. If the two atoms have different charge, this will then correspond to an oscillating dipole. An oscillating dipole also will clearly have strong coupling to the electrons via the Coulomb interaction.

In this case, the magnitude of the electric field is

$$\vec{E} = F\vec{u}, \tag{5.1.25}$$

where F is a constant of proportionality. In other words, the electric field produced by the dipole is directly proportional to the displacement, as opposed to the piezoelectric potential which gives an electric field proportional to the rate of change of the displacement.

As with the piezoelectric interaction, we write

$$H = -e\phi = e\frac{\vec{k} \cdot \vec{E}}{ik^2}, \tag{5.1.26}$$

and quantize the displacement using phonon operators. For optical phonons with frequency ω_{LO}, we must adjust the relation for the displacement amplitude, to

$$\vec{u}(\vec{x}) = \sum_{\vec{k},\lambda} \hat{\eta}_{\vec{k}\lambda} \sqrt{\frac{\hbar}{2m_r N\omega_{LO}}} \left(a_{\vec{k}\lambda} e^{i\vec{k}\cdot\vec{x}} + a_{\vec{k}\lambda}^\dagger e^{-i\vec{k}\cdot\vec{x}}\right), \tag{5.1.27}$$

where we have replaced ρ with m_r/V_{cell} and $V = NV_{\text{cell}}$, where N is the number of unit cells, to account for the fact that the appropriate mass for optical phonons is not the total mass of the unit cell, but the reduced mass relevant to the optical phonon oscillation, deduced in Section 3.1.2, $m_r = m_1 m_2/(m_1 + m_2)$. The $\vec{k} \cdot \vec{E}$ term in (5.1.26) enforces that only longitudinal optical phonons will contribute to this interaction.

As we will deduce later, in Section 7.5.1, the constant of proportionality is equal to

$$F = \omega_{LO}\sqrt{\frac{m_r}{V_{\text{cell}}}} \left(\frac{1}{\epsilon(\infty)} - \frac{1}{\epsilon(0)}\right)^{1/2}, \tag{5.1.28}$$

where $\epsilon(0)$ and $\epsilon(\infty)$ are the low-frequency and high-frequency dielectric constants of the medium, respectively, which, as is discussed in Section 7.5.1, are not independent of the optical phonon oscillations. This leads to the interaction Hamiltonian,

$$\boxed{\begin{aligned} H_{\text{Fr}} &= e\left(\frac{1}{\epsilon(\infty)} - \frac{1}{\epsilon(0)}\right)^{1/2} \sum_{\vec{k},\lambda} \sum_{\vec{k}_1} \frac{\vec{k} \cdot \hat{\eta}_{\vec{k}\lambda}}{k^2} \sqrt{\frac{\hbar\omega_{LO}}{2V}} \\ &\quad \times i\left(a_{\vec{k}\lambda} b_{n,\vec{k}_1+\vec{k}}^\dagger b_{n\vec{k}_1} - a_{\vec{k}\lambda}^\dagger b_{n,\vec{k}_1-\vec{k}}^\dagger b_{n\vec{k}_1}\right). \end{aligned}} \tag{5.1.29}$$

5.1.4 Average Electron–Phonon Scattering Time

Knowing the electron–phonon interaction, we can calculate the average time for phonon emission or absorption by electrons. From Fermi's golden rule, we have, for an electron with wave vector $\vec{q}$,

$$\frac{1}{\tau_{\vec{q}}} = \frac{1}{N_{\vec{q}}} \frac{2\pi}{\hbar} \sum_{f} |\langle \psi_f | H | \psi_i \rangle|^2 \delta(E_f - E_i). \tag{5.1.30}$$

Let us use the Hamiltonian (5.1.10) for the deformation potential for a single phonon mode. The total deformation Hamiltonian for all phonon modes is therefore the sum of this term for all possible $\vec{k}$. Ignoring the effects of anisotropy, we have for the total Hamiltonian,

$$H_{\text{def}} = \sum_{\vec{k},\vec{k}_1} Dk \sqrt{\frac{\hbar}{2\rho V \omega}} \, i \left(a_{\vec{k}} b^{\dagger}_{\vec{k}_1 + \vec{k}} b_{\vec{k}_1} - a^{\dagger}_{\vec{k}} b^{\dagger}_{\vec{k}_1 - \vec{k}} b_{\vec{k}_1} \right). \tag{5.1.31}$$

To get the total rate, we sum over all possible final states. The summation in the Hamiltonian ensures that for any given final state which conserves momentum, there is a term in the Hamiltonian that couples the initial and final states. We define the initial electron wave vector as $\vec{q}$ and sum over all possible final wave vectors of the electron. Following the procedure of Section 4.7, Fermi's golden rule gives us

$$\frac{1}{\tau_{\vec{q}}} = \frac{2\pi}{\hbar} \sum_{\vec{k}} D^2 k^2 \frac{\hbar}{2\rho V \omega} (1 + N^{\text{phon}}_{\vec{k}})(1 - N^{\text{elec}}_{\vec{q}-\vec{k}}) \, \delta\left(E_{\vec{q}-\vec{k}} + \hbar vk - E_{\vec{q}} \right)$$

$$+ \frac{2\pi}{\hbar} \sum_{\vec{k}} D^2 k^2 \frac{\hbar}{2\rho V \omega} N^{\text{phon}}_{\vec{k}} (1 - N^{\text{elec}}_{\vec{q}+\vec{k}}) \, \delta\left(E_{\vec{q}+\vec{k}} - E_{\vec{q}} - \hbar vk \right), \tag{5.1.32}$$

where we have distinguished between the phonon and electron occupation numbers by superscripts. The first term in (5.1.32) corresponds to phonon emission and the second to phonon absorption. Converting the sum to an integral, using the prescription (1.8.1), and assuming that the system is isotropic, we have

$$\frac{1}{\tau_q} = \frac{2\pi}{\hbar} D^2 \frac{\hbar}{2\rho V v} \frac{V}{(2\pi)^3} \int_0^{\infty} k \, d^3k \left[(1 + N^{\text{phon}}_{\vec{k}})(1 - N^{\text{elec}}_{\vec{q}-\vec{k}}) \, \delta\left(E_{\vec{q}-\vec{k}} + \hbar vk - E_{\vec{q}} \right) \right.$$

$$\left. + N^{\text{phon}}_{\vec{k}} (1 - N^{\text{elec}}_{\vec{q}+\vec{k}}) \, \delta\left(E_{\vec{q}+\vec{k}} - E_{\vec{q}} - \hbar vk \right) \right]. \tag{5.1.33}$$

We assume here that the temperature is low enough that we can assume a linear relation $\omega = vk$ for the acoustic phonons and that the Brillouin zone boundary can be replaced by $k = \infty$.

We can make several simplifications. First, if the electron occupation number is much less than 1, that is, the system has high temperature, we can drop the $(1 - N)$ Pauli exclusion terms. If we assume that the electron energy is large compared to the phonon energy $\hbar vk$, then the energy δ-functions are equal to $\delta(\hbar^2 k^2/2m \pm \hbar^2 kq \cos\theta/m)$. We can easily remove the energy δ-functions by performing the integration over θ, the angle between $\vec{k}$ and $\vec{q}$, to obtain

$$\frac{1}{\tau_q} = \frac{D^2}{4\pi\rho v}\left(\frac{m}{\hbar^2 q}\right)\int_0^{2q} k^2 dk \,(1 + 2N_k^{\text{phon}}). \tag{5.1.34}$$

As discussed in Section 4.9.3, in the high-temperature limit, the occupation number of the phonons is approximately

$$N_k^{\text{phon}} \simeq \frac{k_B T}{\hbar v k} \gg 1. \tag{5.1.35}$$

This allows us to write

$$\frac{1}{\tau_q} \simeq \frac{D^2}{4\pi\rho v}\left(\frac{m}{\hbar^2}\right)\frac{k_B T}{\hbar v q}(2q)^2. \tag{5.1.36}$$

For electrons at low density, in a Maxwell–Boltzmann distribution (e.g., free carriers in a semiconductor), the average scattering time is then

$$\frac{1}{\tau_q} \simeq \frac{D^2 m k_B T}{\pi\rho\hbar^3 v^2}\bar{q},$$

$$= \frac{\sqrt{3}D^2 m^{3/2}}{\pi\rho\hbar^4 v^2}(k_B T)^{3/2}, \tag{5.1.37}$$

where $\bar{q}$ is the average wave vector, which is given by $(\hbar\bar{q})^2/2m = \frac{3}{2}k_B T$. For a typical solid with density $\rho = 5$ g/cm^3, $v = 5 \times 10^5$ cm/s, deformation potential $D \sim 1$ eV, and effective carrier mass approximately the vacuum electron mass, this implies an average scattering time at $T = 300$ K of around $\tau = 10^{-11}$ s (i.e., 10 picoseconds). The exact time for electron–phonon scattering can vary widely depending on the exact electron–phonon interaction mechanism and the electron energy distribution. Typical electron–phonon scattering times are in general of the order tens of picoseconds, however.

This time scale is one reason why many scientists are interested in **ultrafast** physics, that is, experiments with time resolution of picoseconds or less. To look at solid state systems far from equilibrium – on time scales short compared to their scattering time – requires time resolution at least this good.

Exercise 5.1.4   Determine the average electron–phonon scattering time for a metal with electrons in a Fermi–Dirac distribution with $\mu = 0.5$ eV above the ground state, and $T = 300$ K, instead of a Maxwell–Boltzmann distribution, and all the other parameters the same as in the example above. In this case, (5.1.34) is not valid, but (5.1.33) is. The phonon occupation number can still be taken in the high-temperature approximation, and you can still assume that the electron energy is large compared to the phonon energy. You will need to use a program like Mathematica for the final weighted average over q. In this case, you cannot use $(\hbar\bar{q})^2/2m = \frac{3}{2}k_B T$.

5.2 Electron–Photon Interactions

The same approach can be used for electron–photon interactions as used for electron–phonon interactions. Again, we need a term in the Hamiltonian which depends on both the photon and electron wave function. We can deduce this from the fundamental electromagnetic theory.

As discussed in many textbooks (e.g., Cohen-Tannoudji *et al.* 1977: appendix III), when there is a magnetic field we must replace the p operator with $\vec{p} - q\vec{A}$. For the energy of an electron. We therefore write

$$\frac{p^2}{2m} \to \frac{|\vec{p} - q\vec{A}|^2}{2m} + U = \frac{(p^2 - q\vec{p} \cdot \vec{A} - q\vec{A} \cdot \vec{p} + q^2 A^2)}{2m} + U. \tag{5.2.1}$$

Using the quantum mechanical definition $\vec{p} = -i\hbar\nabla$, we can write

$$\vec{p} \cdot \vec{A}\psi = -i\hbar(\nabla \cdot \vec{A})\psi - i\hbar\vec{A} \cdot (\nabla\psi). \tag{5.2.2}$$

In the Coulomb gauge, also known as the radiation gauge or transverse gauge, we set $\nabla \cdot \vec{A} = 0$, and therefore $\vec{p} \cdot \vec{A} = \vec{A} \cdot \vec{p}$. We thus have

$$H = \frac{p^2}{2m} + U - \frac{q}{m}\vec{A} \cdot \vec{p} + \frac{q^2}{2m}A^2. \tag{5.2.3}$$

The sum $p^2/2m + U$ is simply the particle energy in the absence of the electromagnetic field. The last term is proportional to A^2, and therefore can be ignored for electromagnetic waves with low amplitude. This leaves us with a new term for the interaction of charged carriers with electromagnetic field,

$$H_{\text{int}} = -\frac{q}{m}\vec{A} \cdot \vec{p}. \tag{5.2.4}$$

The mass m is the vacuum electron mass, not an effective mass of an electron band, because we are starting from scratch with the underlying electrons in the solid.

As we did for the electron–phonon interaction, we replace the classical amplitude with the quantized formula, which for photons is

$$\vec{A}(\vec{x}) = \sum_{\vec{k},\lambda} \hat{\eta}_{\vec{k}\lambda} \sqrt{\frac{\hbar}{2\epsilon V\omega}} \left(a_{\vec{k}\lambda} e^{i\vec{k}\cdot\vec{x}} + a_{\vec{k}\lambda}^{\dagger} e^{-i\vec{k}\cdot\vec{x}} \right), \tag{5.2.5}$$

where we use the appropriate ϵ for the medium. Since we are concerned about transitions between bands, we must define an interaction Hamiltonian matrix as we did for the shear deformation Hamiltonian,

$$H_{n,n'} = \int d^3x \, \Psi_{n'}^{\dagger}(\vec{x}) \, H_{\text{int}}(\vec{x}) \, \Psi_n(\vec{x}). \tag{5.2.6}$$

This yields

$$
\begin{aligned}
H_{n,n'} &= \sum_{\vec{k}_1,\vec{k}_2} \int d^3x \left(\frac{e^{-i\vec{k}_2\cdot\vec{x}}}{\sqrt{V}} u^*_{n'\vec{k}_2}(\vec{x})\, b^\dagger_{n'\vec{k}_2} \right) H_{\mathrm{int}}(\vec{x}) \left(\frac{e^{i\vec{k}_1\cdot\vec{x}}}{\sqrt{V}} u_{n\vec{k}_1}(\vec{x})\, b_{n\vec{k}_1} \right) \\
&= -\frac{q}{m} \sum_{\vec{k}_1,\vec{k}_2} \sum_{\vec{k},\lambda} \sqrt{\frac{\hbar}{2\epsilon V\omega}} \\
&\quad \times \left[a_{\vec{k}\lambda}\, b^\dagger_{n'\vec{k}_2}\, b_{n\vec{k}_1} \left(\frac{1}{V}\int d^3x\, e^{i(-\vec{k}_2+\vec{k})\cdot\vec{x}}\, u^*_{n'\vec{k}_2}(\vec{x})\, (\hat{\eta}_{\vec{k}\lambda}\cdot\vec{p})\, e^{i\vec{k}_1\cdot\vec{x}} u_{n\vec{k}_1}(\vec{x}) \right) \right. \\
&\quad \left. + a^\dagger_{\vec{k}\lambda}\, b^\dagger_{n'\vec{k}_2}\, b_{n\vec{k}_1} \left(\frac{1}{V}\int d^3x\, e^{i(-\vec{k}_2-\vec{k})\cdot\vec{x}}\, u^*_{n'\vec{k}_2}(\vec{x})\, (\hat{\eta}_{\vec{k}\lambda}\cdot\vec{p})\, e^{i\vec{k}_1\cdot\vec{x}} u_{n\vec{k}_1}(\vec{x}) \right) \right].
\end{aligned}
\tag{5.2.7}
$$

Note that the momentum $\vec{p}$ is an operator that acts on the electron states.

Again, as in the case of the electron–phonon interaction, we assume that the wavelengths involved are long compared to the unit cell size, which allows us to write the integral over $\vec{x}$ as the product of a sum over all unit cells times an integral over a single unit cell. Using $q = -e$ for electrons, we finally obtain

$$
\begin{aligned}
H_{\mathrm{e\text{-}phot}} &= \frac{e}{m} \sum_{n,n'} \sum_{\vec{k}_1} \sum_{\vec{k},\lambda} \sqrt{\frac{\hbar}{2\epsilon V\omega}}\, \hat{\eta}_{\vec{k}\lambda}\cdot\langle n'|\vec{p}|n\rangle \\
&\quad \times \left(a_{\vec{k}\lambda}\, b^\dagger_{n',\vec{k}_1+\vec{k}}\, b_{n\vec{k}_1} + a^\dagger_{\vec{k}\lambda}\, b^\dagger_{n',\vec{k}_1-\vec{k}}\, b_{n\vec{k}_1} \right),
\end{aligned}
\tag{5.2.8}
$$

where

$$
\langle n'|\vec{p}|n\rangle = \frac{1}{V_{\mathrm{cell}}} \int_{\mathrm{cell}} d^3x\, u^*_{n'0}(\vec{x})\,(-i\hbar\vec{\nabla})\, u_{n0}(\vec{x}).
\tag{5.2.9}
$$

This interaction term can be used with Fermi's golden rule to determine the transition rate between electronic states due to photon emission and absorption. As in the case of the electron–phonon interaction, it does not conserve the total number of photons, since photons can be either created by the $a^\dagger_{\vec{k}}$ term or destroyed by the $a_{\vec{k}}$ term, but it does conserve the number of electrons.

Exercise 5.2.1   Show that the interaction Hamiltonian (5.2.8) is Hermitian, even if $\langle n'|\vec{p}|n\rangle$ is complex. Hint: The fact that both n and n' are summed over is important.

5.2.1 Optical Transitions Between Semiconductor Bands

Suppose that we have two semiconductor bands as shown in Figure 5.3. If a light wave hits this semiconductor, there can be optical transitions which raise an electron in the valence band to the conduction band.

In this case, the initial state is the unperturbed semiconductor with a full valence band, which we denote as $|0\rangle$, plus a photon field with some number of photons in state $\vec{k}$ with energy $\hbar\omega = \hbar ck$, while the final state has an electron in the conduction band, a hole in the valence band, and one fewer photon.

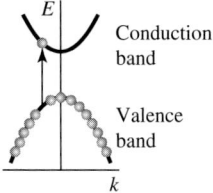

Fig. 5.3 An optical transition in a two-band semiconductor. The solid dots represent electrons in the valence band. A photon can raise an electron into the empty conduction band, leaving behind a hole in the valence band.

As discussed in Section 2.3, the creation of an electron in the conduction band and a hole in the valence band implies in general the creation of an exciton. The final state of the electronic system is therefore not simply an independent electron and hole, but an exciton state. This is true even when the energy of the photon is greater than the band gap of the semiconductor; in this case the final state is an ionized exciton, in which the electron and hole are not bound together but are still correlated. When the photon energy is well above the band gap, however, we can treat the electron and hole as uncorrelated, independent particles. In this case, we consider the initial state of $N_{\vec{k}}$ photons in a single polarization state $\hat{\eta}$, zero conduction electrons, and one electron per valence state, and integrate over all possible final electron states; that is, we sum over all independent electron and hole states. Following the procedure of Section 4.7, using (5.2.8), we have the rate of photon absorption per photon,

$$\frac{1}{\tau} = \frac{1}{N_{\vec{k}}} \frac{2\pi}{\hbar} \sum_{\vec{k}_c} \frac{\hbar e^2}{2\epsilon V \omega m^2} \left| \langle v | \vec{p} | c \rangle \cdot \hat{\eta} \right|^2 N_{\vec{k}} N_{\vec{k}_v} (1 - N_{\vec{k}_c}) \, \delta(E_{\vec{k}_c} - E_{\vec{k}_v} - \hbar\omega)$$

$$= \frac{2\pi}{\hbar} \sum_{\vec{k}_c} \frac{\hbar e^2}{2\epsilon V \omega m^2} \left| \langle v | \vec{p} | c \rangle \cdot \hat{\eta} \right|^2 \delta(E_{\vec{k}_c} - E_{\vec{k}_v} - \hbar\omega), \qquad (5.2.10)$$

where $\vec{k}_c$ and $\vec{k}_v$ are the electron and hole wave vectors, respectively. There is no photon emission term because we assume that $N_{\vec{k}_c} = 0$, and therefore it is not possible to conserve energy with the $a^\dagger b^\dagger b$ term, which creates a photon at the cost of electron energy. The electron and hole wave vectors are constrained by the momentum conservation condition $\vec{k} = \vec{k}_v - \vec{k}_c$.

Converting the sum over $\vec{k}_c$ to an integral over energy $E_{\vec{k}_c}$, assuming that the system is isotropic, we have

$$\frac{1}{\tau} = \frac{2\pi}{\hbar} \int \mathcal{D}(E_c) dE_c \frac{\hbar e^2}{2\epsilon V \omega m^2} \left| \langle v | \vec{p} | c \rangle \cdot \hat{\eta} \right|^2 \delta(E_{\vec{k}_c} - E_{\vec{k}_v} - \hbar\omega). \qquad (5.2.11)$$

In terms of the wave vectors $\vec{k}_c$ and $\vec{k}_v$, the energies $E_{\vec{k}_c}$ and $E_{\vec{k}_v}$ are

$$E_{\vec{k}_c} = E_{\text{gap}} + \frac{\hbar^2 k_c^2}{2m_e}, \quad E_{\vec{k}_v} = -\frac{\hbar^2 k_h^2}{2m_h}, \qquad (5.2.12)$$

where m_e and m_h are the effective masses of the electron and hole, respectively. The difference in energy which appears in the δ-function is therefore

$$
\begin{aligned}
E_{\vec{k}_c} - E_{\vec{k}_v} &= E_{\text{gap}} + \frac{\hbar^2 k_c^2}{2m_e} + \frac{\hbar^2 k_v^2}{2m_h} \\
&= E_{\text{gap}} + \frac{\hbar^2 k_c^2}{2m_e} + \frac{\hbar^2 |\vec{k}_c - \vec{k}|^2}{2m_h} \\
&\simeq E_{\text{gap}} + \frac{\hbar^2 k_c^2}{2m_e} + \frac{\hbar^2 k_c^2}{2m_h} \\
&= E_{\text{gap}} + \frac{\hbar^2 k_c^2}{2m_r},
\end{aligned}
\tag{5.2.13}
$$

where m_r is the exciton reduced mass, $m_r = (1/m_e + 1/m_h)^{-1}$. Here we have made use of the assumption that the photon momentum $\vec{k}$ is negligible compared to the typical electron and hole momentum, that is, small compared to the zone boundary momentum π/a. This is the assumption of **vertical transitions**, in which optical transitions simply move electrons directly vertical on the band diagram.

The energy δ-function can therefore be removed by a change of variables of the integral to $E' = \hbar^2 k_c^2 / 2m_r$. We define the **joint density of states** using m_r in the standard density of states formula (2.2.9), to obtain the rate of absorption per photon,

$$
\begin{aligned}
\frac{1}{\tau} &= \frac{2\pi}{\hbar} \left(\frac{V}{2\pi^2} \frac{\sqrt{2} m_r^{3/2} \sqrt{\hbar\omega - E_{\text{gap}}}}{\hbar^3} \right) \frac{\hbar e^2}{2\epsilon V \omega m^2} \left| \langle v|\vec{p}|c \rangle \cdot \hat{\eta} \right|^2 \\
&= \sqrt{\hbar\omega - E_{\text{gap}}} \frac{e^2 m_r^{3/2}}{\sqrt{2\pi} \hbar^2 \epsilon m^2} \frac{\left| \langle v|\vec{p}|c \rangle \cdot \hat{\eta} \right|^2}{\hbar\omega}.
\end{aligned}
\tag{5.2.14}
$$

Again, this expression is valid only for $\hbar\omega$ well above the band gap, because near the band-gap energy there are significant excitonic corrections to the absorption. In this region, one must write down the exciton states directly instead of free electron and hole states.

Note that we use the free-electron mass m as well as the effective masses m_e and m_h. The free-electron mass enters in through the electromagnetic interaction (5.2.4), while the effective masses enter into the density of states.

The same formalism can be used to calculate an **emission** rate, if the initial state of the system has electrons in the upper, conduction band. In this case, the initial state has $N_{\vec{k}_c}$ electrons in the upper band. We obtain

$$
\begin{aligned}
\frac{1}{\tau} &= \frac{1}{N_{\vec{k}_c}} \frac{2\pi}{\hbar} \sum_{\vec{k}_c} \frac{\hbar e^2}{2\epsilon V \omega m^2} \left| \langle v|\vec{p}|c \rangle \cdot \hat{\eta} \right|^2 (1 + N_{\vec{k}}) N_{\vec{k}_c} (1 - N_{\vec{k}_v}) \delta(E_{\vec{k}_v} + \hbar\omega - E_{\vec{k}_c}) \\
&= \frac{\sqrt{\hbar\omega - E_{\text{gap}}}}{\hbar\omega} \frac{e^2 m_r \sqrt{m_e}}{\sqrt{2\pi} \hbar^2 \epsilon m^2} \left| \langle v|\vec{p}|c \rangle \cdot \hat{\eta} \right|^2 (1 + N_{\vec{k}})(1 - N_{\vec{k}_v}),
\end{aligned}
\tag{5.2.15}
$$

where $\vec{k} = \vec{k}_v - \vec{k}_c$. The $(1 + N_{\vec{k}})$ term gives the spontaneous emission rate when $N_{\vec{k}} = 0$, plus the stimulated emission rate proportional to $N_{\vec{k}}$. The $(1 - N_{\vec{k}_v})$ term gives the Pauli exclusion effect for the final electron state. If the final valence band state is full, there will be Pauli exclusion blocking of the emission.

Exercise 5.2.2   Formula (5.2.14) gives an absorption *rate*. Convert this to an absorption *length*, that is, show that for a constant fluence of photons hitting a surface at $x = 0$, the probability of photon absorption at distance x from the surface is given by

$$P(x) = P(0)e^{-x/l},   \tag{5.2.16}$$

where l is the absorption length, which is inversely proportional to the rate $1/\tau$ given by (5.2.14). Estimate this absorption length for the case $E_{\text{gap}} = 2$ eV, $\epsilon/\epsilon_0 = 10$, and effective masses all of the order of the vacuum electron mass, for $\hbar\omega/E_{\text{gap}} = 1.1$. The matrix element for a semiconductor is typically given in terms of the unitless **oscillator strength**

$$f = \frac{2|\langle p \rangle|^2}{mE_{\text{gap}}},   \tag{5.2.17}$$

which is of the order of unity for typical semiconductors.

Exercise 5.2.3   Show that the spontaneous-emission lifetime is a few hundred femtoseconds for a conduction electron initially in a state with kinetic energy 30 meV falling into an empty valence-band state. Assume the unitless oscillator strength defined in Exercise 5.2.1 is equal to unity, the gap energy $E_{\text{gap}} = 1.5$ eV, the effective masses of both bands are about one-tenth the free-electron mass, and the dielectric constant of the medium is $\epsilon = 10\epsilon_0$. (These are values typical for a III–V semiconductor such as GaAs.)

Exercise 5.2.4   Show that in vacuum, an electron cannot directly recombine with a hole in the Dirac sea, that is, a positron (see Appendix F), by emission of a single photon, because this would violate energy and momentum conservation. Electron–positron recombination requires emission of two photons, unlike electron–hole recombination between bands in semiconductors.

5.2.2 Multipole Expansion

An alternative way of deriving the electron–photon interaction is to start with the Hamiltonian for electrostatic energy in an externally applied potential,

$$H_{\text{int}} = q\phi(\vec{x}).   \tag{5.2.18}$$

Expanding the potential in a Taylor series, we have

$$\phi(\vec{x}) = \phi(0) + \sum_i x_i \frac{\partial}{\partial x_i}\phi(\vec{x}) + \frac{1}{2}\sum_{ij} x_i x_j \frac{\partial}{\partial x_i}\frac{\partial}{\partial x_j}\phi(\vec{x}) + \cdots$$

$$= \phi(0) - \vec{x}\cdot\vec{E} - \frac{1}{2}\sum_{ij} x_i x_j \frac{\partial E_j}{\partial x_i} + \cdots .   \tag{5.2.19}$$

In the standard expansion (see, e.g., Jackson 1999: s. 4.2; Mandel and Wolf 1995: ss. 14.1.2 and 14.1.3), we note that since $\nabla \cdot \vec{E} = 0$, we are free to add a term which is equal to

$\nabla \cdot \vec{E}$ times any constant we like. If we subtract $\frac{1}{6}\nabla \cdot \vec{E}$, we can write the series in terms of the pure dipole and quadrupole operators,

$$H_{\text{int}} = q\phi(0) - \vec{d} \cdot \vec{E} - \frac{1}{6}\sum_{ij} Q_{ij} \frac{\partial E_j}{\partial x_i} + \cdots , \tag{5.2.20}$$

where

$$d_i = q x_i \tag{5.2.21}$$

is the dipole moment operator and

$$Q_{ij} = q(3x_i x_j - x^2 \delta_{ij}) \tag{5.2.22}$$

is the quadrupole moment operator.

Instead of the $\vec{A}$ field, we then have the interaction in terms of the $\vec{E}$ field. From (4.3.20), we have

$$\vec{E}(\vec{x}) = i \sum_{\vec{k},\lambda} \hat{\eta}_{\vec{k}\lambda} \sqrt{\frac{\hbar\omega}{2\epsilon V}} \left(a_{\vec{k}\lambda} e^{i\vec{k}\cdot\vec{x}} - a_{\vec{k}\lambda}^\dagger e^{-i\vec{k}\cdot\vec{x}} \right). \tag{5.2.23}$$

Following the procedure at the beginning of Section 5.2, we then write

$$H_{\text{e–phot}} = \int d^3x \, \Psi_{n'}^\dagger(\vec{x}) \, H_{\text{int}}(\vec{x}) \, \Psi_n(\vec{x}) \tag{5.2.24}$$

which yields, for the dipole term,

$$\begin{aligned} H_{\text{DE}} \;=\; & -iq \sum_{n,n'} \sum_{\vec{k}_1} \sum_{\vec{k},\lambda} \sqrt{\frac{\hbar\omega}{2\epsilon V}} \left(\hat{\eta}_{\vec{k}\lambda} \cdot \langle n'|\vec{x}|n\rangle \right) \\ & \times \left(a_{\vec{k}\lambda} b_{n',\vec{k}_1+\vec{k}}^\dagger b_{n\vec{k}_1} - a_{\vec{k}\lambda}^\dagger b_{n',\vec{k}_1-\vec{k}}^\dagger b_{n\vec{k}_1} \right), \end{aligned} \tag{5.2.25}$$

and for the quadrupole term,

$$\begin{aligned} H_{\text{QE}} \;=\; & -\frac{q}{6} \sum_{n,n'} \sum_{\vec{k}_1} \sum_{\vec{k},\lambda} \sqrt{\frac{\hbar\omega}{2\epsilon V}} \left(\sum_{ij} \eta_{(\vec{k}\lambda)i} k_j \langle n'|(3x_i x_j - x^2 \delta_{ij})|n\rangle \right) \\ & \times \left(a_{\vec{k}\lambda} b_{n',\vec{k}_1+\vec{k}}^\dagger b_{n\vec{k}_1} + a_{\vec{k}\lambda}^\dagger b_{n',\vec{k}_1-\vec{k}}^\dagger b_{n\vec{k}_1} \right), \end{aligned}$$

$$\tag{5.2.26}$$

where the wave vector $\vec{k}$ comes from the derivative of the electric field in (5.2.19).

The dipole interaction term is proportional to $\vec{x}$; that is, it has three terms, proportional to $\eta_x x$, $\eta_y y$, and $\eta_z z$. (We drop the $\vec{k}\lambda$ subscript of the polarization vector $\hat{\eta}$ for the moment.) The quadrupole interaction term at first appears to have nine terms since i and j run from 1 to 3, but in fact there are only five independent terms, corresponding to the same symmetries as the d-orbitals in atomic physics. First, because the x, y, z operators commute, there are only three independent cross terms, proportional to $(\eta_x k_y + \eta_y k_x)xy$, $(\eta_y k_z + \eta_z k_y)yz$, and $(\eta_x k_z + \eta_z k_x)xz$. For the remaining three terms, we note that

$$x^2\eta_x k_x + y^2\eta_y k_y + z^2\eta_z k_z = \frac{1}{6}(2z^2 - x^2 - y^2)(2\eta_z k_z - \eta_x k_x - \eta_y k_y)$$
$$+ \frac{1}{2}(x^2 - y^2)(\eta_x k_x - \eta_y k_y)$$
$$+ \frac{1}{3}(x^2 + y^2 + z^2)(\eta_x k_x + \eta_y k_y + \eta_z k_z). \quad (5.2.27)$$

The last term is proportional to $\hat{\eta}\cdot\vec{k}$, which is zero. There are therefore only five independent terms, which as stated above, have the symmetries of d-orbitals.

The dipole interaction Hamiltonian (5.2.25) is the same as the interaction (5.2.8) which we found earlier. We can write the matrix element $\langle n'|x|n\rangle$ in terms of the quantum mechanical p operator by using the commutation relation $[x, H_0] = [x, p^2/2m + U(x)] = i\hbar p/m$, where H_0 is the Hamiltonian that gives the energy of the electron bands. Therefore

$$\langle n|[x, H_0]|n'\rangle = \langle n|(xH_0 - H_0 x)|n'\rangle = (E_{n'} - E_n)\langle n|x|n'\rangle$$
$$= \frac{i\hbar}{m}\langle n|p|n'\rangle, \quad (5.2.28)$$

and consequently

$$\langle n'|x|n\rangle = -i\frac{\langle n'|p|n\rangle}{m\omega}, \quad (5.2.29)$$

where $\omega = (E_{n'} - E_n)/\hbar$. This relation is exact if $|n\rangle$ and $|n'\rangle$ are eigenstates of the Hamiltonian, while if we use these to represent Bloch cell functions as we did in (5.2.8), it is valid in the long-wavelength approximation, which implies $\psi_{n\vec{k}} \simeq u_{n\vec{k}}$. Substituting (5.2.29) into (5.2.25) gives (5.2.8). For this reason, (5.2.8) is also called the **dipole** interaction.

We could also have derived the quadrupole Hamiltonian using the $\vec{A}$-field. In Section 5.2, when we derived (5.2.8), we made the approximation

$$\langle n'|\vec{p}|n\rangle = \frac{1}{V_{\text{cell}}}\int_{\text{cell}} d^3x\; u^*_{n'\vec{k}_2}(\vec{x})\,\vec{p}\,u_{n\vec{k}_1}(\vec{x}) \approx \frac{1}{V_{\text{cell}}}\int_{\text{cell}} d^3x\; u^*_{n'0}(\vec{x})\,\vec{p}\,u_{n0}(\vec{x}), \quad (5.2.30)$$

which is reasonable when the wavelength $\vec{k} = \vec{k}_2 - \vec{k}_1$ is long compared to the unit cell size. To go one order higher, we can use the approximation (1.6.25) for the k-dependence of the Bloch cell functions,

$$\langle n'|\vec{p}|n\rangle \approx \frac{1}{V_{\text{cell}}}\int_{\text{cell}} d^3x\; \left(u^*_{n'0}(\vec{x})(1 + i\vec{k}_2 \cdot \vec{x})\right)\vec{p}\left(u_{n0}(\vec{x})(1 - i\vec{k}_1 \cdot \vec{x})\right)$$
$$= \frac{1}{V_{\text{cell}}}\int_{\text{cell}} d^3x\; u^*_{n'0}(\vec{x})\,\vec{p}\left(1 + i(\vec{k}_2 - \vec{k}_1)\cdot\vec{x}\right)u_{n0}(\vec{x}). \quad (5.2.31)$$

Since $\vec{k} = \vec{k}_2 - \vec{k}_1$ by momentum conservation, the second-order term of the electron–photon interaction will therefore have a term proportional to $(\hat{\eta}\cdot\vec{p})(\vec{k}\cdot\vec{x})$. We can rewrite this term as the sum of a symmetric and an antisymmetric term,

$$(\vec{p} \cdot \hat{\eta})(\vec{k} \cdot \vec{x}) = \frac{1}{2}\left[(\vec{p} \cdot \hat{\eta})(\vec{k} \cdot \vec{x}) - (\vec{k} \cdot \vec{x})(\vec{p} \cdot \hat{\eta})\right] + \frac{1}{2}\left[(\vec{p} \cdot \hat{\eta})(\vec{k} \cdot \vec{x}) + (\vec{k} \cdot \vec{x})(\vec{p} \cdot \hat{\eta})\right].$$

$$(5.2.32)$$

The first term is the **magnetic dipole** term, and the second is the **electric quadrupole** term. By various algebra manipulations (see, e.g., Bethe and Jackiw 1968: 226), this term gives the same Hamiltonian as (5.2.26).

As we will see in Section 6.10.2, when the first-order dipole interaction is forbidden by symmetry, the quadrupole interaction may be allowed. Since this term is proportional to k, it will be most significant for short-wavelength radiation.

Exercise 5.2.5　Prove that substituting (5.2.29) into (5.2.25) gives (5.2.8). Hint: See Exercise 5.2.1.

5.3 Interactions with Defects: Rayleigh Scattering

In addition to electrons and holes, phonons and photons, quasiparticles in solids can interact with defects in a solid. As discussed in the introduction to this chapter, in the quasiparticle picture we treat a solid as a perfect crystal, which constitutes the vacuum state of the quasiparticles, plus various defects. Quasiparticles such as electrons, phonons, and photons travel without scattering through a perfect crystal, but they scatter off of defects, that is, any imperfections in the crystal.

The defects themselves can therefore be thought of as quasiparticles. Defects can move through the crystal just as other quasiparticles can, but their motion is in general much slower, since it involves displacement of whole atoms. Compared to the electrons and holes, defects have nearly infinite mass. The interaction of the other quasiparticles with defects is therefore usually elastic scattering, that is, no energy is given to or received from a defect. To give energy to a defect, it would have to be excited into a higher energy state, and this usually requires more energy than a typical thermal electron or phonon has.

Phonon-defect scattering. To compute the interaction Hamiltonian for elastic scattering, we use the same approach as in previous sections. We first start with the classical energy, and then use the quantized operators to represent the fields. For phonons, the total energy of a classical wave with frequency ω and amplitude u is

$$H = \frac{1}{2}\rho V \omega^2 u^2,$$

$$(5.3.1)$$

where ρ is the mass density of the crystal, and V is the total volume (cf. (4.4.25) in Section 4.4, where the total energy is $N\hbar\omega$). Since $\omega u = \dot{u}$, we can equally well write this as

$$H = \frac{1}{2}\rho V \dot{u}^2,$$

$$(5.3.2)$$

that is, the energy is just the sum of $\frac{1}{2}m\dot{u}^2$ for all the atoms.

This energy is already accounted for in the Hamiltonian (4.2.14), which gives us the phonon eigenstates. If one part of the crystal has a higher or lower density than the rest of the crystal, however, then we will have an extra energy term

$$H_{\text{def–phon}} = \frac{1}{2}(\rho' - \rho) \int_{V'} d^3x\, |\dot{u}(x)|^2, \tag{5.3.3}$$

where ρ' is the different density and V' is the volume in which it applies. In the quantum mechanical picture, we have, from (4.2.29),

$$\dot{\vec{u}}(\vec{x}) = -i \sum_{\vec{k},\lambda} \hat{\eta}_{\vec{k}\lambda} \sqrt{\frac{\hbar\omega_k}{2\rho V}} \left(a_{\vec{k}\lambda} e^{i\vec{k}\cdot\vec{x}} - a^\dagger_{\vec{k}\lambda} e^{-i\vec{k}\cdot\vec{x}} \right), \tag{5.3.4}$$

where $\hat{\eta}_{\vec{k}\lambda}$ is the unit polarization vector. The appropriate density in this formula is the average density, since we assume that the phonon states are the same, and the defect just introduces a coupling between them. The interaction Hamiltonian is therefore

$$H_{\text{def–phon}} = \int_{V'} d^3x\, \frac{1}{2} \left(\frac{\Delta\rho}{\rho}\right) \sum_{\vec{k}_1,\lambda_1} \sum_{\vec{k}_2,\lambda_2} \left(\hat{\eta}_{\vec{k}_1\lambda_1} \cdot \hat{\eta}_{\vec{k}_2\lambda_2} \right) \frac{\hbar\sqrt{\omega_1\omega_2}}{2V} \tag{5.3.5}$$

$$\times \left(a_{\vec{k}_1\lambda_1} e^{i\vec{k}_1\cdot\vec{x}} - a^\dagger_{\vec{k}_1\lambda_1} e^{-i\vec{k}_1\cdot\vec{x}} \right) \left(a^\dagger_{\vec{k}_2\lambda_2} e^{-i\vec{k}_2\cdot\vec{x}} - a_{\vec{k}_2\lambda_2} e^{i\vec{k}_2\cdot\vec{x}} \right),$$

where $\Delta\rho = \rho' - \rho$. When the two quantized amplitude terms are multiplied together, there will be four products with two creation or destruction operators. The term with two creation operators and the term with two destruction operators do not conserve total energy, which means they will not contribute to first-order transitions, since Fermi's golden rule enforces energy conservation. Therefore, we will drop them in the following.

Let us assume that the defect is a sphere with radius a, as shown in Figure 5.4. We then have

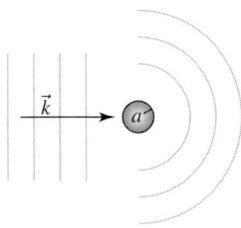

Rayleigh scattering from a spherical defect.

$$H_{\text{def-phon}} = \int_0^a r^2 dr \int_{-1}^1 2\pi \, d(\cos\theta) \frac{1}{2}\left(\frac{\Delta\rho}{\rho}\right) \sum_{\vec{k}_1,\lambda_1} \sum_{\vec{k}_2,\lambda_2} \left(\hat{\eta}_{\vec{k}_1\lambda_1} \cdot \hat{\eta}_{\vec{k}_2\lambda_2}\right) \frac{\hbar\sqrt{\omega_1\omega_2}}{2V}$$

$$\times \left(a_{\vec{k}_1} a_{\vec{k}_2}^\dagger e^{i|\vec{k}_1-\vec{k}_2|r\cos\theta} + a_{\vec{k}_1\lambda_1}^\dagger a_{\vec{k}_2\lambda_2} e^{-i|\vec{k}_1-\vec{k}_2|r\cos\theta}\right)$$

$$= \int_0^a 2\pi r^2 dr \frac{1}{2}\left(\frac{\Delta\rho}{\rho}\right) \sum_{\vec{k}_1,\lambda_1} \sum_{\vec{k}_2,\lambda_2} \left(\hat{\eta}_{\vec{k}_1\lambda_1} \cdot \hat{\eta}_{\vec{k}_2\lambda_2}\right) \frac{\hbar\sqrt{\omega_1\omega_2}}{2V}$$

$$\times \left(a_{\vec{k}_1\lambda_1} a_{\vec{k}_2\lambda_2}^\dagger \frac{2\sin(|\Delta\vec{k}|r)}{|\Delta\vec{k}|r} + a_{\vec{k}_1\lambda_1}^\dagger a_{\vec{k}_2\lambda_2} \frac{2\sin(|\Delta\vec{k}|r)}{|\Delta\vec{k}|r}\right), \tag{5.3.6}$$

where $\Delta k = \vec{k}_1 - \vec{k}_2$.

The integral over r gives

$$\int_0^a 2\pi r \, dr \frac{\sin(|\Delta\vec{k}|r)}{|\Delta\vec{k}|} = \frac{2\pi}{|\Delta\vec{k}|^3}\left(\sin(|\Delta\vec{k}|a) - (|\Delta\vec{k}|a)\cos(|\Delta\vec{k}|a)\right). \tag{5.3.7}$$

In the limit of long wavelength, $\lambda \gg a$, that is, $|\Delta\vec{k}|a \ll 1$ (the **Born** approximation), this becomes

$$\int_0^a 2\pi r \, dr \frac{\sin(|\Delta\vec{k}|r)}{|\Delta\vec{k}|} \simeq \frac{2\pi}{3}a^3. \tag{5.3.8}$$

Substituting this back into (5.3.6), we have

$$H_{\text{def-phon}} = \frac{1}{2}\left(\frac{\Delta\rho}{\rho}\right)\frac{2\pi}{3}a^3 \sum_{\vec{k}_1,\lambda_1} \sum_{\vec{k}_2,\lambda_2} \left(\hat{\eta}_{\vec{k}_1\lambda_1} \cdot \hat{\eta}_{\vec{k}_2\lambda_2}\right) \frac{\hbar\sqrt{\omega_1\omega_2}}{V}$$

$$\times \left(a_{\vec{k}_1\lambda_1} a_{\vec{k}_2\lambda_2}^\dagger + a_{\vec{k}_1\lambda_1}^\dagger a_{\vec{k}_2\lambda_2}\right). \tag{5.3.9}$$

Since the summation is over dummy variables $\vec{k}_1, \lambda_1$ and $\vec{k}_2, \lambda_2$, we can switch these for the sum over the $aa^\dagger$ term, and use the commutation property of the boson operators, to obtain

$$H_{\text{def-phon}} = \frac{1}{2}\left(\frac{\Delta\rho}{\rho}\right)\frac{2\pi}{3}a^3 \sum_{\vec{k}_1,\lambda_1} \sum_{\vec{k}_2,\lambda_2} \left(\hat{\eta}_{\vec{k}_1\lambda_1} \cdot \hat{\eta}_{\vec{k}_2\lambda_2}\right) \frac{\hbar\sqrt{\omega_1\omega_2}}{V}$$

$$\times \left(2a_{\vec{k}_1\lambda_1}^\dagger a_{\vec{k}_2\lambda_2} + \delta_{\vec{k}_1,\vec{k}_2}\right). \tag{5.3.10}$$

The $\delta_{\vec{k}_1,\vec{k}_2}$ term gives an infinite sum, as in the original phonon Hamiltonian, which is a constant that does not affect the interactions of the quasiparticles. We subtract this, since it is a constant, to obtain

$$\boxed{H_{\text{def-phon}} = \left(\frac{\Delta\rho}{\rho}\right)\frac{2\pi}{3}a^3 \sum_{\vec{k}_1,\lambda_1} \sum_{\vec{k}_2,\lambda_2} \left(\hat{\eta}_{\vec{k}_1\lambda_1} \cdot \hat{\eta}_{\vec{k}_2\lambda_2}\right) \frac{\hbar\sqrt{\omega_1\omega_2}}{V} a_{\vec{k}_1\lambda_1}^\dagger a_{\vec{k}_2\lambda_2}.} \tag{5.3.11}$$

This is the interaction for elastic scattering from a single defect, sometimes called **Rayleigh** scattering.

As before, we can insert the interaction Hamiltonian into Fermi's golden rule to find the scattering rate. We start with an initial state of N_i phonons with wave vector $\vec{k}$ and polarization λ, and sum over all possible final states, to obtain the scattering rate per phonon,

$$\frac{1}{\tau} = \frac{1}{N_i} \frac{2\pi}{\hbar} \sum_f |\langle f|H_{\text{int}}|i\rangle|^2 \delta(E_i - E_f)$$

$$= \frac{1}{N_i} \frac{2\pi}{\hbar} \left(\frac{\Delta\rho}{\rho}\right)^2 \left(\frac{2\pi}{3}a^3\right)^2 \sum_{\vec{k}',\lambda'} \left(\hat{\eta}_{\vec{k}\lambda} \cdot \hat{\eta}_{\vec{k}'\lambda'}\right)^2 (1 + N_{\vec{k}})N_i \frac{\hbar^2\omega_{\vec{k}}^2}{V^2} \delta(E_{\vec{k}} - E_{\vec{k}'}).$$

$$(5.3.12)$$

We can convert the sum to an integral using the standard prescription (1.8.1). The energy δ-function forces the magnitude of the outgoing k-vector to be the same as the initial one, so that the integral over final states corresponds to a sum over all possible angles. We then have

$$\frac{1}{\tau} = \frac{2\pi}{\hbar} \left(\frac{\Delta\rho}{\rho}\right)^2 \left(\frac{2\pi}{3}a^3\right)^2 \frac{V}{(2\pi)^3}$$

$$\times \sum_{\lambda'} \int d^3k' (1 + N_k) \frac{\hbar^2\omega_k^2}{V^2} \left(\hat{\eta}_{\vec{k}\lambda} \cdot \hat{\eta}_{\vec{k}'\lambda'}\right)^2 \delta(E_k - E_{k'}). \quad (5.3.13)$$

Note that the momentum is not conserved in the scattering process; in princple, the defect recoils in response to scattering from a quasiparticle, but since its mass is assumed to be much greater, the recoil is negligible.

The integral over the polarizations is tricky; we assume $\hat{k} = \hat{z}$, $\hat{\eta}_{\vec{k}\lambda} = \hat{x}$, and write

$$\hat{k}' = (\sin\theta\cos\phi)\hat{x} + (\sin\theta\sin\phi)\hat{y} + (\cos\theta)\hat{z},$$
$$\hat{\eta}'_1 = (\cos\theta\cos\phi)\hat{x} + (\cos\theta\sin\phi)\hat{y} - (\sin\theta)\hat{z},$$
$$\hat{\eta}'_2 = -(\sin\phi)\hat{x} + (\cos\phi)\hat{y},$$
$$\hat{\eta}'_3 = \hat{k}', \quad (5.3.14)$$

where the $\lambda = 1$ and $\lambda = 2$ polarizations are pure transverse modes, and $\lambda = 3$ is the longitudinal phonon. The integral over angle then gives us a factor of $2\pi/3$ for the transverse polarization 1, a factor of 2π for transverse polarization 2, and a factor of $4\pi/3$ for the longitudinal scattered polarization.

Assuming that we have a single linear phonon branch with $\omega = vk$, when the algebra is all done, the rate of scattering is

$$\frac{1}{\tau} = \frac{4\pi}{9} \left(\frac{\Delta\rho}{\rho}\right)^2 k^4 a^6 (1 + N_k)\frac{v}{V}. \quad (5.3.15)$$

This is the total rate of scattering per phonon from a single defect of radius a in a volume V. If there are N_d defects per volume V, then scattering rate per phonon is just

$$\frac{1}{\tau} = \frac{4\pi}{9} \left(\frac{\Delta\rho}{\rho}\right)^2 k^4 a^6 (1 + N_k)n_d v, \quad (5.3.16)$$

where $n_d = N_d/V$ is the density of defects.

The scattering rate is related to the total cross-section σ for scattering by the relation

$$\frac{1}{\tau} = \sigma n v, \tag{5.3.17}$$

where n is the density of particles, v is the velocity, and σ is the total scattering cross-section. In the limit when there is no stimulated scattering, the total cross-section for Rayleigh scattering from a spherical defect is therefore

$$\sigma = \frac{4\pi}{9} \left(\frac{\Delta\rho}{\rho}\right)^2 k^4 a^6. \tag{5.3.18}$$

The Rayleigh scattering rate is therefore proportional to the fourth power of the frequency. The prefactor, which in this case is $4\pi/9$, depends on the exact geometry of the defect, but it is generally the case that the scattering rate is proportional to the size to the sixth power.

The strong ω^4 frequency dependence of the scattering time for phonons leads to many of the effects we know. Low-frequency sound, in the acoustic frequency range 0–10 kHz, has extremely low scattering, with a mean free path of kilometers in typical materials. Ultrasound, at frequencies of 1–10 MHz, has a mean free path of millimeters to centimeters, while high-frequency phonons, which lie in the terahertz range, which corresponds to room temperature (30 meV = 4.5×10^{13} Hz = 45 THz), have a microscopic mean free path and therefore move only diffusively. This is what we know as **heat flow**, which will be discussed in Section 5.7. As we will see in Section 5.4, there is also a very strong frequency dependence for phonon–phonon scattering.

Exercise 5.3.1   If we assume that scattering of phonons leads to breakup of the coherent properties of sound, then the attenuation of sound in a solid can be computed using the above formulas. Calculate the **attenuation length** of microwave sound with frequency 10 MHz in a solid with speed of sound 2×10^5 cm/s and a defect density of 10^{12} impurities per cm^3, if the size of the impurities is roughly 20 μm, and the impurities have density variation $\Delta\rho/\rho = 0.1$. The attenuation length l is defined by the relation

$$N_{\text{coh}}(x) = N_{\text{coh}}(0)e^{-x/l}, \tag{5.3.19}$$

where $N_{\text{coh}}(0)$ is the initial number of phonons in the coherent wave, and $N_{\text{coh}}(x)$ is the number of unscattered phonons remaining after the wave has traveled a distance x.

Photon-defect scattering. An exactly analogous calculation can be performed for Rayleigh scattering of photons. In this case, the classical energy is, in the MKS system (see, e.g., Jackson 1999: s. 4.7),

$$H = -\frac{1}{2} \int_{V'} d^3x \, \vec{P} \cdot \vec{E}, \tag{5.3.20}$$

where $\vec{P}$ is the polarization induced by the electric field, which for a dielectric sphere of radius a with index of refraction n, surrounded by a vacuum, is equal to

$$\vec{P} = 3\epsilon_0 \left(\frac{n^2 - 1}{n^2 + 2}\right) \vec{E}. \tag{5.3.21}$$

Using the definition (4.3.20) for the electric field in terms of the quantum mechanical creation and destruction operators, we find

$$H_{\text{def–phot}} = -\left(\frac{n^2 - 1}{n^2 + 2}\right) 2\pi a^3 \sum_{\vec{k}_1,\lambda_1} \sum_{\vec{k}_2,\lambda_2} \left(\hat{\eta}_{\vec{k}_1\lambda_1} \cdot \hat{\eta}_{\vec{k}_2\lambda_2}\right) \frac{\hbar\sqrt{\omega_1\omega_2}}{V} a^{\dagger}_{\vec{k}_1\lambda_1} a_{\vec{k}_2\lambda_2}. \tag{5.3.22}$$

Inserting this into Fermi's golden rule, we find the scattering rate, for two transverse polarizations,

$$\frac{1}{\tau} = \frac{8\pi}{3} \left(\frac{n^2 - 1}{n^2 + 2}\right)^2 k^4 a^6 (1 + N_k) n_d c, \tag{5.3.23}$$

and the total cross-section per defect, in the absence of stimulated emission,

$$\sigma = \frac{8\pi}{3} \left(\frac{n^2 - 1}{n^2 + 2}\right)^2 k^4 a^6, \tag{5.3.24}$$

which is the same as the classical Rayleigh scattering cross-section (e.g. Jackson 1999: s. 10.1.B). The strong frequency dependence for Rayleigh scattering of light explains why the sky is blue (when we look at scattered light from the atmosphere) and sunsets are red (when we look at the remaining sunlight after much scattering in the atmosphere).

Exercise 5.3.2   If the Rayleigh scattering of light in the atmosphere is known to be dominated by scattering from a particulate with radius 50 nm and index of refraction $n = 2$, estimate the particulate density if half of the light at wavelength 500 nm is scattered after traveling through 50 km of the atmosphere.

Electron–defect scattering. Last, we can do the same calculation for **electron** waves scattering from defects. In this case, the interaction energy is simply the change of the potential energy at the defect times the probability of finding an electron there. We write

$$H_{\text{def–e}} = \int_{V'} d^3x \, \Psi_n^{\dagger}(\vec{x}) \, \Delta U \, \Psi_n(\vec{x}), \tag{5.3.25}$$

where ΔU is the change in the electron energy in the region of the defect. Using the definition of the Ψ operators, we have

$$H_{\text{def–e}} = \int_{V'} d^3x \sum_{\vec{k}_1,\vec{k}_2} \left(\frac{e^{-i\vec{k}_2\cdot\vec{x}}}{\sqrt{V}} u^*_{n\vec{k}_2}(\vec{x}) \, b^{\dagger}_{n\vec{k}_2}\right) \Delta U \left(\frac{e^{i\vec{k}_1\cdot\vec{x}}}{\sqrt{V}} u_{n\vec{k}_1}(\vec{x}) \, b_{n\vec{k}_1}\right). \tag{5.3.26}$$

Following the procedure of Sections 5.1 and 5.2, we assume that the wavelength is long compared to the unit cell and separate the integration over the unit cell and the long-range integration. In the case of scattering within a single band, the integral over the Bloch functions over the unit cell just gives a factor of unity. For the long-range interaction, we assume that the defect is a sphere of radius a, as above, and assume $|\Delta \vec{k}|a \ll 1$. This gives us

$$H_{\text{def-e}} = \frac{\Delta U}{V} \sum_{\vec{k}_1, \vec{k}_2} \int_0^a r^2 dr \int_{-1}^1 2\pi \, d(\cos \theta) \, e^{i|\vec{k}_1 - \vec{k}_2|r \cos \theta} \, b^\dagger_{n\vec{k}_2} b_{n\vec{k}_1}$$

$$= \frac{\Delta U}{V} \sum_{\vec{k}_1, \vec{k}_2} \int_0^a r^2 dr \, 2\pi \, \frac{2 \sin(|\Delta \vec{k}|r)}{|\Delta \vec{k}|r} \, b^\dagger_{n\vec{k}_2} b_{n\vec{k}_1}, \tag{5.3.27}$$

or

$$H_{\text{def-e}} = \frac{\Delta U}{V} \left(\frac{4\pi a^3}{3} \right) \sum_{\vec{k}_1, \vec{k}_2} b^\dagger_{n\vec{k}_2} b_{n\vec{k}_1}. \tag{5.3.28}$$

In other words, the interaction energy is just a constant, given by the energy times the volume of the defect, for each possible scattering process.

The scattering rate from defects for electrons is found from Fermi's golden rule as usual. The density of states of electrons gives a very different frequency dependence, however. For an initial electron state with $\vec{k}$, we have

$$\frac{1}{\tau} = \frac{2\pi}{\hbar} \left(\frac{\Delta U}{V} \right)^2 \left(\frac{4\pi a^3}{3} \right)^2 \sum_{\vec{k}'} (1 - N_{\vec{k}'}) \delta(E_{\vec{k}'} - E_{\vec{k}})$$

$$= \frac{2\pi}{\hbar} \left(\frac{\Delta U}{V} \right)^2 \left(\frac{4\pi a^3}{3} \right)^2 (1 - N_{\vec{k}}) \mathcal{D}(E_k), \tag{5.3.29}$$

which, for electrons with an effective mass, with density of states (2.2.9), gives a scattering rate proportional to $\sqrt{E - E_0}$, that is, linear with k.

In all of the above, we have assumed that the defect is spherical. The same approach can be used for other types of defects, such as line defects. In this case, one integrates over the appropriate volume V' instead of a sphere of radius a.

Phase factors. In all of the above, we have assumed a defect centered at $\vec{x} = 0$ to calculate the scattering cross-section, and then to treat the case of N defects, we have just multiplied the cross-section of a single defect by N. This procedure actually sweeps under the rug a subtle issue which can be important in some calculations. In the above, we have assumed that the defect potential energy function is given by $U(\vec{x}) = \Delta U \Theta(a - |\vec{x}|)$, where $\Theta(x)$ is the Heaviside function, equal to 1 if $x > 1$ and 0 otherwise. More generally, we can write $U(\vec{x}) = U_{\text{def}}(\vec{x} - \vec{R})$, where U_{def} is a function which is the same for all defects (in the present case, $U_{\text{def}}(\vec{x}) = \Delta U \Theta(a - |\vec{x}|)$), and $\vec{R}$ is the location of the center of the defect. This means that the interaction Hamiltonian for multiple defects is given by, in the case of electrons,

$$H_{\text{def-e}} = \sum_i \int d^3x \, \Psi_n^\dagger(\vec{x}) \, U(\vec{x} - \vec{R}_i) \, \Psi_n(\vec{x}), \tag{5.3.30}$$

where we sum over all defect positions. If we substitute in the definitions of the $\Psi(\vec{x})$ operators in terms of plane wave operators, as we have for each case, as in (5.3.26), then a simple change of variables gives us

$$
H_{\text{def-e}} = \sum_i \sum_{\vec{k}_1, \vec{k}_2} e^{-i(\vec{k}_1 - \vec{k}_2) \cdot \vec{R}_i}\, b_{n\vec{k}_2}^{\dagger}\, b_{n\vec{k}_1}
$$

$$
\times \left(\frac{1}{V} \int d^3 x\, e^{i(\vec{k}_1 - \vec{k}_2) \cdot \vec{x}}\, u_{n\vec{k}_2}^*(\vec{x})\, U(\vec{x})\, u_{n\vec{k}_1}(\vec{x}) \right), \tag{5.3.31}
$$

where the latter term in parentheses does not depend on $\vec{R}_i$. The matrix element for scattering from $\vec{k}$ to $\vec{k}'$ is therefore given by

$$
\langle f | H_{\text{def-e}} | i \rangle = \sum_i e^{-i(\vec{k} - \vec{k}') \cdot \vec{R}_i} \langle f | H_{\text{def-e}}^{(0)} | i \rangle, \tag{5.3.32}
$$

where $\langle f | H_{\text{def-e}}^{(0)} | i \rangle$ is the single-defect matrix element, and the square of the matrix element which enters into Fermi's golden rule is

$$
|\langle f | H_{\text{def-e}}^{(0)} | i \rangle|^2 = \sum_{ij} e^{-i(\vec{k} - \vec{k}') \cdot (\vec{R}_i - \vec{R}_j)} |\langle f | H_{\text{def-e}}^{(0)} | i \rangle|^2. \tag{5.3.33}
$$

The sum of the diagonal terms, when $i = j$, gives just $N |\langle f | H_{\text{def-e}}^{(0)} | i \rangle|^2$ since the phase factors are all equal to unity. If there are a large number of randomly distributed defects, then we can assume that the off-diagonal terms all cancel out, since they have random phase factors which can be both negative and positive. This is consistent with the assumption we made above, that the cross-section of N defects is just N times the cross-section of one defect. If there are only a few defects, however, or if there is a definite correlation of the positions of some defects, for example pairs of defects, then the off-diagonal terms may be important. This is true for phonon and photon scattering as well as electrons.

5.4 Phonon–Phonon Interactions

Besides interactions between electrons and phonons or photons, there is also a self-interaction of phonons with other phonons and of electrons with other electrons. The photon–photon interaction is much weaker, and will be treated in Chapter 7. Here we treat the interaction of phonons with each other. For an extensive review of phonon–phonon interactions, see Srivastava (1990).

In Chapter 3, we introduced the elasticity tensor, which allowed us to write down a generalized Hooke's law. This was justified by modeling atoms in a solid as though they are connected by springs; in other words, as discussed in Section 3.1.1, the potential energy near a minimum can always be approximated as quadratic in the displacement. This is, of course, an approximation. For large displacements from equilibrium, non-quadratic terms in the interaction energy will become important, as illustrated in Figure 5.5. In general, we can write the energy density due to deformation as a Taylor series,

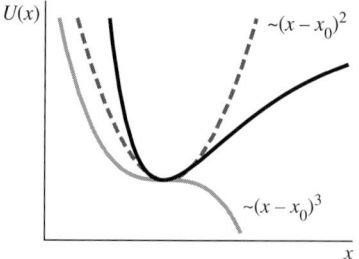

The interatomic potential can be viewed as the sum of a harmonic potential (dashed line) plus a cubic potential (dotted-dashed line) plus higher-order terms.

$$\frac{U}{V} = \frac{U_0}{V} + \sum_{ij} C_{ij}^{(1)} \varepsilon_{ij} + \frac{1}{2} \sum_{ijkl} C_{ijkl}^{(2)} \varepsilon_{ij} \varepsilon_{kl} + \frac{1}{6} \sum_{ijklmn} C_{ijklmn}^{(3)} \varepsilon_{ij} \varepsilon_{kl} \varepsilon_{mn} + \cdots . \tag{5.4.1}$$

The first term is a constant which can be ignored, while the second term, which is linear with strain, vanishes near an equilibrium point. The second-order term is the energy of the elastic-constant Hooke's law, given in (3.4.18). The third-order term, however, can contribute important corrections to the total energy. We can rewrite the series up to the third order as

$$\frac{U}{V} = \frac{1}{2} \sum_{ijkl} \left(C_{ijkl}^{(2)} + \frac{1}{3} \sum_{mn} C_{ijklmn}^{(3)} \varepsilon_{mn} \right) \varepsilon_{ij} \varepsilon_{kl}. \tag{5.4.2}$$

The term with $C_{ijklmn}^{(3)}$ in the parentheses gives a correction to the spring constants, that is, to the elastic-constant tensor. It implies that the spring constant has a correction term which increases linearly with strain. This correction can be parametrized as follows. From Section 4.2, we have the general result that the frequency of a phonon mode is equal to

$$\omega = \sqrt{\frac{C}{\rho}} k, \tag{5.4.3}$$

where C is the effective elastic constant and ρ is the mass density. If the spring constant changes linearly with strain, then we have (dropping subscripts)

$$\Delta \omega = \frac{1}{2} \omega \frac{\Delta C}{C} = \frac{1}{2} \omega \frac{C^{(3)} \varepsilon}{C} \tag{5.4.4}$$

or

$$\gamma = \frac{1}{\omega} \frac{\Delta \omega}{\varepsilon} = \frac{1}{2} \frac{C^{(3)}}{C} = \text{const.} \tag{5.4.5}$$

The unitless constant γ is defined as the **Gruneisen parameter** for a given mode. The anharmonicity of the modes can therefore be characterized in terms of the shift of frequency of the modes with strain. The Gruneisen parameters of the acoustic phonon modes can be measured by measuring the phonon speed for varying static strains.

The anharmonic terms control the phonon–phonon interactions. In Section 4.2, we found the phonons as eigenstates of the quadratic Hamiltonian. We can treat the anharmonic term as a small perturbation of the total energy, and use the perturbation theory of quantum mechanics to give the rate of transition between phonon states using Fermi's golden rule. Again dropping subscripts, we have the interaction energy

$$H_{\text{anharm}} = \int dV \, \frac{1}{3} \gamma C \varepsilon^3. \tag{5.4.6}$$

In general, the ε^3 term will give rise to $(ku)^3$ terms, where, as in Section 5.1,

$$u(\vec{x}) = \sum_{\vec{k}} \sqrt{\frac{\hbar}{2\rho V \omega_{\vec{k}}}} \left(a_{\vec{k}} e^{i\vec{k}\cdot\vec{x}} + a_{\vec{k}}^{\dagger} e^{-i\vec{k}\cdot\vec{x}} \right). \tag{5.4.7}$$

(We ignore the polarization dependence, for simplicity.) It is easy to see that if we substitute these into the Hamiltonian (5.4.6) we will have terms with all possible combinations of three creation and destruction operators. This reflects the fact that phonons are not conserved – a phonon at one frequency can turn into two new ones with lower frequency, two phonons can combine to form a third one, etc. This is a type of inelastic scattering. In terms of Fermi's golden rule, (5.4.6) and (5.4.7) lead to, for a process in which one phonon turns into two others,

$$\frac{1}{\tau_1} = \frac{2\pi}{\hbar} \frac{\gamma^2 C^2}{9} \frac{\hbar^3}{8\rho^3 V v^3} \sum_{\vec{k}_2} k_1 k_2 k_3 (1 + N_{k_3})(1 + N_{k_2}) \, \delta(\hbar v k_1 - \hbar v k_2 - \hbar v k_3), \tag{5.4.8}$$

where we have assumed $\omega_i = v k_i$. The sum is over only one final wave vector because the other is determined by momentum conservation $\vec{k}_1 = \vec{k}_2 + \vec{k}_3$.

Without going into detail on how to calculate the absolute scattering time, we can make several general statements about phonon–phonon scattering based on the form of (5.4.8). If we convert the sum over k_2 to an integral, we will effectively multiply by the density of final states. For acoustic phonons, as we have deduced in Section 4.9.1, the density of states is proportional to ω^2. Therefore, the total rate of inelastic phonon–phonon scattering will in general scale as ω^5. As seen in Section 5.3, the elastic scattering rate with defects scales as ω^4. The strong frequency dependence leads to the experience that sound (low-frequency acoustic waves) and heat (high-frequency acoustic waves) have very different transport behaviors.

The analogy of photons with phonons leads one naturally to expect photon–photon interactions. As we will discuss in Section 7.6, photon–photon interaction is normal in a solid medium, but normally does not occur in vacuum. Actually, at very high intensity the photon states are no longer good eigenstates of the vacuum. This manifests itself as a photon–photon interaction, analogous the phonon–phonon interaction discussed here. Another consequence, however, is that at high amplitude, photons are not eigenstates of the Hamiltonian, just as at high amplitude, the energy minimum in the interaction between two atoms is no longer perfectly harmonic, and the eigenstates of the system are no longer the phonon states deduced in Section 4.2. In this case, one cannot talk of the number of photons in a vacuum in any meaningful way.

Exercise 5.4.1   Show that the scattering rate formula (5.4.8) follows from the form of the Hamiltonian (5.4.6).

5.4.1 Thermal Expansion

The anharmonic terms also control the **thermal expansion** of a material. Since, by definition, we cannot assume constant volume when determining the thermal expansion, we must work a bit at the thermodynamics to come up with the proper equation of state. We begin with the formula for pressure,

$$P = -\frac{\partial F}{\partial V}\bigg|_T, \tag{5.4.9}$$

where $F = U - TS$ is the Helmholtz free energy, and write the entropy as

$$S = \int_0^T \frac{1}{T'}\frac{\partial U}{\partial T'}dT', \tag{5.4.10}$$

where $S = 0$ at $T = 0$. Writing the total energy as

$$U = \sum_{\vec{k}} \hbar\omega_{\vec{k}}n_{\vec{k}}, \tag{5.4.11}$$

we have

$$TS = T\int_0^T \frac{1}{T'}\sum_{\vec{k}} \hbar\omega_{\vec{k}}\frac{\partial n_{\vec{k}}}{\partial T'}dT'. \tag{5.4.12}$$

Using the Planck occupation number $N_k = (e^{\hbar\omega_{\vec{k}}/k_BT} - 1)^{-1}$, converting the variable of integration from T' to $x = \hbar\omega_{\vec{k}}/k_BT'$ and integrating by parts, we obtain, after some mathematics,

$$TS = U + k_BT\sum_{\vec{k}} \int_{\frac{\hbar\omega_{\vec{k}}}{k_BT}}^{\infty} \frac{1}{e^x - 1}dx, \tag{5.4.13}$$

which implies, carrying through the math,

$$P = -\frac{\partial}{\partial V}(U - TS)$$

$$= -\sum_{\vec{k}} \hbar\frac{\partial\omega_{\vec{k}}}{\partial V}N_{\vec{k}}. \tag{5.4.14}$$

Converting the sum to an integral using the density of states as in (4.9.2), assuming that the system is isotropic, we have

$$P = -\int \frac{\partial E}{\partial V}f(E)\mathcal{D}(E)dE, \tag{5.4.15}$$

where $E = \hbar\omega$. This is the **equation of state** for this system.

The frequency shift with volume is given by the Gruneisen parameter, according to definition (5.4.5),

$$\gamma(\omega) = -\frac{V}{\omega}\frac{\partial\omega}{\partial V}, \tag{5.4.16}$$

where we have used the hydrostatic strain $\epsilon = \partial V / V$, and we allow for the possibility that γ is different for different ω. Therefore, the equation of state is

$$P = -\frac{1}{V} \int \gamma(E) \, Ef(E)\mathcal{D}(E)dE. \tag{5.4.17}$$

We can now use this equation of state to determine the thermal expansion. By implicit differentiation,

$$dV(T,P) = \frac{\partial V}{\partial T}\bigg|_P dT + \frac{\partial V}{\partial P}\bigg|_T dP, \tag{5.4.18}$$

and therefore,

$$\frac{1}{V}\frac{\partial V}{\partial T}\bigg|_P = -\frac{\dfrac{\partial P}{\partial T}\bigg|_V}{V \dfrac{\partial P}{\partial V}\bigg|_T}. \tag{5.4.19}$$

The denominator on the right-hand side is simply the bulk modulus, which relates the hydrostatic stress to the hydrostatic strain, and is given in terms of the elastic constants for an isotropic solid by $B = \frac{1}{3}(C_{11} + 2C_{12})$. From the equation of state (5.4.17), we have

$$\frac{\partial P}{\partial T}\bigg|_V = \frac{1}{V} \int \gamma(E) \, E\frac{\partial f}{\partial T}\mathcal{D}(E)dE. \tag{5.4.20}$$

Apart from the $\gamma(E)$ factor, the integral is just the equation for the specific heat of the phonons, discussed in Section 4.9.3. We therefore define the average Gruneisen parameter as

$$\gamma = \frac{\displaystyle\int \gamma(E)c_V(E)dE}{\displaystyle\int c_V(E)dE} = -\frac{1}{C_V}\frac{\partial P}{\partial T}\bigg|_V, \tag{5.4.21}$$

where $c_V(E)dE$ is the contribution to the specific heat by a given energy range of phonon states. We then have, for the coefficient of thermal expansion,

$$\frac{1}{V}\frac{\partial V}{\partial T}\bigg|_P = \frac{\gamma C_V}{B}. \tag{5.4.22}$$

For an isotropic system, the volume expansion is related simply to the linear expansion by defining $V = l^3$, which yields

$$\frac{1}{l}\frac{\partial l}{\partial T} = \frac{1}{l}\frac{\partial l}{\partial V}\frac{\partial V}{\partial T} = \frac{1}{3V}\frac{\partial V}{\partial T}, \tag{5.4.23}$$

and therefore,

$$\alpha = \frac{1}{l}\frac{\partial l}{\partial T}\bigg|_P = \frac{\gamma C_V}{3B}. \tag{5.4.24}$$

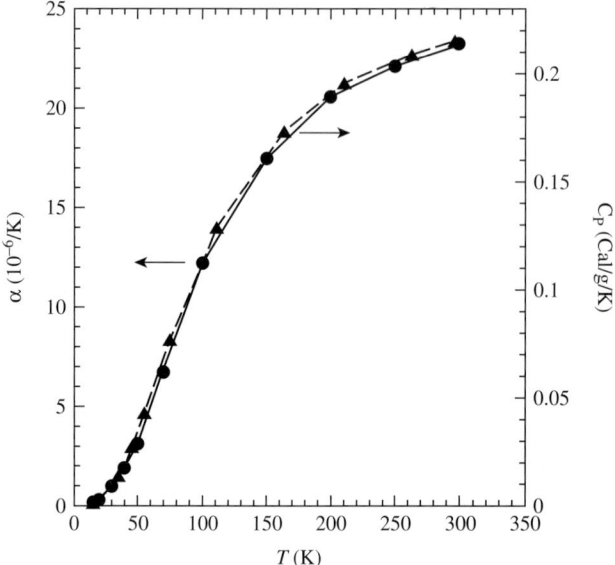

Fig. 5.6 Comparison of volume expansion data and specific heat for aluminum. Data from Fraser and Hallett (1961) and Giaque and Meads (1941).

Figure 5.6 shows a comparison of volume expansion data and specific heat for aluminum. The average Gruneisen parameter has been obtained by fitting the volume expansion data to the specific heat data by an overall multiplicative factor.

Exercise 5.4.2 Fill in the missing mathematical steps from (5.4.12) to (5.4.14).

5.4.2 Crystal Phase Transitions

Figure 5.7 shows a possible set of vibrational modes in a two-atom linear chain model. Recall that in Section 5.4 we defined the Gruneisen parameter

$$\gamma = \frac{1}{\omega}\frac{\Delta\omega}{\varepsilon}, \tag{5.4.25}$$

which gives the shift of frequency of a phonon mode for a given strain ε, due to the anharmonic terms in the vibrational Hamiltonian. It is quite common for the Gruneisen parameters of some phonon modes to be negative. (This is called **softening** of a phonon mode, rather than red shift, since sound doesn't have color!) What would happen if we put the crystal under strain, shifting the phonon mode at the zone boundary downward until it reaches zero frequency? In this case, there would be no energy penalty for having a new long-range ordering with wave vector k. If the phonon mode continued to shift downward with increasing strain, it would actually become energetically favorable for the crystal to go into a state with this new long-range order.

This type of pressure-driven phase transition between different crystal symmetries has been seen experimentally. The pressures needed are very high, since solids are not

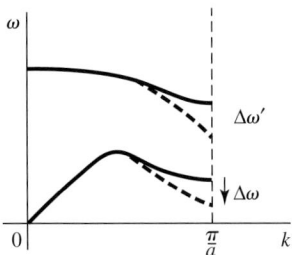

A phonon dispersion relation with phonon mode softening, leading to a structural phase transition under pressure.

very compressible, as discussed in Section 2.4.1. We can estimate the pressure needed by noting that a typical value of a Young's modulus (see Section 3.4.1) is 100 kbar. For $\Delta\omega$ to be an appreciable fraction of ω, we expect that the strain ε should be an appreciable fraction of unity. This means we expect that pressures around 100 kbar are needed.

The same anharmonic effects come into the topic of phase transitions when temperature is changed (e.g., melting). As discussed in Section 5.4, the anharmonic terms become important only at high strain. This can be static strain due to pressure, or it can be strain due to high-amplitude thermal vibrations. It is a reasonable assumption that when the amplitude of the vibrations is a significant fraction of the lattice constant a of a crystal, the spring constant may soften so much that there is essentially no restoring force on an atom – it can leave and never come back. In this case, the crystal is melted.

From (4.10.5) in Section 4.10, the average amplitude of the displacement is connected to the average occupation number of a phonon mode:

$$\langle x^2 \rangle = \frac{2\hbar}{\rho V \omega_k} \sum_k N_k$$

$$= \frac{2\hbar}{M \omega_k} \left(\frac{1}{N} \sum_k N_k \right), \tag{5.4.26}$$

where $M = \rho V_{\text{cell}}$ is the mass of the unit cell. In the high-temperature limit, the occupation number N_k of the phonon mode is given by

$$\frac{1}{e^{\hbar\omega/k_B T} - 1} \approx \frac{k_B T}{\hbar\omega}. \tag{5.4.27}$$

Let us pick a characteristic vibrational frequency of the solid at high temperature, for example $\hbar\omega = k_B \Theta$, where Θ is Debye temperature defined in Section 4.9.3, roughly equal to the acoustic phonon frequency at the zone boundary, or to the optical phonon frequency.

If we assume that melting occurs when the rms average of x is some fixed fraction f of the lattice constant a, then we have the condition at the melting temperature T_m

$$N_k = \frac{k_B T_m}{k_B \Theta} = \frac{M\omega(fa)^2}{2\hbar} \tag{5.4.28}$$

or

$$T_m = f^2 \frac{M k_B \Theta^2 a^2}{2\hbar^2}. \tag{5.4.29}$$

This is known as the **Lindemann melting criterion**, which relates the melting temperature to the Debye temperature. It agrees reasonably well with experimental data, for f of about 10%.

As we will discuss in Section 2.9, near a phase transition there are large fluctuations. In the case of a softening phonon mode, the fluctuations correspond simply to vibrational modes, which are excited thermally when the energy of the phonons has been shifted down near to zero. In general, the theory of crystal symmetry phase transitions has much in common with the theory of other classical phase transitions discussed in Chapter 10. Phase transitions between different crystal symmetries are discussed at length by Toledano and Toledano (1987).

Exercise 5.4.3   Estimate the melting temperature of a solid with lattice constant 6 Å, optical phonon energy 30 meV, and effective mass of the unit cell given by the mass of a silicon atom. Does your answer seem reasonable?

5.5 Electron–Electron Interactions

By now, the method of writing down a quantum mechanical interaction Hamiltonian for quasiparticles should be familiar. We can do the same for electron–electron interactions, starting with the known, classical potential.

The Coulomb energy of interaction between two particles can be written as

$$U(\vec{r}_1 - \vec{r}_2) = \frac{e^2}{4\pi\epsilon|\vec{r}_1 - \vec{r}_2|} \Psi^\dagger_{ns}(\vec{r}_1)\Psi^\dagger_{n's'}(\vec{r}_2)\Psi_{n's'}(\vec{r}_2)\Psi_{ns}(\vec{r}_1), \tag{5.5.1}$$

where n, n' and s, s' label the band and the spins of the electrons. The order of the field operators is important – the four field operators are a **correlation function**, that is, the Coulomb potential is proportional to the probability of finding a particle at $\vec{r}_2$ given a particle already at $\vec{r}_1$. For simplicity, in the following we drop the Bloch cell functions and band and spin indices, and treat the only case of identical electrons scattering within a single band.

In this case, using the field operators (4.6.4) for the electrons gives the total energy of interaction in the system

$$\begin{aligned}
H_{\text{int}} &= \frac{1}{2}\int d^3r_1 d^3r_2\, U(\vec{r}_1 - \vec{r}_2) \\
&= \frac{1}{2V^2}\int d^3r_1 d^3r_2 \sum_{\vec{k}_1,\vec{k}_2,\vec{k}_3,\vec{k}_4} \frac{e^2}{4\pi\epsilon|\vec{r}_1 - \vec{r}_2|} e^{i(-\vec{k}_4\cdot\vec{r}_1 - \vec{k}_3\cdot\vec{r}_2 + \vec{k}_2\cdot\vec{r}_2 + \vec{k}_1\cdot\vec{r}_1)} \\
&\quad \times b^\dagger_{\vec{k}_4} b^\dagger_{\vec{k}_3} b_{\vec{k}_2} b_{\vec{k}_1},
\end{aligned} \tag{5.5.2}$$

where, once again, we have used the long-wavelength approximation to separate out the integral of the Bloch functions over the unit cell, which just gives a factor of unity for scattering within a single band. The factor of $\frac{1}{2}$ adjusts for the double counting in the integral over $\vec{r}_1$ and $\vec{r}_2$.

The integral over $\vec{r}_1$ and $\vec{r}_2$ is a Fourier transform which can be computed. Changing variables to $\vec{R} = (\vec{r}_1 + \vec{r}_2)/2$ and $\vec{r} = \vec{r}_1 - \vec{r}_2$, we write

$$
\frac{1}{2V^2} \int d^3R\, d^3r \frac{e^2}{4\pi\epsilon r} e^{i(\vec{k}_1+\vec{k}_2-\vec{k}_3-\vec{k}_4)\cdot\vec{R}} e^{i(\vec{k}_1-\vec{k}_2+\vec{k}_3-\vec{k}_4)\cdot\vec{r}/2}
$$

$$
= \frac{(2\pi)^3}{V^2} \delta^3(\vec{k}_1+\vec{k}_2-\vec{k}_3-\vec{k}_4) \int 2\pi r^2 dr\, d(\cos\theta) \frac{e^2}{4\pi\epsilon r} e^{i|\Delta\vec{k}|r\cos\theta}
$$

$$
= \frac{(2\pi)^3}{V^2} \delta^3(\vec{k}_1+\vec{k}_2-\vec{k}_3-\vec{k}_4) \frac{e^2}{\epsilon} \int_0^\infty dr\, \frac{\sin(|\Delta\vec{k}|r)}{|\Delta\vec{k}|}, \tag{5.5.3}
$$

where we have defined $\Delta\vec{k} = (\vec{k}_1 - \vec{k}_4) = (\vec{k}_3 - \vec{k}_2)$, which can be viewed as the momentum exchanged in a collision. The δ-function arises from using the mathematical relation

$$
\int_{-\infty}^\infty dx\, e^{ikx} = 2\pi\,\delta(k), \tag{5.5.4}
$$

and ensures momentum conservation $\vec{k}_1 + \vec{k}_2 = \vec{k}_3 + \vec{k}_4$ in a collision.

The final integral is ill-defined for the upper limit $r = \infty$, but we can solve this problem if we make a small adjustment to the form of the Coulomb potential. If we replace the Coulomb potential with the **screened** Coulomb potential,

$$
\frac{e^2}{4\pi\epsilon r} \rightarrow \frac{e^2}{4\pi\epsilon r} e^{-\kappa r}, \tag{5.5.5}
$$

then we have

$$
H_{\text{e-e}} = \frac{1}{2} \sum_{\vec{k}_1,\vec{k}_2,\vec{k}_3,\vec{k}_4} \frac{(2\pi)^3}{V^2} \delta^3(\vec{k}_1+\vec{k}_2-\vec{k}_3-\vec{k}_4) \frac{e^2}{\epsilon} \left(\int_0^\infty dr\, \frac{\sin(|\Delta\vec{k}|r)}{|\Delta\vec{k}|} e^{-\kappa r} \right)
$$

$$
\times b_{\vec{k}_4}^\dagger b_{\vec{k}_3}^\dagger b_{\vec{k}_2} b_{\vec{k}_1}, \tag{5.5.6}
$$

where the integral in the parentheses is now well defined. The momentum-conserving δ-function eliminates the summation over one momentum variable (after converting the sum to an integral), so that after we perform the integral, we finally have

$$
\boxed{ H_{\text{e-e}} = \frac{1}{2V} \sum_{\vec{k}_1,\vec{k}_2,\vec{k}_3} \frac{e^2/\epsilon}{|\Delta\vec{k}|^2 + \kappa^2} b_{\vec{k}_4}^\dagger b_{\vec{k}_3}^\dagger b_{\vec{k}_2} b_{\vec{k}_1}, } \tag{5.5.7}
$$

where $\vec{k}_4$ is determined by momentum conservation. The parameter κ is known as the **screening** parameter; that is, it implies that at lengths greater than $1/\kappa$, the Coulomb potential is screened out. This calculation shows that the Coulomb interaction between particles is problematic unless we assume there is at least some small amount of screening. We can, of course, take the limit $\kappa \rightarrow 0$, to obtain the interaction Hamiltonian for the unscreened

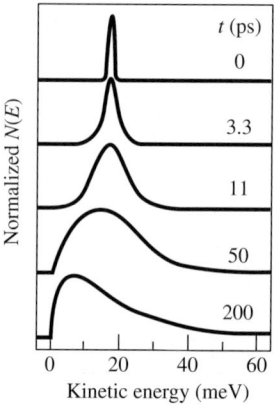

Fig. 5.8   Evolution of the energy distribution of electrons via electron–electron scattering, found by numerical solution of the quantum Boltzmann equation (4.8.16), for electron density $n = 4 \times 10^{14}$ cm^{-3} and screening parameter $\kappa = 10^5$ cm^{-1}. From Snoke *et al.* (1992).

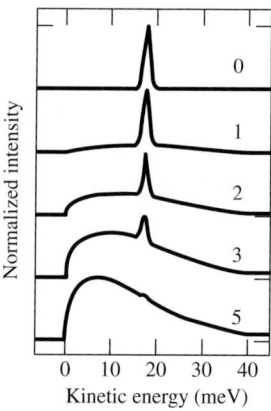

Fig. 5.9   Evolution of the energy distribution of electrons, found by numerical solution of the quantum Boltzmann equation (4.8.16) for the same conditions as for Figure 5.8, but with a short-range interaction. The curves are labeled by the average number of scattering events per particle.

Coulomb potential, but this will be problematic in the limit $\Delta \vec{k} \to 0$, that is, for collisions with small momentum exchange, known as **forward scattering**.

As discussed above, the interaction (5.5.7) has a natural interpretation in terms of the momentum creation and destruction operators. We remove two particles from states $\vec{k}_1$ and $\vec{k}_2$, and put them back in states $\vec{k}_3$ and $\vec{k}_4$. The interaction depends on the momentum lost by one particle and gained by the other, $\vec{k}_1 - \vec{k}_3$.

Even in the presence of screening, electron–electron scattering is strongly enhanced for low-momentum exchange, that is, forward scattering. Figure 5.8 shows a numerical solution of the evolution of a gas of electrons at low density using the quantum Boltzmann equation method discussed in Section 4.8. By contrast, Figure 5.9 shows the evolution in the case when the electron–electron scattering cross-section is a constant independent of momentum exchange (which is characteristic of short-range, or hard sphere scattering). Although both end up in a Maxwell–Boltzmann distribution, the path to the final equilibrium state is quite different.

Exercise 5.5.1   Determine the interaction Hamiltonian for Coulomb scattering of electrons confined to move in only two dimensions. Use the screened Coulomb interaction (5.5.5). For the actual two-dimensional screened interaction, see the last part of Section 5.5.1.

5.5.1 Semiclassical Estimation of Screening Length

The screening length used to make the Coulomb interaction manageable can be approximated by means of a simple, semiclassical calculation. Let us assume that the potential energy felt by the electrons varies slowly. Then in a given region, we can assume that number of electrons at energy E is just the standard density of states for a three-dimensional gas,

$$N(E)dE = f(E)\mathcal{D}(E)dE. \tag{5.5.8}$$

We assume that the distribution function $f(E)$ is the same everywhere.

Suppose that at some small region of space, the electric potential is slightly higher, but constant over that whole region. The density of states is measured relative to the new ground state. In thermal equilibrium, however, the occupation number $f(E)$ is unchanged. Therefore, we have, for a potential energy U,

$$N'(E)dE = f(E)\mathcal{D}(E - U)dE. \tag{5.5.9}$$

To get the total number of electrons, we just sum this over all energies.

A local change in the potential energy will therefore lead to a change in the number of electrons in a given region of space. This will cause a change in the charge density. Relative to an electron gas at $U = 0$, the net charge density, for electrons with charge $-e$, is

$$\begin{aligned}
\rho(r) = \frac{Q}{V} &= -\frac{e}{V}\int_{-\infty}^{\infty} \left(N'(E) - N(E)\right)dE \\
&= -\frac{e}{V}\int_{-\infty}^{\infty} f(E)\left(\mathcal{D}(E - U(r)) - \mathcal{D}(E)\right)dE,
\end{aligned} \tag{5.5.10}$$

where V is the volume. By a change of variables, this is the same as

$$\rho(r) = -\frac{e}{V}\int_{-\infty}^{\infty} \left(f(E + U(r)) - f(E)\right)\mathcal{D}(E)dE. \tag{5.5.11}$$

If we assume that U is a small perturbation, then this can also be written as

$$\rho(r) = -\frac{e}{V}\int_{-\infty}^{\infty} \frac{\partial f}{\partial E}U(r)\,\mathcal{D}(E)dE. \tag{5.5.12}$$

This charge difference in turn has an effect on the potential energy, since a net positive charge lowers the potential energy for electrons. Writing the potential energy as $U(r) = -eV(r)$, where $V(r)$ is the electric potential, we have

$$\rho(r) = \frac{e^2}{V} \int_{-\infty}^{\infty} \frac{\partial f}{\partial E} V(r) \, \mathcal{D}(E) dE. \tag{5.5.13}$$

Poisson's equation from classical electrodynamics gives us the effect of this extra charge on the electric potential,

$$\nabla^2 V(r) = -\frac{\rho(r)}{\epsilon}$$

$$= -\frac{e^2}{\epsilon} \left(\frac{1}{V} \int_{-\infty}^{\infty} \frac{\partial f}{\partial E} \, \mathcal{D}(E) dE \right) V(r). \tag{5.5.14}$$

In the case of spherical symmetry, we can write this as

$$\frac{1}{r} \frac{\partial^2}{\partial r^2} (rV(r)) = \kappa^2 V(r), \tag{5.5.15}$$

where

$$\boxed{\kappa^2 = -\frac{e^2}{\epsilon} \left(\frac{1}{V} \int_{-\infty}^{\infty} \frac{\partial f}{\partial E} \, \mathcal{D}(E) dE \right).} \tag{5.5.16}$$

In the limit $\kappa \to 0$, we know that the potential between two charges of $-e$ is

$$U(r) = \frac{e^2}{4\pi\epsilon r}. \tag{5.5.17}$$

When $\kappa \neq 0$, the solution of (5.5.15) for $V(r)$ at $r = 0$ should approach this. Using this as a boundary condition, we obtain the solution

$$U(r) = \frac{e^2}{4\pi\epsilon r} e^{-\kappa r}. \tag{5.5.18}$$

The screening length given by (5.5.16) depends on the temperature of the electrons through the temperature dependence of the distribution function $f(E)$. It can be shown that in the case of a Maxwell–Boltzmann distribution of electrons,

$$\kappa^2 = \frac{e^2 n}{\epsilon k_B T}, \tag{5.5.19}$$

where n is the electron density. This is known as the **Debye** screening approximation. The screening is less efficient at high temperature, since there is less fractional change of the energy of the electrons due to the presence of another electron.

Physically, we understand screening to come about from the fact that electrons tend to avoid each other because of the Coulomb repulsion. Therefore, there tends to be a region of net positive charge around each electron which partially cancels out its negative charge, and which completely cancels it out at long distances.

Although we used a semiclassical approximation to obtain (5.5.16), when we do a full many-body calculation in Section 8.11, we will see that this is the correct answer in the limit of low momentum and low energy exchange in electron–electron scattering.

Exercise 5.5.2   Show that for a three-dimensional isotropic electron gas, (5.5.16) becomes (5.5.19) when $f(E) \propto e^{-E/k_B T}$, that is, when the electrons have a Maxwell–Boltzmann distribution.

Screening in a two-dimensional gas. As we have seen in Chapter 2, it is possible to create a two-dimensional electron gas in a solid system, for example, in a quantum well structure. To calculate the screening for this type of system, we must first realize that we are dealing with a two-dimensional electron gas embedded in a three-dimensional universe, not an electron gas in a two-dimensional universe.

Unlike the three-dimensional derivation above, we do not have spherical symmetry; instead we have cylindrical symmetry. The Poisson equation in this case is

$$\nabla^2 V(\vec{r}, z) = \nabla_r^2 V + \frac{\partial^2 V}{\partial z^2} = -\frac{\sigma(\vec{r})}{\epsilon} \delta(z), \tag{5.5.20}$$

where $\vec{r}$ is in the plane of the electron gas. Following the argument above, the induced charge density is

$$\sigma_{\mathrm{ind}}(\vec{r}) = -\frac{e^2}{\epsilon} \left(\frac{1}{A} \int_{-\infty}^{\infty} \frac{\partial f}{\partial E} D(E) \, dE \right) V(\vec{r}, 0) \equiv 2\kappa V(\vec{r}, 0). \tag{5.5.21}$$

The factor 2 in the definition of κ is for convenience, as we will see below. For a test charge at $\vec{r} = 0$, the total charge density in the plane is then $\sigma = \sigma_{\mathrm{ind}} + e\delta^2(\vec{r})$. (In the three-dimensional derivation above, we could also have included a point test charge with density $e\delta^3(\vec{r})$, but we did not do this since we knew the form of the solution in the $r \to 0$ limit.)

The solution can be found by switching to the Fourier transform for the plane. We write

$$V(\vec{r}, z) = \frac{1}{(2\pi)^2} \int d^3q \, e^{-i\vec{q}\cdot\vec{r}} V(\vec{q}, z), \tag{5.5.22}$$

in which case the Poisson equation becomes

$$\left(-q^2 + \frac{\partial^2}{\partial z^2} \right) V(\vec{q}, z) = \left(-\frac{e}{\epsilon} + 2\kappa V(\vec{q}, 0) \right) \delta(z). \tag{5.5.23}$$

This is a one-dimensional differential Green's function equation in z, with the general form of solution

$$V(q, z) = \frac{C}{q} e^{-q|z|}. \tag{5.5.24}$$

Setting the discontinuity in the slope at $z = 0$ equal to the integral over the factor with $\delta(z)$ yields

$$V(q, z) = -\frac{1}{2q} e^{-q|z|} \left(-\frac{e}{\epsilon} + 2\kappa V(\vec{q}, 0) \right). \tag{5.5.25}$$

Setting $z = 0$ and solving for $V(q, 0)$ we finally obtain

$$V(q, 0) = \left(\frac{e}{2\epsilon} \right) \frac{1}{q + \kappa}. \tag{5.5.26}$$

The interaction energy between two particles with charge e is therefore

$$U(q) = \left(\frac{e^2}{2\epsilon} \right) \frac{1}{q + \kappa}. \tag{5.5.27}$$

The parameter κ prevents a divergence at $q = 0$ just as in the three-dimensional case. We can reverse Fourier transform back to get $U(r)$ in real space, in which case we obtain

$$U(r) = \left(\frac{\pi e^2}{\epsilon}\right) \int_0^\infty \frac{q\,dq}{q + \kappa} J_0(\sqrt{2}qr), \tag{5.5.28}$$

where J_0 is a Bessel function. This integral diverges as $1/r$ near $r = 0$ and decays exponentially at large r with length scale of order $1/\kappa$, as expected.

From (5.5.21), we have

$$\kappa = -\frac{e^2}{2\epsilon}\frac{1}{A}\int_{-\infty}^\infty \frac{\partial f}{\partial E}\mathcal{D}(E)dE. \tag{5.5.29}$$

For a low-density Maxwellian gas, this is

$$\kappa = \frac{e^2}{2\epsilon}\frac{n}{k_B T}, \tag{5.5.30}$$

where n is the area density of the gas.

Exercise 5.5.3 Compare the screening length for a three-dimensional electron gas at room temperature, in the Maxwell–Boltzmann limit with density $n = 10^{18}$ cm^{-3}, with that of a two-dimensional electron gas with the same density, in a quantum well with width 10 nm, that is, an area density of 10^{12} cm^{-2}, also in the Maxwell–Boltzmann limit. Can you justify using the Maxwell–Boltzmann limit in both of these cases?

5.5.2 Average Electron–Electron Scattering Time

Just as we did for other scattering mechanisms, we can calculate the rate of scattering for Coulomb scattering. We run into a problem, however, because of the high rate of forward scattering, that is, scattering with very little change of the electron momentum. The total scattering rate may not be very important for many physical processes, because it includes a huge number of scattering events in which almost nothing happens to the electron. We can therefore talk about three different "scattering times" for Coulomb scattering.

First, the **total** scattering rate is

$$\frac{1}{\tau_{\text{tot}}(\vec{k})} = \frac{2\pi}{\hbar}\frac{1}{(2\pi)^6}\int d^3k'\,d^3K\,N_{\vec{k}'}(1 - N_{\vec{k}+\vec{K}})(1 - N_{\vec{k}'-\vec{K}})\frac{(e^2/\epsilon)^2}{(K^2 + \kappa^2)^2}$$
$$\times \delta(E_k + E_{k'} - E_{\vec{k}+\vec{K}} - E_{\vec{k}'-\vec{K}}). \tag{5.5.31}$$

This can also be called the **dephasing** time, because any scattering process disturbs the phase of the electron wave function. We will discuss the significance of dephasing times, called T_2 times, in Chapter 9.

Second, we could weight the integral by K, the momentum change of an electron in a collision. This can be called the **momentum** relaxation time, and is the most appropriate time scale for transport measurements, in which the randomization of the direction of the motion of the electrons is important.

Third, we could also weight the integral by $\Delta E = |E_{\vec{k}+\vec{K}} - E_{\vec{k}}|$, the energy exchange in a collision, which is proportional to $K^2 + 2\vec{k} \cdot \vec{K}$. This can be called the **energy** relaxation time, or the **thermalization** time, since the speed at which an electron gas will reach a thermal distribution depends on how fast it can redistribute kinetic energy.

For scattering matrix elements which do not depend strongly on $\vec{k}$, such as deformation potential scattering or elastic scattering from a neutral defect, these three all will have approximate the same time scale. For screened Coulomb scattering, however, these three can have very different behaviors.

For example, consider the low-density limit, when the Pauli exclusion factors $(1 - N_k)$ can be ignored. In this case, eliminating the energy δ-function gives

$$
\begin{aligned}
\frac{1}{\tau_{\text{tot}}(\vec{k})} &= \frac{1}{(2\pi)^4} \frac{e^4}{\epsilon^2} \frac{m}{\hbar^3} \int d^3k' \, N_{\vec{k}'} \frac{1}{k_r} \int_0^{2k_r} K dK \frac{1}{(K^2 + \kappa^2)^2} \\
&= \frac{1}{(2\pi)^4} \frac{e^4}{\epsilon^2} \frac{m}{\hbar^3} \int d^3k' \, N_{\vec{k}'} \frac{1}{k_r} \left(\frac{2k_r^2}{\kappa^4 + 4\kappa^2 k_r^2} \right),
\end{aligned}
\tag{5.5.32}
$$

where $\vec{k}_r = (\vec{k} - \vec{k}')/2$. If the density is low enough that $\kappa \ll k_r$, the term in the parentheses becomes just $1/2\kappa^2$, and therefore the integral will be proportional to n/κ^2, where n is the electron density. The Debye screening formula (5.5.19), and also the more general formula (5.5.16), implies that κ^2 is proportional to n, however. This gives the counterintuitive result that the total scattering rate is independent of density in the low-density limit. Although there are fewer particles to scatter with, the cross-section for scattering becomes larger and larger as the screening drops off at low density (assuming that the size of the system is larger than the screening length at all densities). Here we have ignored the contribution of electron exchange, which can become important at low densities, as discussed in Section 10.8.

On the other hand, when the integral is weighted by the momentum exchange K, in the limit $\kappa \ll k_r$ the integral is proportional to $1/\kappa$, so that the total momentum relaxation rate decreases as $n^{1/2}$ with decreasing density, as one might expect. The energy relaxation rate also decreases as density decreases.

In the high-density limit, the Pauli exclusion factors change the scattering rate substantially. The need to satisfy energy and momentum conservation in the scattering process, as well as to find empty states into which both particles can scatter, severely limits the number of possible scattering processes. It is easy to see that the scattering rate approaches zero as the momentum $\vec{k}$ approaches the Fermi momentum k_F from above. An electron scattering with electrons in the Fermi sea cannot *gain* energy, because that would imply one of the other electrons loses energy, but that is forbidden by Pauli exclusion. On the other hand, it can only lose energy equal to the difference between its initial energy and the Fermi energy We will return to these considerations in Section 8.15.

Exercise 5.5.4   Use (5.5.32) to show that in the limit of low density, the total scattering time (dephasing time) for an electron gas in vacuum due to Coulomb scattering at room temperature is independent of density and of the order of a femtosecond. This involves the following steps:

(a) Taking the limit $k_r \gg \kappa$, the integral in (5.5.32) becomes

$$
\frac{1}{\kappa^2} \int_0^\infty 4\pi k'^2 dk' \, N(E_{k'}) \int_{-1}^1 d(\cos\theta) \frac{1}{\sqrt{k^2 + k'^2 + 2kk' \cos\theta}}.
\tag{5.5.33}
$$

Assuming a Maxwell–Boltzmann distribution of electrons, convert the integral to a constant times a unitless integral, and perform the integral over angle and then over k', for k determined by the condition $E_k = k_B T$.

(b) Using the answer from part (a), calculate the rate $1/\tau_{\text{tot}}$ using the vacuum electron mass and the permittivity of the vacuum ϵ_0, and using the Debye formula (5.5.19) for κ^2.

The time scale for electron–electron dephasing is typically shorter than the time scales of all other dynamic processes in solids, even at very low electron density. Time scales this short can be studied using sub-picosecond laser pulses to excite the electrons.

5.6 The Relaxation-Time Approximation and the Diffusion Equation

We now have a palette of different interactions with which to work in understanding the dynamics of quasiparticles. A greatly simplifying approximation which is often made is to assign to each type of quasiparticle a single **relaxation time**, which is equal to the average scattering time for all processes, and then to assume that all of the quasiparticles of this type scatter with exactly this relaxation time. From the previous sections, we can see that this approximation is more reasonable for some processes than others. For elastic scattering of electrons from defects, it is very reasonable, and for electron–phonon scattering it is also fairly reasonable, while for phonon scattering with k^4 or k^5 dependence, or electron scattering with $1/k$ dependence, it seems much less reasonable. Nevertheless, the relaxation-time approximation has been used successfully for a number of different effects, and also helps us to have an intuitive understanding of things like heat flow and electrical conductivity.

Suppose we have a gas of quasiparticles, as shown in Figure 5.10. In a homogeneous system, when the motion of the particles is random, in equilibrium, there will be no net flow of particles in any one direction. If there is a gradient of the density, however, then more particles will tend to flow from a region of high density to a region of low density, leading to a net velocity of the particles. This flow is described by the diffusion equation,

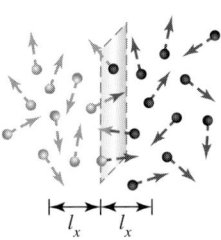

Fig. 5.10   Diffusion of a gas through an imaginary surface.

which we will derive here. In this derivation, we must distinguish carefully between the **root-mean-squared speed**,

$$\bar{v} = \sqrt{\sum_i v_i^2}, \tag{5.6.1}$$

which is proportional to the temperature in a classical gas, and the **average velocity**,

$$\vec{v} = \sum_i \vec{v}_i, \tag{5.6.2}$$

which is zero when there is no density gradient, no matter what the temperature, and in general is much lower than $\bar{v}$. Diffusion due to a density gradient leads to an average velocity in one direction, which as we will see, depends on the root-mean-squared speed.

Figure 5.10 shows a surface in the gas, normal to the x-direction. We define the **mean free path** of the quasiparticles in the x-direction, l_x, as the average distance traveled by a particle in the x-direction before it collides with another quasiparticle. In the relaxation time approximation, this distance is just

$$l_x = \bar{v}_x \tau, \tag{5.6.3}$$

where τ is the relaxation time, that is, the time to hit another particle, and v_x is the root-mean-squared speed in the x-direction.

On average, particles within a distance l_x from the surface, and moving toward it, will pass through without scattering, while particles further away will scatter before they get there. We define the current density $\vec{g}$ as the density of particles times their average velocity. The microscopic current density passing through the surface from the left to the right is then given by

$$g_x^+ = \frac{1}{2} n_{x-l_x} \bar{v}_x, \tag{5.6.4}$$

where we approximate that the density of particles over a distance of l_x is constant, equal to n_{x-l_x}, and one half of the particles are moving toward the surface while the other half are moving away. In the same way, the current density passing through from the right to the left is given by

$$g_x^- = \frac{1}{2} n_{x+l_x} \bar{v}_x. \tag{5.6.5}$$

The net current density in the x-direction, normal to the surface, is therefore

$$g_x = \frac{1}{2}(n_{x-l_x} - n_{x+l_x}) \bar{v}_x. \tag{5.6.6}$$

Multiplying and dividing by l_x, we have

$$g_x = \frac{(n_{x-l_x} - n_{x+l_x})}{2l_x} l_x \bar{v}_x = \frac{(n_x - n_{x+dx})}{dx} l_x \bar{v}_x$$

$$= -\frac{\partial n}{\partial x} l_x \bar{v}_x = -\frac{\partial n}{\partial x} \bar{v}_x^2 \tau, \tag{5.6.7}$$

where we have taken the limit that the mean free path is small compared to the distances over which the density n changes significantly. This is the key assumption of the diffusion equation, which we can call the **diffusive approximation**.

This can be rewritten more generally as

$$\vec{g} = -D\vec{\nabla}n, \tag{5.6.8}$$

or, writing $\vec{g} = n\vec{v}$, where $\vec{v}$ is the average velocity of the particles in a local region,

$$\vec{v} = -D\frac{\vec{\nabla}n}{n}. \tag{5.6.9}$$

Here we have defined $D = \bar{v}_x^2 \tau$ as the **diffusion constant** of the gas. The root-mean-squared speed in the x-direction is related to the total root-mean-squared speed $\bar{v}^2$ according to $\bar{v}^2 = \bar{v}_x^2 + \bar{v}_y^2 + \bar{v}_z^2 = 3\bar{v}_x^2$. Here we assume the net speed $|\vec{v}|$ in any one direction is negligible compared to the root-mean-squared speed, so that the three spatial directions are equivalent. We then have

$$D = \frac{1}{3}\bar{v}^2\tau. \tag{5.6.10}$$

In general, in a system with density gradients, we can define the **velocity field**, $\vec{v}(\vec{x})$, which gives the average velocity of the particles in some small region around the point $\vec{x}$. The velocity field is subject to the constraint that the total number of particles in the system is constant, if we ignore processes which destroy particles (e.g., recombination of carriers by photon emission). Consider the surface shown in Figure 5.11. The only way for the total number of particles inside the surface to change is for a current to flow in or out through the surface. We can therefore equate the two,

$$\frac{\partial}{\partial t}\left(\int_V d^3x\, n\right) = -\int_S d\vec{a} \cdot \vec{g}, \tag{5.6.11}$$

where $\vec{g} = \vec{v}n$ is the current density of the particles. By Gauss' law, the surface integral is equal to a volume integral of the divergence,

$$\int_V d^3x\, \frac{\partial n}{\partial t} = -\int_V d^3x\, \vec{\nabla} \cdot \vec{g}, \tag{5.6.12}$$

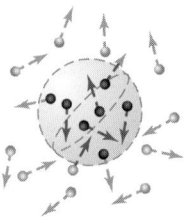

Fig. 5.11 A closed surface used in the derivation of the continuity equation.

or

$$\frac{\partial n}{\partial t} = -\vec{\nabla} \cdot \vec{g}. \tag{5.6.13}$$

This is known as the **mass conservation law**, or, since not all quasiparticles have an effective mass, it can be better called the **continuity equation**.

Substituting (5.6.8) into (5.6.13), we have

$$\frac{\partial n}{\partial t} = D\nabla^2 n. \tag{5.6.14}$$

This is the three-dimensional **diffusion equation**.

The diffusion equation gives rise to evolution of the gas in which sharp features are smeared out over time. For example, suppose that at $t = 0$ the spatial distribution is $n(\vec{x}) = \delta(\vec{x})$. The solution for all later times is

$$n(\vec{x}, t) = \frac{1}{(4\pi Dt)^{3/2}} e^{-|\vec{x}|^2/4Dt}. \tag{5.6.15}$$

In one dimension, the solution is similar. For $n(x) = \delta(x)$ at $t = 0$, the time evolution of the distribution is given by

$$n(x, t) = \frac{1}{(4\pi Dt)^{1/2}} e^{-x^2/4Dt}. \tag{5.6.16}$$

This solution for various times is shown in Figure 5.12.

As discussed above, the derivation of the diffusion equation is based on the diffusive approximation $l_x \ll dx$. In other words, we assume that each particle scatters many times before going whatever distance dx which is relevant to our measurements. If the particles do not have a mean free path short compared to the length scales of interest, we say that their motion is **ballistic**. These two regimes of motion have very different behaviors. As seen in the above solutions of the diffusion equation, the average distance traveled by a particle in the diffusive regime is proportional to $t^{1/2}$. By contrast, particles moving ballistically

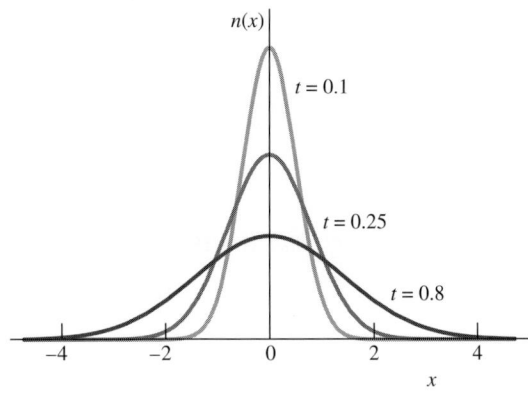

Fig. 5.12   Solutions of the diffusion equation for $n(x, 0) = \delta(x)$ at successive times, for $D = 1$.

move at nearly constant velocity, and therefore the distance they travel is proportional to t.

Exercise 5.6.1 Prove that (5.6.15) is a solution to the diffusion equation (5.6.14) with the initial condition $n(\vec{x}, 0) = \delta(\vec{x})$. Show that it conserves the total number of particles.

What is the full width at half maximum of this distribution as a function of time?

In this section, we have used a semiclassical picture of the quasiparticles to deduce the diffusion equation. Does this imply that the wave picture of electrons or phonons is incorrect in this case? No, it is just easier to use the particle picture. In the quantum mechanical wave picture, the scattering of the wave from many random influences leads to **dephasing**, or **decoherence**. This will be discussed in detail in Chapter 9. In this case, the phase of the wave is not a good observable, which means that the number of particles is (by number-phase uncertainty, as discussed in Section 4.4). It is therefore natural to think in terms of scattering of single particles in this case. But as we saw in the first part of this chapter, the wave nature of the quasiparticles is essential for determining the scattering cross-sections.

5.7 Thermal Conductivity

The diffusion equation applies to a great number of different phenomena. We can apply it to heat flow as well. In general, of course, the number of phonons is not conserved. If the variation in temperature is not too great, however, we can make a linear assumption,

$$du = \left(\frac{\partial u}{\partial n}\right)\bigg|_T dn \equiv \bar{E} dn, \tag{5.7.1}$$

where u is the energy density and $\bar{E}$ is the average energy per particle. Then (5.6.14) becomes

$$\frac{\partial u}{\partial t} = D\nabla^2 u. \tag{5.7.2}$$

Equivalently, we can use the heat capacity defined in Section 4.9.3,

$$C = \frac{\partial U}{\partial T} = V\frac{\partial u}{\partial T}, \tag{5.7.3}$$

which implies $dT = (V/C)du$, to write the diffusion equation in terms of temperature,

$$\frac{\partial T}{\partial t} = D\nabla^2 T. \tag{5.7.4}$$

As in Section 5.6, $D = \frac{1}{3}\bar{v}^2\tau$, but here $\bar{v}$ is the speed of sound in the medium, which is a constant for all acoustic phonons on same branch. In most cases, we can use an average value of the speed of sound for all the branches.

The diffusion constant for heat can be written in terms of other constants which are found from macroscopic measurements. As we found in Section 5.6, there is a current associated with a gradient in particle density. For heat energy density u, (5.6.9) becomes

$$\vec{h} = u\vec{v} = -D\vec{\nabla}u,\tag{5.7.5}$$

where $\vec{h}$ is the heat flow in units of energy per area per second. In standard heat flow measurements, however, we define the **thermal conductivity**, K, by the relation

$$\vec{h} = -K\vec{\nabla}T.\tag{5.7.6}$$

Using the relation $dT = (V/C)du$, we therefore have $\vec{h} = -(KV/C)\vec{\nabla}u$, or, comparing this to (5.7.5),

$$\boxed{K = \frac{DC}{V}.}\tag{5.7.7}$$

Figure 5.13 shows a typical thermal-conductivity curve for a solid. At low temperature, the mean free path l due to phonon–phonon scattering or phonon–defect scattering is much longer than the typical size of the system. In this case, the thermal conductivity is simply proportional to the heat capacity, which as deduced in Section 4.9.3 is proportional to T^3. At high temperature, the relaxation time τ of the phonons due to phonon–phonon scattering decreases as $1/\omega^5$, as shown in Section 5.4 (i.e., as $1/T^5$). Since D is proportional to the scattering time, the thermal conductivity decreases rapidly at high temperature in all materials. As seen in this figure, the elastic scattering with defects also plays an important role in the scattering; isotopically pure Ge has many fewer defects, and therefore has much higher thermal conductivity.

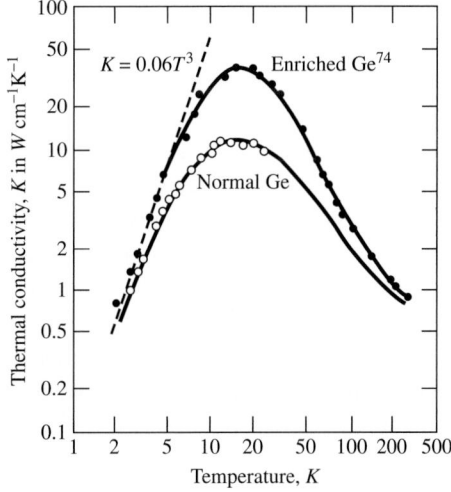

Fig. 5.13   Thermal conductivity for germanium. From Geballe and Hull (1958).

We have seen that many different effects are related to the same anharmonic terms in the phonon energy Hamiltonian. The Gruneisen parameter (introduced in Section 5.4) determines the change of the speed of sound in the medium with strain, and also determines the phonon–phonon scattering rate, and therefore also the thermal conductivity.[2] The anharmonic terms can also lead to structural phase transitions and melting, as we have seen in Section 5.4.

In metals, free electrons can also carry heat. This is why metals tend to be good heat conductors. At high temperature, the scattering rate of electrons increases much less slowly; for example, the scattering rate for electron-impurity scattering increases as $k \sim \sqrt{T}$, according to our calculation in Section 5.3. Therefore, at high temperature the electron contribution can dominate. We will discuss heat conductivity by electrons further in Sections 5.8 and 5.9.

Exercise 5.7.1   If there is a heat source within the medium, the heat diffusion equation (5.7.2) is altered to

$$\frac{\partial u}{\partial t} = D\nabla^2 u + G, \tag{5.7.8}$$

where G is a generation term which can depend on $\vec{x}$ and t. Solve this equation in one dimension for the steady-state temperature profile of a metal bar with specific heat per unit mass $C/\rho V = 400$ J/kg-K, heat conductivity $K = 1$ W/K-cm, and density $\rho = 10$ g/cm^3, if the bar is 1 meter long, with cross-section 10 cm^2, which is heated everywhere along its length with $G = 20$ mW/cm^3, and clamped at each end to a highly conducting medium with temperature held fixed at $T = 300$ K. Show that the heat flow out the ends of the bar is equal to the total heat generated in the medium. What is the maximum T reached?

5.8 Electrical Conductivity

In the case of electrons, diffusion also occurs, but we must account for another effect which is usually more important, namely **drift**, which is average motion in response to a force. This force may be either an externally applied force or a force of the particles on each other, that is, a pressure. In the diffusion equation, motion occurs just through the random scattering processes. This applies well to phonons. The electrons, however, have charge and therefore respond strongly to the potential generated by other electrons and to externally applied fields. In general, we ignore diffusion for electrons and worry only about drift when we are dealing with electrical circuits, because the force of the electrons on each other due to the Coulomb interaction dominates the transport behavior.

---

[2] Some have argued that umklapp scattering processes (introduced in Section 2.2) are necessary for thermal conductivity, but there is no sharp cutoff between the contribution of normal and umklapp processes in the phonon scattering which controls thermal conduction. See Maznev and Wright (2014).

In the **Drude** approximation for electron transport, we assume that each electron accelerates under a force, which produces an acceleration equal to

$$\vec{a} = \frac{d\vec{v}}{dt} = \frac{\vec{F}}{m},$$
(5.8.1)

where $\vec{F}$ is called the **drift force**, and then after each electron has traveled a time τ, the relaxation time due to scattering, it undegoes a collision that randomizes the direction of its velocity. If we assume that the average velocity after each collision is zero, then the average velocity of the particles is just given by the velocity they reach by acceleration in the time since the last collision,

$$\vec{v} = \frac{\vec{F}}{m}\tau.$$
(5.8.2)

The average velocity $\vec{v}$ is called the **drift velocity**.

The electrical current density is equal to the charge per particle times the average velocity times the average density of electrons,

$$\vec{J} = q\vec{v}n = q\frac{\vec{F}}{m}\tau n.$$
(5.8.3)

The force is determined by the electric field $\vec{E}$, which implies

$$\vec{J} = q\frac{q\vec{E}}{m}\tau n = q^2\frac{\tau}{m}n\vec{E}.$$
(5.8.4)

This is **Ohm's law;** in other words, the electrical current is linear with the electric field. We write

$$\vec{J} = \sigma\vec{E},$$
(5.8.5)

where σ is the electrical conductivity,

$$\sigma = q^2\frac{\tau}{m}n = |q|\mu n,$$
(5.8.6)

and μ is the **mobility**,

$$\boxed{\mu = \frac{|q|\tau}{m}.}$$
(5.8.7)

This last parameter is useful because it depends on the properties of a single electron and not on the electron density, which can vary.

Exercise 5.8.1 (a) Prove the assumption made above, that if the probability of a collision per unit time is dt/τ, then the average time since the last collision is τ. This follows a similar procedure as Exercise 1.5.3. The probability of a last collision in time interval t is equal to the probability of no collision in all intervals from 0 to t, times the probability of a collision exactly in the range $(t, t+dt)$. Show that this implies that the probability $P(t)$ of no collision in time interval t is equal to $e^{-t/\tau}$, and therefore the average time since the last collision is τ.

(b) The same argument can equally well be applied to the time until the *next* collision. Therefore, the mean time between collisions is 2τ, and the average energy gained by an electron between collisions is $\frac{1}{2}m|v|^2$ with $|v| = (F/m)(2\tau)$. Show that this implies that the average rate of energy loss in a resistor with cross-sectional area A and length l is equal to $P = IV = V^2/R$, where $R = l/A\sigma$ is the total resistance.

(This exercise was inspired by a problem originally in Ashcroft and Mermin (1976).)

Exercise 5.8.2 For a typical resistance $R = 100\ \Omega$ in a resistor of length $l = 2$ mm and cross-sectional area $A = 1$ mm^2, where $R = l/A\sigma$, calculate the average scattering time τ for an electron, for a free carrier density of 10^{14} cm^{-3} and effective mass of $0.1m_0$.

The mobility and the diffusion constant D are clearly not independent. Recalling the definition of D, we have, for a Maxwell–Boltzmann distribution of electrons in three dimensions,

$$\begin{aligned}
D &= \frac{1}{3}\bar{v}^2\tau \\
&= \frac{1}{3}\left(\frac{3k_B T}{m}\right)\tau \\
&= k_B T\left(\frac{\tau}{m}\right)
\end{aligned} \tag{5.8.8}$$

or

$$\boxed{D = \frac{\mu k_B T}{|q|}.} \tag{5.8.9}$$

This is known as the **Einstein relation**.

As we have seen in Section 5.1.4, the electron–phonon scattering time in the low-density limit decreases as $T^{-3/2}$. At low temperature, this time becomes very long, and therefore the mobility of electrons will be dominated by scattering with defects, which has a nearly constant rate, as discussed in Section 5.3. At high temperature, however, the electron mobility will be dominated by electron–phonon scattering. This is seen, for example, in Figure 5.14, which shows the temperature dependence of the electrical mobility in a typical semiconductor. At high temperature, the mobility decreases as $T^{-3/2}$.

Since the mobility diverges as $T \to 0$, one might expect that all metals are good conductors in the limit of $T = 0$. It turns out that this is wrong. The reason is that in our discussion so far, we have looked at the transport in terms of the electrons undergoing a sequence of independent scattering events. In other words, we have ignored the phase of the electrons. This is usually reasonable when there is a strong inelastic scattering rate. At very low temperature, however, when the dephasing due to inelastic scattering is much less, it can be proven that conductors with even a low amount of disorder can become insulators due to **Anderson localization** in the limit $T \to 0$. We will discuss this effect in Chapter 9.

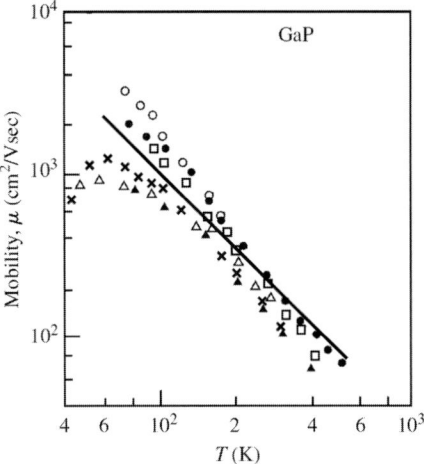

Fig. 5.14   Electron mobility as a function of temperature for the semiconductor GaP. The different symbols are different samples of various purities. The solid curve is proportional to $T^{-3/2}$. From Rode (1972).

Combining the results of the average velocity due to drift and the velocity due to diffusion, which we found in Section 5.6, we have

$$\vec{v} = -D\frac{\vec{\nabla}n}{n} + \frac{\tau}{m}\vec{F}. \tag{5.8.10}$$

Multiplying by n and taking the divergence of both sides, we have

$$\vec{\nabla}\cdot(\vec{v}n) = -D\vec{\nabla}\cdot\vec{\nabla}n + \frac{\tau}{m}\vec{\nabla}\cdot(\vec{F}n). \tag{5.8.11}$$

From the number conservation condition (5.6.13), the left-hand side is equal to $-\partial n/\partial t$. This gives us

$$\frac{\partial n}{\partial t} = D\nabla^2 n - \frac{\tau}{m}\vec{\nabla}\cdot(\vec{F}n). \tag{5.8.12}$$

This is the **drift–diffusion** equation. If we assume that $n(\vec{x}) = n_0 e^{\mu(\vec{x})/k_B T}$ for a classical gas, where $\mu(\vec{x})$ is the local chemical potential, we can also rewrite (5.8.10) as

$$\vec{v} = \frac{\tau}{m}(-\nabla\mu(\vec{x}) + \vec{F}) = -\frac{\tau}{m}\nabla(\mu(\vec{x}) + U(\vec{x})), \tag{5.8.13}$$

where $U(\vec{x})$ is the potential energy that gives rise to the force $\vec{F}$. This shows why the chemical potential is called a potential: Diffusion can be viewed as motion in response to the force arising from the gradient of chemical potential, just as drift is motion in response to a gradient of the electrical potential.

Diffusion constant of a degenerate Fermi gas. In Section 5.6, we assumed that the particles had a Maxwell–Boltzmann distribution of velocities. In the case of a degenerate

Fermi gas, we must adjust these formulas. We start with the relation of the diffusion constant to the net velocity in (5.6.9),

$$\vec{v} = -D\frac{\nabla n}{n},\tag{5.8.14}$$

and compute $(\nabla n)/n$ for a Fermi distribution using (2.4.2) in the $T = 0$ approximation. Setting all the constants in (2.4.2) equal to C, this gives us

$$n = C\mu^{3/2}$$
$$\rightarrow \frac{\partial n}{\partial \mu} = \frac{3}{2}\mu^{1/2}C = \frac{3}{2}\frac{n}{\mu}.\tag{5.8.15}$$

Therefore

$$\frac{\nabla n}{n} = \frac{1}{n}\frac{\partial n}{\partial \mu}\nabla\mu = \frac{3}{2}\frac{\nabla\mu}{\mu}.\tag{5.8.16}$$

On the other hand, we have the velocity due to a force, which can be equated with a chemical potential gradient,

$$\vec{v} = \frac{F}{m}\tau = -\nabla\mu\frac{\tau}{m}.\tag{5.8.17}$$

Putting these into (5.8.14) gives us

$$-\nabla\mu\frac{\tau}{m} = -D\frac{3}{2}\frac{\nabla\mu}{\mu},\tag{5.8.18}$$

or

$$D = \frac{2}{3}E_F\frac{\tau}{m}.\tag{5.8.19}$$

Equating this with definition (5.6.10),

$$D = \frac{1}{3}v_{\text{eff}}^2\tau,\tag{5.8.20}$$

we obtain $\frac{1}{2}mv_{\text{eff}}^2 = E_F$. In other words, the gas acts as though only electrons at the Fermi surface contribute to transport.[3]

Wiedemann–Franz law. It should come as no surprise that the heat conductivity due to free electrons is proportional to the electrical conductivity. The energy carried by electrons is, in the relaxation time approximation, $\vec{h} = n\bar{E}\vec{v}$, where $\bar{E}$ is the average energy per electron. Since the same scattering processes control the drift velocity of electrons whether we are talking about heat or electrical current, the two conductivities are directly proportional. Using the Sommerfeld result for the heat capacity of an electron gas in a Fermi distribution, it can be shown that

$$K = \frac{\pi^2}{3}\frac{k_B^2 T}{e^2}\sigma.\tag{5.8.21}$$

This is known as the **Wiedemann–Franz** law.

---

[3] For the general formula for the diffusion constant of a Fermi gas going between the Maxwell–Boltzmann limit and the degenerate Fermi gas case, see Jyegal (2017).

Exercise 5.8.3   Derive the Wiedemann–Franz law (5.8.21) using the definitions of the electrical conductivity, the thermal conductivity, and the diffusion constant given in the previous sections, along with the Sommerfeld relation (4.9.31) for the heat capacity of the electron gas, and the diffusion constant for a degenerate Fermi gas.

Exercise 5.8.4   Show that the electric field term dominates over the diffusion term in (5.8.10) in a typical situation. Compute D using $\bar{v}$ given by the root-mean-squared speed of electrons in a Maxwell–Boltzmann distribution at room temperature and a constant scattering time of $\tau = 10^{-11}$ s, and assume the carrier density drops by a factor of two over a distance of $100\ \mu$m. Over the same distance there is a voltage drop of 1 V. Compare the diffusion velocity and the drift velocity terms in (5.8.10) under these circumstances.

5.9 Thermoelectricity: Drift and Diffusion of a Fermi Gas

Suppose that we have a gas of electrons in a material with a thermal gradient. Because the diffusion constant depends on the temperature, the electrons in the hotter region will diffuse faster than those in the colder region. Therefore, extra charge will build up in the cold region. The imbalance of the diffusion will end when the extra charge in the cold region generates an electric field strong enough to repel new electrons from coming in. This will lead to a voltage difference across the region. This is known as the **Seebeck effect**.

   To determine this effect quantitatively, we must go back to first principles to derive the drift and diffusion laws for a Fermi gas. From (5.6.7), we have the net current due to spatial variation

$$J_x = qg_x = -q\frac{\partial n}{\partial x}v_x^2\tau, \tag{5.9.1}$$

where τ is the average relaxation time. We can generalize this for the set of carriers in a small energy range $(E, E + dE)$ in a volume V, as

$$J_x(E)dE = -q\left(\frac{1}{V}\frac{\partial f}{\partial x}\mathcal{D}(E)dE\right)D(E), \tag{5.9.2}$$

where $f(E)$ is the Fermi–Dirac distribution function, $\mathcal{D}(E)$ is the density of states at energy E, and $D(E)$ is the energy-dependent diffusion constant.

   When there is a thermal gradient, (5.9.2) can be expanded as

$$J_x(E)dE = -\frac{q}{V}\left(\left.\frac{\partial f}{\partial T}\right|_n\frac{\partial T}{\partial x} + \left.\frac{\partial f}{\partial n}\right|_T\frac{\partial n}{\partial x}\right)D(E)\mathcal{D}(E)dE, \tag{5.9.3}$$

where μ is the chemical potential, which depends on the values of T and the total density n. The second term in the parentheses gives rise to the diffusion equation at constant temperature already deduced for a classical gas in Section 5.6. The temperature-gradient term is

$$\frac{\partial f}{\partial T}\bigg|_n = \frac{\partial f}{\partial T} + \frac{\partial f}{\partial \mu}\frac{\partial \mu}{\partial T}\bigg|_n, \tag{5.9.4}$$

where the second term gives the shift of the chemical potential with temperature at constant density. For a degenerate Fermi gas, this is negligible, but it is an important correction term at low particle densities.

For the Fermi–Dirac function,

$$\frac{\partial f}{\partial T} = -\frac{\partial f}{\partial E}\frac{E - \mu}{T}. \tag{5.9.5}$$

Keeping just this term, for the case of a degenerate Fermi gas, the net current due to the thermal gradient is then found by integrating all the contributions at different energies:

$$\vec{J}_{\text{therm}} = \vec{\nabla}T\frac{qk_B}{V}\int_0^\infty \frac{E - \mu}{k_B T}\frac{\partial f}{\partial E}D(E)\mathcal{D}(E)dE. \tag{5.9.6}$$

In the same way, we can deduce the net drift current for a gradient of potential energy U. Following a similar argument to that which gave us the formula (5.5.12), we write

$$J_x(E)dE = \frac{q}{V}\frac{\partial f}{\partial E}\frac{\partial U}{\partial x}D(E)\mathcal{D}(E)dE. \tag{5.9.7}$$

For positive q, the gradient of U is given by the electric field, $\vec{\nabla}U = q\vec{\nabla}V = -q\vec{E}$. The total drift current is then

$$\vec{J}_{\text{drift}} = -\vec{E}\frac{q^2}{V}\int_0^\infty \frac{\partial f}{\partial E}D(E)\mathcal{D}(E)dE. \tag{5.9.8}$$

This implies the general formula for the conductivity defined by $\vec{J}_{\text{drift}} = \sigma\vec{E}$,

$$\sigma = -\frac{q^2}{V}\int_0^\infty \frac{\partial f}{\partial E}D(E)\mathcal{D}(E)dE. \tag{5.9.9}$$

For a gas with a Maxwellian distribution of the electrons $f(E) = e^{-(E-\mu)/k_B T}$, $D(E) = \frac{1}{3}v^2\tau = \frac{2}{3}(E/m)$, and density of states $\mathcal{D}(E) \propto E^{1/2}$, this becomes

$$\sigma = q^2\frac{\tau}{m}n, \tag{5.9.10}$$

where $n = (1/V)\int f(E)\mathcal{D}(E)dE$. This recovers the Drude formula (5.8.6).

In steady state, the two currents (5.9.6) and (5.9.8) will be equal and opposite. Setting them equal, we have

$$\vec{E} = \left(\frac{k_B}{q}\right)\frac{\displaystyle\int_0^\infty \frac{E - \mu}{k_B T}\frac{\partial f}{\partial E}D(E)\mathcal{D}(E)dE}{\displaystyle\int_0^\infty \frac{\partial f}{\partial E}D(E)\mathcal{D}(E)dE}\vec{\nabla}T \tag{5.9.11}$$

$$\equiv S\vec{\nabla}T,$$

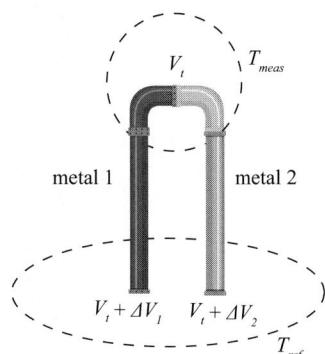

Fig. 5.15   Basic structure of a thermocouple, using two different conductors with different Seebeck coefficients.

where S is the Seebeck coefficient, also known as the **thermoelectric power** or **thermopower**. As discussed above, a temperature gradient leads to an electric field, that is, an electric potential difference. Note that this result depends on the sign of the charge of the majority carriers – when the majority carriers are electrons with negative charge, a hotter region will become positively charged as electrons move away, while if holes are the majority carriers, a hot region will become more negatively charged.

The Seebeck effect underlies the operation of a **thermocouple**, in which a voltage is generated proportional to the temperature. As illustrated in Figure 5.15, when two conductors with different Seebeck coefficients are connected at a junction, they must have a common voltage V_t there. If there is an equal temperature difference across the two conductors, between the reference temperature T_{ref} and the measured temperature T_{meas}, the voltage difference between the other two ends of the conductors will be nonzero and will be proportional to the temperature. This voltage difference can drive a current. The power to drive this current comes ultimately from the thermal gradient between the two regions of different temperature, which transfers heat flow via moving charge carriers.

Sommerfeld expansion. Since the Fermi–Dirac function $f(E)$ varies rapidly only near $E = \mu$, we can use the Sommerfeld expansion method, as we did in Section 4.9.4.

For the integrand of the numerator of (5.9.11), we write

$$K(E) = \mathcal{D}(E)D(E) = K(\mu) + (E - \mu)\left.\frac{\partial K}{\partial E}\right|_{E=\mu} + \cdots \qquad (5.9.12)$$

which gives

$$\int \frac{E-\mu}{k_B T} K(E) \frac{\partial f}{\partial E} dE = K(\mu) \int_{-\infty}^{\infty} \varepsilon \frac{\partial f}{\partial \varepsilon} d\varepsilon + k_B T \left.\frac{\partial K}{\partial E}\right|_{E=\mu} \int_{-\infty}^{\infty} \varepsilon^2 \frac{\partial f}{\partial \varepsilon} d\varepsilon + \cdots ,$$

$$(5.9.13)$$

with $\varepsilon = (E - \mu)/k_B T$. The lowest-order integral vanishes, and the next integral is equal to $-\pi^2/3$.

Table 5.1 Seebeck coefficient for various metals	
Metal	Seebeck coefficient (μV/K)
Aluminum	-1.8
Copper	$+1.8$
Gold	$+1.9$
Lead	-1.1
Platinum	-5.3
Silver	$+1.5$

For the conductivity, we take the lowest order of the Sommerfeld expansion,

$$\sigma \simeq -\frac{q^2}{V}K(\mu)\int_{-\infty}^{\infty}\frac{\partial f}{\partial \varepsilon}d\varepsilon = \frac{q^2}{V}\mathcal{D}(\mu)D(\mu). \tag{5.9.14}$$

The Seebeck coefficient is then

$$S = \frac{\pi^2}{3}\frac{k_B^2 T}{q}\frac{1}{K(\mu)}\frac{\partial K}{\partial E}\bigg|_{E=\mu}. \tag{5.9.15}$$

This is known as the **Mott formula**. It gives values in fair agreement with measured magnitudes in metals (see Table 5.1, from Solyom 2009).

Exercise 5.9.1   Estimate the Seebeck coefficient for a metal with a Fermi level of 0.5 eV, at $T = 300$ K. To do this, you will need to approximate $D(E)$ as a peaked function at $E = E_F$, according to the discussion at the end of Section 5.8. Your answer should not depend on the electron mass, to lowest order.

Classical limit. In the case of a classical gas with a Maxwellian distribution $f(E) = e^{-(E-\mu)/k_B T}$, we can perform the integrals in (5.9.11) directly. However, in this case we cannot ignore the term for shift of the chemical potential with temperature in (5.9.4), which we treated as negligible for a degenerate Fermi gas. The density, which we want to keep constant, is

$$n = \frac{1}{V}\int_0^{\infty}\mathcal{D}(E)e^{-(E-\mu)/k_B T}dE$$
$$= \frac{C}{V}e^{\mu/k_B T}(k_B T)^{3/2}, \tag{5.9.16}$$

which we can solve for μ,

$$\mu = k_B T \ln[n/C(k_B T)^{3/2}], \tag{5.9.17}$$

for a three-dimensional gas. Therefore,

$$\frac{\partial \mu}{\partial T}\bigg|_n = k_B \ln[n/C(k_B T)^{3/2}] - \frac{3}{2}k_B$$
$$= \frac{\mu}{T} - \frac{3}{2}k_B, \tag{5.9.18}$$

and

$$\frac{\partial f}{\partial T} + \frac{\partial f}{\partial \mu}\frac{\partial \mu}{\partial T}\Big|_n = -\frac{\partial f}{\partial E}\frac{E-\mu}{T} - \frac{\partial f}{\partial E}\left(\frac{\mu}{T} - \frac{3}{2}k_B\right)$$

$$= -\frac{\partial f}{\partial E}\left(\frac{E}{T} - \frac{3}{2}k_B\right). \tag{5.9.19}$$

The Seebeck coefficient is then

$$S = \left(\frac{k_B}{q}\right)\frac{\dfrac{1}{(k_B T)^2}\displaystyle\int_0^\infty (E - \tfrac{3}{2}k_B T)f(E)\mathcal{D}(E)D(E)dE}{\dfrac{1}{k_B T}\displaystyle\int_0^\infty f(E)\mathcal{D}(E)D(E)dE}$$

$$= \frac{k_B}{q}. \tag{5.9.20}$$

For a classical gas, the magnitude of the Seebeck coefficient is universal, equal simply to k_B/e. The sign still depends on the sign of the charge of the carriers, however.

Onsager relations. The thermoelectric effect works in reverse also – if a voltage gradient is used to drive a current, the electrons will carry heat with them, leading to a temperature gradient. This is known as the **Peltier effect**, and is the basis of solid state cooling devices with no moving parts.

The Seebeck and Peltier effects can be viewed as two implications of the general relations between the heat flow $\vec{h}$ carried by electrons and the electrical current density $\vec{J}$, which are an example of **Onsager relations**:

$$\vec{J} = -\frac{L_{11}}{T}\vec{\nabla}(\mu + qV) + \frac{L_{12}}{T^2}\vec{\nabla}T$$

$$\vec{h} = \frac{L_{21}}{T}\vec{\nabla}(\mu + qV) - \frac{L_{22}}{T^2}\vec{\nabla}T. \tag{5.9.21}$$

We can immediately read off that $L_{11}q/T$ is the electrical conductivity (which is related to the diffusion constant by the Einstein relation), L_{22}/T^2 is the heat conductivity, and $L_{12}/L_{11}qT$ is the Seebeck coefficient we have just derived. L_{11} and L_{22} are related by the Wiedemann–Franz law (5.8.21) discussed in Section 5.8. As discussed in Section 9.8, the basic principle of time-reversal symmetry implies that $L_{12} = L_{21}$.

The Peltier coefficient Π is defined by the relation

$$\vec{h} = \Pi\vec{J}. \tag{5.9.22}$$

This implies that $\Pi = L_{12}/L_{11}q = ST$.

Phonon wind. The above relations have been derived assuming that electron motion is the only relevant transport. In most real systems, however, the interaction of phonons and electrons must be taken into account. In computing the electrical conductivity in Section 5.8, we allowed for random scattering of electrons and phonons with no net force in any direction; this contributed to the drag force on the electrons characterized by the scattering time τ. It is also possible also for phonons to have a net drift in one direction, known as a **phonon wind**. Phonons diffusing from a hot region can push carriers selectively in one direction due to the electron–phonon interaction, leading to a net electrical current. This can lead to altered values for the measured thermoelectric coefficients.

5.10 Magnetoresistance

So far we have used a single constant for the conductivity of electrons in a medium. In general, however, we can write down a conductivity tensor relating the current density and the electric field in all possible directions. We have already seen in Section 2.9.3 that there can be a Hall voltage perpendicular to the current flow. If there is a magnetic field, there will be force in the direction perpendicular to the drift velocity of the electron:

$$\vec{v} = \frac{\tau}{m}(q\vec{E} + q\vec{v} \times \vec{B}). \tag{5.10.1}$$

For a magnetic field in the z-direction, we can solve this for $\vec{v}$ to obtain

$$v_x = \frac{q\tau}{m}E_x + \frac{qB\tau}{m}v_y$$

$$v_y = \frac{q\tau}{m}E_y - \frac{qB\tau}{m}v_x$$

$$v_z = \frac{q\tau}{m}E_z. \tag{5.10.2}$$

Solving for v_x and v_y, we find

$$\vec{J} = q\vec{v}n = \tilde{\sigma}\vec{E}, \tag{5.10.3}$$

where

$$\tilde{\sigma} = \begin{pmatrix} \sigma_{xx} & \sigma_{xy} & 0 \\ -\sigma_{xy} & \sigma_{xx} & 0 \\ 0 & 0 & \sigma_0 \end{pmatrix}, \tag{5.10.4}$$

with

$$\sigma_{xx} = \frac{\sigma_0}{1 + (\omega_c\tau)^2}$$

$$\sigma_{xy} = \frac{\sigma_0}{1 + (\omega_c\tau)^2}\frac{q\tau B}{m}. \tag{5.10.5}$$

Here, $\omega_c = |q|B/m$ is the cyclotron frequency introduced in Section 2.9, and $\sigma_0 = q^2(\tau/m)n$ is the standard conductivity found in Section 5.8.

The off-diagonal term σ_{xy} is the **Hall conductivity**. Another way of understanding the quantum Hall effect, discussed in Section 2.9.3, is to hypothesize that $\tau \to \infty$ when there is a full Landau level; in other words, the scattering rate goes to zero because there are no free states to scatter into, in a full level. In this case, the (5.10.5) becomes

$$\sigma_{xx} = 0$$

$$\sigma_{xy} = \frac{|q|n}{B}. \tag{5.10.6}$$

From Section 2.9, the density n for a full Landau level is

$$n = \frac{N}{A} = \frac{|q|B}{h}. \tag{5.10.7}$$

Therefore, the Hall conductivity is

$$\sigma_{xy} = \frac{q}{B} \frac{|q|B}{h} = \frac{q|q|}{h}, \tag{5.10.8}$$

which is what we found in Section 2.9.3.

The transport matrix (5.10.4) is antisymmetric. This is another example of an Onsager relation. In general, the Onsager reciprocity relation says that the transport coefficient matrices must be symmetric under time reversal. For the case of magnetic field, which switches sign on time reversal, we have

$$\sigma_{ij}(B) = \sigma_{ji}(-B). \tag{5.10.9}$$

The Onsager reciprocity relation is a basic result of thermodynamics for dissipative systems, as we will discuss in Section 9.8.

Exercise 5.10.1   For an electronic system with constant scattering time $\tau = 10^{-11}$ s and effective electron mass one-tenth the vacuum electron mass, determine the magnetic field at which magnetoresistance effects become important, that is, when σ_{xy} is equal to σ_{xx}.

Exercise 5.10.2   Show that not only σ_{xx}, but also ρ_{xx} vanishes in the limit $\tau \to \infty$, where ρ_{xx} is the diagonal resistivity defined by

$$\tilde{\rho} = \begin{pmatrix} \rho_{xx} & -\rho_{xy} & 0 \\ \rho_{xy} & \rho_{xx} & 0 \\ 0 & 0 & \rho_0 \end{pmatrix} = \tilde{\sigma}^{-1}. \tag{5.10.10}$$

5.11 The Boltzmann Transport Equation

The diffusion equation is actually a special case of a more general equation for transport, known as the **Boltzmann transport equation**, or simply the Boltzmann equation. In general, as we have seen, the scattering time can depend on the momentum $\vec{k}$. Therefore, in many cases we would like an equation for $n_{\vec{k}}$ instead of the total density n.

Starting with the function $n_{\vec{k}}$ which depends on $\vec{k}, t$, and $\vec{x}$, we write down the implicit derivative

$$\frac{dn_{\vec{k}}}{dt} = \frac{\partial n_{\vec{k}}}{\partial t} + \left(\vec{\nabla}_{\vec{k}} \, n_{\vec{k}} \right) \cdot \frac{\partial \vec{k}}{\partial t} + \left(\vec{\nabla}_{\vec{x}} \, n_{\vec{k}} \right) \cdot \frac{\partial \vec{x}}{\partial t}. \tag{5.11.1}$$

The total change in the particle distribution can only come about by scattering processes such as we have studied in this chapter. But we already have a formula for the redistribution of the particle distribution by scattering, namely, the quantum Boltzmann equation discussed in Section 4.8. Therefore, we can equate the two as

$$\frac{\partial n_{\vec{k}}}{\partial t} + \left(\vec{\nabla}_{\vec{k}} \, n_{\vec{k}} \right) \cdot \frac{\partial \vec{k}}{\partial t} + \left(\vec{\nabla}_{\vec{x}} \, n_{\vec{k}} \right) \cdot \frac{\partial \vec{x}}{\partial t} = \frac{\partial n_{\vec{k}}}{\partial t} \bigg|_{\text{coll}}, \tag{5.11.2}$$

where the term on the right-hand side, known as the collision term, is from the quantum Boltzmann equation for the appropriate particle interaction; for example, for elastic particle–particle scattering the collision term is given by (4.8.16). The first term on the left-hand side is an explicit time dependence of the density, for example, a recombination term for carriers in a semiconductor conduction band returning to the valence band by photon emission. If the number of particles is conserved, this term is dropped.

The time derivative of $\vec{k}$ is given by the force $\vec{F} = \partial\vec{p}/\partial t = \partial(\hbar\vec{k})/\partial t$, while the time derivative of $\vec{x}$ is just the velocity $\hbar\vec{k}/m$. Therefore, we have

$$\frac{\partial n_{\vec{k}}}{\partial t} + \left(\vec{\nabla}_{\vec{k}}\, n_{\vec{k}}\right) \cdot \frac{\vec{F}}{\hbar} + \left(\vec{\nabla}_{\vec{x}}\, n_{\vec{k}}\right) \cdot \frac{\hbar\vec{k}}{m} = \left.\frac{\partial n_{\vec{k}}}{\partial t}\right|_{\text{coll}}. \tag{5.11.3}$$

This is the Boltzmann equation. In the case of an isotropic effective mass, we can rewrite the gradient with respect to $\vec{k}$ as

$$\left(\vec{\nabla}_{\vec{k}}\, n_{\vec{k}}\right) = \frac{\partial n_{\vec{k}}}{\partial E}\left(\vec{\nabla}_{\vec{k}}\, E\right) = -\frac{n_{\vec{k}}}{k_B T}\frac{\hbar^2\vec{k}}{m}, \tag{5.11.4}$$

where we have assumed a Maxwellian distribution of the particles, that is, the kinetic energy E is much greater than the chemical potential of the particles. The Boltzmann equation then becomes

$$\frac{\partial n_{\vec{k}}}{\partial t} - \frac{n_{\vec{k}}}{k_B T}\frac{\hbar^2\vec{k}}{m} \cdot \frac{\vec{F}}{\hbar} + \vec{\nabla}n_{\vec{k}} \cdot \frac{\hbar\vec{k}}{m} = \left.\frac{\partial n_{\vec{k}}}{\partial t}\right|_{\text{coll}}. \tag{5.11.5}$$

Notice that in the Boltzmann equation we treat $n_{\vec{k}}$ as a function of both $\vec{k}$ and $\vec{r}$ and do not take into account the uncertainty principle which says that we cannot exactly define the position and momentum of a particle, even though we use a quantum mechanical calculation in the quantum Boltzmann equation for the collision term. This is legitimate as long as the wavelength of the particles is short compared to the characteristic length scale, which in this case is the mean free path. From the definition of the mean free path in Section 5.6, we have the requirement

$$\lambda \ll v\tau, \tag{5.11.6}$$

which, since $p = h/\lambda$, is the same as

$$\frac{1}{\tau} \ll \frac{pv}{h}$$
$$= \frac{mv^2}{h} \sim \frac{k_B T}{h}. \tag{5.11.7}$$

In other words, the rate of change of the population should be much less than the typical energy scale. This is just the same as the condition for the validity of Fermi's golden rule, discussed in Section 4.7, which is also the condition of validity of the quantum Boltzmann equation. When this condition is not satisfied, the wave nature of the particle wave function becomes important. This is the case in Anderson localization at low temperatures, which we will discuss in Section 9.11. It is also the case for superconductors, discussed in Chapter 11.

The Boltzmann equation reduces to the drift–diffusion equation (5.8.12) when we make the relaxation-time approximation, which was used to deduce the diffusion equation. We can see this in the following.

We assume that the distribution $n_{\vec{k}}$ is just slightly different from the equilibrium distribution $n_{\vec{k}}^0$, and is equal to the equilibrium distribution shifted by the drift velocity $\vec{v} = \hbar \vec{k}_{\vec{v}}/m$,

$$
\begin{aligned}
n_{\vec{k}} = n_{\vec{k}-\vec{k}_v}^0 &\simeq n_{\vec{k}}^0 + \vec{\nabla}_{\vec{k}} n_{\vec{k}}^0 \cdot \vec{k}_{\vec{v}} \\
&= n_{\vec{k}}^0 - \frac{n_{\vec{k}}^0}{k_B T} \frac{\hbar^2}{m} (\vec{k} \cdot \vec{k}_{\vec{v}}) \\
&= n_{\vec{k}}^0 - \frac{n_{\vec{k}}^0}{k_B T} \vec{v} \cdot \hbar \vec{k}.
\end{aligned}
\tag{5.11.8}
$$

The relaxation-time approximation says that the distribution $n_{\vec{k}}$ relaxes toward the equilibrium distribution $n_{\vec{k}}^0$ with a time constant τ. Therefore, assuming the total number of particles is conserved, so that we can drop the explicit dependence on t, the Boltzmann equation (5.11.5) becomes

$$
-\frac{n_{\vec{k}}}{k_B T} \frac{\hbar^2 \vec{k}}{m} \cdot \frac{\vec{F}}{\hbar} + \vec{\nabla} n_{\vec{k}} \cdot \frac{\hbar \vec{k}}{m} = \frac{n_{\vec{k}} - n_{\vec{k}}^0}{\tau}
$$

$$
= -\frac{1}{\tau} \frac{n_{\vec{k}}^0}{k_B T} \vec{v} \cdot \hbar \vec{k}.
\tag{5.11.9}
$$

Note that all three terms have a dot product with $\vec{k}$. As long as $\vec{k} \neq 0$, this implies

$$
-\frac{1}{k_B T} \frac{\hbar}{m} n_{\vec{k}} \vec{F} + \vec{\nabla} n_{\vec{k}} \frac{\hbar}{m} = -\frac{1}{\tau} \frac{\hbar}{k_B T} n_{\vec{k}}^0 \vec{v}.
\tag{5.11.10}
$$

Integrating over all $\vec{k}$, and multiplying through by the various constants, we have

$$
-\frac{\tau}{m} n \vec{F} + \frac{k_B T}{m} \tau \vec{\nabla} n = -n \vec{v}.
\tag{5.11.11}
$$

Recall that $k_B T = \frac{1}{3} m \bar{v}^2$, which means that $(k_B T/m)\tau = \frac{1}{3}\bar{v}^2\tau = D$. Therefore, (5.11.11) is just the same as (5.8.10), which gives us the drift–diffusion equation.

The Boltzmann equation implies that in the presence of a drift force, the momentum distribution of the particles is not a pure equilibrium distribution. Instead, we have the "drifted" distribution implied by (5.11.8). For a Maxwell–Boltzmann distribution, this is

$$
n_{\vec{k}} = n_0 e^{-\hbar^2 |\vec{k}-\vec{k}_v|^2/mk_B T}.
\tag{5.11.12}
$$

Typical drift velocities in solids are around 10^5 cm/s, while the thermal root-mean-squared speed $\bar{v} = \sqrt{3k_B T/m}$ is around 10^7 cm/s at room temperature, for an effective mass of the order of the free-electron mass. Therefore, the drifted distribution is hardly any different from the equilibrium distribution.

Exercise 5.11.1   (a) Show that a typical drift velocity is of the order of 10^5 cm/s for electrons in a solid subject to an electric field of 5 V over a distance of 1 cm, for τ around 10^{-11} s as calculated in Section 5.1.4.

    (b) Plot the equilibrium momentum distribution and the drifted momentum distribution as functions of the magnitude k for $T = 300$ K.

5.12 Drift of Defects and Dislocations: Plasticity

So far in this chapter, we have treated defects in the crystal as static objects. As mentioned in the introduction to this chapter, however, these defects can, in general, move through a crystal and interact with each other.

The study of defects and dislocations is a large subject on its own; for a lengthy treatment see Kovacs and Zsoldos (1973), and Chaikin and Lubensky (1995: ch. 9). Here we summarize just a few of the basic effects.

Imperfections need not only be point defects. It is also common to have a **line defect**, also known as a **dislocation** line. Defects can also occur as two-dimensional surfaces, or **domain walls**, between different regions of a crystal, which we will discuss in Section 10.7.

Figure 5.16 shows an example of how a line defect can be created. In Figure 5.16(a), the lattice is undistorted. In (b), we imagine a plane, indicated by the dashed line, and suppose that the atomic bonds on one side of a line on the plane are cut and replaced with bonds to atoms one site to the left. After the lattice has had a chance to relax, it will look like Figure 5.16(c). A line of dangling bonds is created inside the crystal. Note that the dislocation creates a built-in strain field in the crystal, as seen in the distortion of the unit cells nearby.

A general result of the study of the motion of dislocations is that the strength of materials is largely defined by the motion of dislocations and not by the elastic constants introduced in Section 3.4.1. It is easy to see why. Figure 5.16(d) shows the same crystal as in (b), subject to a shear strain. The dashed line indicates where a new bond can form, as the dangling bond of the dislocation moves over to the next atom, causing the adjacent bond to be broken. Since the bond lengths are nearly the same, there is little difference in energy if the bond is in one position or the other. As seen in Figure 5.16(e), this corresponds to motion of the dislocation. This is known as **slip**. Instead of having any one unit cell undergo substantial additional shear deformation, the unit cells are simply rearranged without substantial change of the total energy of the crystal. For this reason, line defects, which define slip planes, are the most important consideration in determining the plasticity of materials.

Defects and dislocations can move either by diffusion or by drift. A stress field puts a force on defects which causes them to move. Defects and dislocations also create stress fields by their presence, which means that they interact with each other. We can

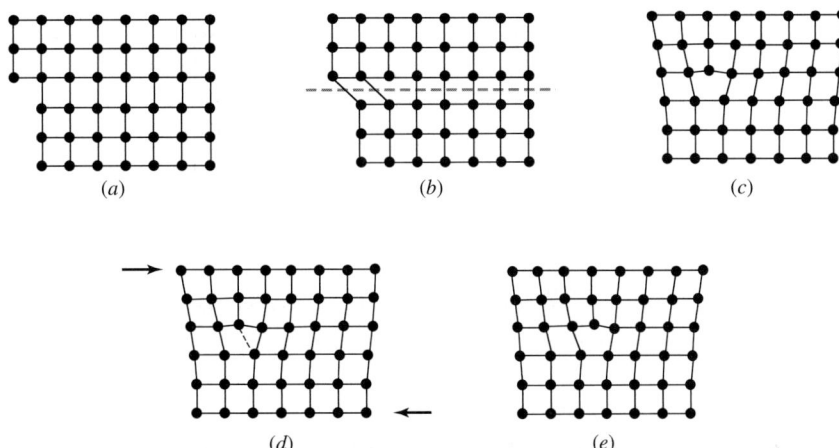

Fig. 5.16 (a) A crystal with an edge dislocation. (b) A line dislocation is created by defining a slip plane (indicated by the dashed line) and moving all the bonds to the left of a line (perpendicular to the plane of the page) by one place. (c) The crystal after relaxation from the process of (b). (d) The same crystal subject to a shear strain. The dashed line indicates a new bond which is energetically close to the one next to it. (e) Deformation of the crystal by slip, which corresponds to motion of the dislocation line.

determine the effect of stress on a line defect using the stress and strain formalism of Chapter 3.

In general, the energy density of a static stress field can be written as

$$\frac{U}{V} = \frac{1}{2}\sum_{ij}\sigma_{ij}\varepsilon_{ij}, \tag{5.12.1}$$

where σ_{ij} and ε_{ij} are the stress and strain fields defined in Section 3.4.1. This equation is analogous to the linear energy density of a spring obeying Hooke's law: $U = \frac{1}{2}kx^2 = -\frac{1}{2}Fx$.

As given in (3.4.8), the strain is related to the displacement by the relation

$$\varepsilon_{ij} = \frac{1}{2}\left(\frac{\partial u_i}{\partial x_j} + \frac{\partial u_j}{\partial x_i}\right), \tag{5.12.2}$$

which implies

$$\frac{U}{V} = \frac{1}{4}\sum_{ij}\sigma_{ij}\left(\frac{\partial u_i}{\partial x_j} + \frac{\partial u_j}{\partial x_i}\right)$$

$$= \frac{1}{2}\sum_{ij}\sigma_{ij}\frac{\partial u_i}{\partial x_j}. \tag{5.12.3}$$

By the chain rule of differentiation, this is the same as

$$\frac{U}{V} = \frac{1}{2} \sum_{ij} \left(\frac{\partial}{\partial x_j} (\sigma_{ij} u_i) - u_i \frac{\partial \sigma_{ij}}{\partial x_j} \right)$$

$$= \frac{1}{2} \left(\vec{\nabla} \cdot (\tilde{\sigma} \cdot \vec{u}) - \sum_i u_i \sum_j \frac{\partial \sigma_{ij}}{\partial x_j} \right). \tag{5.12.4}$$

If there are no body forces, that is, the only externally applied forces are on the surface of the crystal, then for static loading (3.4.23) implies the second term vanishes. We therefore have

$$U = \frac{1}{2} \int_V d^3x \, \vec{\nabla} \cdot (\tilde{\sigma} \cdot \vec{u}). \tag{5.12.5}$$

By Gauss' theorem, this can be converted to a surface integral,

$$U = \frac{1}{2} \int_S (\tilde{\sigma} \cdot \vec{u}) \cdot d\vec{a}. \tag{5.12.6}$$

This is only possible if the volume which is integrated does not have a discontinuity in $\tilde{\sigma} \cdot \vec{u}$. A dislocation is just that. We therefore want to define a surface which cuts out the dislocation. As shown in Figure 5.17(a), we can define a surface which includes the whole volume of the crystal but cuts out a thin slice that contains a line dislocation. The volume inside this slice can be made negligible if it is thin enough.

As seen in Figure 5.18, if we start at a site above a dislocation and count four unit cells to the left, then four cells down, then four cells to the right, and four cells up, we do not come back to our starting point. The vector which points from the end point to the starting point is known as the **Burgers vector**. No matter what path we take around the dislocation, we will always find the same Burgers vector, which defines the discontinuity of the medium.

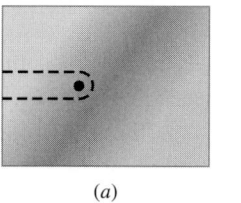 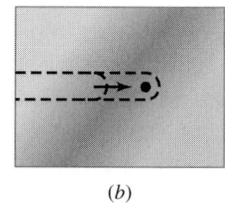

(a)   (b)

Fig. 5.17   (a) A surface which cuts out a line dislocation. (b) After the line dislocation has moved.

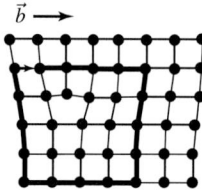

Fig. 5.18   The Burgers vector is defined as the discontinuity in a path which follows an equal number of unit cells in each direction.

If we compare the displacement $\vec{u}$ on the upper side of the thin slice in Figure 5.17(a) to the displacement on the lower side, the difference is just equal to the Burgers vector. Therefore, we can write

$$U = \frac{1}{2}\vec{b} \cdot \int_S \tilde{\sigma} \cdot d\vec{a}. \tag{5.12.7}$$

Imagine now that we move the line defect by a distance dr, as shown in Figure 5.17(b). To keep the line defect cut out, an additional surface must be added to both the top and bottom of our surface of integration. For a line segment $d\vec{l}$, the added area is $2d\vec{r} \times d\vec{l}$. The change of the energy is therefore

$$\begin{aligned} dU &= (\vec{b} \cdot \tilde{\sigma}) \cdot (d\vec{r} \times d\vec{l}) \\ &= -(\vec{b} \cdot \tilde{\sigma}) \cdot (d\vec{l} \times d\vec{r}). \end{aligned} \tag{5.12.8}$$

By the vector identity $(\vec{A} \times \vec{B}) \cdot \vec{C} = (\vec{C} \times \vec{A}) \cdot \vec{B}$, this becomes

$$dU = -\left((\tilde{\sigma} \cdot \vec{b}) \times d\vec{l}\right) \cdot d\vec{r}. \tag{5.12.9}$$

Since $\vec{F} = -\vec{\nabla}U = \partial U/\partial \vec{r}$, this gives the force per element of the line defect as

$$\boxed{d\vec{F} = (\tilde{\sigma} \cdot \vec{b}) \times d\vec{l}.} \tag{5.12.10}$$

Thus, a stress field puts a force on a dislocation. Bending or stressing a material causes the dislocations to move in such a way as to minimize the energy. This is the origin of slip, which determines the real strength of materials.

One of the implications of the theory of defects and dislocations is the well-known effect of **work hardening**. In the standard view, the dislocation lines in a medium are like long polymer chains. A dislocation line can only terminate either at a surface of the crystal, at a junction with another line defect, or in a closed loop. When the medium is worked, these lines can become entangled, similar to the way that polymers can become entangled. In this case, the defects become pinned, and the material cannot easily undergo slip.

Exercise 5.12.1   Prove that the force law (3.4.23) follows from the definition of the energy density (5.12.1) and the energy conservation law

$$\frac{\partial}{\partial t}\left(\frac{1}{2}\rho|\vec{v}|^2 + \frac{U}{V}\right) = 0, \tag{5.12.11}$$

where $\vec{v} = \partial \vec{u}/\partial t$.

References

N.W. Ashcroft and N.D. Mermin, *Solid State Physics* (Holt, Rinehart, and Winston, 1976).

H.A. Bethe and R. Jackiw, *Intermediate Quantum Mechanics* (Benjamin Cummings, 1968).

P.M. Chaikin and T.C. Lubensky, *Principles of Condensed Matter Physics* (Cambridge University Press, 1995).

C. Cohen-Tannoudji, B. Diu, and F. Laloë, *Quantum Mechanics* (Wiley, 1977).

D.B. Fraser and A.D. Hallett, "The coefficient of linear expansion and Gruneisen gamma of Cu, Ag, Au, Fe, Ni, and Al from 4 to 300 K," *Proceedings of the 7th International Conference on Low Temperature Physics*, 689 (1961).

T.H. Geballe and G.W. Hull, "Isotopic and other types of thermal resistance in germanium," *Physical Review* **110**, 773 (1958).

W.F. Giaque and P.F. Meads, "The heat capacities and entropies of aluminum and copper from 15 to 300° K," *Journal of the American Chemical Society* **63**, 1897 (1941).

J.D. Jackson, *Classical Electrodynamics*, 3rd ed. (Wiley, 1999).

J. Jyegal, "Thermal energy diffusion incorporating generalized Einstein relation for degenerate semiconductors," *Applied Sciences* **7**, 773 (2017).

I. Kovacs and L. Zsoldos, *Dislocations and Plastic Deformation* (Pergamon, 1973).

L. Mandel and E. Wolf, *Optical Coherence and Quantum Optics* (Cambridge University Press, 1995).

A.A. Maznev and O.B. Wright, "Demystifying umklapp vs normal scattering in lattice thermal conductivity," *American Journal of Physics* **82**, 1062 (2014).

D.L. Rode, "Electron mobility in Ge, Si, and GaP," *Physica Status Solidi B* **53**, 245 (1972).

D.W. Snoke, W.W. Rühle, Y. Lu, and E. Bauser, "Evolution of a nonthermal electron energy distribution in GaAs," *Physical Review* B **45**, 10979 (1992).

J. Solyom, *Fundamentals of the Physics of Solids*, Vol. II (Springer, 2009).

G.P. Srivastava, *The Physics of Phonons* (Adam Hilger, 1990).

J.C. Toledano and P. Toledano, *The Landau Theory of Phase Transitions* (World Scientific, 1987).

6 Group Theory

Group theory is one of the most powerful methods of modern physics. In systems with known symmetry, many properties can be deduced using group theory without lengthy calculations. Crystals, of course, with their periodic lattices of known symmetry, lend themselves naturally to the methods of group theory. Many molecules also have well-defined symmetry.

Solid state physicists seem to fall into one of two camps – those who feel all solid state physics begins with group theory, and those who feel group theory can be completely ignored. Unfortunately, many physicists never learn group theory because it involves many mathematical theorems, and courses in group theory often take a whole semester to get through the proofs. It is not necessary, however, to take such an in-depth approach. The aim of this chapter is to present many of the techniques of group theory, using the theorems as given, without proof. For the details of the proofs and theorems, the reader may refer to, for example, Tinkham (1964), Bir and Pikus (1974), or Bassani and Parravicini (1975); for a modern review of group theory in condensed matter, see Dresselhaus *et al.* (2008).

6.1 Definition of a Group

Group theory starts with the definition of a **group**. A group is any set of things (called **elements**) which have the following properties:

- **Closure**. There is some operation ∘, which we will call **multiplication**, but which can be any action that applies to two of the elements of the group, which has the property that if A and B belong to the group, then $A \circ B$ is also a member of the group. We typically drop the symbol ∘ and just write AB for the multiplication of A and B.
- **Associativity**. The order of multiplication does not matter: $(AB)C = A(BC)$.
- **Identity**. One of the elements of the group, which we will call E, has the property that $AE = A$.
- **Inverse**. For any element A, there is another element A^{-1} such that $A^{-1}A = E$.

Much of the power of group theory is that the elements of a group can be just about anything, as long as they have these properties. Group theory can therefore be applied to all kinds of systems. Note that commutativity, the property that $AB = BA$, is *not* a necessary property. Groups that are commutative are known as **Abelian** groups.

In solid state theory, the group of interest is a *set of real-space operations which map a crystal* (or some other object) *into itself.* These are called **symmetry operations**. Examples are rotations, reflections, and inversion. Translations (shifts of the whole crystal in one direction or another) are not normally included as elements of the group defining the crystal symmetry.

In the case of real-space symmetry operations, the multiplication operation of the group of real space operators which corresponds to AB is "do operation A after operation B." For example, two real-space symmetry operations might be right-handed rotation by 90° around the z-axis and right-handed rotation by 180° around the z-axis. Let us call the first operation A and the second operation B. Then in group theory notation we can say that $AA = B$; in other words, doing a right-handed 90° rotation about the z-axis twice is the same as a right-handed 180° rotation about the same axis.

Note that we are now talking about an operation on an operation. The multiplication operation is *not* one of the members of the group. It is an operation on the members of the group, which in this case are real-space operations on the crystal. It is also not simple mathematical multiplication. The multiplication operation of a group is simply defined as a way of taking two group members and getting a third. In the case of crystal symmetry operations, we get the third symmetry operation by performing two successive symmetry operations one after the other.

The group of operations which can map a crystal into itself defines the symmetry of the crystal. Figure 6.1 gives an example of a set of symmetry operations which map a crystal into itself. A crystal with *high* symmetry has many operations which map it into itself. Others which are not very symmetric have just a few. A sphere has perfect symmetry – all rotations, reflections, and inversions map it into itself.

The power of group theory in solid state physics comes from a surprising fact: In three dimensions, only a *finite number* of possible symmetries can be used to fill all of space in a periodic structure (see Table 6.1). If the repeated basis of a lattice (defined in Section 1.4) has point-like (spherical) symmetry, then there are only *seven* possible periodic symmetries. (Therefore, the number 7, often viewed as a magic number, does have fundamental importance in the Universe.) If other types of basis with lower symmetry are allowed, then there are only 32 ways to fill space. These are called the 32 **point groups**, so named because one point in the crystal remains invariant under any symmetry operation. These are the groups that will concern us in this chapter. Table 6.2 gives a list of the point groups. A complete tabulation of the group theory tables of these point groups is given by Koster *et al.* (1963), and at the website associated with this book (www.cambridge.org/snoke).

One could also, in principle, expand the set of operations to allow translations, so that no single point remains invariant. This requires an infinite system, since at the boundaries

Table 6.1 The number of possible symmetries that can fill space

	Basis of spherical symmetry	Arbitrary basis
Point groups	7	32
Space groups	14	230

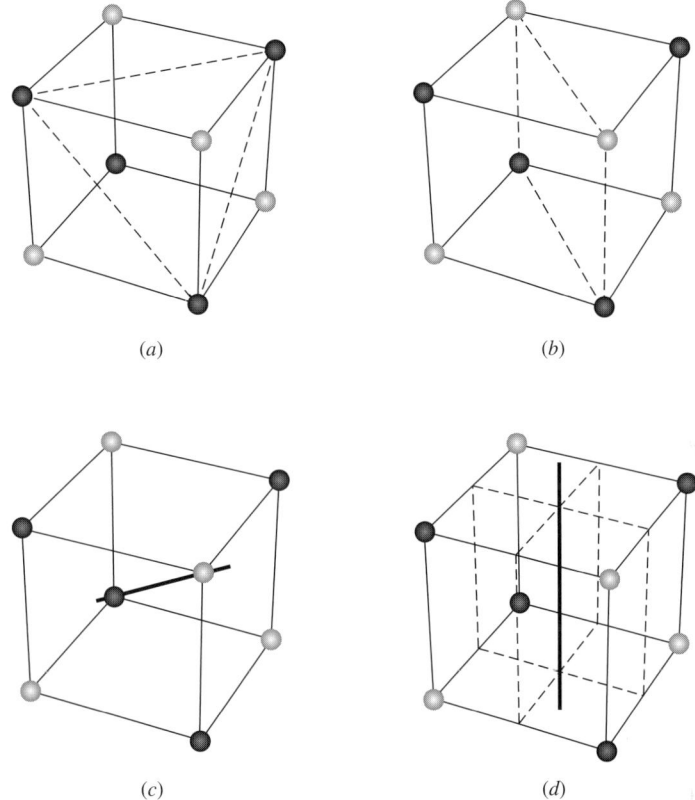

(a)

(b)

(c)

(d)

Fig. 6.1 A structure with T_d symmetry. This is a cubic lattice (zincblende) which is also a tetrahedral symmetry, since a tetrahedron can be constructed of four planes between identical atoms like the one shown in (a). There are six reflections across planes like that shown in (b) which map the crystal back into itself. There are also eight 120° rotations around axes like that shown in (c) (i.e., four such axes, and a left rotation and right rotation around each). Next, there are six operations in which the crystal is rotated around an axis like that shown in (d), and then reflected across one of the planes shown (there are three such axes, with rotations to the left and right). Last, there are three 180° rotations around the same axes.

of a finite medium, a translation would take some point outside. Symmetry groups with these operations are called **space groups**. In this case, there are 14 possible symmetries for a basis with spherical symmetry, and 230 with an arbitrary basis. Even with this expanded set of symmetry operations, there is still a *finite* number of possibilities. In solid state physics, however, we are mostly concerned with the 32 point groups.

6.2 Representations

As discussed above, the elements of the group are the **real space operations** which map a crystal into itself. We can *represent* these operations with square matrices, however. A

Table 6.2 Names of the 32 point groups			
Schönflies notation	Hermann–Maughin	Description	Example lattices
Cubic (three lattice vectors equal length, orthogonal)			
O_h	$(m3m)$	Octahedral, with inversion	sc, fcc, bcc, Diamond
O	(432)	Octahedral	
T_d	$(\bar{4}3m)$	Tetrahedral with four-fold rotation/reflection	Zincblende
T_h	$(m3)$	Tetrahedral, with inversion	
T	(23)	Tetrahedral	
Tetragonal (two lattice vectors equal length, all orthogonal)			
D_{4h}	$(\frac{4}{m}mm)$	One four-fold rotation axis, with two-fold rotation, inversion	Rutile
D_4	(422)	One four-fold rotation axis, with two-fold rotation	
D_{2d}	$(\bar{4}2m)$	One four-fold rotation/reflection axis, with reflection	
S_4	$(\bar{4})$	One four-fold rotation/reflection axis	
C_{4h}	$(\frac{4}{m})$	One four-fold rotation axis, with inversion	
C_{4v}	$(4mm)$	One four-fold rotation axis, with reflection	
C_4	(4)	One four-fold rotation axis	
Orthorhombic (lattice vectors three unequal lengths, all orthogonal)			
D_{2h}	$(\frac{2}{m}mm)$	Three two-fold rotation axes, with inversion	
D_2	(222)	Three two-fold rotation axes	
C_{2v}	$(2mm)$	One two-fold rotation axis, with reflection	
Hexagonal (two lattice vectors equal length, 120° angle)			
D_{6h}	$(\frac{6}{m}mm)$	One six-fold rotation axis, with two-fold rotation, inversion	sh
D_6	(622)	One six-fold rotation axis, with two-fold rotation	
D_{3h}	$(\bar{6}2m)$	One three-fold rotation axis, with two-fold rotation, inversion	hcp
C_{6h}	$(\frac{6}{m})$	One six-fold rotation axis, with inversion	
C_{6v}	$(6mm)$	One six-fold rotation axis, with reflection	Wurtzite
C_6	(6)	One six-fold rotation axis	
C_{3h}	$(\bar{6})$	One three-fold rotation axis, with inversion	
Rhombohedral/trigonal (three lattice vectors equal length, equal angles but not 90°)			
D_{3d}	$(\bar{3}m)$	One three-fold rotation/reflection axis	Calcite
D_3	(32)	One three-fold rotation axis, with two-fold rotation	
$S_6(C_{3i})$	$(\bar{3})$	One six-fold rotation/reflection axis	Quartz
C_{3v}	$(3m)$	One three-fold rotation axis, with reflection	
C_3	(3)	One three-fold rotation axis	

Table 6.2 *(cont.)*

Schönflies notation	Hermann–Maughin	Description	Example lattices
Monoclinic (lattice vectors three unequal lengths, one orthogonal to plane of the other two)			
C_{2h}	$(\frac{2}{m})$	One two-fold rotation axis, with inversion	
C_2	(2)	One two-fold rotation axis	
$C_s(C_{1h})$	$(m, \bar{2})$	One reflection plane	
Triclinic (lattice vectors three unequal lengths, no right angles)			
$C_i(S_2)$	$(\bar{1})$	Only inversion	
C_1	(1)	No real-space symmetry operations	

representation of a symmetry group is any set of square matrices which have the exact same multiplication properties as the real-space operations. For example, in the previous section we said that the real-space operation of right-handed rotation by $90°$ around the z-axis and the real-space operation of right-handed rotation by $180°$ around the z-axis satisfied the relation $AA = B$. We came to this conclusion based on our experience of the real world. With matrices, we could represent operation A by

$$A = \begin{pmatrix} 0 & 1 & 0 \\ -1 & 0 & 0 \\ 0 & 0 & 1 \end{pmatrix} \tag{6.2.1}$$

and operation B by

$$B = \begin{pmatrix} -1 & 0 & 0 \\ 0 & -1 & 0 \\ 0 & 0 & 1 \end{pmatrix}. \tag{6.2.2}$$

It is easy to verify that these matrices have the property $AA = B$ using the normal rules of matrix multiplication. Here we have represented the group multiplication operation, which is "do real-space operation A after real-space operation B," by mathematical matrix multiplication. The relations of the real space operations have been mapped to the relations of a group consisting of matrices undergoing regular mathematical multiplication.

In the above example, we used 3×3 matrices, but there is no rule for how many rows and columns a matrix representation can have. A representation might have 100 rows and columns. A powerful theorem of group theory tells us, however, that we do not need such large matrices to represent the point groups for crystal symmetries. Large matrices can always be reduced to block diagonal matrices made of smaller matrices by the proper choice of **similarity transformation**. A block diagonal matrix consists of smaller matrices along the diagonal, as illustrated in Figure 6.2.

We define a similarity transformation as follows. Given a unitary matrix T of the same dimension as the representation matrices, with the inverse matrix T^{-1}, we transform one matrix representation into another matrix representation (with the same multiplication properties) by transforming *every* matrix in the group according to the rule $A' = T^{-1}AT$,

$$\begin{pmatrix} \boxed{1} & 0 & 0 & 0 & 0 & 0 & 0 & 0 \\ 0 & \boxed{\begin{matrix} 1 & 1 \\ 1 & 1 \end{matrix}} & 0 & 0 & 0 & 0 & 0 \\ 0 & & 0 & 0 & 0 & 0 & 0 \\ 0 & 0 & 0 & \boxed{\begin{matrix} 0 & 2 \\ 2 & 0 \end{matrix}} & 0 & 0 & 0 \\ 0 & 0 & 0 & & 0 & 0 & 0 \\ 0 & 0 & 0 & 0 & 0 & \boxed{\begin{matrix} 0 & 1 & 1 \\ 1 & 0 & 1 \\ 1 & 1 & 0 \end{matrix}} \\ 0 & 0 & 0 & 0 & 0 & & \\ 0 & 0 & 0 & 0 & 0 & & \end{pmatrix}$$

Fig. 6.2 A block diagonal matrix.

$B' = T^{-1}BT$, etc., using the same matrix T to transform each matrix in the group. It is easy to see that the transformed matrices also satisfy the group properties: If $AB = C$, then $(T^{-1}AT)(T^{-1}BT) = T^{-1}ABT = T^{-1}CT$, in other words, $A'B' = C'$.

With the proper choice of similarity transformation, every matrix in the group can be block diagonalized so that the blocks of all the matrices are the same size. If all the members of the group have blocks of the same size, then matrix multiplication will only multiply one block by another block of the same size in the same position in the matrix. Therefore, each set of blocks must satisfy the multiplication rules of the group, and consequently any one set of blocks can be a representation of the group by itself.

This process of transforming a matrix representation into block diagonal matrices is called **reducing** a matrix representation. A set of matrices which cannot be further reduced into blocks of smaller dimension is called an **irreducible** representation. (Other matrix representations are called **reducible**.) For each of the 32 point groups, the number of possible irreducible representations is finite. Surprisingly, the largest matrix you will ever need for an irreducible representation of a point group is a 4×4 matrix. Every other representation of any point group can be block diagonalized into blocks with at most four rows.

We can summarize this in the following theorem:

All possible matrix representations of a point group can be block diagonalized by a similarity transformation into blocks with finite dimension, and which belong to a finite set of matrices for a particular point group. These are called the **irreducible representations** of the group.

This will allow us to use a finite set of tables which list the irreducible representations of all 32 point groups.

Each set of matrices in a representation operates on a common set of **basis functions**, also called basis states. The basis functions of a representation are the things which the rows of the matrix stand for. In the example above, of 3×3 matrices used to represent $90°$ and $180°$ rotation about the z-axis, the matrices operated on the x, y, and z components of a vector in three-dimensional space. The x, y, and z unit vectors are therefore the three basis functions of this representation.

The choice of basis functions determines the exact form of the matrices in the representation, but for a given representation, there is more than one possible choice of basis functions. In our example, we could have chosen the components of a vector projected onto three other perpendicular axes instead of the x, y, and z axes. This would change the specific numbers in the matrices, but it would not change the multiplication properties of the matrices, since two real-space 90° rotations still equal a single 180° rotation, no matter how we represent these operations with matrices.

Another way of describing a similarity transformation, defined above, is as a **change of basis** of the matrices. The process of block diagonalizing a representation through a similarity transformation is the same as finding a set of basis functions that has subsets such that a member of one subset only transforms to other members of the same subset when acted on by the operations in the group.

Exercise 6.2.1 (a) In cubic symmetry, there are six C_4 rotations (90° rotations about the x, y, and z axes), eight C_3 rotations (120° rotations about the [111] directions, or cube corners), and three C_2 rotations (180° rotations about the x, y, and z axes), among other symmetry operations. Write down 3×3 matrix representations of all of these operations, using the x, y, and z unit vectors as the basis functions of the matrix representations.

(b) Does this set of matrices, along with the identity matrix, form a group? Is multiplication of these matrices commutative in any or all cases?

(c) Can you come up with a 2×2 representation which has the same multiplication properties? It does not matter what your matrices are, as long as they have the same multiplication properties. Hint: There is a trivial solution, which is reducible.

How about a 1×1 representation? What is the simplest possible matrix representation that has the same multiplication properties?

6.3 Character Tables

We have already simplified the task of representing a group by knowing that there is a finite set of irreducible representations of the group. This means that we can write down a finite set of matrices to represent all the group members, which in the case of a crystal are the real-space symmetry operations which transform the crystal back into itself. It turns out in most cases, however, that life is even simpler. Often we never even need to write down these matrices! All we need to know for many applications are the traces of the matrices, that is, the sum of the diagonal elements. The trace of a matrix member of a representation is called its **character**. A general theorem is that a similarity transformation does not change the characters of the representation matrices; in other words, the trace of a matrix is conserved even on a change of the basis functions. Therefore, we can write down the character of each representation matrix even without defining the basis functions of the representation.

A further simplification is possible by separating the members of the group into **classes**. A class is formally defined as the set of all **conjugates** of a given member, where a conjugate is defined as a member of the group generated by the transformation

$$C = X^{-1}AX, \tag{6.3.1}$$

where X is any other member of the whole group. In other words, a class is a set of members of the group that can be transformed into each other by symmetry operations. All of the members of the whole group naturally fall into separate classes. For example, in a cubic crystal, the set of all 90° rotations form a class, and the set of all 120° rotations form another class.

The advantage of collecting the members of the group into classes like this is that another general theorem says that all members of a class have the same character. We do not need to write out every member in a table, we only need to write down the characters for each class.

The symmetry of a crystal can therefore be represented by a **character table**. The character tables for different crystal symmetries are given on the website which accompanies this book. Table 6.3 shows an example for a crystal with cubic symmetry. The rows are the irreducible representations, and the columns are the classes. Another general theorem says that the number of irreducible representations is equal to the number of classes. The irreducible representations are typically labeled with the symbol Γ_i, where i is the number of the representation. The classes are labeled by the number of symmetry operations in the class and a symbol for the symmetry operations. Table 6.4 gives standard notation for the various symmetry operations which make up the classes of the group. These can include both **proper** and **improper** rotations, as illustrated in Figure 6.3. Improper rotations enter in only when spin is taken into account, as discussed below.

The math of group theory allows us to make some general statements about all character tables. First, the identity operator E always forms a class by itself, and since it is always represented by 1 in every location on the diagonal, its character is always simply the dimension of the matrices of the irreducible representation.

Table 6.3 The character table of the T_d cubic symmetry double group

	E	$\bar{E}$	$8C_3$	$8\bar{C}_3$	$3C_2$ $3\bar{C}_2$	$6S_4$ $6S_4$	$6\bar{S}_4$ $6\bar{S}_4$	$6\sigma_d$ $6\bar{\sigma}_d$	Basis functions
Γ_1	1	1	1	1	1	1	1	1	$R = x^2 + y^2 + z^2$
Γ_2	1	1	1	1	1	-1	-1	-1	$L_x L_y L_z$
Γ_3	2	2	-1	-1	2	0	0	0	$(2z^2 - x^2 - y^2), \sqrt{3}(x^2 - y^2)$
Γ_4	3	3	0	0	-1	1	1	-1	L_x, L_y, L_z
Γ_5	3	3	0	0	-1	-1	-1	1	x, y, z
Γ_6	2	-2	1	-1	0	$\sqrt{2}$	$-\sqrt{2}$	0	$\phi_{1/2,-1/2}, \phi_{1/2,1/2}$
Γ_7	2	-2	1	-1	0	$-\sqrt{2}$	$\sqrt{2}$	0	$\Gamma_6 \times \Gamma_2$
Γ_8	4	-4	-1	1	0	0	0	0	$\phi_{3/2,-3/2}, \phi_{3/2,-1/2},$ $\phi_{3/2,1/2}, \phi_{3/2,3/2}$

Table 6.4 Notation for symmetry operations

E	Identity
C_6, C_4 etc.	six-fold, four-fold rotation, etc. (proper rotation)
I	Inversion
σ	Reflection
S_6, S_4 etc.	Rotation + reflection across plane perpendicular to the rotation axis
bar over regular operators	*Improper* rotation = change sign of spinor
$\bar{E}$	Simple spin flip

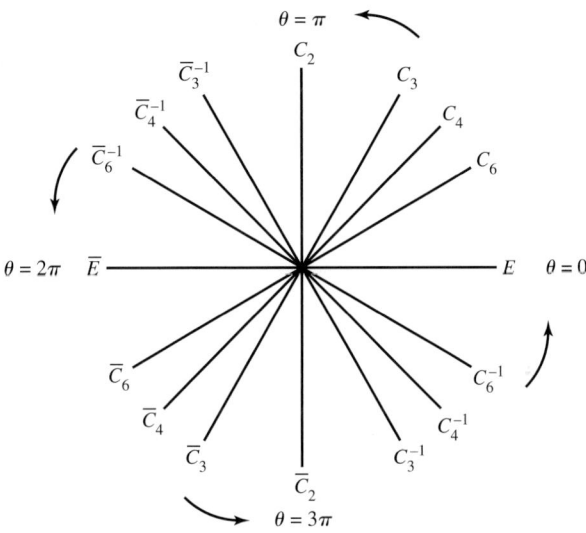

Fig. 6.3   Proper and improper rotations. Note that a full rotation in this diagram corresponds to a rotation of 4π radians, not 2π radians.

Also, there is always one irreducible representation, which is typically called Γ_1, which is 1×1 and which has every member equal to the number 1. Since $1 \times 1 = 1$, this representation always preserves the multiplication properties of the group.

Unfortunately, there are several different notations for labeling the irreducible representations of a given group. Table 6.5 gives some of the standard labels in the case of cubic symmetry. In this book, we will use the Koster symmetry notation in which the irreducible representations are simply numbered in order. This has the advantage of simplicity, but the disadvantage that similar representations in different symmetry groups can have different numbers, and the notation gives no indication of the similarity. A common mistake is to assume that the Γ_4 representation in one group is similar to the Γ_4 representation in another group, but this is often not the case. For example, in the O_h cubic point group (which includes the symmetry operation of inversion), the 3×3 representation which has basis functions of x, y, and z is called Γ_4, while in the T_d cubic symmetry group (without inversion), the representation with the same basis is called Γ_5.

Table 6.5 Common notations in the T_d point group (tetrahedral symmetry, or cubic without inversion)

Koster	BSW ("Wigner")	Molecular
Γ_1	Γ_1	A_1
Γ_2	Γ_2	A_2
Γ_3	Γ_{12}	E
Γ_4	Γ_{25}	T_1
Γ_5	Γ_{15}	T_2

Some groups include the inversion operator I. In this case, the irreducible representations in the Koster notation are labeled by their parity, either $+$ or $-$. The characters of the negative parity representations are the same as those of the positive parity representations except they have the opposite sign. In molecular chemistry notation, positive parity representations are labeled with a "g" (for "gerade" in German) and negative parity representations are labeled with "u" (for "ungerade").

Each character table usually also gives typical examples of the basis functions of the matrix representations. As discussed above, these are not unique; different choices of the basis functions are possible which will not change the numbers in the character table. In Table 6.3, two special notations are used for the basis functions: L_x, L_y, and L_z transform as x, y, and z but do not change sign on inversion; these can be represented as the classical angular momentum components, for example, $L_x = yp_z - zp_y$. The basis state $\phi_{l,m}$ is the eigenstate of the quantum angular momentum operator J with total angular momentum l and projection m along the z-axis.

As seen in Table 6.3, there are two sets of irreducible representations, those in which spin does not enter the basis functions, and those in which it does. The full group which includes spin flip is called the **double group**. If spin flip (rotation by 2π radians) is not included in the symmetry operations, then three of the classes of this symmetry group will be removed, as well as the three representations which have spin-related basis functions, leaving a new group described by a 5×5 character table. As stated above, a general theorem says that the number of classes and the number of irreducible representations in a group must be the same.

Exercise 6.3.1   Verify the characters in Table 6.3 for the irreducible representation Γ_5 which has basis functions which are the three components of a vector in real space along x, y, and z, by explicitly constructing a matrix representation for one operator in each class.

6.4 Equating Physical States with the Basis States of Representations

The usefulness of group theory in solid state physics comes from the fact that the electronic states and vibrational states in symmetric crystals can be used as the basis states of the irreducible representations of the symmetry group of the crystal.

Let R be a symmetry operation which belongs to the symmetry group which transforms the Hamiltonian into itself. In other words,

$$R^{-1}HR = H, \qquad (6.4.1)$$

or $[R, H] = 0$, since RR^{-1} is equal to the identity operator. Therefore, for any eigenstate $|\psi_i\rangle$ of H,

$$H|\psi_i\rangle = E_i|\psi_i\rangle$$

$$(R^{-1}HR)|\psi_i\rangle = E_i|\psi_i\rangle$$

$$H(R|\psi_i\rangle) = E_i(R|\psi_i\rangle), \qquad (6.4.2)$$

where E_i is the energy of state $|\psi_i\rangle$. This means that if $|\psi_i\rangle$ is an eigenstate of H, then $R|\psi_i\rangle$ is also an eigenstate of H, for any operation R in the symmetry group. We can therefore write

$$R|\psi_i\rangle = \alpha_{ij}|\psi_j\rangle, \qquad (6.4.3)$$

where $|\psi_j\rangle$ is some other eigenstate, and α_{ij} is a multiplicative constant. Multiplying both sides by $\langle\psi_j|$, we have

$$\langle\psi_j|R|\psi_i\rangle = \langle\psi_j|\alpha_{ij}|\psi_j\rangle$$

$$= \alpha_{ij}\langle\psi_j|\psi_j\rangle = \alpha_{ij}. \qquad (6.4.4)$$

Therefore, we can write R in matrix form, using as its basis the eigenstates of the Hamiltonian, with matrix elements $\langle\psi_j|R|\psi_i\rangle$.

The matrix representation defined on all the eigenfunctions of the Hamiltonian can always be *reduced* to block diagonal form, as discussed in Section 6.2. The Hamiltonian will be reduced to blocks of the same dimension defined on the same basis functions.

The physics implied by this reduction is that the eigenfunctions that make up the basis functions of a single irreducible representation must be degenerate in energy. We can prove this by the following. First, multiplying (6.4.3) by R^{-1} implies

$$|\psi_i\rangle = R^{-1}\alpha_{ij}|\psi_j\rangle, \qquad (6.4.5)$$

or

$$R^{-1}|\psi_j\rangle = \frac{1}{\alpha_{ij}}|\psi_i\rangle. \qquad (6.4.6)$$

Then, multiplying both sides of (6.4.3) by $R^{-1}H$, we have

$$(R^{-1}H)R|\psi_i\rangle = (R^{-1}H)\alpha_{ij}|\psi_j\rangle$$

$$H|\psi_i\rangle = E_j\alpha_{ij}R^{-1}|\psi_j\rangle$$

$$E_i|\psi_i\rangle = E_j\alpha_{ij}\frac{1}{\alpha_{ij}}|\psi_i\rangle$$

$$\Rightarrow E_i = E_j. \qquad (6.4.7)$$

In other words, if any symmetry operation R transforms $|\psi_i\rangle$ into $|\psi_j\rangle$, then these states must have the same energy. If no symmetry operation transforms one state into the other, then by definition they belong to different irreducible representations. Since an irreducible matrix representation of R transforms its basis states into each other, it follows that all the eigenstates of the Hamiltonian that are basis states of an irreducible representation must have the same energy. This is called an **essential degeneracy**. We can therefore label the electron and phonon states at a given energy by the irreducible representation to which they belong.

Eigenfunctions of the Hamiltonian which are basis functions of different irreducible representations can have different energies. Of course, they still could have the same energy, if no term exists in the Hamiltonian to split them. If they happen to have the same energy, but do not belong to the same irreducible representation, this is called an **accidental degeneracy**.

Exercise 6.4.1 Show that a symmetry operator R must be **unitary**, that is, its adjoint is equal to its inverse; for a matrix representation this means that its inverse is equal to the complex conjugate of the transposed matrix. There are several steps to showing this.

(a) Show that $|\alpha_{ij}|^2 = 1$.

(b) Inserting a sum over a complete set of states into $\langle \psi_i | R^\dagger R | \psi_i \rangle$ gives

$$\sum_k \langle \psi_i | R^\dagger | \psi_k \rangle \langle \psi_k | R | \psi_i \rangle.$$

By evaluating this, show that $\alpha_{ji}^* = \alpha_{ij}^{-1}$, and therefore R is unitary. Note that this analysis also implies that only one row in any column is nonzero, and vice versa.

Figure 6.4 gives an example of the symmetry of the electronic states in the bands of the semiconductor GaAs, which has T_d symmetry. The points in the Brillouin zone are labeled with letters according to the fcc convention shown in Figure 1.25(b). The symmetry of the bands changes at different points in the Brillouin zone. As we saw in Section 1.9.4, the Hamiltonian for Bloch waves can be written as

$$H = \frac{p^2}{2m} + U(\vec{r}) + \frac{\hbar}{m}\vec{k}\cdot\vec{p} + \frac{\hbar^2 k^2}{2m}, \tag{6.4.8}$$

where $\vec{k}$ is the wave vector in the Brillouin zone and $\vec{p}$ is the momentum operator acting on the electronic state. The potential energy $U(\vec{r})$ has exactly the same symmetry as the crystal, and the p^2 and k^2 terms are invariant under the symmetry operations. Therefore, the Bloch wave functions at zone center ($\vec{k} = 0$) are the basis functions of the irreducible representations of the symmetry group of the crystal itself. This is why the symbol Γ, which labels the center of the Brillouin zone as shown in Figure 1.25(a), is used for the irreducible representations in the Koster character tables.

Since the linear term $\vec{k}\cdot\vec{p}$ is not rotationally invariant, when $\vec{k} \neq 0$ an extra term in the Hamiltonian exists which can make the Hamiltonian belong to a different symmetry group from the crystal as a whole. One can think of the new symmetry group in this case as that of the original crystal plus an arrow pointing in the direction of $\vec{k}$. The direction of this arrow must be preserved under any symmetry transformations in the new group;

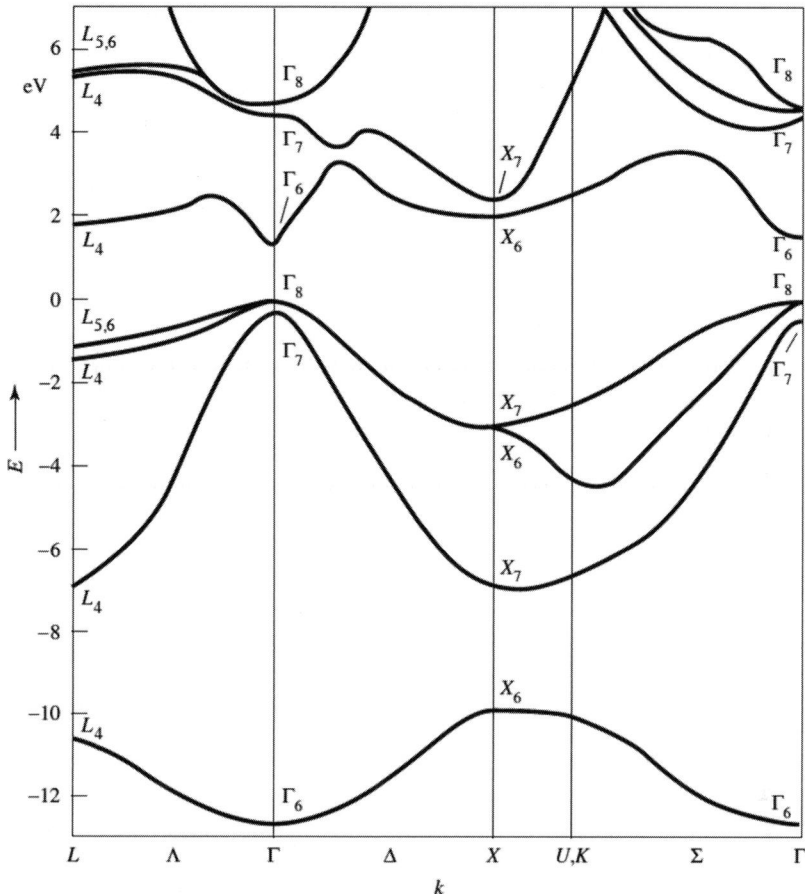

Fig. 6.4 Band structure of the semiconductor GaAs, from a numerical calculation, with the critical points labeled by the irreducible representations of the Koster notion used here. From Chelikowksy and Cohen (1976).

therefore some of the symmetry operations in the original group must be removed. As we will discuss in Section 6.8, this reduced symmetry can lead to energy splitting of some bands away from zone center. This can be seen in Figure 6.4. Since the symmetry group changes, the total number of irreducible representations may change, and consequently the numbering of the representations can change. When the k-vector reaches a critical point on the Brillouin zone boundary, the symmetry can change once again, as some symmetry operations may be added back.

In Figure 6.4, the electronic states at the critical points are labeled by the irreducible representations for the appropriate symmetry group at each point. It is common to label the irreducible representations at points away from the zone center by the symbols for the symmetry points, for example, $X_1, X_2, \ldots, L_1, L_2, \ldots$, instead of using the subscripted symbol Γ for the irreducible representations everywhere.

In general, to determine the splitting of bands for $k \neq 0$, energies one can use degenerate-state perturbation theory to construct a matrix for the energies, using the degenerate states

Table 6.6	Symmetry notation associated with different levels of description of the system	
Example	Name	Meaning
0_h	**Group**	Set of *operations* in real space on crystal as a whole; defines symmetry of crystal as a whole.
Γ_5	**Representation** (irreducible)	Set of *matrices* which is a group with the same properties as the real-space group; defines symmetry of electronic bands, vibrational bands, and quantum mechanical operators.
$\lvert u_x^5 \rangle$	**Basis function**	Single electronic Bloch wave function or vibrational polarization state; defines rows and columns of representation matrices.

as the basis states. Up to second order, the general matrix for Löwdin degenerate-state perturbation theory (reviewed in Appendix E) is

$$E_{nm} = \langle m | H_{\text{int}} | n \rangle + \sum_l \frac{\langle m | H_{\text{int}} | l \rangle \langle l | H_{\text{int}} | n \rangle}{E_n - E_l}, \tag{6.4.9}$$

where states n and m belong to the set of states which are degenerate in the absence of the perturbation term H_{int}. Diagonalizing this matrix to get the new eigenstates and their energies corresponds to reducing the original representation to a set of irreducible representations in the new symmetry that includes the perturbation term. We will return to an example of this in Section 6.11.1.

The vibrational eigenstates of a crystal also comprise a basis for a matrix representation of the symmetry operations, since the Hamiltonian of interatomic forces also must remain invariant under the symmetry operations of the crystal symmetry group. In Chapter 3, we used a classical approach to find the vibrational modes, with force equations based on Hooke's law, but force equations are just another way of solving for the eigenstates of the Hamiltonian, and therefore the above analysis also applies to eigenstates found in this way. The basis functions belonging to a single irreducible representation correspond to degenerate vibrational modes.

As noted in Section 3.1.3, the vibrational modes of crystals are typically labeled by group-theory notation for the symmetry of the modes, such as in Figure 3.8. Like the Bloch wave functions, the modes at zone center (the Γ-point) have the full symmetry of the crystal, but vibrational modes with a finite $\vec{k}$ in some direction, that is, further out in the Brillouin zone, belong to a group with different symmetry.

Table 6.6 lists the three different levels of symmetry description for the quantum mechanics states in a crystal. It is important to not confuse these three different levels of description.

6.5 Reducing Representations

As discussed in Section 6.2, all matrix representations of a group can be reduced by block diagonalization to a finite set of irreducible matrices. How do we perform this task in practice? That is, how do we find the smaller matrices into which the larger matrices can be block diagonalized, or reduced? A large amount of mathematical technology has been developed to solve this problem. In this section, we simply state the most useful results without proof. These theorems may seem tedious, but the beauty is that they involve entirely straightforward summation – there is no guessing involved.

As we have discussed, there is a finite number of irreducible representations, which are tabulated for each point group. If we know the point group, a formula tells us how to find the irreducible representations into which a matrix representation can be block diagonalized. This formula is

$$n_{\Gamma_i} = \frac{1}{h} \sum_R \chi_{\Gamma_i}(R)\chi(R). \tag{6.5.1}$$

Here, n_{Γ_i} is the number of times the irreducible representation Γ_i appears in the blocks of the matrix when it is reduced, h is the total number of operations in the group, $\chi_{\Gamma_i}(R)$ is the character of the operation R in the irreducible representation Γ_i, and $\chi(R)$ is the character (trace) of the matrix representation of operation R in the representation which is to be reduced. If we know the character, that is the diagonal elements, of some matrix representation, we can therefore find all of the irreducible representations by seeing which n_{Γ_i} are nonzero according to the above formula. The diagonal elements of a matrix representation are easy to determine, since we need only ask which basis states are mapped into themselves by a given operation R.

Formula (6.5.1) tells us what irreducible representations will occur in the block diagonalization, but it does not tell us how to write those irreducible blocks in terms of the original basis states of the unreduced matrix. To do that, we need to use another theorem of group theory.

Any basis state $|\phi\rangle$ of the original, unreduced representation can be written as a linear combination of the basis states of the irreducible representations:

$$|\phi\rangle = \sum_i \sum_{n=1}^{d_i} \alpha_n^i |\phi_n^i\rangle, \tag{6.5.2}$$

where d_i is the dimension of irreducible representation i, $|\phi_n^i\rangle$ is the nth basis function of the irreducible representation i, and α_n^i is a constant. The new states $|\phi_n^i\rangle$ are obtained by the formula

$$\alpha_n^i |\phi_n^i\rangle = P_{nn}(\Gamma_i)|\phi\rangle, \tag{6.5.3}$$

where $P_{nm}(\Gamma_i)$ is the **projection operator**, defined as

$$P_{mn}(\Gamma_i) = \left(\frac{d_i}{h}\right) \sum_R T_{mn}^*(R)O(R), \tag{6.5.4}$$

where $T_{mn}(R)$ is the matrix representation of operation R in the irreducible representation i and $O(R)$ is the operation R in the original, unreduced representation.

In other words, to find the new basis states, one follows the following recipe:

- Start with a basis function $|\phi\rangle$ in the original, unreduced representation.
- Find the transformation of this basis function under the action of each operation R in the symmetry group.
- Sum up all these transformed basis functions $O(R)|\phi\rangle$, each weighted by the appropriate diagonal element $T_{nn}^*(R)$ of the irreducible matrix representation i for which one wants to find the basis functions.
- The resulting linear superposition of the original basis functions is the nth basis function of the representation i, apart from an overall multiplicative factor.

For this procedure, it doesn't matter which basis function $|\phi\rangle$ one chooses in the original representation, as along as $|\phi\rangle$ is not orthogonal to the new basis state one wants to find, that is, $P_{nn}(\Gamma_i)|\phi\rangle \neq 0$.

Formulas (6.5.3) and (6.5.4) clearly require an explicit choice for the matrices of the irreducible representations. As discussed above, these are not unique; we can in general choose different sets of matrices for an irreducible representation of a given group, although the traces (characters) of these matrices must always be the same. For example, a three-dimensional representation might have x, y, and z as basis functions $|\phi_1\rangle$, $|\phi_2\rangle$, and $|\phi_3\rangle$, respectively, but of course we could change the order to $|\phi_1\rangle = y$, $|\phi_3\rangle = x$, and $|\phi_3\rangle = z$, which would change the representation matrices but not the traces nor the multiplication properties of the group.

Often one does not need to go through all the work of the above formulas, which, though they have the advantage of requiring nothing but straightforward summation, can be time consuming. It is also possible to deduce the reduction into irreducible representations by simple inspection of the character table. The sum of the characters of the irreducible representations in each class must equal the character for that class of the original representation. This is because no similarity transformation can change the trace of a matrix. If some selection of representations can be found in which the characters add up properly, it must be the correct reduction, because this reduction is unique. For simple representations, we can often simply guess the proper reduction this way.

Example. To see how all this group theory technology works, let us present an example from molecular chemistry. Figure 6.5 shows an example of a molecule with cubic symmetry and eight degenerate s-orbitals. We can use symmetry arguments to find the eigenstates of the molecular states in the LCAO approximation. In this approximation, the energy states can be written as linear superpositions of the atomic orbitals, which are labeled 1–8 in Figure 6.5.

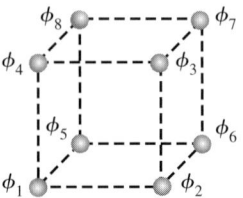

Fig. 6.5 A cubic molecule.

Table 6.7 The character table of the O_h cubic symmetry single group

	E	$8C_3$	$3C_2$	$6C_4$	$6C_2'$	I	$8S_6$	$3\sigma_h$	$6S_4$	$6\sigma_d$	Basis functions
Γ_1^+	1	1	1	1	1	1	1	1	1	1	$r = x^2 + y^2 + z^2$
Γ_2^+	1	1	1	-1	-1	1	1	1	-1	-1	$(x^2 - y^2)(y^2 - z^2)(z^2 - x^2)$
Γ_3^+	2	-1	2	0	0	2	-1	2	0	0	$(2z^2 - x^2 - y^2), \sqrt{3}(x^2 - y^2)$
Γ_4^+	3	0	-1	1	-1	3	0	-1	1	-1	L_x, L_y, L_z
Γ_5^+	3	0	-1	-1	1	3	0	-1	-1	1	yz, xz, xy
Γ_1^-	1	1	1	1	1	-1	-1	-1	-1	-1	$\Gamma_2^+ \times xyz$
Γ_2^-	1	1	1	-1	-1	-1	-1	-1	1	1	xyz
Γ_3^-	2	-1	2	0	0	-2	1	-2	0	0	$\Gamma_3^+ \times xyz$
Γ_4^-	3	0	-1	1	-1	-3	0	1	-1	1	x, y, z
Γ_5^-	3	0	-1	-1	1	-3	0	1	1	-1	$\Gamma_5^+ \times xyz$

The first step is to determine the symmetry group of the system by asking what operations transform it back into itself. In this case, there are six four-fold (C_4) rotations (left and right about each face of the cube), three two-fold (C_2) rotations (180° rotation about each face of the cube), another set of six two-fold (C_2) rotations (about the cube edges), and eight three-fold (C_3) rotations (about the cube corners) which map the system into itself, plus the inversion operation. This puts the system in the O_h symmetry group, which has the character table shown in Table 6.7 (only the single group is shown, neglecting spin). The second group of operators starting with I are just the same as the first five classes, but with inversion.

Next, we write the characters of this group in the representation using the basis of the atomic states. These are as follows:

	E	$8C_3$	$3C_2$	$6C_4$	$6C_2'$	I	$8S_6$	$3\sigma_h$	$6S_4$	$6\sigma_d$
Γ_ϕ	8	2	0	0	0	0	0	0	0	4

It may seem difficult to get these characters, but actually it is simple – the character of a given symmetry operation is equal to the number of atoms that are mapped back into themselves by the operation.

This representation can then be *reduced* using (6.5.1). This yields

$$\Gamma_\phi = \Gamma_1^+ \oplus \Gamma_2^- \oplus \Gamma_4^- \oplus \Gamma_5^+, \tag{6.5.5}$$

where we have used the symbol "$\oplus$" to indicate the reduction of the larger matrix into a sum of smaller matrices as blocks along the diagonal. This is sometimes called an **outer sum**.

Is is easy to verify that the sum of the characters of the irreducible representations equals the characters of the original representation. This is seen in the following sums:

	E	$8C_3$	$3C_2$	$6C_4$	$6C_2'$	I	$8S_6$	$3\sigma_h$	$6S_4$	$6\sigma_d$
Γ_1^+	1	1	1	1	1	1	1	1	1	1
Γ_2^-	1	1	1	−1	−1	−1	−1	−1	1	1
Γ_4^-	3	0	−1	1	−1	−3	0	1	−1	1
Γ_5^+	3	0	−1	−1	1	3	0	−1	−1	1
Γ_ϕ	8	2	0	0	0	0	0	0	0	4

.

We have already deduced, without actually solving the Schrödinger equation, that the eigenstates of this molecule fall into four groups of degenerate states. To find the linear combinations of the original states that correspond to these eigenstates, we must construct explicit representations using a specific choice of basis.

For the one-dimensional Γ_1^+ and Γ_2^- representations, the matrices are identical to the characters listed in Table 6.7. For example, for Γ_1^+, starting with the basis function $|\phi_1\rangle$, we use (6.5.4) to get

$$
\begin{aligned}
P_{xx}(\Gamma_1^+)|\phi_1\rangle = \tfrac{1}{48} [\, &1 \cdot (|\phi_1\rangle) & E \\
+&1 \cdot (2|\phi_1\rangle + 2|\phi_3\rangle + 2|\phi_6\rangle + 2|\phi_8\rangle)) & 8\,C_3 \\
+&1 \cdot (|\phi_3\rangle + |\phi_6\rangle + |\phi_8\rangle)) & 3\,C_2 \\
+&1 \cdot (2|\phi_2\rangle + 2|\phi_4\rangle + 2|\phi_5\rangle)) & 6\,C_4 \\
+&1 \cdot (3|\phi_7\rangle + |\phi_2\rangle + |\phi_4\rangle + |\phi_5\rangle)) & 6\,C_2' \\
+&1 \cdot (|\phi_7\rangle)) & I \\
+&1 \cdot (2|\phi_7\rangle + 2|\phi_5\rangle + 2|\phi_4\rangle + 2|\phi_2\rangle)) & 8\,S_6 \\
+&1 \cdot (|\phi_5\rangle + |\phi_4\rangle + |\phi_2\rangle)) & 3\,\sigma_h \\
+&1 \cdot (2|\phi_8\rangle + 2|\phi_6\rangle + 2|\phi_3\rangle)) & 6\,S_4 \\
+&1 \cdot (3|\phi_1\rangle + |\phi_8\rangle + |\phi_6\rangle + |\phi_3\rangle))\,] & 6\,\sigma_d
\end{aligned}
$$

$$
= \tfrac{1}{8} (|\phi_1\rangle + |\phi_2\rangle + |\phi_3\rangle + |\phi_4\rangle + |\phi_5\rangle + |\phi_6\rangle \\
+ |\phi_7\rangle + |\phi_8\rangle).
$$

(6.5.6)

For each operator R in the sum, we have found the basis state $R\phi_1$ into which $|\phi_1\rangle$ is transformed in the symmetry operation, and multiplied by the character for R in the Γ_1 representation. The result in this case is a completely symmetric (s-like) linear combination of the atomic states.

For the Γ_4^- states, we need to use an explicit representation of the symmetry operations. We will use the standard basis functions given in Table 6.7, which are x, y, and z. As discussed above, we do not need to construct each matrix representation completely – we only need to find the diagonal components of each matrix, that is, to find the basis function for x we count only operations which map the x-axis back into itself. For example, a C_4 rotation around the x-axis is represented by

$$
\begin{pmatrix} 1 & 0 & 0 \\ 0 & 0 & 1 \\ 0 & -1 & 0 \end{pmatrix},
$$

which leaves the x-axis unchanged, and therefore contributes a term $T_{xx}^*(C_{4x}) = 1$ to (6.5.4), while a C_2 rotation about the z-axis is represented by

$$\begin{pmatrix} -1 & 0 & 0 \\ 0 & -1 & 0 \\ 0 & 0 & 1 \end{pmatrix}$$

and contributes a term $T_{xx}^*(C_{2z}) = -1$. A C_3 rotation about a corner of the cube does not contribute any term because none of the axes is mapped into itself.

The task is made easier by the fact that we only really need to calculate the matrices for four types of rotation; the last five transformations in Table 6.7 are just the same rotations plus inversion. These have the same representations but opposite sign, if the representation has negative parity. Again starting with the $|\phi_1\rangle$ basis state of the original representation, (6.5.4) therefore gives for the ϕ_x basis function of the Γ_4^- representation:

$$
\begin{aligned}
P_{xx}(\Gamma_4^-)|\phi_1\rangle = \tfrac{3}{48} \big[&|\phi_1\rangle & E \\
&- |\phi_3\rangle - |\phi_6\rangle + |\phi_8\rangle & 3\,C_2 \\
&+ |\phi_4\rangle + |\phi_5\rangle & 2\,C_4 \\
&- |\phi_7\rangle - |\phi_2\rangle & 2\,C_2' \\
&- |\phi_7\rangle & I \\
&+ |\phi_5\rangle + |\phi_4\rangle - |\phi_2\rangle & 3\,\sigma_h \\
&- |\phi_6\rangle - |\phi_3\rangle & 2\,S_4 \\
&+ |\phi_1\rangle + |\phi_8\rangle \big] & 2\,\sigma_d
\end{aligned}
$$

$$
= \tfrac{1}{8}(|\phi_1\rangle - |\phi_2\rangle - |\phi_3\rangle + |\phi_4\rangle + |\phi_5\rangle - |\phi_6\rangle \\
- |\phi_7\rangle + |\phi_8\rangle).
$$

$$(6.5.7)$$

This state is shown in Figure 6.6. As expected, it has negative parity in the x-direction.

In this way, group theory allows us to construct the eigenstates of the Hamiltonian as linear combinations of the atomic orbitals simply by knowing the symmetry properties of the system, without solving the Schrödinger equation.

Exercise 6.5.1   (a) Find the basis state of the Γ_2^- irreducible representation for this molecule as a linear superposition of the original atomic states.

(b) Working with an explicit representation for the Γ_5^+ irreducible representation, find one of the basis states in terms of the original atomic states.

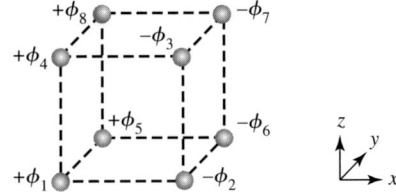

Fig. 6.6 Γ_4^- orbital of a cubic molecule.

6.6 Multiplication Rules for Outer Products

If two terms in the total quantum mechanical Hamiltonian commute, then the eigenstates of the total Hamiltonian can be written simply as products of the eigenstates of the individual terms.

It often happens that different terms in a Hamiltonian have eigenstates that are the basis states of different irreducible representations of the crystal symmetry group. For example, the wave function of an electron may have the spatial symmetry of a p-orbital of an atom and also a spin state. In the cubic T_d point group, the p-orbitals form the basis of the Γ_5 representation, since this representation has a basis that transforms like the x, y, and z components of a vector, while the two spin states form the basis of the Γ_6 representation. In the p-orbitals, there are therefore six different orbitals for the electron, each of which is the product of a spatial state times a spin state.

This type of multiplication is different from the multiplication process we defined in Section 6.1. That process, which was written as $\circ$, defines the relationships of the elements of a group. It is represented by mathematical multiplication of the square matrices of a representation of the group. By the definition of a group, multiplying two members of the group always gives another member of the same group.

When we multiply the basis states of different representations, we are performing an **outer product**, also known as a **direct product** (usually written $\otimes$), which multiplies any two representation matrices for the same symmetry operation, even if they have different dimensions. To form the outer product, we multiply each element of one matrix by the entire other matrix. For example, let A be a 3×3 representation of one of the operations in a group, and let B be a 2×2 representation of the same operation. Then the outer product $A \otimes B$ is equal to

$$
\begin{pmatrix}
A_{11}\begin{pmatrix} B_{11} & B_{12} \\ B_{21} & B_{22} \end{pmatrix} & A_{12}\begin{pmatrix} B_{11} & B_{12} \\ B_{21} & B_{22} \end{pmatrix} & A_{13}\begin{pmatrix} B_{11} & B_{12} \\ B_{21} & B_{22} \end{pmatrix} \\
A_{21}\begin{pmatrix} B_{11} & B_{12} \\ B_{21} & B_{22} \end{pmatrix} & A_{22}\begin{pmatrix} B_{11} & B_{12} \\ B_{21} & B_{22} \end{pmatrix} & A_{23}\begin{pmatrix} B_{11} & B_{12} \\ B_{21} & B_{22} \end{pmatrix} \\
A_{31}\begin{pmatrix} B_{11} & B_{12} \\ B_{21} & B_{22} \end{pmatrix} & A_{32}\begin{pmatrix} B_{11} & B_{12} \\ B_{21} & B_{22} \end{pmatrix} & A_{33}\begin{pmatrix} B_{11} & B_{12} \\ B_{21} & B_{22} \end{pmatrix}
\end{pmatrix}
$$

$$
= \begin{pmatrix}
A_{11}B_{11} & A_{11}B_{12} & A_{12}B_{11} & A_{12}B_{12} & A_{13}B_{11} & A_{13}B_{12} \\
A_{11}B_{21} & A_{11}B_{22} & A_{12}B_{21} & A_{12}B_{22} & A_{13}B_{11} & A_{13}B_{12} \\
A_{21}B_{11} & A_{21}B_{12} & A_{22}B_{11} & A_{22}B_{12} & A_{23}B_{11} & A_{23}B_{12} \\
A_{21}B_{21} & A_{21}B_{22} & A_{22}B_{21} & A_{22}B_{22} & A_{23}B_{21} & A_{23}B_{22} \\
A_{31}B_{11} & A_{31}B_{12} & A_{32}B_{11} & A_{32}B_{12} & A_{33}B_{11} & A_{33}B_{12} \\
A_{31}B_{21} & A_{31}B_{22} & A_{32}B_{21} & A_{32}B_{22} & A_{33}B_{21} & A_{33}B_{22}
\end{pmatrix} .
$$

$$\tag{6.6.1}$$

If we compute this outer product for each operation in the group, we will have a set of 6×6 matrices which is also a representation of the group. This is a general theorem: The outer product of two matrix representations of a group is also a matrix representation of the group. The set of basis functions for this new representation consists of all possible combinations of the two different sets of basis functions. Note that since the terms $A_{11}B_{11}$, etc., are simple mathematical products, the outer product is commutative, even though the products of the elements of the group, that is, products of the matrices within a single group representation, are generally not commutative.

As discussed in Section 6.5, this representation is *reducible* by a proper transformation to block diagonal form, in which all of the blocks are irreducible representations. One can use the formulas given in Section 6.5, but in the case of an outer product of two irreducible representations, it is often easier to deduce the reduction of an outer product by inspection of the character table. It is easy to see from the above example that the character (trace) of an outer product must be equal to the product of the traces of the two matrices which are multiplied. As discussed in Section 6.5, the sum of the characters of the irreducible representations in any given class must add up to the character of the original, reducible representation. Therefore, the outer product of two irreducible representations can be found by writing down the product of the characters of the two representations for each class, and then finding by inspection a set of irreducible representations in the same symmetry group which have characters which add up to those products, for each class, as we did in the example at the end of Section 6.5.

In the example of the Hamiltonian with two terms discussed above, the new irreducible blocks give the essential degeneracies of the full Hamiltonian. As discussed in Section 6.4, there can be no term in the Hamiltonian that leaves the symmetry unchanged and also splits the energy degeneracy of states in the same irreducible block. States that belong to different irreducible representations can have different energies.

Since there is only a finite number of irreducible representations in a given group, there is only a finite number of possible outer products of irreducible representations. This allows us to write down a multiplication table which contains the rule for each possible outer product of the irreducible representations of a group. (Multiplication tables are used only for outer products of irreducible representations of the same symmetry group; it makes no sense to multiply representation matrices from different groups, since there may not even be the same number of symmetry operation matrices.) Table 6.8 gives the multiplication table for the T_d and O groups. (The group O_h has the same multiplication table, with the additional rule that the sign of the parity must be multiplied.) In general, a given irreducible representation may occur more than once when a representation is reduced. An irreducible representation that occurs twice is indicated in Table 6.8 by a 2 in front of a representation. Each time the same irreducible representation appears, it represents a different block in the reduced representation, which corresponds to a different set of basis functions. Thus, for example, the basis functions for one irreducible representation may correspond to electron states with different energy from those in another irreducible representation, even though they happen to have the same symmetry.

The basis functions of the new irreducible representations in terms of the original, multiplied representations can also be tabulated, using (6.5.1) and (6.5.3). The coefficients give the relative weight of each original component. In general, we write

$$|w_l^{r\alpha}\rangle = \sum_{jk} C_{jk,l}^{pq,r\alpha} |u_j^p\rangle |v_k^q\rangle, \tag{6.6.2}$$

Table 6.8 Multiplication table for the groups O and T_d

$\otimes$	Γ_1	Γ_2	Γ_3	Γ_4	Γ_5	Γ_6	Γ_7	Γ_8
Γ_1	Γ_1	Γ_2	Γ_3	Γ_4	Γ_5	Γ_6	Γ_7	Γ_8
Γ_2	Γ_2	Γ_1	Γ_3	Γ_5	Γ_4	Γ_7	Γ_6	Γ_8
Γ_3	Γ_3	Γ_3	$\Gamma_1 \oplus \Gamma_2 \oplus \Gamma_3$	$\Gamma_4 \oplus \Gamma_5$	$\Gamma_4 \oplus \Gamma_5$	Γ_8	Γ_8	$\Gamma_6 \oplus \Gamma_7 \oplus \Gamma_8$
Γ_4	Γ_4	Γ_5	$\Gamma_4 \oplus \Gamma_5$	$\Gamma_1 \oplus \Gamma_3 \oplus \Gamma_4 \oplus \Gamma_5$	$\Gamma_2 \oplus \Gamma_3 \oplus \Gamma_4 \oplus \Gamma_5$	$\Gamma_6 \oplus \Gamma_8$	$\Gamma_7 \oplus \Gamma_8$	$\Gamma_6 \oplus \Gamma_7 \oplus 2\Gamma_8$
Γ_5	Γ_5	Γ_4	$\Gamma_4 \oplus \Gamma_5$	$\Gamma_2 \oplus \Gamma_3 \oplus \Gamma_4 \oplus \Gamma_5$	$\Gamma_1 \oplus \Gamma_3 \oplus \Gamma_4 \oplus \Gamma_5$	$\Gamma_7 \oplus \Gamma_8$	$\Gamma_6 \oplus \Gamma_8$	$\Gamma_6 \oplus \Gamma_7 \oplus 2\Gamma_8$
Γ_6	Γ_6	Γ_7	Γ_8	$\Gamma_6 \oplus \Gamma_8$	$\Gamma_7 \oplus \Gamma_8$	$\Gamma_1 \oplus \Gamma_4$	$\Gamma_2 \oplus \Gamma_5$	$\Gamma_3 \oplus \Gamma_4 \oplus \Gamma_5$
Γ_7	Γ_7	Γ_6	Γ_8	$\Gamma_7 \oplus \Gamma_8$	$\Gamma_6 \oplus \Gamma_8$	$\Gamma_2 \oplus \Gamma_5$	$\Gamma_1 \oplus \Gamma_4$	$\Gamma_3 \oplus \Gamma_4 \oplus \Gamma_5$
Γ_8	Γ_8	Γ_8	$\Gamma_6 \oplus \Gamma_7 \oplus \Gamma_8$	$\Gamma_6 \oplus \Gamma_7 \oplus 2\Gamma_8$	$\Gamma_6 \oplus \Gamma_7 \oplus 2\Gamma_8$	$\Gamma_3 \oplus \Gamma_4 \oplus \Gamma_5$	$\Gamma_3 \oplus \Gamma_4 \oplus \Gamma_5$	$\Gamma_1 \oplus \Gamma_2 \oplus \Gamma_3 \oplus 2\Gamma_4 \oplus 2\Gamma_5$

Table 6.9 Coupling coefficients for the outer product $\Gamma_5 \otimes \Gamma_6$ in the T_d group

	$|w_1^7\rangle$	$|w_2^7\rangle$	$|w_1^8\rangle$	$|w_2^8\rangle$	$|w_3^8\rangle$	$|w_4^8\rangle$
$|u_1^5\rangle|v_1^6\rangle$	0	$-i/\sqrt{3}$	$-i/\sqrt{6}$	0	$-i/\sqrt{2}$	0
$|u_1^5\rangle|v_2^6\rangle$	$-i/\sqrt{3}$	0	0	$i/\sqrt{2}$	0	$i/\sqrt{6}$
$|u_2^5\rangle|v_1^6\rangle$	0	$1/\sqrt{3}$	$1/\sqrt{6}$	0	$-1/\sqrt{2}$	0
$|u_2^5\rangle|v_2^6\rangle$	$-1/\sqrt{3}$	0	0	$-1/\sqrt{2}$	0	$1/\sqrt{6}$
$|u_3^5\rangle|v_1^6\rangle$	$i/\sqrt{3}$	0	0	0	0	$i\sqrt{2}/\sqrt{3}$
$|u_3^5\rangle|v_2^6\rangle$	0	$-i/\sqrt{3}$	$i\sqrt{2}/\sqrt{3}$	0	0	0

where $|u_j^p\rangle$ and $|v_k^q\rangle$ are basis functions in the original two irreducible representations Γ_q and Γ_p that are multiplied together, and $|w_l^{r\alpha}\rangle$ is a basis function of one of the new irreducible representations Γ_r found by block diagonalizing the product representation. The superscript α is used to allow for the possibility that in the block diagonalization, two or more blocks with the same Γ_r symmetry are obtained. The coefficients $C_{jk,l}^{pq,r\alpha}$ are called the **coupling coefficients** or **Clebsch–Gordan coefficients**. Tables 6.9 and 6.10 give some examples.

Exercise 6.6.1 Verify the multiplication rules given in Table 6.8,

$$\Gamma_6 \otimes \Gamma_5 = \Gamma_7 \oplus \Gamma_8, \tag{6.6.3}$$

Table 6.10 Coupling coefficients for the outer product $\Gamma_4 \otimes \Gamma_4$ in the T_d group

	$\lvert w_1^1\rangle$	$\lvert w_1^3\rangle$	$\lvert w_2^3\rangle$	$\lvert w_1^4\rangle$	$\lvert w_2^4\rangle$	$\lvert w_3^4\rangle$	$\lvert w_1^5\rangle$	$\lvert w_2^5\rangle$	$\lvert w_3^5\rangle$
$\lvert u_1^4\rangle\lvert v_1^4\rangle$	$1/\sqrt{3}$	$-1/\sqrt{6}$	$1/\sqrt{2}$	0	0	0	0	0	0
$\lvert u_1^4\rangle\lvert v_2^4\rangle$	0	0	0	0	0	$1/\sqrt{2}$	0	0	$1/\sqrt{2}$
$\lvert u_1^4\rangle\lvert v_3^4\rangle$	0	0	0	0	$-1/\sqrt{2}$	0	0	$1/\sqrt{2}$	0
$\lvert u_2^4\rangle\lvert v_1^4\rangle$	0	0	0	0	0	$-1/\sqrt{2}$	0	0	$1/\sqrt{2}$
$\lvert u_2^4\rangle\lvert v_2^4\rangle$	$1/\sqrt{3}$	$-1/\sqrt{6}$	$-1/\sqrt{2}$	0	0	0	0	0	0
$\lvert u_2^4\rangle\lvert v_3^4\rangle$	0	0	0	$1/\sqrt{2}$	0	0	$1/\sqrt{2}$	0	0
$\lvert u_3^4\rangle\lvert v_1^4\rangle$	0	0	0	0	$1/\sqrt{2}$	0	0	$1/\sqrt{2}$	0
$\lvert u_3^4\rangle\lvert v_2^4\rangle$	0	0	0	$-1/\sqrt{2}$	0	0	$1/\sqrt{2}$	0	0
$\lvert u_3^4\rangle\lvert v_3^4\rangle$	$1/\sqrt{3}$	$\sqrt{2}/\sqrt{3}$	0	0	0	0	0	0	0

and

$$\Gamma_6 \otimes \Gamma_8 = \Gamma_3 \oplus \Gamma_4 \oplus \Gamma_5, \tag{6.6.4}$$

by using (6.5.1) and showing that for each operator in the T_d group, the sum of the characters of the representations in the outer sum add up to the product of the characters of the original representations.

Spin–orbit splitting. As an example, let us look at the valence bands of the semiconductor GaAs, which has T_d symmetry. The topmost valence band of this material arises from the atomic p-orbital states, which have symmetry x, y, z. From Table 6.3, these states transform as the Γ_5 irreducible representation of T_d.

An electron in each orbital can have two spin states, which are the basis states of the Γ_6 representation of the T_d group. The electronic states of the s-orbital conduction band are therefore represented by the product $\Gamma_6 \otimes \Gamma_1$, which is just Γ_6 (any representation multiplied by Γ_1 is unchanged). The electronic states of the p-orbitals are given by the product $\Gamma_6 \otimes \Gamma_5$, which according to the multiplication table is $\Gamma_6 \otimes \Gamma_5 = \Gamma_7 \oplus \Gamma_8$. In other words, the valence band is split into two essential degeneracies in T_d symmetry. This means that the Hamiltonian can have a term that splits these groups of states.

Such a term does exist, namely the spin–orbit coupling term; it is derived in Appendix F, and we used this term in Section 1.13, where we found that p-like states are split by spin–orbit coupling into two degenerate states (Γ_7 in our group theory notation) and four degenerate states (Γ_8 in our group theory notation). The spin–orbit interaction Hamiltonian, introduced in Section 1.13, can be written as

$$
\begin{aligned}
H_{SO} &= \xi \vec{L} \cdot \vec{S} \\
&= \xi (L_x S_x + L_y S_y + L_z S_z) \\
&= \frac{\xi}{4}\big((L_+ + L_-)(S_+ + S_-) - (L_+ - L_-)(S_+ - S_-)\big) + \xi L_z S_z \\
&= \frac{\xi}{2}(L_+ S_- + L_- S_+ + 2L_z S_z),
\end{aligned}
\tag{6.6.5}
$$

where the L operators have the properties

$$L_\pm|lm\rangle = \hbar\sqrt{l(l+1) - m(m\pm 1)}|l, m\pm 1\rangle$$
$$L_z|lm\rangle = \hbar m|lm\rangle, \tag{6.6.6}$$

and the same for the S operators (see, e.g., Cohen-Tannoudji *et al.* 1977: s. VI.C.)

From the coupling coefficients in Table 6.9, we can write the six valence band states in terms of the original p and spin states. For example, three of the six are

$$|w_1^7\rangle = -\frac{i}{\sqrt{3}}|x\rangle|\uparrow\rangle - \frac{1}{\sqrt{3}}|y\rangle|\uparrow\rangle + \frac{i}{\sqrt{3}}|z\rangle|\downarrow\rangle$$

$$|w_1^8\rangle = -\frac{i}{\sqrt{6}}|x\rangle|\downarrow\rangle + \frac{1}{\sqrt{6}}|y\rangle|\downarrow\rangle + \frac{i\sqrt{2}}{\sqrt{3}}|z\rangle|\uparrow\rangle$$

$$|w_2^8\rangle = \frac{i}{\sqrt{2}}|x\rangle|\uparrow\rangle - \frac{1}{\sqrt{2}}|y\rangle|\uparrow\rangle, \tag{6.6.7}$$

where $|\uparrow\rangle$ and $|\downarrow\rangle$ are the two spin states. We can rewrite these using the definitions of the angular momentum eigenstates for $l = 1$, $m = -1, 0, 1$:

$$|1\rangle = -\frac{1}{\sqrt{2}}(|x\rangle + i|y\rangle)$$

$$|0\rangle = |z\rangle$$

$$|-1\rangle = \frac{1}{\sqrt{2}}(|x\rangle - i|y\rangle). \tag{6.6.8}$$

Note a subtlety here: These are eigenstates of the L_z operator, but they form basis states of the Γ_5 irreducible representation, not Γ_4, in the T_d character table shown in Table 6.3. The basis functions of Γ_4 in T_d transform as the set of *operators* L_x, L_y, L_z; that is, as $xp_y - yp_x$, etc.

In the basis of (6.6.8), the product states (6.6.7) become

$$|w_1^7\rangle = \sqrt{\frac{2}{3}}|-1\rangle|\uparrow\rangle + \frac{i}{\sqrt{3}}|0\rangle|\downarrow\rangle$$

$$|w_1^8\rangle = \frac{i}{\sqrt{3}}|1\rangle|\downarrow\rangle + \frac{i\sqrt{2}}{\sqrt{3}}|0\rangle|\uparrow\rangle$$

$$|w_2^8\rangle = -|1\rangle|\uparrow\rangle. \tag{6.6.9}$$

The states have definite values of total angular momentum; in the cases here, $m_J = -\frac{1}{2}$, $\frac{1}{2}$, and $\frac{3}{2}$, respectively. It is easy to show that the states are grouped into two degenerate sets with different values of the spin–orbit energy (6.6.5), namely a doublet with $j = \frac{1}{2}$ and four degenerate states with $j = \frac{3}{2}$.

Exercise 6.6.2 Verify, using the coupling coefficients for the $\Gamma_6 \otimes \Gamma_5$ product states in Table 6.9, that the interaction (6.6.5) acting on the product states gives the same energy shift for both Γ_7 states, and a common energy shift for the four Γ_8 states.

Excitons. Going further with our example of GaAs, an exciton formed from an electron in the Γ_6 conduction band and a hole in the Γ_8 valence band will have states represented by

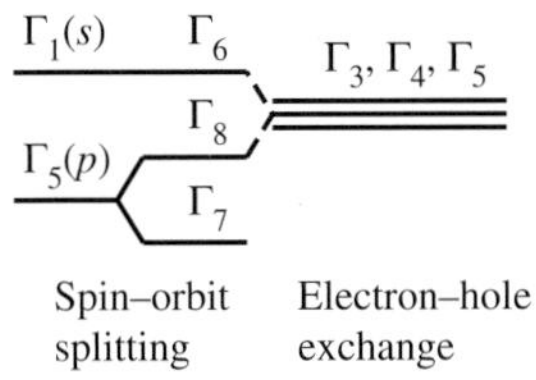

Spin–orbit　　　Electron–hole
splitting　　　　exchange

Fig. 6.7　Spin–orbit interaction in a crystal with T_d symmetry, such as, GaAs, leads to splitting of the p-like valence band. Excitons formed from conduction-band electrons and the topmost valence-band holes are split by spin–spin exchange into three levels, according to the group theory, as discussed in the text.

the product $\Gamma_6 \otimes \Gamma_8 = \Gamma_3 \oplus \Gamma_4 \oplus \Gamma_5$. This means that the exciton states formed from these bands will be split into a doublet and two triplets. Figure 6.7 illustrates the level splitting in this case. The term in the Hamiltonian that can split these states is known as **electron–hole exchange**, which has the general form

$$H_{eh} = \xi_{eh} \vec{S}_e \cdot \vec{S}_h, \tag{6.6.10}$$

where $\vec{S}_e$ and $\vec{S}_h$ are the electron and hole spin operators, respectively, and ξ_{eh} is an interaction constant which depends on the details of the atomic orbitals. We will discuss the origin of electron–hole exchange in Section 10.8.3.

By the same approach, an exciton formed from an electron in the Γ_6 conduction band and the Γ_7 valence band of a GaAs will be split according to $\Gamma_6 \otimes \Gamma_7 = \Gamma_2 \oplus \Gamma_5$; that is, the excitons formed from these bands will split into a singlet and triplet state.

Exercise 6.6.3　In many semiconductors, a **biexciton** can be formed as a bound state of two excitons, analogous to a hydrogen molecule formed from two hydrogen atoms. For the Γ_6 conduction band and Γ_7 and Γ_8 valence bands discussed in the above example in T_d symmetry, find the possible symmetries of the biexciton states by computing all the possible outer products of two exciton states.

6.7 Review of Types of Operators

We have now defined several types of operators. This can make group theory confusing to learn. Table 6.11 summarizes the different types of operators used.

Symmetry operations act on the crystal itself. These operations form the elements of the group that defines the symmetry of the system.

We also have *operators that act on these operators* – the multiplication operator, the outer product, and the outer sum. The elements of the group (which in this case are symmetry operations) are treated as mathematical entities (which can be represented by square matrices) with a defined set of multiplication rules. The multiplication operator, outer product, and outer sum are operators in the algebra of group theory.

Last, we also have *quantum mechanical operators*. These act on quantum mechanical states. They do not directly act on the symmetry elements at all. These quantum mechanical

Table 6.11 Four types of operators	
E, I, C_n, σ_i, etc.	Elements of the group, operate on the Hamiltonian itself
$\circ$	Multiplication, operates on elements of the group (symmetry operations) within a single irreducible representation; the result is also a member of the same irreducible representation.
$\otimes, \oplus$	Outer product, outer sum; relate elements in different irreducible representations within a group.
p_i, x_i, etc.	Quantum mechanical operators, act on quantum states. The eigenstates of a quantum mechanical operator can be used as the basis functions of representations of the R operations. Sets of operators can also be used as basis functions.

operators can be used to form the basis functions for representations of group elements, however.

The *basis functions* of a representation are any things that are transformed properly by the matrices of the representation. As discussed above, the eigenstates of the Hamiltonian can form the basis functions of a representation. A basis function need not be a wave function, however. A set of quantum mechanical operators can also be a basis for a representation. For example, the components of the momentum operator p_x, p_y, p_z can be the three basis functions of the Γ_5 representation in T_d symmetry, since they transform in the same way as x, y, and z.

It is possible to have confusion about the two types of "basis" in use when we talk about crystals. The *basis functions* define the representations of the symmetry operations in the group. We also have, in the definition of a Bravais lattice, the *basis* of a lattice, which is the thing that is actually repeated at each lattice site, such as an atom or molecule. This lattice basis has nothing to do with the basis functions of a representation.

Note also that we speak of "a" representation, but each representation is a *set* of square matrices that have the same multiplication properties as the group of symmetry operations.

6.8 Effects of Lowering Symmetry

One of the uses of group theory is to determine when and how a set of degenerate states is split when the symmetry of the system is lowered; for example, if stress or magnetic field is applied to a crystal, or, as discussed in Section 6.4, if k-states not at the center of the Brillouin zone are in view.

The first step in determining the new symmetry of the states is to determine which symmetry operations are no longer valid, and find the new symmetry group to which the system belongs. For example, as illustrated in Figure 6.8, if uniaxial stress is applied to a cubic crystal along the z-axis, only 2 C_4 rotations (about the z-axis) will map the system back into

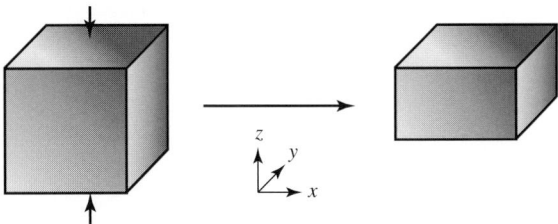

Unit cell of cubic symmetry O_h reduced to D_{4h}.

itself, and no C_3 rotations. Looking at the list of operator classes in the symmetry tables, we can see that the O_h symmetry group without these elements becomes the D_{4h} group.

In this new group, an irreducible representation in the old group may become reducible. If that is the case, the reduction can be performed using (6.5.1) or by inspection, knowing that the characters of the new representations must add up to the characters of the old representation for the symmetry operations that are common to both groups. The following table shows this breakdown for Γ_4^- representation going from O_h symmetry to D_{4h} symmetry, which would occur, for example, if uniaxial stress is applied along a [100] direction:

		E	C_2	$2C_4$	C_2'	I	σ_n	$2S_4$	$\bar{\sigma}_d$
O_h	Γ_4^-	3	−1	1	−1	−3	1	−1	1
D_{4h}	Γ_2^-	1	1	1	−1	−1	−1	−1	1
	Γ_5^-	2	−2	0	0	−2	2	0	0

Reducing the symmetry of a system in general has the effect of splitting the essential degeneracy of states. The splitting of each representation upon reduction is listed in the **compatibility table** of a group (e.g., Table 6.12). It is important to notice that even if a representation does not split into smaller irreducible representations upon lowering the symmetry, the name of the representation with the same basis functions may change from one group to another. For example, the negative parity representations in O_h do not keep the same name in T_d. The representation Γ_4^- in O_h, which has basis functions x, y, and z, becomes Γ_5 in T_d.

Reduction of atomic symmetry. As an example of reduction of symmetry, let us examine how crystal symmetry can affect the splitting of bands. As discussed in Section 1.1.2, in the LCAO approximation, the bands of crystals arise from linear combinations of the underlying atomic states.

An isolated atom can be rotated or reflected in any direction. Therefore, atomic orbits in isolation belong to the **full rotation group**, which has an infinite number of classes and irreducible representations. The character table for the full rotation group is shown in Table 6.13. There is an infinite number of possible symmetry operations and representations; only a few are shown. This group is not a point group – it cannot be used for a Bravais lattice which fills all space. We can find how the atomic orbitals will split when arranged

Table 6.12 Compatibility table for the group O_h, showing the change of the names of the representations under transformation to T_d and D_{4h}

O_h	Γ_1^+	Γ_2^+	Γ_3^+	Γ_4^+	Γ_5^+	Γ_6^+	Γ_7^+	Γ_8^+
T_d	Γ_1	Γ_2	Γ_3	Γ_4	Γ_5	Γ_6	Γ_7	Γ_8
D_{4h}	Γ_1^+	Γ_3^+	$\Gamma_1^+ \oplus \Gamma_3^+$	$\Gamma_2^+ \oplus \Gamma_5^+$	$\Gamma_4^+ \oplus \Gamma_5^+$	Γ_6^+	Γ_7^+	$\Gamma_6^+ \oplus \Gamma_7^+$

O_h	Γ_1^-	Γ_2^-	Γ_3^-	Γ_4^-	Γ_5^-	Γ_6^-	Γ_7^-	Γ_8^-
T_d	Γ_2	Γ_1	Γ_3	Γ_5	Γ_4	Γ_7	Γ_6	Γ_8
D_{4h}	Γ_1^-	Γ_3^-	$\Gamma_1^- \oplus \Gamma_3^-$	$\Gamma_2^- \oplus \Gamma_5^-$	$\Gamma_4^- \oplus \Gamma_5^-$	Γ_6^-	Γ_7^-	$\Gamma_6^- \oplus \Gamma_7^-$

Table 6.13 Character table for the full rotation group

	E	C_6	C_4	C_3	C_2	I	S_3	S_4	S_6	σ
$D_0^{\pm}$	1	1	1	1	1	± 1	± 1	± 1	± 1	± 1
$D_1^{\pm}$	3	2	1	0	-1	± 3	± 2	± 1	0	∓ 1
$D_2^{\pm}$	5	1	-1	-1	1	± 5	± 1	∓ 1	0	± 1
$D_3^{\pm}$	7	-1	-1	1	-1	± 7	∓ 1	∓ 1	± 1	∓ 1
$D_4^{\pm}$	9	-2	1	0	1	± 9	∓ 2	± 1	0	± 1
$D_5^{\pm}$	11	-1	1	-1	-1	± 11	∓ 1	± 1	∓ 1	∓ 1
$D_{1/2}^{\pm}$	2	$\sqrt{3}$	$\sqrt{2}$	1	0	± 2	$\pm\sqrt{3}$	$\pm\sqrt{2}$	± 1	0
$D_{3/2}^{\pm}$	4	$\sqrt{3}$	0	-1	0	± 4	$\pm\sqrt{3}$	0	∓ 1	0
$D_{5/2}^{\pm}$	6	0	$-\sqrt{2}$	0	0	± 6	0	$\mp\sqrt{2}$	0	0
$D_{7/2}^{\pm}$	8	$-\sqrt{3}$	0	1	0	± 8	$\mp\sqrt{3}$	0	± 1	0
$D_{9/2}^{\pm}$	10	$-\sqrt{3}$	$-\sqrt{2}$	-1	0	± 10	$\mp\sqrt{3}$	$\mp\sqrt{2}$	∓ 1	0
$D_{11/2}^{\pm}$	12	0	0	0	0	± 12	0	0	0	0

in a crystal, however, by finding the splitting of the full rotation group on lowering the symmetry to the point group of the crystal. Again, we can do this by inspection.

Going from full rotation symmetry to T_d symmetry, a p-orbital of an atom (D_1^-) will become Γ_5, as can be seen by comparison of characters of Tables 6.13 and 6.3, while a d-orbital (D_2^+) will be split into $\Gamma_3 \oplus \Gamma_5$, as seen in the following comparison.

	E	C_3	C_2	S_4	σ
D_2^+	5	-1	1	-1	1
Γ_3	2	-1	2	0	0
Γ_5	3	0	-1	-1	1

The five d-orbitals of atoms will therefore be split into a doublet and triplet, by the so-called **crystal field**.

Exercise 6.8.1   Determine how the Γ_4 states in a crystal with T_d symmetry will be split if uniaxial stress is applied along the [111] direction. To do this requires access to a set of symmetry tables, which can be found at the website associated with this book, (www.cambridge.org/snoke).

Exercise 6.8.2   In Figure 6.4, the states at the critical points X and L, in the [100] and [111] directions, respectively, correspond to different symmetries from the T_d symmetry of the crystal as a whole. Determine what these symmetries are, by explicitly determining which rotations are no longer included as operations that map the system, including the k-vector, into itself. Are the splittings of some of the Γ-point degeneracies seen in the band diagram in the [111] direction consistent with your assignment of the symmetry in that direction? To do this requires access to a set of symmetry tables, which can be found at the website associated with this book, (www.cambridge.org/snoke).

6.9 Spin and Time Reversal Symmetry

Besides the spatial symmetry operations of a group, a physical system can have another type of symmetry: time reversal symmetry. A static object is obviously invariant under time reversal, but a particle with momentum reverses direction on time reversal, and a particle with spin changes to the opposite spin under time reversal. Reviews of time reversal in quantum mechanics are given in numerous textbooks (e.g., Koster *et al.* 1963; Schiff 1968; Bir and Pikus 1974).

For a spinless particle, time reversal corresponds to complex conjugation of the wave function, since this reverses momenta. We define the complex conjugation operator K such that $K\psi = \psi^*$. Then it is easy to see that if we take the expectation value of the momentum,

$$\langle \vec{p} \rangle = \int d^3r \, \psi^* \frac{\hbar}{i} \nabla \psi, \tag{6.9.1}$$

the momentum of the complex conjugate of the wave function will be

$$\int d^3r \, (K\psi)^* \frac{\hbar}{i} \nabla (K\psi) = \int d^3r \, \psi \frac{\hbar}{i} \nabla \psi^*$$
$$= -\int d^3r \left(\psi^* \frac{\hbar}{i} \nabla \psi \right)^*$$
$$= -\langle \vec{p} \rangle. \tag{6.9.2}$$

To take into account the spin of particles, we need the time-reversal operator to also act to flip spins, because all angular momenta are reversed under time reversal. We therefore can write generally the time-reversal operator as $T = UK$, where U is a unitary operator acting on the spin, and K is the complex conjugate operator. We can deduce U by the following argument.

The spin operator $\vec{S}$ is written in terms of the Pauli matrices, namely $\vec{S} = \frac{1}{2}\hbar\vec{\sigma}$, with

$$\sigma_z = \begin{pmatrix} 1 & 0 \\ 0 & -1 \end{pmatrix}, \quad \sigma_x = \begin{pmatrix} 0 & 1 \\ 1 & 0 \end{pmatrix}, \quad \sigma_y = \begin{pmatrix} 0 & -i \\ i & 0 \end{pmatrix}. \quad (6.9.3)$$

The operator $\vec{S}$ transforms a wave function ψ into some new function ψ', i.e., $S\psi = \psi'$. If we time-reverse both sides of this equation, the spin operator must give the opposite sign. Therefore, we have

$$\vec{S}(T\psi) = -(T\psi') = -T(S\psi)$$
$$\Rightarrow \vec{S}T = -T\vec{S}. \quad (6.9.4)$$

For each component i of the spin, we then have

$$S_i T\psi = -TS_i\psi$$
$$S_i UK\psi = -UK(S_i\psi)$$
$$S_i U\psi^* = -US_i^*\psi^*$$
$$\Rightarrow S_i U = -US_i^*. \quad (6.9.5)$$

It is easy to show that this is satisfied for $U = e^{i\phi}\sigma_y$, where the σ_y is the Pauli matrix for the y-component given above, and $e^{i\phi}$ is an arbitrary phase factor. We write

$$T = \begin{pmatrix} 0 & -1 \\ 1 & 0 \end{pmatrix} K. \quad (6.9.6)$$

It immediately follows that $T^2 = -E$, where E is the identity operator.

Exercise 6.9.1   Prove the statement above, that $U = e^{i\phi}\sigma_y$ satisfies the requirement (6.9.5). You will need to use the anticommutation rule $\{\sigma_i, \sigma_j\} = 2\delta_{ij}$ and $\sigma_y^* = -\sigma_y$.

Kramers' theorem. Suppose that we have a Hamiltonian which is invariant under time reversal, with a set of eigenstates ψ_i. For each state, the time-reversed eigenstate $T\psi_i$ is also an eigenstate of H, with the same eigenvalue. This follows because if the Hamiltonian is invariant under time reversal, it commutes with the time-reversal operator,

$$T^{-1}HT = H \Rightarrow HT = TH, \quad (6.9.7)$$

and therefore

$$H(T\psi_i) = TH\psi_i = TE_i\psi_i = E_i(T\psi_i). \quad (6.9.8)$$

In other words, the time-reversed state has the same energy as the original state. In Section 1.6, we derived the result (1.6.16) from Bloch function considerations, but here we can see that the same relation is implied because $T\psi_{\vec{k}} = \psi_{-\vec{k}}$.

When spin and angular momentum are taken into account, Kramers' theorem becomes

$$\boxed{E_{\pm m_J}(-\vec{k}) = E_{\mp m_J}(\vec{k}),} \quad (6.9.9)$$

since both spin and momentum are reversed by time reversal. One possibility, of course, is that $T\psi_i = \psi_i$, so there is just one state, and the statement that the eigenvalue of the time-reversed state is the same as that of the original state is trivial. If the number of electrons in a system is odd, however, then there must be a degeneracy of the energy eigenstates. We can see this by the following argument.

The time-reversal operator for a system with multiple electrons is the product

$$T_n = \left(\prod_{i=1}^{n} U_i\right) K \equiv U_n K, \tag{6.9.10}$$

where i is an index for the spin of each of the electrons. Then, for our definition (6.9.6),

$$T_n^2 = U_n^2, \tag{6.9.11}$$

since $K^2 = 1$. Each pair of U_i operators gives a factor of -1. Therefore, if there is an odd number of spin-$\frac{1}{2}$ particles, then $T_n^2 = -E$, while for an even number of electrons, $T_n^2 = E$.

We now write the inner product integral

$$\int d^3r (T_n^2 \psi)^* (T_n \psi) = \int d^3r \, (U_n K T_n \psi)^* \, (U_n K \psi)$$

$$= \int d^3r (U_n (T_n \psi)^*)^* U_n \psi^*$$

$$= \int d^3r (U_n U_n^* \psi)^* (T_n \psi)$$

$$= \int d^3r \, \psi^* (T_n \psi), \tag{6.9.12}$$

where we have used $U^\dagger U = E$. At the same time, we have for an odd number of electrons

$$\int d^3r (T_n^2 \psi)^* (T_n \psi) = -\int d^3r \, \psi^* (T_n \psi). \tag{6.9.13}$$

Comparing (6.9.12) and (6.9.13), the inner product of ψ and $T\psi$ is equal to its negative, and therefore must be zero, which means that ψ and $T\psi$ are orthogonal. There must then be at least two states with the same energy. This is known as a **Kramers degeneracy.**

The band structures that we calculated in Chapter 1 are single-particle energy states, so they will have this type of Kramers degeneracy. This implies that at zone center, each electron band has a twofold spin degeneracy (or any even-number degeneracy), and no terms in the Hamiltonian will lift this degeneracy, except for a term that removes the time-reversal invariance, such as a magnetic field. At other points in the Brillouin zone, the degeneracy of the spin states may be lifted, because the Kramers degeneracy applies to states with opposite $\vec{k}$, not the same $\vec{k}$. In this case, the k-dependent band energy terms act like an effective magnetic field in the crystal. Terms that lead to spin splitting are also responsible for spin flip processes. We will return to discuss this in Section 10.9.

Time reversal and coupling coefficients. Suppose that a set of degenerate eigenstates of H are the basis functions for an irreducible representation of the group of symmetry operators that commute with the Hamiltonian. Then the time-reversed states must also comprise the basis functions for an irreducible representation, with the same eigenvalue. It can be shown that for a single electron wave function, there are three possible relationships between the representation in the original basis functions and the representation in the time-reversed basis functions.

	E	$\bar{E}$	C_2	$\bar{C}_2$	Basis functions
					Table 6.14 The character table of the C_2 double group
Γ_1	1	1	1	1	L_z
Γ_2	1	1	-1	-1	L_x or L_y
Γ_3	1	-1	i	$-i$	$\phi_{1/2,1/2}$
Γ_4	1	-1	$-i$	i	$\phi_{1/2,-1/2}$

- The time-reversed basis functions equal the original functions; the representation is real.
- The time-reversed basis functions are not equal to the original functions, but the representation in the new basis is the same as the original.
- The time-reversed functions are not the same as the original ones, and the representation in the new basis is a different irreducible representation from the original.

In this last case, we must think carefully about the coupling coefficients for multiplication of representations. For example, consider the point group C_2, which has just the symmetry operation of the identity E, twofold rotation C_2, and each of these with spin flip. It has the character table and basis functions shown in Table 6.14. The Γ_3 and Γ_4 representations have the basis functions spin up and spin down, respectively. Multiplication of $\Gamma_3 \otimes \Gamma_4$ gives Γ_1, as can be verified by inspection of the character table. Since all three representations are singly degenerate, we might expect that the basis function of the product state is just $|u_3\rangle|v_4\rangle = |\uparrow\downarrow\rangle$. However, this state is not time-reversal invariant.

If the Hamiltonian is time-reversal invariant, then we would like a basis that is time-reversal invariant. We can therefore write down a time-reversal invariant basis for the Γ_1 representation which is a linear superposition of the product state and its time-reversed state,

$$|w_1^1\rangle = \frac{1}{\sqrt{2}}(|u_3\rangle|v_4\rangle + T(|u_3\rangle|v_4\rangle)) = \frac{1}{\sqrt{2}}(|\uparrow\downarrow\rangle - |\uparrow\downarrow\rangle). \tag{6.9.14}$$

Alternatively, we could have used the linear combination

$$|w_1^1\rangle = \frac{1}{\sqrt{2}}(|u_3\rangle|v_4\rangle - T(|u_3\rangle|v_4\rangle)) = \frac{1}{\sqrt{2}}(|\uparrow\downarrow\rangle + |\uparrow\downarrow\rangle). \tag{6.9.15}$$

There are therefore two linearly independent superpositions of states that can both be time-reversal invariant basis states for the Γ_1 product state.

In the standard tables, therefore, given in the website associated with this book, the coupling coefficients for groups with this property (essentially, groups that do not have an essential spin degeneracy) give linear combinations for time-invariant basis states. Table 6.15 gives the coupling coefficient table for the group C_2. Of course, if the Hamiltonian is not invariant under time reversal (for example, when there is a magnetic field that splits the spin degeneracy) then we do not need to be concerned about creating time-reversal-invariant basis states.

Table 6.15 Coupling coefficients for time-reversal invariant basis states of the group C_2

	$\lvert w_1^1\rangle$	$\lvert w_1^1\rangle$		$\lvert w_1^2\rangle$	$\lvert w_1^2\rangle$
$\lvert u_1^3\rangle\lvert v_1^4\rangle$	$1/\sqrt{2}$	$i/\sqrt{2}$	$\lvert u_1^3\rangle\lvert v_1^3\rangle$	$1/\sqrt{2}$	$i/\sqrt{2}$
$\lvert u_1^4\rangle\lvert v_1^3\rangle$	$-1/\sqrt{2}$	$i/\sqrt{2}$	$\lvert u_1^4\rangle\lvert v_1^4\rangle$	$1/\sqrt{2}$	$-i/\sqrt{2}$

6.10 Allowed and Forbidden Transitions

In quantum mechanics, we often want to compute a matrix element of the form $\langle\psi_l^r\lvert Q_k^q\rvert\phi_j^p\rangle$, where $\lvert\phi_j^p\rangle$ is a quantum eigenstate which is the jth basis function for a representation p, $\lvert\psi_l^r\rangle$ is a quantum eigenstate which is the lth basis function for a different representation r, and Q_k^q is a quantum mechanical operator which can also be the kth basis function of a representation q. For example, Fermi's golden rule (Section 4.7) uses matrix elements of this type to compute the transition rate between states,

$$\Gamma_{i\to f} = \frac{2\pi}{\hbar}\lvert\langle f\lvert H_I\rvert i\rangle\rvert^2\mathcal{D}(E_f)dE, \tag{6.10.1}$$

where H_I is an interaction term in the Hamiltonian that causes transitions between initial state $\lvert i\rangle$ and final state $\lvert f\rangle$.

Group theory allows us to easily determine whether this matrix element is nonzero, that is, whether the transition is allowed or forbidden according to Fermi's golden rule, without actually computing the integral in terms of the wave functions. This is one of the most common uses of group theory, to find **selection rules**.

The trick in computing selection rules is to realize that the product $Q_k^q\lvert\phi_j^p\rangle$ is an outer product, and the set of all the states $Q_1^q\lvert\phi_1^p\rangle, Q_2^q\lvert\phi_1^p\rangle, Q_1^q\lvert\phi_2^p\rangle$, etc., form the basis for a new representation. This new representation can then be reduced into irreducible representations using the rules of Section 6.5 (typically, by looking at the multiplication table of a group).

If none of the new irreducible representations in the outer product of $\Gamma_q\otimes\Gamma_p$ is the same as the representation Γ_r of the final state, then the matrix element must be zero. This is essentially the same type of argument as used often in physics to show that some integrals must vanish by symmetry. For example, an odd function (that is, a function with odd parity in a system with inversion symmetry) integrated over the range $-\infty$ to $+\infty$ must vanish; the product of two odd functions integrated over the same range need not vanish. Group theory simply extends this kind of symmetry argument to systems with more symmetry operations than just inversion.

The basic theorem used to determine selection rules is the **Wigner–Eckart** theorem:

$$\boxed{\langle\psi_l^r\lvert Q_k^q\rvert\phi_j^p\rangle = \sum_\alpha (C_{jk,l}^{pq,r\alpha})^*\langle r\lVert Q^q\rVert p\rangle_\alpha.} \tag{6.10.2}$$

Here, $(C_{jk,l}^{pq,r\alpha})^*$ is the complex conjugate of the Clebsch–Gordan coefficient in the transformation from the product basis functions of $\Gamma_p \otimes \Gamma_q$ to the basis functions of the new, reduced representation Γ_r, defined by (6.6.2):

$$|w_l^{r\alpha}\rangle = \sum_{jk} C_{jk,l}^{pq,r\alpha} |u_j^p\rangle |v_k^q\rangle, \tag{6.10.3}$$

where $|u_j^p\rangle$, $|v_k^q\rangle$, and $|w_l^{r\alpha}\rangle$ are the basis functions of the irreducible representations Γ_q, Γ_p, and Γ_r, respectively. If Γ_r does not appear in the reduction of the outer product $\Gamma_q \otimes \Gamma_p$, then of course $(C_{jk,l}^{pq,r\alpha})^*$ is zero and the transition is forbidden. The Wigner–Eckart theorem goes further, however, to tell us that if the transition is not forbidden, we only need to do one integral, because all the various possible matrix elements between the j, k, and l components are equal to a single number $\langle r|Q^q|p\rangle_\alpha$, which depends only on the irreducible representations involved and not on j, k, and l, times the proper Clebsch–Gordan coefficient. We use $|r\rangle$ and $|p\rangle$ to stand for any basis state of the r and p representations, respectively. The Wigner–Eckart theorem says that it does not matter which one we use, in cases when the representation has more than one basis state. If we compute the matrix element for any one basis state in the r representation and any one basis state in the p representation, we will know all the matrix elements for all the different basis states, because they are all equal to a constant $\langle r|Q^q|p\rangle$ times the proper Clebsch–Gordan coefficient.

The index α in these sums is the index for the number of times irreducible representation r appears in block diagonal reduction of the outer product. In general, there can be more than one block of the same irreducible representation in the reduction of a large matrix; for example, as seen in Table 6.8, the outer product $\Gamma_8 \otimes \Gamma_8$ in T_d symmetry leads to a 16×16 matrix which is reduced into seven blocks, which include two Γ_4 blocks and two Γ_5 blocks.

Example. To see how this works, let us deduce some optical selection rules. As we saw in Section 5.2, the optical interaction matrix element is proportional to the momentum $\vec{p}$ operator. In O_h symmetry (cubic, with inversion) this operator belongs to the Γ_4^- symmetry representation. Suppose we want to know if an optical transition between a valence band with Γ_7^+ and a conduction band with Γ_6^+ symmetry is allowed, that is, whether the matrix element $\langle \psi_i^{6+}|\vec{p}|\phi_j^{7+}\rangle$ is nonzero for any states i and j.

To decide this, we first multiply $\Gamma_4^- \otimes \Gamma_7^+$. The multiplication table for the O_h group is the same as for groups O and T_d, shown in Table 6.8. This gives us $\Gamma_4^- \otimes \Gamma_7^+ = \Gamma_7^- \oplus \Gamma_8^-$. The parity of the final representations is just the product of the two parities.

Since the final product does not include the final state representation, Γ_6^+, the transition is not allowed. Actually, we did not need to work that hard in this case, since one could already see that the transition is forbidden by parity.

Suppose that we want to know if the same optical transition is allowed in a crystal with T_d symmetry. In this case, the momentum operator belongs to the Γ_5 representation. Therefore, we perform the outer product $\Gamma_5 \otimes \Gamma_7 = \Gamma_6 \oplus \Gamma_8$. The final state representation Γ_6 is included in the product, so the transition is allowed.

Exercise 6.10.1 Determine all the dipole-allowed optical transitions between states in a crystal with T_d symmetry, for electronic bands with Γ_6, Γ_7, and Γ_8 symmetry.

Exercise 6.10.2   For an optical transition in a crystal with T_d symmetry between a valence band with Γ_8 symmetry and a conduction band with Γ_6 symmetry, which states are coupled to each other by left-circular and right-circular polarized light along the z-axis? To answer this, use the Wigner–Eckart theorem; you can use either Table 6.9 or the symmetry tables at the website associated with this book.

6.10.1 Second-Order Transitions

As shown in Section 5.2, the rate of optical transitions is proportional to the square of the matrix element $\langle u_{f0}|\vec{p}|u_{i0}\rangle$, where $\vec{p}$ is the momentum and $|u_{i0}\rangle$ and $|u_{f0}\rangle$ are the initial and final Bloch functions, respectively. As we have seen above, the symmetry of the crystal gives selection rules which tell us whether this matrix element is zero for a given pair of bands.

Suppose that the selection rules of a crystal give $\langle u_{f0}|\vec{p}|u_{i0}\rangle = 0$ at zone center. Then the optical transitions are forbidden. Are they really completely forbidden, though? Not exactly. It may be the case that the transition is forbidden at zone center, but is nonzero for $\vec{k}$ away from zone center. The $k\cdot p$ theory presented in Section 1.9.4 implies that the symmetry changes slightly as k increases, so that the transition can become slightly allowed. What is the transition rate then?

As deduced in Section 1.9.4, from $k \cdot p$ theory

$$|u_{nk}\rangle = |u_{n0}\rangle + \frac{\hbar}{m}\sum_{m\neq n}\frac{\vec{k}\cdot\langle u_{m0}|\vec{p}|u_{n0}\rangle|u_{m0}\rangle}{E_{n0} - E_{m0}}. \tag{6.10.4}$$

Instead of using only the matrix element $\langle u_{f0}|\vec{p}|u_{i0}\rangle$, we can therefore write, to lowest order in $\vec{k}$,

$$\langle u_{f\vec{k}}|\vec{p}|u_{i\vec{k}}\rangle = \left(\langle u_{f0}| + \frac{\hbar}{m}\sum_{m\neq f}\frac{\vec{k}\cdot\langle u_{f0}|\vec{p}|u_{m0}\rangle\langle u_{m0}|}{E_{f0} - E_{m0}}\right)$$

$$\times\vec{p}\left(|u_{i0}\rangle + \frac{\hbar}{m}\sum_{m'\neq i}\frac{\vec{k}\cdot\langle u_{m'0}|\vec{p}|u_{i0}\rangle|u_{m'0}\rangle}{E_{i0} - E_{m'0}}\right)$$

$$\simeq \frac{\hbar}{m}\sum_{m\neq f}\frac{\left(\vec{k}\cdot\langle u_{f0}|\vec{p}|u_{m0}\rangle\right)\langle u_{m0}|\vec{p}|u_{i0}\rangle}{E_{f0} - E_{m0}}$$

$$+\frac{\hbar}{m}\sum_{m\neq i}\frac{\left(\vec{k}\cdot\langle u_{m0}|\vec{p}|u_{i0}\rangle\right)\langle u_{f0}|\vec{p}|u_{m0}\rangle}{E_{i0} - E_{m0}}. \tag{6.10.5}$$

Therefore, if there is some other band m that has an allowed transition to both bands, then the transition can occur in second-order perturbation theory. Figure 6.9 illustrates the second-order process, which we can think of as a virtual excitation to a higher band. The energy denominator means that the further away the intermediate band is, the less effective it will be in acting as an intermediate state. This can be viewed as energy uncertainty in the virtual excitation process; the larger the virtual energy needed, the less likely the process is.

Fig. 6.9 A second-order transition, with a virtual intermediate state.

As seen here, the second-order matrix elements are proportional to k. Therefore, the square of the matrix element, which goes into Fermi's golden rule, will be proportional to k^2. In Section 5.2.1, we found that the rate of optical transitions for an allowed transition is proportional to $\sqrt{E}$, that is, proportional to k. This factor came from the density of states of the electron bands with their effective masses. That factor will still be present in a second-order transition. Therefore, the total rate of second-order optical transitions between two semiconductor bands will be proportional to k^3, that is, $E^{3/2}$. The energy dependence of the absorption spectrum thus becomes a way of determining whether a transition is allowed or forbidden in first order.

Exercise 6.10.3 (a) In Chapter 5, we did not calculate any interaction Hamiltonian terms that would couple a photon directly to a phonon. However, a second-order process is possible in which an electronic transition to a virtual state occurs by dipole photon interaction, and then the electron returns to its original state by a phonon emission, leading effectively to the creation of a phonon by a photon.

The semiconductor Cu_2O has O_h symmetry (Tables 6.7 and 6.8), electron bands with symmetries $\Gamma_6^\pm$, $\Gamma_7^\pm$, and $\Gamma_8^\pm$, and optical phonons with symmetries $\Gamma_2^-, \Gamma_3^-, \Gamma_4^+$, and Γ_5^+ (all at zone center). For which optical phonons is this process allowed, at zone center?

(b) In Section 7.8, we will introduce **Raman scattering**, in which a photon scatters from a crystal, leaving behind a phonon and continuing on its way with its energy and momentum shifted by the loss of the phonon. One type is resonant Raman scattering, a third-order process with three multiplied matrix elements, $\langle i|H_{\text{dipole}}|n\rangle\langle n|H_{\text{phon}}|m\rangle\langle m|H_{\text{dipole}}|i\rangle$, where $|i\rangle$, $|m\rangle$, and $|n\rangle$ are electronic states. (We will also return to this in Chapter 8.) Find the optical phonons for which this process is allowed in Cu_2O at zone center, and show that they do not also participate in the absorption process of part (a). This is generally true, that phonons active in infrared absorption are not active for Raman scattering, and vice versa.

6.10.2 Quadrupole Transitions

In Section 6.10.1, we looked at second-order transitions, meaning transitions that involve second-order perturbation theory with intermediate states. There is another type of "higher-order" transition. This is one that is higher order in the matrix element, namely, going from

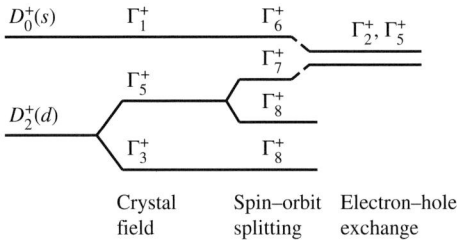

Fig. 6.10 Level splitting in Cu_2O, a crystal with O_h symmetry.

dipole to quadrupole (and possibly higher) moments. The derivation of these higher-order terms was presented in Section 5.2.2. In this section, we will work an extended example of quadrupole-allowed transitions using the semiconductor Cu_2O, which is a naturally occurring crystal with the highest possible crystal symmetry, namely the O_h group. Note that the second-order transitions and quadrupole transitions are both *single-photon* transitions. There is yet another type of "higher-order" transition, namely a transition involving two or more photons in the excitation of a single electron. This involves nonlinear terms in the Hamiltonian, which will be discussed in Section 7.6.

The highest occupied atomic state of Cu_2O is the $3d$ orbital of the copper atoms, and the lowest unoccupied orbital arises from the $4s$ orbitals. In the crystal field, the s-like conduction band has Γ_1^+ representation in O_h, while the d-like valence band becomes $\Gamma_3^+ \oplus \Gamma_5^+$ (see Section 6.8 for a discussion of how the atomic states split in a periodic crystal). The highest of these is found, from experiments, to be the Γ_5^+ band. When spin is included, the conduction band becomes $\Gamma_1^+ \otimes \Gamma_6^+ = \Gamma_6^+$, and the highest valence band becomes $\Gamma_5^+ \otimes \Gamma_6^+ = \Gamma_7^+ \oplus \Gamma_8^+$, according to Table 6.8. This splitting is due to the spin–orbit interaction, which causes the Γ_7^+ band to be the highest of these two.

Exciton states can be created from the product of electron and hole states. Multiplying the lowest conduction band with the highest valence band yields $\Gamma_7^+ \otimes \Gamma_6^+ = \Gamma_2^+ \oplus \Gamma_5^+$. The splitting between these states is due to spin-exchange discussed in detail in Section 10.8.3. The higher, triply degenerate state is called the **orthoexciton** state, while the lower, singlet state is called the **paraexciton** state, by analogy with the ortho and para states of hydrogen. Figure 6.10 illustrates the level splitting in this crystal.

Suppose that we want to create excitons in this material by direct absorption of photons. The state with no excitons has Γ_1^+ symmetry, that is, perfect symmetry since there is nothing there. The transition matrix elements for single-photon dipole transitions to the paraexciton and orthoexciton states can therefore be written as $\langle \Gamma_1^+ | \Gamma_4^- | \Gamma_2^+ \rangle$ and $\langle \Gamma_1^+ | \Gamma_4^- | \Gamma_5^+ \rangle$, respectively, since the p_x, p_y, p_z operators in O_h are basis functions of the Γ_4^- representation. These transitions are both clearly forbidden by parity.

As discussed in Section 5.2.2, however, we can still possibly have transitions via the quadrupole operator, which has terms proportional to

$$(x^2 - y^2)(\eta_x k_x - \eta_y k_y), \quad \tfrac{1}{3}(2z^2 - x^2 - y^2)(2\eta_z k_z - \eta_x k_x - \eta_y k_y),$$
$$3(\eta_x k_y + \eta_y k_x)xy, \quad 3(\eta_y k_z + \eta_z k_y)yz, \quad 3(\eta_x k_z + \eta_z k_x)xz, \tag{6.10.6}$$

where $\vec{\eta}$ is the polarization vector and $\vec{k}$ is the propagation wave vector. From Table 6.7, we see that the functions of x, y, and z are the basis functions of the Γ_3^+ (upper line) and Γ_5^+ (lower line) representations.

To determine the selection rules for the quadrupole transitions to the Γ_2^+ paraexciton states in Cu_2O, we therefore multiply, using Table 6.8,

$$\Gamma_3^+ \otimes \Gamma_2^+ = \Gamma_3^+$$
$$\Gamma_5^+ \otimes \Gamma_2^+ = \Gamma_4^+,$$
(6.10.7)

and for the Γ_5^+ orthoexciton states,

$$\Gamma_3^+ \otimes \Gamma_5^+ = \Gamma_4^+ \oplus \Gamma_5^+$$
$$\Gamma_5^+ \otimes \Gamma_5^+ = \Gamma_1^+ \oplus \Gamma_3^+ \oplus \Gamma_4^+ \oplus \Gamma_5^+.$$
(6.10.8)

This tells us that only the orthoexciton state has an allowed quadrupole transition, because the Γ_1^+ representation only appears for this case.

We can go further to work out the specific polarizations of light that give allowed transitions. For this, we need to look at the basis functions. The orthoexciton Γ_5^+ states transform as xy, yz, and xz. Looking at the associated $\eta_i k_j$ functions in (6.10.6), we can see that for light directed along the [100] direction, the polarization vector must point along [010] or [001]. We define [001] as $\theta = 0$. Then for $\vec{k}$ along [100], the quadrupole operator is proportional to $xz \cos\theta + xy \sin\theta$. When the polarization is along [001] (i.e., $\theta = 0$), the light couples to the orthoexciton state with xz symmetry, while for $\theta = 90°$, the light excites the xy state. The orthoexciton state with yz symmetry is not coupled to light with $\vec{k}$ along this direction. We can see this by noting that the matrix element $\langle 1|xz|xz \rangle$ corresponds to the integral

$$\int_{-L}^{L} dx \int_{-L}^{L} dy \int_{-L}^{L} dz (1)(xz)(xz) = \int_{-L}^{L} dx \int_{-L}^{L} dy \int_{-L}^{L} dz \, x^2 z^2,$$
(6.10.9)

which is nonzero, while integrals such as

$$\int_{-L}^{L} dx \int_{-L}^{L} dy \int_{-L}^{L} dz \, (1)(xz)(yz)$$
(6.10.10)

are odd in at least one variable and therefore vanish in the integration. The same result could be deduced using the Wigner–Eckart formula (6.10.2) and the coupling constant table for the product $\Gamma_5^+ \otimes \Gamma_5^+$ in the O_h group.

Exercise 6.10.4 Determine the polarization-angle dependence of the matrix element for light absorption by each of the orthoexciton states in the crystal Cu_2O, for light directed along the [111] axis.

Reduction of symmetry with stress. Suppose now that we add stress to the crystal along the [001] direction, so that the crystal symmetry now becomes D_{4h}. We can use the methods of Section 6.8 to deduce the change of the selection rules in this case.

Table 6.16 Character table of the D_{4h} symmetry single group

	E	$2C_4$	C_2	$2C_2'$	$2C_2''$	I	$2S_4$	σ_h	$2\sigma_v$	$2\sigma_d$	Basis functions
Γ_1^+	1	1	1	1	1	1	1	1	1	1	R
Γ_2^+	1	1	1	-1	-1	1	1	1	-1	-1	S_z
Γ_3^+	1	-1	1	1	-1	1	-1	1	1	-1	$(x^2 - y^2)$
Γ_4^+	1	-1	1	-1	1	1	-1	1	-1	1	xy
Γ_5^+	2	0	-2	0	0	2	0	-2	0	0	L_x, L_y
Γ_1^-	1	1	1	1	1	-1	-1	-1	-1	-1	$(x^2 - y^2)xyz$
Γ_2^-	1	1	1	-1	-1	-1	-1	-1	1	1	z
Γ_3^-	1	-1	1	1	-1	-1	1	-1	-1	1	xyz
Γ_4^-	1	-1	1	-1	1	-1	1	-1	1	1	$(x^2 - y^2)z$
Γ_5^-	2	0	-2	0	0	-2	0	2	0	0	x, y

Table 6.17 Multiplication table for the group D_{4h}

$\otimes$	Γ_1	Γ_2	Γ_3	Γ_4	Γ_5	Γ_6	Γ_7
Γ_1	Γ_1	Γ_2	Γ_3	Γ_4	Γ_5	Γ_6	Γ_7
Γ_2	Γ_2	Γ_1	Γ_4	Γ_3	Γ_5	Γ_6	Γ_7
Γ_3	Γ_3	Γ_4	Γ_1	Γ_2	Γ_5	Γ_7	Γ_6
Γ_4	Γ_4	Γ_3	Γ_2	Γ_1	Γ_5	Γ_7	Γ_6
Γ_5	Γ_5	Γ_5	Γ_5	Γ_5	$\Gamma_1 \oplus \Gamma_2 \oplus \Gamma_3 \oplus \Gamma_4$	$\Gamma_6 \oplus \Gamma_7$	$\Gamma_6 \oplus \Gamma_7$
Γ_6	Γ_6	Γ_6	Γ_7	Γ_7	$\Gamma_6 \oplus \Gamma_7$	$\Gamma_1 \oplus \Gamma_2 \oplus \Gamma_5$	$\Gamma_3 \oplus \Gamma_4 \oplus \Gamma_5$
Γ_7	Γ_7	Γ_7	Γ_6	Γ_6	$\Gamma_6 \oplus \Gamma_7$	$\Gamma_3 \oplus \Gamma_4 \oplus \Gamma_5$	$\Gamma_1 \oplus \Gamma_2 \oplus \Gamma_5$

Table 6.16 gives the character table of the new symmetry group, and Table 6.17 gives the multiplication table. From the O_h compatibility table given in Table 6.12, the Γ_2^+ paraexciton state becomes Γ_3^+, and the orthoexciton splits into Γ_4^+ and Γ_5^+ symmetry. The Γ_4^+ state transforms as xy, while the Γ_5^+ is a doublet with basis states that transform as xz and yz. Note that in Table 6.16 the basis functions of the Γ_5^+ representation are listed as L_x and L_y. This is because for the symmetry operations included in the D_{4h} group, xy and yz transform exactly as L_x and L_y.

In D_{4h} symmetry, the quadrupole operators are represented by $\Gamma_1^+, \Gamma_3^+, \Gamma_4^+$, and Γ_5^+. The product of the Γ_3^+ paraexciton state and the Γ_3^+ quadrupole operator now gives

$$\Gamma_3^+ \otimes \Gamma_3^+ = \Gamma_1^+,$$

which means that the matrix element $\langle \Gamma_1^+ | \Gamma_3^+ | \Gamma_3^+ \rangle$ is nonzero, and optical transitions to the paraexciton state are allowed. The stress has made a forbidden transition into an allowed one.

Exercise 6.10.5 Determine the polarization dependence of the dipole and quadrupole optical transitions to paraexciton and orthoexciton states in Cu_2O if the direction of the propagation of the light is along [110], if the crystal is placed under stress so that it has D_{4h} symmetry, with the [001] direction as the selected axis.

Exercise 6.10.6 The change of basis function for the orthoexcitons from xz, yz in O_h symmetry to L_x, L_y in D_{4h} was allowable because these functions transform the same under all D_{4h} symmetry operations. The paraexciton basis function changes from $(x^2 - y^2)(y^2 - z^2)(z^2 - x^2)$ in O_h to just $(x^2 - y^2)$ in D_{4h}. Show that if you keep the full, original basis function for paraexcitons in the O_h group, it does not change the selection rules in D_{4h} symmetry for $\vec{k}$ along [100] or [110].

Note that when the symmetry is changed, matrix element integrals of the form (6.10.9) must have the integration over z changed to going from $-L'$ to L', where L' is different from L in the other two directions. Show that for the paraexciton state, if $L' = L$, the quadrupole matrix elements to this state vanish, in agreement with the selection rule deduced for O_h symmetry.

6.11 Perturbation Methods

One very useful application of group theory is to write down the Hamiltonians of systems under small perturbations. In one approach, this can be connected directly to the $\vec{k} \cdot \vec{p}$ perturbation theory introduced in Section 1.9.4. In a second, more general approach, we can simply make an assumption of linearity, that changes in the Hamiltonian are linear with the perturbations.

6.11.1 Group Theory in $\boldsymbol{k} \cdot \boldsymbol{p}$ Theory

Recall from Section 1.9.4 that the energy of an electron band away from zone center depends on the matrix elements of the $\vec{p}$ operator, as follows:

$$E_n(\vec{k}) = E_n(0) + \frac{\hbar^2 k^2}{2m} + \frac{\hbar^2}{m^2} \sum_{m \neq n} \frac{|\vec{k} \cdot \langle u_{m0}|\vec{p}|u_{n0}\rangle|^2}{E_n(0) - E_m(0)}. \tag{6.11.1}$$

Suppose we have a band that has degenerate states at some point in the Brillouin zone; that is, they comprise the basis states of an irreducible representation of the crystal symmetry group. To adapt $k \cdot p$ theory to this case, we need to use Löwdin degenerate-state perturbation theory, summarized in Appendix E. This gives the prescription that the energies of the states are the eigenvalues of the following matrix:

$$H_{n,ij}(\vec{k}) = E_n(0) + \frac{\hbar^2 k^2}{2m} + \frac{\hbar^2}{m^2} \sum_{m \neq n} \frac{(\vec{k} \cdot \langle u_{n0,i}|\vec{p}|u_{m0}\rangle)(\langle u_{m0}|\vec{p}|u_{n0,j}\rangle \cdot \vec{k})}{E_n(0) - E_m(0)},$$

$$\tag{6.11.2}$$

where i and j run over the degenerate states in band n, and m runs over all other electron states at zone center.

Let us take again the example of a III-V semiconductor with T_d symmetry, and consider the valence band states arising from three degenerate p-like states. Since the valence band is full, we ignore all terms that couple a valence band state to another valence band state, and because of the energy denominator in (6.11.2), which suppresses the contribution of states far away, we only worry about the lowest conduction band, which is s-like. In principle, there are nine different matrix elements to calculate, but group theory allows us to set some to zero using the coupling coefficients. The coupling coefficients for $\Gamma_5 \otimes \Gamma_5 = \Gamma_1 \oplus \Gamma_3 \oplus \Gamma_4 \oplus \Gamma_5$ in T_d symmetry are given in the website associated with this book, but it turns out that these coupling coefficients are exactly the same as the coefficients for $\Gamma_4 \otimes \Gamma_4$ in T_d, which is also the product of 3×3 representations. These are given in Table 6.10. For the coupling to the Γ_1 representation, for the s-like conduction band, the first column of this table gives us

$$|w_1^1\rangle = \tfrac{1}{\sqrt{3}}(p_x|u_x\rangle + p_y|u_y\rangle + p_z|u_z\rangle), \tag{6.11.3}$$

where $|u_x\rangle$, $|u_y\rangle$, and $|u_z\rangle$ are the p-like valence band states, and p_x, p_y, and p_z are the momentum operator components. This then gives us the nonzero terms $\langle s|p_x|u_x\rangle k_x$, $\langle s|p_y|u_y\rangle k_y$, and $\langle s|p_z|u_z\rangle k_z$ in (6.11.2), where $|s\rangle$ is the conduction-band state. By the cubic symmetry of the system, the matrix elements are all equal. This then gives the Hamiltonian matrix

$$H_n = E_n(0) + \frac{\hbar^2 k^2}{2m} + \frac{\hbar^2}{m^2 E_g}|\langle s|p_x|u_x\rangle|^2 \begin{pmatrix} k_x^2 & k_x k_y & k_x k_z \\ k_x k_y & k_y^2 & k_y k_z \\ k_x k_z & k_y k_z & k_z^2 \end{pmatrix}, \tag{6.11.4}$$

where E_g is the gap energy, or more generally, since $k^2 = k_x^2 + k_y^2 + k_z^2$, the k-dependent matrix is

$$H_{LK} = \begin{pmatrix} lk_x^2 + m(k_y^2 + k_z^2) & nk_x k_y & nk_x k_z \\ nk_x k_y & lk_y^2 + m(k_x^2 + k_z^2) & nk_y k_z \\ nk_x k_z & nk_y k_z & lk_z^2 + m(k_x^2 + k_y^2) \end{pmatrix}, \tag{6.11.5}$$

where l, m, and n are constants. This is known as the **Luttinger–Kohn** Hamiltonian.

We now can introduce the electron spin. We have already discussed in Section 6.6 how the spin–orbit interaction splits the p-like valence bands in T_d symmetry. Using the coupling coefficients, we write

$$\langle w_{l'}^{r\alpha}|H|w_l^{r\alpha}\rangle = \left(\sum_{j'k'} (C_{j'k',l'}^{pq,r\alpha})^* \langle u_{j'}^p|\langle v_{k'}^q|\right) H \left(\sum_{jk} C_{jk,l}^{pq,r\alpha}|u_j^p\rangle|v_k^q\rangle\right). \tag{6.11.6}$$

The Hamiltonian (6.11.5) couples different p-states, but does not flip spins; therefore only terms with the same electron spin will couple to each other. For the case of the Γ_8 valence band in GaAs, with T_d symmetry, we can use the coupling coefficients of Table 6.9 to obtain the matrix elements. Two examples are:

$$\langle w^8_{1/2}|H_{LK}|w^8_{1/2}\rangle = \left(\frac{i}{\sqrt{6}}\langle x|\langle\downarrow| + \frac{1}{\sqrt{6}}\langle y|\langle\downarrow| - \frac{i\sqrt{2}}{\sqrt{3}}\langle z|\langle\uparrow|\right)H_{LK}$$

$$\times \left(-\frac{i}{\sqrt{6}}|x\rangle|\downarrow\rangle + \frac{1}{\sqrt{6}}|y\rangle|\downarrow\rangle + \frac{i\sqrt{2}}{\sqrt{3}}|z\rangle|\uparrow\rangle\right)$$

$$= \frac{1}{6}(lk^2_x + m(k^2_y + k^2_z)) + \frac{1}{6}(lk^2_y + m(k^2_x + k^2_z)) + \frac{2}{3}(lk^2_z + m(k^2_x + k^2_y))$$

$$= \frac{l+2m}{3}(k^2_x + k^2_y + k^2_z) + \frac{l-m}{6}(k^2_x + k^2_y - 2k^2_z)$$

$$\tag{6.11.7}$$

and

$$\langle w^8_{3/2}|H_{LK}|w^8_{1/2}\rangle = \left(-\frac{i}{\sqrt{2}}\langle x|\langle\uparrow| - \frac{1}{\sqrt{2}}\langle y|\langle\uparrow|\right)H_{LK}$$

$$\times \left(-\frac{i}{\sqrt{6}}|x\rangle|\downarrow\rangle + \frac{1}{\sqrt{6}}|y\rangle|\downarrow\rangle + \frac{i\sqrt{2}}{\sqrt{3}}|z\rangle|\uparrow\rangle\right)$$

$$= \frac{1}{\sqrt{3}}n(k_xk_z - ik_yk_z). \tag{6.11.8}$$

When all the matrix elements are calculated, and the spin–orbit energy discussed in Section 6.6 is included, we obtain the Luttinger–Kohn Hamiltonian with spin:

$$H_{LK} = \begin{pmatrix} P+Q & S & R & 0 & -\frac{1}{\sqrt{2}}S & -\sqrt{2}R \\ S^* & P-Q & 0 & R & \sqrt{2}Q & -\sqrt{\frac{3}{2}}S \\ R^* & 0 & P-Q & -S & -\sqrt{\frac{3}{2}}S^* & -\sqrt{2}Q \\ 0 & R^* & -S^* & P+Q & \sqrt{2}R^* & \frac{1}{\sqrt{2}}S^* \\ -\frac{1}{\sqrt{2}}S^* & \sqrt{2}Q & -\sqrt{\frac{3}{2}}S & \sqrt{2}R & P+\Delta_{so} & 0 \\ -\sqrt{2}R^* & -\sqrt{\frac{3}{2}}S^* & -\sqrt{2}Q & \frac{1}{\sqrt{2}}S & 0 & P+\Delta_{so} \end{pmatrix},$$

$$\tag{6.11.9}$$

defined for the six states in the order $|w^8_{3/2,3/2}\rangle$, $|w^8_{3/2,1/2}\rangle$, $|w^8_{3/2,-1/2}\rangle$, $|w^8_{3/2,-3/2}\rangle$, $|w^7_{1/2,1/2}\rangle$, $|w^7_{1/2,-1/2}\rangle$. (Note that these differ from the eigenstates found in Section 1.13 by arbitrary phase factors.) Here,

$$P = \frac{l+2m}{3}(k^2_x + k^2_y + k^2_z)$$

$$Q = \frac{l-m}{6}(k^2_x + k^2_y - 2k^2_z)$$

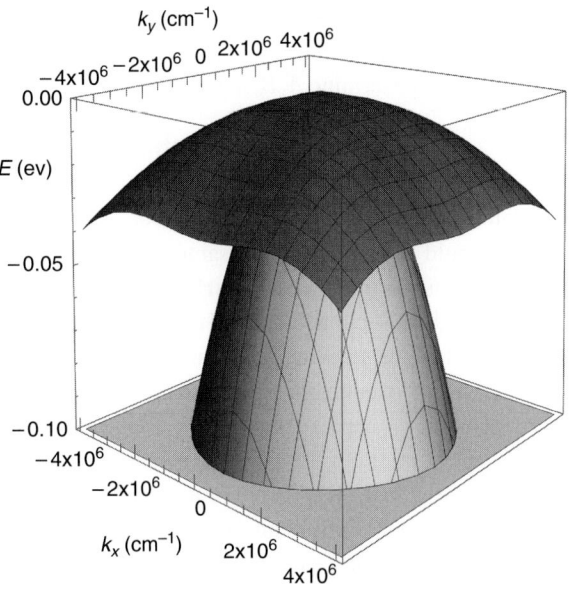

Fig. 6.11 Topmost valence bands (Γ_8) for realistic parameters for the semiconductor GaAs, for $k_z = 0$, from (6.11.11). The upper, darker surface is the heavy hole, which has substantial "warping," and the lower surface is the light hole.

$$R = \frac{l-m}{6}\sqrt{3}(k_x^2 - k_y^2) - \frac{i}{\sqrt{3}}nk_xk_y$$

$$S = \frac{1}{\sqrt{3}}n(k_xk_z - ik_yk_z), \tag{6.11.10}$$

and Δ_{so} is the energy splitting between the Γ_7 and Γ_8 states.

When the splitting Δ_{so} between the Γ_7 and Γ_8 states is large and the magnitude of k is small, we can ignore the coupling between the Γ_7 and Γ_8 states, and treat the Γ_8 states just by the 4×4 block in the upper left corner of (6.11.9) coupling the Γ_8 states to each other. Diagonalization of this 4×4 block gives eigenvalues of the form

$$E_{hh} = -Ak^2 \pm \sqrt{B^2k^4 + C^2(k_x^2k_y^2 + k_y^2k_z^2 + k_z^2k_x^2)}. \tag{6.11.11}$$

This is the standard form for the "warping" of the uppermost Γ_8 valence band in III–V semiconductors like GaAs; Figure 6.11 shows the shape. In practice, the values of A, B, and C are determined from experiments, such as optical band-to-band absorption spectroscopy. The band with the $-$ sign is typically called the **light hole** (since it has lower effective mass near $k = 0$) and the band with the $+$ sign is called the **heavy hole**. The Γ_7 states are known as the **split–off** valence band.

Exercise 6.11.1   Verify the values of P, Q, R, and S in (6.11.10) by evaluating four of the terms in (6.11.9) above, using the coupling coefficients of Table 6.9.

Exercise 6.11.2   Use a program like Mathematica to diagonalize the 4×4 block for the Γ_8 valence band states Hamiltonian (6.11.9) and verify the form of the solution (6.11.11). Determine A, B, and C in terms of l, m, and n.

6.11.2 Method of Invariants

In Section 6.11.1, we used $\vec{k} \cdot \vec{p}$ theory to deduce the variation of bands away from zone center. We could have obtained the same result by another method, which is more general. This is known as the **method of invariants**. For a full treatment, see Bir and Pikus (1974).

As discussed in Section 6.4, the Hamiltonian of a system can be written in terms of blocks that act on degenerate eigenstates, and these degenerate eigenstates can act as the basis functions of irreducible representations. Let us suppose that we add a perturbation to the system that couples these states. This perturbation can in general be written in terms of a set of matrices for all possible couplings. For example, for the three p-orbitals discussed in Section 6.11.1, there are nine possible couplings. The matrices must be Hermitian, because the Hamiltonian must be Hermitian. We can construct nine independent Hermitian coupling matrices of the following forms:

$$\sigma_{xx} = \begin{pmatrix} 1 & 0 & 0 \\ 0 & 0 & 0 \\ 0 & 0 & 0 \end{pmatrix}, \quad \sigma_{xy}^{+} = \frac{1}{2} \begin{pmatrix} 0 & 1 & 0 \\ 1 & 0 & 0 \\ 0 & 0 & 0 \end{pmatrix},$$

$$\sigma_{xy}^{-} = \frac{1}{2} \begin{pmatrix} 0 & -i & 0 \\ i & 0 & 0 \\ 0 & 0 & 0 \end{pmatrix}, \quad \text{etc.} \tag{6.11.12}$$

In addition, suppose that the perturbation Hamiltonian depends linearly on some set of parameters. These parameters could be, for example, the three terms k_x, k_y, and k_z which are first order in k, or they could be the six terms which are second order in k, namely k_x^2, k_y^2, k_z^2, $k_x k_y$, $k_y k_z$, and $k_x k_z$. Other examples of first-order parameters include the three electric field components or the three magnetic field components; other examples of second-order parameters include the strain ε_{ij}, introduced in Section 3.4.1. In general, the Hamiltonian will have terms that are the products of the coupling matrices σ_{ij} and the various parameters; for example, for our p-orbital system, we will have terms of the form $k_x \sigma_{xy}$, $k_y \sigma_{zz}$, etc., in first order, and terms of the form $k_x^2 \sigma_{yz}$, $k_x k_y \sigma_{zz}$, etc., in second order.

This allows for a large number of possible terms, but we can greatly simplify things by using group theory. It can be shown that the set of all the matrices σ_{ij} can themselves be used as a set of basis functions for a representation of the symmetry group of the Hamiltonian. In the example given here, this will give a 9×9 representation. This representation can then be reduced to irreducible blocks using the methods we have already encountered in this chapter.

The 9×9 representation can be reduced by using the fact that the matrices arise from the product of two sets of states with x, y, z symmetry, In T_d symmetry, these states are the basis functions of the Γ_5 representation. From the T_d multiplication table, we have $\Gamma_5 \otimes \Gamma_5 = \Gamma_1 \oplus \Gamma_3 \oplus \Gamma_4 \oplus \Gamma_5$. We can therefore create four irreducible representations by adding the σ_{ij} matrices in the same combinations as the basis functions of these four representations, which are listed in Table 6.3. For the Γ_1 representation, the basis function is $x^2 + y^2 + z^2$; the sum of matrices that transforms the same way is

$$\sigma_{xx} + \sigma_{yy} + \sigma_{zz} = \begin{pmatrix} 1 & 0 & 0 \\ 0 & 1 & 0 \\ 0 & 0 & 1 \end{pmatrix}. \tag{6.11.13}$$

For the Γ_3 representation, the basis functions are $2z^2 - x^2 - y^2$ and $\sqrt{3}(x^2 - y^3)$. The matrices that transform the same way are

$$2\sigma_{zz} - \sigma_{xx} - \sigma_{yy} = \begin{pmatrix} -1 & 0 & 0 \\ 0 & -1 & 0 \\ 0 & 0 & 2 \end{pmatrix},$$

$$\sqrt{3}(\sigma_{xx} - \sigma_{yy}) = \sqrt{3} \begin{pmatrix} 1 & 0 & 0 \\ 0 & -1 & 0 \\ 0 & 0 & 0 \end{pmatrix}. \tag{6.11.14}$$

Finally, we have for the Γ_4 representation,

$$\sigma_{xy}^- = \frac{1}{2} \begin{pmatrix} 0 & -i & 0 \\ i & 0 & 0 \\ 0 & 0 & 0 \end{pmatrix}, \quad \sigma_{yz}^- = \frac{1}{2} \begin{pmatrix} 0 & 0 & -i \\ 0 & 0 & 0 \\ i & 0 & 0 \end{pmatrix},$$

$$\sigma_{xz}^- = \frac{1}{2} \begin{pmatrix} 0 & 0 & 0 \\ 0 & 0 & -i \\ 0 & i & 0 \end{pmatrix}, \tag{6.11.15}$$

and for the Γ_5 representation,

$$\sigma_{xy}^+ = \frac{1}{2} \begin{pmatrix} 0 & 1 & 0 \\ 1 & 0 & 0 \\ 0 & 0 & 0 \end{pmatrix}, \quad \sigma_{yz}^+ = \frac{1}{2} \begin{pmatrix} 0 & 0 & 1 \\ 0 & 0 & 0 \\ 1 & 0 & 0 \end{pmatrix},$$

$$\sigma_{xz}^+ = \frac{1}{2} \begin{pmatrix} 0 & 0 & 0 \\ 0 & 0 & 1 \\ 0 & 1 & 0 \end{pmatrix}. \tag{6.11.16}$$

The six parameters k_x^2, $k_x k_y$, etc., form the basis functions of a 6×6 representation, which can also be reduced. We can reduce this representation by noting that it also arises from a $\Gamma_5 \otimes \Gamma_5$ product, but the basis functions of the Γ_4 irreducible representation vanish, since the k-components commute. This leaves us with $\Gamma_1 \oplus \Gamma_3 \oplus \Gamma_5$. Formally, we can prove this by constructing the character table for the full 6×6 representation, by determining which parameters are mapped to themselves under each symmetry operation, as we did in Section 6.5. Then we find by inspection that the characters of the irreducible representations add up to $\Gamma_1 \oplus \Gamma_3 \oplus \Gamma_5$:

For the products of the parameters and the coupling matrices to transform consistently under the symmetry operations, only products of parameters and matrices that transform the same way can occur. Thus, for example, the Hamiltonian will include a term in the Γ_5 representation proportional to

$$k_x k_y \begin{pmatrix} 0 & 1 & 0 \\ 1 & 0 & 0 \\ 0 & 0 & 0 \end{pmatrix}. \tag{6.11.17}$$

	E	$8C_3$	$3C_2$	$6S_4$	$6\sigma_d$
Γ_1	1	1	1	1	1
Γ_3	2	-1	2	0	0
Γ_5	3	0	-1	-1	1
$\Gamma_{6\times6}$	6	0	2	0	2

Because there is no Γ_4 representation for the $k_i k_j$ parameters, there will be no Γ_4 term in the Hamiltonian using the σ_{ij} matrices. (Note that in Table 6.3, the basis functions for Γ_5 are listed as x, y, z, but in T_d symmetry the basis functions yz, xz, xy transform the same way. That is not the case in some symmetry groups, such as O_h.)

Since the basis functions of each irreducible representation transform into each other, they must give the same energy in the Hamiltonian, and therefore must have the same overall multiplier. We therefore have, for the three irreducible representations,

$$H = a(k_x^2 + k_y^2 + k_z^2) \begin{pmatrix} 1 & 0 & 0 \\ 0 & 1 & 0 \\ 0 & 0 & 1 \end{pmatrix}$$

$$+ \frac{b}{2}(2k_z^2 - k_x^2 - k_y^2) \begin{pmatrix} -1 & 0 & 0 \\ 0 & -1 & 0 \\ 0 & 0 & 2 \end{pmatrix} + \frac{3b}{2}(k_x^2 - k_y^2) \begin{pmatrix} 1 & 0 & 0 \\ 0 & -1 & 0 \\ 0 & 0 & 0 \end{pmatrix}$$

$$+ d \begin{pmatrix} 0 & k_x k_y & k_x k_z \\ k_x k_y & 0 & k_y k_z \\ k_x k_z & k_y k_z & 0 \end{pmatrix} \tag{6.11.18}$$

$$= a(k_x^2 + k_y^2 + k_z^2)$$

$$+ \begin{pmatrix} b(2k_x^2 - k_y^2 - k_z^2) & dk_x k_y & dk_x k_z \\ dk_x k_y & b(2k_y^2 - k_x^2 - k_z^2) & dk_y k_z \\ dk_x k_z & dk_y k_z & b(2k_z^2 - k_x^2 - k_y^2) \end{pmatrix},$$

where a, b, and d are constants that must be determined by comparison to experiment. (The Γ_3 terms are multiplied by $\frac{1}{2}$, while the Γ_5 terms are not, for convenience, since we can define the constants as multiplied by any factor.) It is easy to see that this has the same form as the Luttinger–Kohn Hamiltonian (6.11.5).

It may not seem that we have saved much work by using the method of invariants, but we have derived this general form without referring to any of the particular dipole transition matrix elements used in $\vec{k} \cdot \vec{p}$ theory. Thus, we can say that the Hamiltonian must have this general form because of symmetry considerations, no matter how many other bands we include in the $\vec{k} \cdot \vec{p}$ summation. In Section 6.11.1, we included only one other band, the

s-like conduction band, but we can see now that even if we included others, it would not change the overall form of the Hamiltonian acting on p-like states.

Furthermore, we can generalize the same argument to other second-order parameters. In particular, the strain tensors ε_{xx}, ε_{xy}, etc., have the same symmetry properties as the parameters $k_x^2, k_x k_y$, etc. Therefore, we can immediately write down the Hamiltonian for strain acting on three degenerate p-orbitals in cubic T_d symmetry, without needing to look at the details of how the strain affects the electron bands:

$$H_{PB} = a(\varepsilon_{xx} + \varepsilon_{yy} + \varepsilon_{zz})$$

$$+ \begin{pmatrix} b(2\varepsilon_{xx} - \varepsilon_{yy} - \varepsilon_{zz}) & d\varepsilon_{xy} & d\varepsilon_{xz} \\ d\varepsilon_{xy} & b(2\varepsilon_{yy} - \varepsilon_{xx} - \varepsilon_{zz}) & d\varepsilon_{yz} \\ d\varepsilon_{xz} & d\varepsilon_{yz} & b(2\varepsilon_{zz} - \varepsilon_{xx} - \varepsilon_{yy}) \end{pmatrix}.$$

$$(6.11.19)$$

This is known as the **Pikus–Bir** Hamiltonian. Notice that the a term corresponds to the effect of hydrostatic strain while the b and d terms correspond to the effects of traceless shear strains. As in Section 6.11.1, the effect of spin–orbit interaction can be included, giving a 6×6 matrix for the Γ_7 and Γ_8 states, which can be diagonalized to get the band energies.

We have used the method of invariants here for the specific case of three degenerate p-orbitals in cubic symmetry, but the method is quite general. It can be used whenever the Hamiltonian can be assumed to be linear in some parameters.

Exercise 6.11.3   For couplings between spins, we can construct four independent Hermitian coupling matrices,

$$\begin{pmatrix} 1 & 0 \\ 0 & 1 \end{pmatrix}, \quad \begin{pmatrix} 1 & 0 \\ 0 & -1 \end{pmatrix}, \quad \begin{pmatrix} 0 & 1 \\ 1 & 0 \end{pmatrix}, \quad \begin{pmatrix} 0 & -i \\ i & 0 \end{pmatrix}.$$

$$(6.11.20)$$

Show explicitly, by calculating the characters for the C_2 and C_4 rotations, that one of these matrices can act as the basis of a D_0^+ representation of the full rotation symmetry group, and the other three can act as the basis functions for a D_1^+ representation, that is, they transform as L_x, L_y, L_z. The rotation matrix acting on a spinor, for a rotation of angle θ around an axis $\hat{n}$, is

$$R(\theta) = \begin{pmatrix} \cos\theta/2 - in_z\sin\theta/2 & (-in_x - n_y)\sin\theta/2 \\ (-in_x + n_y)\sin\theta/2 & \cos\theta/2 + in_z\sin\theta/2 \end{pmatrix}.$$

$$(6.11.21)$$

By the method of invariants, we can then multiply each by the magnetic field component that transforms the same way, to get the Hamiltonian for the effect of magnetic field on a particle with spin. Write down this full Hamiltonian. Note that spin, angular momentum, and magnetic field are all even under an inversion transformation I.

References

F. Bassani and G.P. Parravicini, *Electronic States and Optical Transitions in Solids*, R. A. Ballinger, ed. (Oxford University Press, 1975).

G.L. Bir and G.E. Pikus, *Symmetry and Strain-Induced Effects in Semiconductors* (Wiley, 1974).

J.R. Chelikowsky and M.L. Cohen, "Nonlocal pseudopotential calculations for the electronic structure of eleven diamond and zinc-blende semiconductors," *Physical Review* B **14**, 556 (1976).

Cohen-Tannoudji, B. Diu, and F. Laloë, *Quantum Mechanics* (Wiley, 1977).

M.S. Dresselhaus, G. Dresselhaus, and A. Jorio, *Group Theory: Application to the Physics of Condensed Matter* (Springer, 2008).

G.F. Koster, J.O. Dimmock, R.G. Wheeler, and H. Statz, *Properties of the Thirty-Two Point Groups* (MIT Press, 1963).

L.I. Schiff, *Quantum Mechanics,* 3rd ed. (McGraw-Hill, 1968).

M. Tinkham, *Group Theory and Quantum Mechanics* (McGraw-Hill, 1964).

7

The Complex Susceptibility

Most of the topics in this chapter relate directly to optics. The general approach, however, is actually that of the general theory of **response functions**, that is, the response of a system to a driving field. The most obvious example is the response of a charged particle to an electromagnetic field, which controls the optical properties of a condensed matter system, but there are other examples, such as the response of a spin system to a magnetic field, or a resonant acoustic response to a sound wave.

The theoretical tool introduced in this chapter is the use of complex analysis to keep track of the phase of the response relative to the driving field. As we will see, the complex susceptibility contains all of the information needed to describe the interaction of the driving field with the medium, including reflection, absorption, nonlinear effects, etc.

7.1 A Microscopic View of the Dielectric Constant

In Section 3.5.1, we deduced the effect of a medium on light due to the polarization of charge in the medium. At that point, we simply invoked a constant of proportionality, the susceptibility χ, to describe the polarization due to the electric field,

$$\vec{P} = \chi \epsilon_0 \vec{E}. \tag{7.1.1}$$

Is this a reasonable assumption? Where does χ come from, physically? To understand the connection between the microscopic structure and the macroscopic properties, we can make a model of a solid as consisting of a number of charged harmonic oscillators, as illustrated in Figure 7.1. For each oscillator, we can write the equation of motion $F = m\ddot{x}$ as

$$m\ddot{x} = qE_0 e^{-i\omega t} - m\Gamma\dot{x} - m\omega_0^2 x, \tag{7.1.2}$$

where the three terms on the right give the three forces acting on the charge, namely: the driving force due to the oscillating electric field, which we write as $E_0 e^{-i\omega t}$, acting on the charge q; the damping force proportional to the velocity $\dot{x}$; and the spring force $F = -Kx$, where we have used the natural frequency $\omega_0 = \sqrt{K/m}$. Γ is a constant of proportionality for the local damping, or dissipative term. We ignore anisotropy and drop the vector notation at this point, although it is easy to see that, as in the case of lattice vibrations, anisotropy will lead to a spring-constant tensor instead of a single spring constant.

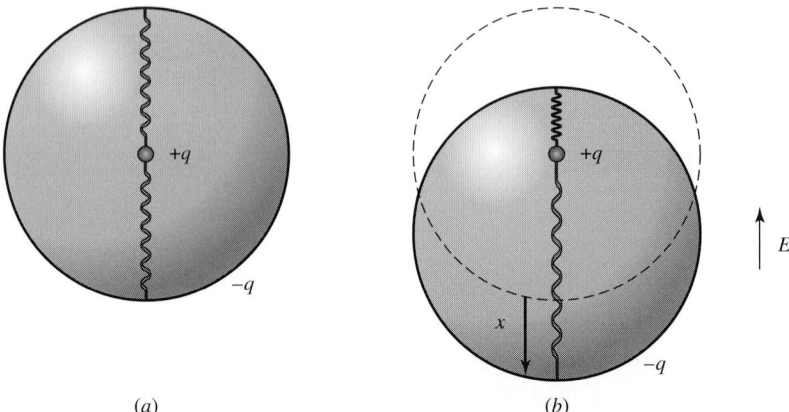

Fig. 7.1 Model of a charged oscillator. (a) In equilibrium, there is no dipole moment. (b) in the presence of electric field, a dipole moment qx arises. The central positive charge (e.g., an atomic nucleus) is assumed to not move.

As we have done before, we solve this by guessing a solution. We suppose

$$x(t) = x_0 e^{-i(\omega t + \delta)}, \tag{7.1.3}$$

where x_0 is the amplitude of the oscillation and δ is a phase shift to take into account the possibility that the oscillator is not in phase with the driving force. The first time derivative is therefore

$$\dot{x} = -i\omega x_0 e^{-i(\omega t + \delta)} \tag{7.1.4}$$

and the second time derivative is

$$\ddot{x}(t) = -\omega^2 x_0 e^{-i(\omega t + \delta)}. \tag{7.1.5}$$

Substituting these into (7.1.2), we have

$$- m\omega^2 x_0 e^{-i\delta} = qE_0 + im\Gamma\omega x_0 e^{-i\delta} - m\omega_0^2 x_0 e^{-i\delta}. \tag{7.1.6}$$

This gives us two equations, since the real parts and the imaginary parts on each side of the equation must be equal, and two unknowns, x_0 and δ. After some algebra, we find

$$x_0 = \frac{qE_0}{m} \frac{1}{\sqrt{(\omega_0^2 - \omega^2)^2 + \Gamma^2\omega^2}},$$

$$\tan\delta = -\frac{\Gamma\omega}{\omega_0^2 - \omega^2}. \tag{7.1.7}$$

Figure 7.2 illustrates the behavior of these variables. The phase shift δ is zero at low frequency, that is, the charge moves in phase with the driving field; at $\omega = \omega_0$, it is 90° out of phase, and eventually at high frequency it moves exactly opposite of the driving field.

As discussed in Section 3.5.1, the polarization is equal to the dipole moment per volume. The dipole moment of any single oscillator is just qx, which means that the polarization is

$$P = qx\frac{N}{V}, \tag{7.1.8}$$

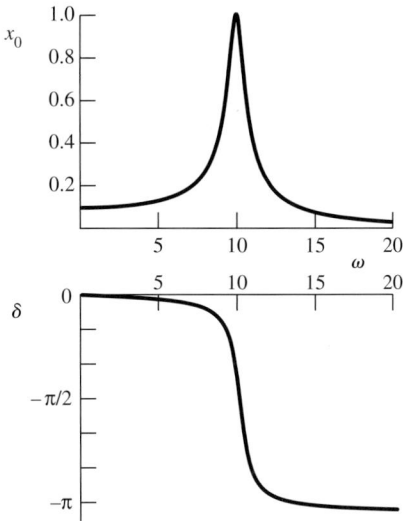

Fig. 7.2   The amplitude and phase shift of the oscillator response as functions of frequency, for $\omega_0 = 10$ and $\Gamma = 1$.

where N is the total number of oscillators and V is the total volume. Substituting in our solution for x, we have

$$P = q \frac{N}{V} \left(\frac{qE_0}{m} \frac{e^{-i(\omega t + \delta)}}{\sqrt{(\omega_0^2 - \omega^2)^2 + \Gamma^2 \omega^2}} \right). \tag{7.1.9}$$

Equating this with (7.1.1), we have

$$\chi = \frac{q^2 N}{\epsilon_0 m V} \frac{e^{-i\delta}}{\sqrt{(\omega_0^2 - \omega^2)^2 + \Gamma^2 \omega^2}}, \tag{7.1.10}$$

which after solving for $\cos \delta$ and $\sin \delta$ can be rearranged as

$$\chi = \frac{q^2 N}{\epsilon_0 m V} \frac{(\omega_0^2 - \omega^2) + i\Gamma\omega}{(\omega_0^2 - \omega^2)^2 + \Gamma^2 \omega^2}. \tag{7.1.11}$$

In other words, χ has a real and an imaginary part. The real part corresponds to a polarization that is in phase with the driving electric field, and is responsible for the index of refraction, as we saw in Section 3.5.1. The imaginary part gives a polarization that is 90° out of phase with the driving field. What does this correspond to, physically?

As already derived in Section 3.5.1, Maxwell's (isotropic) wave equation in one dimension is

$$\frac{\partial^2 E}{\partial x^2} = \frac{(1 + \chi)}{c^2} \frac{\partial^2 \vec{E}}{\partial t^2}. \tag{7.1.12}$$

The solution of this is $E = E_0 e^{i(kx - \omega t)}$. If χ has an imaginary part, however, then we must allow for the possibility of k being complex. We write $\chi = \chi_R + i\chi_I$ and $k = k_R + ik_I$, in which case the wave equation becomes

$$-(k_R + ik_I)^2 E = -(1 + \chi_R + i\chi_I)\frac{\omega^2}{c^2}E \tag{7.1.13}$$

or

$$k_I^2 - k_R^2 - 2ik_Rk_I = -(1 + \chi_R + i\chi_I)\frac{\omega^2}{c^2}. \tag{7.1.14}$$

If we match the real and imaginary parts, we find

$$k_R^2 - k_I^2 = (1 + \chi_R)\frac{\omega^2}{c^2}$$

$$2k_Rk_I = \chi_I\frac{\omega^2}{c^2}. \tag{7.1.15}$$

The latter equation implies that if χ_I is nonzero, and if ω and k_R are nonzero (as for a propagating wave), then k_I cannot be nonzero. This therefore implies that if χ has an imaginary part, then the solutions have the form

$$E = E_0 e^{i(k_Rx - \omega t)}e^{-k_Ix}. \tag{7.1.16}$$

Our result (7.1.11) gives $\chi_I > 0$ for all frequencies; therefore (7.1.15) implies that k_R and k_I must have the same signs. For example, if a wave is propagating with $k_R > 0$ toward positive x, then k_I will be positive, leading to decay.

The χ_I term therefore corresponds to **absorption** of the signal. This is not surprising if we note, from (7.1.11), that the imaginary part of χ is proportional to Γ, the damping constant, which gives the strength of the dissipative term. We can also note that the real part of χ corresponds to polarization that is *in phase* with the driving electric field, which means that the current,

$$\vec{J} = \frac{\partial \vec{P}}{\partial t} = i\omega\vec{P}, \tag{7.1.17}$$

is 90° *out of phase* with the driving field. The imaginary part, conversely, corresponds to a current in phase with the driving field, which is the source of resistive energy loss; as we are familiar with from electronics, the power loss is given by the ohmic term in which the voltage and current are in phase with each other.

Figure 7.3 shows curves for χ_I and χ_R. Several observations stand out. First, a finite, nonzero susceptibility is seen at frequencies well below ω_0, and therefore the fact that the index of refraction is different from unity is due to the presence of absorption in the system, even in ranges of frequency where χ_I is negligible. If Γ were zero, n would equal 1 everywhere except $\omega = \omega_0$, at which point it would be infinite. In general, the absorption properties and the index of refraction of a material are not independent parameters; both arise from the same underlying resonances.

Also, the existence of dispersion, that is, the fact that $n(\omega)$ is not constant when ω varies, is not surprising, but a necessary consequence of the resonance that exists due to a damped oscillator.

Last, the existence of **anomalous dispersion**, that is, a region of frequency above ω_0 at which the index of refraction decreases with increasing ω, is also not surprising, but part of the same resonance behavior. The index of refraction can even become less than unity, which implies a phase velocity faster than c. Typically, one does not find a sub-unity index

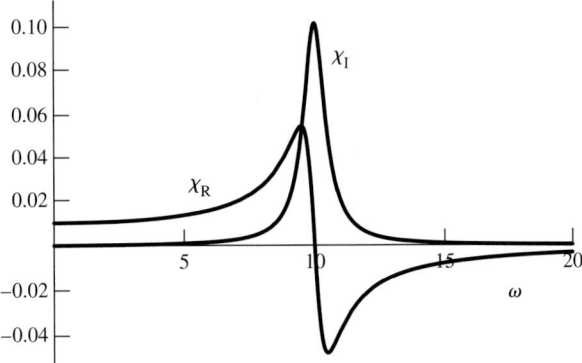

Fig. 7.3 The real and the imaginary susceptibility as functions of frequency, for the same classical oscillator as in Figure 7.2.

of refraction because more than one resonance exists. The susceptibilities of the different resonances add up, so that a positive χ_R below one resonance cancels the negative χ_R above a lower resonance. In a system with a single, isolated resonance, however, it is possible to have an index of refraction less than unity. Even stranger, it is possible to design materials with negative index of refraction when there is a magnetic resonance as well as an electric dipole resonance, allowing both μ and ϵ to be negative. We will return to discuss this in Section 7.3.

Local field corrections. In many solids, it can be argued that the electric field felt by a given atom is quite different from the average electric field in the medium. This leads to a correction for the macroscopic susceptibility in terms of the microscopic polarizability.

The need for such a correction can be seen by considering a simple example: an atom in the center of a crystal with cubic symmetry. Each atom is surrounded by six nearest neighbors, each of which is a dipole aligned with the externally applied electric field, which we can take as along the z-direction. The z-component of the electric field generated by each dipole is, in the far-field limit,

$$E_z = \frac{qd}{4\pi\epsilon_0} \frac{2z^2 - x^2 - y^2}{r^5}, \tag{7.1.18}$$

where qd is the electric dipole moment for a dipole oriented along the $\hat{z}$-direction. Dipoles above and below a given atom give an electric field in the positive direction, while atoms to the side give a contribution in the negative direction. If all the neighboring dipoles are the same distance away, and there are two neighbors in the z-direction and four neighbors in the x–y plane, as in a crystal with cubic symmetry, then the contributions from the nearest neighbors exactly cancel. If the atom sits at the center of a symmetric object, such as a cube or sphere, then the same argument can be made for all the other atoms as well – for each electric field contribution in one direction, there is an equal and opposite contribution. The atom at the center of the medium should therefore feel no polarization field at all, only the externally applied field. This will be larger than the average electric field, which is equal to the externally applied field minus the average polarization field. The same argument applies for an atom in the center of a homogeneous disordered medium, in which the contributions of the neighboring atoms will cancel out on average.

The standard picture breaks down in this case because we can only talk about the average electric field on length scales large compared to the size of an atom. One way to model the local inhomogeneity is to break the medium into two parts: the medium very far from the atom of interest, which can be treated as having the standard average electric field, and a sphere centered around the atom of interest, with some radius much larger than the atomic spacing, in which we calculate the local field for the specific symmetry of the neighboring atoms. In the **Clausius–Mossotti** model (see, e.g., Jackson 1999: s. 4.5), one assumes that the field generated by the local induced dipoles in this sphere is exactly zero due to the cancellation by symmetry discussed in the previous paragraph, leaving only the externally applied field. Such models do not actually fit experimental data for solids very well. We can in general assume, though, that the electric field felt by an atom locally is equal to the average electric field in the medium times some enhancement factor due to the local field corrections.

Exercise 7.1.1   Suppose that instead of a harmonic oscillator, we have a metal in which the electrons move directly in response to the electric field, that is, there is a damping constant but no spring constant, so that $\omega_0 = 0$. What is the susceptibility $\chi(\omega)$ in the limit $\omega \to 0$? Using the formula (3.8.7) for the reflectivity, what does this imply about the infrared reflectance of a metal?

Show that if $\Gamma \equiv 1/\tau$, the solution is consistent with the formula $J = \sigma E$, where σ is given by (5.8.6).

Exercise 7.1.2   Using the result of the previous exercise, calculate the **skin depth** $\lambda = 1/k_I$ for the penetration of an electromagnetic wave into a metal surface, as a function of frequency and the material parameters n (the electron density), m (the electron mass) and τ (the average electron scattering time). For realistic parameters for a metal, that is, $n = N/V = 10^{21}$ cm^{-3} and $\tau = 1$ ps, what is the skin depth of a 10 MHz electromagnetic wave?

7.1.1 Fresnel Equations for the Complex Dielectric Function

Recall from Section 3.8.1 that the reflectivity of a medium depends on its index of refraction. To deduce the reflectivity when there is a complex susceptibility, we need to go back to the derivation of the Fresnel equations. A key step in the derivation in Section 3.8.1 was to use $B_0 = (n/c)E_0$ for the field amplitudes in a plane wave. More generally, we can write, from the Maxwell equation $\nabla \times \vec{E} = -\partial \vec{B}/\partial t$,

$$ i\vec{k} \times \vec{E}_0 = i\omega \vec{B}_0, \qquad (7.1.19) $$

which for the field directions assigned in Figure 3.25 implies $B_0 = E_0(k_R + ik_I)/\omega$. Solving (7.1.15) for k_R and k_I gives

$$
k_R = \pm \frac{\omega}{\sqrt{2}c} \left(\sqrt{(1 + \chi_R)^2 + \chi_I^2} + (1 + \chi_R) \right)^{1/2}
$$

$$
k_I = \pm \frac{\omega}{\sqrt{2}c} \left(\sqrt{(1 + \chi_R)^2 + \chi_I^2} - (1 + \chi_R) \right)^{1/2}. \tag{7.1.20}
$$

As discussed above, (7.1.15) implies that if χ_I is positive, k_R and k_I must have the same signs. In the Fresnel equations (3.8.5) and (3.8.6), one should therefore use the index of refraction

$$
n_{\text{eff}} = \frac{c(k_R + ik_I)}{\omega}
$$

$$
= \left(\frac{\sqrt{(1 + \chi_R)^2 + \chi_I^2} + (1 + \chi_R)}{2} \right)^{1/2}
$$

$$
+ i \left(\frac{\sqrt{(1 + \chi_R)^2 + \chi_I^2} - (1 + \chi_R)}{2} \right)^{1/2}. \tag{7.1.21}
$$

The will in general lead to reflectivity and transmission coefficients r and t that are complex, which is not a problem, as it simply means that there can be phase shifts of the reflected and transmitted light relative to the incident light.

The phase velocity of the waves is given by $v = \omega/k_R$, which is not equal to $c/|n_{\text{eff}}|$. Thus, in Snell's law for refraction, which is derived from equating the spacing of the wave fronts along both sides of an interface (see Section 3.5.2), only the real part of n_{eff} should be used. We can write $n_{\text{eff}} = n + i\alpha$, where n is the index of refraction used in Snell's law, and $\alpha = c|k_I|/\omega$ is the imaginary term giving the absorption.

For normal incidence, Fresnel equations for a complex index of refraction give the simple equation for the reflectivity,

$$
|r|^2 = \frac{(n_1 - n_2)^2 + (\alpha_1 - \alpha_2)^2}{(n_1 + n_2)^2 + (\alpha_1 + \alpha_2)^2}. \tag{7.1.22}
$$

Exercise 7.1.3   Surprisingly, near a resonance, not only the phase velocity, but also the **group velocity** can exceed the speed of light. This still does not violate the theory of relativity, because in the region near the resonance, the signal is strongly absorbed, so all that really happens is that a wave pulse is selectively absorbed on its trailing edge, which makes the leading edge appear to move faster than the speed of light.

To see this effect, use (7.1.20) to obtain a relation for k_R as a function of ω. Then take the derivative of this function with respect to ω. The inverse of this derivative, $d\omega/dk$, is the group velocity. Plot the group velocity vs ω for $c = 1$, $\omega_0 = 10$, $\Gamma = 1$, and $q^2 N/\epsilon_0 mV = 1$. You should see that away from the resonance on either side, the group velocity is less than 1, but very near the resonance, in the region of strong absorption, the group velocity becomes larger than $c\ (= 1)$. This is easily done using a program like Mathematica. This effect has been seen in atomic systems with a sharp, strong resonance.

Exercise 7.1.4 The complex Fresnel equations allow a powerful method of materials analysis known as **ellipsometry**. In general, linearly polarized light reflected at an oblique angle will become elliptically polarized, with linear and circular polarization components that depend on the real and imaginary parts of χ. Measuring the reflectivity of both components of polarization therefore gives the real and imaginary parts of χ.

Plot the reflectivity spectrum for s-polarized light for an ensemble of N identical, isolated damped oscillators in a medium, as a function of frequency for several angles of incidence. (The s and p polarizations are defined in Section 3.8.1.) Pick $\omega_0 = 1$, $\Gamma = 0.1$, and $e^2 N/mV = 1$.

Exercise 7.1.5 As discussed in Exercise 7.1.2, we can model a metal as an ensemble of classical oscillators with $\omega_0 = 0$. Calculate the absolute reflectivity r for an infrared electromagnetic wave, with wavelength $10\ \mu m$, incident normal to the surface of a metal with electron density $n = 10^{21}\ cm^{-3}$, electron scattering time $\tau = 1$ ps, and electron mass equal to the vacuum electron mass.

7.1.2 Fano Resonances

So far, we have considered an isolated resonance. Consider the case of an oscillator coupled to a second oscillator. In this case, we have two coupled equations, for the position x of the first charge, which is driven by an electric field, and the position y of a second charge, which is not directly affected the electric field:

$$m\ddot{x} = qE_0 e^{-i\omega t} - m\Gamma_1 \dot{x} - m\omega_1^2 x + Ky$$
$$m\ddot{y} = -m\Gamma_2 \dot{y} - m\omega_1^2 y + Kx. \qquad (7.1.23)$$

The coupling constant K, which gives the force of one oscillator on the other, must be the same in both equations, by Newton's third law. As before, we assume that the driving field is given by $E_0 e^{-i\omega t}$ and guess solutions of the form $x = x_0 e^{-i(\omega t + \delta_1)}$ and $y = y_0 e^{-i(\omega t + \delta_2)}$. We must then solve the equations

$$-m\omega^2 x_0 e^{-i\delta_1} = qE_0 + im\Gamma_1 \omega x_0 e^{-i\delta_1} - m\omega_1^2 x_0 e^{-i\delta_1} + Ky_0 e^{-i\delta_2}$$
$$-m\omega^2 y_0 e^{-i\delta_2} = im\Gamma_2 \omega y_0 e^{-i\delta_2} - m\omega_2^2 y_0 e^{-i\delta_2} + Kx_0 e^{-i\delta_1} \qquad (7.1.24)$$

for the amplitude and phase of each oscillator, equating the real and imaginary parts of each equation, as before.

Figure 7.4 shows the response of the first oscillator to the driving field, as a function of frequency. Although the second oscillator is not coupled to the driving field, it affects the response of the first oscillator through the coupling. As seen in this figure, if the second resonance is sharp, with low damping constant Γ_2, it can drive the response of the first oscillator all the way to zero. This type of asymmetric response is known as a **Fano resonance**. Effectively, there is an interference between the responses of the oscillators that cancels out. Fano resonances are common in many systems, whenever there is a sharp resonance, such as an isolated dopant atom coupled to a broad response function.

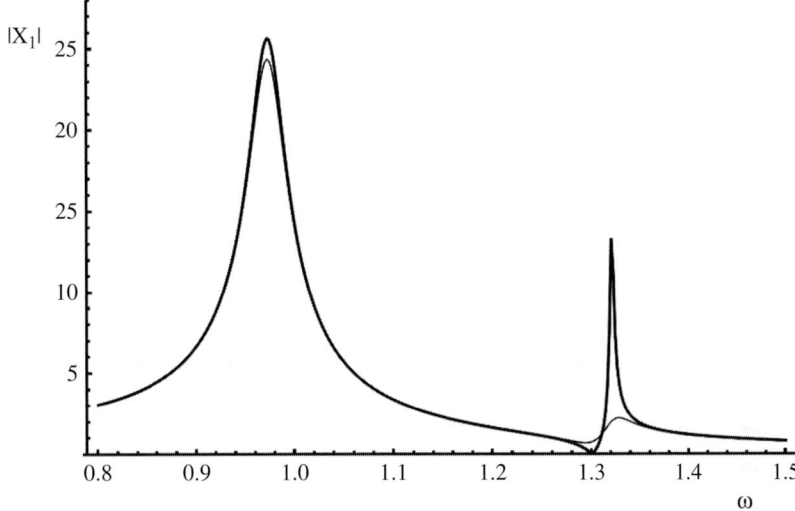

Fig. 7.4 The susceptibility of an oscillator (proportional to x_0) coupled to a second oscillator, from (7.1.24) in the text, for $\omega_1 = 1, \omega_2 = 1.3, \Gamma_1 = 0.04, m = 1$ and $K = 0.2$, with two values of the damping of the second resonance: thin line: $\Gamma_2 = \Gamma_1/2$; heavy line: $\Gamma_2 = \Gamma_1/40$.

Exercise 7.1.6 Use a program like Mathematica to solve (7.1.24) for $x_0, y_0, \cos\delta_1, \sin\delta_1,$ $\cos\delta_2,$ and $\sin\delta_2$ (subject to the constraints $\cos^2\delta + \sin^2\delta = 1$) and plot the response for $qE_0 = 1, m = 1$, your own choice of parameters ω_1, ω_2 and the damping constants.

7.2 Kramers–Kronig Relations

Since we have found that the susceptibility χ is a complex variable, it makes sense to apply the mathematics of complex analysis to this case. One reason why we may want to do this is because in a typical solid system, instead of just a single isolated resonance, a large number of resonances all add together to give the total dielectric function.

We will not review all of the results of complex analysis here, but just recall two theorems of complex calculus that are especially useful. (For a review of complex calculus, see, e.g., Wyld 1976.) The first is Cauchy's residue theorem that if we define a counterclockwise closed path in the complex plane, then for a function $f(z)$ that is analytic everywhere inside the path,

$$\oint \frac{f(z)}{z - z'}dz = \begin{cases} 2\pi i f(z'), & \text{if } z' \text{ lies inside the closed path,} \\ 0, & \text{otherwise.} \end{cases} \tag{7.2.1}$$

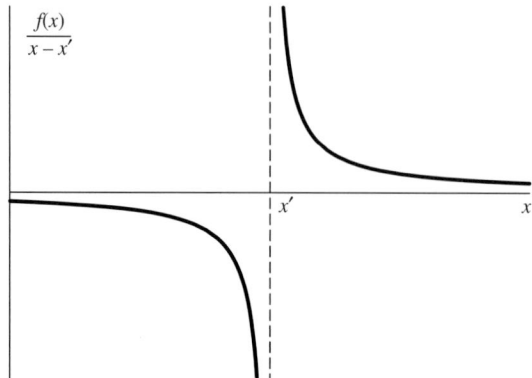

Fig. 7.5 Illustration of the cancellation that allows the principal value of the integral over a pole to be finite.

The function $f(z')$ is known as the residue. For a clockwise path, the sign of the integral will be the opposite.

The second useful formula is the Dirac equality,

$$\frac{1}{x - x' \pm i\epsilon} = \mathcal{P}\frac{1}{x - x'} \mp i\pi\delta(x - x'), \tag{7.2.2}$$

where ϵ is an infinitesimal, real number, and $\mathcal{P}$ is the principal value. The principal value of the integral means that we integrate x on the real axis over all points except $x = x'$. Since $1/x$ is an odd function, this integral will be finite, because for every place where the function is large and positive, there is another place where the function has equal and opposite negative value, as illustrated in Figure 7.5.

Response of a single resonance to an impulse. Let us take a close look at the susceptibility we deduced in Section 7.1. The response (7.1.11) can be rewritten as

$$\chi(\omega) = C\frac{1}{(\omega_0^2 - \omega^2) - i\Gamma\omega}, \tag{7.2.3}$$

where $C = q^2 N/\epsilon_0 m V$. If Γ is small enough, this is strongly peaked near $\omega = \omega_0$. In this case, we can approximate $|\omega - \omega_0| \ll \omega_0$, so that we have

$$\chi(\omega) = C\frac{1}{(\omega_0 - \omega)(\omega_0 + \omega) - i\Gamma\omega}$$

$$\approx \left(\frac{C}{2\omega_0}\right)\frac{1}{\omega_0 - \omega - i\Gamma/2}. \tag{7.2.4}$$

This has a pole at $\omega = \omega_0 - i\Gamma/2$, in the lower half of the complex plane. This pole in frequency space corresponds to a time response with decay. We can see this by the following calculation.

To obtain the real-time response, we recall that $\vec{P}(\omega) = \chi(\omega)\epsilon_0\vec{E}(\omega)$. Suppose that $\vec{E}$ is given by a constant times $\delta(t)$, that is, a short impulse. The Fourier transform of this will be equal to a constant (see Appendix B), which we will call $\vec{E}_0$. Taking the inverse Fourier transform of $\vec{P}(\omega)$ to get $\vec{P}(t)$ gives us

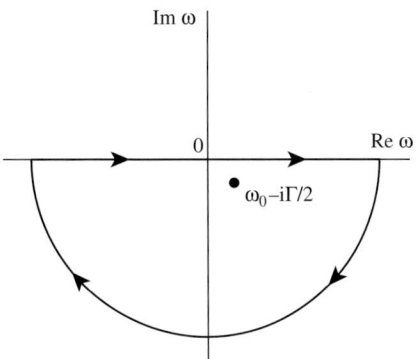

Fig. 7.6 The path in the complex plane used for the integration discussed in the text.

$$\vec{P}(t) = \frac{C\epsilon_0 \vec{E}_0}{4\pi\omega_0} \int_{-\infty}^{\infty} d\omega \, e^{-i\omega t} \frac{1}{\omega_0 - \omega - i\Gamma/2}. \tag{7.2.5}$$

We can calculate this using the prescription of (7.2.1). When $t > 0$, and ω has a large, negative imaginary component, the factor $e^{-i\omega t} \to 0$ as $|\omega| \to \infty$. This allows us to use a closed path that includes the entire real axis and then loops back on a clockwise path at $|\omega| = \infty$ in the lower half of the complex plane. Figure 7.6 shows this path. Since this closed path includes the pole, the residue is nonzero. On the other hand, when $t < 0$, the factor $e^{-i\omega t} \to 0$ as $|\omega| \to \infty$ only in the upper half plane. Since there is no pole there, the residue is zero. We therefore have

$$\vec{P}(t) = \left(\frac{C\epsilon_0 \vec{E}_0}{2\omega_0}\right) i\Theta(t) e^{-i\omega_0 t - (\Gamma/2)t}, \tag{7.2.6}$$

where $\Theta(t)$ is the Heaviside function, equal to 1 for $t > 0$ and equal to 0 for $t < 0$. This response is the standard "ring-down" response to an impulse, illustrated in Figure 7.7, decaying with a rate given by the damping Γ.

Derivation of the Kramers–Kronig relations. The total susceptibility $\chi(\omega)$ of a medium is simply equal to the sum of the susceptibilities of all the individual resonances in a system, which can have different relative weight factors C. If we assume that the resonance frequencies are distributed over a finite range, then $|\chi(\omega)|$ decreases at least as fast as $|1/\omega|$ as $|\omega| \to \infty$, since any single resonance is a pole of the form (7.1.11).

Note that if the sign of the imaginary term in the denominator of (7.2.4) was the opposite, that is, if the pole was in the upper half of the complex plane, then the response function would be nonzero for $t < 0$, before the impulse at $t = 0$. We therefore conclude that causality requires that any susceptibility function have only poles in the lower half of the complex plane.

We define the function

$$f(\omega) = \frac{\chi(\omega)}{\omega' - \omega - i\epsilon}, \tag{7.2.7}$$

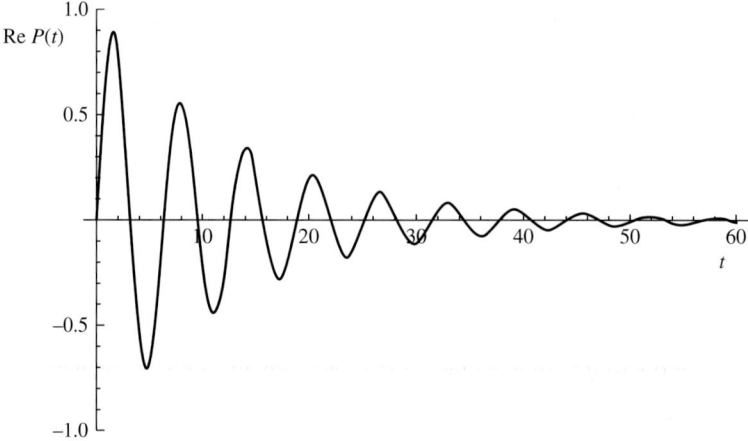

Fig. 7.7   The response of the polarization of a simple oscillator to a $\delta(t)$ impulse, for $\omega = 1$ and $\Gamma = 0.15$.

which is analytic in the entire upper half plane, where ϵ is an infinitesimal real, positive number. Since $\chi(\omega)$ decreases at least as fast as $1/\omega$, we know that $|f(\omega)|$ decreases at least as fast as $1/|\omega|^2$. Therefore, the contribution of a loop in the upper half plane with $|\omega| \to \infty$ will vanish. We therefore create a closed path consisting of the real axis and a loop back in the upper half plane, which contains no poles. We can then use (7.2.1) to conclude

$$\int_{-\infty}^{\infty} d\omega \, \frac{\chi(\omega)}{\omega' - \omega - i\epsilon} = 0. \tag{7.2.8}$$

(If we took a loop in the lower half plane, we would have to deal with multiple poles, including the one introduced at $\omega' - i\epsilon$ and others in $\chi(\omega)$. Since this loop must give the same answer as the loop in the upper half plane, we know that the contributions from these poles must cancel to zero.)

By the Dirac formula (7.2.2), we then have

$$\mathcal{P} \int_{-\infty}^{\infty} d\omega \, \frac{\chi(\omega)}{\omega' - \omega} + i\pi \chi(\omega') = 0, \tag{7.2.9}$$

or, switching the definitions of ω and ω',

$$\chi(\omega) = \frac{1}{i\pi} \mathcal{P} \int_{-\infty}^{\infty} d\omega' \, \frac{\chi(\omega')}{\omega' - \omega}. \tag{7.2.10}$$

Equating the real and imaginary parts on each side immediately gives

$$
\begin{aligned}
\mathrm{Re}\,\chi(\omega) &= \frac{1}{\pi} \mathcal{P} \int_{-\infty}^{\infty} d\omega' \, \frac{\mathrm{Im}\,\chi(\omega')}{\omega' - \omega} \\[2mm]
\mathrm{Im}\,\chi(\omega) &= -\frac{1}{\pi} \mathcal{P} \int_{-\infty}^{\infty} d\omega' \, \frac{\mathrm{Re}\,\chi(\omega')}{\omega' - \omega}.
\end{aligned}
\tag{7.2.11}
$$

This is one version of the **Kramers–Kronig relations**. The relation between Re χ and Im χ is known mathematically as a **Hilbert transform**, analogous to a Fourier transform but with $1/(x - y)$ in the place of e^{ixy}.

We can put the Kramers–Kronig relations in another useful form if we assume that $\chi(\omega)$ has the following symmetry properties:

$$\text{Re } \chi(-\omega) = \text{Re } \chi(\omega)$$

$$\text{Im } \chi(-\omega) = -\text{Im } \chi(\omega), \tag{7.2.12}$$

in other words, Re χ is an even function of ω and Im χ is an odd function of ω. The standard form of a resonance (7.1.11) has these properties: the imaginary part of the resonance is proportional to ω while the real part contains only terms with ω^2. Any susceptibility that is a sum of resonances of this type will therefore have these properties.

We break the integration over the real axis in (7.2.10) into two parts, as follows:

$$\chi(\omega) = \frac{1}{i\pi} \mathcal{P} \int_0^\infty d\omega' \frac{\chi(\omega')}{\omega' - \omega} + \frac{1}{i\pi} \mathcal{P} \int_{-\infty}^0 d\omega' \frac{\chi(\omega')}{\omega' - \omega}. \tag{7.2.13}$$

Using the properties that Re $\chi(\omega)$ is even and Im $\chi(\omega)$ is odd, we have

$$\text{Re } \chi(\omega) + i \text{Im } \chi(\omega) = \frac{1}{i\pi} \mathcal{P} \int_0^\infty d\omega' \text{ Re } \chi(\omega') \left(\frac{1}{\omega' - \omega} - \frac{1}{\omega' + \omega} \right)$$

$$+ \frac{1}{\pi} \mathcal{P} \int_0^\infty d\omega' \text{ Im } \chi(\omega') \left(\frac{1}{\omega' - \omega} + \frac{1}{\omega' + \omega} \right). \tag{7.2.14}$$

Adding the terms over a common denominator, we have

$$\text{Re } \chi(\omega) + i \text{Im } \chi(\omega) = \frac{1}{i\pi} \mathcal{P} \int_0^\infty d\omega' \text{ Re } \chi(\omega') \frac{2\omega}{\omega'^2 - \omega^2}$$

$$+ \frac{1}{\pi} \mathcal{P} \int_0^\infty d\omega' \text{ Im } \chi(\omega') \frac{2\omega'}{\omega'^2 - \omega^2}. \tag{7.2.15}$$

Finally, equating the real and imaginary parts on each side, we have

$$\boxed{\begin{aligned} \text{Re } \chi(\omega) &= \frac{2}{\pi} \mathcal{P} \int_0^\infty d\omega' \frac{\text{Im } \chi(\omega')\omega'}{\omega'^2 - \omega^2} \\ \text{Im } \chi(\omega) &= -\frac{2}{\pi} \mathcal{P} \int_0^\infty d\omega' \frac{\text{Re } \chi(\omega')\omega}{\omega'^2 - \omega^2}. \end{aligned}} \tag{7.2.16}$$

A surprising implication of the Kramers–Kronig relations is that if one knows the real part of the susceptibility over the entire frequency range, one can automatically calculate the entire imaginary part of the susceptibility over the whole range without any further information, and vice versa. In physical terms, this means that if we know the absorption spectrum of a material, we can compute its index of refraction over the entire wavelength range without any additional measurements, and on the other hand, if we know the index of refraction over the whole range, we can calculate the absorption spectrum without any additional information. This comes from the fundamental result deduced in Section 7.1, that absorption and refraction are not independent properties of a material, but come from the same underlying physical mechanism. As discussed in Section 7.1, the index of refraction can be deduced from reflectivity measurements. Therefore, one can equally well say that the absorption spectrum and the reflectivity spectrum are not independent, but can be calculated from each other.

Of course, using the Kramers–Kronig relations to deduce the reflectivity spectrum from the absorption spectrum assumes that we have one set of spectrum over the entire frequency range $(0, \infty)$. Experimentally, this is never the case, so there is always some degree of uncertainty in deriving one from the other. Nevertheless, it is a powerful method for analyzing optical data. In particular, if the absorption is too strong, the transmission through a thick slab may be too weak to measure, but the reflectivity can still be measured, and the absorption spectrum can be deduced from that.

Exercise 7.2.1 The dielectric function is subject to various **sum rules**. For example, show that the Kramers–Kronig relations imply

$$\int_0^\infty |\mathrm{Im}\ \epsilon(\omega)|\ \omega d\omega = \frac{\pi e^2 N}{2mV}, \tag{7.2.17}$$

for a set of single oscillators, no matter what their frequency distribution is.

7.3 Negative Index of Refraction: Metamaterials

As discussed in Section 7.1, the index of refraction can in principle be less than 1, or even negative. To see this, we need to redo the derivation of the Maxwell wave equation, presented in Section 3.5.1. We start with Maxwell's equations, and assume a homogeneous and charge-neutral medium:

$$\nabla \cdot \vec{E} = 0$$
$$\nabla \cdot \vec{B} = 0$$
$$\nabla \times \vec{E} = -\frac{\partial \vec{B}}{\partial t}$$
$$\nabla \times \vec{B} = \mu_0\epsilon_0\frac{\partial \vec{E}}{\partial t} + \mu_0\vec{J}. \tag{7.3.1}$$

We now allow for the current to arise not only from an electric polarization $\vec{P}$, but also from a magnetization $\vec{M}$ that is assumed to be linearly proportional to the $\vec{H}$-field:

$$\vec{P} = \chi \epsilon_0 \vec{E}$$
$$\vec{M} = \chi_m \vec{H}, \tag{7.3.2}$$

where χ_m is a unitless parameter, called the magnetic susceptibility, and $\vec{H}$ is defined by

$$\vec{B} = \mu_0(\vec{H} + \vec{M}) = \mu_0(1 + \chi_m)\vec{H} \equiv \mu\vec{H}. \tag{7.3.3}$$

The total current arises from the time dependence of the electric polarization and the circulating current due to the magnetization:

$$\vec{J} = \frac{\partial \vec{P}}{\partial t} + \nabla \times \vec{M}. \tag{7.3.4}$$

Substituting this and the definition of $\vec{H}$ in (7.3.3) into the fourth Maxwell equation, we obtain

$$\mu_0 \nabla \times (\vec{H} + \vec{M}) = \mu_0 \epsilon_0 \frac{\partial \vec{E}}{\partial t} + \chi \mu_0 \epsilon_0 \frac{\partial \vec{E}}{\partial t} + \mu_0 \nabla \times \vec{M}, \tag{7.3.5}$$

which simplifies to

$$\nabla \times \vec{H} = \epsilon \frac{\partial \vec{E}}{\partial t}, \tag{7.3.6}$$

with $\epsilon = \epsilon_0 (1 + \chi)$, and finally by (7.3.3),

$$\nabla \times \vec{B} = \mu \epsilon \frac{\partial \vec{E}}{\partial t}. \tag{7.3.7}$$

We deduce a wave equation in the same way that we did in Section 3.5.1, by taking the time derivative of both sides of (7.3.7) and simplifying using the second and third Maxwell equations:

$$\nabla \times \frac{\partial \vec{B}}{\partial t} = \mu \epsilon \frac{\partial^2 \vec{E}}{\partial t^2}$$

$$\Rightarrow \nabla^2 \vec{E} = \mu \epsilon \frac{\partial^2 \vec{E}}{\partial t^2}. \tag{7.3.8}$$

The real parts of μ and ϵ can both be negative. As we have seen for electric dipoles, the susceptibility of a single oscillator is

$$\chi(\omega) = \frac{C}{(\omega_0^2 - \omega^2) - i\Gamma\omega}. \tag{7.3.9}$$

As shown in Figure 7.8, for a resonance that is narrow enough and strong enough, the factor $(1 + \chi_R)$ can dip below zero on the high side of the resonance. In the same way, a magnetic resonance can have the same feature. To get a material with a magnetic resonance is non-trivial; methods will be discussed at the end of this section. In both the electric and magnetic resonances, one wants a sharp resonance with frequency below the range of electromagnetic wave frequencies of interest.

If the real parts of ϵ and μ are both negative, then (7.3.8) will still allow traveling wave solutions of the form $E_0 e^{i(k_R x - \omega t)} e^{-k_I x}$. However, (7.3.6) implies that $\vec{H}$ will point in the opposite direction as $\vec{B}$. This affects the Poynting vector, defined as

$$\vec{S} = \vec{E}^* \times \vec{H}, \tag{7.3.10}$$

which gives the direction of the energy flow. Since $\vec{E}$, $\vec{B}$, and $\vec{k}$ still form a right-handed system, this implies that energy will flow in the opposite direction to the $\vec{k}$-vector. In other

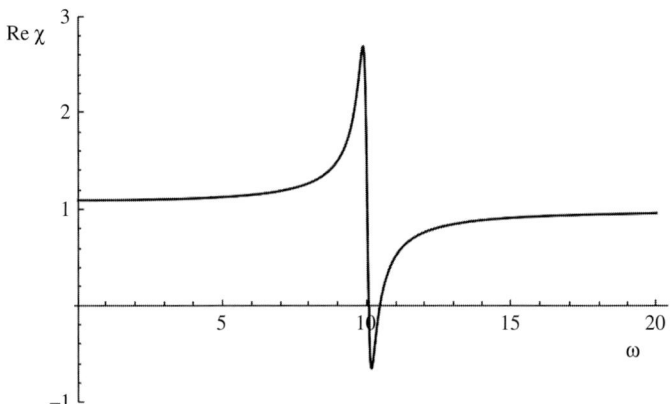

Fig. 7.8   The real part of the susceptibility for an isolated, sharp resonance, with $\omega = 10$, $\Gamma = 0.3$, and $C = 10$.

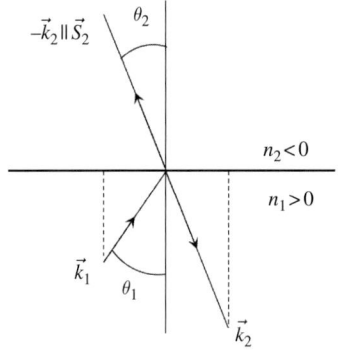

Fig. 7.9   Rays for refraction at the interface of a negative-index material, for an incident ray along $\vec{k}_1$.

words, the wave fronts move backward through the wave. Causality is still maintained, however. When magnetic resonance is taken into account, the equations in (7.1.15) become

$$k_R^2 - k_I^2 = [(1 + \chi_R)(1 + \chi_{mR}) - \chi_I \chi_{mI}] \frac{\omega^2}{c^2}, \tag{7.3.11}$$

$$2k_R k_I = [(1 + \chi_R)\chi_{mI} + (1 + \chi_{mR})\chi_I] \frac{\omega^2}{c^2}, \tag{7.3.12}$$

where χ_{mR} and χ_{mR} are the real and imaginary parts of the magnetic susceptibility, respectively. When ϵ and μ are negative, the right side of (7.3.12) is negative, since χ_I and χ_{mI} are always positive. This implies that k_R and k_I must have the opposite sign. For S positive, corresponding to a wave moving in the $+x$ direction, k_R will be negative, which then implies that k_I is positive, giving decay in the forward direction.

Moving to a full three-dimensional description, Figure 7.9 shows the case of refraction between a medium with Re $\mu\epsilon > 0$ and one with Im $\mu\epsilon < 0$. Conservation of the in-plane components of the k-vectors (cf. Section 3.5.2) implies that the real part of $\vec{k}_2$ must point

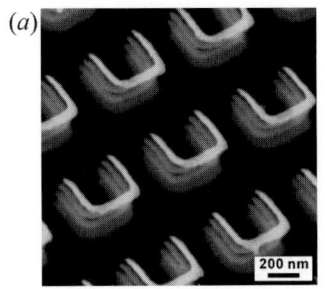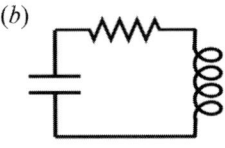

Fig. 7.10   (a) Image of stacked split rings used to create a metamaterial with negative index of refraction. From the Liu *et al.* (2008). (b) Equivalent circuit.

to the right. This implies that $\vec{S}_2$, which is antiparallel to $\vec{k}_2$, will point up and to the left. The conservation of in-plane k-components then gives

$$k_{R1}\sin\theta_1 = -k_{R2}\sin\theta_2. \qquad (7.3.13)$$

Since we write the index of refraction that goes into Snell's law as $n_{\mathrm{eff}} = k_R(c/\omega)$, this means that the second medium acts as if it has negative index of refraction.

Metamaterials. To make a material with a strong magnetic resonance leading to negative index of refraction, a set of split rings can be fabricated that have small size compared to the wavelength of the electromagnetic radiation but large compared to the atoms that make up the material. Figure 7.10(a) shows an example. The equivalent circuit of a resistive split ring is a resonant "tank" circuit, shown in Figure 7.10(b). Charge will oscillate in response to an electromagnetic wave, moving around the ring from one side of the gap to the other, with a natural frequency equal to $\omega_0 = 1/\sqrt{LC}$, leading to an oscillating magnetic field through the ring.

This type of material, in which fabrication is used to make patterned structures with macroscopic properties that differ from the underlying properties, is known as a **metamaterial**. Apart from negative index of refraction, many other novel effects have been demonstrated. For a general review, see Liu and Zhang (2011).

7.4 The Quantum Dipole Oscillator

In Section 7.1. we derived the dielectric response of a classical oscillator. The electrons in a solid belong to bands described by quantum mechanics, however.

Let us consider a quantum mechanical two-level picture for the electrons, as shown in Figure 7.11. These could be any two electron states, but we will label these as v and c for a valence band state and conduction band state, because this model can also work for two electron bands separated by a gap. We will continue to view the electric field as a classical

A quantum mechanical two-level electron system. Where $|v\rangle$ is the valence state (initially full) and $|c\rangle$ is the conduction state (initially empty).

field, however. As discussed in Section 4.4, this is reasonable because a coherent state can be constructed from the bosonic quantum mechanical operators for the electromagnetic field, and a coherent state is equivalent to a classical field.

As in Section 7.1, the polarization of the medium is given by (7.1.8)

$$P = \frac{N}{V} q x, \tag{7.4.1}$$

where x is the displacement of the charge q (we ignore the vector nature of x and P here, but now we treat both as quantum mechanical operators). The polarization at any moment in time is therefore written as

$$\langle P \rangle = \frac{qN}{V} \langle \psi_t | x | \psi_t \rangle, \tag{7.4.2}$$

where we assume there are N identical electrons, and $|\psi_t\rangle$ is the state of each electron at time t. As in Section 7.7, we switch to the interaction representation $|\psi(t)\rangle = e^{iH_0 t/\hbar} |\psi_t\rangle$, where H_0 is the unperturbed Hamiltonian of the system. In the interaction representation, the polarization is

$$\langle P \rangle = \frac{qN}{V} \langle \psi(t) | e^{iH_0 t/\hbar} \, x \, e^{-iH_0 t/\hbar} | \psi(t) \rangle. \tag{7.4.3}$$

Inserting a sum over a complete set of states on each side of the x operator, but restricting ourselves to just two electronic states in our two-level system, we have

$$
\begin{aligned}
\langle P \rangle &= \frac{qN}{V} \langle \psi(t) | e^{iH_0 t/\hbar} \sum_i |i\rangle\langle i| \, x \sum_j |j\rangle\langle j| e^{-iH_0 t/\hbar} | \psi(t) \rangle \\
&= \frac{qN}{V} \Big[\langle \psi(t) | e^{iH_0 t/\hbar} |v\rangle\langle v|x|c\rangle\langle c| e^{-iH_0 t/\hbar} | \psi(t) \rangle \\
&\qquad + \langle \psi(t) | e^{iH_0 t/\hbar} |c\rangle\langle c|x|v\rangle\langle v| e^{-iH_0 t/\hbar} | \psi(t) \rangle \Big] \\
&= \frac{qN}{V} \Big[\langle \psi(t)|v\rangle\langle v|x|c\rangle\langle c|\psi(t)\rangle e^{-i(E_c - E_v)t/\hbar} + \text{c.c.} \Big], \tag{7.4.4}
\end{aligned}
$$

where "c.c." means the complex conjugate of the first term. Here we have used the assumption that $\langle c|x|c\rangle = \langle v|x|v\rangle = 0$, by symmetry; that is, the individual band states are not polarized.

Finding the polarization thus reduces to finding the coefficients $\langle\psi(t)|v\rangle$ and $\langle c|\psi(t)\rangle$ as the electron wave function evolves as a superposition over the two different states. As we have before (cf. Section 4.7), we use time-dependent perturbation theory, defining

$$H = H_0 + V_{int},\tag{7.4.5}$$

where V_{int} is the electron–electromagnetic field interaction Hamiltonian, introduced in Section 5.2.2,

$$V_{int} = -qxE,\tag{7.4.6}$$

ignoring the direction of the electric field. For a coherent driving field, we write $E = E_0\cos\omega t$, or

$$E = \frac{E_0}{2}\left(e^{i\omega t} + e^{-i\omega t}\right).\tag{7.4.7}$$

Note that if we want a Hermitian Hamiltonian, we cannot use the notation $E = E_0 e^{-i\omega t}$.

If we assume the electron starts in the lower state $|v\rangle$, time-dependent perturbation theory then gives, to first order,

$$|\psi(t)\rangle = |\psi(0)\rangle + \frac{1}{i\hbar}\int_{t_0}^{t}dt'\, V_{int}(t')|\psi(0)\rangle,$$

$$= |v\rangle + \frac{1}{i\hbar}\int_{t_0}^{t}dt'\, e^{iH_0 t'/\hbar}\left(-qx\frac{E_0}{2}\left(e^{i\omega t'} + e^{-i\omega t'}\right)\right)e^{-iH_0 t'/\hbar}|v\rangle$$

$$\tag{7.4.8}$$

and therefore, to first order in E_0,

$$\langle v|\psi(t)\rangle = 1,$$

$$\langle c|\psi(t)\rangle = -\frac{1}{2i\hbar}\langle c|x|v\rangle\, qE_0\int_{t_0}^{t}dt'\left(e^{i\omega t'} + e^{-i\omega t'}\right)e^{i(E_c-E_v)t'/\hbar}.$$

$$\tag{7.4.9}$$

In other words, the electron wave function remains mostly that of the valence band, with a little mixing in of the conduction band state.

We want to find the steady-state solution that the system reaches after a long time, and eliminate the effects of memory of the initial state. We therefore assume that the electric field has been turned only slowly, proportional to $e^{\epsilon t}$, and take the limit $\epsilon \to 0$. If we set $t_0 = -\infty$, this kills all memory of the initial state. Writing $E_c - E_v = \hbar\omega_0$, the integral over time is then equal to

$$\langle c|\psi(t)\rangle = \frac{q}{2\hbar}\langle c|x|v\rangle\, E_0\left(\frac{1}{\omega_0 + \omega - i\epsilon}e^{i(\omega_0+\omega)t} + \frac{1}{\omega_0 - \omega - i\epsilon}e^{i(\omega_0-\omega)t}\right).$$

$$\tag{7.4.10}$$

Substituting this into (7.4.4), the polarization is therefore

$$\langle P\rangle = \frac{q^2 N}{V}\frac{|\langle c|x|v\rangle|^2}{2\hbar}E_0\left(\frac{e^{i\omega t}}{\omega_0 + \omega - i\epsilon} + \frac{e^{-i\omega t}}{\omega_0 - \omega - i\epsilon}\right) + \text{c.c.}$$

$$\tag{7.4.11}$$

Note that the oscillating term $e^{-i(E_c - E_v)t/\hbar} = e^{-i\omega_0 t}$ in (7.4.4) exactly cancels the $e^{i\omega_0 t}$ factor in (7.4.10), so that we are left with just terms at oscillation frequency ω.

We can write the matrix element $\langle c|x|v \rangle$ in terms of the quantum mechanical p operator by using the relation (5.2.29) that we deduced in Chapter 5, namely

$$\langle c|x|v \rangle = -i\frac{\langle c|p|v \rangle}{m\omega_0}. \tag{7.4.12}$$

Using the Dirac formula (7.2.2), we then obtain

$$\langle P \rangle = \frac{q^2 N}{mV} \left(\frac{2|\langle c|p|v \rangle|^2}{m\hbar\omega_0} \right) \frac{1}{\omega_0^2 - \omega^2} E_0 \cos \omega t \tag{7.4.13}$$

$$+ \frac{\pi}{2\omega_0} \frac{q^2 N}{mV} \left(\frac{2|\langle c|p|v \rangle|^2}{m\hbar\omega_0} \right) [\delta(\omega_0 - \omega) - \delta(\omega_0 + \omega)]E_0 \sin \omega t.$$

We see that this quantum mechanical system acts as an **oscillator**, with charge motion in response to an electric field. We sometimes fall into thinking of electrons as little billiard balls which must be in one state or the other, but in the presence of a coherent driving field, the electron wave function will exhibit a coherent oscillatory response in which a fraction of the wave function is in each of the two available states. The oscillation of the wave function corresponds to a real motion of the charge, since the spatial distribution of the charge in the two electron states is different. Figure 7.12 shows how the charge moves for a transition between two states, one which has s-like symmetry, and the other which has p-like symmetry. As seen in this figure, when the electron is fully in either one state or the other, the charge distribution is symmetric and there is no net polarization of the atom. When the electron is in an equal superposition $\frac{1}{\sqrt{2}}(\varphi_s + \varphi_p)$ of the two states, however, then the charge has moved to be mostly on the right. The orthogonal superposition of the two states $\frac{1}{\sqrt{2}}(\varphi_s - \varphi_p)$ will give charge mostly on the left. Even if the electron remains mostly in one of the two states, as we have assumed in our calculation here, the introduction of a small superposition of the other state will lead to a net asymmetry of the charge distribution, which then will oscillate in time.

Note that if the two states did not have different symmetries, there would not be an oscillating polarization, because then the $\vec{p}$ operator would not couple the two states.

Comparison to the classical oscillator. We can directly compare the result for the quantum mechanical oscillator with the classical case discussed in Section 7.1 by looking at the response of the classical oscillator to the same form of input electric field, namely $E(t) = E_0 \cos \omega t$. This can be rewritten as

$$E(t) = E_0 \frac{e^{-i\omega t} + e^{i\omega t}}{2}. \tag{7.4.14}$$

In other words, we just need to add a signal with negative frequency to the input of the classical oscillator discussed in Section 7.1. This negative frequency part is just a bookkeeping tool for including the complex conjugate to give a real field.

The classical response is then written as

$$P(t) = \chi \epsilon_0 E = \frac{\epsilon_0 E_0}{2} \left(\chi_R(\omega)e^{-i\omega t} + i\chi_I(\omega)e^{-i\omega t} + \chi_R(-\omega)e^{i\omega t} + i\chi_I(-\omega)e^{i\omega t} \right). \tag{7.4.15}$$

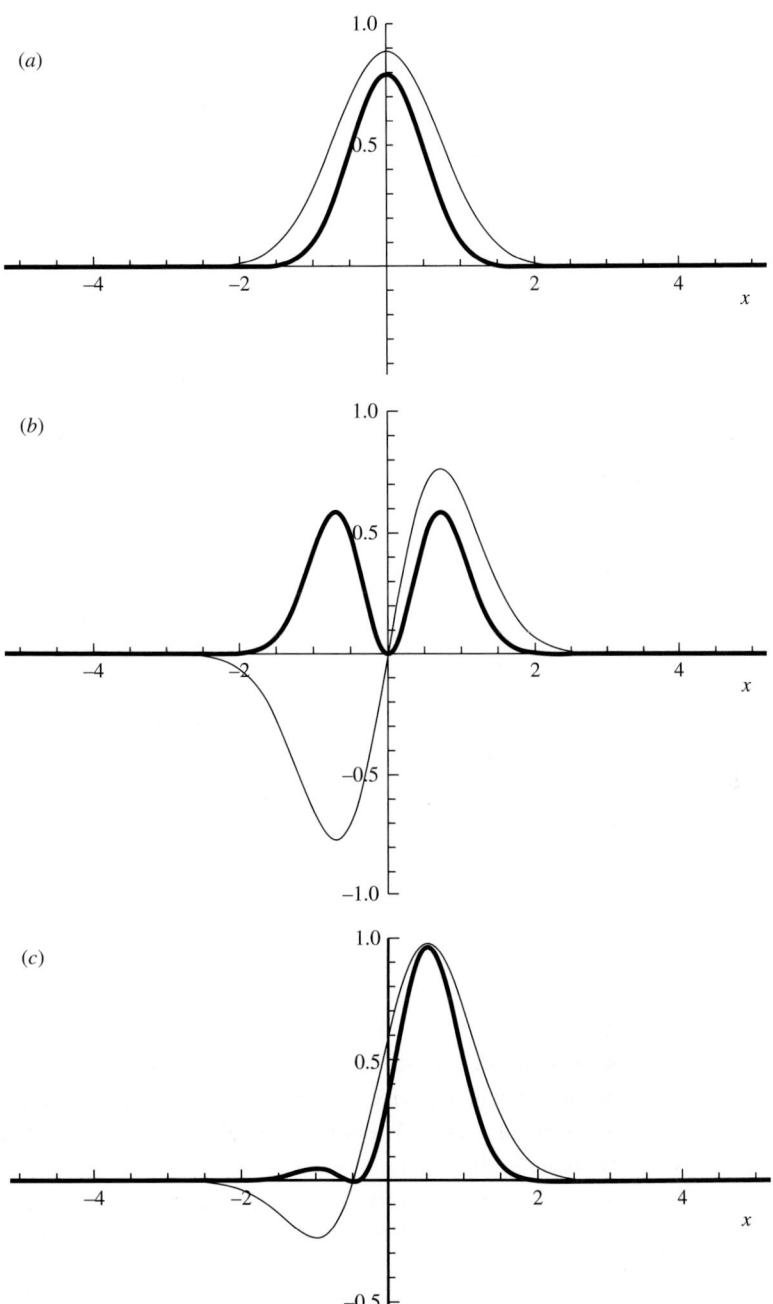

Fig. 7.12 (a) Wave function (thin lines) and charge probability distribution (thick lines) for an s-like atomic orbital ψ_s. (b) Wave function and charge probability distribution for a p-like atomic orbital φ_p. (c) Wave function and charge probability distribution for the superposition $\frac{1}{\sqrt{2}}(\varphi_s + \varphi_p)$.

Using the symmetries $\chi_R(-\omega) = \chi_R(\omega)$ and $\chi_I(-\omega) = -\chi_I(\omega)$, this becomes

$$P(t) = \chi_R(\omega)\epsilon_0 E_0 \cos \omega t + \chi_I(\omega)\epsilon_0 E_0 \sin \omega t. \tag{7.4.16}$$

We see immediately that the two terms of (7.4.13) correspond to the real and imaginary parts of the susceptibility introduced in Section 7.1. We therefore write for the quantum oscillator

$$\chi = \frac{q^2 N}{\epsilon_0 m V} \left(\frac{2|\langle c|p|v \rangle|^2}{m \hbar \omega_0} \right) \frac{1}{\omega_0^2 - \omega^2} \tag{7.4.17}$$

$$+ i \frac{\pi}{2\omega_0} \frac{q^2 N}{\epsilon_0 m V} \left(\frac{2|\langle c|p|v \rangle|^2}{m \hbar \omega_0} \right) (\delta(\omega_0 - \omega) - \delta(\omega_0 + \omega)).$$

The classical result (7.1.11) gives us, in the limit $\Gamma \to 0$,

$$\chi_R = \frac{q^2 N}{\epsilon_0 m V} \frac{1}{\omega_0^2 - \omega^2}. \tag{7.4.18}$$

Comparing this to (7.4.13), we see that the real parts are the same except for the dimensionless term $2|\langle c|\vec{p}|v \rangle|^2/m\hbar\omega_0$, which we have already encountered before, in Sections 1.9.4 and 5.2.1. We see now why it is called the **oscillator strength**; it is the ratio of the response of the quantum oscillator to that of a classical oscillator with the same electron mass and resonance frequency.

The imaginary susceptibility of (7.4.13) also is the same as the classical result times the oscillator strength, in the limit $\Gamma \to 0$. We can see this by using the identity (C.2) in Appendix C, to write

$$\delta(\omega_0 - \omega) - \delta(\omega_0 + \omega) = \lim_{\Gamma \to 0} \frac{1}{\pi} \left[\frac{\Gamma/2}{(\omega_0 - \omega)^2 + (\Gamma/2)^2} - \frac{\Gamma/2}{(\omega_0 + \omega)^2 + (\Gamma/2)^2} \right]$$

$$= \lim_{\Gamma \to 0} \frac{1}{\pi} \frac{\Gamma\omega(2\omega_0)}{(\omega_0^2 - \omega^2)^2 + \Gamma^2(2\omega_0^2 + 2\omega^2)/4 + \Gamma^4/16}$$

$$\simeq \lim_{\Gamma \to 0} \frac{2\omega_0}{\pi} \frac{\Gamma\omega}{(\omega_0^2 - \omega^2)^2 + \Gamma^2\omega_0^2}, \tag{7.4.19}$$

where in the last line we have approximated $\omega^2 \sim \omega_0^2$ for a sharply peaked function. Substituting this into (7.4.13) immediately gives the classical result from (7.1.11) times the oscillator strength.

The imaginary susceptibility, which corresponds to absorption, is just the same as deduced from Fermi's golden rule. We can see this easily by rewriting the positive-frequency term of (7.4.17) as

$$\chi_I = \frac{1}{\omega_0} \left(\frac{2\pi}{\hbar} \frac{q^2 N}{2\epsilon_0 V \omega_0 m^2} |\langle c|p|v \rangle|^2 \, \delta(\omega_0 - \omega) \right). \tag{7.4.20}$$

We define the absorption length in terms of the decay of the intensity as a function of distance, proportional to $e^{-x/l}$. Then, from (7.1.15), the absorption length for intensity is given by $l = (2k_I)^{-1} = (k_R\chi_I)^{-1}$, for $k_R = \omega/c$. For linear absorption, the absorption length is just the speed of light times τ, the absorption time deduced from Fermi's golden rule, and therefore $\chi_I = 1/\omega\tau$. Equation (7.4.20) is therefore equivalent to

$$\frac{1}{\tau} = \omega\chi_I = \frac{2\pi}{\hbar} \frac{q^2 N}{2\epsilon_0 V \omega_0 m^2} |\langle c|p|v\rangle|^2 \, \delta(\omega_0 - \omega). \tag{7.4.21}$$

By comparison, (5.2.10), which we deduced using Fermi's golden rule for the rate of absorption per photon, is in our simplified notation

$$\frac{1}{\tau} = \frac{2\pi}{\hbar} \sum_{\vec{k}} \frac{\hbar e^2}{2\epsilon_0 V \omega m^2} |\langle c|\vec{p}|v\rangle|^2 \, \delta(\hbar\omega_0 - \hbar\omega). \tag{7.4.22}$$

This is the same, assuming that we can treat the band states as N identical oscillators, as we have throughout this section.

Damped oscillators. In all of the above, we have assumed that damping of the quantum transition is negligible. It is also possible to account for damping in a quantum mechanical system. In Section 8.1, we will see how damping can be introduced in terms of an imaginary self-energy of a quantum state, and in Section 8.4 we will see how this leads to line broadening. An imaginary term in the energy will give a decaying term in the time dependence, which will then give a Lorentzian shape of the resonance spectrum.

Exercise 7.4.1   The optical response of a semiconductor can be treated as a set of two-level oscillators, with the number of oscillators at a given frequency equal to the joint density of states (see Section 5.2.1) at that frequency. Compute the frequency dependence of the index of refraction in the spectral range of the band gap of a three-dimensional, direct-gap semiconductor (ignore excitonic effects) and plot this function using a program like Mathematica, using arbitrary constants for the material parameters.

Exercise 7.4.2   Section 8.1 of Chapter 8 will show that we can write an imaginary part of the energy of an electron state, corresponding to $\hbar/\tau$, where τ is the time to scatter out of that state. Put in $E_c - i\Gamma$ for the conduction state energy in (7.4.9), and show that in this case the susceptibility has the same form as that of the classical damped oscillator, found in Section 7.1, in the limit of $\Gamma \ll \omega_0$.

General polarization operator. Above, we wrote the polarization as

$$\langle P \rangle = \frac{qN}{V} \langle \psi_t | x | \psi_t \rangle, \tag{7.4.23}$$

where $|\psi_t\rangle$ is the state of a single oscillator at time t, and we assumed that there were N identical oscillators acting the same way. More generally, we may want an operator for a system in which the different oscillators are not all acting the same, for example due to spatial variation of the electric field, or random scattering processes.

Following the approach of Section 5.1, we write the displacement $\vec{x}$ as the weighted average:

$$\vec{P} = \frac{qN}{V} \int d^3x \, \Psi^\dagger(\vec{x}) \, \vec{x} \, \Psi(\vec{x}), \tag{7.4.24}$$

with

$$\Psi^\dagger(\vec{x}) = \sum_n \varphi_n^*(\vec{x}) b_n^\dagger, \tag{7.4.25}$$

where $\varphi_n(\vec{x})$ is the wave function of the state n in real space and $b_n^\dagger$ is the fermion creation operator for an electron in that state. The states n here can be any type of states, for example, Bloch states labeled by k-vector and a band index, or localized states.

For the model used earlier in this section, we assume that we have an ensemble of identical, localized two-level oscillators of the type shown in Figure 7.11. We then identify the states by their site i and their level, which we limit to two possibilities v or c. The polarization operator then becomes

$$\vec{P} = \frac{Nq}{V} \sum_{ij} \int d^3x \, \vec{x} \, \left(\varphi_{iv}^*(\vec{x}) \varphi_{jc}(\vec{x}) b_{iv}^\dagger b_{jc} + \varphi_{ic}^*(\vec{x}) \varphi_{jv}(\vec{x}) b_{ic}^\dagger b_{jv} + \varphi_{ic}^*(\vec{x}) \varphi_{jc}(\vec{x}) b_{ic}^\dagger b_{jc} \right.$$
$$\left. + \, \varphi_{iv}^*(\vec{x}) \varphi_{jv}(\vec{x}) b_{iv}^\dagger b_{jv} \right). \tag{7.4.26}$$

The wave functions at different sites are assumed to be isolated and therefore orthogonal, so that this becomes

$$\vec{P} = \frac{Nq}{V} \sum_i \left[\left(\int d^3x \, \vec{x} \, \varphi_{iv}^*(\vec{x}) \varphi_{ic}(x) \right) b_{iv}^\dagger b_{ic} + \left(\int d^3x \, x \, \varphi_{ic}^*(\vec{x}) \varphi_{iv}(\vec{x}) \right) b_{ic}^\dagger b_{iv} \right.$$
$$\left. + \left(\int d^3x \, x \, \varphi_{ic}^*(\vec{x}) \varphi_{ic}(\vec{x}) \right) b_{ic}^\dagger b_{ic} + \left(\int d^3x \, x \, \varphi_{iv}^*(\vec{x}) \varphi_{iv}(\vec{x}) \right) b_{iv}^\dagger b_{iv} \right]$$
$$\equiv \frac{Nq}{V} \sum_i \left[\langle c|\vec{x}|v \rangle b_{iv}^\dagger b_{ic} + \langle v|\vec{x}|c \rangle b_{ic}^\dagger b_{iv} + \langle c|\vec{x}|c \rangle b_{ic}^\dagger b_{ic} + \langle v|\vec{x}|v \rangle b_{iv}^\dagger b_{iv} \right].$$
$$\tag{7.4.27}$$

We now assume, as we did before, that the individual states are not polarized, so that only terms coupling different states are nonzero. Thus, we finally have the polarization operator

$$\vec{P} = \frac{Nq}{V} \sum_i \left(\langle c|\vec{x}|v \rangle b_{iv}^\dagger b_{ic} + \langle v|\vec{x}|c \rangle b_{ic}^\dagger b_{iv} \right). \tag{7.4.28}$$

Since the sum is over the different sites i, we can write a local polarization operator for a single site,

$$\vec{P}(\vec{r}_i) = \frac{Nq}{V} \left(\langle c|\vec{x}|v \rangle b_{iv}^\dagger b_{ic} + \langle v|\vec{x}|c \rangle b_{ic}^\dagger b_{iv} \right). \tag{7.4.29}$$

This operator will be useful later in the book as a direct measure of the polarization of any general state of electrons in the system, for example in Sections 7.5.3 and 9.3.

7.5 Polaritons

In addition to transitions between the electron bands, there are other natural oscillators in a solid. We have already seen two of these – optical phonons and excitons. These oscillators will therefore naturally have an effect on the dielectric constant of the medium. The effect on the dielectric constant can lead to new states in the system, known as **polaritons**.

7.5.1 Phonon-Polaritons

As discussed in Section 3.1.2, typical solids with two or more atoms per unit cell have optical phonon modes that correspond, at $k = 0$, to an internal oscillation of the unit cell. The oscillation of two atoms relative to each other clearly will involve the oscillation of charge, so it is natural to expect that the optical phonons will contribute to the electric field susceptibility.

Neglecting the effect of damping, we can write the susceptibility for the optical phonon using (7.1.11) for a charged harmonic oscillator, as

$$\chi_{TO} = \frac{q^2 N}{\epsilon_0 m_r V} \frac{1}{\omega_T^2 - \omega^2}, \tag{7.5.1}$$

where we use the reduced mass m_r as the appropriate mass for a two-atom oscillation. Here ω_T is the optical phonon frequency, and q is the effective charge that participates in the oscillation. (This need not be exactly one electron charge, since the electron cloud around an atom or molecule can become only partially polarized.) The dielectric constant is therefore

$$\epsilon = \epsilon_0 (1 + \chi_{TO} + \chi'), \tag{7.5.2}$$

where χ' is the susceptibility due to all other resonances in the system. We assume that these are all at much higher frequency than the optical phonon (e.g., electronic resonances in a typical semiconductor have energy of a few eV, while optical phonons have typical energies of tens of meV.) Then we can write

$$\epsilon(\omega) = \epsilon(\infty) + \frac{q^2 N}{m_r V} \frac{1}{\omega_T^2 - \omega^2}, \tag{7.5.3}$$

where $\epsilon(\infty)$ is the high-frequency dielectric constant, which includes all terms other than the phonon term. In the range of frequency of the phonons, we assume that $\epsilon(\infty)$ is constant.

Gauss' law says that $\nabla \cdot \vec{D} = 0$, which for a plane wave with $\vec{E} = \vec{E}_0 e^{i(\vec{k} \cdot x - \omega t)}$ becomes

$$\epsilon(\vec{k} \cdot \vec{E}) = 0 \tag{7.5.4}$$

for an isotropic medium. This is automatically satisfied for a transverse wave, but it can also be satisfied if $\vec{k}$ is parallel to $\vec{E}$ but $\epsilon = 0$. This can occur if $(\omega_T^2 - \omega^2) < 0$, in which case the phonon term can cancel the high-frequency term of (7.5.3). When this occurs, the electric field is longitudinal and will couple to longitudinal optical (LO) phonons with frequency ω_L. We define this frequency by the condition

$$0 = \epsilon(\infty) + \frac{q^2 N}{m_r V} \frac{1}{\omega_T^2 - \omega_L^2}, \qquad (7.5.5)$$

which becomes

$$\frac{q^2 N}{m_r V} = \epsilon(\infty)(\omega_L^2 - \omega_T^2). \qquad (7.5.6)$$

Setting this back into (7.5.3), we have

$$\epsilon(\omega) = \epsilon(\infty)\left(1 + \frac{\omega_L^2 - \omega_T^2}{\omega_T^2 - \omega^2}\right). \qquad (7.5.7)$$

At $\omega = 0$, this implies

$$\frac{\epsilon(0)}{\epsilon(\infty)} = \frac{\omega_L^2}{\omega_T^2}. \qquad (7.5.8)$$

This is the **Lyddane–Sachs–Teller** relation. The two optical constants, $\epsilon(0)$ and $\epsilon(\infty)$, are directly related to the optical phonon frequencies, with no other inputs needed. This relation assumes, of course, that the optical phonon is the lowest-frequency resonance that contributes to the dielectric constant, but this is a good assumption in many materials. From group theory considerations, we could already have determined that the longitudinal and transverse optical phonons could have energy splitting. The physical cause of this splitting is the interaction with the electric field, which we have deduced here.

Given the formula (7.5.7), we can determine how the optical modes of the system are affected by the phonon modes. The standard isotropic relation for photons, $\omega = ck/n$, can be written as

$$\omega^2 = \frac{c^2 k^2}{\epsilon(\omega)/\epsilon_0}, \qquad (7.5.9)$$

which becomes

$$\omega^2 = \frac{c^2 k^2}{\epsilon(\infty)/\epsilon_0}\left(\frac{\omega_T^2 - \omega^2}{\omega_L^2 - \omega^2}\right), \qquad (7.5.10)$$

which can then be solved for ω as a function of k. Figure 7.13 shows the two branches of the solution. As seen in this figure, there is a mixing of the photon and phonon states. The lower branch changes from a photon-like to a phonon-like character, while the upper branch changes from a phonon-like to a photon-like character. In the intermediate mixing region, the states have a mixed character. These are known as **polaritons**.

This takes us back to the philosophy of quasiparticles. Photons, phonons, and polaritons are all excitations of the system and none has more fundamental character than the others. If we quantize these states, all three will be bosons with operators that follow the same rules. One thing to keep in mind, however, is that in deducing the dispersion laws we

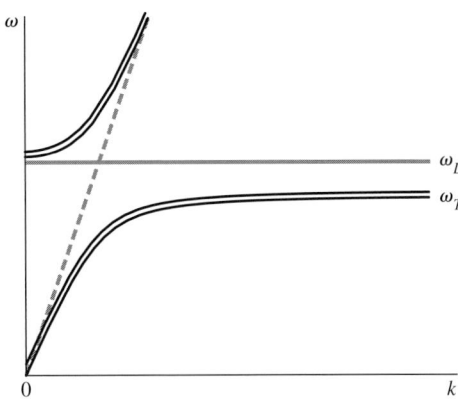

Fig. 7.13   The phonon-polariton dispersion. Degenerate states are shown slightly separated for illustration.

have assumed that the Fermi statistics of the oscillating electrons do not play a role. If the electrons are so dense that the Fermi statistics come into play, then the simple calculation presented here must be adjusted.

The region of polariton mixing only occurs for momenta comparable to a photon momentum, which is very small compared to typical momenta of electrons at thermal velocities. At any wave vector k, there must be a total of five modes, since in the limit of zero mixing there are two transverse photon modes, two transverse phonon modes, and one longitudinal phonon mode. This is shown in Figure 7.13. The two transverse photon modes at low frequency evolve into the two transverse phonon modes, while a new branch of transverse modes appears at ω_L. The longitudinal phonon branch does not mix with the photon modes and is degenerate with the transverse modes at $k = 0$.

Exercise 7.5.1   Use a program like Mathematica to solve (7.5.10) for $\omega(k)$. Setting $c = 1$ and $\epsilon(\infty) = 1$, plot both branches of $\omega(k)$ for various values of the splitting $\omega_L^2 - \omega_T^2$, and show that in the limit of zero splitting there is one single photon branch with $\omega = ck/n$.

Fröhlich interaction. The Lyddane–Sachs–Teller relation allows us to write down a simple form of the matrix element for the Fröhlich electron–phonon interaction discussed in Section 5.1.3. We solve (7.5.6) for q, the effective charge,

$$
\begin{aligned}
|q| &= \left(\frac{\epsilon(\infty) m_r V}{N} \right)^{1/2} \left(\omega_L^2 - \omega_T^2 \right)^{1/2} \\
&= \left(\frac{\epsilon(\infty) m_r V}{N} \right)^{1/2} \left(1 - \frac{\epsilon(\infty)}{\epsilon(0)} \right)^{1/2} \omega_L \\
&= \epsilon(\infty) \left(\frac{m_r V}{N} \right)^{1/2} \left(\frac{1}{\epsilon(\infty)} - \frac{1}{\epsilon(0)} \right)^{1/2} \omega_L.
\end{aligned}
\tag{7.5.11}
$$

The Fröhlich electron–phonon interaction arises because of the electric field created by the polarization of the unit cell during an oscillation. The electric field induced by the oscillation is related to the polarization by the relation $\vec{P} = \chi_{TO}\epsilon_0\vec{E}$. If $\epsilon = 0$, for the case of a longitudinal electric field, then since $\epsilon = \epsilon(\infty) + \chi_{TO}\epsilon_0$, we have $\chi_{TO}\epsilon_0 = -\epsilon(\infty)$, and $\vec{E} = -\vec{P}/\epsilon(\infty)$. The polarization is the effective charge times the displacement, times the number of oscillators per volume, as given in (7.1.8). We therefore have the electric field induced by an optical phonon

$$
\begin{aligned}
\vec{E} &= \frac{1}{\epsilon(\infty)}\frac{N}{V}|q|\vec{u} \\
&= \frac{N}{V}\left(\frac{m_r V}{N}\right)^{1/2}\left(\frac{1}{\epsilon(\infty)} - \frac{1}{\epsilon(0)}\right)^{1/2}\omega_L\vec{u} \\
&= \left(\frac{m_r}{V_{\text{cell}}}\right)^{1/2}\left(\frac{1}{\epsilon(\infty)} - \frac{1}{\epsilon(0)}\right)^{1/2}\omega_L\vec{u},
\end{aligned}
\tag{7.5.12}
$$

where $\vec{u}$ is the atomic displacement and $V_{\text{cell}} = V/N$ is the unit cell volume. We therefore have a constant of proportionality between the atomic displacement and the electric field generated by the displacement, used in Section 5.1.3.

7.5.2 Exciton-Polaritons

Just as phonon motion can correspond to a polarizable oscillator, so also can electronic excitation. All types of electronic excitations can therefore make polaritons, inlcuding plasmons, which will be introduced in Section 8.11.1. In this section, we examine exciton-polaritons. Remember that the exciton is the fundamental quantum of excitation of the electrons and holes. As we saw in Section 7.4, the transition from one band to another of different symmetry corresponds to a time variation of the polarization of the medium.

Just as we did for the optical phonon, we write the dielectric constant in terms of the exciton resonance frequency, ignoring damping, as

$$
\epsilon(\omega) = \epsilon(\infty) + \frac{q^2 N}{mV}\frac{1}{\omega_{\text{ex}}^2 - \omega^2}.
\tag{7.5.13}
$$

In this case, however, the resonance frequency is not a constant. In the Wannier limit, the exciton energy is equal to the semiconductor band-gap energy minus the exciton binding energy, plus the exciton kinetic energy,

$$
\hbar\omega_{\text{ex}} = E_{\text{gap}} - \text{Ry}_{\text{ex}} + \frac{\hbar^2 k^2}{2m_{\text{ex}}}.
\tag{7.5.14}
$$

As in the case of the phonon-polariton, we write

$$
\omega^2 = \frac{c^2 k^2}{\epsilon(\omega)/\epsilon_0},
\tag{7.5.15}
$$

which we can solve for $\omega(k)$. Figure 7.14 shows the two branches of the solution. The exciton states are mixed with the photon states in an exciton-polariton. As in the case of a phonon-polariton, the interaction with the electric field induces a splitting between **transverse** and **longitudinal** excitons.

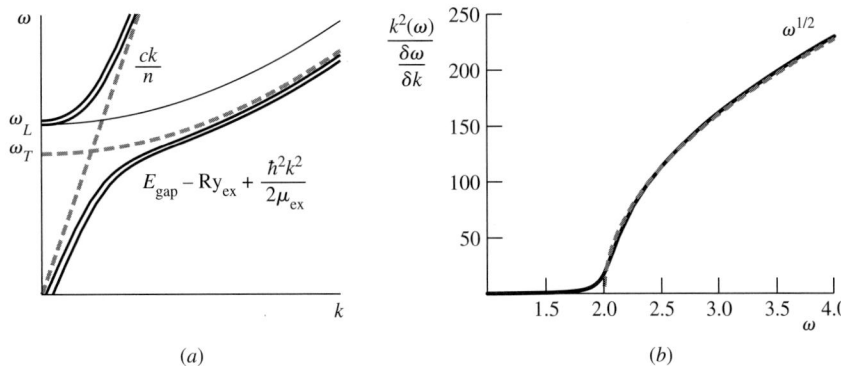

(a)

(b)

Fig. 7.14 (a) The exciton-polariton dispersion. (b) Density of states for the lower exciton-polariton branch.

At first glance, it might seem that all excitons can lose energy by moving down the dispersion curve from the polariton region into the photon-like region. This does not happen so readily, however, because of the effect known as the **polariton bottleneck**. If we plot the density of states as a function of ω, as shown in Figure 7.14(b), we see that the density of states drops dramatically at the bottom of the mixing region. Excitons therefore cannot easily convert into polaritons by phonon emission.

As in the case of phonon-polaritons, there is not a hard-and-fast distinction between what is a photon, what is an exciton, and what is a polariton. All three are bosonic excitations of the medium. In the excitonic range of the spectrum, however, the interactions between the quasiparticles are much stronger. Photons are nearly noninteracting, except for the high-intensity nonlinear effects discussed in Section 7.6.

In Section 5.2.1, we determined the lifetime of electrons in the conduction band due to photon emission using Fermi's golden rule. There, we assumed that the electrons and holes had energy well above the band bottom, so that we could distinguish between the electron and hole states and the photon states as separate eigenstates of the Hamiltonian, and view the coupling between them as a small perturbation of the total Hamiltonian.

In the energy range of exciton-polaritons, however, the coupling between the electron–hole pairs and the photons is strong. The exciton-polaritons are the proper eigenstates of the system including both the photon and electron Hamiltonian, and therefore do not decay into other states, unless there is an additional interaction such as phonon scattering. At the surface of the medium, however, exciton-polaritons couple to photons outside. The coupling is determined by Snell's law using (7.5.13) for the index of refraction in the medium, and the Fresnel equations (3.8.7) for the transmission coefficient, as discussed below. The lifetime of the polariton states is therefore determined by their path length in the medium, that is, the size of the crystal, and the fraction of energy coupled out to photons each time the polaritons reflect off the surface.

Exercise 7.5.2 Use a program like Mathematica to solve equations (7.5.13)–(7.5.15) for $\omega(k)$. Setting $c = 1$, $\epsilon(\infty)/\epsilon_0 = 2$, and $E_g - \mathrm{Ry}_{ex} = 1$, plot both branches of $\omega(k)$ for various values of the coupling constant $q^2 N/mV$ and the exciton mass, and show the mixing region. For a case of reasonably strong mixing, plot the three-dimensional density of states for the lower branch using (1.8.5), and show that the density of states for the photon-like region is well below that of the exciton-like region. To do this, you must solve for $k^2(\omega)$.

Coupling of polaritons to external photons. Every photon inside a transparent medium technically belongs to a polariton branch, since the index of refraction comes from the coupling of light to electronic resonances. There is thus no such thing as "photon emission" from an exciton inside the material, since the coupling to photons is already taken into account by the polariton effect. Photon emission from excitons technically only occurs at surfaces.

To determine the coupling of polariton states inside the medium to light outside, we recall from Section 3.5.2 that we match the components of the k-vector parallel to the surface. The condition for emission of an external photon with energy and momentum conservation is

$$\omega(k^2) = \omega(k_\parallel^2 + k_{\perp i}^2) = \hbar c \sqrt{k_\parallel^2 + k_{\perp o}^2}, \qquad (7.5.16)$$

where $\omega(k^2)$ is the polariton dispersion and $k_{\perp i}$ and $k_{\perp o}$ are the k-components perpendicular to the surface on the inside and outside, respectively. The k-component parallel to the surface, $k_\parallel$, must be the same inside and outside. For a polariton traveling normal to the surface, with $k_\parallel = 0$, this condition can always be satisfied, but for a nonzero angle of incidence this condition gives a severe restriction, so that only a small cone of internal angle of incidence can couple to external photons; polaritons with angle of incidence outside this cone are internally reflected.

The determination of the angle of emission for the external photon can be visualized using Figure 7.15. A polariton traveling at a given angle relative to a surface has well-defined values of $k_\perp$ and $k_\parallel$. The heavy solid lines show the frequencies of the upper and lower polariton branches for a fixed value of $k_\perp$; the heavy dot indicates the lower polariton value for a specific value of $k_\parallel$. To match this energy and momentum, we can vary the external photon $k_\perp$ to slide the photon dispersion (dashed line) up and down until it hits the same spot.

The external photon dispersion curve for a nonzero value of $k_\perp$ always lies above the dispersion for $k_\perp = 0$, which is simply $\omega = c|k_\parallel|$ on this type of plot. There is thus a limited range of values of k for the polaritons that can couple to external photons, known as the "light cone." Polaritons with angle of incidence outside this cone are internally reflected.

Exercise 7.5.3   Determine the maximum internal angle of incidence at which a lower polariton can couple to an external photon, as a function of frequency, for the parameters of an exciton-polariton given in Figure 7.15, using a numerical plotting package such as Mathematica.

7.5.3 Quantum Mechanical Formulation of Polaritons

Although the classical calculations of the previous sections give the correct basic picture for polaritons, there are important quantum-mechanical corrections. Also, the quantum

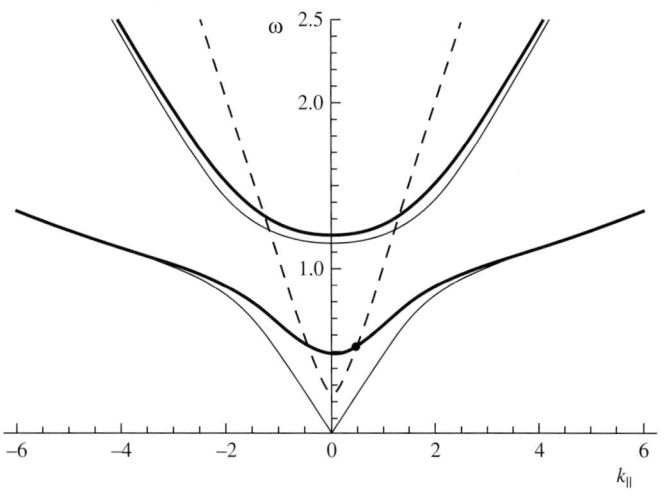

Fig. 7.15 Graphical solution of the energy-momentum conservation condition for photon emission from polaritons. The exciton-polariton dispersion is shown for $E_{\text{gap}} - \text{Ry}_{\text{ex}} = 1$, $\hbar^2/2m = 0.01$, $q^2 N/\epsilon_0 m V = 1$, and $\epsilon(\infty)/\epsilon_0 = 3$. Heavy line: $k_\perp = 1$; thin line: $k_\perp = 0$. The dashed line shows the dispersion for photons outside the medium, with index of refraction $n = 1$, for finite $k_\perp = 0.25$.

mechanical formulation allows us to include selection rules for the transitions between different states.

Phonon-polaritons. We start with the interaction energy of a dipole in an external electric field, $V_{\text{int}} = -q\vec{x}\cdot\vec{E}$, introduced in Section 5.2.2. For a polarization field $\vec{P} = (N/V)q\vec{x}$, the total interaction energy is then

$$V_{\text{int}} = -\int d^3r\, \vec{P}\cdot\vec{E}.$$ (7.5.17)

This formula differs from (5.3.20) because the polarization field here is assumed to be fixed, while in (5.3.20) the polarization is assumed to be induced by the electric field in a linear dielectric medium. The extra factor of $\frac{1}{2}$ in (5.3.20) comes from integration, effectively $\int E\,dE = \frac{1}{2}E^2$. (See Jackson 1999: s. 4.7.)

The polarization, which is proportional to the local displacement, and the electric field can be written in terms of the phonon and photon operators using the formulas from Chapter 4. The electric field, from (4.3.19), is

$$E(\vec{r}) = i\sum_{\vec{k}} \sqrt{\frac{\hbar\omega_k}{2\epsilon_\infty V}} \left(a_{\vec{k}} e^{i\vec{k}\cdot\vec{r}} - a_{\vec{k}}^\dagger e^{-i\vec{k}\cdot\vec{r}} \right),$$ (7.5.18)

where ϵ_∞ is the dielectric constant of the medium, not counting the contribution of the optical phonons.

The polarization field due to the phonons is, using (4.2.27),

$$P(\vec{r}) = \frac{N}{V}qx(r) = \frac{qN}{V}\sqrt{\frac{\hbar}{2m_r N\omega_0}} \sum_{\vec{k}} \left(c_{\vec{k}} e^{i\vec{k}\cdot\vec{r}} + c_{\vec{k}}^\dagger e^{-i\vec{k}\cdot\vec{r}} \right),$$ (7.5.19)

where we have used creation and destruction operators $c_k^\dagger$ and c_k for the phonons to distinguish them from the photon operators, and we assume an optical phonon with constant frequency ω_0. This may be written as

$$P(\vec{r}) = \sqrt{\frac{\hbar \epsilon_\infty \Omega^2}{2V\omega_0}} \sum_{\vec{k}} \left(c_{\vec{k}} e^{i\vec{k}\cdot\vec{r}} + c_{\vec{k}}^\dagger e^{-i\vec{k}\cdot\vec{r}} \right), \tag{7.5.20}$$

where

$$\Omega = \sqrt{\frac{q^2 N}{\epsilon_\infty m_r V}}. \tag{7.5.21}$$

Substituting these formulas for E and P into (7.5.17), we obtain

$$V_{\text{int}} = -i\frac{\hbar\Omega}{2}\sqrt{\frac{\omega_k}{\omega_0}}\frac{1}{V}\sum_{\vec{k},\vec{k}'}\int d^3r \left(a_{\vec{k}} e^{i\vec{k}\cdot\vec{r}} - a_{\vec{k}}^\dagger e^{-i\vec{k}\cdot\vec{r}} \right) \left(c_{\vec{k}'} e^{i\vec{k}'\cdot\vec{r}} + c_{\vec{k}'}^\dagger e^{-i\vec{k}'\cdot\vec{r}} \right). \tag{7.5.22}$$

The integral of the exponential factors over $\vec{r}$ gives us $\delta_{\vec{k},\vec{k}'}$, which eliminates one momentum sum. Adding in the energy of the photon and phonon states, we have the total Hamiltonian

$$H = \sum_{\vec{k}} \left[\hbar\omega_k a_{\vec{k}}^\dagger a_{\vec{k}} + \hbar\omega_0 c_{\vec{k}}^\dagger c_{\vec{k}} - \frac{i}{2}\hbar\Omega\sqrt{\frac{\omega_k}{\omega_0}} \left(a_{\vec{k}} c_{-\vec{k}} + a_{\vec{k}} c_{\vec{k}}^\dagger - a_{\vec{k}}^\dagger c_{\vec{k}} - a_{\vec{k}}^\dagger c_{-\vec{k}}^\dagger \right) \right]. \tag{7.5.23}$$

This Hamiltonian can be simplified by defining new operators that are linear superpositions of the creation and destruction operators that appear in the Hamiltonian. We define the new bosonic operator $\xi_{\vec{k}} = \alpha_k a_{\vec{k}} + \beta_{\vec{k}} c_{\vec{k}} + \gamma_k a_{-\vec{k}}^\dagger + \delta_k c_{-\vec{k}}^\dagger$ and its Hermitian conjugate for the creation operator, with the coefficients $\alpha_k, \dots, \delta_k$ chosen such that

$$H = \sum_{\vec{k}} \hbar\omega\, \xi_{\vec{k}}^\dagger \xi_{\vec{k}}, \tag{7.5.24}$$

which implies

$$[\xi_{\vec{k}}, H] = \hbar\omega \xi_{\vec{k}}. \tag{7.5.25}$$

This is known as **diagonalizing** the Hamiltonian to produce new eigenstates. We will use this method again when we discuss Bose–Einstein condensates in Chapter 11.

The condition (7.5.25) is equivalent to the matrix equation

$$\begin{pmatrix} \omega_k & \frac{i}{2}\Omega\sqrt{\frac{\omega_k}{\omega_0}} & 0 & -\frac{i}{2}\Omega\sqrt{\frac{\omega_k}{\omega_0}} \\ -\frac{i}{2}\Omega\sqrt{\frac{\omega_k}{\omega_0}} & \omega_0 & -\frac{i}{2}\Omega\sqrt{\frac{\omega_k}{\omega_0}} & 0 \\ 0 & -\frac{i}{2}\Omega\sqrt{\frac{\omega_k}{\omega_0}} & -\omega_k & \frac{i}{2}\Omega\sqrt{\frac{\omega_k}{\omega_0}} \\ -\frac{i}{2}\Omega\sqrt{\frac{\omega_k}{\omega_0}} & 0 & -\frac{i}{2}\Omega\sqrt{\frac{\omega_k}{\omega_0}} & -\omega_0 \end{pmatrix} \begin{pmatrix} \alpha_k \\ \beta_k \\ \gamma_k \\ \delta_k \end{pmatrix} = \omega \begin{pmatrix} \alpha_k \\ \beta_k \\ \gamma_k \\ \delta_k \end{pmatrix}. \tag{7.5.26}$$

Following the standard diagonalization procedure of setting the determinant to zero, we write

$$\omega^4 - \omega^2(\omega_0^2 + \omega_k^2) + \omega_k^2\omega_0^2 - \Omega^2\omega_k^2 = 0. \tag{7.5.27}$$

This is equivalent to (7.5.10), with the bare photon frequency $\omega_k = ck/\sqrt{\epsilon_\infty}$ and $\omega_T^2 = \omega_L^2 - \Omega^2$, using (7.5.6) and our definition of Ω above. There are two positive-frequency solutions, which are the same as shown in Figure 7.13.

When $\omega_k = \omega_0$, it is easy to show that in the limit $\omega_0 \gg \Omega$, the energies of the eigenstates are $\hbar\omega = \hbar\omega_0 \pm \hbar\Omega/2$, and the corresponding eigenstates are $\xi_k = (a_k \mp ic_k)/\sqrt{2}$. The energy $\hbar\Omega$ is known as the **Rabi splitting**. Away from the crossover region, when either $\omega_k \ll \omega_0$ or $\omega_k \gg \omega_0$, the eigenmodes correspond to nearly pure photon a_k and phonon c_k operators.

Exciton-polaritons: Frenkel picture. To treat excitons quantum mechanically, the simplest approach is to imagine each exciton as a simple two-level electron oscillator, of the type studied in Section 7.4, localized at some discrete location i. This is equivalent to the Frenkel limit for excitons, discussed in Section 2.3, in which the binding between the electron and hole is so tight that they both lie on the same atom. In other words, the exciton consists of an excitation of a single atom; the excitation can hop from one atom to another.

From Section 7.4, the polarization per unit volume at a location $\vec{r}_i$, due to excitation of an electron state, is

$$\vec{P}(\vec{r}_i) = \frac{Nq}{V}\vec{x}_i = \frac{Nq}{V}\left(\langle c|\vec{x}|v\rangle b_{iv}^\dagger b_{ic} + \langle v|\vec{x}|c\rangle b_{ic}^\dagger b_{iv}\right), \tag{7.5.28}$$

where the $b^\dagger$ and b operators are fermion creation and destruction operators. Note that $\vec{x}_i$ is the displacement of the local oscillator i, and $\vec{r}_i$ is the location of that oscillator in the medium.

The product $b_{ic}^\dagger b_{iv}$ is equivalent to the creation of an exciton at site i, since an electron is removed from the valence band state and put in the conduction band state. We can define the Fourier transform of the exciton operator, that is, an exciton in a plane-wave state with momentum $\vec{k}$, by

$$c_{\vec{k}}^\dagger = \frac{1}{\sqrt{N}}\sum_i e^{i\vec{k}\cdot\vec{r}_i} b_{ic}^\dagger b_{iv}. \tag{7.5.29}$$

The inverse Fourier transform is

$$b_{ic}^\dagger b_{iv} = \frac{1}{\sqrt{N}}\sum_k e^{-i\vec{k}\cdot\vec{r}_i} c_{\vec{k}}^\dagger. \tag{7.5.30}$$

Substituting this into (7.5.28), we obtain

$$\vec{P}(\vec{r}_i) = -i\frac{Nq}{V}\frac{1}{m\omega_0\sqrt{N}}\sum_{\vec{k}}\left(p_{cv}c_{\vec{k}}e^{i\vec{k}\cdot\vec{r}_i} - p_{cv}^*c_{\vec{k}}^\dagger e^{-i\vec{k}\cdot\vec{r}_i}\right), \tag{7.5.31}$$

where we have invoked another relation used in Section 7.4, namely $p_{cv} = \langle c|p|v\rangle = im\omega_0\langle c|x|v\rangle$.

The polarization (7.5.31) has a form similar to the polarization due to phonons (7.5.19), and can be used the same way to derive a polariton spectrum. Although the exciton operators are not purely bosonic, they are approximately so (see Section 11.6), and so in the following we will treat them as pure bosons.

We now drop the subscript i from the spatial coordinate, and treat the medium as continuous. We then have

$$V_{\text{int}} = -\int d^3r\, \vec{P} \cdot \vec{E}$$

$$= \sqrt{\frac{\hbar \omega_k}{2\epsilon V}}\, \frac{\sqrt{N}q}{m\omega_0 V} \sum_{\vec{k},\vec{k}'} \int d^3r \left(a_{\vec{k}}e^{i\vec{k}\cdot\vec{r}} - a_{\vec{k}}^\dagger e^{-i\vec{k}\cdot\vec{r}} \right) \left(p_{cv}^* c_{\vec{k}'}^\dagger e^{-i\vec{k}'\cdot\vec{r}} - p_{cv}c_{\vec{k}'}e^{i\vec{k}'\cdot\vec{r}} \right)$$

$$= \frac{\hbar}{2}\sqrt{\frac{q^2 N}{\epsilon m V}}\sqrt{\frac{2}{m\hbar\omega_0}}\sqrt{\frac{\omega_k}{\omega_0}} \sum_k \left(-p_{cv}a_{\vec{k}}c_{-\vec{k}} + p_{cv}^* a_{\vec{k}}c_{\vec{k}}^\dagger + p_{cv}a_{\vec{k}}^\dagger c_{\vec{k}} - p_{cv}^* a_{\vec{k}}^\dagger c_{-\vec{k}}^\dagger \right).$$

(7.5.32)

As we did for the phonon-polaritons, we diagonalize the total Hamiltonian, writing $\xi_k = \alpha_k a_k + \beta_k c_k + \gamma_k a_{-k}^\dagger + \delta_k c_{-k}^\dagger$. Then

$$H = \sum_k \hbar\omega \xi_k^\dagger \xi_k = \sum_k (\hbar\omega_k a_k^\dagger a_k + \hbar\omega_0 c_k^\dagger c_k) + V_{\text{int}}$$

(7.5.33)

and

$$[\xi_k, H] = \hbar\omega\xi_k.$$

(7.5.34)

Expanding out this commutation relation using bosonic commutation of the operators gives us the matrix equation

$$\begin{pmatrix} \omega_k & \frac{1}{2}\Omega^*\sqrt{\frac{\omega_k}{\omega_0}} & 0 & \frac{1}{2}\Omega\sqrt{\frac{\omega_k}{\omega_0}} \\ \frac{1}{2}\Omega\sqrt{\frac{\omega_k}{\omega_0}} & \omega_0 & \frac{1}{2}\Omega\sqrt{\frac{\omega_k}{\omega_0}} & 0 \\ 0 & -\frac{1}{2}\Omega^*\sqrt{\frac{\omega_k}{\omega_0}} & -\omega_k & -\frac{1}{2}\Omega\sqrt{\frac{\omega_k}{\omega_0}} \\ -\frac{1}{2}\Omega^*\sqrt{\frac{\omega_k}{\omega_0}} & 0 & -\frac{1}{2}\Omega^*\sqrt{\frac{\omega_k}{\omega_0}} & -\omega_0 \end{pmatrix} \begin{pmatrix} \alpha_k \\ \beta_k \\ \gamma_k \\ \delta_k \end{pmatrix} = \omega \begin{pmatrix} \alpha_k \\ \beta_k \\ \gamma_k \\ \delta_k \end{pmatrix},$$

(7.5.35)

with

$$\Omega = \sqrt{\frac{q^2 N}{\epsilon m V}}\, F$$

(7.5.36)

and $F = p_{cv}\sqrt{2/m\hbar\omega_0}$, and $|F|^2$ is the oscillator strength. The determinant is

$$\omega^4 - \omega^2(\omega_0^2 + \omega_k^2) + \omega_k^2\omega_0^2 - |\Omega|^2\omega_k^2 = 0.$$

(7.5.37)

This is equivalent to the standard polariton equation

$$\omega^2 = \frac{c^2 k^2}{\epsilon_\infty} \left(\frac{\omega_T^2 - \omega^2}{\omega_0^2 - \omega^2} \right), \tag{7.5.38}$$

with $\omega_T^2 = \omega_0^2 - \Omega^2$ and $\omega_k = ck/\sqrt{\epsilon_\infty}$. The Frenkel exciton dispersion relation looks just like that for optical phonons, because we have assumed a constant energy for all the excitons.

Exciton-polaritons: Wannier picture. The Wannier limit of excitons consists of the case in which the electron and hole are no longer confined to the same atom, but instead, the electron and hole orbit each other at some distance, with the electron and hole on different atoms that could be quite far from each other. In this case, as discussed in Section 2.3, there will be a Coulomb attraction between the free electron and the hole, which gives rises to bound states of the free electron and hole that are exactly the same as the Rydberg bound states of a hydrogen atom, but with the energy scaled by the dielectric constant of the medium. The mixing with photons in this case is harder to calculate; we cannot do an exact diagonalization as we did for optical phonons and Frenkel excitons. Instead, we will do an approximation that is valid at the relevant optical frequencies.

For this calculation, we use the interaction Hamiltonian in k-space for electron–photon interaction, which we derived in Section 5.2. Dropping the explicit accounting of the photon polarization vector, the interaction Hamiltonian (5.2.8) is

$$V_{\text{int}} = -\frac{q}{m} \sum_{\vec{k},\vec{k}'} \sqrt{\frac{\hbar}{2\epsilon V \omega_k}} \left[\langle c|p|v \rangle \left(a_{\vec{k}} b^\dagger_{c,\vec{k}'+\vec{k}} b_{v,\vec{k}'} + a^\dagger_{\vec{k}} b^\dagger_{c,\vec{k}'-\vec{k}} b_{v,\vec{k}'} \right) \right.$$

$$\left. + \langle c|\vec{p}|v \rangle^* \left(a_{\vec{k}} b^\dagger_{v,\vec{k}'+\vec{k}} b_{c,\vec{k}'} + a^\dagger_{\vec{k}} b^\dagger_{v,\vec{k}'-\vec{k}} b_{c,\vec{k}'} \right) \right]. \tag{7.5.39}$$

To account for the orbit of the electron and hole, the exciton creation operator in the Wannier case includes the wave function of the relative motion of the two particles (see Section 11.6):

$$c^\dagger_{\vec{k}} = \sum_{\vec{k}'} \phi(|\vec{k}/2 - \vec{k}'|) b^\dagger_{c,\vec{k}-\vec{k}'} b_{v,-\vec{k}'}, \tag{7.5.40}$$

where $\phi(k)$ is the Fourier transform of the exciton 1s Bohr wave function,

$$\phi(k) = \frac{1}{\sqrt{V}} \frac{8\sqrt{\pi a^3}}{(1 + a^2 k^2)^2}, \tag{7.5.41}$$

and a is the exciton Bohr radius.

We cannot invert this to write the Hamiltonian in terms of the exciton operators, as we did for Frenkel excitons. Therefore, instead of an exact diagonalization, we write down a matrix on the states $|\,\text{ex} \rangle = c^\dagger_{\vec{k}}|0\rangle$ and $|\text{phot}\rangle = a^\dagger_{\vec{k}}|0\rangle$, of the form

$$\tilde{M} = \begin{pmatrix} \langle \text{ex}|H|\text{ex} \rangle & \langle \text{ex}|H|\text{phot} \rangle \\ \langle \text{phot}|H|\text{ex} \rangle & \langle \text{phot}|H|\text{phot} \rangle \end{pmatrix}, \tag{7.5.42}$$

and diagonalize this matrix.

The off-diagonal term is

$$
\langle \text{ex}|H|\text{phot}\rangle = -\langle 0| \sum_{\vec{k}'''} \phi(|\vec{k}/2 - \vec{k}'''|) b^\dagger_{v,-\vec{k}'''} b_{c,\vec{k}-\vec{k}'''} \frac{q}{m} \sum_{\vec{k}'',\vec{k}'} \sqrt{\frac{\hbar}{2\epsilon V \omega_{k''}}}
$$

$$
\times \left[\langle c|\vec{p}|v\rangle \left(a_{\vec{k}''} b^\dagger_{c,\vec{k}'+\vec{k}''} b_{v,\vec{k}'} + a^\dagger_{\vec{k}''} b^\dagger_{c,\vec{k}'-\vec{k}''} b_{v,\vec{k}'} \right) \right.
$$

$$
\left. + \langle v|\vec{p}|c\rangle \left(a_{\vec{k}''} b^\dagger_{v,\vec{k}'+\vec{k}''} b_{c,\vec{k}'} + a^\dagger_{\vec{k}''} b^\dagger_{v,\vec{k}'-\vec{k}''} b_{c,\vec{k}'} \right) \right] a^\dagger_{\vec{k}} |0\rangle
$$

$$
= -\sum_{\vec{k}'} \phi(|\vec{k}/2 + \vec{k}'|) \frac{q}{m} \sqrt{\frac{\hbar}{2\epsilon V \omega_k}} \langle c|\vec{p}|v\rangle. \tag{7.5.43}
$$

Treating ω_k as nearly constant, we can resolve the sum over k' by converting it to an integral of the exciton wave function:

$$
\frac{V}{(2\pi)^3} \int 2\pi k'^2 dk' d(\cos\theta) \frac{1}{\sqrt{V}} \frac{8\sqrt{\pi a^3}}{(1 + a^2|\vec{k}/2 + \vec{k}'|^2)^2} = \frac{\sqrt{V}}{\sqrt{\pi a^3}}. \tag{7.5.44}
$$

We then have the matrix element

$$
\langle \text{ex}|H|\text{phot}\rangle = \frac{\hbar}{2} \frac{1}{\sqrt{\pi a^3}} \sqrt{\frac{q^2}{\epsilon m}} \sqrt{\frac{\omega_0}{\omega_k}} \langle c|\vec{p}|v\rangle \sqrt{\frac{2}{m\hbar\omega_0}}
$$

$$
= \frac{\hbar}{2} \Omega \sqrt{\frac{\omega_0}{\omega_k}}, \tag{7.5.45}
$$

where

$$
\Omega = \sqrt{\frac{q^2}{\epsilon m \pi a^3}} F, \tag{7.5.46}
$$

and F is defined as above, for Frenkel excitons, in terms of the oscillator strength.

Thus, writing $|\text{polariton}\rangle = \alpha|\text{phot}\rangle + \beta|\text{ex}\rangle$, we have the matrix equation

$$
\begin{pmatrix} \omega_0 & \frac{1}{2}\Omega\sqrt{\omega_0/\omega_k} \\ \frac{1}{2}\Omega^*\sqrt{\omega_0/\omega_k} & \omega_k \end{pmatrix} \begin{pmatrix} \alpha \\ \beta \end{pmatrix} = \omega \begin{pmatrix} \alpha \\ \beta \end{pmatrix} \tag{7.5.47}
$$

which has the determinant equation

$$
\omega_0\omega_k - \omega(\omega_k + \omega_0) + \omega^2 - \frac{1}{4}|\Omega|^2\omega_0/\omega_k = 0. \tag{7.5.48}
$$

This has the same behavior as the Frenkel determinant equation (7.5.37), except near $\omega_k = 0$, where it breaks down.

Note that the Rabi coupling term (7.5.46) depends on the exciton Bohr radius a through the volume factor πa^3. The Frenkel limit can be viewed as the same as the Wannier picture in the case when the exciton Bohr radius becomes equal to the unit cell size of the underlying crystal. In the Frenkel limit, the volume becomes the unit cell size $a_L^3 = V/N$, in which case the Rabi frequency (7.5.46) of the Wannier limit becomes exactly the same as (7.5.36) in the Frenkel limit.

Exercise 7.5.4   Plot the solutions of (7.5.37) and (7.5.48) as functions of ω_k for the same values of $\Omega = 0.1$ and $\omega_0 = 1$, and show that they have the same behavior near the crossing of the photon and exciton branches, but differ far from this crossover point.

7.6 Nonlinear Optics and Photon–Photon Interactions

As discussed in Section 3.5.1, the susceptibility χ is not a constant, but depends on many different factors. In Section 3.6, we allowed for the possibility that χ depends on the electric field, and expanded the polarization as a series in powers of E, as follows:

$$P = \chi \epsilon_0 E + 2\chi^{(2)} E^2 + 4\chi^{(3)} E^3 + \cdots . \tag{7.6.1}$$

Linear response theory truncates this series with the first term. **Nonlinear** theory deals with the higher-order terms. We have already seen in Section 3.5.1 that the higher-order terms are the basis of electro-optics, in which DC electric fields produce a change in the index of refraction of a material. The higher-order terms above, however, also mean that the response can depend on the intensity of the electric field. These higher-order terms are the basis of the active field of nonlinear optics. This field is amply discussed by Yariv (1989). Here we present the main effects that can occur.

The general method used here is called **coupled wave analysis**. We assume that we are given one wave that is dominant, and add additional waves that are perturbations of this main wave in order to find the solution of the wave equation in the medium.

7.6.1 Second-Harmonic Generation and Three-Wave Mixing

For simplicity, let us consider a wave propagating in one dimension, and ignore any effects of anisotropy. Recall, from Section 3.5.1, that Maxwell's wave equation in a neutral medium is (ignoring the direction of the polarization)

$$\frac{\partial^2 E}{\partial x^2} = \epsilon_0 \mu_0 \frac{\partial^2 E}{\partial t^2} + \mu_0 \frac{\partial J}{\partial t}. \tag{7.6.2}$$

Using the expansion (7.6.1), and keeping only terms up to second-order, the current term is given by

$$J = \frac{\partial P}{\partial t} = \frac{\partial}{\partial t} \left(\epsilon_0 \chi E + 2\chi^{(2)} E^2 \right). \tag{7.6.3}$$

Substituting this into the wave equation, we have

$$\frac{\partial^2 E}{\partial x^2} = \frac{(1 + \chi)}{c^2} \frac{\partial^2 E}{\partial t^2} + 4\mu_0 \chi^{(2)} \left(\left(\frac{\partial E}{\partial t} \right)^2 + E \frac{\partial^2 E}{\partial t^2} \right). \tag{7.6.4}$$

Suppose that we assume a typical plane-wave solution of the form $E = E_0 e^{i(kx - \omega t)}$. If we substitute this into (7.6.4), we find that the terms with a second derivative of E have time dependence of $e^{-i\omega t}$, while the square of the first derivative, which appears on the right-hand side, has a time dependence of $e^{-2i\omega t}$. Therefore, we cannot solve the equation for all t with a solution of this form. The nonlinear current term generates terms with frequency 2ω.

Let us instead assume that the solution has the form

$$E = E_0(x)e^{i(kx-\omega t)} + E_2(x)e^{i(k_2 x - \omega_2 t)}, \tag{7.6.5}$$

where $\omega_2 = 2\omega$ and $E_2 \ll E_0$. Since $\chi^{(2)}$ is also small, we drop terms on the right-hand side of (7.6.4) that are proportional to $E_2\chi^{(2)}$. Equation (7.6.4) therefore becomes

$$\frac{\partial^2 E}{\partial x^2} = -\frac{n^2}{c^2}\omega^2 E_0 e^{i(kx-\omega t)} - \frac{n^2}{c^2}\omega_2^2 E_2 e^{i(k_2 x - \omega_2 t)} + 4\mu_0 \chi^{(2)}\left(-2\omega^2 E_0^2 e^{2i(kx-\omega t)}\right). \tag{7.6.6}$$

The spatial second derivative is

$$\frac{\partial^2 E}{\partial x^2} = \frac{\partial^2 E_0}{\partial x^2}e^{i(kx-\omega t)} + 2\frac{\partial E_0}{\partial x}ike^{i(kx-\omega t)} - k^2 E_0 e^{i(kx-\omega t)}$$
$$+ \frac{\partial^2 E_2}{\partial x^2}e^{i(k_2 x - \omega_2 t)} + 2\frac{\partial E_2}{\partial x}ik_2 e^{i(k_2 x - \omega_2 t)} - k_2^2 E_2 e^{i(k_2 x - \omega_2 t)}. \tag{7.6.7}$$

We assume that the overall intensity changes slowly on length scales of the wavelength, that is,

$$k\frac{\partial E}{\partial x} \gg \frac{\partial^2 E}{\partial x^2}, \tag{7.6.8}$$

and therefore we drop the terms with the second derivative of the amplitudes $E_0(x)$ and $E_2(x)$. Then Maxwell's wave equation (7.6.4) is

$$2\frac{\partial E_0}{\partial x}ike^{i(kx-\omega t)} + \left(\frac{n^2}{c^2}\omega^2 - k^2\right)E_0 e^{i(kx-\omega t)} + 2\frac{\partial E_2}{\partial x}ik_2 e^{i(k_2 x - \omega_2 t)}$$
$$+ \left(\frac{n^2}{c^2}\omega_2^2 - k_2^2\right)E_2 e^{i(k_2 x - \omega_2 t)} = 4\mu_0 \chi^{(2)}\left(-2\omega^2 E_0^2 e^{2i(kx-\omega t)}\right). \tag{7.6.9}$$

From linear, first-order optics, we have $\omega = (c/n)k$ and $\omega_2 = (c/n)k_2$, so two terms drop out. There is no wave term with a frequency to match the first term, and therefore we must have $\partial E_0/\partial x \approx 0$ and

$$2\frac{\partial E_2}{\partial x}ik_2 e^{i(k_2 x - \omega_2 t)} = 4\mu_0 \chi^{(2)}\left(-2\omega^2 E_0^2 e^{2i(kx-\omega t)}\right), \tag{7.6.10}$$

or

$$\frac{\partial E_2}{\partial x} = i\sqrt{\frac{\mu_0}{\epsilon}}\chi^{(2)}(2\omega)E_0^2 e^{i(2k-k_2)x}, \tag{7.6.11}$$

where we have identified $\omega_2 = 2\omega$.

If we assume that monochromatic light with frequency ω is sent into a crystal, we have the boundary condition $E_2 = 0$ at $x = 0$. For a crystal of length L, we then have the solution

$$E_2(x) = \sqrt{\frac{\mu_0}{\epsilon}}\chi^{(2)}(2\omega)E_0^2 \frac{e^{i\Delta kL} - 1}{\Delta k}, \tag{7.6.12}$$

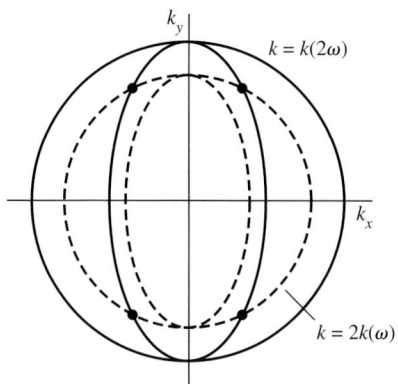

Fig. 7.16   Phase matching in a uniaxial crystal. The solid lines are the slowness surfaces (see Section 3.5.2) for ω and 2ω. The dashed line corresponds to doubling the slowness surface $\vec{k}(\omega)$. In general, phase matching with $\vec{k}(2\omega) = 2\vec{k}(\omega)$ can only occur at certain angles, labeled by the dark points.

where $\Delta k = 2k - k_2$. The square of the amplitude is therefore

$$|E_2(x)|^2 = \frac{\mu_0}{\epsilon}|\chi^{(2)}|^2(2\omega)^2 E_0^4 L^2 \frac{\sin^2 \Delta k L/2}{(\Delta k L/2)^2}. \tag{7.6.13}$$

If $k_2 = 2k$ exactly, the intensity of the frequency-doubled wave grows as L^2. If $\Delta k \neq 0$, the intensity oscillates with period $1/\Delta k$. A mismatch of k can occur if the index of refraction of the frequency-doubled light is not the same as the index of refraction of the input light (which is usually the case.) In certain anisotropic crystals, the angle can be tuned to a special angle at which the indices in a particular $\vec{k}$-direction nearly match, as illustrated in Figure 7.16. This process of getting Δk to approach zero is known as **phase matching**.

This effect of frequency doubling is a generic effect of nonlinear optics, used in numerous laser devices. It is actually a special case of the more general effect of **three-wave mixing**. Instead of the wave solution (7.6.5), we can in general allow solutions with three different frequencies,

$$E = E_1(x)e^{i(k_1 x - \omega_1 t)} + E_2(x)e^{i(k_2 x - \omega_2 t)} + E_3(x)e^{i(k_3 x - \omega_3 t)}. \tag{7.6.14}$$

It is easy to show that for input light with two different frequencies ω_1 and ω_2, light is generated with sum and difference frequencies $\omega_3 = |\omega_1 \pm \omega_2|$. The corresponding phase matching condition is $k_3 = |k_1 \pm k_2|$. If we allow the system to be fully three-dimensional, it is easy to show that the phase matching condition will be $\vec{k}_3 = \pm\vec{k}_1 \pm \vec{k}_2$. In terms of the photon picture, one can view this as a photon–photon interaction in which photon energy $\hbar\omega$ and momentum $\hbar\vec{k}$ must be conserved inside the crystal. The intensity of the output light is proportional to the product of the input intensities, which is equivalent to saying that the rate of the process is proportional to the probability of two input photons colliding.

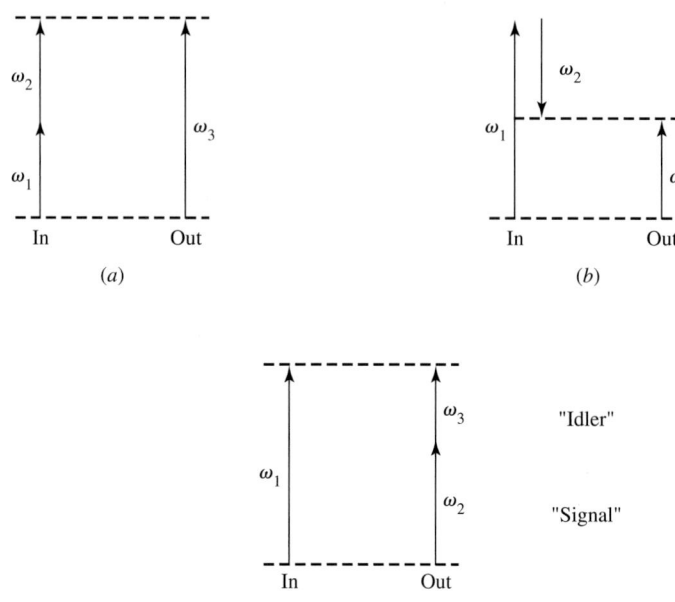

Fig. 7.17 Typical three-wave mixing processes. (a) Sum frequency generation. (b) Difference frequency generation. (c) Parametric down conversion. The output with higher frequency is known as the *signal*, and the other output as the *idler*.

Exercise 7.6.1   (a) Show that the wave equation (7.6.4) allows sum and difference frequency generation in the case of two input waves with different frequencies. Show that the amplitude of the sum wave with frequency $\omega_3 = \omega_1 + \omega_2$ is proportional to the product $E_1 E_2$.

   (b) Show that in the case of three-dimensional waves, the phase matching condition is $\vec{k}_3 = \pm \vec{k}_1 \pm \vec{k}_2$.

   Figure 7.17 illustrates possible three-wave mixing processes. It is also possible to send in light of a single frequency and have it converted into two output beams of lower frequency, such that the two output frequencies sum to the input frequency. This is known as **parametric down conversion**. In this case, the exact breakdown of frequencies will depend on the phase matching condition. Since, in general, phase matching will occur only for one specific angle of an anisotropic crystal, the angle of a crystal can be tuned to allow parametric down conversion only for one specific pair of output frequencies. In this way, a light source with tunable wavelength can be created.

   Second-harmonic generation, and three-wave mixing, can only occur in crystals of proper symmetry. In a centrosymmetric crystal in which all representations have a defined parity, the initial photon state has negative parity, while the final state, as a product of two photon states, must necessarily have positive parity. Therefore, the three-wave mixing process is forbidden. At the surface of a crystal, however, the centrosymmetric symmetry

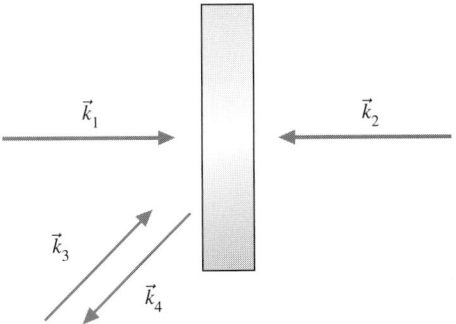

Fig. 7.18   Optical phase conjugation, or a "time-reversing mirror."

will be broken. Therefore, second-harmonic generation is often used as a sensitive probe of surface effects.

7.6.2 Higher-Order Effects

It should not be hard to imagine that if we allow a polarization term $4\chi^{(3)}E^3$ as in (7.6.1), that the same type of coupled wave analysis used in Section 7.6.1 will give rise to *four*-wave mixing, just as the $\chi^{(2)}$ term gave rise to three-wave mixing. Four-wave mixing processes, like three-wave processes, can be viewed as photon–photon interactions in which three photons convert into a fourth photon that conserves energy and momentum inside the medium. Figure 7.18 illustrates the special case of phase conjugation. If two waves are sent into a medium that are time-reversed replicas of each other, e.g., beams $\vec{k}_1$ and $\vec{k}_2$, then a third beam $\vec{k}_3$ hitting this medium will generate a beam $\vec{k}_4$ that is the time-reversed copy of itself. In other words, instead of Snell's law reflection, the medium will exhibit time-reversed reflection, in which the reflected beam travels back along the orginal path.

As we will see in Section 9.4, four-wave mixing is an important tool for understanding scattering processes in condensed matter systems. Another example of a four-wave mixing processes is a **transient grating process**, illustrated in Figure 7.19. Two of the beams interfere with each other, leading to an interference pattern. The nonlinearities of the medium lead to changes of the index of refraction of the medium with period corresponding to that of the interference pattern. This leads to an effective diffraction grating inside the medium. A third beam then diffracts from this diffraction grating.

This picture works because the $\chi^{(3)}$ term in the polarizability can be viewed as an intensity-dependent index of refraction. From the definition of the index of refraction (3.5.13),

$$n(E) = \sqrt{1 + \chi + 2\frac{\chi^{(2)}}{\epsilon_0}E + 4\frac{\chi^{(3)}}{\epsilon_0}E^2 + \cdots}.$$
(7.6.15)

Identifying the low-intensity index of refraction as $n(0) = \sqrt{1 + \chi}$, and expanding the square root as a Taylor series, assuming the higher-order terms are small, we have

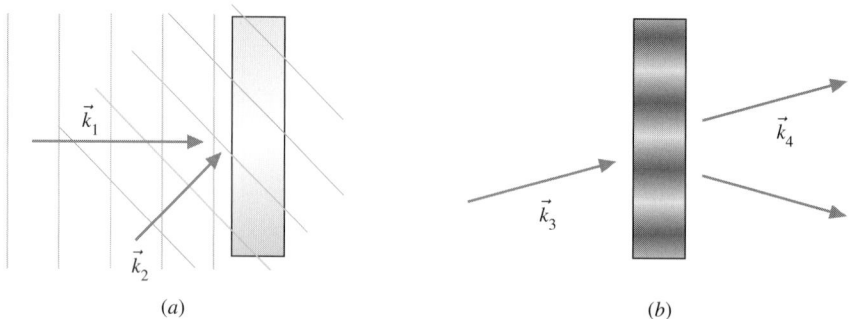

Fig. 7.19   The transient diffraction grating model of four-wave mixing. (a) Two beams entering a medium create an interference pattern. (b) The interference pattern leads to a periodic modulation of the optical properties of the medium. This acts as a grating from which a third beam diffracts.

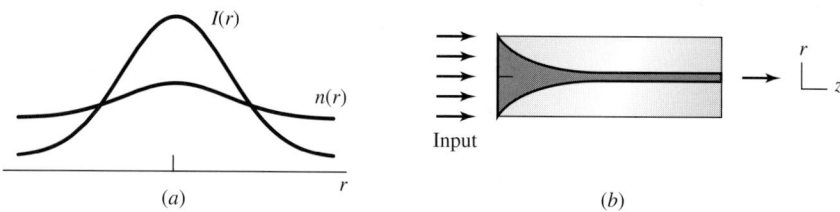

Fig. 7.20   Formation of a spatial soliton by self-focusing. (a) Intensity profile and consequent index of refraction profile, for a beam with Gaussian radial distribution. (b) Self-focusing in a nonlinear medium, which can lead to very tight spatial confinement, in a **filament**. In general, the path of a filament is very unstable to perturbations, and self-focusing can lead to multiple filaments with a chaotic path, like lightning.

$$n(E) \simeq n(0) \left(1 + \frac{\chi^{(2)}}{\epsilon} E + \frac{2\chi^{(3)}}{\epsilon} E^2 + \cdots \right). \tag{7.6.16}$$

As discussed above, in many materials $\chi^{(2)}$ is zero, by symmetry. In this case, since the intensity of electromagnetic radiation is proportional to E^2, it implies

$$n(E) = n(0) + AI, \tag{7.6.17}$$

where I is the intensity and A is a constant. The transient grating picture does not work for all four-wave mixing processes, however; for example, optical phase conjugation is not well described in this model.

The intensity dependence of the index of refraction, which is a generally occurring effect, leads to all manner of nonlinear effects. For example, it can lead to **self focusing** and **spatial solitons**. As illustrated in Figure 7.20, if an optical beam has a Gaussian intensity profile, and the index of refraction increases with increasing intensity, then the index of refraction, and therefore the phase delay, will be largest in the center of the beam. This is exactly the condition for a converging focus, that is, a graded-index lens. The beam will contract, leading to even higher power in the center, and therefore more lensing, in a runaway effect. Very high-power lasers often break up into thin filaments due to this effect.

Exercise 7.6.2   Show explicitly that the diffraction grating made by two beams interfering in a medium will give diffracted output of a third beam consistent with momentum conservation of the photons. In other words, first calculate the interference pattern for two beams with the same wavelength and wave vectors $\vec{k}_1$ and $\vec{k}_2$ interfering in a planar medium. Assume that the medium then has index of refraction proportional to the intensity of the interference pattern in the plane. Last, calculate the Fraunhofer (far-field) diffraction pattern for a third beam with the same wavelength, and wave vector $\vec{k}_3$ impinging on this medium. Show that the output pattern corresponds to beams with $\vec{k}_4 = \pm(\vec{k}_1 - \vec{k}_2) + \vec{k}_3$.

One can also show this for non-equal wavelengths, but then one must perform a volume interference pattern instead of just an interference pattern in a two-dimensional plane.

7.7 Acousto-Optics and Photon–Phonon Interactions

In Section 7.6.1, we allowed χ to depend on the electric field $\vec{E}$ and expanded the polarization as a Taylor series in powers of E. The susceptibility can also depend on the local atomic displacement, that is, on the strain in a material. We can expand the polarization as a series in powers of the displacement u,

$$P = \epsilon_0 \chi E + \epsilon_0 \frac{\partial \chi}{\partial u} u E + \cdots . \tag{7.7.1}$$

Let us assume that the effect of strain on the polarization is linear, and set $\partial \chi / \partial u = C = $ constant. If there is a coherent sound wave in the system, we have

$$u = u_0 e^{i(k_A x - \omega_A t)}, \tag{7.7.2}$$

which means that the one-dimensional Maxwell's wave equation (7.6.2) becomes, to first order in u,

$$\begin{aligned}
\frac{\partial^2 E}{\partial x^2} &= \epsilon_0 \mu_0 \frac{\partial^2 E}{\partial t^2} + \mu_0 \frac{\partial^2 P}{\partial t^2} \\
&= \epsilon_0 \mu_0 (1 + \chi) \frac{\partial^2 E}{\partial t^2} + \mu_0 \epsilon_0 C \left(u \frac{\partial^2 E}{\partial t^2} + 2 \frac{\partial u}{\partial t} \frac{\partial E}{\partial t} + \frac{\partial^2 u}{\partial t^2} E \right) \\
&= \frac{(1 + \chi)}{c^2} \frac{\partial^2 E}{\partial t^2} + \frac{C}{c^2} \frac{1}{2} u_0 e^{i(k_A x - \omega_A t)} \left(-\omega_A^2 E - 2i\omega_A \frac{\partial E}{\partial t} + \frac{\partial^2 E}{\partial t^2} \right).
\end{aligned} \tag{7.7.3}$$

It is easy to see that, as in the case of second-harmomic generation discussed in Section 7.6.1, we cannot find a nontrivial solution of the form $E = E_0 e^{i(kx - \omega t)}$, because we will have terms on the right-hand side that are proportional to $e^{-i(\omega \pm \omega_A)t}$. Instead, we can use the same approach of coupled wave analysis used in Section 7.6.1, and guess a solution of the form

$$E = E_0(x)e^{i(kx-\omega t)} + E_2(x)e^{i(k_2 x - \omega_2)t}. \tag{7.7.4}$$

Following the same procedure as in Section 7.6.1, we obtain the one-dimensional Maxwell's wave equation

$$2\frac{\partial E_0}{\partial x}ike^{i(kx-\omega t)} + \left(\frac{n^2}{c^2}\omega^2 - k^2\right)E_0 e^{i(kx-\omega t)} + 2\frac{\partial E_2}{\partial x}ik_2 e^{i(k_2 x - \omega_2 t)}$$

$$+ \left(\frac{n^2}{c^2}\omega_2^2 - k_2^2\right)E_2 e^{i(k_2 x - \omega_2 t)} = -\frac{C}{c^2}u_0 E_0(\omega_A + \omega)^2 e^{i((k+k_A)x - (\omega+\omega_A)t)}, \tag{7.7.5}$$

which, setting $\omega = (c/n)k$ and $\omega_2 = (c/n)k_2$, gives

$$2\frac{\partial E_2}{\partial x}ik_2 e^{i(k_2 x - \omega_2 t)} = -\frac{C}{c^2}u_0 E_0(\omega_A + \omega)^2 e^{i((k+k_A)x - (\omega+\omega_A)t)} \tag{7.7.6}$$

and

$$|E_2(x)|^2 = \frac{|C|^2}{c^2}\omega_2^2 u_0^2 E_0^2 L^2 \frac{\sin^2 \Delta kL/2}{(\Delta kL/2)^2}, \tag{7.7.7}$$

for the initial condition $E_2 = 0$. As in the case of second-harmonic generation, this implies the existence of a growing electromagnetic wave, with $\omega_2 = \omega + \omega_A$ and the phase matching condition $\Delta k = k + k_A - k_2 = 0$. The intensity of the outgoing wave is proportional to both the sound intensity and the input light intensity. This process is known as **acousto-optic** interaction, and also as **Brillouin scattering**. In a general, three-dimensional system, we will have the phase matching condition $\Delta \vec{k} = \vec{k} + \vec{k}_A - \vec{k}_2 = 0$. Phase matching is generally not a problem in acousto-optic processes, because the frequency of the outgoing electromagnetic wave is nearly the same as the frequency of the input wave, in contrast to the case of frequency doubling, in which the outgoing wave has twice the frequency of the input one.

There are several ways of viewing this physically. One is to see the sound wave as effectively creating a moving diffraction grating due to the effect of strain on the index of refraction, and the outgoing wave as the light diffracted from this grating, as illustrated in Figure 7.21. If we had included the complex conjugate of the acoustic wave, that is,

$$u = u_0 \frac{1}{2}\left[e^{i(\vec{k}_A \cdot \vec{x} - \omega_A t)} + e^{-i(\vec{k}_A \cdot \vec{x} - \omega_A t)}\right], \tag{7.7.8}$$

to make a real-valued wave, then we would have two outgoing electromagnetic waves, corresponding to $\omega_2 = \omega \pm \omega_A$ and $\vec{k}_2 = \vec{k} \pm \vec{k}_A$. By comparison, consider a diffraction grating that consists of medium imprinted with varying index of refraction, that is,

$$\chi = A\cos k_A x = \frac{1}{2}A(e^{i\vec{k}_A \cdot \vec{x}} + e^{-i\vec{k}_A \cdot \vec{x}}). \tag{7.7.9}$$

By the same coupled wave analysis, we will obtain two outgoing waves with $\vec{k}_2 = \vec{k} \pm \vec{k}_A$, but no frequency shift. In the case of the diffraction grating created by the sound wave, there is also a frequency shift, which can viewed as the Doppler shift of the light due to

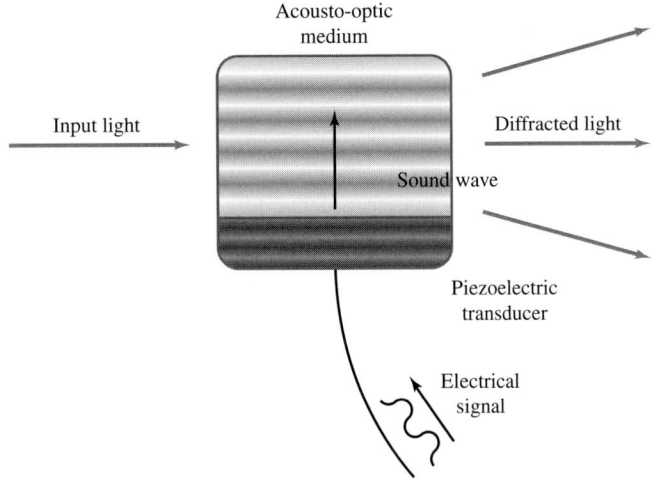

Fig. 7.21　The acousto-optic effect as diffraction from a moving diffraction grating.

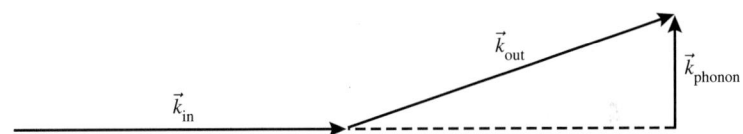

Fig. 7.22　Phonon absorption in the acousto-optic effect.

reflecting from a moving grating. This analysis shows that a standard diffraction grating can also be viewed as a printed-in sound wave. Standard diffraction gratings produce a series of diffracted outgoing waves, not just two outgoing waves, because they are not usually printed as pure sine waves. There will be two outgoing electromagnetic waves for each Fourier component of the printed-in pattern.

Alternatively, we can view the acousto-optic effect as a photon–phonon interaction in which a photon either absorbs or emits a phonon, with the consequent energy and momentum conservation conditions $\hbar\omega_2 = \hbar\omega \pm \hbar\omega_A$ and $\hbar\vec{k}_2 = \hbar\vec{k} \pm \hbar\vec{k}_A$, as illustrated in Figure 7.22. This picture is perhaps the most convenient for calculations.

The acousto-optic interaction is the basis of many modern devices. For example, an acousto-optic device can be used as a frequency shifter for light, since the outgoing, new light beam will have its frequency shifted by ω_A, which can be controlled by a sound generator. Alternatively, an acousto-optic device can be used to shunt a light beam into a new direction, if the phonon momentum is perpendicular to the incoming light beam, as in Figure 7.22. Conservation of momentum implies that the outgoing light beam must have a component of $\vec{k}$ in the direction of the phonon that was absorbed.

Exercise 7.7.1 For a solid with index of refraction 1.6 and speed of sound 5×10^5 cm/s, determine the sound frequency needed to deflect a light beam of $\lambda = 600$ nm by an angle of $30°$.

In the above, we have simply used a constant C to parametrize the effect of strain on the medium. More generally, the effect of strain of the index of refraction is written in terms of the **photoelastic tensor** $\tilde{p}$, defined by the relation

$$\eta_{ij} = \eta_{ij}^{(0)} + \sum_{kl} p_{ijkl}\, \varepsilon_{kl}, \tag{7.7.10}$$

where $\tilde{\eta}$ is the index ellipsoid tensor defined in Section 3.5.3 and $\tilde{\varepsilon}$ is the strain tensor. As with the electro-optic and piezoelectric tensors, the number of independent elements of the photoelastic tensor is determined by the symmetry of the medium.

The photoelastic constant is the standard way of parametrizing the effect of the strain on the optical susceptibility, rather than the constant C we used above. We can obtain an equation equivalent to (7.7.7) by expanding the polarizability in terms of the strain, similar to the way we did in terms of the atomic displacement in (7.7.1). Assuming that we have only a hydrostatic strain ε and a single photoelastic constant p, we write

$$P = \epsilon_0 \chi E + \epsilon_0 \frac{\partial \chi}{\partial \varepsilon} \varepsilon E + \cdots. \tag{7.7.11}$$

From Section 3.5.3 we have

$$\frac{1}{\eta} = 1 + \chi, \tag{7.7.12}$$

and therefore

$$\frac{\partial \chi}{\partial \varepsilon} = -\frac{1}{\eta^2} \frac{\partial \eta}{\partial \varepsilon} = -n^4 \text{p}, \tag{7.7.13}$$

which implies, following the same procedure that led to (7.7.7),

$$|E_2(x)|^2 = \frac{n^8 \text{p}^2}{c^2} \omega_2{}^2 \varepsilon_0^2 E_0^2 L^2 \frac{\sin^2 \Delta kL/2}{(\Delta kL/2)^2}. \tag{7.7.14}$$

We thus have an equation for the intensity of the light diffracted from the acoustic wave as a function of the square of the strain amplitude, that is, the diffracted light intensity is proportional to the acoustic intensity. The acoustic intensity was defined quantitatively in Section 3.8.2 by (3.8.23). We have to be careful about this definition, however, because it depends on how we define the wave. For a real wave, such as defined in (7.7.8), $S_A = \frac{1}{2} Z \omega_A^2 u_0^2 = \frac{1}{2} \rho v^3 \varepsilon_0^2$. For the complex wave defined in (7.7.2), $S_A = \rho v^3 \varepsilon_0^2$.

It is standard to define an acousto-optic figure of merit of a medium,

$$\mathcal{M} = \frac{\text{p}^2 n^6}{\rho v^3}, \tag{7.7.15}$$

where ρ is the mass density of the medium and v is the speed of sound. In terms of this figure of merit, for phase-matched conditions, the light output of an acousto-optic cell then obeys the simple equation

$$\frac{|E_2|^2}{|E_0|^2} = \mathcal{M}(k_2 L)^2 S_A. \tag{7.7.16}$$

Exercise 7.7.2 What acousto-optic figure of merit is needed to get 50% of an input light beam to be redirected into a diffracted beam, if the acoustic intensity is 5 W/cm^2? Use the following parameters: The input light has wavelength 600 nm, the index of refraction of the medium is 1.8, the acousto-optic cell size is 1 mm, the speed of sound in the medium is 5×10^5 cm/s, and the density is 5 g/cm^3. What value does this imply for the photoelastic constant?

7.8 Raman Scattering

Suppose that we do not have a macroscopic coherent sound wave introduced in a medium, but instead just have the thermal vibrations of the medium, that is, a Planck distribution of phonons. We still have the possibility of a photon–phonon interaction in a medium. This is known as **Raman scattering**. This process has been used in a great number of solid state physics studies, and is well reviewed by Cardona and Gunterodt (1982).

We begin with the same expansion (7.7.1),

$$P = \epsilon_0 \chi E + \epsilon_0 \frac{\partial \chi}{\partial u} u E + \cdots , \tag{7.8.1}$$

and again assume that the change of the polarizability with displacement is a constant $C = \partial \chi / \partial u$. In terms of the quantum mechanical operators, the displacement amplitude is given by the formula (4.2.27) (ignoring the polarization angle),

$$u(\vec{x}) = \sum_{\vec{k}} \sqrt{\frac{\hbar}{2\rho V \omega_k}} \left(a_{\vec{k}} e^{i\vec{k}\cdot\vec{x}} + a_{\vec{k}}^\dagger e^{-i\vec{k}\cdot\vec{x}} \right), \tag{7.8.2}$$

and the electric field is given by (4.3.19),

$$E(\vec{x}) = i \sum_{\vec{p}} \sqrt{\frac{\hbar \omega_p}{2\epsilon_0 V}} \left(a_{\vec{p}} e^{i\vec{p}\cdot\vec{x}} - a_{\vec{p}}^\dagger e^{-i\vec{p}\cdot\vec{x}} \right). \tag{7.8.3}$$

(We will use the notation $\vec{p}$ for the photon wave vector to distinguish photon operators from phonon operators with wave vector $\vec{k}$.) Following the standard procedure of Chapter 5, we can determine the rate of phonon absorption or emission by photons, using the interaction Hamiltonian and Fermi's golden rule.

We write the net energy due to polarization (7.5.17),

$$H = -\int_V d^3x \, \vec{P} \cdot \vec{E}, \tag{7.8.4}$$

which yields the interaction Hamiltonian (ignoring the polarization angle)

$$H_{\text{Raman}} = \frac{1}{2} \int_V d^3x \, \epsilon_0 \frac{\partial \chi}{\partial u} \sum_{\vec{k}} \sqrt{\frac{\hbar}{2\rho V \omega_k}} \left(a_{\vec{k}} e^{i\vec{k}\cdot\vec{x}} + a_{\vec{k}}^\dagger e^{-i\vec{k}\cdot\vec{x}} \right)$$

$$\times \sum_{p,p'} \frac{\hbar \sqrt{\omega_p \omega_{p'}}}{2\epsilon_0 V} \left(a_{\vec{p}} e^{i\vec{p}\cdot\vec{x}} - a_{\vec{p}}^\dagger e^{-i\vec{p}\cdot\vec{x}} \right) \left(a_{\vec{p}'} e^{i\vec{p}'\cdot\vec{x}} - a_{\vec{p}'}^\dagger e^{-i\vec{p}'\cdot\vec{x}} \right). \tag{7.8.5}$$

As in the standard procedure of Chapter 5, the integration over x yields a momentum-conserving δ-function that eliminates one of the summations.

If the initial state is a photon with wave vector $\vec{q}$ along with a Planck distribution of phonons, the Raman interaction Hamiltonian will couple this state to a final state with a single photon at a different energy and momentum, which has either emitted or absorbed a phonon from the Planck distribution of thermal phonons. For phonon emission, we write the scattering rate according to Fermi's golden rule (4.7.16), summing over final states with a photon with wave vector $\vec{q}'$ and an extra phonon with wave vector $\vec{k} = \vec{q} - \vec{q}'$, following the same procedure as in Chapter 5 (e.g., Section 5.1.4),

$$
\begin{aligned}
\frac{1}{\tau} &= \frac{2\pi}{\hbar} \sum_f |\langle f|H_{\text{Raman}}|i\rangle|^2 \, \delta(E_f - E_i) \\
&= \frac{2\pi}{\hbar} \left|\frac{\partial \chi}{\partial u}\right|^2 \frac{\hbar}{2\rho\omega_k} \frac{(\hbar\omega_{q'})^2}{4V}(1 + N_k) \, \mathcal{D}(E_{q'}) \\
&= \frac{1}{2\pi} \left|\frac{\partial \chi}{\partial u}\right|^2 \frac{\hbar}{2\rho\omega_k} \frac{\omega_{q'}^4}{(c/n)^3} (1 + N_k),
\end{aligned} \tag{7.8.6}
$$

where we have used the density of states for photons (4.9.7), with n the index of refraction in the medium, and we have assumed that $\omega_q \simeq \omega_{q'}$, that is, the phonon energy is negligible compared to the photon energy. The term $(1 + N_k)$ gives the effects of spontaneous and stimulated emission of phonons and arises from the phonon creation operator $a_{\vec{k}}^\dagger$ in the Hamiltonian (7.8.5).

This phonon emission process is known as **Stokes** Raman scattering. There will also be a light-scattering process corresponding to phonon absorption by a photon, which is known as **Antistokes** Raman scattering, directly proportional to the number of phonons $N_{\vec{q}-\vec{q}'}$. For example, suppose that the relevant phonon is an optical phonon with $\omega_{\vec{k}} = \omega_{TO} \simeq$ constant. The scattered photon frequency for Stokes scattering will just be $\omega_{q'} = \omega_q - \omega_{TO}$, while the frequency for Antistokes scattered light will be $\omega_{q'} = \omega_q + \omega_{TO}$, as illustrated in Figure 7.23. Since the energy and momentum of the photon change, Raman scattering is an inelastic scattering process for photons, in contrast to elastic scattering of photons from impurities, discussed in Section 5.3.

Note that the ratio of the Stokes and the Antistokes Raman emission intensities depends on the temperature. We write

$$
\frac{I_A}{I_S} = \frac{N_{TO}}{1 + N_{TO}} = \frac{\dfrac{1}{e^{\hbar\omega_{TO}/k_B T} - 1}}{1 + \dfrac{1}{e^{\hbar\omega_{TO}/k_B T} - 1}} = e^{-\hbar\omega_{TO}/k_B T}. \tag{7.8.7}
$$

Raman scattering therefore provides an optical thermometer.

In the case of an optical phonon, it is actually not proper to use the mass density ρ in the denominator of (7.8.6). As discussed in Section 5.1.3, the relevant mass is the effective mass of the two-atom phonon oscillation. We should then substitute $\rho \to M_{\text{eff}}/V_{\text{cell}}$.

The Raman scattering rate can be expressed as a cross-section through the relation

$$
\frac{1}{\tau} = \sigma \left(\frac{N}{V}\right) v, \tag{7.8.8}
$$

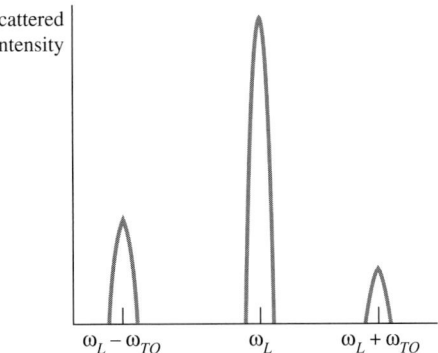

Fig. 7.23 The Raman light scattering spectrum including Stokes and Antistokes Raman scattering, for a single optical phonon. The main peak is elastically scattered input light, which is typically much more intense than the Raman inelastically scattered light.

where σ is the cross-section, N/V is the photon density, and $v = c/n$. We have considered just one photon in the above calculation of (7.8.6), which implies

$$\sigma = \frac{V}{2\pi} \left| \frac{\partial \chi}{\partial u} \right|^2 \frac{\hbar V_{\text{cell}}}{2M_{\text{eff}} \omega_{\vec{q}-\vec{q}'}} |q'|^4 \left(1 + N_{\vec{q}-\vec{q}'} \right) \tag{7.8.9}$$

for Stokes Raman scattering.

Exercise 7.8.1 Derive the Raman scattering cross-section corresponding to the above process, when polarization of the incoming and outgoing photons are taken into account.

Resonant Raman scattering. So far, we have viewed the parameter $C = \partial \chi / \partial u$ as a constant. In principle, however, if there is a distortion of the lattice, then the electronic bands can be shifted, and χ can also depend on the electronic states.

Recall from Section 7.1 that for an ensemble of single oscillators, neglecting damping,

$$\chi = \frac{q^2 N}{MV} \frac{1}{\omega_0^2 - \omega^2}, \tag{7.8.10}$$

where N is the number of oscillators. Both q, the effective charge, and ω_0, the resonance frequency, can depend on u. Therefore, we can write

$$\frac{\partial \chi}{\partial u} = -2 \frac{q^2 N}{MV} \frac{\omega_0}{(\omega_0^2 - \omega^2)^2} \frac{\partial \omega_0}{\partial u} + \frac{N}{MV} \frac{1}{\omega_0^2 - \omega^2} \frac{\partial (q^2)}{\partial u}. \tag{7.8.11}$$

Notice that the resonance in the electronic oscillator gives rise to a resonance in the Raman scattering cross-section. Raman scattering can therefore be a sensitive probe of the electronic states, and not just the phonon states of a system. This fact has been exploited in numerous experiments, reviewed by Cardona and Gunterodt (1982). Figure 7.24 shows an example of resonant Raman scattering.

As seen in Section 7.4, electron bands act as resonators. If a semiconductor is excited by light with photon energy greater than the band gap, one expects Raman lines to be stronger.

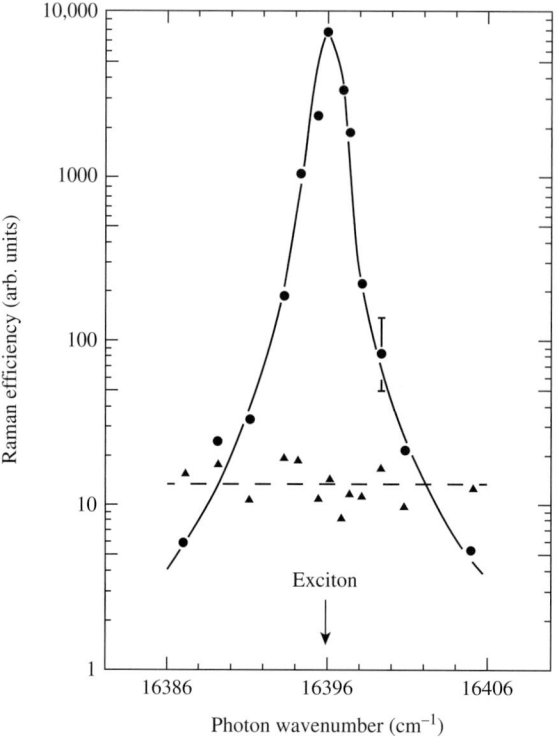

Raman efficiency (arb. units)

Photon wavenumber (cm^{-1})

Fig. 7.24   Circles: resonant Raman scattering efficiency as a function of photon wavenumber (inverse wavelength) for a phonon emission process in the semiconductor cuprous oxide with an allowed optical transition to an exciton state. Triangles: Raman efficiency for a phonon for which the electronic transition is forbidden by symmetry. From Compaan and Cummins (1973).

On the other hand, if a light source has photon energy greater than the band gap, real transitions can occur in which electrons are promoted from the valence band to the conduction band. These free carriers can emit phonons by the normal electron–phonon interaction, dropping down to one optical phonon energy below the laser photon energy. If these free carriers emit light at this energy, the light emission will be hard to tell apart from Raman-scattered light. There is longstanding ambiguity in the literature between Raman scattering in semiconductors and **hot luminescence**, that is, normal light emission from carriers excited directly across the band gap that have not yet equilibrated. Figure 4.10 shows an example of hot luminescence at various times after a laser pulse. These lines have been interpreted as resonant Raman scattering lines sitting on top of normal luminescence, but when the spectrum is time-resolved, as in Figure 4.10, the emission is clearly seen to arise from a single, non-equilibrium distribution. The fit to the theory in Figure 4.10 is the solution to a quantum Boltzmann equation, as discussed in Section 4.8.1, using the appropriate deformation potential matrix elements to describe the exciton–phonon scattering.

In general, Raman scattering is subject to selection rules, discussed in Section 6.10, when the symmetries of the electron bands and phonons are taken into account. In

Section 8.5, we will see how the electron band transition matrix elements enter into the resonant Raman scattering rates.

Raman scattering has great importance for materials research because it can be used to study practically any system. One does not need a transparent material or one that emits luminescence – one can simply bounce enough light off the material and learn about the electronic and vibrational states from the Raman scattering.

References

A. Compaan and H.Z. Cummins, "Resonant quadrupole–dipole Raman scattering at the 1S yellow exciton in Cu_2O," *Phys. Rev. Lett.* **31**, 41 (1973).

M. Cardona and G. Gunterodt, eds. *Light Scattering in Solids*, Vols. 1 and 2, (Springer, 1982).

J.D. Jackson, *Classical Electrodynamics*, 3rd ed. (Wiley, 1999).

N. Liu, H. Guo, L. Fu, S. Kaiser, H. Schweizer, and H. Giessen, "Three-dimensional photonic metamaterials at optical frequencies," *Nature Materials* **7**, 31 (2008).

Yongmin Liu and Xiang Zhang, "Metamaterials: A new frontier of science and technology," *Chem. Soc. Rev.* **40**, 2494 (2011).

H.W. Wyld, *Mathematical Methods for Physics* (W.A. Benjamin, 1976).

A. Yariv, *Quantum Electronics*, 3rd ed. (Wiley, 1989).

8

Many-Body Perturbation Theory

The defining feature of solid state systems is that they involve the interactions of a vast number of particles in such a way that we cannot simply treat one or two particles as an isolated system, as in the case of atomic or particle physics. Since we cannot compute the trajectories of millions of particles at all times, we must find other ways of predicting their behavior.

In Chapter 2, we found that one way to simplify a system is to find its ground state, and then deal only with excitations out of this state. These excitations can then be treated as single particles, known as quasiparticles. As we discussed in Chapter 2, we can hypothesize that perhaps every particle we know is simply an excitation of an underlying field. Are there truly fundamental particles? It would be hard to insist on this for empirical reasons alone, because every quasiparticle acts like a real particle.

In the past few chapters, we have treated these quasiparticles (electrons, phonons, and photons) as eigenstates of the system that are unchanged by interactions. If we allow for interactions, however, then these states cannot be pure eigenstates. When the interactions are weak, we can use perturbation theory and talk in terms of transitions between the original eigenstates, that is, Fermi's golden rule. If the interactions are strong, however, this picture breaks down.

One way to treat a strongly interacting system is to find the proper eigenstates of the whole system. Often, however, it is not possible to find the exact eigenstates, because the mathematics are too complicated. Many-body perturbation theory gives us methods of saying meaningful things about the system even when we do not know the exact eigenstates.

8.1 Higher-Order Time-Dependent Perturbation Theory

In Section 4.7, we deduced a general result of time-dependent perturbation theory,

$$|\psi(t)\rangle = S(t,0)|\psi(0)\rangle = e^{-(i/\hbar)\int_{t_0}^{t} V_{int}(t')dt'/\hbar}|\psi(0)\rangle \tag{8.1.1}$$

$$= \left(1 + \frac{1}{i\hbar}\int_0^t dt' \, V_{int}(t') + \left(\frac{1}{i\hbar}\right)^2 \int_0^t dt' \int_0^{t'} dt'' \, V_{int}(t')V_{int}(t'') + \cdots\right)|\psi(0)\rangle,$$

where $|\psi(t)\rangle$ is the state of the system at time t in the interaction representation. (The full state at time t as determined by the Schrödinger equation is written in our notation as

$|\psi_t\rangle$, and is related to the interaction representation by the relation $|\psi_t\rangle = e^{-iH_0t/\hbar}|\psi(t)\rangle$.) We used the result (8.1.1) to deduce Fermi's golden rule, restricting our attention to the first-order term.

Suppose that we want to include higher orders of perturbation theory. Once again, following the dicussion of Section 4.7, we assume that the Hamiltonian consists of a main part H_0 plus a perturbation term V_{int},

$$H = H_0 + V_{\mathrm{int}}, \tag{8.1.2}$$

in which we assume that the eigenstates of H_0 are known, defined by $H_0|n\rangle = E_n|n\rangle$.

Recalling that $V_{\mathrm{int}}(t) = e^{iH_0t/\hbar}V_{\mathrm{int}}e^{-iH_0t/\hbar}$, and writing the initial state at $t_0 = 0$ as $|i\rangle$, we then have

$$
\begin{aligned}
\langle n|\psi(t)\rangle = \langle n|i\rangle &+ \frac{1}{i\hbar}\int_0^t \langle n|V_{\mathrm{int}}|i\rangle e^{(i/\hbar)(E_n-E_i)t'}\,dt' \\
&+ \left(\frac{1}{i\hbar}\right)^2 \int_0^t dt' \int_0^{t'} dt'' \sum_m \langle n|V_{\mathrm{int}}|m\rangle \langle m|V_{\mathrm{int}}|i\rangle \\
&\times e^{(i/\hbar)(E_n-E_m)t'}\, e^{(i/\hbar)(E_m-E_i)t''} \\
&+ \cdots .
\end{aligned}
\tag{8.1.3}
$$

Here we have inserted a sum over the complete set of eigenstates, $\sum |m\rangle\langle m| = 1$.

Depletion of the initial state. The case when $|n\rangle = |i\rangle$ requires special attention. In this case, assuming V_{int} is time-independent, performing the integrals over time gives the expansion

$$
\begin{aligned}
\langle i|\psi(t)\rangle = 1 &+ \frac{1}{i\hbar}\langle i|V_{\mathrm{int}}|i\rangle t + \frac{1}{2}\frac{1}{(i\hbar)^2}|\langle i|V_{\mathrm{int}}|i\rangle|^2 t^2 \\
&+ \frac{1}{i\hbar}\sum_{m\neq i}|\langle m|V_{\mathrm{int}}|i\rangle|^2 \int_0^t dt'\, e^{(i/\hbar)(E_i-E_m)t'}\frac{\left(e^{(i/\hbar)(E_m-E_i)t'}-1\right)}{E_i - E_m} + \cdots \\
= 1 &+ \frac{1}{i\hbar}\langle i|V_{\mathrm{int}}|i\rangle t + \frac{1}{2}\frac{1}{(i\hbar)^2}|\langle i|V_{\mathrm{int}}|i\rangle|^2 t^2 \\
&+ \frac{1}{i\hbar}\sum_{m\neq i}\frac{|\langle m|V_{\mathrm{int}}|i\rangle|^2}{E_i - E_m}\left(t - \frac{1}{(i/\hbar)(E_i - E_m)}\left(e^{(i/\hbar)(E_i-E_m)t}-1\right)\right) + \cdots .
\end{aligned}
$$

$$\tag{8.1.4}$$

Here we have separated out the terms in the sum over m corresponding to $m = i$ from the rest of the m states in the second-order term.

As in Section 4.7, we assume that we can convert the sum over states m to an integral, with a density of states $\mathcal{D}(E_m)$ nearly constant over a range of energy ΔE_m around E_i. Taking the long-time limit $t \gg \hbar/\Delta E_m$, the time-dependent factor in the last term can be simplified as follows. When $|E_i - E_m|$ is large, the oscillations of the exponential function cancel out in the integral over E_m, and therefore we can neglect the second term in the parentheses, and set the time-dependent factor equal to just t. On the other hand, when

$\epsilon_m \equiv E_i - E_m$ is small, we can expand the exponential term in powers of ϵ_m, which gives us the following factor in the integral:

$$\frac{1}{\epsilon_m}\left(t - \frac{1}{i\epsilon_m/\hbar}\left(1 + (i\epsilon_m/\hbar)t + \frac{1}{2}(i\epsilon_m/\hbar)^2 t^2 + \cdots - 1\right)\right).$$

$$(8.1.5)$$

The leading-order term of this expansion in this case is $-it^2/2\hbar$, which we can rewrite as $-(i/\eta)t$, where we define $\eta = 2\hbar/t$. In the limit $\epsilon_m \to 0$, the real part of this factor is therefore equal to zero, while the imaginary part is large; for all other ϵ_m the factor vanishes. We can then use

$$\delta(\epsilon) = \lim_{\eta \to 0} \frac{1}{\pi}\frac{\eta}{\epsilon^2 + \eta^2},$$

$$(8.1.6)$$

which implies $1/\eta \simeq \pi\delta(\epsilon)$, to write this lowest order term as $i\pi\delta(\epsilon_m)$.

Taking into account the two cases $\epsilon_m = 0$ and $\epsilon_m \neq 0$, we can then rewrite (8.1.5) as

$$\left(\mathcal{P}\frac{1}{\epsilon_m} - i\pi\delta(\epsilon_m)\right)t,$$

$$(8.1.7)$$

where $\mathcal{P}$ is the principal value function. Using the Dirac formula

$$\frac{1}{x + i\eta} = \mathcal{P}\frac{1}{x} - i\pi\delta(x),$$

$$(8.1.8)$$

this is the same as rewriting (8.1.4) as

$$\langle i|\psi(t)\rangle = 1 + \frac{1}{i\hbar}\langle i|V_{\text{int}}|i\rangle t + \frac{1}{2}\frac{1}{(i\hbar)^2}|\langle i|V_{\text{int}}|i\rangle|^2 t^2 + \frac{1}{i\hbar}\sum_{m\neq i}\frac{|\langle m|V_{\text{int}}|i\rangle|^2}{E_i - E_m + i\eta}t + \cdots,$$

$$(8.1.9)$$

where η is vanishingly small when t is large.

If we keep expanding this series, we find that it is the expansion of the exponential,

$$\langle i|\psi(t)\rangle = \exp\left[-(i/\hbar)\left(\langle i|V_{\text{int}}|i\rangle + \sum_{m\neq i}\frac{|\langle m|V_{\text{int}}|i\rangle|^2}{E_i - E_m + i\eta} + \cdots\right)t\right],$$

$$(8.1.10)$$

or, if we return to the Schrödinger representation,

$$\langle i|\psi_t\rangle = \exp\left[-(i/\hbar)\left(E_i + \langle i|V_{\text{int}}|i\rangle + \sum_{m\neq i}\frac{|\langle m|V_{\text{int}}|i\rangle|^2}{E_i - E_m + i\eta} + \cdots\right)t\right].$$

$$(8.1.11)$$

Looking at this result, we see that the effect of the interactions is to give corrections to the unperturbed energy E_i of the state $|i\rangle$. The correction to the energy of a state due to its interactions is known as its **self-energy**. The first-order correction to the energy,

$\Delta^{(1)} = \langle i|V_{\text{int}}|i\rangle$, is called the **mean-field energy**, and has no imaginary part. Using the Dirac formula to rewrite the second-order term, we have

$$\sum_{m\neq i} \frac{|\langle m|V_{\text{int}}|i\rangle|^2}{E_i - E_m + i\eta} = \mathcal{P}\left(\sum_{m\neq i} \frac{|\langle m|V_{\text{int}}|i\rangle|^2}{E_i - E_m}\right) - i\pi \sum_{m\neq i} |\langle m|V_{\text{int}}|i\rangle|^2 \delta(E_i - E_m)$$

$$\equiv \Delta^{(2)} - i\Gamma^{(2)}. \tag{8.1.12}$$

We define $\Delta^{(2)}$ as the real, second-order self-energy of the state $|i\rangle$, and $\Gamma^{(2)}$ as the second-order imaginary self-energy of the state $|i\rangle$.

The second-order imaginary self-energy has the same form as the total scattering rate determined by Fermi's golden rule, from Section 4.7,

$$\frac{1}{\tau} = \frac{2\pi}{\hbar} \sum_{m\neq i} |\langle m|V_{\text{int}}|i\rangle|^2 \delta(E_i - E_m). \tag{8.1.13}$$

We can therefore write, to second order,

$$\langle i|\psi_t\rangle = e^{-(i/\hbar)(E_i + \Delta^{(1)} + \Delta^{(2)})t - \Gamma^{(2)}t/\hbar}$$

$$= e^{-(i/\hbar)E_i't} e^{-t/2\tau}, \tag{8.1.14}$$

where E_i' is the adjusted real energy of the state, and

$$|\langle i|\psi_t\rangle|^2 = e^{-t/\tau}, \tag{8.1.15}$$

which is consistent with the interpretation of Fermi's golden rule that the total rate of out-scattering gives the total depletion of the initial state. It may seem odd to introduce the idea of imaginary energy, which comes about because of the infinitesimal term $i\eta$ in the denominator of the second-order term, but without this term, we would not have self-consistency, since we know from Fermi's golden rule that the initial state $|i\rangle$ is depleted, but we only get depletion of state $|i\rangle$ in (8.1.15) if there is an imaginary self-energy term. To put it another way, the state $|i\rangle$ is no longer an eigenstate of the full system when interactions are taken into account, and therefore it does not have a real eigenvalue.

Comparing the real and imaginary terms of (8.1.12), we see that both involve the same scattering matrix elements. The only difference is that the imaginary part includes only energy-conserving transitions, which we call **real** transitions, while the real part is derived from transitions that violate energy conservation, which we call **virtual** transitions. Again: *real* self-energy corrections come from *virtual* transitions, and **imaginary** self-energy comes from *real* transistions, in second order!

In virtual transitions, there is an energy denominator that acts to suppress violation of energy conservation but does not strictly prevent it. Because there is an energy uncertainty principle $\Delta E \delta t \geq \hbar$, we can visualize this type of process as one in which energy conservation is violated just for a short time during a quick transition to some state and back from it.

Transitions to other states. We can now use the same approach to generalize Fermi's golden rule to higher orders of perturbation theory. Going back to the expansion (8.1.3), and taking now the case when $n \neq i$, we perform the integrals over t'' in each term, to obtain

$$
\langle n|\psi(t)\rangle = \frac{1}{i\hbar}\langle n|V_{\text{int}}|i\rangle \int_0^t dt'\, e^{(i/\hbar)(E_n-E_i)t'} \left(1 - \frac{i}{\hbar}\langle i|V_{\text{int}}|i\rangle t' + \cdots\right)
$$
$$
+\frac{1}{i\hbar} \sum_{m\neq i} \langle n|V_{\text{int}}|m\rangle \langle m|V_{\text{int}}|i\rangle
$$
$$
\times \int_0^t dt'\, e^{(i/\hbar)(E_n-E_m)t'} \frac{\left(e^{(i/\hbar)(E_m-E_i)t'}-1\right)}{E_i-E_m}
$$
$$
+\cdots, \tag{8.1.16}
$$

where we have once again treated the $m = i$ term in the sum separately. The mean-field term $\langle i|V|i\rangle$ and higher-order terms with this mean-field energy sum up to give an exponential factor corresponding to a corrected self-energy for the state $|i\rangle$, just as above. Ignoring these corrections, either because they are small or because they are already taken into account in the calculation of the energy E_i, the integral over t' gives

$$
\langle n|\psi(t)\rangle = \langle n|V_{\text{int}}|i\rangle \frac{\left(e^{(i/\hbar)(E_n-E_i)t}-1\right)}{E_i-E_n}
$$
$$
+ \sum_{m\neq i} \frac{\langle n|V_{\text{int}}|m\rangle \langle m|V_{\text{int}}|i\rangle}{E_i-E_m}
$$
$$
\times \left(\frac{\left(e^{(i/\hbar)(E_n-E_i)t}-1\right)}{E_i-E_n} - \frac{\left(e^{(i/\hbar)(E_n-E_m)t}-1\right)}{E_m-E_n}\right)
$$
$$
+\cdots. \tag{8.1.17}
$$

We can rewrite the last time-dependent factor to simplify the second-order term in the same way we did for the self-energy calculation above. We first factor out the term that does not depend on E_m, to obtain

$$
\frac{1}{E_i-E_m}\left(1 - \frac{\left(e^{(i/\hbar)(E_n-E_m)t}-1\right)}{E_m-E_n}\frac{E_i-E_n}{\left(e^{(i/\hbar)(E_n-E_i)t}-1\right)}\right)\frac{\left(e^{(i/\hbar)(E_n-E_i)t}-1\right)}{E_i-E_n}.
$$
$$
\tag{8.1.18}
$$

As we have already seen in deducing Fermi's golden rule in Section 4.7, for large t the factor

$$
\frac{\left(e^{(i/\hbar)(E_n-E_i)t}-1\right)}{E_i-E_n} \tag{8.1.19}
$$

is strongly peaked near $E_n = E_i$, where it is equal to $t/i\hbar$. Assuming $E_n \approx E_i$, the factor that depends on E_m therefore becomes, for small $\epsilon_m = E_i - E_m$,

$$\frac{1}{\epsilon_m}\left(1 + \frac{\left(e^{(i/\hbar)\epsilon_m t} - 1\right)}{\epsilon_m}\frac{1}{t/i\hbar}\right)$$

$$= \frac{1}{\epsilon_m}\left(1 + \frac{\left(1 + (i/\hbar)\epsilon_m t + \frac{1}{2}((i/\hbar)\epsilon_m t)^2 + \cdots - 1\right)}{\epsilon_m}\frac{1}{t/i\hbar}\right),$$

$$(8.1.20)$$

which, to leading order, is equal to $-it/2\hbar = -i/\eta$, using our definition of η above. For large $|\epsilon_m|$, the term vanishes due to the oscillating exponential. Following the same procedure as for the self-energy calculation above, we therefore have

$$\langle n|\psi(t)\rangle = \left(\langle n|V_{\text{int}}|i\rangle + \sum_{m\neq i}\frac{\langle n|V_{\text{int}}|m\rangle\langle m|V_{\text{int}}|i\rangle}{E_i - E_m + i\eta} + \cdots\right)\frac{e^{(i/\hbar)(E_n - E_i)t} - 1}{E_i - E_n}.$$

$$(8.1.21)$$

Switching back to the time-dependent Schrödinger representation, using $\langle n|\psi_t\rangle = \langle n|e^{-iH_0 t}|\psi(t)\rangle = e^{-iE_n t}\langle n|\psi_t\rangle$, and using the same assumptions as used in Section 4.7 for deducing Fermi's golden rule, we deduce the rate for transitions from $|i\rangle$ to $|n\rangle$,

$$\boxed{\frac{\partial}{\partial t}|\langle n|\psi_t\rangle|^2 = \frac{2\pi}{\hbar}\left|\langle n|V_{\text{int}}|i\rangle + \sum_{m\neq i}\frac{\langle n|V_{\text{int}}|m\rangle\langle m|V_{\text{int}}|i\rangle}{E_i - E_m + i\eta} + \cdots\right|^2 \delta(E_i - E_n).}$$

$$(8.1.22)$$

This is Fermi's golden rule with second-order correction. Once again, transitions involving virtual intermediate states come in to play. For example, for the second-order amplitude in (8.1.22),

$$\mathcal{P}\sum_{m\neq i}\frac{\langle n|V_{\text{int}}|m\rangle\langle m|V_{\text{int}}|i\rangle}{E_i - E_m},$$

$$(8.1.23)$$

we view the system as virtually jumping to an intermediate state $|m\rangle$ in violation of energy conservation but quickly returning to another state $|n\rangle$ that does satisfy energy conservation, as illustrated in Figure 8.1.

Connection to time-independent perturbation theory. Although we have worked fairly hard to justify the time-dependent formulas (8.1.11) and (8.1.22), we find that when all the dust settles, we only need to use the results of time-*independent* perturbation theory (see Appendix E), and add a small term $i\eta$ in the denominators. This is also known as **Rayleigh–Schrödinger** perturbation theory. (For an alternative derivation of these results, see Combescot 2001.)

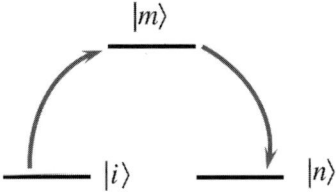

Fig. 8.1 The second-order transition as a virtual transition to an intermediate state.

In other words, the transition rate (8.1.22) is the same as the standard Fermi's golden rule, with the matrix element

$$\langle n|\psi_t\rangle = \langle n|V_{\text{int}}|i'\rangle, \tag{8.1.24}$$

where $|i'\rangle$ is the adjusted eigenstate from time-independent perturbation theory,

$$|i'\rangle = |i\rangle + \sum_{m\neq i} \frac{|m\rangle\langle m|V_{\text{int}}|i\rangle}{E_i - E_m + i\eta} + \cdots. \tag{8.1.25}$$

In the same way, the inner product $\langle i|\psi_t\rangle$ is given by

$$\langle i|\psi_t\rangle = e^{-i\tilde{E}_i t/\hbar}, \tag{8.1.26}$$

where the energy $\tilde{E}_i$ is given by the standard time-independent perturbation series

$$\tilde{E}_i = E_i + \langle i|V_{\text{int}}|i\rangle + \sum_{m\neq i} \frac{|\langle m|V_{\text{int}}|i\rangle|^2}{E_i - E_m + i\eta} + \cdots. \tag{8.1.27}$$

Ultimately, it turns out that in time-dependent perturbation theory one can adopt a very simple method: drop terms that arise from the lower bound of all the time integrals except the final one, and add a small imaginary term to any energy denominators. This is equivalent to dropping the memory of the initial state of the wave function, as we did in Section 7.4, but there it was justified differently, in what is sometimes called the **adiabatic** approximation. We could not adopt that approach here because we started with a definite initial state of the system at $t = 0$; if we dropped the memory of the initial state everywhere, we would have to drop it in the factor (8.1.19) also, in which case we would not obtain the correct Fermi's golden rule. Note also that the sign of the small imaginary term in the denominator is important here; if it were negative the population in the initial state would grow, according to (8.1.12)–(8.1.15).

Exercise 8.1.1   As discussed above, in time-dependent perturbation theory one can simply drop terms from the lower bounds of all time integrals except the final one, and add a small imaginary term in the denominator consistent with causality. Show that following this procedure for the real part of the third order of the expansion (8.1.4) gives the same third-order term as time-independent Rayleigh–Schrödinger theory,

$$\Delta_n^{(3)} = \sum_{m,m'\neq i} \langle i|V_{\text{int}}|m\rangle \frac{\langle m|V_{\text{int}}|m'\rangle\langle m'|V_{\text{int}}|i\rangle}{(E_i - E_m)(E_i - E_{m'})}$$

$$- \sum_{m\neq i} \langle i|V_{\text{int}}|m\rangle \frac{\langle m|V_{\text{int}}|i\rangle\langle i|V_{\text{int}}|i\rangle}{(E_i - E_m)^2}. \tag{8.1.28}$$

Exercise 8.1.2   Deduce the existence of the second-order susceptibility $\chi^{(2)}$ in (7.6.1), which gives a response proportional to the square of the electric field E^2, following the procedure of Section 7.4 for $P \propto \langle\psi_t|x|\psi_t\rangle$, but using the second-order perturbation theory method introduced here to obtain $\langle c|\psi(t)\rangle$, for a transition from state $|v\rangle$ to state $|c\rangle$ via one intermediate state $|m\rangle$. Assume that transitions between all three states are dipole-allowed (which means there cannot be inversion symmetry), and for simplicity, assume $E = E_0 e^{-i\omega t}$.

8.2 Polarons

Let us now use a specific interaction term from Chapter 5 for the perturbation energy V_{int} in the above formalism. We have from Section 5.1.3 for the electron–phonon Fröhlich interaction, neglecting polarization,

$$H_{\text{Fr}} = \frac{C_{\text{Fr}}}{\sqrt{V}} \sum_{\vec{k}} \sum_{\vec{k}_1} \frac{1}{k} \; i \left(a_{\vec{k}} b^{\dagger}_{n,\vec{k}_1+\vec{k}} b_{n\vec{k}_1} - a^{\dagger}_{\vec{k}} b^{\dagger}_{n,\vec{k}_1-\vec{k}} b_{n\vec{k}_1} \right), \tag{8.2.1}$$

where C_{Fr} is a constant which depends on the phonon frequency and the dielectric function.

If the initial state $|i\rangle$ is a Fock state with an electron in band n at wave vector $\vec{k}$ and $n_{\vec{p}}$ phonons at wave vector $\vec{p}$, and the final state $|m\rangle$ has an electron at wave vector $\vec{k} + \vec{p}$ and one less phonon, the matrix element $\langle m | V_{\text{int}} | i \rangle$ is

$$\langle m | V_{\text{int}} | i \rangle = i \frac{C_{\text{Fr}}}{\sqrt{V}} \frac{\sqrt{N_{\vec{p}}}}{p}. \tag{8.2.2}$$

If a phonon is emitted, we have the same matrix element except that we must multiply by $\sqrt{1 + N_{\vec{p}}}$. The second-order correction of the real self-energy due to the electron–phonon Fröhlich interaction is therefore

$$\Delta_{\vec{k}}^{(2)} = \frac{1}{V} \sum_{\vec{p}} \frac{C_{\text{Fr}}^2 N_{\vec{p}}/p^2}{E_{\vec{k}} + \hbar\omega_{LO} - E_{\vec{k}+\vec{p}}} + \frac{1}{V} \sum_{\vec{p}} \frac{C_{\text{Fr}}^2 (1 + N_{\vec{p}})/p^2}{E_{\vec{k}} - \hbar\omega_{LO} - E_{\vec{k}-\vec{p}}}, \tag{8.2.3}$$

where the first term accounts for phonon absorption and the second term accounts for phonon emission. For a thermal distribution of phonons, the occupation number N_p is just given by the Planck formula,

$$\bar{N}_p = \frac{1}{e^{\hbar\omega_p/k_B T} - 1}. \tag{8.2.4}$$

There is no first-order or third-order correction due to electron–phonon scattering, because for every phonon creation operator there must be a phonon destruction operator to return the system to its initial state $|i\rangle$.

At low temperature (i.e., $k_B T \ll \hbar\omega_{LO}$), we can approximate $N_p \approx 0$, which means we can ignore phonon absorption and stimulated emission of phonons. Expanding the denominator of the second term, to account for spontaneous phonon emission, gives

$$E_{\vec{k}} - \hbar\omega_{LO} - E_{\vec{k}-\vec{p}} = \frac{\hbar^2 k^2}{2m} - \hbar\omega_{LO} - \left(\frac{\hbar^2 k^2}{2m} - \frac{\hbar^2 \vec{k} \cdot \vec{p}}{m} + \frac{\hbar^2 p^2}{2m} \right)$$

$$= \frac{\hbar^2 \vec{k} \cdot \vec{p}}{m} - \frac{\hbar^2 p^2}{2m} - \hbar\omega_{LO}, \tag{8.2.5}$$

where m is the effective mass of the electrons. When $k = 0$, the denominator is therefore negative, and therefore the second-order correction to the energy is negative. The electron ground state is pushed lower (i.e., red-shifted), by emission of virtual phonons with negative energy which couple the ground state to higher electronic states.

When $k \neq 0$, there will still be a correction to the electron energy. Converting the sum over $\vec{p}$ to an integral, we have

$$\Delta_{\vec{k}}^{(2)} = \frac{1}{(2\pi)^2} \int dp \, d(\cos\theta) \frac{C_{Fr}^2}{(kp\hbar^2/m)\cos\theta - \hbar^2 p^2/2m - \hbar\omega_{LO}}. \tag{8.2.6}$$

The optical phonon energy $\hbar\omega_{LO}$ is generally much larger than the electron energy, which is comparable to $k_B T$, at low temperature. We can therefore write the integral over angle as

$$\int \frac{1}{A\cos\theta - B} d(\cos\theta), \tag{8.2.7}$$

where $A = kp(\hbar^2/m) \ll B$. This equals

$$\int_{-1}^{1} \frac{1}{Ax - B} dx = \frac{1}{A} \log\left(\frac{B-A}{B+A}\right), \tag{8.2.8}$$

which can be expanded as a Taylor series,

$$\frac{1}{A} \log\left(\frac{B-A}{B+A}\right) = -\frac{2}{B} - \frac{2}{3B^3} A^2 + \cdots. \tag{8.2.9}$$

The A^2 term, which is proportional to k^2, has the same sign as the $k = 0$ energy correction. The red shift is therefore *greater* for larger k, as illustrated in Figure 8.2. Since the correction is proportional to k^2, it changes the effective mass of the electrons, to make them have greater mass than one would expect from the band curvature.

This effect is known as the **polaron** effect. The electron energy and mass are **renormalized** to different values due to the interaction with the phonons. Physically, one imagines a cloud of virtual optical phonons surrounding each electron, which is another way of saying that the lattice is distorted due to the polarization caused by the charge of the electron. In principle, the renormalization of the electron energy can be so great that it has nearly infinite effective mass, that is, it becomes localized and unable to move.

It is often difficult to determine from experiments, such as the cyclotron experiment discussed in Section 2.2, what the real effective mass of the electrons is. The polaron mass changes with temperature, since as seen in (8.2.3), the phonon occupation number enters into the electron energy renormalization, but even at $T = 0$ there is a polaron effect which cannot be eliminated.

In general, electron–phonon interactions can lead to large changes of the intrinsic electron behavior. As we will see in Section 11.7, the electron–phonon interaction gives Cooper pairing of electrons, which leads to superconductivity.

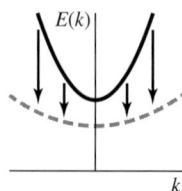

Fig. 8.2   The polaron effect.

Exercise 8.2.1   How large an effect should **acoustic** phonons produce? Estimate the polaron mass of electrons at zero temperature due to acoustic phonons, if the band effective mass is equal to the free electron mass, the electron–phonon deformation potential energy is $D = 1$ eV, the speed of sound is $v = 5 \times 10^5$ cm/s, the density is 5 g/cm^3, and the Debye temperature is 30 meV.

8.3 Shift of Bands with Temperature

The electron–phonon interaction discussed in Section 8.2 also leads to another ubiquitous effect, which is the shift of the electron bands with temperature. Looking again at (8.2.3), we see that when $\vec{k} = 0$, the total second-order energy correction is

$$E_0^{(2)} = \frac{1}{V} \sum_{\vec{p}} \frac{C_{\mathrm{Fr}}^2}{p^2} \left(\frac{N_{\vec{p}}}{\hbar\omega_{LO} - E_{\vec{p}}} + \frac{N_{\vec{p}}}{-\hbar\omega_{LO} - E_{\vec{p}}} + \frac{1}{-\hbar\omega_{LO} - E_{\vec{p}}} \right). \tag{8.3.1}$$

As discussed in Section 8.2, the spontaneous emission term always gives a negative correction. The second term, corresponding to stimulated emission, is also always negative. The first term, corresponding to absorption, can be positive or negative. The sum over $\vec{p}$, however, includes values over the entire Brillouin zone. The electron energy $E_{\vec{p}}$ for all possible transitions typically has values of the order of eV, while the optical phonon energy $\hbar\omega_{LO}$ is typically a few tens of meV. Therefore, the net contribution of this term is also negative. The second-order energy correction therefore includes a term which is proportional to the phonon occupation number and negative.

The number of optical phonons N_p just depends on the phonon energy,

$$\langle N_p \rangle = \frac{1}{e^{\hbar\omega_{LO}/k_B T} - 1}. \tag{8.3.2}$$

To first order, it does not depend on $\vec{p}$, and therefore can be taken outside the sum as a constant that depends only on the temperature.

For semiconductors, the same argument can also be made for *holes*, which can undergo virtual excitations to higher states (corresponding to electron states lower in the valence band). The band-gap energy can therefore be written as

$$E_g(T) = E_g(0) - A \frac{1}{e^{\hbar\omega_{LO}/k_B T} - 1}. \tag{8.3.3}$$

For $k_B T \gg \hbar\omega_{LO}$, the phonon occupation number is approximately

$$\frac{1}{e^{\hbar\omega_{LO}/k_B T} - 1} = \frac{1}{1 + \hbar\omega_{LO}/k_B T + (\hbar\omega_{LO}/k_B T)^2 + \cdots - 1}$$

$$\simeq \frac{(k_B T)^2}{(\hbar\omega_{LO})k_B T + (\hbar\omega_{LO})^2} \tag{8.3.4}$$

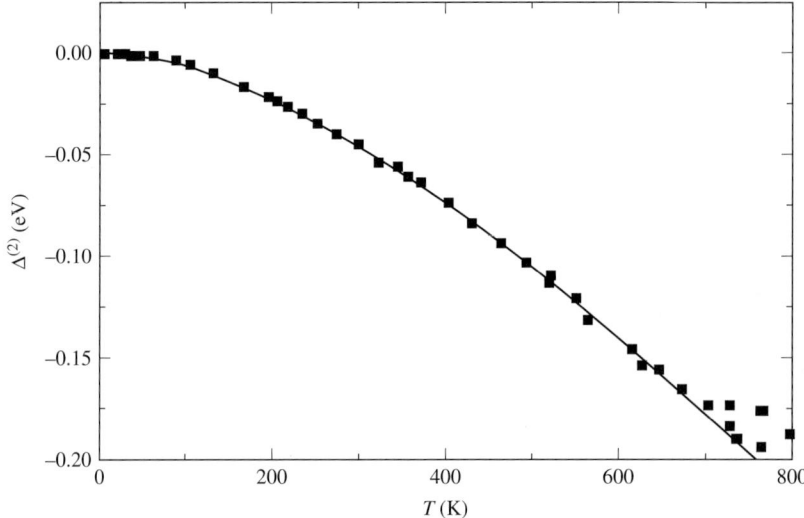

Fig. 8.3 Squares: experimentally measured energy gap shift of silicon as a function of temperature. Solid line: fit to the Varshni formula with $E_0 = 1.17$ eV, $\alpha = 4.8 \times 10^{-4}$ eV/K, and $\beta = 655$ K. From Alex *et al.* (1996).

Equation (8.3.3) can therefore be rewritten as the **Varshni** formula,

$$E_g(T) = E_g(0) - \frac{\alpha T^2}{T + \beta}, \tag{8.3.5}$$

where α and β are constants.

Figure 8.3 shows an example of the measured band gap of a semiconductor as a function of temperature, with a fit to the Varshni formula. This is a strong confirmation of the theory of the renormalization of the electron states by the phonon interaction. Surprisingly, the formula works well even at low temperature. A proper calculation would need to take into account not only a single optical phonon frequency, but all phonons, including low-frequency acoustic phonons.

Exercise 8.3.1   Estimate the band shift with temperature due to electron–phonon Fröhlich interaction, for an electron band with effective mass equal to the free electron mass, optical phonon energy $\hbar\omega_{LO} = 30$ meV, and $\epsilon(0)/\epsilon_0 = 10$ and $\epsilon(\infty)/\epsilon_0 = 1$. What is the magnitude of the band shift due to the electron–phonon interaction at zero temperature? Even at $T = 0$, calculation of the electron band energies needs to take into account the vibrations of the lattice.

8.4 Line Broadening

Figure 8.4 shows the photon emission spectrum from a semiconductor material as the temperature is raised. One effect that is obvious is the red shift of the band gap with temperature, discussed in Section 8.3. Another effect, which may not be so obvious at

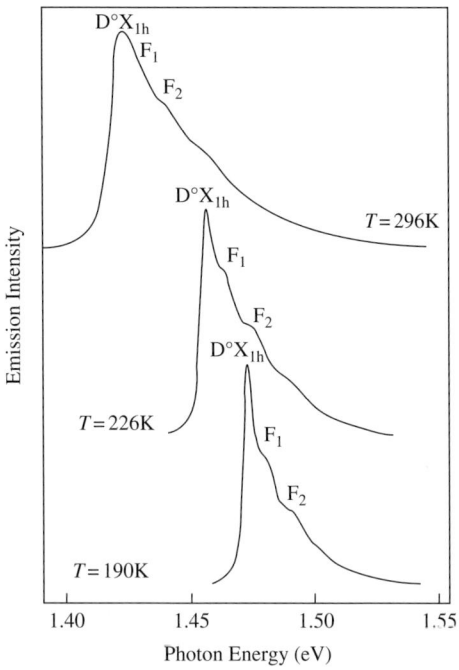

Fig. 8.4   Photoluminescence from GaAs quantum wells (with $Al_{0.25}Ga_{0.75}As$ barriers) at various temperatures. The line labeled D^0X_{1h} arises from a donor-bound hole, while the lines labeled F_1 and F_2 arise from transitions from higher-energy quantized states in the well (see Section 2.8). From Yu *et al.* (1985).

first glance, is the increase of the spectral *broadening* of the emission. As the temperature is increased, sharp features in the luminescence are smeared out. This is an example of a common effect, generally called **homogeneous broadening**. Although we will study it here in the context of photon emission, it also occurs in other contexts, such as angle-resolved photoemission of electrons (see Section 1.10).

Consider the Hamiltonian

$$H = H_0 + V_{int} + V'_{int}, \tag{8.4.1}$$

where V_{int} is an "internal" interaction of a system, and V'_{int} is a "probe" of the system. For example, suppose that V_{int} is the electron–phonon interaction and V'_{int} is the electron–photon interaction, and we measure single photons emitted from the system in definite k-states. We will assume that the probe is weak, so that we can take V'_{int} only in the lowest order.

Keeping track of both perturbation terms, the probability amplitude for transitions from initial state $|i\rangle$ to a final state $|n\rangle$ is

$$\langle n|\psi(t)\rangle = \langle n|e^{-(i/\hbar)\int_0^t (V'_{int}(t')+V_{int}(t'))dt'}|i\rangle$$

$$= \langle n|\left(1 + \frac{1}{i\hbar}\int_0^t dt' V'_{int}(t') + \frac{1}{i\hbar}\int_0^t dt' V_{int}(t')\right.$$

$$+ \frac{1}{(i\hbar)^2} \int_0^t dt' dt'' \left(V'_{\text{int}}(t') V'_{\text{int}}(t'') + V_{\text{int}}(t') V'_{\text{int}}(t'') \right.$$

$$\left. + V'_{\text{int}}(t') V_{\text{int}}(t'') + V_{\text{int}}(t') V_{\text{int}}(t'') \right) + \cdots \Big) |i\rangle. \tag{8.4.2}$$

We assume that state $|i\rangle$ is coupled to state $|n\rangle$ only by V'_{int}, and not by V_{int}; for example, state $|n\rangle$ has a photon, while state $|i\rangle$ doesn't, and V_{int} has no photon operators at all. Terms with only V_{int} therefore vanish. We also drop all terms that are second order or higher in V'_{int}. Inserting a sum over a complete set of states $|m\rangle$ in the second-order terms, we obtain

$$\langle n|\psi(t)\rangle = \frac{1}{i\hbar} \int_0^t dt' \langle n|V'_{\text{int}}(t')|i\rangle + \sum_m \frac{1}{i\hbar} \int_0^t dt' \int_0^{t'} dt'' \langle n|V_{\text{int}}(t')|m\rangle \langle m|V'_{\text{int}}(t'')|i\rangle$$

$$+ \frac{1}{i\hbar} \int_0^t dt' \int_0^{t'} dt'' \langle n|V'_{\text{int}}(t')|m\rangle \langle m|V_{\text{int}}(t'')|i\rangle + \cdots . \tag{8.4.3}$$

As above, we assume that V_{int} does not couple to the final state $|n\rangle$, and therefore $\langle n|V_{\text{int}}(t')|m\rangle$ is zero, and the first second-order term vanishes. Applying a similar logic to the higher-order terms, we then have

$$\langle n|\psi(t)\rangle = \sum_m \left(\frac{1}{i\hbar} \int_0^t dt' \langle n|V'_{\text{int}}(t')|m\rangle \right) \langle m| \left(1 + \frac{1}{i\hbar} \int_0^{t'} dt'' V_{\text{int}}(t'') + \cdots \right) |i\rangle$$

$$= \sum_m \frac{1}{i\hbar} \int_0^t dt' \langle n|V'_{\text{int}}|m\rangle e^{(i/\hbar)(E_n - E_m)t'} \langle m|e^{-(i/\hbar)\int_0^{t'} V_{\text{int}}(t'')dt''}|i\rangle$$

$$= \sum_m \frac{1}{i\hbar} \int_0^t dt' \langle n|V'_{\text{int}}|m\rangle e^{(i/\hbar)(E_n - E_m)t'} \langle m|\psi_i(t')\rangle, \tag{8.4.4}$$

where $|\psi_i(t')\rangle$ is the state to which $|i\rangle$ would evolve if only the effect of V_{int} were taken into account.

We then break the sum over states m into two parts,

$$\langle n|\psi(t)\rangle = \frac{1}{i\hbar} \int_0^t dt' \langle n|V'_{\text{int}}|i\rangle e^{(i/\hbar)(E_n - E_i)t'} \langle i|\psi_i(t')\rangle$$

$$+ \sum_{m \neq i} \frac{1}{i\hbar} \int_0^t dt' \langle n|V'_{\text{int}}|m\rangle e^{-(i/\hbar)(E_n - E_m)t'} \langle m|\psi_i(t')\rangle$$

$$+ \cdots . \tag{8.4.5}$$

The first term depends on the amplitude $\langle i|\psi_i(t')\rangle$, while the second term depends on the amplitude $\langle m|\psi_i(t')\rangle$. We have computed both of these aleady in Section 8.1. We will look at the second term first.

Using (8.1.21), the second term of (8.4.5) is

$$\sum_{m \neq i} \frac{1}{i\hbar} \int_0^t dt' \langle n|V'_{\text{int}}|m\rangle e^{-(i/\hbar)(E_n - E_m)t'} \langle m|V_{\text{int}}|i\rangle \left(\frac{e^{(i/\hbar)(E_m - E_i)t'} - 1}{E_i - E_m} \right).$$

$$\tag{8.4.6}$$

Following the same procedure as in Section 8.1, performing this time integral gives

$$\sum_{m \neq i} \frac{\langle n|V'_{\text{int}}|m\rangle\langle m|V_{\text{int}}|i\rangle}{E_i - E_m + i\eta} \left(\frac{e^{(i/\hbar)(E_n - E_i)t'} - 1}{E_i - E_n} \right). \tag{8.4.7}$$

Comparing this to (8.1.22), we see that this is the amplitude for real transitions from $|i\rangle$ to $|n\rangle$ via a second-order transition process, in which energy is conserved – squaring this amplitude will turn the term in parentheses into an energy-conserving $\delta(E_i - E_n)$. We could obtain higher-order terms for real transition processes as well, by keeping terms with larger powers of V_{int} and V'_{int}.

Assuming that we can neglect these higher-order processes, we are left with the first term of (8.4.5), with $\langle i|\psi_i(t')\rangle$. Using (8.1.10), along with (8.1.12), the first term in (8.4.5) becomes

$$\begin{aligned} \langle n|\psi(t)\rangle &= \frac{1}{i\hbar} \int_0^t dt' \, \langle n|V'_{\text{int}}|i\rangle e^{(i/\hbar)(E_n - E_i)t'} \langle i|\psi_i(t')\rangle \\ &= \frac{1}{i\hbar} \int_0^t dt' \, \langle n|V'_{\text{int}}|i\rangle e^{(i/\hbar)(E_n - E'_i + i\Gamma)t'}, \end{aligned} \tag{8.4.8}$$

where $E'_i = E_i + \Delta^{(1)} + \Delta^{(2)}$ is the renormalized energy of the state $|i\rangle$, and $\Gamma = \Gamma^{(2)}$ is the imaginary self-energy, found from the total scattering rate out of state $|i\rangle$ due to V_{int}. Completing the time integral, we have

$$\langle n|\psi(t)\rangle = \langle n|V'_{\text{int}}|i\rangle \frac{e^{(i/\hbar)(E_n - E'_i + i\Gamma)t} - 1}{E'_i - E_n - i\Gamma}. \tag{8.4.9}$$

The imaginary self-energy term Γ means that when we square this term, we will not just be left with energy-conserving transitions. Switching back to the Schrödinger representation, using $\langle n|\psi_t\rangle = e^{-iE_n t}\langle n|\psi(t)\rangle$, the probability of the system being in the final state $|n\rangle$, after a long time, is

$$\begin{aligned} |\langle n|\psi_\infty\rangle|^2 &= |\langle n|V'_{\text{int}}|i\rangle|^2 \left| \frac{-1}{(E'_i - E_n) - i\Gamma} \right|^2 \\ &= |\langle n|V'_{\text{int}}|i\rangle|^2 \frac{1}{(E'_i - E_n)^2 + \Gamma^2}. \end{aligned} \tag{8.4.10}$$

To convert this to a transition rate, we must divide by the total amount of time spent in the initial state $|i\rangle$,

$$\begin{aligned} \int_0^\infty dt \, |\langle i|\psi_t\rangle|^2 &= \int_0^\infty dt \left| e^{-(i/\hbar)(E'_i - i\Gamma)t} \right|^2 \\ &= \int_0^\infty dt \, e^{-2\Gamma t/\hbar} \\ &= \frac{\hbar}{2\Gamma}, \end{aligned} \tag{8.4.11}$$

which yields

$$\boxed{\frac{1}{\tau} = |\langle n|V'_{\text{int}}|i\rangle|^2 \frac{2\Gamma/\hbar}{(E'_i - E_n)^2 + \Gamma^2}.} \tag{8.4.12}$$

In the limit $\Gamma \to 0$, this becomes

$$\frac{1}{\tau} = \frac{2\pi}{\hbar} |\langle n|V'_{\text{int}}|i\rangle|^2 \delta(E'_i - E_n), \qquad (8.4.13)$$

which is just the same as we found in Section 4.7 for Fermi's golden rule.

In other words, the effect of imaginary self-energy due to interactions, that is, depletion of the initial state, is that transitions from $|i\rangle$ to other states are broadened, with a Lorentzian spectrum instead of strict energy conservation. One way to view this is as an effect of the uncertainty principle – if the particle stays in a state only for some time Δt, then its energy will be uncertain by the amount $\Delta E = \hbar/\Delta t$. Recall that the self-energy Γ is given by the rate of scattering out of state $|i\rangle$. Alternatively, one can view the broadening as coming about because during a transition, energy can be lost or gained to other particles which are interacting with the particle in the initial state $|i\rangle$. The extra energy is given to or taken from collective excitations of the interacting system.

Note that to get the Lorentzian spectrum, we had to take the limit $t \to \infty$ in (8.4.10), which means that the result (8.4.12) is valid only for times longer than $T_2 = \hbar/\Gamma$. T_2 here is the time for depletion of state $|i\rangle$. This is often known as the **dephasing time** (we will discuss this further in Section 9.5). This is to be distinguished from the **population lifetime**, which is the time for depletion of all states with energy equal to E_i. In many cases, the population lifetime can be much longer than the dephasing time because particles can move rapidly back and forth between states with the same energy.

Lorentzian line broadening is a ubiquitous effect, both in photon emission and electron photoemission (see Section 1.10). The same broadening also occurs in photon absorption spectra. It is a generic effect that the Fourier transform of a sine wave with decay is a broadened Lorentzian (cf. Section 7.2). The broadening energy Γ can come from any out-scattering process, such as electron–phonon scattering, electron–electron scattering, etc. Anything that shortens the lifetime for the quasiparticle to stay in a single quantum state will contribute to line broadening.

Lorentzian line broadening is called **homogeneous** broadening to distinguish it from **inhomogeneous** line broadening. Recall from Section 1.8.2 that long-wavelength disorder leads to random shifts of the energy bands in a solid. A random Gaussian distribution of energies due to the inhomogeneities will therefore lead to a Gaussian line shape of light emitted from a solid. Inhomogeneous broadening reflects this built-in uncertainty in the energy due to disorder in the system. Homogeneous broadening, on the other hand, reflects the strength of interactions in a system. The homogeneous broadening linewidth therefore can vary depending on the nature of the interactions; for example, line broadening due to electron–phonon interaction will depend on the temperature, which controls the number of phonons. Figure 8.4 shows an example of the increase of the line broadening with increasing temperature. If $I_0(E)$ is the unbroadened emission spectrum, the homogeneously broadened spectrum is given by the convolution

$$I(E) = \int_{-\infty}^{\infty} dE' \, I_0(E') \frac{2\Gamma}{(E' - E)^2 + \Gamma^2}, \qquad (8.4.14)$$

assuming Γ does not depend on E. As shown in Section 5.1.4, the electron–phonon scattering rate is proportional to the average number of phonons. Since the number of phonons increases with increasing T, therefore the line broadening of transitions involving electrons will also increase.

The electron-photon interaction can also lead to broadening of electron–photon transitions. To see this, imagine that V_{int} is the full electron–photon interaction Hamiltonian (5.2.8), except for one term that corresponds to photon emission along a certain direction with a certain polarization. This one, excised term, we will call V'_{int}. Then all of the analysis at the beginning of this section still applies. If there are no other, stronger interaction processes, then the photon emission process itself will lead to broadening of the photon emission lines, an effect known as **lifetime broadening**. This is a commonly observed effect in atomic physics, in which single atoms do not strongly interact with each other.

Exercise 8.4.1   What does the transition spectrum look like on short time scales? Instead of taking the limit $t \to \infty$, determine the probability $|\langle f | \psi(t) \rangle|^2$, and use a program like Mathematica to plot it as a function of final state energy E_f, for various times t. You should see that at early times the spectrum oscillates rapidly, but that it converges to a Lorentzian on time scales long compared to $\hbar / \Gamma$.

Exercise 8.4.2   In Section 7.5.3 we defined the exciton creation operator $c_k^\dagger$ and the exciton–photon interaction

$$V'_{int} \simeq \frac{\hbar}{2} \sqrt{\frac{q^2 N}{\epsilon m V}} \sqrt{\frac{2}{m\hbar\omega_0}} \sqrt{\frac{\omega_k}{\omega_0}} \sum_{\vec{k}} \left(p_{cv}^* a_k c_k^\dagger + p_{cv} a_k^\dagger c_k \right), \qquad (8.4.15)$$

where a_k is the photon operator, and the constants in front of the sum are defined in Section 7.5.3.

Suppose that excitons exist in a semiconductor with energy gap $\hbar\omega_0$ and exciton binding energy Δ, and the initial state is an ensemble of single excitons with identical wave vector $\vec{K}$, and they have an average scattering time τ with phonons in the medium. Show that this system will yield a photon spectrum with a Lorentzian energy distribution, and determine the width of the Lorentzian.

8.5 Diagram Rules for Rayleigh–Schrödinger Perturbation Theory

As seen in Section 8.1, the Rayleigh–Schrödinger perturbation method gives an infinite series of terms. Typically, one keeps only the lowest few terms, but in many cases, such as when the lower-order terms are symmetry forbidden, or when several quasiparticles are involved in a single interaction, we must consider higher-order terms.

Going to higher orders can become a bookkeeping nightmare. One way to keep track of how to write down the higher-order integrals is to use a diagrammatic approach. The rules for writing down matrix elements using Rayleigh–Schrödinger perturbation theory are as follows:

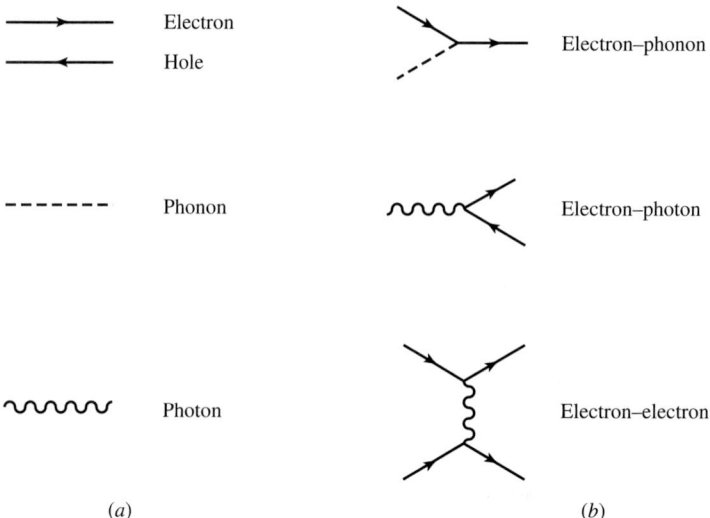

Fig. 8.5 (a) Lines used for various quasiparticles in the diagrammatic approach. (b) Vertices for solid state interactions. Electrons are represented by solid lines with arrows, phonons are represented by dashed lines, and photons are represented by wavey lines.

- Pin down the external "legs," that is, the incoming and outgoing quasiparticles for the process of interest, assigning a momentum to each. This involves defining a direction of time for the diagram – usually time goes from left to right. An incoming hole is counted as an outgoing electron, and vice versa, that is, a hole is an electron moving backwards in time. Figure 8.5(a) shows the lines typically used to represent different quasiparticles.
- Draw all possible diagrams that connect those legs using vertices for the possible interactions. Depending on the interaction, some internal legs may be produced. These must be terminated either as external legs or in the legs of other vertices. Figure 8.5(b) shows some examples of vertices for interactions we have computed in Chapter 5.

   Note that the squiggly line used for photons is also used for the Coulomb interaction vertex. This represents the instantaneous electric interaction in the Coulomb gauge. The photon line in this case is always drawn perpendicular to the time axis, in other words, vertical, for time going left to right.
- Conserve momentum strictly at each vertex. Momentum is counted as positive in the direction of the arrow of an electron line, which means for holes it is the opposite of the actual hole momentum. In the case of elastic scattering from a defect, momentum is not conserved, and one must multiply by a phase factor for the momentum change.
- For each vertex, multiply by the appropriate matrix element. For example, from Chapter 5 we have the matrix elements (neglecting polarization vectors):

electron–phonon (deformation) $M_k^{\text{Def}} = Dk\sqrt{\dfrac{\hbar}{2\rho V \omega_k}}$

electron–phonon (Fröhlich) $M_k^{\text{Fr}} = \dfrac{e}{k}\left(\dfrac{1}{\epsilon(\infty)} - \dfrac{1}{\epsilon(0)}\right)^{1/2}\sqrt{\dfrac{\hbar \omega_{LO}}{2V}}$

electron–photon $M_k^{\text{dipole}} = \dfrac{e}{m}\langle p\rangle\sqrt{\dfrac{\hbar}{2\epsilon V \omega_k}}$

electron–electron $M_{\Delta k}^{\text{Coul}} = \dfrac{1}{V}\dfrac{e^2/\epsilon}{|\vec{\Delta k}|^2 + \kappa^2}\,.$

Each of these matrix elements has units of energy. Note that the electron–electron interaction accounts for two vertices.

- For each internal state, multiply by a term

$$\frac{1}{E_i - E_n + i\eta},$$

where E_i is the initial energy, including all external incoming legs, and E_n is the energy of the internal state. Note that with Rayleigh–Schrödinger diagrams, unlike the Feynman diagrams which will be introduced in the last half of this chapter, we do not associate a factor with each internal *line*; rather, we have a factor for each internal *state*, which can have multiple lines. States are viewed as occurring sequentially in time, with a new state starting at each vertex. The time-ordering of the vertices (interaction events) must therefore be tracked explicitly.

- For each incoming external leg entering a vertex, multiply by $\sqrt{N_{\vec{p}}}$, where $N_{\vec{p}}$ is the occupation number of the quasiparticle in state $\vec{p}$, and for each outgoing external leg, multiply by $\sqrt{1 \pm N_{\vec{p}}'}$ (+ for bosons, − for fermions), where $N_{\vec{p}}'$ is the occupation number of state $\vec{p}$ *after* taking into account the depletion of that state when incoming legs are terminated at vertices. If the field is coherent (e.g., a laser beam), multiply instead by $A_p e^{i\omega t}$ for an incoming wave and $A_p e^{-i\omega t}$ for an outgoing wave, where A_p is given by, as shown in Section 4.4,

phonons $A_p = \sqrt{\dfrac{\rho V \omega_p}{2\hbar}}\, x_0$

photons $A_p = \sqrt{\dfrac{\epsilon_0 V}{2\hbar \omega_p}}\, E_0.$

- Sum over all internal states, that is, all momenta that are not determined by momentum conservation and the momenta of the external legs.
- For each crossing of fermion lines, multiply by -1.

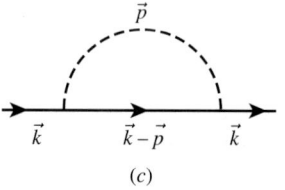

Fig. 8.6 (a) Diagram for the electron-phonon absorption self-energy. (b) Diagram for the electron–phonon emission self-energy. (c) Diagram for electron–phonon self-energy due to emission and reabsorption.

Example. The electron–phonon self-energy which leads to electron energy renormalization, discussed in Section 8.3, can be written as the three diagrams shown in Figure 8.6. In each case, the self-energy is found by returning the system to its initial state. Each of the two vertices gives a contribution M_p^{Fr}. Diagram (a) has two external phonon legs, which yield two factors $\sqrt{N_{\vec{p}}}$, since $N_{\vec{p}}$ is depleted by 1 after a phonon is absorbed, so that the factor for the outgoing leg is $\sqrt{1 + N_{\vec{p}}'} = \sqrt{1 + N_{\vec{p}} - 1} = \sqrt{N_{\vec{p}}}$. The initial energy is just the electron energy $E_{\vec{k}}$ plus the phonon energy $\hbar\omega_{LO}$, while the intermediate energy is $E_{\vec{k}+\vec{p}}$. Summing over all possible external phonon legs gives for diagram (a), we have the self-energy

$$\langle i|H^{(2)}|i\rangle^{(a)} = \sum_{\vec{p}} \frac{(M_p^{\text{Fr}})^2 N_{\vec{p}}}{E_{\vec{k}} + \hbar\omega_{LO} - E_{\vec{k}+\vec{p}} + i\eta}. \tag{8.5.1}$$

The sum over $\vec{p}$ will allow us to get the real part of the self-energy by taking the principal value with $\eta = 0$.

For diagram (b), the incoming phonon contributes an energy $\hbar\omega_{LO}$ to both the initial and intermediate state, which therefore cancels out in the difference of the two energies. The energy of the electron and the other phonon in the intermediate state is $E_{\vec{k}-\vec{p}} + \hbar\omega_{LO}$, which gives

$$\langle i|H^{(2)}|i\rangle^{(b)} = \sum_{\vec{p}} \frac{(M_p^{\text{Fr}})^2 N_{\vec{p}}}{E_{\vec{k}} - \hbar\omega_{LO} - E_{\vec{k}-\vec{p}} + i\eta}. \tag{8.5.2}$$

Last, diagram (c) involves no external phonon legs, but requires a sum over all internal momentum states,

$$\langle i|H^{(2)}|i\rangle^{(c)} = \sum_{\vec{p}} \frac{(M_p^{\text{Fr}})^2}{E_{\vec{k}} - \hbar\omega_{LO} - E_{\vec{k}-\vec{p}} + i\eta}. \tag{8.5.3}$$

It is easy to see that the sum of these three diagrams is the same as (8.2.3), which we deduced in Section 8.2. Diagram (c) is the spontaneous emission term, which is always present even if there are no phonons in the initial state.

Raman scattering. As an example of a more complicated diagram, consider the matrix element for resonant Raman scattering from a semiconductor. Resonant Raman scattering was introduced in Section 7.8. For the case shown in Figure 8.7(a), the incoming photon creates an electron–hole pair (holes are shown as electrons moving backward in time), and the electron or the hole can emit a phonon. The initial state is a photon with momentum $\vec{q}$, and the final state is a photon with momentum $\vec{q} - \vec{p}$ plus a phonon with momentum $\vec{p}$. The internal momentum $\vec{k}$ is not fixed by momentum conservation and the external legs, which means we must sum over it, and there are two internal states, which will give us two energy denominators. For electron–phonon Fröhlich scattering, the result for this diagram is

$$\langle f|H^{(3)}|i\rangle = M_{\vec{q}}^{\text{dipole}} M_{\vec{q}-\vec{p}}^{\text{dipole}} M_{\vec{p}}^{\text{Fr}} \sum_{\vec{k}} \frac{1}{\hbar\omega_{\text{phot}} - E_{c,\vec{k}} + E_{v,\vec{k}-\vec{q}} + i\eta}$$

$$\times \frac{1}{\hbar\omega_{\text{phot}} - E_{c,\vec{k}-\vec{p}} + E_{v,\vec{k}-\vec{q}} - \hbar\omega_{LO} + i\eta} \sqrt{1 + N_{\vec{p}}}, \qquad (8.5.4)$$

where the energy of a conduction electron is $E_g + E_{c,k}$ and the energy of the valence electron is $-E_{v,k}$. We have assumed that none of the other photon states is initially occupied. If the phonon state is unoccupied, then the square root factor for its occupation will also just be unity. The two factors in the denominators vanish at different values of $\vec{k}$, so we can once again find the real part by taking the principal value, setting $\eta = 0$.

There are many other diagrams that one can write down. For example, Figure 8.7(b) shows another possible diagram with the same external legs, in which an electron-hole pair is created by *emission* of a photon, and finally recombines by *absorption* of a photon. In this case, the matrix element is strongly suppressed by the denominator for the internal

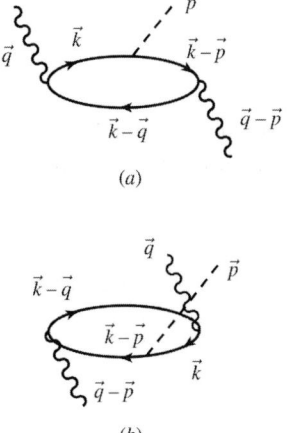

(a)

(b)

Fig. 8.7   (a) Diagram for a resonant Raman scattering process. (b) Another resonant Raman scattering process with the same initial and final states.

states, since the internal state energy is of the order of twice the band-gap energy, because both the incoming and outgoing photons exist at the same time.

It is important to remember that this procedure gives an *amplitude*, that is, a matrix element with units of energy. To find the scattering rate using Fermi's golden rule, we must square this amplitude and multiply by $2\pi/\hbar$ and the final density of states. If more then one diagram is involved, we must first add the amplitudes and then square the sum.

This diagrammatic method is common in solid state physics, especially optics, but should not be confused with the standard methods of Feynman diagrams, which are discussed in the rest of this chapter. Examples of texts using the Rayleigh–Schrödinger approach in solid state physics are those by Yariv (1989) and Yu and Cardona (1996).

Exercise 8.5.1 The resonant Raman scattering diagram shown in Figure 8.7(a) can be compared to the results of Section 7.8 by taking the initial state $|i\rangle$ as N_i photons in state $\vec{q}$ and the final state as $N_i - 1$ photons in state $\vec{q}$ and a phonon in state $\vec{p}$.

(a) Calculate the rate of transitions for the matrix element (8.5.4), and show that it is independent of the volume V and proportional to the total number of incoming photons (assuming no stimulated photon emission).

(b) Using (7.8.8), calculate the Raman scattering cross-section for this process. Verify that your answer has the units of area, and is proportional to the volume of the illuminated scattering medium.

Exercise 8.5.2 Write down the integral for the diagram shown in Figure 8.7(b).

Exercise 8.5.3 Write down all the possible third-order Rayleigh–Schrödinger diagrams that have an initial state of a photon and a final state of an emitted phonon and a photon, for two electron bands c and v. If you apply a selection rule that phonons do not cause electrons to change bands, which diagrams are eliminated?

8.6 Feynman Perturbation Theory

So far in this chapter, we have used Rayleigh–Schrödinger perturbation theory to determine the time-dependent states of a system. This has the advantage of simplicity, but it has a significant problem. In this method. we must assign a sequence of times for the interactions and write down the energy of the state at each time. In some cases, it is not so clear what the sequence of times is. For example, the diagram in Figure 8.8 has several phonon lines interacting with electron lines. Is the last phonon leg on the upper electron line emitted before or after the phonon is emitted by the hole? Before or after the absorption of the phonon? Does it matter?

We would like a method that has more general applicability. This leads us to the general formalism commonly known as Feynman perturbation theory. This is the standard approach in many-body physics; the Rayleigh–Schrödinger theory is mainly used only for optics, where the electron–photon interaction can be taken as small compared to all other perturbations. For general textbooks on many-body Feynman perturbation theory, see Mahan (2000), Fetter and Walecka (1971), or Abrikosov *et al.* (1975). Our purpose

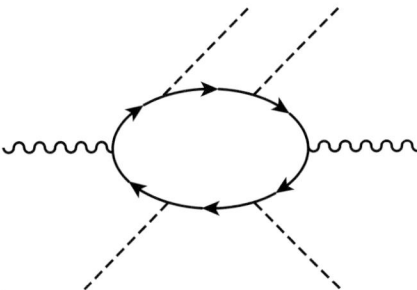

Fig. 8.8   A complicated scattering diagram.

here is not to present a full discussion of many-particle theory, but to present the basic concepts so that Feynman diagrams are not a mystery. The first few pages that justify the approach are slow-going, but once these are mastered, the Feynman diagrammatic method is easy to use.

Once again, we start with the general form for the time evolution of any given wave function,

$$|\psi(t)\rangle = e^{-(i/\hbar)\int V_{\text{int}}(t)dt}|\psi(0)\rangle$$

$$= \Bigg(1 + (1/i\hbar)\int_0^t dt'\, V_{\text{int}}(t')$$

$$+ (1/i\hbar)^2 \int_0^t dt' \int_0^{t'} dt''\, V_{\text{int}}(t')V_{\text{int}}(t'') + \cdots\Bigg)|\psi(0)\rangle, \qquad (8.6.1)$$

which we first derived in Section 4.7. The operator $S(t,t_0) = e^{-(i/\hbar)\int_{t_0}^t V_{\text{int}}(t')dt'}$ is known as the **S-matrix**. As we saw in Section 8.1, this operator determines not only the rate of scattering from one state into other states, through $\langle n|S|i\rangle$, it also tells us the energy renormalization of a given state, through $\langle i|S|i\rangle$.

The S-matrix can in general be written as

$$S(t,t_0) = 1 + \sum_{n=1}^{\infty} (1/i\hbar)^n \int_{t_0}^t dt_1 \int_{t_0}^{t_1} dt_2 \cdots \int_{t_0}^{t_{n-1}} dt_n\, V_{\text{int}}(t_1)V_{\text{int}}(t_2)\cdots V_{\text{int}}(t_n). \quad (8.6.2)$$

The upper limits of the time integrals are all different. We can rewrite the S-matrix with the same upper limit for all the time integrals as follows:

$$S(t,t_0) = 1 + \sum_{n=1}^{\infty} \frac{(1/i\hbar)^n}{n!} \int_{t_0}^t dt_1 \int_{t_0}^t dt_2 \cdots \int_{t_0}^t dt_n\, \mathsf{T}(V_{\text{int}}(t_1)V_{\text{int}}(t_2)\cdots V_{\text{int}}(t_n)),$$

$$(8.6.3)$$

in which $\mathsf{T}(V_{\text{int}}(t_1)V_{\text{int}}(t_2)\cdots V_{\text{int}}(t_n))$ is the **time-ordered product** of the operators $V_{\text{int}}(t_1), V_{\text{int}}(t_2)$, etc. In the time-ordered product, the operators are rearranged such that $t_1 > t_2 > \cdots > t_n$.

To see that (8.6.3) is equivalent to (8.6.2), consider just the second-order term,

$$\int_{t_0}^{t} dt_1 \int_{t_0}^{t} dt_2 \, \mathsf{T}(V_{\text{int}}(t_1)V_{\text{int}}(t_2)) = \int_{t_0}^{t} dt_1 \int_{t_0}^{t_1} dt_2 \, V_{\text{int}}(t_1)V_{\text{int}}(t_2)$$
$$+ \int_{t_0}^{t} dt_2 \int_{t_0}^{t_2} dt_1 \, V_{\text{int}}(t_2)V_{\text{int}}(t_1).$$

$$(8.6.4)$$

By the simple change of variables, $t_1 \to t_2, t_2 \to t_1$, the second integral is equivalent to the first. By normalizing the left-hand side by a factor of $\frac{1}{2}$, we obtain the second-order term of (8.6.2). The same procedure works for all orders of the expansion.

Wick's theorem. In Chapter 5, the interaction terms we wrote down were all written in terms of boson and fermion creation and destruction operators $a_k, a_k^\dagger, b_k, b_k^\dagger$, etc. We would like to convert the time-ordered product of the interactions into a **normal-ordered product** of the creation and destruction operators, that is, a product in which all creation operators are to the left of all destruction operators. (Note that the form (8.6.2) is equivalent to (8.6.3) only if there are an even number of fermion operators in the interaction V_{int}; for example, if for every fermion created, another is destroyed, conserving the number of fermions.)

The key theorem of many-body theory that allows us to do this is **Wick's theorem**, which says, for operators $A, B, C, \ldots, X, Y, Z$,

$$\mathsf{T}(ABC\ldots YZ) = \mathsf{N}(ABCD\cdots YZ) + \mathsf{N}(\overbrace{A}B C D \cdots YZ)$$
$$+ \mathsf{N}(A\overbrace{B}C D \ldots YZ) + \mathsf{N}(A\overbrace{B}\overbrace{C}D \cdots YZ)$$
$$+ \mathsf{N}(A\overbrace{BC}D \ldots YZ) + \cdots,$$

$$(8.6.5)$$

where $\overbrace{AB}$ is a complex number (a "c-number") known as the **contraction** of operators A and B, defined by

$$\overbrace{AB} = \mathsf{T}(AB) - \mathsf{N}(AB),$$

$$(8.6.6)$$

and $\mathsf{N}(ABC\ldots)$ is the normal-ordered product of any number of operators. The normal-ordered product is obtained by putting all the creation operators to the left of all the destruction operators. (The order of the creation operators or the destruction operators within each set doesn't matter.) If some of the operators are fermionic, both the time-ordered product and the normal-ordered product include additionally a factor $(-1)^F$, where F is the number of interchanges of neighboring fermion operators needed to obtain the product from the original ordering. Note that the creation operator for a *hole* is an electron *destruction* operator acting on a state below the Fermi level; such an operator will be moved to the left in normal-ordering.

Wick's theorem says that the time-ordered product is equal to the sum of all the possible normal-ordered products, for every possible contraction of two operators. On one hand, we have increased the complexity, since we now have a large sum of products instead of just one, but on the other hand, normal-ordered products and contractions are much easier to handle.

Wick's theorem for two operators amounts to just the statement that the contraction of our standard creation and destruction operators $a_{\vec{k}}^{\dagger}(t)$ and $a_{\vec{k}}(t)$ is a c-number. To show that, we must first determine the commutator of these two operators.

In the interaction picture, for any operator $A(t)$, we can differentiate with respect to time to obtain

$$i\hbar\frac{\partial}{\partial t}A(t) = e^{iH_0 t/\hbar}(AH_0 - H_0 A)e^{-iH_0 t/\hbar}$$

$$= [A(t), H_0]. \tag{8.6.7}$$

The unperturbed Hamiltonian H_0 by definition is diagonal in the number operators, that is,

$$H_0 = \sum_{\vec{k}} E_{\vec{k}}\left(\hat{N}_{\vec{k}} + \frac{1}{2}\right) = \sum_{\vec{k}} \hbar\omega_k\left(a_{\vec{k}}^{\dagger}a_{\vec{k}} + \frac{1}{2}\right), \tag{8.6.8}$$

and therefore

$$[a_{\vec{k}}, H_0] = E_{\vec{k}}[a_{\vec{k}}, \hat{N}_{\vec{k}}] = \hbar\omega_{\vec{k}}a_{\vec{k}}, \tag{8.6.9}$$

$$[a_{\vec{k}}^{\dagger}, H_0] = E_{\vec{k}}[a_{\vec{k}}^{\dagger}, \hat{N}_{\vec{k}}] = -\hbar\omega_{\vec{k}}a_{\vec{k}}^{\dagger}, \tag{8.6.10}$$

using the commutation relations (4.8.11) from Section 4.8. Then

$$i\hbar\frac{\partial}{\partial t}a_{\vec{k}}(t) = e^{iH_0 t/\hbar}[a_{\vec{k}}, H_0]e^{-iH_0 t/\hbar}$$

$$= \hbar\omega_{\vec{k}}a_{\vec{k}}(t) \tag{8.6.11}$$

and therefore, solving the differential equation with initial condition $a_k(0) = a_k$,

$$\boxed{a_{\vec{k}}(t) = e^{-i\omega_{\vec{k}}t}a_{\vec{k}}.} \tag{8.6.12}$$

All of the above applies to both fermion and boson operators. Using these results it is easy to show that

$$\boxed{\begin{aligned}[a_{\vec{k}}(t_1), a_{\vec{k}}^{\dagger}(t_2)] &= e^{-i\omega_{\vec{k}}(t_1-t_2)} \quad \text{for bosons,} \\ \{b_{\vec{k}}(t_1), b_{\vec{k}}^{\dagger}(t_2)\} &= e^{-i\omega_{\vec{k}}(t_1-t_2)} \quad \text{for fermions,}\end{aligned}} \tag{8.6.13}$$

where the square brackets are the commutator and the curly brackets are the anticommutator (cf. (4.6.1)).

Let us now compute the contraction for two boson operators. If $t_1 > t_2$, we have

$$T(a_{\vec{k}}(t_1)a_{\vec{k}}^{\dagger}(t_2)) - N(a_{\vec{k}}(t_1)a_{\vec{k}}^{\dagger}(t_2)) = a_{\vec{k}}(t_1)a_{\vec{k}}^{\dagger}(t_2) - a_{\vec{k}}^{\dagger}(t_2)a_{\vec{k}}(t_1)$$

$$= [a_{\vec{k}}(t_1), a_{\vec{k}}^{\dagger}(t_2)]$$

$$= e^{-i\omega_{\vec{k}}(t_1-t_2)}. \tag{8.6.14}$$

If $t_1 < t_2$, then

$$T(a_{\vec{k}}(t_1)a_{\vec{k}}^{\dagger}(t_2)) = a_{\vec{k}}^{\dagger}(t_2)a_{\vec{k}}(t_1) = N(a_{\vec{k}}(t_1)a_{\vec{k}}^{\dagger}(t_2)), \tag{8.6.15}$$

and the contraction is equal to zero. Putting both possibilities $t_1 > t_2$ and $t_2 > t_1$ together, we have

$$
\begin{aligned}
T(a_{\vec{k}}(t_1)a_{\vec{k}}^\dagger(t_2)) - N(a_{\vec{k}}(t_1)a_{\vec{k}}^\dagger(t_2)) &= \overbrace{a_{\vec{k}}(t_1)a_{\vec{k}}^\dagger}(t_2) \\
&= \overbrace{a_{\vec{k}}^\dagger(t_2)a_{\vec{k}}}(t_1) \\
&= e^{-i\omega_{\vec{k}}(t_1-t_2)}\Theta(t_1 - t_2),
\end{aligned} \tag{8.6.16}
$$

where $\Theta(t)$ is the Heaviside function, equal to 1 if $t > 0$, and equal to 0 if $t < 0$.

Equation (8.6.16) is Wick's theorem in the case of two boson operators. As advertised, the contraction is a c-number regardless of the ordering of t_1 and t_2. The contraction is simply zero for the case of two creation operators or for two destruction operators, since they commute.

It should not be too hard to see that Wick's theorem will work for larger numbers of operators. The time-ordered product is converted to a normal-ordered product by switching the order of two operators, one pair at a time. Each time two operators are commuted, a commutator (which is a c-number) is generated, that is, a contraction.

Exercise 8.6.1   Verify Wick's theorem explicitly for two fermion operators acting on the same state $\vec{k}$. Show also that

$$
\overbrace{b_k(t_1)b_k^\dagger}(t_2) = -\overbrace{b_k^\dagger(t_2)b_k}(t_1). \tag{8.6.17}
$$

Exercise 8.6.2   Verify Wick's theorem explicitly for the case of four operators, namely two creation operators and two destruction operators for bosons in the same state $\vec{k}$. There are $4! = 24$ possible time orderings, but just pick two possible orderings to verify.

Green's functions. It is convenient to write the contraction of creation and destruction operators in terms of a **Green's function**,[1] defined as follows:

$$
G_{\vec{k}}(t_1 - t_2) \equiv -i\overbrace{a_k(t_1)a_k^\dagger}(t_2). \tag{8.6.18}
$$

For the case above, this definition implies

$$
G_{\vec{k}}(t) = -ie^{-i\omega_k t}\Theta(t), \tag{8.6.19}
$$

which has the Fourier transform

$$
\begin{aligned}
G(\vec{k}, \omega) &= -i\int_{-\infty}^{\infty} dt\, e^{i\omega t}\, e^{-i\omega_k t}\Theta(t) \\
&= \lim_{\epsilon \to 0} -i\int_0^{\infty} dt\, e^{i(\omega-\omega_k)t}e^{-\epsilon t} \\
&= \frac{1}{\omega - \omega_k + i\epsilon}.
\end{aligned} \tag{8.6.20}
$$

---

[1]   Some people have objected to the non-grammatical nature of this term – properly it should be just "Green's function" or "the Green function." Some have also tried to define the term "Greenian." These seem awkward; in deference to common usage, we stick to the term "the Green's function."

The Green's function can also be expressed in terms of the vacuum expectation value. We write

$$G_{\vec{k}}(t_1 - t_2) \equiv -i\langle \text{vac}|T(a_k(t_1)a_k^\dagger(t_2))|\text{vac}\rangle. \tag{8.6.21}$$

This automatically gives the right behavior for both cases $t_1 > t_2$ and $t_1 < t_2$ because a destruction operator in the rightmost position acting on vacuum always gives zero. The vacuum state here does not necessarily mean a true vacuum, but the state in which no quasiparticles of interest exist. (As we have discussed before, for example, in Section 2.1, we can always move to a quasiparticle picture in which the ground state of the system becomes our new vacuum.)

We will see below that the Green's functions have a central role in the perturbation theory; namely, they are the lines in the Feynman diagrams.

Calculating an S-matrix. Wick's theorem gives us a recipe for determining the S-matrix, that is, the time evolution of a system, in any order of perturbation theory. Here we work out an example explicitly. Although it may seem tedious, it brings out all the elements of the theory that we will summarize in the simple diagram rules in the next section.

Suppose that we have an initial state with one electron with momentum $\vec{q}_1$, that is, $|\psi(0)\rangle = |i\rangle = b_{q_1}^\dagger |\text{vac}\rangle$, and we want to know the probability of ending in a final state with an electron with momentum $\vec{q}_2$ and a phonon with momentum $\vec{q}_3$, that is, $|f\rangle = a_{q_3}^\dagger b_{q_2}^\dagger |\text{vac}\rangle$, due to an electron–phonon interaction potential of the form

$$V_{\text{int}} = \sum_{\vec{k},\vec{k}_1} M_{\vec{k}}\, i \left(a_{\vec{k}} b_{\vec{k}_1+\vec{k}}^\dagger b_{\vec{k}_1} - a_{\vec{k}}^\dagger b_{\vec{k}_1-\vec{k}}^\dagger b_{\vec{k}_1} \right), \tag{8.6.22}$$

where $M_{\vec{k}}$ is the appropriate matrix element for the electron–phonon interaction of interest. The S-matrix is given by

$$S(t, t_0) = 1 + \frac{1}{i\hbar} \int_{t_0}^t dt_1\, V_{\text{int}}(t_1) + \frac{1}{2} \frac{1}{(i\hbar)^2} \int_{t_0}^t dt_1 \int_{t_0}^t dt_2\, T(V_{\text{int}}(t_1)V_{\text{int}}(t_2)) + \cdots, \tag{8.6.23}$$

where

$$V_{\text{int}}(t) = e^{iH_0 t/\hbar} V_{\text{int}} e^{-iH_0 t/\hbar} \tag{8.6.24}$$
$$= \sum_{\vec{k},\vec{k}_1} M_{\vec{k}}\, i \left(a_{\vec{k}}(t) b_{\vec{k}_1+\vec{k}}^\dagger(t) b_{\vec{k}_1}(t) - a_{\vec{k}}^\dagger(t) b_{\vec{k}_1-\vec{k}}^\dagger(t) b_{\vec{k}_1}(t) \right).$$

Wick's theorem converts the time-ordered products to sums of normal-ordered products. We therefore just have to pick out each term in the S-matrix expansion that has a set of creation and destruction operators that turn the initial state into the final state.

For the first-order term in the S-matrix expansion above, we do not use any contractions, because we need each of the operators in V_{int}, one to kill the initial state and two to create the final state. Using (8.6.12) for the creation and destruction operators in the interaction representation, the only term from the sum in V_{int} that does not vanish after acting on the initial state is

$$S^{(1)}(t, t_0) = -iM_{\vec{q}_3} a^{\dagger}_{\vec{q}_3} b^{\dagger}_{\vec{q}_2} b_{\vec{q}_1} \frac{1}{i\hbar} \int_{t_0}^{t} dt_1 \, e^{-i(\omega_1 - \omega_2 - \omega_3)t_1}, \qquad (8.6.25)$$

where $\vec{q}_3 = \vec{q}_1 - \vec{q}_2$, and ω_1, ω_2, and ω_3 are the associated energies of the particles with these momenta. In the limit $t - t_0 \to \infty$, the integration of the exponential factor over time gives rise to an overall energy-conserving factor $\delta(\hbar\omega_1 - \hbar\omega_2 - \hbar\omega_3)$ when converting this to a scattering rate in the form of Fermi's golden rule, as shown in Section 4.7. The creation and destruction operators will in general give the appropriate factors of $\sqrt{N_{\vec{q}_1}}$, $\sqrt{1 + N_{\vec{q}_3}}$, etc.; here, they are just equal to 1 for our choice of the initial and final states. The contribution to the matrix element in Fermi's golden rule for this term is then just $\langle f | H^{(1)} | i \rangle = -iM_{\vec{q}_3}$.

There are no nonzero second-order terms in the S-matrix expansion, because there will be an extra phonon creation or destruction operator that does not correspond to any particle in the initial or final state. In the third-order term, however, contractions will convert pairs of operators into c-numbers which will not change any of the states. An example of a nonzero term is

$$S^{(3)}(t, t_0) = \frac{1}{(i\hbar)^3} \int_{t_0}^{t} dt_1 \int_{t_0}^{t} dt_2 \int_{t_0}^{t} dt_3 \, (-iM_{\vec{q}_3} M_{\vec{k}}^2)$$

$$\times \sum_{\vec{k}} \left(a^{\dagger}_{\vec{q}_3}(t_1) b^{\dagger}_{\vec{q}_2}(t_1) b_{\vec{q}_1}(t_1) a_{\vec{k}}(t_2) b^{\dagger}_{\vec{q}_1}(t_2) b_{\vec{q}_1 - \vec{k}}(t_2) a^{\dagger}_{\vec{k}}(t_3) b^{\dagger}_{\vec{q}_1 - \vec{k}}(t_3) b_{\vec{q}_1}(t_3) \right.$$

$$\left. + a^{\dagger}_{\vec{q}_3}(t_1) b^{\dagger}_{\vec{q}_2}(t_1) b_{\vec{q}_1}(t_1) a^{\dagger}_{-\vec{k}}(t_2) b^{\dagger}_{\vec{q}_1}(t_2) b_{\vec{q}_1 - \vec{k}}(t_2) a_{-\vec{k}}(t_3) b^{\dagger}_{\vec{q}_1 - \vec{k}}(t_3) b_{\vec{q}_1}(t_3) \right)$$

$$= \frac{1}{(i\hbar)^3} \int_{t_0}^{t} dt_1 \int_{t_0}^{t} dt_2 \int_{t_0}^{t} dt_3 \, (-iM_{\vec{q}_3} M_{\vec{k}}^2) \, e^{-i\omega_1 t_3 + i(\omega_2 + \omega_3)t_1} \, a^{\dagger}_{\vec{q}_3} b^{\dagger}_{\vec{q}_2} b_{\vec{q}_1}$$

$$\times \sum_{\vec{k}} \left[iG^{\text{phon}}_{\vec{k}}(t_2 - t_3) + iG^{\text{phon}}_{-\vec{k}}(t_3 - t_2) \right] iG^{e}_{\vec{q}_1}(t_1 - t_2) iG^{e}_{\vec{q}_1 - \vec{k}}(t_2 - t_3),$$

$$(8.6.26)$$

among others. The sum is only over $\vec{k}$ because the contractions are only nonzero for creation and destruction operators with the same $\vec{k}$, and only momentum-conserving possibilities are included in the sum of the interaction V_{int}. The factor $1/6 = 1/3!$ for the third-order perturbation expansion does not appear because there are six terms in the sums over all the momenta in the interaction Hamiltonians that give identical contributions (to see this, just imagine all possible orderings of the times t_1, t_2, and t_3). In general, the factor $1/n!$ in the S-matrix expansion (8.6.3) will not appear, because accounting for the redundant, identical terms in the S-matrix expansion exactly cancels this factor.

We can now switch to the Fourier transforms of the Green's functions, using

$$G_{\vec{k}}(t) = \frac{1}{2\pi} \int_{-\infty}^{\infty} d\omega \, e^{-i\omega t} G(\vec{k}, \omega). \qquad (8.6.27)$$

The pair of phonon Green's functions that appears in (8.6.26) commonly occurs in the computation of the S-matrix, because of the form of the interaction Hamiltonian (8.6.22).

We can simplify this to

$$G_{\vec{k}}(t) + G_{-\vec{k}}(-t) = \frac{1}{2\pi} \int_{-\infty}^{\infty} d\omega \left(e^{-i\omega t} G(\vec{k}, \omega) + e^{i\omega t} G(-\vec{k}, \omega) \right)$$

$$= \frac{1}{2\pi} \int_{-\infty}^{\infty} d\omega \, e^{-i\omega t} \left(G(\vec{k}, \omega) + G(-\vec{k}, -\omega) \right)$$

$$= \frac{1}{2\pi} \int_{-\infty}^{\infty} d\omega \, e^{-i\omega t} \tilde{G}(\vec{k}, \omega), \qquad (8.6.28)$$

where

$$\tilde{G}(\vec{k}, \omega) = \frac{1}{\omega - \omega_k + i\epsilon} + \frac{1}{-\omega - \omega_{-k} + i\epsilon}$$

$$= \frac{2\omega_k - 2i\epsilon}{\omega^2 - \omega_k^2 + 2i\omega_k\epsilon + \epsilon^2}, \qquad (8.6.29)$$

assuming $\omega_k = \omega_{-k}$, which follows from time-reversal symmetry (see Section 6.9). Since ϵ is an infinitesimal, we can rewrite this as

$$\tilde{G}(\vec{k}, \omega) = \frac{2\omega_k}{\omega^2 - \omega_k^2 + i\epsilon}. \qquad (8.6.30)$$

Equation (8.6.26) then becomes

$$S^{(3)}(t, t_0) = \frac{1}{(\hbar)^3} \int_{t_0}^{t} dt_1 \int_{t_0}^{t} dt_2 \int_{t_0}^{t} dt_3 \, (-iM_{\vec{q}_3} M_{\vec{k}}^2) \, e^{-i\omega_1 t_3 + i(\omega_2 + \omega_3)t_1} \, a_{\vec{q}_3}^\dagger b_{\vec{q}_2}^\dagger b_{\vec{q}_1}$$

$$\times \sum_{\vec{k}} \frac{1}{(2\pi)^3} \int_{-\infty}^{\infty} d\omega \int_{-\infty}^{\infty} d\omega' \int_{-\infty}^{\infty} d\omega'' \, e^{-i\omega(t_2 - t_3)} e^{-i\omega'(t_1 - t_2)} e^{-i\omega''(t_2 - t_3)}$$

$$\times \tilde{G}^{\text{phon}}(\vec{k}, \omega) G^e(\vec{q}_1, \omega') G^e(\vec{q}_1 - \vec{k}, \omega'').$$

As before, if t and t_0 are far enough apart, the integral over t_3 will converge to an energy-conserving δ-function, since then we can approximate the integral as

$$\int_{-\infty}^{\infty} dt_3 \, e^{-i(\omega_1 - \omega - \omega'')t_3} = 2\pi \delta(\omega_1 - \omega - \omega''). \qquad (8.6.31)$$

This can then be used to eliminate the integration over $\omega'' = \omega_1 - \omega$. The same thing can be done to eliminate the integration over t_2 and ω'. We then have

$$S^{(3)}(t, t_0) = \frac{1}{(\hbar)^3} (-iM_{\vec{q}_3} M_{\vec{k}}^2) a_{\vec{q}_3}^\dagger b_{\vec{q}_2}^\dagger b_{\vec{q}_1} \int_{t_0}^{t} dt_1 \, e^{-i(\omega_1 - \omega_2 - \omega_3)t_1}$$

$$\times \sum_{k} \frac{1}{(2\pi)} \int_{-\infty}^{\infty} d\omega \, \tilde{G}^{\text{phon}}(k, \omega) G^e(\vec{q}_1, \omega_1) G^e(\vec{q}_1 - \vec{k}, \omega_1 - \omega).$$

$$(8.6.32)$$

The last integration over t_1 leads to an overall energy-conserving δ-function through the same procedure used to deduce Fermi's golden rule, just as in the first-order term (8.6.25). We write

$$\lim_{(t - t_0) \to \infty} |\langle f | S^{(3)}(t, t_0) | i \rangle|^2 = \frac{2\pi}{\hbar} |\langle f | H^{(3)} | i \rangle|^2 \delta(E_i - E_f)(t - t_0), \qquad (8.6.33)$$

where

$$\langle f|H^{(3)}|i\rangle =$$

$$(-iM_{\vec{q}_3}M_{\vec{k}}^2)\sum_{\vec{k}}\frac{i}{(2\pi)\hbar^3}\int_{-\infty}^{\infty}d(\hbar\omega)\,\tilde{G}^{\text{phon}}(\vec{k},\omega)G^e(\vec{q}_1,\omega_1)G^e(\vec{q}_1-\vec{k},\omega_1-\omega)$$

$$(8.6.34)$$

has units of energy. As we did before for Fermi's golden rule, we then divide by the time interval to obtain the transition rate

$$\frac{\partial}{\partial t}|\langle f|\psi_t\rangle|^2 = \frac{\partial}{\partial t}\left[\lim_{(t-t_0)\to\infty}|\langle f|S^{(3)}(t,t_0)|i\rangle|^2\right]$$

$$= \frac{2\pi}{\hbar}|\langle f|H^{(3)}|i\rangle|^2\delta(E_i-E_f).\qquad(8.6.35)$$

The term (8.6.34) is a higher-order matrix element for the transitions between states $|i\rangle$ and $|f\rangle$. To get the transition rates properly, one must first add the matrix elements for all orders and *then* square the sum. This is possible because all of the S-matrix terms between the initial and final states have the same integral over time, leading to the same energy-conserving δ-function.

Going back, one can see that the Wick's theorem expansion of the S-matrix contains many other possible sets of contractions beside the ones we made in (8.6.26). This might seem to make the calculation of higher-order terms very complicated. Luckily, there is a diagrammatic method, like that of Rayleigh–Schrödinger perturbation theory, which allows us to easily do the bookkeeping, as we will discuss in Section 8.7.

Exercise 8.6.3   Suppose that an initial state consists of two electrons with the same spin in states $\vec{q}_1$ and $\vec{q}_2$, and the final state consists of electrons in states $\vec{q}_3$ and $\vec{q}_4$. Deduce the scattering matrix elements that are first and second order in the Coulomb interaction

$$\frac{1}{2}\sum_{\vec{k}_1,\vec{k}_2,\vec{k}_3}M_{|\Delta\vec{k}|}^{\text{Coul}}\,b_{\vec{k}_4}^{\dagger}\,b_{\vec{k}_3}^{\dagger}\,b_{\vec{k}_2}\,b_{\vec{k}_1},\qquad(8.6.36)$$

following the same procedure as used to deduce (8.6.34). Show that your answers have units of energy.

8.7 Diagram Rules for Feynman Perturbation Theory

Based on the theory of the previous section, we can write down a set of rules for Feynman diagrams for writing matrix elements that contribute to $S(t,t_0)$ in the limit $(t-t_0)\to\infty$. Note that these rules are quite different from those of Section 8.5, although there are some similarities. These Feynman diagram rules are not unique. In general, there are many different approaches to many-body theory in which different diagram rules are used, depending on the system under study. The ones presented here are convenient for solid state calculations.

- Pin down the external "legs" for incoming and outgoing particles in the process of interest. Assign a momentum and energy for each, and, if spin or polarization are being taken into account, assign a spin state or a polarization state to each external leg.
- Draw all the topologically distinct diagrams that connect the external legs, using the vertices of Figure 8.5. Assign a momentum and direction to each internal line, and an energy. The direction of electron lines, which is indicated by the arrow, must be continuous.
- Conserve both momentum and energy at each vertex. An outgoing hole at a vertex is counted as an incoming electron, that is, an electron moving backwards in time. (Elastic scattering from a defect is an exception; instead of conserving momentum one multiplies by a phase factor for each vertex.)
- Sum over each momentum that is not determined by momentum conservation. If spin, polarization, or interband transitions are being taken into account, sum over all possible spin, polarization states, and different bands that are not determined by the external legs. Converting sums over momentum to integrals gives an extra factor of $V/(2\pi)^3$ for each sum.
- Integrate over each energy not determined by energy conservation, and multiply by $i/2\pi$ for each integration over energy.
- For each internal line, write down the frequency-domain Green's function for the appropriate particle. These are also called **propagators** because they connect the vertices. The basic form of the electron propagators is

$$G(\vec{k}, E) = \frac{1}{E - E_{\vec{k}} + i\epsilon}, \tag{8.7.1}$$

where $E_{\vec{k}}$ is the energy of the particle as calculated from the momentum $\vec{k}$. Here we have switched the Green's function to a function of energy $E = \hbar\omega$ by dividing by $\hbar$. Recall that the vacuum state corresponds to the ground state of the system, that is, electrons filling all states up to the Fermi level. Therefore, the energy $E_{\vec{k}}$ is measured relative to the Fermi level E_F. When the state $\vec{k}$ has energy less than E_F, then the definition of the Green's function (8.6.18) implies $G_{\vec{k}}(t) = ie^{-i\omega_{\vec{k}}t}\Theta(-t)$, because as mentioned in Section 8.6, destruction operators that create holes are treated as creation operators, and moved to the left in the normal-ordered product. The hole propagator is then

$$G(\vec{k}, E) = \frac{1}{E - E_{\vec{k}} - i\epsilon}, \tag{8.7.2}$$

where the sign of the ϵ term must be the opposite in order to handle the time-dependent Green's function in the limit $t \to -\infty$.

For phonons or photons, we write

$$G(\vec{k}, E) = \frac{2E_{\vec{k}}}{E^2 - E_{\vec{k}}^2 + i\epsilon}, \tag{8.7.3}$$

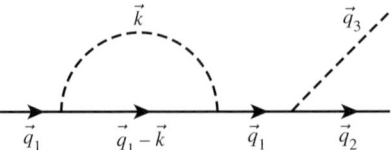

Fig. 8.9 The diagram for the electron–phonon interaction example that gives (8.6.34).

to take into account the two terms that appear in the interaction Hamiltonians for absorption and emission, as for example in (8.6.26), leading to (8.6.30). Here, we again neglect polarization.

- For each vertex, multiply by the appropriate constant. These are the same as from Section 8.5.
- For each crossing of fermion lines, multiply by -1. When working with external legs, do not cross fermion lines if there is a topologically equivalent diagram with uncrossed lines. Additionally, for each internal fermion loop, multiply by -1. (This was not the case for Rayleigh–Schrödinger theory.)
- As in Rayleigh–Schrödinger theory, for each external incoming leg that connects to a vertex, multiply by $\sqrt{N_{\vec{p}}}$, where $N_{\vec{p}}$ is the occupation number of the state $\vec{p}$, and for each outgoing leg from a vertex multiply by $\sqrt{1 \pm N'_{\vec{p}}}$, where $N'_{\vec{p}}$ is the occupation number of state $\vec{p}$ after taking into account the depletion of that state by incoming legs.

It is easy to see using these rules that the third-order electron–phonon interaction term we calculated in Section 8.6 is represented by the diagram in Figure 8.9.

Raman scattering. For an example, let us return to the case of resonant Raman scattering, for the three-vertex diagram shown in Figure 8.7(a). Using the above rules, we have, for an initial state with one photon and no phonons,

$$\langle f|H^{(3)}|i\rangle = -\frac{i}{2\pi} M_{\vec{q}}^{\text{dipole}} M_{\vec{q}-\vec{p}}^{\text{dipole}} M_{\vec{p}}^{\text{Fr}}$$

$$\times \int_{-\infty}^{\infty} dE \sum_{\vec{k}} \left(\frac{1}{E - \hbar\omega_{\text{phot}} - (E_{v,\vec{k}-\vec{q}} - E_F) - i\epsilon} \right) \qquad (8.7.4)$$

$$\times \left(\frac{1}{E - (E_{c,\vec{k}} - E_F) + i\epsilon} \right) \left(\frac{1}{E - \hbar\omega_{LO} - (E_{c,\vec{k}-\vec{p}} - E_F) + i\epsilon} \right).$$

The integral over E can be resolved using analytical calculus, viewing the real axis as part of a closed loop in the complex plane. There are two poles in the lower half plane, and one pole in the upper half plane. Since the integrand of (8.7.4) vanishes as $1/E^3$ at large magnitude of E, we can equate the integral over the real axis with the integral over the loop over the upper half plane. Cauchy's residue formula (7.2.1) gives

$$\oint \frac{f(z)}{z - z'} dz = 2\pi i f(z'), \qquad (8.7.5)$$

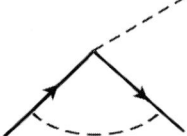

Fig. 8.10   Electron–phonon vertex with a higher-order correction.

for z' inside the closed loop. The pole in the upper half plane occurs at $E = \hbar\omega_{\text{phot}} + (E_{v,\vec{k}-\vec{q}} - E_F)$. Setting E equal to this value in the other two terms in the denominator to find the residue, we find

$$\langle f|H^{(3)}|i\rangle = M_{\vec{q}}^{\text{dipole}} M_{\vec{q}-\vec{p}}^{\text{dipole}} M_{\vec{p}}^{\text{Fr}} \sum_{\vec{k}} \frac{1}{E_{c,\vec{k}} - \hbar\omega_{\text{phot}} - E_{v,\vec{k}-\vec{q}} + i\epsilon}$$

$$\times \frac{1}{E_{c,\vec{k}-\vec{p}} + \hbar\omega_{LO} - \hbar\omega_{\text{phot}} - E_{v,\vec{k}-\vec{q}} + i\epsilon}, \tag{8.7.6}$$

which corresponds to the integral (8.5.4) written earlier using Rayleigh–Schrödinger theory. The integral for Figure 8.7(b) is actually topologically the same as Figure 8.7(a), so we do not need to write a separate diagram. The integral for Figure 8.7(b) can be obtained simply by remembering that we should sum over all electron bands, which we did not do in writing down (8.7.4).

Exercise 8.7.1   Show explicitly that (8.7.4) is equal to (8.7.6), using the procedure outlined in this section.

Exercise 8.7.2   Assign momenta for the external legs and internal propagators, and write down the integral for the vertex interaction shown in Figure 8.10, using the diagram rules above. You do not need to evaluate the integral.

8.8 Self-Energy

The diagrammatic method allows us to easily write down equations with series of terms. For example, consider the diagram equation shown in Figure 8.11. In the S-matrix expansion, there is an infinite number of terms. If we represent a single propagator as $1/A$ and the inner diagram as B, then this equation corresponds to the sum

$$S = \frac{1}{A} + \frac{1}{A}B\frac{1}{A} + \frac{1}{A}B\frac{1}{A}B\frac{1}{A} + \cdots. \tag{8.8.1}$$

Simple algebra leads to the result

$$S = \frac{1}{A - B}, \tag{8.8.2}$$

which can be verified easily by multiplying S by $(A - B)$.

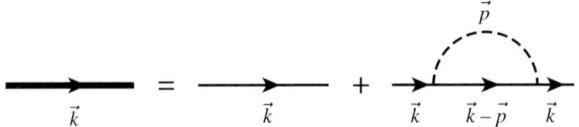

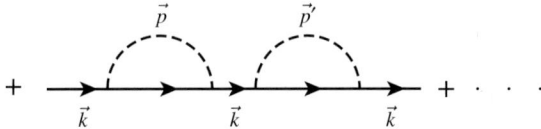

Infinite series of diagrams for the self-energy calculation.

In general, if the original propagator $1/A$ has the form

$$G(\vec{k}, \omega) = \frac{1}{\hbar\omega - E_{\vec{k}} + i\epsilon} \tag{8.8.3}$$

then S will be a new, *renormalized* propagator of the form

$$G'(\vec{k}, \omega) = \frac{1}{\hbar\omega - E_{\vec{k}} - \Sigma(\vec{k}, \omega) + i\epsilon}. \tag{8.8.4}$$

The term $E_{\vec{k}} + \Sigma(\vec{k}, \omega)$ is the renormalized energy of the quasiparticle. $\Sigma(\vec{k}, \omega)$ is the self-energy of the particle due to interactions, which plays the same role as the self-energy we introduced in Section 8.1. We can rewrite this more generally as

$$G' = G + G\Sigma G'. \tag{8.8.5}$$

This is known as a **Dyson equation**.

For example, for an electron–phonon Fröhlich interaction in the zero-temperature limit, the self-energy, given by the one-loop diagram in Figure 8.11 minus the external legs, will be

$$\Sigma(\vec{k}, \omega) = \frac{i}{2\pi} \int_{-\infty}^{\infty} dE \sum_{\vec{p}} (M_{\vec{p}}^{\mathrm{Fr}})^2 \frac{2\hbar\omega_{LO}}{(E^2 - (\hbar\omega_{LO})^2 + i\epsilon)} \frac{1}{(\hbar\omega - E - E_{\vec{k}-\vec{p}} + i\epsilon)}$$

$$= \frac{i}{2\pi} \int_{-\infty}^{\infty} dE \sum_{\vec{p}} (M_{\vec{p}}^{\mathrm{Fr}})^2 \left[\frac{1}{(E - \hbar\omega_{LO} + i\epsilon)} \frac{1}{(\hbar\omega - E - E_{\vec{k}-\vec{p}} + i\epsilon)} \right.$$

$$\left. + \frac{1}{(-E - \hbar\omega_{LO} + i\epsilon)} \frac{1}{(\hbar\omega - E - E_{\vec{k}-\vec{p}} + i\epsilon)} \right]. \tag{8.8.6}$$

Here we have used E for the energy of the phonon, and $\hbar\omega$ for the energy of the particle in state $\vec{k}$.

One can show that the second term vanishes, while the first of these two terms can be computed using Cauchy's residue formula in the upper half plane as in the Raman scattering example of Section 8.7. We obtain

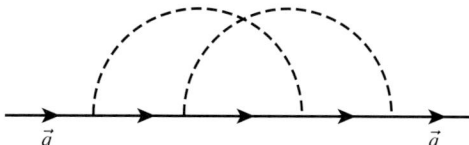

Fig. 8.12 Another self-energy diagram for electron–phonon interaction.

$$\text{Re } \Sigma_{\vec{k}} = \sum_{\vec{p}} \frac{(M_{\vec{p}}^{\text{Fr}})^2}{(E_{\vec{k}} - \hbar\omega_{LO} - E_{\vec{k}-\vec{p}})}, \tag{8.8.7}$$

where we have set $\hbar\omega = E_{\vec{k}}$ for the external legs. This is the polaron correction discussed in Section 8.2, for the zero-temperature diagram of Figure 8.6(c).

The self-energy summation above means that if there is an interaction that leads to a self-energy term, this can be simply incorporated into a new propagator with an additional energy term. We have renormalized the energy of the quasiparticles. Just as we found in Rayleigh–Schrödinger theory, the new energy term can have three effects: first, a shift of the ground state energy, due to a constant term in the self-energy; second, an adjustment of the effective mass of the particles, if there is a term proportional to k^2; and third, an imaginary self-energy term that corresponds to broadening of the state. The renormalized propagators can then be used in new diagrams. The Feynman diagram method therefore allows us to build up diagrams in a heirarchy of renormalized propagators.

In general, there is an infinite number of possible diagrams, and any sum will account for only a subset of these. One of the tricks of many-body theory is to be able to identify which are the important diagrams.

Exercise 8.8.1   Show that the second of the two terms in (8.8.6) vanishes.

Exercise 8.8.2   Write down the self-energy integral for the diagram shown in Figure 8.12. How does this compare to the two-phonon term in Figure 8.11?

Photon self-energy and the dielectric constant. In the same way that we can calculate an electron self-energy, we can also calculate a photon self-energy due to interaction with electrons. Let us assume that the photon interacts with a semiconductor, with electrons in two bands, a filled valence band and an empty conduction band. Then the Feynman rules for the diagram shown in Figure 8.13 give (summing over all possible electron bands)

$$\Sigma(\vec{k}, \omega) = \frac{\hbar}{2\epsilon_0 \omega_{\vec{k}} V} \frac{e^2 p_{cv}^2}{m^2} \frac{-i}{2\pi} \int_{-\infty}^{\infty} dE$$

$$\times \sum_{\vec{p}} \left[\frac{1}{(E - (E_{c,\vec{p}} - E_F) + i\epsilon)} \frac{1}{(E - \hbar\omega - (E_{v,\vec{p}-\vec{k}} - E_F) - i\epsilon)} \right.$$

$$\left. + \frac{1}{(E - (E_{v,\vec{p}} - E_F) + i\epsilon)} \frac{1}{(E - \hbar\omega - (E_{c,\vec{p}-\vec{k}} - E_F) - i\epsilon)} \right], \tag{8.8.8}$$

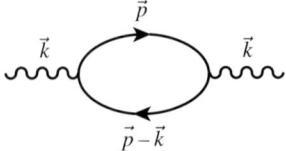

Diagram for the photon self-energy, that is, the polarizability of a medium.

where we have two terms, because in the sum over internal electron states, we restrict ourselves to terms that couple the valence and conduction band through the p_{cv} matrix element. Using Cauchy's residue formula for the upper half plane, we obtain

$$\Sigma(\vec{k}, \omega) = \frac{\hbar}{2\epsilon_0 \omega_{\vec{k}} V} \frac{e^2 p_{cv}^2}{m^2} \sum_{\vec{p}} \left[\frac{1}{\hbar\omega - E_{c,\vec{p}} + E_{v,\vec{p}-\vec{k}} + i\epsilon} + \frac{1}{\hbar\omega - E_{v,\vec{p}} + E_{c,\vec{p}-\vec{k}} + i\epsilon} \right].$$

(8.8.9)

Approximating $E_{c,\vec{k}} - E_{v,\vec{k}-\vec{p}} \approx \hbar\omega_0$, in other words, treating the bands as flat, with fixed energy, we then have

$$\Sigma \approx \frac{1}{2\epsilon_0 \omega} \frac{e^2 p_{cv}^2}{m^2} \frac{N_s}{V} \left[\frac{1}{\omega - \omega_0 + i\epsilon} + \frac{1}{\omega + \omega_0 + i\epsilon} \right]$$

$$= -\frac{e^2 p_{cv}^2}{\epsilon_0 m^2} \frac{N_s}{V} \frac{1}{\omega_0^2 - \omega^2 - i\epsilon},$$

(8.8.10)

where N_s is the number of states in the band. We once again apply the rule (8.8.2), with $1/A$ equal to the photon propagator (8.7.3) and B equal to this self-energy. We obtain

$$G'(\vec{k}, E) = \frac{2E_{\vec{k}}}{E^2 - E_{\vec{k}}^2 - 2E_{\vec{k}}\Sigma + i\epsilon}.$$

(8.8.11)

Renormalized vertices. Note that the vertex used for photons, given in Section 8.5,

$$M_{\vec{k}}^{\text{dipole}} = \frac{e}{m} p_{cv} \sqrt{\frac{\hbar}{2\epsilon V \omega_{\vec{k}}}},$$

(8.8.12)

contains the permittivity ϵ. In the calculation of the photon self-energy above, we assumed that this was equal to the vacuum value, ϵ_0. For consistency, we would like a set of propagators and vertices that both account for the change of ϵ due to the electron–photon interaction.

Since each end of a photon propagator will be connected to an electron–photon interaction vertex (ignoring any direct photon–phonon interactions), we can absorb part of the photon self-energy into a renormalized dielectric constant used in the vertices. Figure 8.14 shows a diagram that corresponds to the following equation, namely two vertices and a renormalized propagator:

$$M_{\vec{k}}^{\text{dipole}} G'(\vec{k}, E) M_{\vec{k}}^{\text{dipole}} = \frac{e^2 p_{cv}^2}{m^2} \frac{\hbar^2}{2\epsilon_0 V E_{\vec{k}}} \frac{2E_{\vec{k}}}{(E^2 - E_{\vec{k}}^2 - 2E_{\vec{k}}\Sigma + i\epsilon)}.$$

(8.8.13)

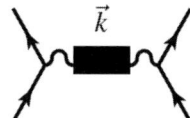

Fig. 8.14   A renormalized photon propagator between two electron–hole vertices.

We want to rewrite this as

$$\tilde{M}_{\vec{k}}^{\text{dipole}} \tilde{G}(\vec{k}, E) \tilde{M}_{\vec{k}}^{\text{dipole}} = \frac{e^2 p_{cv}^2}{m^2} \frac{\hbar^2}{2\epsilon V \tilde{E}_{\vec{k}}} \frac{2\tilde{E}_{\vec{k}}}{(E^2 - \tilde{E}_{\vec{k}}^2 + i\epsilon)}, \tag{8.8.14}$$

where $\tilde{E}_{\vec{k}} = \hbar ck/n$, and $n = \sqrt{\epsilon/\epsilon_0}$. Setting the two terms equal implies

$$E^2 - E_{\vec{k}}^2 - 2E_{\vec{k}}\Sigma = (\epsilon/\epsilon_0)(E^2 - \tilde{E}_{\vec{k}}^2) \tag{8.8.15}$$

or

$$2E_{\vec{k}}\Sigma = (1 - n^2)E^2 = -\chi E^2, \tag{8.8.16}$$

where χ is the optical susceptibility we have used before, e.g., in Section 7.1. For this reason, the self-energy is often called the **polarizability**, not only for photons but for other quasiparticles.

From (8.8.16) we have

$$\chi = -\frac{2E_{\vec{k}}\Sigma}{E^2} = \frac{e^2 N_s}{m\epsilon_0 V} \left(\frac{2p_{cv}^2}{m\hbar\omega}\right) \frac{1}{\omega_0^2 - \omega^2 - i\epsilon}, \tag{8.8.17}$$

where we set $E = E_{\vec{k}}$. Comparing to (7.2.3) and (7.4.17) from Section 7.4, we see that once again we recover the standard form for the susceptibility of the two-level electronic oscillator.

Note that in the above we absorbed the effect of the interactions into a renormalized dielectric constant in the factors for the vertices, but we could have instead left the vertices untouched and generated a renormalized propagator. Because of this arbitrariness, the above renormalization procedure is by no means universal; one must be careful to follow how the renormalization is done in different textbooks.

Exercise 8.8.3   Comparing (8.8.17) to (5.2.10) of Chapter 5, show that this result gives the same absorption coefficient, if the electron bands are assumed to have constant energy.

8.9 Physical Meaning of the Green's Functions

We now have "bare" Green's functions and "renormalized" Green's functions. What do the Green's functions represent, physically?

Consider a state prepared as a single particle created in a vacuum. The initial state corresponds to a single particle created in a $\vec{k}$-state at time t_0:

$$|i\rangle = a_{\vec{k}}^{\dagger}|\text{vac}\rangle. \tag{8.9.1}$$

Suppose we would like to know the probability of finding a single particle in the same $\vec{k}$-state after some time t. We evolve this state forward in time, to time t:

$$\begin{aligned}|\psi_t\rangle &= e^{-iH_0 t/\hbar}|i\rangle \\ &= e^{-iH_0 t/\hbar}a_{\vec{k}}^{\dagger}|\text{vac}\rangle.\end{aligned}$$

The probability amplitude for finding the particle in the original state is then given by

$$\begin{aligned}\langle i|\psi_t\rangle &= \left(\langle\text{vac}|a_{\vec{k}}\right)\left(e^{-iH_0 t/\hbar}a_{\vec{k}}^{\dagger}|\text{vac}\rangle\right) \\ &= \langle\text{vac}|a_{\vec{k}}(t)a_{\vec{k}}^{\dagger}(0)|\text{vac}\rangle,\end{aligned} \tag{8.9.2}$$

where we have used the fact that the energy of the vacuum is zero, by definition, so that $|\text{vac}\rangle = e^{-iH_0 t/\hbar}|\text{vac}\rangle$. If t is less than zero, this would correspond to asking the probability of finding a particle in state $\vec{k}$ before it had been created. Setting this to zero probability, we have

$$\begin{aligned}\langle i|\psi_t\rangle &= \langle\text{vac}|a_{\vec{k}}(t)a_{\vec{k}}^{\dagger}(0)|\text{vac}\rangle\,\Theta(t) \\ &= \langle\text{vac}|T(a_{\vec{k}}(t)a_{\vec{k}}^{\dagger}(0))|\text{vac}\rangle \\ &\equiv iG_{\vec{k}}(t).\end{aligned} \tag{8.9.3}$$

This is the Green's function we introduced in Section 8.6 for a particle in state $\vec{k}$. In other words, the Green's function gives us the probability of creating a particle in state $\vec{k}$, evolving it forward in time, and finding it in the same state.

If we include interactions, we must use the full interacting Hamiltonian to evolve the system in time. We generalize to

$$iG_{\vec{k}}'(t) = \langle\text{vac}|e^{iHt/\hbar}a_{\vec{k}}e^{-iHt/\hbar}a_{\vec{k}}^{\dagger}|\text{vac}\rangle\Theta(t), \tag{8.9.4}$$

where $H = H_0 + V_{\text{int}}$ is the full Hamiltonian. As we have seen, the time evolution under the full Hamiltonian is given by $e^{-iHt/\hbar} = e^{-iH_0 t/\hbar}S(t, 0)$. We therefore have

$$\begin{aligned}iG_{\vec{k}}'(t) &= \left(\langle\text{vac}|S(0,t)e^{iH_0 t/\hbar}\right)\left(a_{\vec{k}}e^{-iH_0 t/\hbar}S(t,0)a_{\vec{k}}^{\dagger}|\text{vac}\rangle\right)\Theta(t) \\ &= \langle\text{vac}|S(0,t)a_{\vec{k}}(t)S(t,0)a_{\vec{k}}^{\dagger}(0)|\text{vac}\rangle\Theta(t).\end{aligned} \tag{8.9.5}$$

We can simplify this by making the assumption that the action of the S-matrix on the vacuum state just gives a phase factor as $t \to \infty$, such that $S(t,0)|\text{vac}\rangle = \alpha_0(t)|\text{vac}\rangle = e^{i\theta(t)}|\text{vac}\rangle$. In that case,

$$\begin{aligned}iG_{\vec{k}}'(t) &= \langle\text{vac}|e^{-i\theta(t)}a_{\vec{k}}(t)S(t,0)a_{\vec{k}}^{\dagger}(0)|\text{vac}\rangle\Theta(t) \\ &= \frac{\langle\text{vac}|a_{\vec{k}}(t)S(t,0)a_{\vec{k}}^{\dagger}(0)|\text{vac}\rangle\Theta(t)}{\langle\text{vac}|S(t,0)|\text{vac}\rangle}.\end{aligned} \tag{8.9.6}$$

The numerator here corresponds to the renormalized Green's function of Section 8.8, in which the operators and the S-matrix are time-ordered, generating a whole set of diagrams via Wick's theorem. The denominator gives the vacuum renormalization factor, which will be discussed below.

The assumption that the effect of the interaction on the vacuum is just a phase factor originated in the theory of scattering of single particles, in which it can be assumed that at late times, the interaction drops to zero as the particles move apart, so that the eigenstates return to those of the unperturbed vacuum. Another approach is to argue that even in the presence of interactions, there can be no long-term, permanent creation of new quasiparticles. Because the wave function is normalized, we can write

$$(\langle \text{vac}|S(0,t))\,(S(t,0)|\text{vac}\rangle) = 1 \tag{8.9.7}$$
$$= \sum_n \langle \text{vac}|S(0,t)|n\rangle\langle n|S(t,0)|\text{vac}\rangle = \sum_n |\langle n|S(t,0)|\text{vac}\rangle|^2,$$

where the sum is over the complete set of eigenstates. In this sum, all of the terms $|\langle n|S(t,0)|\text{vac}\rangle|^2$ are exactly what we calculate in Fermi's golden rule, discussed in Section 8.1. Since the states $|n\rangle$ all have energy higher than the ground state, we know that after a sufficiently long time, transitions to these states will be suppressed by the energy-conserving δ-function that occurs for real transitions. Thus, for large enough t we are left with only the ground state term, $|\langle \text{vac}|S(t,0)|\text{vac}\rangle|^2 = 1$, which implies that the only effect of $S(t,0)$ on the ground state is a phase factor.

Disconnected diagrams and vacuum renormalization. So far we have ignored an odd possibility that arises in our diagrammatic method. This is the case of **disconnected diagrams**, as shown in Figure 8.15. For a given order of the perturbation theory, that is, for a given number of vertices, we can have valid diagrams that have "bubbles" not connected to the propagator of interest. The diagram shown in Figure 8.15 is a fourth-order electron–phonon interaction diagram. Do we need to account for all possible disconnected diagrams in calculating self-energies?

It turns out that these diagrams are eliminated by accounting for the normalization of the wave function. In (8.9.6), we defined the renormalized Green's function, which had a denominator that accounted for the effect of the interactions on the vacuum. This denominator can be computed in a perturbation expansion just as we have done for the numerator,

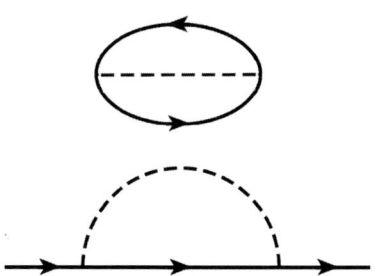

Fig. 8.15   A disconnected diagram in the calculation of a renormalized propagator.

using the same diagrammatic methods. The expansion will consist of all the same disconnected diagrams as can be found for the numerator, and only these. We know that when the interactions are neglected, the normalization factor will be exactly unity. We therefore have

$$\frac{1}{\langle \text{vac}|S(t,0)|\text{vac}\rangle} = \frac{1}{(1+\Lambda)},$$ (8.9.8)

where Λ represents the contributions from the disconnected diagrams.

At the same time, since each term in the numerator can be accompanied by any number of disconnected diagrams, we can write

$$\langle \text{vac}|a_{\vec{k}}(t)S(t,0)a_{\vec{k}}^{\dagger}(0)|\text{vac}\rangle = \Pi_{\vec{k}}(1+\Lambda),$$ (8.9.9)

where $\Pi_{\vec{k}}$ is the contribution of the standard set of diagrams, and Λ is the contribution of disconnected diagrams. The $(1 + \Lambda)$ terms will exactly cancel. Thus, in one step we eliminate both the need to worry about the disconnected diagrams and the need to worry about the vacuum state evolution when we compute renormalized propagators.

Vacuum energy. Although the effects of disconnected diagrams cancel out in calculations like those above, we can ask what they represent. In the S-matrix expansion using Wick's theorem acting on the vacuum state, there are any number of diagrams, such as those shown in Figure 8.16. These essentially correspond to particles being created out of vacuum for a very short time and then annihilating themselves again. (As we have done throughout this book, we define the 'vacuum' as the ground state of the system.) By the uncertainty principle, if these quasiparticles excited out of the vacuum live for a short enough time, there is no limit on how many or how energetic they may be.

As an example of one such term, consider the first of the diagrams shown in Figure 8.16. There are two internal momenta that must be summed over. From the rules in Section 8.7, we have the energy

$$\Delta = \frac{-1}{(2\pi)^2} \int_{-\infty}^{\infty} dE \sum_{\vec{k}} \int_{-\infty}^{\infty} dE' \sum_{\vec{p}} M_k^2 \frac{2\hbar c k}{E^2 - (\hbar c k)^2 + i\epsilon}$$

$$\times \frac{1}{(E' - (E_{c,\vec{p}} - E_F) + i\epsilon)} \frac{1}{(E + E' - (E_{v,\vec{p}+\vec{k}} - E_F) - i\epsilon)}.$$ (8.9.10)

Fig. 8.16   Examples of the infinite series of all possible disconnected diagrams.

The integral over E' can be resolved by taking a loop in the upper half of the complex plane, giving

$$\frac{-2\pi i}{(E_{c,\vec{p}} - E_F) - (E_{v,\vec{p}+\vec{k}} - E_F) - E + i\epsilon} = \frac{-2\pi i}{E_g + \dfrac{\hbar^2 p^2}{2m_r} + \dfrac{\hbar^2 \vec{k} \cdot \vec{p}}{m_v} + \dfrac{\hbar^2 k^2}{2m_v} - E + i\epsilon},$$

(8.9.11)

where m_r is the reduced mass. The integral over E is similarly resolved by taking a loop in the upper half of the complex plane. We then have

$$\Delta = \sum_{\vec{k},\vec{p}} M_k^2 \frac{1}{\hbar ck + E_g + \dfrac{\hbar^2 p^2}{2m_r} + \dfrac{\hbar^2 \vec{k} \cdot \vec{p}}{m_v} + \dfrac{\hbar^2 k^2}{2m_v} - i\epsilon}.$$

(8.9.12)

For a finite solid, this term can be computed and is nonzero. More strangely, the same diagram exists in a pure vacuum; in this case the "gap" energy E_g is the energy gap between the matter and antimatter bands, $2mc^2$ (see Appendix F). In this case, there is no upper bound to the sums over $\vec{k}$ and $\vec{p}$; when the relativistic energy–momentum relation is taken into account, the integral that corresponds to (8.9.12) will diverge. (See, e.g., Mandl and Shaw 1984 for a fully relativistic Feymann perturbation theory.) This would seem to present a problem, but as discussed in Section 2.1, the idea of a renormalized vacuum means that we can view the ground state of any system as a new vacuum, no matter what its energy. We simply move to a new picture in which the only thing that matters is excitations out of the ground state, that is, vacuum, and the energy of the ground state is irrelevant.

This means that the vacuum state is active, full of particles popping into existence and disappearing at all times. This is true not just of a solid state system but also of empty space, which has various vacuum diagrams involving matter–antimatter pairs appearing and disappearing. In philosophical terms, the vacuum is not "nothing," but an active field. This lends support to the idea that every particle is simply a consequence of quantizing the underlying field, and we cannot treat particles in vacuum as more real than quasiparticles in a solid; for example, photons as more real than phonons or electrons as more real than holes.

The vacuum energy leads to a measurable effect. Between two parallel conducting plates, the boundary conditions will allow virtual photons to exist if they have a wavelength equal to the separation between the plates divided by an integer. If this separation is small, then a significant fraction of states will be forbidden. This will lead to a reduction of the vacuum energy density between the plates, which implies an attractive force between the plates. This is known as the **Casimir effect**. The Casimir effect has been measured in experiments in recent years (for a review see Imry 2005).

Exercise 8.9.1   Write down the integral for the second diagram in Figure 8.16 and show that it is nonzero.

The spectral function. Recall, from the discussion of line broadening in Rayleigh–Schrödinger theory in Section 8.4 that the probability amplitude for a transition from a state $|i\rangle$ to state $|n\rangle$ via an interaction V' is given by (8.4.8),

$$\langle n|\psi(t)\rangle = \frac{1}{i\hbar} \int_0^t dt' \; \langle n|V'_{\text{int}}|i\rangle \; e^{(i/\hbar)(E_n - E_i)t'} \langle i|\psi(t')\rangle. \tag{8.9.13}$$

As we have seen at the beginning of this section, the probability amplitude $\langle i|\psi(t)\rangle$ which was so important in Rayleigh–Schrödinger perturbation theory corresponds to the Feynman Green's function in the interaction representation. Switching to the Heisenberg representation, using the relation $|\psi(t)\rangle = e^{iH_0 t}|\psi_t\rangle$, allows us to express (8.9.13) in terms of the Feynman Green's function:

$$\langle n|\psi(t)\rangle = \frac{1}{i\hbar} \int_0^t dt' \; \langle n|V'_{\text{int}}|i\rangle \; e^{(i/\hbar)E_n t'} \langle i|\psi_{t'}\rangle$$

$$= \frac{1}{\hbar} \int_0^t dt' \; \langle n|V'_{\text{int}}|i\rangle \; e^{(i/\hbar)E_n t'} \; G_{\vec{k}}(t'). \tag{8.9.14}$$

Suppose that the Green's function has a self-energy correction $\Delta_{\vec{k}} - i\Gamma_{\vec{k}}$, and $E'_{\vec{k}} = E_{\vec{k}} + \Delta_{\vec{k}}$. Then the energy-dependent Green's function is given by

$$G(\vec{k}, E) = \frac{1}{E - E'_{\vec{k}} + i\Gamma_{\vec{k}}}, \tag{8.9.15}$$

which, it can be verified, is the Fourier transform of

$$G_{\vec{k}}(t) = -ie^{-i(E'_{\vec{k}} - i\Gamma_{\vec{k}})t/\hbar} \; \Theta(t). \tag{8.9.16}$$

Substituting this into (8.9.14), we have the same results as in Section 8.4,

$$\frac{1}{\tau} = \frac{|\langle n|\psi_\infty\rangle|^2}{\int_0^\infty dt\, |\langle i|\psi_t\rangle|^2} = |\langle n|V'_{\text{int}}|i\rangle|^2 \frac{2\Gamma/\hbar}{(E_n - E'_{\vec{k}})^2 + \Gamma^2}. \tag{8.9.17}$$

The imaginary self-energy Γ gives the spread of energies of the state due to the interactions in the system. In other words, because a pure momentum state is no longer an eigenstate, its energy cannot be exactly defined. The stronger the interactions, the less time the particle spends in a given momentum state, and therefore, by the uncertainty principle, the broader its energy spectrum.

The spectral broadening in this case is directly proportional to the **spectral function**,

$$A(\vec{k}, E) \equiv -2\text{Im}\, G(\vec{k}, E) = iG(\vec{k}, E) - iG^*(\vec{k}, E), \tag{8.9.18}$$

which for the case of a particle with renormalized self-energy, discussed in Section 8.8, is

$$A(\vec{k}, E) = i\left(\frac{1}{E - E'_{\vec{k}} + i\Gamma} - \frac{1}{E - E'_{\vec{k}} - i\Gamma} \right)$$

$$= \frac{2\Gamma}{(E - E'_{\vec{k}})^2 + \Gamma^2}. \tag{8.9.19}$$

The spectral function can be seen as giving the probability that a particle with momentum $\vec{k}$ has energy E, given that the k-states are not eigenstates in the presence of interactions.

The spectral function need not always be Lorentzian. In general, the self-energy terms Δ and Γ can depend on E as well as $\vec{k}$. When finite temperature is taken into account, we will define a revised spectral function, as discussed in Section 8.14.

Exercise 8.9.2   Prove that (8.9.16) is the Fourier transform of (8.9.15), when the integral over time is divided by $\hbar$ to give units of inverse energy.

8.10 Finite Temperature Diagrams

The Feynman perturbation theory we have developed so far has been applied to a zero-temperature state, that is, a system in its ground state, or effective vacuum state, to which a few extra particles are added as incoming legs. There is no reason, however, why we cannot have any number of particles in the initial state. In particular, we can account for a thermal background of particles by creating an initial state with some distribution of particles, and then performing the thermal average over all such initial states.

We can compute the thermal average at a given temperature T and chemical potential μ as

$$\langle\ldots\rangle_T \equiv \frac{1}{Z_0}\sum_j e^{-(E_j-\mu N_j)/k_B T}\langle j|\ldots|j\rangle, \tag{8.10.1}$$

where E_j is the total energy and N_j is the total number of particles of state $|j\rangle$, according to all of the individual occupation numbers $N_{p1}, N_{p2}, \ldots$ that define that state; the partition function is

$$Z_0 = \sum_j e^{-(E_j-\mu N_j)/k_B T}. \tag{8.10.2}$$

In principle, the energy E_j that enters into the weighting should include the effects of interactions and not just H_0. In Section 8.13, we will introduce a diagrammatic method for doing this. In many cases, however, it is sufficient to do the thermal average only in lowest order, keeping track only of H_0. In this case, the Feynman diagram method we have introduced can be modified quite simply to account for finite temperature.

We first recall that in the expansion of the S-matrix using Wick's theorem (8.6.5), there is no rule that every quasiparticle in the initial state must interact with the expansion term of interest. Some quasiparticles in the initial state may be left untouched. These will be treated the same as disconnected diagrams, discussed in Section 8.9, and ignored. If there are particles that are created or destroyed by an interaction, however, then creation and destruction operators will give extra factors of $\sqrt{N_p}$ or $\sqrt{1 \pm N_p}$. In the case of a self-energy diagram, all of the incoming external legs must be put back where they were via outgoing external legs in the same states. This will always turn these square roots into just N_p or $(1 \pm N_p)$. In the case of a scattering amplitude, the incoming external legs do not have to be the same as the outgoing external legs, so that there may be unmatched square roots. However, when a scattering rate is calculated, Fermi's golden rule dictates that the scattering amplitude is squared, so that once again the square roots will be turned into linear terms.

Assuming an incoherent thermal population, we therefore can add the following set of rules for Feynman diagrams at finite temperatures:

- Define the initial and final states as Fock states of the form $|j\rangle = |\ldots, N_{p_1}, N_{p_2}, \ldots\rangle$, which assigns an occupation number to each momentum state of the background thermal population. In principle, this corresponds to a large number of external legs, most of which will not intersect with the diagram and therefore will be irrelevant. Therefore, just include external legs from this background that intersect with the diagram of interest.
- Sum over all the momenta of these external legs.
- In addition, add explicitly any external legs that represent particles of interest in the presence of this background.
- After all other calculations have been made, perform the thermal average, that is, perform the average (8.10.1). For terms linear in $N_{\vec{p}}$, this amounts to replacing $N_{\vec{p}}$ with its thermal average value $\langle N_{\vec{p}}\rangle_T$ anywhere it occurs. In general, when the system does not have phase coherence, products of different occupation numbers are assumed to be factorizable, i.e., $\langle N_{p_1} N_{p_2}\rangle_T = \langle N_{p_1}\rangle_T \langle N_{p_2}\rangle_T$.

  For scattering rates, first square the amplitude (the matrix element), obtaining terms of the form $\langle j| \ldots |i\rangle\langle i| \ldots |j\rangle$, and *then* take the thermal average over states $|j\rangle$.

This diagrammatic method can also be used for nonequilibrium distributions. The instantaneous state of the particles can be represented as a Fock state with distribution function $N_{\vec{p}}$, as we did for the quantum Boltzmann equation (see Section 4.8). This approach will be valid under the same conditions as the quantum Boltzmann equation, that is, the same conditions under which Fermi's golden rule applies.

The approach here of representing the thermal background as an average over Fock states presumes that only the occupation numbers of the states are relevant. As discussed in Section 4.8, this amounts to assuming that there is no macroscopic coherent phase. This is known as the **random phase approximation** (RPA). This term is sometimes used in a much more restricted sense to refer to the screening calculation done in Section 8.11.

Effective finite-temperature propagators. The above approach, of treating thermal background particles as external legs, works well when there are not too many such legs intersecting a diagram. For higher orders of the perturbation theory, accounting for these legs can get confusing. Therefore, it is sometimes convenient to define an effective Green's function that accounts for the thermal background without external legs.

In the presence of the thermal background, we define an effective Green's function which is the same as our previous one (given in (8.9.3) of Section 8.9), but instead of a vacuum expectation value, we take the average over all possible initial states,

$$\tilde{G}_{\vec{k}}(t_1 - t_2) = -i\langle \mathrm{T}(a_{\vec{k}}(t_1) a_{\vec{k}}^\dagger(t_2))\rangle_T \tag{8.10.3}$$

$$= -i\Bigg[\frac{1}{Z_0} \sum_j e^{-\beta(E_j - \mu N_j)} \langle j| a_{\vec{k}}(t_1) a_{\vec{k}}^\dagger(t_2) |j\rangle \Theta(t_1 - t_2)$$

$$+ \frac{1}{Z_0} \sum_j e^{-\beta(E_j - \mu N_j)} \langle j| a_{\vec{k}}^\dagger(t_2) a_{\vec{k}}(t_1) |j\rangle \Theta(t_2 - t_1) \Bigg],$$

where $\beta = 1/k_B T$. In this case, a destruction operator acting on the state $|j\rangle$ does not automatically remove that term, since the initial state may be occupied.

We take each of the time-dependent operators and act on the state $|j\rangle$. This gives

$$\tilde{G}_{\vec{k}}(t_1 - t_2) = -i\frac{1}{Z_0} \sum_j e^{-iE_{\vec{k}}(t_1 - t_2)/\hbar} \langle j|(1 + N_{\vec{k}})|j\rangle e^{-\beta(E_j - \mu N_j)} \Theta(t_1 - t_2)$$

$$+ \frac{1}{Z_0} \sum_j e^{-iE_{\vec{k}}(t_1 - t_2)/\hbar} \langle j|N_{\vec{k}}|j\rangle e^{-\beta(E_j - \mu N_j)} \Theta(t_2 - t_1)$$

$$= -ie^{-iE_{\vec{k}}(t_1 - t_2)/\hbar} \left((1 + \langle N_{\vec{k}}\rangle_T)\Theta(t_1 - t_2) + \langle N_{\vec{k}}\rangle_T \Theta(t_2 - t_1) \right).$$

$$(8.10.4)$$

Because there is thermal occupation of the initial state, there is a probability proportional to $\langle N_{\vec{k}}\rangle_T$ that the destruction operator will act on an occupied state. In the same way, the probability of creating a particle is stimulated by the familiar $(1 + \langle N_{\vec{k}}\rangle_T)$ factor. A similar calculation for fermions gives

$$\tilde{G}_{\vec{k}}(t_1 - t_2) = -ie^{-iE_{\vec{k}}(t_1 - t_2)/\hbar} \left((1 - \langle N_{\vec{k}}\rangle_T)\Theta(t_1 - t_2) - \langle N_{\vec{k}}\rangle_T \Theta(t_2 - t_1) \right).$$

$$(8.10.5)$$

Both of these become our previously introduced Green's function in the limit $\langle N_{\vec{k}}\rangle_T \to 0$, in particular, in the $T = 0$ limit for a boson with a Planck distribution, and for free electrons above the Fermi level at $T = 0$.

The Fourier transform of (8.10.4) for bosons is then

$$\tilde{G}(\vec{k}, \omega) = -i \int_{-\infty}^{\infty} dt\, e^{i\omega t} \tilde{G}_{\vec{k}}(t)$$

$$= \frac{1}{\omega - \omega_{\vec{k}} + i\epsilon}(1 + \langle N_{\vec{k}}\rangle_T) - \frac{1}{\omega - \omega_{\vec{k}} - i\epsilon}\langle N_{\vec{k}}\rangle_T. \qquad (8.10.6)$$

As discussed in Section 8.6, phonons and photons have parallel terms for $-\omega$ and $-\vec{k}$. When these are added in, the Green's function becomes

$$\tilde{G}(\vec{k}, \omega) = \frac{2\omega_k}{\omega^2 - \omega_{\vec{k}}^2 + i\epsilon}(1 + \langle N_{\vec{k}}\rangle_T) - \frac{2\omega_{\vec{k}}}{\omega^2 - \omega_{\vec{k}}^2 - i\epsilon}\langle N_{\vec{k}}\rangle_T.$$

$$(8.10.7)$$

The Fourier transform for fermions is

$$\tilde{G}(\vec{k}, \omega) = \frac{1}{\omega - \omega_{\vec{k}} + i\epsilon}(1 - \langle N_{\vec{k}}\rangle_T) + \frac{1}{\omega - \omega_{\vec{k}} - i\epsilon}\langle N_{\vec{k}}\rangle_T. \qquad (8.10.8)$$

This has a natural interpretation: We use the electron propagator according to the probability that the state $\vec{k}$ is empty, allowing us to create an electron in that state, and the hole propagator according to the probability that the state $\vec{k}$ is full, allowing us to create a hole. (Note that these finite-temperature Green's functions do *not* generate spectral functions subject to the sum rules discussed in Section 8.14, except at $T = 0$).

We can use these finite-temperature Green's functions just like our previous ones, under the conditions of the random-phase approximation, that is, when the thermal average of a product is equal to the product of the individual averages. That allows us to use Wick's theorem as before to write the S-matrix in terms of contractions, and then take the thermal average of each contraction as done above.

Because these effective Green's functions are the sum of two propagators, we have to be careful how we use a Dyson equation to get a renormalized propagator: These effective Green's functions do not have the simple form $1/A$ invoked in the summation process of (8.8.1). However, we can write a Dyson equation for each of the two propagators in (8.10.4) and (8.10.5) separately, and then weight each of them by the appropriate thermal population factor. This means that we can calculate a self-energy using the full finite-temperature Green's functions, and then use this self-energy to adjust the denominator of each propagator just as we did for the zero-temperature propagators. The diagram rule is to use the finite-temperature, effective propagators given here for internal lines whenever the method discussed at the beginning of this section would bring in additional external lines sampling background particles, and to use the simple, bare propagator whenever continuity with the original external legs is needed, when these legs are selected out, and not considered part of the thermal background.

Example: energy renormalization due to phonons. If we use this form of the propagator in our standard Feynman diagram method, we will recover the same results that we obtained previously using our original propagators and a set of external legs to account for background particles at finite temperature. For example, Figure 8.6 of Section 8.5 gives three diagrams for the renormalization of the energy of a free electron due to phonon interactions. These gave the self-energy

$$\sum_{\vec{p}} \frac{(M_{\vec{p}}^{\mathrm{Fr}})^2 (1 + \langle N_{\vec{p}} \rangle_T)}{E_{\vec{k}} - \hbar\omega_{\vec{p}} - E_{\vec{k}-\vec{p}} + i\epsilon} + \sum_{\vec{p}} \frac{(M_{\vec{p}}^{\mathrm{Fr}})^2 \langle N_{\vec{p}} \rangle_T}{E_{\vec{k}} + \hbar\omega_{\vec{p}} - E_{\vec{k}+\vec{p}} + i\epsilon}. \tag{8.10.9}$$

Using the present approach, we write only the diagram shown in Figure 8.17. From the Feynman rules with our finite-temperature propagator, we have

$$\sum_{\vec{p}} (M_{\vec{p}}^{\mathrm{Fr}})^2 \frac{i}{2\pi} \int_{-\infty}^{\infty} dE \left(\frac{2\hbar\omega_{\vec{p}}}{E^2 - (\hbar\omega_{\vec{p}})^2 + i\epsilon} (1 + \langle N_{\vec{p}} \rangle_T) - \frac{2\hbar\omega_{\vec{p}}}{E^2 - (\hbar\omega_{\vec{p}})^2 - i\epsilon} \langle N_{\vec{p}} \rangle_T \right)$$

$$\times \frac{1}{E_{\vec{k}} - E - E_{\vec{k}-\vec{p}} + i\epsilon}. \tag{8.10.10}$$

In using Cauchy's residue theorem to resolve the integrals over energy, terms with two poles on the same side of the complex plane will vanish. (See Exercise 8.8.1.) There are just two terms with poles in both the upper and lower half planes, giving

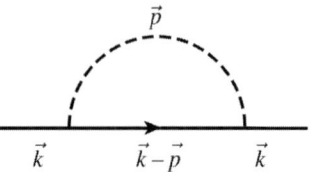

Fig. 8.17   Diagram for the self-energy correction of a free electron due to phonon interaction.

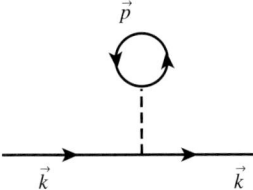

Fig. 8.18 Another diagram for the self-energy correction of a free electron due to phonon interaction.

$$\sum_{\vec{p}} (M_{\vec{p}}^{\text{Fr}})^2 \left(\frac{1}{E_{\vec{k}} - \hbar\omega_{\vec{p}} - E_{\vec{k}-\vec{p}} + i\epsilon}(1 + \langle N_{\vec{p}}\rangle_T) + \frac{1}{E_{\vec{k}} + \hbar\omega_{\vec{p}} - E_{\vec{k}-\vec{p}} + i\epsilon} \langle N_{\vec{p}}\rangle_T \right)$$

$$(8.10.11)$$

which is the same as (8.10.9), since we assume time-reversal symmetry and can switch the summation for the second integral from $\vec{p}$ to $-\vec{p}$.

Exercise 8.10.1   Using the finite-temperature Green's functions given in this section, compute the real self-energy correction for a free electron that corresponds to the diagram shown in Figure 8.18, known as a "tadpole" diagram, for optical phonons with constant energy $\hbar\omega_{LO}$. To have a sensible answer, you will need to multiply your integral by a factor $e^{i\eta E}$, to make the integrand vanish in the limit of large imaginary E in the upper half of the complex plane, and take the limit $\eta \to 0$.

Exercise 8.10.2   Use the finite-temperature Green's functions method of this section to compute the diagram shown in Figure 8.32, used in the Fermi liquid theory at the end of this chapter. Show that the result for the imaginary self-energy has the same two terms as in (8.15.17), which are also the same as the rates found by the quantum Boltzmann equation for two-body scattering of fermions. Hint: Note that Cauchy's theorem for a loop around the complex plane will give a zero result unless there are poles on both the top and bottom halves of the plane.

8.11 Screening and Plasmons

Just as the propagators can be renormalized by interactions, so can the vertices. Figure 8.19 shows a diagrammatic equation for the Coulomb interaction between two electrons. Just as the self-energy diagrams discussed in Section 8.8 put the system back into its original state, the additional loops in this series of diagrams all leave the momentum exchanged in the Coulomb interaction the same. We can therefore write down a renormalized Coulomb interaction that takes into account all of these diagrams for a given momentum exchange.

Starting with bare Coulomb interaction

$$\frac{1}{V} \frac{e^2/\epsilon}{|\Delta\vec{k}|^2},$$

$$(8.11.1)$$

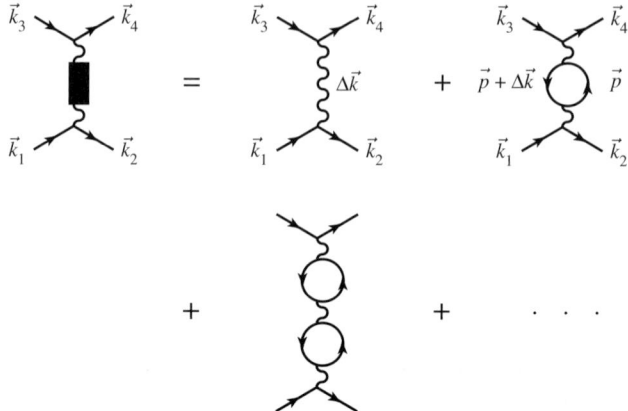

Fig. 8.19   Diagram for Thomas–Fermi screening at zero temperature.

the infinite series shown in Figure 8.19 gives us a sum of the form

$$S = \frac{1}{A} + \frac{1}{A}B\frac{1}{A} + \frac{1}{A}B\frac{1}{A}B\frac{1}{A} + \cdots, \tag{8.11.2}$$

where $A = \epsilon V |\Delta \vec{k}|^2/e^2$, and B is the contribution from the propagators in a single loop, which we will call $\Pi_{\Delta \vec{k}}$, which depends on the momentum exchange in the interaction. Since the sum of the series $S = 1/(A - B)$, as discussed at the beginning of Section 8.8, we have

$$S = \frac{1}{V}\frac{e^2/\epsilon}{|\Delta \vec{k}|^2 - (1/V)(e^2/\epsilon)\Pi_{\Delta \vec{k}}}. \tag{8.11.3}$$

In other words, the bare Coulomb interaction is renormalized to the form

$$S = \frac{1}{V}\frac{e^2/\epsilon}{|\Delta \vec{k}|^2 + \kappa^2}, \tag{8.11.4}$$

where $\kappa^2 = -(1/V)(e^2/\epsilon)\Pi_{\Delta \vec{k}}$, which is just the form we assumed in Section 5.5. Instead of making an ad hoc assumption about the value of the screening constant, however, we can calculate it now using our Feynman diagram method by determining the contribution of the loop that appears in Figure 8.19.

Zero-temperature Fermi gas. In Section 8.8, the loops represented electrons excited in a semiconductor from the valence band to the conduction band. Here we consider the case of a Fermi gas at zero temperature, in which the electrons are in a single, isotropic electron band, with a Fermi energy E_F, as shown in Figure 8.20.

Holes and free electrons are defined relative to this energy. In the single-loop diagram shown in Figure 8.19, the momentum $\vec{p}$ for the hole cannot have magnitude greater than the Fermi momentum p_F corresponding to E_F, and the electron cannot have momentum less than this value. This gives us

$$\Pi_{\Delta \vec{k}} = -\frac{i}{2\pi}\int_{-\infty}^{\infty} dE \sum_{\vec{p}} \frac{\Theta(E_{\vec{p}+\Delta \vec{k}} - E_F)}{(E + \Delta E - (E_{\vec{p}+\Delta \vec{k}} - E_F) + i\epsilon)} \frac{\Theta(E_F - E_{\vec{p}})}{(E - (E_{\vec{p}} - E_F) - i\epsilon)}, \tag{8.11.5}$$

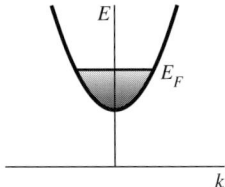

Fig. 8.20   A Fermi level in a simple band.

where ΔE is the energy exchanged in the collision, $\Delta \vec{k}$ is the momentum exchanged, and $\Theta(E)$ is the Heaviside function. Using the Cauchy residue formula for a loop in the upper half of the complex plane, as we have done in the previous sections, gives

$$\Pi_{\Delta \vec{k}} = \sum_{\vec{p}} \frac{\Theta(E_{\vec{p}+\Delta \vec{k}} - E_F)\Theta(E_F - E_{\vec{p}})}{\Delta E + E_{\vec{p}} - E_{\vec{p}+\Delta \vec{k}} + i\epsilon}$$

$$= \sum_{\vec{p}} \frac{\Theta(E_F - E_{\vec{p}}) - \Theta(E_F - E_{\vec{p}+\Delta \vec{k}})}{\Delta E + E_{\vec{p}} - E_{\vec{p}+\Delta \vec{k}} + i\epsilon}, \tag{8.11.6}$$

where we have used the properties $\Theta(-E) = 1 - \Theta(E)$ and $\Theta(E - a)\Theta(E - b) = \Theta(E - \max(a, b))$.

Assuming that the energy exchanged is small (i.e., $\Delta E \to 0$), and the long-wavelength limit, $|\Delta \vec{k}| \ll |\vec{p}|$, the principal value of the ratio in (8.11.6) is just the derivative of the Heaviside function. This is equal to $-\delta(E)$. The total screening parameter is therefore equal to

$$\kappa^2 = \frac{1}{V} \frac{e^2}{\epsilon} \sum_{\vec{p}} \delta(E_{\vec{p}} - E_F)$$

$$= \frac{1}{V} \frac{e^2}{\epsilon} \mathcal{D}(E_F). \tag{8.11.7}$$

If we take into account the two spin states of electrons, we then obtain

$$\boxed{\kappa^2 = \frac{3e^2 n}{2\epsilon E_F},} \tag{8.11.8}$$

where n is the electron density. This is known as the **Thomas–Fermi** screening approximation.

Finite temperature screening. We can use a similar approach at finite temperature. Using the method introduced in Section 8.10, we define the vacuum as the empty band, and treat all the electrons as new particles created in this vacuum, that is, as external legs. No holes exist in this case. We will have a hierarchy of diagrams with increasing number of external legs that enter into the diagram, just as we had a hierarchy of number of loops in the above case at $T = 0$. We sum an infinite series of diagrams with external legs, as shown in Figure 8.21. In each of these diagrams, the momentum exchange is not affected, since for each incoming external leg we must have the outgoing external leg with the same momentum, so that this remains an interaction vertex between only two electrons.

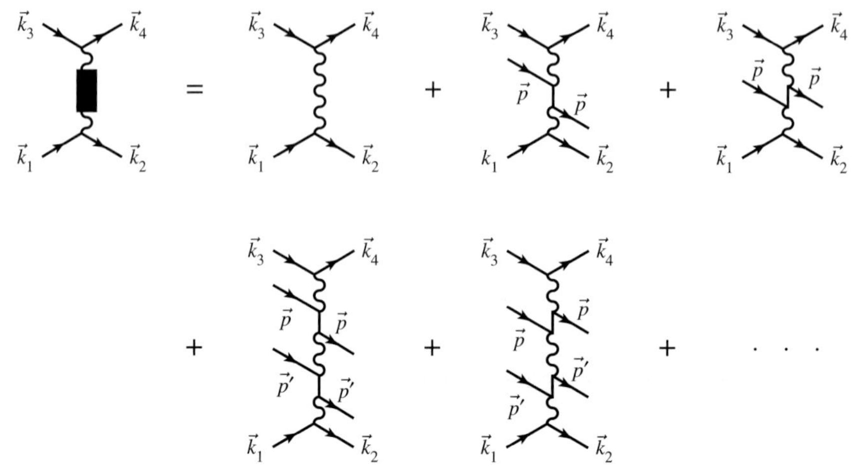

Fig. 8.21   Infinite series of diagrams for the interaction of two electrons.

The B term in the infinite series is then, from inspection of the first two diagrams with extra external legs in Figure 8.21, using the finite-temperature rules of Section 8.10,

$$\Pi_{\Delta \vec{k}} = \sum_{\vec{p}} \left[\frac{\bar{N}_{\vec{p}}}{(E_{\vec{p}} + \Delta E) - E_{\vec{p}+\Delta \vec{k}} + i\epsilon} + \frac{\bar{N}_{\vec{p}}}{(E_{\vec{p}} - \Delta E) - E_{\vec{p}-\Delta \vec{k}} - i\epsilon} \right],$$

(8.11.9)

where ΔE is the energy exchanged in the collision, and $\Delta \vec{k}$ is the momentum exchanged. The $N_{\vec{p}}$ factors arise because an incoming leg contributes a factor $\sqrt{N_{\vec{p}}}$; if $N_{\vec{p}} = 1$, the outgoing leg at the same momentum contributes $\sqrt{1 - (1 - N_{\vec{p}})} = \sqrt{N_{\vec{p}}}$, while if $N_p = 0$, the whole term is zero. Note that we only replace $N_{\vec{p}}$ with its thermal average *after* computing these factors.

We then change the dummy variable in the sum from $\vec{p}$ to $\vec{p} - \Delta \vec{k}$, for the second term, to obtain

$$\Pi_{\Delta \vec{k}} = \sum_{\vec{p}} \left[\frac{\bar{N}_{\vec{p}}}{\Delta E + E_{\vec{p}} - E_{\vec{p}+\Delta \vec{k}} + i\epsilon} + \frac{\bar{N}_{\vec{p}+\Delta \vec{k}}}{-\Delta E + E_{\vec{p}+\Delta \vec{k}} - E_{\vec{p}} - i\epsilon} \right]$$

$$= \sum_{\vec{p}} \frac{\bar{N}_{\vec{p}} - \bar{N}_{\vec{p}+\Delta \vec{k}}}{\Delta E + E_{\vec{p}} - E_{\vec{p}+\Delta \vec{k}} + i\epsilon}.$$

(8.11.10)

This is known as the **Lindhard** formula.

The real part of (8.11.10) is found by taking the principal value of the integral, using the Dirac formula (7.2.2). Taking the limit $\Delta E \to 0$, and subsequently the long-wavelength approximation $|\Delta \vec{k}| \ll |\vec{p}|$, we obtain

$$\text{Re} \, \Pi_{\Delta \vec{k}} = \sum_{\vec{p}} \frac{\partial N(E_{\vec{p}})}{\partial E_{\vec{p}}},$$

(8.11.11)

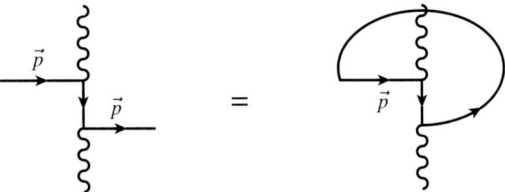

The correlation function, or polarizability, diagrams for a system with preexisting particles viewed as a loop diagram.

where we treat the average occupation number $\bar{N}_{\vec{p}}$ as a continuous function $N(E_{\vec{p}})$. In other words, for an isotropic system,

$$\kappa^2 = -\frac{1}{V}\frac{e^2}{\epsilon}\int dE\,\mathcal{D}(E)\frac{\partial N(E)}{\partial E}, \tag{8.11.12}$$

which is the same as (8.11.16). It is straightforward to show that this result is the same as (8.11.8) in the zero-temperature limit. In the high-temperature limit, when the electron distribution function is a Maxwell–Boltzmann function, it is equal to

$$\boxed{\kappa^2 = \frac{e^2 n}{\epsilon k_B T}.} \tag{8.11.13}$$

This is known as the **Debye** screening approximation. Thus, we see that the semiclassical approach we used for the screening length in Section 5.5.1 actually gives us accurately the low-frequency, long-wavelength limit of the many-body vertex correction.

As with the renormalization of the propagators, we can use the renormalized vertex to calculate new, higher-order corrections to the vertex in a hierarchy of different levels of renormalization.

Exercise 8.11.1   Show that (8.11.8) and (8.11.13) both follow from (8.11.12) for an electron gas with two degenerate spin states, in the two limits of a $T = 0$ electron gas with a Fermi level, and in the high-temperature, Maxwell–Boltzmann limit.

Exercise 8.11.2   Use the finite-temperature propagators of Section 8.10 for fermions for the single-loop diagram shown in Figure 8.19, and show that you obtain the same Lindhard formula (8.11.10).

Actually, a two-external-leg diagram can also be viewed as a loop, as illustrated in Figure 8.22, if we view the departing particle as the *same* particle as the incoming particle.

8.11.1 Plasmons

Since the occupation number $f(E)$ decreases with increasing energy, the derivative $\partial f(E)/\partial E$ is negative, and therefore the screening constant κ^2 given by (8.11.13) is positive. If we do not make the low-frequency approximation $\Delta E \to 0$, however, then there is the possibility that the $\Pi_{\Delta \vec{k}}$ term may cancel the $|\Delta \vec{k}|^2$ term in (8.11.3). Writing out the kinetic energy terms in the denominator of (8.11.9), we have

$$\mathrm{Re}\,\Pi_{\Delta \vec{k}} = \sum_{\vec{p}} \left[\frac{\bar{N}_{\vec{p}}}{\Delta E - \hbar^2 \vec{p} \cdot \Delta \vec{k}/m - \hbar^2 |\Delta \vec{k}|^2/2m} \right.$$

$$\left. + \frac{\bar{N}_{\vec{p}}}{-\Delta E + \hbar^2 \vec{p} \cdot \Delta \vec{k}/m - \hbar^2 |\Delta \vec{k}|^2/2m} \right]$$

$$= \sum_{\vec{p}} \frac{\bar{N}_{\vec{p}}\, (\hbar^2 |\Delta \vec{k}|^2/m)}{-(\hbar^2 |\Delta \vec{k}|^2/2m)^2 + (\Delta E - \hbar^2 \vec{p} \cdot \Delta \vec{k}/m)^2}. \tag{8.11.14}$$

In the long-wavelength limit, $|\Delta \vec{k}| \ll p$, this becomes simply

$$\mathrm{Re}\,\Pi_{\Delta \vec{k}} = \frac{N\hbar^2 |\Delta \vec{k}|^2}{m(\Delta E)^2}, \tag{8.11.15}$$

where $N = \sum \bar{N}_p$ is the total number of particles. The term (8.11.15) will lead to cancellation of the $|\Delta \vec{k}|^2$ term in the denominator of (8.11.3) when

$$\frac{N}{V} \frac{e^2}{\epsilon} \frac{\hbar^2}{m(\Delta E)^2} = 1, \tag{8.11.16}$$

or

$$\boxed{\hbar \omega_{\mathrm{pl}} = \Delta E = \sqrt{\frac{ne^2 \hbar^2}{\epsilon m}},} \tag{8.11.17}$$

where $n = N/V$ is the particle density. This is known as the **plasmon frequency**, and the zero in the denominator of the Coulomb vertex is known as the plasmon pole. The same result is obtained in the $T = 0$ Thomas–Fermi limit.

Classical model. We can get a basic understanding of how the plasmon resonance arises by imagining a classical gas of electrons, as shown in Figure 8.23. If there is a displacement x of the gas with particle density n relative to a positive background of the same charge density, there will be a force pulling the electron gas back which is proportional to the distance, which can be determined by viewing the region of net positive charge and the region of net negative charge as a set of charged capacitor plates. The energy of a capacitor is

$$U = \frac{1}{2} \frac{Q^2}{C}, \tag{8.11.18}$$

where $C = \epsilon A/d$ for parallel plates. The charge Q depends on the amount of the displacement of the electron gas,

$$Q = enAx, \tag{8.11.19}$$

and therefore the restoring force per particle is

$$F = -\frac{1}{N} \frac{\partial U}{\partial x} = -\frac{1}{N} \frac{e^2 n^2 Axd}{\epsilon} = -\frac{e^2 nx}{\epsilon}, \tag{8.11.20}$$

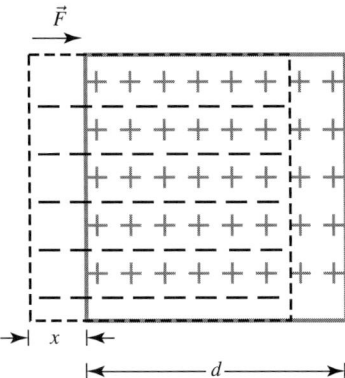

The plasmon resonance visualized as the oscillation of a charged, homogeneous gas relative to a positive background.

since $N = nV = nAd$. In other words, the electrons feel a restoring force with a spring constant $K = e^2 n/\epsilon$. The natural frequency of the spring is $\omega_p = \sqrt{K/m} = \sqrt{e^2 n/\epsilon m}$, which is just the frequency deduced from the many-body theory above.

The plasmon as a new quasiparticle. Consider the diagram in Figure 8.24. We can think of this as a **two-particle propagator** in which the electron–hole pair has a net momentum $\vec{k}$. The lowest-order diagram is just the same as the loop that we called Π_k in (8.11.6); this has units of inverse energy, as do all propagators. The ground state of the system is the Fermi sea, and this is the propagator for an excitation out of that ground state.

When we take into account all the higher order diagrams shown in Figure 8.24, we will have a sum of the form

$$S = B + B\frac{1}{A}B + B\frac{1}{A}B\frac{1}{A}B + \cdots, \tag{8.11.21}$$

which sums to $S = B/(1 - B/A)$, or, in terms of the Coulomb vertex $M_k = 1/A = e^2/\epsilon V k^2$ and Π_k given by (8.11.15), with ΔE given by (8.11.17),

$$S = \frac{\Pi_k}{1 - M_k \Pi_k}$$
$$= \frac{2NE_k}{(\Delta E)^2 - (\hbar\omega_{pl})^2}. \tag{8.11.22}$$

This has the form of a phonon-like propagator, with a pole at which the denominator vanishes, like the standard pole in the bare Green's functions. The numerator has the number of electrons N which is extensive with the size of the system, but this will be controlled by the volume factor that appears in the denominator of the vertices linking this propagator to other diagrams. The number N could in principle be absorbed into a new definition of the vertices, similar to the procedure for photon vertices in Section 8.8.

From the discussion of Section 8.9, we know that the poles in the frequency-dependent Green's functions correspond to undamped oscillations in the time-dependent Green's functions, which in turn correspond to quasiparticles with infinite lifetime. If the interactions give an imaginary self-energy that is not too large, the renormalized propagator corresponds to a state that is nearly an eigenstate. We are therefore justified in treating

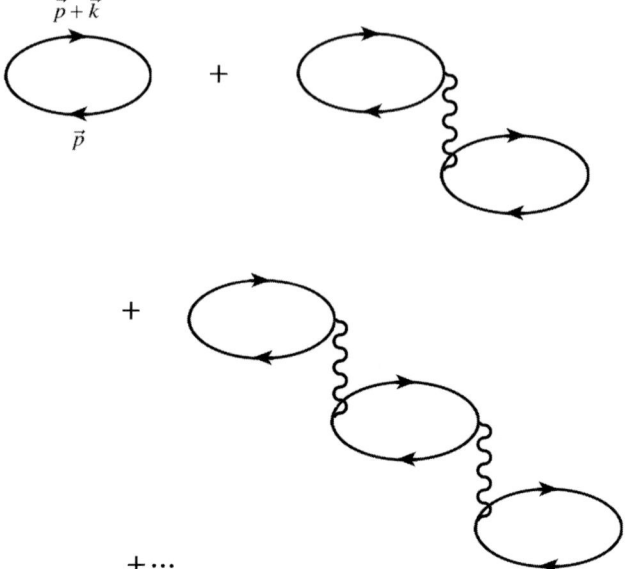

Fig. 8.24 Infinite series for the renormalized two-particle propagator for a plasmon.

this as a new quasiparticle state in the system, which we call a **plasmon**. Physically, it comes about because there is a natural oscillation in the medium with frequency ω_p, which is quantized just like the oscillating phonon or photon fields. The appearance of plasmons thus illustrates a general point: *Quasiparticles are equated with poles in the Green's functions.*

As we have done so often before, we have moved to a new picture in which the equilibrium electron gas becomes the new vacuum, and the oscillations of the medium are characterized as quasiparticles excited out of this vacuum, just like phonons. These plasmons are like phonons in many ways.

Damping of plasmons. So far we have ignored the imaginary part of $\Pi_{\Delta\vec{k}}$ in (8.11.10). Let us now look at this imaginary term. Using the Dirac formula (7.2.2), we have

$$\text{Im } \Pi_{\Delta\vec{k}} = -\pi \sum_{\vec{p}} \left(\bar{N}_{\vec{p}} - \bar{N}_{\vec{p}'+\Delta\vec{k}} \right) \delta(\Delta E + E_{\vec{p}} - E_{\vec{p}+\Delta\vec{k}}). \qquad (8.11.23)$$

The energy-conserving δ-function implies

$$\Delta E = E_{\vec{p}+\Delta\vec{k}} - E_{\vec{p}}. \qquad (8.11.24)$$

In other words, when the plasmon energy $\Delta E = \hbar\omega_{pl}$ can be lost to the kinetic energy of a free electron, there will be decay of the plasmons. This is known as **Landau damping** of plasmons.

Suppose that the electrons are in a Fermi sea from $p = 0$ to $p = p_F$. The minimum allowed value of $\Delta\vec{k}$ then occurs when $|\vec{p}| = 0$, in which case $\Delta E = E_{\Delta\vec{k}}$. The maximum occurs when $|\vec{p}| = p_F$, in which case

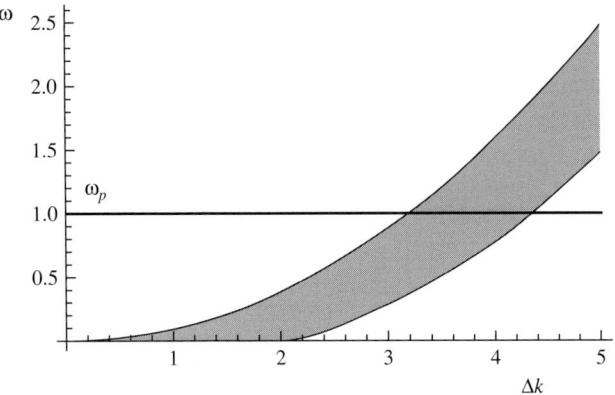

Fig. 8.25 The plasmon resonance and the range of energies and momenta that lead to Landau damping (shown as the gray region), in arbitrary units.

$$\Delta E = \frac{\hbar^2 p_F |\Delta \vec{k}| \cos\theta}{m} + \frac{\hbar^2 |\Delta \vec{k}|^2}{2m}, \qquad (8.11.25)$$

where θ is the angle between $\vec{p}$ and $\Delta\vec{k}$. For a given value of ΔE, the maximum value of $|\Delta\vec{k}|$ occurs when $\cos\theta = -1$. Figure 8.25 illustrates the range of $|\Delta\vec{k}|$ that contributes to Landau damping of the plasmons, in the case of a Fermi sea. As seen in this figure, at the plasmon pole, there will be no damping at low momentum, but at higher momentum, the plasmons will decay.

Exercise 8.11.3   Determine the $k = 0$ plasmon frequency for a gas of electrons with mass equal to the free electron mass, in a system with low-frequency dielectric constant $\epsilon = 2\epsilon_0$, and density 10^{19} cm^{-3}. A convenient unit conversion is $e^2/4\pi\epsilon_0 = 1.44 \times 10^{-7}$ eV-cm.

Exercise 8.11.4   Determine the plasmon frequency to quadratic order in Δk, assuming an isotropic, Maxwell–Boltzmann distribution of electrons.

8.11.2 The Conductor–Insulator Transition and Screening

The previous discussion assumed a Fermi gas of just one type of fermion, namely electrons. The situation is considerably more complicated when there are two types of fermions with opposite charge, for example, in the case of an excited semiconductor with a neutral plasma composed of equal numbers of free electrons and holes. In that case, there is also the possibility of pairing of particles with opposite charge into bound states. These bound states, namely the excitons introduced in Section 2.3, can have very different contributions to screening.

In Section 2.5.4, we discussed the Mott transition from conductor to insulator in terms of the wave function overlap of the bound pairs. When the pairs are packed close together, they no longer have an individual identity, and a conducting plasma appears.

Significant screening can change this picture substantially. When screening reduces the strength of the Coulomb interaction, this leads to a reduction of the binding energy of the pair states. From numerical calculations (e.g., Smith 1964), the binding energy as a function of screening is approximately given by

$$
\text{Ry}(\kappa) = \begin{cases} \text{Ry}(0)\left(1 - \dfrac{2}{1 + (\kappa a)^{-1}}\right), & \kappa a < 1 \\ 0, & \kappa a \geq 1, \end{cases} \tag{8.11.26}
$$

where $\text{Ry}(0)$ is the standard hydrogenic Rydberg in the case of no screening, and a is the Bohr radius in the absence of screening. In other words, when the screening length $1/\kappa$ is shorter than the orbital radius of the bound pair, then bound states are not possible.

Suppose that we start with a neutral gas of positive and negative free particles (e.g., electrons and holes in a semiconductor), such that the screening is strong enough to prevent the appearance of bound states. Now imagine that we reduce the total number of particles slowly, at the same temperature and volume. At some point, the screening length will grow longer than the Bohr radius, and at that point there will be a sudden transition to an insulating state of mostly bound pairs. The condition for this transition is $\kappa = 1/a$, or

$$
n = \frac{a^2 \epsilon k_B T}{e^2}, \tag{8.11.27}
$$

using the Debye screening formula (8.11.13).

Now imagine that we start with this same gas of bound pairs, and increase the density. The bound states do not contribute to screening on length scales longer than their Bohr radius, to first order, since they are charge-neutral. However, ionization can occur due to collisions of the pairs. The higher the density, the more likely these collisions. Eventually, the number of free particles due to ionization will be enough to screen the Coulomb interaction. As the screening increases, the binding energy decreases, making ionization even more likely. Eventually the screening will prevent the existence of bound states, switching the gas to a conducting state. This is known as an **ionization catastrophe**. The condition is $\kappa = 1/a$, but κ depends on the density of free particles, given by the equilibrium value from the appropriate mass-action equation. From Section 2.5.3, we have

$$
\frac{n_e^2}{n_{ex}} = \frac{n_Q^{(e)} n_Q^{(h)}}{n_Q^{(ex)}} e^{-\text{Ry}/k_B T} \equiv n_Q e^{-\text{Ry}/k_B T}, \tag{8.11.28}
$$

where n_{ex} is the pair (exciton) density and n_Q is the effective density of states, which depends on the temperature and the effective masses of the particles.

Assuming Ry is constant and $n_e \ll n_{ex}$, this means the critical density for the transition is approximately

$$
n = \left(\frac{a^2 \epsilon k_B T}{e^2}\right)^2 \frac{e^{\text{Ry}/k_B T}}{n_Q}. \tag{8.11.29}
$$

This clearly gives a very different threshold than (8.11.27) above. For example, at low temperature, very few pairs will ionize and therefore the critical density becomes exponentially large. The difference between the two critical densities means that there can be **hysteresis**

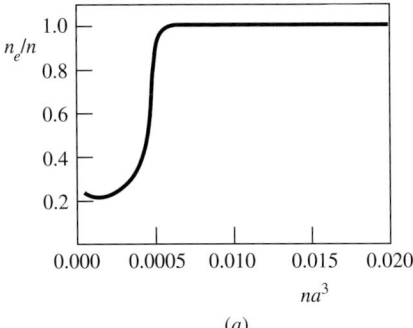 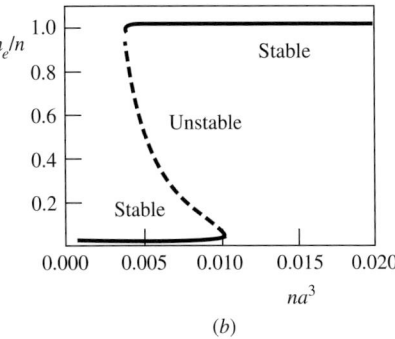

Fig. 8.26   Steady-state solutions of n_e/n as a function of na^3, for (a) high temperature, $a^2/\lambda_D^2 = 0.024$, and (b) low temperature, $a^2/\lambda_D^2 = 0.0016$, where λ_D is the thermal DeBroglie wavelength and a is the exciton Bohr radius. From Snoke and Crawford (1995).

in the system, in which the state of the system depends on its history and not just on the total particle number.

Figure 8.26 shows the results of a self-consistent numerical solution for the fraction of electrons in a semiconductor at constant temperature, taking into account the effect of screening on the exciton binding energy. In the ionized fraction as a function of total pair density at a fixed temperature, a classic hysteresis curve appears. The upper and lower solutions for the ionized fraction are stable against fluctuations, while the middle one is unstable.

This analysis applies to any insulator–plasma transition, for example electrons and protons in outer space. The binding energy of hydrogen atoms is so large, however, that unless the free particles are at very high density, the screening length will not be small enough to prevent formation of bound states. By contrast, photo-generated excitons in semiconductors can easily have densities such that the interparticle spacing is comparable to the screening length.

The electron–hole gas is normally metastable, since electrons can recombine with holes, but systems can be designed with very long recombination lifetime due to a small matrix element for photon emission. On time scales short compared to the lifetime of the electrons and holes to recombine by photon emission, we can treat the electron–hole system as a gas of particles with equal and opposite charge, just like protons and electrons. Depending on the various parameters of temperature, total pair density, effective mass of the particles, and so on, there are a number of different possible phases. Besides the exciton and plasma phases discussed above, if there is an attractive interaction betwen the excitons, as with atomic hydrogen, they may form molecular states, known as **biexcitons**. These biexcitons can be attracted to each other, like molecules in a gas, to form the **electron–hole liquid**, a conducting Fermi liquid state analogous to liquid mercury. It is also possible for excitons to undergo Bose–Einstein condensation at low temperature, in a state similar in many ways to BCS superconductivity, discussed in Chapter 11, but with Coulomb attraction of the electrons and holes as the pairing mechanism instead of the phonon-assisted mechanism of Cooper pairing, as discussed in Section 11.13.

Some readers may find it hard to imagine that all of these effects are occurring in a gas of carriers in a solid – we still tend to expect that electrons in a solid are bound to the atoms. The quasiparticle picture of electrons and holes in extended Bloch states, however, allows us to view a semiconductor with free electrons and holes essentially as the same as a gas of free particles in a box. On time scales short compared to the recombination lifetime, the system is no different in character from a gas of electrons and protons. In this context, plasma, atomic gas, molecular gas, and liquid states of the carriers are to be expected. The physics of liquid–gas phase transitions in electron–hole systems is treated at length in Rice *et al.* (1977) and Keldysh *et al.* (1987).

Exercise 8.11.5 (a) Use a program like Mathematica to plot the phase boundary (8.11.29) in the n–T plane, for the ionization catastrophe, and also the phase boundary (8.11.27), for exciton radius $a = 50$ Å, exciton binding energy $\Delta = 10$ meV, $\epsilon/\epsilon_0 = 10$, and effective electron and hole masses both one-tenth the vacuum electron mass.

(b) Plot the same two curves for the case of hydrogen.

8.12 Ground State Energy of the Fermi Sea: Density Functional Theory

We have already discussed in Section 2.4.1 the case of the noninteracting Fermi gas. There we assumed that the energy of interaction in the ground state of the electrons had already been taken into account in computing the electronic band structure of the system. Suppose we want to calculate that band structure, however. We then would want to start with the unrenormalized electrons and take into account the interactions of the electrons.

To start, let us take the case of an infinite, homogeneous electron gas, in a single parabolic band. We want to know the ground state of this system. The first-order contribution to the self-energy includes the diagram shown in Figure 8.27(a). This corresponds to the self-energy

$$\Sigma_{\text{Hartree}} = \frac{1}{V} \sum_{\vec{p}} N_{\vec{p}} \frac{e^2/\epsilon}{|\Delta \vec{k}|^2}, \tag{8.12.1}$$

where, as seen in the diagram, $\Delta \vec{k} = 0$. This is known as the **Hartree** term, or **direct** interaction term, and corresponds to the classical Coulomb repulsion between all the electrons. The denominator $|\Delta \vec{k}|^2 = 0$ means that this term diverges to infinity. We could handle this by including screening of the Coulomb interaction, but in a solid we can also remember that there is a positive background (from the atomic nuclei) with the same average charge density, which also has a direct Coulomb interaction with the electrons, of the opposite

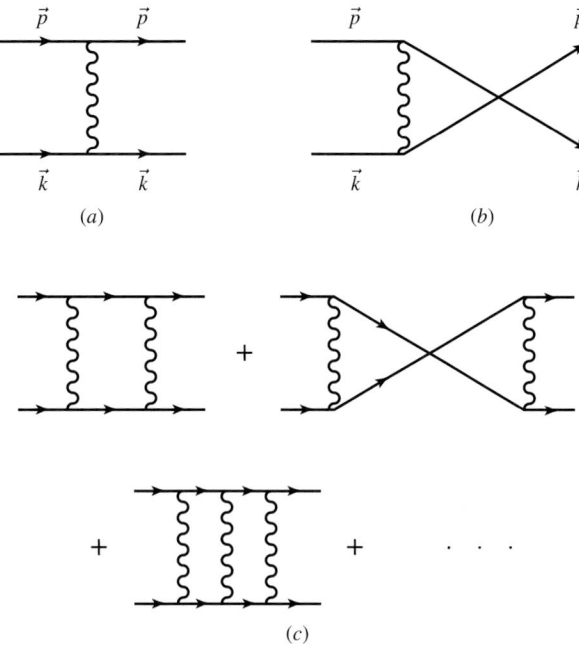

Diagrams for Coulomb interactions in a Fermi gas.

sign. Therefore, in a homogeneous gas, we assume that these terms cancel out, and ignore the Hartree term in the self-energy of the electrons.

For electrons with the same spin, we can write another diagram with one vertex, as shown in Figure 8.27(b). This corresponds to

$$\Sigma_{\text{Fock}}(\vec{k}) = -\frac{1}{V} \sum_{\vec{p}} N_{\vec{p}} \frac{e^2/\epsilon}{|\vec{k} - \vec{p}|^2}. \qquad (8.12.2)$$

This is known as the **Fock**, or **exchange** energy, and is negative. In general, exchange energy has the opposite sign of the direct interaction. The exchange self-energy, like the direct self-energy, increases with increasing particle density. Neglecting screening, in the case of electrons at $T = 0$ with a Fermi level E_F, corresponding to maximum momentum $\hbar k_F$, this integral gives

$$\Sigma_{\text{Fock}}(k) = -\frac{e^2 k_F}{4\pi^2 \epsilon} \left(1 + \frac{1 - (k/k_F)^2}{2k/k_F} \ln \left| \frac{1 + k/k_F}{1 - k/k_F} \right| \right). \qquad (8.12.3)$$

This function is shown in Figure 8.28. Exactly at the point $k = k_F$, the slope diverges, although the function does not. This can lead to interesting consequences for the effective mass. At finite temperature, the Fermi edge is smeared out, so this divergence does not occur.

In addition to the two first-order self-energy diagrams, there is an infinite number of higher-order diagrams such as those shown in Figure 8.27(c). All of these higher-order self-energy diagrams are known as the **correlation energy**.

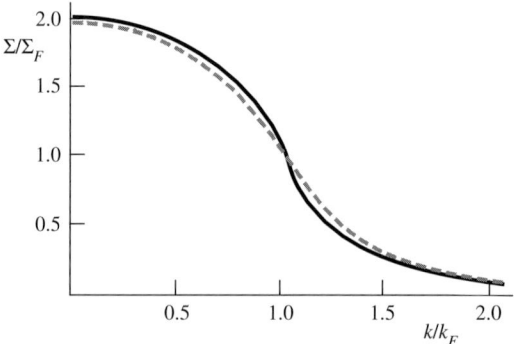

Fig. 8.28   Unscreened exchange energy for an electron gas at $T = 0$ (heavy line), and at finite T, with $k_B T = 0.2\mu$ (thin line).

Exercise 8.12.1   (a) Verify that (8.12.3) follows from (8.12.2) for the case of a three-dimensional electron gas with a Fermi level E_F.

(b) Use a program like Mathematica to calculate numerically the unscreened exchange energy from (8.12.2) for the case of a Maxwell–Boltzmann distribution of electrons, that is, the high-temperature limit.

Density functional theory. We now want to move to the more general case of an inhomogeneous electron gas, to model the electron bands in a spatially varying potential, in particular the lattice of atoms in a crystal. **Density functional theory** has greatly boosted the accuracy of band structure calculations. This method follows from a powerful theorem, known as the Hohenberg–Kohn theorem, which says that the total many-body electron energy in any system is a unique functional of the local charge density $n(\vec{r})$. It is not obvious at first glance that this must be true, since the many-body self-energy depends on interactions with all of the particles in the gas, even ones very far away. Nevertheless, this can be proven, as follows.

Suppose that we have an electron gas subject to an external potential $U(\vec{r})$, with a total many-particle ground state $|\Psi\rangle$ that corresponds to the density function $\rho(\vec{r})$. Imagine that there is some other external potential $U'(\vec{r})$ with a different ground state, $|\Psi'\rangle$, which also corresponds to the *same* density function $\rho(\vec{r})$, for the same number of particles. Since $|\Psi'\rangle$ is different from $|\Psi\rangle$, it is not the ground state of the Hamiltonian with $U(\vec{r})$, and therefore

$$\langle\Psi|H|\Psi\rangle < \langle\Psi'|H|\Psi'\rangle$$
$$= \langle\Psi'|H' + U(\vec{r}) - U'(\vec{r})|\Psi'\rangle, \tag{8.12.4}$$

where H and H' are the total Hamiltonians of the system with the two different external potentials. Defining $E = \langle\Psi|H|\Psi\rangle$ and $E' = \langle\Psi'|H'|\Psi'\rangle$, we then have

$$E < \langle\Psi'|H' + U(\vec{r}) - U'(\vec{r})|\Psi'\rangle$$
$$= E' + \int d^3r \left(U(\vec{r}) - U'(\vec{r}) \right) \rho(\vec{r}). \tag{8.12.5}$$

On the other hand, $|\Psi\rangle$ is not the ground state of the Hamiltonian H', and therefore we must also have

$$\langle\Psi'|H'|\Psi'\rangle < \langle\Psi|H'|\Psi\rangle, \tag{8.12.6}$$

and

$$E' < E + \int d^3r \left(U'(\vec{r}) - U(\vec{r}) \right) \rho(\vec{r}). \tag{8.12.7}$$

Subtracting (8.12.7) from (8.12.5) implies

$$E - E' < E' - E, \tag{8.12.8}$$

which cannot be true. By argument *reductio ad absurdum*, the premise cannot be true, namely, that another Hamiltonian could have a different ground state with the same density function $\rho(\vec{r})$. Therefore, the ground state wave function must be a unique functional of $\rho(\vec{r})$, for a given number of electrons.

One can therefore search for the ground state of the many-body system by guessing a form for the function $\rho(\vec{r})$, computing the total energy including the correlation energy to the best approximation, and then varying $\rho(\vec{r})$ and recomputing the total energy, until one converges on a density function $\rho(\vec{r})$ that minimizes the energy. Handling a function of a single variable $\vec{r}$ is far easier than handling the complete many-body wave function $\Psi(\vec{r}_1, \vec{r}_2, \dots \vec{r}_N)$.

One approximation that further simplifies the procedure is the **local density approximation**. In this approximation, one assumes that the external potential $U(\vec{r})$ is slowly varying, and therefore the exchange and correlation energies of the electron gas at every point in space can be taken to be the same as those of an infinite, homogeneous gas with density equal to the local density $\rho(\vec{r})$. The total energy can therefore be written as

$$E[\rho] = T[\rho] + \int d^3r \, U(\vec{r})\rho(\vec{r}) + \frac{1}{2}\frac{e^2}{4\pi\epsilon}\int d^3r \, d^3r' \frac{\rho(\vec{r})\rho(\vec{r}')}{|\vec{r} - \vec{r}'|} + E_{xc}[\rho], \tag{8.12.9}$$

where $T[\rho]$ is the total kinetic energy, and $E_{xc}[\rho]$ is the total of the exchange and correlation energies.

The kinetic energy can be calculated self-consistently from the effective potential felt by the electrons. This is be done using the **Kohn–Sham** equations. In this approach, we find the eigenstates of the Schrödinger equation

$$\left(-\frac{\hbar^2 \nabla^2}{2m} + U_{KS}(\vec{r}) \right) \phi_j(\vec{r}) = E_j \phi_j(\vec{r}), \tag{8.12.10}$$

with

$$U_{KS}(\vec{r}) = U(\vec{r}) + \frac{1}{2}\frac{e^2}{4\pi\epsilon}\int d^3r' \frac{\rho(\vec{r}')}{|\vec{r} - \vec{r}'|} + \frac{1}{V}\frac{\partial E_{xc}[\rho]}{\partial \rho}\bigg|_{\rho=\rho(\vec{r})}. \tag{8.12.11}$$

The eigenfunctions in turn imply a density function,

$$\rho(\vec{r}) = \sum_{j=1}^{N} |\phi_j(\vec{r})|^2, \tag{8.12.12}$$

which must be consistent with the density function used for the rest of the calculation, when the computation finally converges.

Then the total kinetic energy is equal to

$$T[\rho(\vec{r})] = \sum_{j=1}^{N} \int d^3r \, \phi_j^*(\vec{r}) \left(E_j - U_{KS}(\vec{r}) \right) \phi_j(\vec{r})$$

$$= \sum_{j=1}^{N} E_j - \int d^3r \, \rho(\vec{r}) U_{KS}(\vec{r}). \tag{8.12.13}$$

In this approach, we use the exchange and correlation energies of a homogeneous gas. These are known from standard calculations. Numerical calculations (e.g., Ceperley and Adler 1980; Perdew and Zunger 1983) give

$$E_{xc} = E_{\text{exch}} + E_{\text{corr}}, \tag{8.12.14}$$

where

$$E_{\text{exch}} = -\frac{0.4582}{r_s}, \tag{8.12.15}$$

where $r_s = (3\rho/4\pi)^{-1/3}$ (see Exercise 2.5.3 of Chapter 2), and

$$E_{\text{corr}} = \begin{cases} -\dfrac{0.1423}{1 + 1.9529\sqrt{r_s} + 0.3334r_s}, & r_s \geq 1 \\[4mm] -0.0480 + 0.0311 \ln r_s - 0.0116 r_s + 0.002 r_s \ln r_s, & r_s < 1. \end{cases} \tag{8.12.16}$$

The units of the energy here are Hartrees, where one Hartree is equal to $e^4 m/\epsilon^2\hbar^2$.

Density functional theory has become a successful tool for predicting the properties of materials, and there have been many refinements. For a basic review of density functional theory, see Kohn and Vashishta (1983).

Exercise 8.12.2   Show the consistency of the exchange energy formula (8.12.15) and the result of integrating (8.12.3).

8.13 The Imaginary-Time Method for Finite Temperature

In Section 8.10, we introduced thermal averages of the type

$$\langle \ldots \rangle_T = \frac{1}{Z_0} \sum_j e^{-\beta(E_j - \mu N_j)} \langle j| \ldots |j\rangle, \tag{8.13.1}$$

where $\beta = 1/k_B T$. So far, we have treated this as a thermal average over the unperturbed eigenstate energies E_j, which is valid in low orders of the perturbation theory. But to be self-consistent, we should account for the effects of the interactions on the states $|j\rangle$. As we have seen, the interactions lead to shifts of the energies of the states, and the energies of the states affect the probability of them being occupied. The self-consistent thermal average should then be

$$\langle \ldots \rangle_T = \frac{1}{Z} \sum_j \langle j | \ldots e^{-\beta(H - \mu \hat{N})} | j \rangle, \tag{8.13.2}$$

where $H = H_0 + V_{\text{int}}$ is the full Hamiltonian, and $\hat{N}$ is the total number operator, which commutes with H_0. The same problem arises here as in the time-evolution operator, however, that the interaction term V_{int} does not commute with H_0, and therefore the effect of the exponential factor on $|j\rangle$ will be nontrivial.

It turns out, however, that by a simple trick, we can use all of the formalism we have already developed for the time evolution, with just a few alterations. The trick is to write the exponential factor in terms of an **imaginary time** $\tilde{t} = -i\tau$, with $\tau = \beta\hbar$. Then the statistical weight factor will be the operator

$$e^{-\beta(H - \mu \hat{N})} = e^{-i(H - \mu \hat{N})\tilde{t}/\hbar}, \tag{8.13.3}$$

which has exactly the same form as the time-evolution operator we have used all through this chapter. We can therefore use all the same formalism, including defining an interaction representation, expanding an S-matrix in powers of V_{int}, using a version of Wick's theorem with contractions, and defining Green's functions. This is known as the **Matsubara** method. General treatments can be found in Kittel (1963), Fetter and Walecka (1971), Mattuck (1976), and Mahan (2000).

Wick's theorem and contractions. To have a Feynman-type method, we must find an equivalent of Wick's theorem and define the contraction of two operators. We start by defining the interaction representation. Since $\hat{N}$ commutes with H_0, we can group it with H_0 and write $K_0 = H_0 - \mu\hat{N}$, which amounts to just subtracting μ from all energies. This then allows us to write the interaction representation of any operator A as

$$A(\tau) = e^{K_0 \tau/\hbar} A e^{-K_0 \tau/\hbar}. \tag{8.13.4}$$

In this representation, the time evolution equation becomes, analogous to (8.6.7),

$$-\hbar \frac{\partial}{\partial \tau} A(t) = e^{K_0 \tau/\hbar}(AK_0 - K_0 A)e^{-K_0 \tau/\hbar}$$
$$= [A(\tau), K_0]. \tag{8.13.5}$$

The commutator of two operators then follows from the same logic as (8.6.13), namely

$$[a_{\vec{k}}(\tau_1), a_{\vec{k}}^{\dagger}(\tau_2)] = e^{-(E_{\vec{k}} - \mu)(\tau_1 - \tau_2)/\hbar} \tag{8.13.6}$$

for bosons, and the similar anticommutator for fermions.

In the same way that we defined K_0, we define $K = H - \mu\hat{N}$ for the full Hamiltonian with interactions. We can then write $e^{-\beta K} = e^{-\beta K_0} S(\beta\hbar, 0)$, analogous to how we wrote $e^{-iHt/\hbar} = e^{-\beta H_0} S(t, 0)$ in the Feynman method, and expand the new S-matrix $S(\tau, 0)$ as a sum of integrals of time-ordered products of operators in the interaction representation, for imaginary time τ. In the following, we will distinguish between the variables of integration, which we will write as τ with various subscripted indices, and the fixed value $\beta\hbar$, which will always be given by the actual temperature of the system. The integrals in the S-matrix expansion will therefore correspond to varying the temperature so that each τ ranges up to $\beta\hbar$. It may be hard to visualize what is meant physically by these integrals, but it is not important; for our purposes, it is simply the proper way to do the S-matrix expansion. As in the Feynman approach, each term in the expansion is written in terms of a time-ordered product, which can then be simplified using a variant of Wick's theorem.

In the standard formulation of Feynman perturbation theory we have used so far, we took the initial state and final states as the vacuum state acted on by some set of creation operators to give a definite set of particles in the initial and final states (the external legs). We then used Wick's theorem to take any given term in the time-ordered expansion and reduce it to a product of contractions and leftover, normal-ordered creation and destruction operators, which then canceled out the operators that created the particles in the initial and final states. In the present case, we average over a wide variety of initial states, so we cannot assume that a destruction operator will give zero when acting on the initial state. This means that we cannot formulate Wick's theorem as an operator identity, with free, leftover operators. We can formulate a version of Wick's theorem, however, which applies to averages. The revised Wick's theorem is

$$\langle \mathrm{T}(ABCD\ldots YZ)\rangle_T = (\overset{\sqcap\;\sqcap}{ABCD\ldots YZ}) + (\overset{\sqcap\;\overset{\sqcap\sqcap}{}}{ABCD\ldots YZ}) + \ldots,$$

$$(8.13.7)$$

where the brackets indicate a thermal average over the eigenstates of H_0, and the terms on the right are all possible **fully contracted** terms, in which every operator is contracted to another one. Clearly, this requires that there be an equal number of creation and destruction operators.

The contraction here is defined by Wick's theorem for just two operators. For $\tau_1 > \tau_2$, we have, for bosons,

$$\overline{a_{\vec{k}}(\tau_1)a_{\vec{k}}^{\dagger}(\tau_2)} \equiv \langle \mathrm{T}(a_{\vec{k}}(\tau_1)a_{\vec{k}}^{\dagger}(\tau_2))\rangle_T$$

$$= \langle a_{\vec{k}}(\tau_1)a_{\vec{k}}^{\dagger}(\tau_2)\rangle_T \qquad (8.13.8)$$

$$= \frac{1}{Z_0}\sum_j \langle j|a_{\vec{k}}(\tau_1)a_{\vec{k}}^{\dagger}(\tau_2)e^{-\beta K_0}|j\rangle,$$

where Z_0 is the partition function for all eigenstates of H_0. We can pass the operator $a_{\vec{k}}^{\dagger}(\tau_2)$ through the operator $e^{-K_0\tau/\hbar}$ by considering the action of each on the state $|j\rangle$. If the creation operator acts on $|j\rangle$ first, there will be one more particle in the state with energy $E_{\vec{k}}$ when K_0 acts on it. Therefore, we have

$$\langle a_{\vec{k}}(\tau_1)a_{\vec{k}}^{\dagger}(\tau_2)\rangle_T = \frac{1}{Z_0}\sum_j \langle j|a_{\vec{k}}(\tau_1)e^{-\beta K_0}e^{\beta(E_{\vec{k}}-\mu)}a_{\vec{k}}^{\dagger}(\tau_2)|j\rangle$$

$$= \frac{1}{Z_0}\sum_j \langle j|a_{\vec{k}}^{\dagger}(\tau_2)a_{\vec{k}}(\tau_1)e^{-\beta K_0}e^{\beta(E_{\vec{k}}-\mu)}|j\rangle, \qquad (8.13.9)$$

where in the last line we have used the general cyclic property of the trace, namely

$$\text{Tr}(ABC\ldots XYZ) = \sum_j \langle j|ABC\ldots XYZ|j\rangle = \sum_j \langle j|ZABC\ldots XY|j\rangle.$$

$$(8.13.10)$$

Using the commutation relation (8.13.6), we then have

$$\langle a_{\vec{k}}(\tau_1)a_{\vec{k}}^\dagger(\tau_2)\rangle_T = \langle a_{\vec{k}}(\tau_1)a_{\vec{k}}^\dagger(\tau_2) + [a_{\vec{k}}^\dagger(\tau_2), a_{\vec{k}}(\tau_1)]\rangle_T \, e^{\beta(E_{\vec{k}}-\mu)}$$

$$= \langle a_{\vec{k}}(\tau_1)a_{\vec{k}}^\dagger(\tau_2)\rangle_T \, e^{\beta(E_{\vec{k}}-\mu)} - e^{-(E_{\vec{k}}-\mu)(\tau_1-\tau_2)/\hbar} e^{\beta(E_{\vec{k}}-\mu)}.$$

$$(8.13.11)$$

Solving for $\langle a_{\vec{k}}^\dagger(\tau_1)a_{\vec{k}}(\tau_2)\rangle_T$, we obtain

$$\langle a_{\vec{k}}(\tau_1)a_{\vec{k}}^\dagger(\tau_2)\rangle_T = e^{-(E_{\vec{k}}-\mu)(\tau_1-\tau_2)/\hbar} \frac{-e^{\beta(E_{\vec{k}}-\mu)}}{1 - e^{\beta(E_{\vec{k}}-\mu)}}.$$

$$(8.13.12)$$

The last factor is equal to $1 + \langle N_{\vec{k}}\rangle_T$, where $\langle N_{\vec{k}}\rangle_T$ is the standard boson equilibrium distribution,

$$\langle N_{\vec{k}}\rangle_T = \frac{1}{e^{\beta(E_{\vec{k}}-\mu)} - 1}.$$

$$(8.13.13)$$

A similar calculation can be done for the case $\tau_2 > \tau_1$, so that we finally have

$$\overbrace{a_{\vec{k}}(\tau_1)a_{\vec{k}}^\dagger(\tau_2)} = \langle T(a_{\vec{k}}(\tau_1)a_{\vec{k}}^\dagger(\tau_2))\rangle_T$$

$$(8.13.14)$$

$$= e^{-(E_{\vec{k}}-\mu)(\tau_1-\tau_2)/\hbar}\left((1 + \langle N_{\vec{k}}\rangle_T)\Theta(\tau_1 - \tau_2) + \langle N_{\vec{k}}\rangle_T\, \Theta(\tau_2 - \tau_1)\right).$$

Like the contractions we had before, this is a c-number. This has exactly the same form as (8.10.4), except that now the exponential is the thermal weight factor. We define the Matsubara Green's functions as

$$\mathcal{G}_{\vec{k}}(\tau) = -\langle T(a_{\vec{k}}(\tau)a_{\vec{k}}^\dagger(0))\rangle_T,$$

$$(8.13.15)$$

analogous to our previous time-dependent Green's functions.

The same approach for fermions gives

$$\langle T(b_{\vec{k}}(\tau_1)b_{\vec{k}}^\dagger(\tau_2))\rangle_T = e^{-(E_{\vec{k}}-\mu)(\tau_1-\tau_2)/\hbar}$$

$$(8.13.16)$$

$$\times \left((1 - \langle N_{\vec{k}}\rangle_T)\Theta(\tau_1 - \tau_2) - \langle N_{\vec{k}}\rangle_T\, \Theta(\tau_2 - \tau_1)\right).$$

When $\langle N_{\vec{k}}\rangle_T \to 0$, for example, for a Planck distribution of bosons at $T = 0$, or for fermions above the Fermi energy of a Fermi sea, the functions (8.13.14) and (8.13.16) have the same form as the ones we had before for the vacuum state.

To obtain the finite-temperature Wick's theorem (8.13.7) for larger numbers of operators, we follow the same approach that we did here for two operators. In each case, we commute one operator through all the rest of the operators and then use cyclic permutation to bring it back to the beginning. We continue doing this until all of the operators have been contracted (see, e.g., Fetter and Walecka 1971: s. 24).

Exercise 8.13.1 Verify the revised Wick's theorem (8.13.7) explicitly for the thermal average of four operators, $\langle a_{\vec{k}}(\tau_1)b_{\vec{q}}^\dagger(\tau_1)b_{\vec{q}}(\tau_2)a_{\vec{k}}^\dagger(\tau_2)\rangle_T$.

Fourier series for frequency. In our original Feynman approach, we converted the time-dependent Green's functions to functions of energy, by taking the Fourier transforms. There is an additional complication here, that when we take a Fourier transform, we cannot do an integral from $\tau = -\infty$ to $\tau = +\infty$, as we did for the real time integrals, because the Green's functions are not well defined for $\tau < 0$ or $\tau > \hbar\beta$. We can see this if we rewrite (8.13.8), for $\tau_1 > \tau_2$, as

$$
\begin{aligned}
\langle a_{\vec{k}}(\tau_1) a_{\vec{k}}^{\dagger}(\tau_2)\rangle_T &= \frac{1}{Z_0} \sum_j \langle j | e^{-\beta K_0} a_{\vec{k}}(\tau_1) a_{\vec{k}}^{\dagger}(\tau_2) | j \rangle \\
&= \frac{1}{Z_0} \sum_{jj'} \langle j | e^{-\beta K_0} e^{K_0 \tau_1/\hbar} a_{\vec{k}} e^{-K_0 \tau_1/\hbar} | j'\rangle \langle j' | e^{K_0 \tau_2/\hbar} a_{\vec{k}}^{\dagger} e^{-K_0 \tau_2/\hbar} | j \rangle \\
&= \frac{1}{Z_0} \sum_{jj'} e^{-(\beta - \tau_1/\hbar)(E_j - \mu)} \langle j | a_{\vec{k}} | j'\rangle e^{-(\tau_1 - \tau_2)(E_{j'} - \mu)/\hbar} \langle j' | a_{\vec{k}}^{\dagger} | j \rangle e^{-\tau_2(E_j - \mu)/\hbar}.
\end{aligned}
$$

$$(8.13.17)$$

Since there is no upper bound for the total energies of the states E_j and $E_{j'}$, this will only be well controlled if $\tau > 0$ and $\tau < \beta\hbar$.

Since τ is only defined over a finite interval, we cannot use a Fourier transform that integrates τ from $-\infty$ to ∞. But we can use a Fourier *series* to switch functions of τ over a finite range to a sum of frequencies. We write

$$
\mathcal{G}_{\vec{k}}(\tau) = \frac{1}{\beta\hbar} \sum_{n=-\infty}^{\infty} e^{-i\omega_n \tau} G(\vec{k}, \omega_n), \tag{8.13.18}
$$

with

$$
\mathcal{G}(\vec{k}, \omega_n) = \frac{1}{2} \int_{-\beta\hbar}^{\beta\hbar} d\tau \, e^{i\omega_n \tau} G_{\vec{k}}(\tau), \tag{8.13.19}
$$

and $\omega_n = \pi n/\beta\hbar$, where n is an integer. The integral is from $-\beta\hbar$ to $\beta\hbar$ because when τ ranges from 0 to $\beta\hbar$, the difference $\tau_1 - \tau_2$ can be as low as $-\beta\hbar$.

We can find the frequency-dependent, bare Green's function by substituting (8.13.14) into (8.13.19). For bosons,

$$
\begin{aligned}
\mathcal{G}(\vec{k}, \omega_n) &= -\frac{1}{2}\left(1 + \frac{1}{e^{\beta(E_{\vec{k}} - \mu)} - 1}\right) \int_0^{\beta\hbar} d\tau \, e^{i\omega_n \tau} e^{-(E_{\vec{k}} - \mu)\tau/\hbar} \\
&\quad - \frac{1}{2}\left(\frac{1}{e^{\beta(E_{\vec{k}} - \mu)} - 1}\right) \int_{-\beta\hbar}^0 d\tau \, e^{i\omega_n \tau} e^{-(E_{\vec{k}} - \mu)\tau/\hbar}.
\end{aligned} \tag{8.13.20}
$$

The integrals will have factors $e^{\pm i\omega_n(\beta\hbar)}$, which from the condition $\omega_n = \pi n/\beta\hbar$ will be either 1 or -1. When these are taken into account, the Green's function (8.13.20) will have the simple form

$$
\mathcal{G}(\vec{k}, \omega_n) =
\begin{cases}
\dfrac{1}{i\omega_n - (E_{\vec{k}} - \mu)/\hbar}, & n \text{ even} \\[3mm]
0, & n \text{ odd.}
\end{cases} \tag{8.13.21}
$$

For fermions, we will have the same result, but with the opposite assignment of the integers n: The value will be 0 for even n, and the propagator given above for n odd.

Exercise 8.13.2   Show that the Matsubara Green's function is periodic, that is, for $G(\tau_1 - \tau_2) = -i\langle T(a_{\vec{k}}(\tau_1)a_{\vec{k}}^{\dagger}(\tau_2))\rangle_T$, prove

$$G_{\vec{k}}(\tau_1 - \tau_2 + \hbar\beta) = G_{\vec{k}}(\tau_1 - \tau_2), \tag{8.13.22}$$

for $0 < \tau_1, \tau_2 < \hbar\beta$. Hint: Use the cyclic property of traces (8.13.10), above.

Exercise 8.13.3   Verify that (8.13.21) follows from (8.13.20) for bosons, and derive the comparable relation for fermions.

Rules for Matsubara diagrams. We can now write down a set of diagram rules similar to those we used for Feynman diagrams. The rules are the same as for our Feynman diagrams, including assigning momenta and energies to the internal lines which must be conserved at each vertex, summing over unconstrained momenta, -1 factors for fermion loops and crossings, and associating the same matrix elements we used before with the vertices, except for the following differences:

- For each internal line, associate a propagator of the form (8.13.21). Switching to energy units, we write

$$\mathcal{G}(\vec{k}, E_n) = \frac{1}{iE_n - E_{\vec{k}} + \mu}, \tag{8.13.23}$$

with $E_n = \hbar\omega_n$. As before, we can write a combined phonon or photon propagator that accounts for both emission and absorption. For a Planck distribution, the chemical potential is zero, so that we have

$$\mathcal{G}(\vec{k}, E_n) = \frac{2E_{\vec{k}}}{(iE_n)^2 - E_{\vec{k}}^2}$$

$$= -\frac{2E_{\vec{k}}}{(E_n)^2 + E_{\vec{k}}^2}. \tag{8.13.24}$$

- Instead of integrating over energy, perform a sum over frequencies $\omega_n = \pi n/\hbar\beta$ for any unconstrained energy, with n even for bosons and odd for fermions. In general, calculating these sums may require use of various sum rules or numerical methods, but one can often use a simple analytical trick, discussed in Section 8.15, to resolve the sums. Multiply each sum over frequencies by a factor $-1/\beta$.

  Just as energy was conserved previously, now the integer for the frequency sum must be conserved. At each vertex, the sum of the integers of the incoming lines must add up to the sum for the outgoing lines. For example, an incoming fermion line with an odd integer frequency going to an outgoing fermion line also with an odd integer frequency can only do so by a jump of an even integer frequency, which corresponds to a boson line entering or leaving the vertex.

Example: interaction self-energy. As an example, consider the correction to the single-particle propagator shown in Figure 8.29. This type of self-energy diagram also exists for the zero-temperature Feynman formalism we have used.

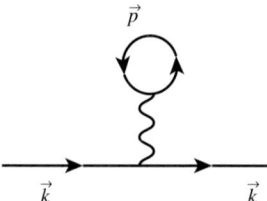

Fig. 8.29   Tadpole self-energy diagram for the electron propagator for the direct Coulomb interaction.

Since the momentum out is equal to the momentum in, the momentum transfer in the Coulomb vertex must be zero. The central self-energy correction is therefore, for the screened Coulomb potential,

$$\Sigma_0 = \frac{e^2/\epsilon}{V\kappa^2} \sum_{\vec{p}} \frac{1}{\beta} \sum_{n=-\infty}^{\infty} \frac{1}{i\hbar\omega_n - E_{\vec{p}} + \mu}. \tag{8.13.25}$$

This sum as it stands will not converge. However, we notice that this sum is the same as that which appears in the Fourier series formula (8.13.18) for $\tau = 0$. If we multiply by a factor $e^{-i\omega_n \eta}$, for $\eta \to 0^-$, we will recover the time-dependent Green's function $G_{\vec{k}}(\tau)$ for $\tau = 0$ coming from the negative side, which according to (8.13.14) is just $\langle N_{\vec{p}} \rangle$. (A negative sign is picked up in the reverse time-ordering for fermions.) We then have

$$\Sigma_0 = \frac{e^2/\epsilon}{V\kappa^2} \sum_{\vec{p}} \langle N_{\vec{p}} \rangle = \frac{e^2/\epsilon}{\kappa^2} \frac{N_{\text{tot}}}{V}. \tag{8.13.26}$$

This diagram is equivalent to that shown in Figure 8.27(a) for the direct Coulomb contribution to the electron self-energy, discussed in Section 8.12. As discussed in that section, this term is often assumed to be canceled by the interaction of the electrons with a positive background of equal and opposite charge.

Exercise 8.13.4   (a) Calculate the tadpole diagram for a zero-temperature Fermi gas using the Feynman method (treating the vacuum as the ground state of the Fermi sea) and show that this gives the same answer as the $T = 0$ limit of the Matsubara calculation given here. Hint: You will need to add a similar exponential factor to have the integral over energy for the single-propagator loop give a meaningful answer.

(b) What Matsubara self-energy diagram (and self-energy diagram for a $T = 0$ Fermi sea) corresponds to the exchange self-energy diagram of Figure 8.27(b)? Show that this diagram gives an energy of opposite sign to the direct Coulomb energy diagram.

Green's function renormalization. In (8.13.15), we defined the bare Matsubara Green's function in terms of the creation and destruction operators, analogous to the real-time-dependent Green's functions. We can gain some insight about what this Green's function means physically by writing it out explicitly for $\tau = \beta\hbar$:

$$\mathcal{G}_{\vec{k}}(\beta\hbar) = -\frac{1}{Z_0}\sum_j \langle j|e^{\beta K_0} a_{\vec{k}} e^{-\beta K_0} a_{\vec{k}}^\dagger e^{-\beta K_0}|j\rangle$$

$$= -\frac{1}{Z_0}\sum_j \langle j|a_{\vec{k}} e^{-\beta K_0} a_{\vec{k}}^\dagger|j\rangle \qquad (8.13.27)$$

$$= -\frac{1}{Z_0}\sum_j \langle j|a_{\vec{k}}^\dagger a_{\vec{k}} e^{-\beta K_0}|j\rangle \equiv -\frac{1}{Z_0}\sum_j \langle j|\hat{N}_{\vec{k}} e^{-\beta K_0}|j\rangle.$$

In each of the last two lines we have used the property of cyclic permutation discussed above. The bare Matsubara Green's function at $\tau = \beta\hbar$ is thus the thermal average of the occupation number of state $\vec{k}$.

For the fully interacting system, we replace H_0 with the full H, which here means replacing K_0 with K. We thus write

$$\mathcal{G}'_{\vec{k}}(\beta\hbar) = -\frac{1}{Z}\sum_j \langle j|a_{\vec{k}} e^{-\beta K} a_{\vec{k}}^\dagger|j\rangle$$

$$= -\frac{1}{Z}\sum_j \langle j|a_{\vec{k}} e^{-\beta K_0} \mathcal{S}(\beta\hbar,0) a_{\vec{k}}^\dagger|j\rangle$$

$$= -\frac{1}{Z}\sum_j \langle j|e^{\beta K_0} a_{\vec{k}} e^{-\beta K_0} \mathcal{S}(\beta\hbar,0) a_{\vec{k}}^\dagger e^{-\beta K_0}|j\rangle$$

$$= -\frac{\langle a_{\vec{k}}(\tau)\mathcal{S}(\beta\hbar,0) a_{\vec{k}}^\dagger(0)\rangle_T}{Z/Z_0}, \qquad (8.13.28)$$

where the thermal average is defined for the eigenstates of K_0, with the partition function Z_0, and we have again used the rule of cyclic permutation of a trace. This is analogous to the way we wrote the renormalized Green's function (8.9.6) in Section 8.9 We can do the same trick that we did in Section 8.9 to eliminate the denominator while also getting rid of disconnected diagrams. We write

$$\frac{Z}{Z_0} = \frac{1}{Z_0}\sum_j \langle j|e^{-\beta K}|j\rangle$$

$$= \frac{1}{Z_0}\sum_j \langle j|e^{-\beta K_0}\mathcal{S}(\beta\hbar,0)|j\rangle = \langle \mathcal{S}(\beta\hbar,0)\rangle_T, \qquad (8.13.29)$$

where we calculate $\langle \mathcal{S}(\tau,0)\rangle_T = 1 + \Lambda$ using the Matsubara diagrammatic method, where Λ stands for all the disconnected diagrams. We then have

$$\frac{Z}{Z_0} = 1 + \Lambda. \qquad (8.13.30)$$

Since the disconnected diagrams in the expansion of $\mathcal{S}(\beta\hbar,0)$ in the numerator give us the same factor, we once again have the result that the factor in the denominator exactly cancels out the disconnected diagrams multiplying the terms in the numerator.

8.14 Symmetrized Green's Functions

So far in this chapter, we have used two kinds of Feynman Green's functions: the time-ordered Green's functions introduced in Section 8.7, with their variant at finite temperature given in Section 8.10, and the Matsubara Green's functions for thermal averages. There is another type of Green's function that is also quite useful for thermal calculations, which allows us to make connections between the Feynman and Matsubara methods. This is generally known as a "retarded Green's function" in the literature. More accurately, these new Green's functions can be called "symmetrized" Green's functions.

We defined the symmetrized, retarded Green's functions as

$$G_{\vec{k}}^{S}(t_1 - t_2) = -i\langle [a_{\vec{k}}(t_1), a_{\vec{k}}^{\dagger}(t_2)]\rangle_T \, \Theta(t_1 - t_2)$$

$$= -i\langle (a_{\vec{k}}(t_1)a_{\vec{k}}^{\dagger}(t_2) - a_{\vec{k}}^{\dagger}(t_2)a_{\vec{k}}(t_1))\rangle_T \, \Theta(t_1 - t_2) \quad (8.14.1)$$

for bosons, and the same but with the anticommutator for fermions. For a non-interacting system, the commutation relations (8.6.13) immediately give

$$G_{\vec{k}}^{S}(t_1 - t_2) = -i e^{-iE_{\vec{k}}(t_1-t_2)/\hbar} \Theta(t_1 - t_2). \quad (8.14.2)$$

Comparing this to our previous time-ordered Green's function (8.10.4) for finite temperature, we can see that although we have a thermal average, the population factors $\langle N_{\vec{k}}\rangle_T$ have been stripped off, so that we have just a "bare" propagator. We also see that the symmetrized Green's functions are equal to our previous time-dependent Green's functions in the $T \to 0$ limit.

For the interacting system, we generalize this to the renormalized Green's function.

$$G_{\vec{k}}^{S'}(t_1 - t_2) = -i\langle [e^{iHt_1/\hbar}a_{\vec{k}}e^{-iHt_1/\hbar}, e^{iHt_2/\hbar}a_{\vec{k}}^{\dagger}e^{iHt_2/\hbar}]\rangle_T \, \Theta(t_1 - t_2),$$

$$(8.14.3)$$

where $H = H_0 + V_{\text{int}}$ is the full Hamiltonian including interactions. As we will see below, it generates a spectral function, which we introduced in Section 8.9. We will be able to calculate this spectral function by connecting it to the results obtained from the Matsubara diagrammatic method.

The Lehmann representation. Since $G_{\vec{k}}^{S}(t)$ involves a thermal average, we can average over any complete set of states that we want. So far, we have averaged over the eigenstates of H_0. But we can just as well suppose that we have found the true eigenstates of the full interacting Hamiltonian H. This is a mathematical trick, because we never actually need to find the true eigenstates of the full many-body system. We average over these:

$$G_{\vec{k}}^{S}(t) = -i\frac{1}{Z}\sum_{m} e^{-\beta\varepsilon_m} \langle m|[e^{iHt/\hbar}a_{\vec{k}}e^{-iHt/\hbar}, a_{\vec{k}}^{\dagger}]|m\rangle \, \Theta(t), \quad (8.14.4)$$

where $\varepsilon_m = \langle m|(H - \mu\hat{N})|m\rangle$. We insert a complete set of the same states:

$$G^S_{\vec{k}}(t) = -i\frac{1}{Z}\sum_{m,n} e^{-\beta\varepsilon_m}\left(\langle m|e^{iHt/\hbar}a_{\vec{k}}e^{-iHt/\hbar}|n\rangle\langle n|a^\dagger_{\vec{k}}|m\rangle\right. \tag{8.14.5}$$

$$\left. - \langle m|a^\dagger_{\vec{k}}|n\rangle\langle n|e^{iHt/\hbar}a_{\vec{k}}e^{-iHt/\hbar}|m\rangle\right)\Theta(t)$$

$$= -i\frac{1}{Z}\sum_{m,n} e^{-\beta\varepsilon_m}\left(e^{i(E_m - E_n)t/\hbar}|\langle m|a_{\vec{k}}|n\rangle|^2 - e^{i(E_n - E_m)t/\hbar}|\langle n|a_{\vec{k}}|m\rangle|^2\right)\Theta(t).$$

Note that $a^\dagger_{\vec{k}}$ and $a_{\vec{k}}$ do not act as simple creation and destruction operators to take one eigenstate $|n\rangle$ to another, in general.

Since m and n are both summed over, we can switch them in the sum for the second term, to get

$$G^S_{\vec{k}}(t) = -i\frac{1}{Z}\sum_{m,n} e^{-i\omega_{nm}t}\left(e^{-\beta\varepsilon_m} - e^{-\beta\varepsilon_n}\right)|\langle m|a_{\vec{k}}|n\rangle|^2\,\Theta(t), \tag{8.14.6}$$

where we have written $\hbar\omega_{nm} = E_n - E_m$.

The Fourier transform is then

$$G^S(\vec{k},\omega) = -i\int_0^\infty dt\, e^{i\omega t}\frac{1}{Z}\sum_{m,n} e^{-i\omega_{nm}t}\left(e^{-\beta\varepsilon_m} - e^{-\beta\varepsilon_n}\right)|\langle m|a_{\vec{k}}|n\rangle|^2$$

$$= \frac{1}{Z}\sum_{m,n}\frac{e^{-\beta\varepsilon_m} - e^{-\beta\varepsilon_n}}{\omega - \omega_{nm} + i\epsilon}|\langle m|a_{\vec{k}}|n\rangle|^2, \tag{8.14.7}$$

where we once again have made the integral tractable by adding a term $e^{-\epsilon t}$. A similar calculation can be done for fermions, which have an anticommutator, so that we finally have

$$G^S(\vec{k},\omega) = \frac{1}{Z}\sum_{m,n}\frac{e^{-\beta\varepsilon_m} \mp e^{-\beta\varepsilon_n}}{\omega - \omega_{nm} + i\epsilon}|\langle m|a_{\vec{k}}|n\rangle|^2, \tag{8.14.8}$$

where the $-$ sign is for bosons and the $+$ sign for fermions.

This is known as the **Lehmann representation** for the Green's functions. It is not useful for computing anything directly, since we don't know the exact eigenstates $|n\rangle$ and $|m\rangle$, but it will be useful for the sum rules we will prove below.

Sum rules for the spectral function. The spectral function of these Green's functions, as defined in Section 8.9, can be found using the Dirac formula (7.2.2); it is

$$A^S(\vec{k},\omega) = -2\mathrm{Im}\, G^S(\vec{k},\omega) = 2\pi\frac{1}{Z}\sum_{m,n}(e^{-\beta\varepsilon_m} \mp e^{-\beta\varepsilon_n})|\langle m|a_{\vec{k}}|n\rangle|^2\delta(\omega - \omega_{nm}).$$

$$\tag{8.14.9}$$

If we integrate over ω, we obtain

$$\frac{1}{2\pi}\int_{-\infty}^\infty d\omega A^S(\vec{k},\omega) = \frac{1}{Z}\sum_{m,n}(e^{-\beta\varepsilon_m} \mp e^{-\beta\varepsilon_n})|\langle m|a_{\vec{k}}|n\rangle|^2.$$

$$\tag{8.14.10}$$

Again noting that we can switch the dummy summation variables n and m, we find

$$\frac{1}{2\pi}\int_{-\infty}^{\infty}d\omega A^S(\vec{k},\omega) = \frac{1}{Z}\sum_{m,n}\left(e^{-\beta\varepsilon_m}\langle m|a_{\vec{k}}|n\rangle\langle n|a_{\vec{k}}^\dagger|m\rangle \mp e^{-\beta\varepsilon_n}\langle m|a_{\vec{k}}^\dagger|n\rangle\langle n|a_{\vec{k}}|m\rangle\right)$$

$$= \frac{1}{Z}\sum_{m}e^{-\beta\varepsilon_m}\langle m|a_{\vec{k}}a_{\vec{k}}^\dagger \mp a_{\vec{k}}^\dagger a_{\vec{k}}|m\rangle. \tag{8.14.11}$$

Using the standard commutation relations for fermions and bosons, and the definition of the partition function, we obtain

$$\boxed{\frac{1}{2\pi}\int_{-\infty}^{\infty}d\omega A^S(\vec{k},\omega) = 1.} \tag{8.14.12}$$

We thus see that the spectral function is normalized, as the spectral function for the standard time-ordered Green's function was, given in Section 8.9.

We can also weight the integral over ω by the equilibrium occupation number $N(\omega)$. In this case, we have the integral

$$\frac{1}{2\pi}\int_{-\infty}^{\infty}d\omega\, N(\omega)A^S(\vec{k},\omega) \tag{8.14.13}$$

$$= \int_{-\infty}^{\infty}d\omega\frac{1}{(e^{\beta(\hbar\omega-\mu)}\mp 1)}\frac{1}{Z}\sum_{m,n}(e^{-\beta\varepsilon_m}\mp e^{-\beta\varepsilon_n})|\langle m|a_{\vec{k}}|n\rangle|^2\delta(\omega-\omega_{nm})$$

$$= \frac{1}{Z}\sum_{m,n}\frac{1}{(e^{\beta(\hbar\omega_{nm}-\mu)}\mp 1)}e^{-\beta\varepsilon_n}\left(e^{\beta(\varepsilon_n-\varepsilon_m)}\mp 1\right)\langle n|a_{\vec{k}}^\dagger|m\rangle\langle m|a_{\vec{k}}|n\rangle.$$

The difference $\varepsilon_n - \varepsilon_m \simeq E_n - E_m - \mu \equiv \hbar\omega_{nm} - \mu$, because the $a_{\vec{k}}$ operator that takes $|n\rangle$ to $|m\rangle$ leaves one fewer particle in state $|m\rangle$. Even though the two states are not necessarily eigenstates of $\hat{N}$, the action of the $a_{\vec{k}}$ operator will still reduce the expectation value of $\hat{N}$ by approximately one particle. The two factors in the denominator and numerator of (8.14.13) therefore cancel. We can then remove the sum over states $|m\rangle$ since no other terms depend on m, to obtain another sum rule,

$$\boxed{\frac{1}{2\pi}\int_{-\infty}^{\infty}d\omega\, N(\omega)A^S(\vec{k},\omega) = \langle N_{\vec{k}}\rangle_T,} \tag{8.14.14}$$

where $\langle N_{\vec{k}}\rangle_T$ is the occupation number of the $\vec{k}$-states taking into account all of the interactions, since the states $|n\rangle$ are the eigenstates of the full Hamiltonian.

This sum rule, along with the normalization condition (8.14.12), leads to the following physical picture. We suppose that the particles are distributed in a thermal distribution of their true eigenstates. For each eigenstate, there is a probability given by the spectral function that the particle is in the plane wave state $\vec{k}$. The sum rule (8.14.14) then tells us that to find the average occupation number of a given $\vec{k}$-state, we use $A^S(\vec{k},\omega)$ as a convolution function, to get the contribution of each ω state to that $\vec{k}$.

The sum rule (8.14.14) also implies

$$A^S(\vec{k},-\omega) = -A^S(\vec{k},\omega) \tag{8.14.15}$$

for bosons, because when the Bose distribution $N(\omega)$ becomes negative for $\omega < 0$ we must get a positive value for $\langle N_{\vec{k}} \rangle_T$.

Connection to the Matsubara Green's functions. Recall that we defined the Matsubara Green's functions as

$$\mathcal{G}_{\vec{k}}(\tau) = -\langle T(a_{\vec{k}}(\tau)a_{\vec{k}}^{\dagger}(0))\rangle_T$$

$$= -\langle a_{\vec{k}}(\tau)a_{\vec{k}}^{\dagger}(0)\rangle_T \theta(\tau) \mp \langle a_{\vec{k}}^{\dagger}(0)a_{\vec{k}}(\tau)\rangle_T \theta(-\tau). \qquad (8.14.16)$$

As we saw in Section 8.13, in the case of an interacting system, we substitute the full K for K_0, to get the renormalized Green's function. But because we assume we know the true eigenstates, we can continue to use the same notation, treating the full K as we did K_0 before. We switch to the Lehmann representation by inserting a complete set of the (unknown) true eigenstates of K:

$$\mathcal{G}_{\vec{k}}(\tau) = -\frac{1}{Z}\sum_{m,n} e^{-\beta\varepsilon_m} \langle m|a_{\vec{k}}(\tau)|n\rangle \langle n|a_{\vec{k}}^{\dagger}(0)|m\rangle \theta(\tau) \qquad (8.14.17)$$

$$\mp \frac{1}{Z}\sum_{m,n} e^{-\beta\varepsilon_m} \langle m|a_{\vec{k}}^{\dagger}(0)|n\rangle \langle n|a_{\vec{k}}(\tau)|m\rangle \theta(-\tau)$$

$$= -\frac{1}{Z}\sum_{m,n} e^{-(\varepsilon_n-\varepsilon_m)\tau/\hbar}\left(e^{-\beta\varepsilon_m}\theta(\tau) \pm e^{-\beta\varepsilon_n}\theta(-\tau)\right)|\langle m|a_{\vec{k}}|n\rangle|^2,$$

where we once again have used the ability to switch n and m. The Fourier transform is then

$$\mathcal{G}(\vec{k},\omega_l) = \frac{1}{2}\int_{-\beta\hbar}^{\beta\hbar} d\tau \, e^{i\omega_l\tau}\mathcal{G}_{\vec{k}}(\tau)$$

$$= \frac{1}{Z}\sum_{m,n} \frac{(e^{-\beta\varepsilon_m} \mp e^{-\beta\varepsilon_n})}{i\omega_l - \omega_{nm}}|\langle m|a_{\vec{k}}|n\rangle|^2, \qquad (8.14.18)$$

where l is an integer used for a discrete frequency, as discussed in Section 8.13. This is exactly the same as the symmetrized, retarded Green's function (8.14.8) if we substitute $i\omega_n \to \omega + i\epsilon$, a process known as **analytic continuation**. Thus, even without ever calculating a Green's function in the Lehmann representation, we know that if we calculate the renormalized Green's function using the Matsubara diagram method, and hence its spectral function, then we immediately know the symmetrized Green's function and its spectral function by simply substituting for $i\omega_n$ wherever it occurs. This allows us, for example, to use the sum rule (8.14.14) given above to calculate the redistribution of $N_{\vec{k}}$ due to interactions, which we will do in Section 8.15 for an electron gas. Also, since we know that the symmetrized Green's functions are equal to the non-symmetrized Green's functions in the $T \to 0$ limit, the Matsubara result for any diagram will be equal to the Feynman result for the same diagram as when we set $T = 0$. This is a useful way to check whether a calculation is correct.

Exercise 8.14.1 Prove that

$$N(\omega)A^S(\vec{k},\omega) = 2\pi\frac{1}{Z}\sum_{m,n} e^{-\beta\varepsilon_n}|\langle m|a_{\vec{k}}|n\rangle|^2\delta(\omega - \omega_{nm}). \qquad (8.14.19)$$

Exercise 8.14.2   Show that in the limit of a noninteracting gas, the spectral function $A^S(\vec{k}, \omega)$ is proportional to $\delta(\omega - \omega_{\vec{k}})$.

8.15 Matsubara Calculations for the Electron Gas

In this section, we will do two extended examples to illustrate how to use the Matsubara diagram method, with the primary application to a gas of electrons. First, we redo the screening calculation of Section 8.11.

Screening at finite temperature. Figure 8.30 shows a single loop diagram that plays the same role as the loop diagram we introduced for electrons at zero temperature, shown in Figure 8.19, which we gave the name $\Pi_{\vec{k}}$. As discussed in Section 8.11.1, this can be viewed as a renormalized bosonic propagator carrying momentum $\vec{k}$.

Following the rules for Matsubara diagrams given in Section 8.13, we associate the energy $E_m = \hbar\omega_m$ with the internal momentum $\vec{p}$ and the energy E_n with the propagator momentum $\vec{k}$. Using the Matsubara diagram rules, we write down two propagators for the fermions, conserve energy and momentum at the vertices, and sum over both, to obtain

$$\Pi_{\vec{k},n} = \sum_{\vec{p}} \frac{1}{\beta} \sum_{m=-\infty}^{\infty} \frac{1}{(i\hbar\omega_m - E_{\vec{p}} + \mu)} \frac{1}{(i\hbar\omega_m + i\hbar\omega_n - E_{\vec{p}+\vec{k}} + \mu)}.$$

$$(8.15.1)$$

The integers m must be odd, because they correspond to the fermionic propagator with momentum $\vec{p}$, while the integers n must be even, that is, bosonic, since the other propagator is also fermionic, and must have frequencies with odd integers for the total of $\omega_m + \omega_n$.

The sum over m can be done using an analytical trick that is generally useful for Matsubara summations. We start by noting that the sum over odd m of any function $F(i\omega)$ can be written as

$$-\frac{1}{\beta} \sum_{m=-\infty}^{\infty} F(i\omega_n) = \frac{\hbar}{2\pi i} \oint_{C_1} d\omega \, F(\omega) N_F(\omega),$$

$$(8.15.2)$$

where $N_F(\omega) = 1/(e^{\beta\hbar\omega} + 1)$ is the Fermi–Dirac distribution, and the contour integral is the loop around the whole imaginary axis of the complex plane, shown as curve C_1 in Figure 8.31. The reason is that the Fermi–Dirac function has poles that act like $-1/(\omega - \omega_m)$

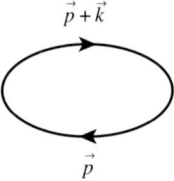

Fig. 8.30   Self-energy loop for finite temperature screening.

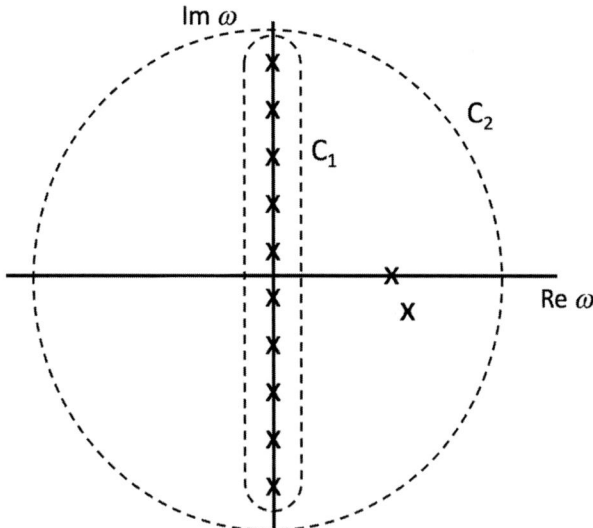

Fig. 8.31 Paths in the complex plane for computing the infinite sum over odd n in (8.15.1). The crosses mark the locations of poles.

every time $e^{\beta\hbar\omega} = -1$, which occurs whenever $\beta\hbar\omega = i(\pi + 2\pi m)$. This implies poles when $\omega = i\pi m/\beta\hbar \equiv i\omega_m$ for all odd m, exactly the same set of values of ω_m we have used in the Matsubara sum for a fermion energy. By Cauchy's residue theorem, the contour integral contributes a residue $F(i\omega_m)$ for each of these poles.

Suppose now that $F(\omega)$ has some set of poles at ω_i, so that at any given pole, it can be written $F(\omega) = f(\omega)/(\omega - \omega_i)$. Then a second loop C_2 can be drawn around the entire complex plane to include these. Cauchy's residue theorem says that the integral for this loop is just the sum for C_1 plus the residues of these extra poles. We can therefore write

$$\frac{\hbar}{2\pi i}\oint_{C_2} d\omega\, F(\omega)N_F(\omega) = \frac{\hbar}{2\pi i}\oint_{C_1} d\omega\, F(\omega)N_F(\omega) + \hbar\sum_i f(\omega_i)N_F(\omega_i).$$

(8.15.3)

If $F(\omega)$ decreases for large $|\omega|$ at least as fast as $1/|\omega|^2$, then the contour integral over C_2 vanishes.[2] Therefore, we have

$$\frac{\hbar}{2\pi i}\oint_{C_1} d\omega\, F(\omega)N_F(\omega) = -\hbar\sum_i f(\omega_i)N_F(\omega_i),$$

(8.15.4)

and

$$\boxed{\frac{1}{\beta}\sum_{m=-\infty}^{\infty} F(i\omega_m) = \hbar\sum_i f(\omega_i)N_F(\omega_i).}\quad \text{(fermions)}$$

(8.15.5)

---

[2] Because the series of poles on the vertical axis continue to infinity, one might wonder if the contribution of C_2 vanishes when it crosses that axis. It turns out that the integrand is odd in ω as it crosses this axis in the horizontal direction, so that the contributions of the poles on opposite sides of this axis cancel out.

We thus end up with a simple rule, as long as $F(\omega)$ falls fast enough at $|\omega| \to \infty$, that the infinite sum over frequencies can be replaced by just a finite sum of the residues at the poles of $F(\omega)$. The same trick can be used to sum over the energies of bosons, using a series of poles for even n, giving the same formula but with the boson distribution $N_B(\omega) = 1/(e^{\beta\hbar\omega} - 1)$, and the opposite sign:

$$\boxed{\frac{1}{\beta} \sum_{m=-\infty}^{\infty} F(i\omega_m) = -\hbar \sum_i f(\omega_i) N_B(\omega_i).} \quad \text{(bosons)} \qquad (8.15.6)$$

We now return to the sum over n in (8.15.1). There are two poles of $F(\omega)$, at $i\hbar\omega_m = E_{\vec{p}} - \mu$ and $i\hbar\omega_m = E_{\vec{p}+\vec{k}} - \mu - i\hbar\omega_n$, as illustrated in Figure 8.31. Using (8.15.5), the sum over m gives us

$$\Pi_{\vec{k},n} = \sum_{\vec{p}} \left(\frac{N_F(E_{\vec{p}} - \mu)}{i\hbar\omega_n + E_{\vec{p}} - E_{\vec{p}+\vec{k}}} + \frac{N_F(E_{\vec{p}+\vec{k}} - \mu - i\hbar\omega_n)}{E_{\vec{p}+\vec{k}} - i\hbar\omega_n - E_{\vec{p}}} \right). \qquad (8.15.7)$$

The $\hbar$ in (8.15.5) has been absorbed in converting the integral over frequency ω into an integral over energy.

The Fermi–Dirac distribution has the property $N_F(E_{\vec{k}} + i\hbar\omega_n) = N_F(E_{\vec{k}})$ for $\omega_n = i\pi m/\beta\hbar$, because ω_n is a bosonic frequency corresponding to even integers n, and therefore $e^{-\beta\hbar\omega_n} = 1$. Therefore, (8.15.7) becomes

$$\Pi_{\vec{k},n} = \sum_{\vec{p}} \frac{N_F(E_{\vec{p}} - \mu) - N_F(E_{\vec{p}+\vec{k}} - \mu)}{i\hbar\omega_n + E_{\vec{p}} - E_{\vec{p}+\vec{k}}}. \qquad (8.15.8)$$

Using analytic continuation, as discussed in Section 8.14, we replace $i\omega_m$ with $\omega + i\eta$. In the long-wavelength limit $\omega \to 0$, we then obtain

$$\Pi_{\vec{k}} = \sum_{\vec{p}} \frac{\partial N}{\partial E}\bigg|_{E_{\vec{p}}=\mu}. \qquad (8.15.9)$$

We thus obtain the same result as (8.11.11), which we found using our earlier finite temperature method.

Second order self-energy of a Fermi system. Figure 8.32 shows the direct second-order diagram for the Coulomb interaction renormalization of a free electron. The imaginary part of the self-energy diagram corresponds to the collision rate for two-body interactions given by the quantum Boltzmann equation in Section 4.8. We assign the energy E_l to the incoming leg with momentum $\vec{q}$, and E_m and E_n to $\vec{p}$ and $\vec{k}$, respectively. Note that the integers l and m for the frequency sums are odd, since they correspond to fermion energies, while the integer n corresponding to the bosonic virtual photon in the Coulomb interaction is even; n must be even because the fermion propagator with $iE_l - iE_n$ must be fermionic, and l is odd. The Matsubara integral for this diagram is then

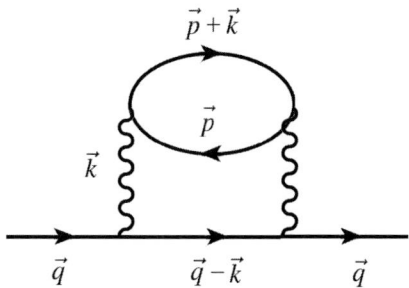

Fig. 8.32   Diagram for the second-order self-energy of a Fermi liquid.

$$\Sigma_{\vec{q},l} = -\frac{1}{\beta^2} \sum_{m,n} \sum_{\vec{k},\vec{p}} M_{\vec{k}}^2 \frac{1}{(iE_m - E_{\vec{p}} + \mu)} \frac{1}{(iE_m + iE_n - E_{\vec{p}+\vec{k}} + \mu)}$$

$$\times \frac{1}{(iE_l - iE_n - E_{\vec{q}-\vec{k}} + \mu)}. \tag{8.15.10}$$

The sum over $\vec{p}$ and m of the first two factors is just the screening bubble we computed above. We therefore have

$$\Sigma_{\vec{q},l} = -\frac{1}{\beta} \sum_{\vec{k},n} M_{\vec{k}}^2 \left(\sum_{\vec{p}} \frac{N_F(E_{\vec{p}} - \mu) - N_F(E_{\vec{p}+\vec{k}} - \mu)}{iE_n + E_{\vec{p}} - E_{\vec{p}+\vec{k}}} \right) \frac{1}{iE_l - iE_n - E_{\vec{q}-\vec{k}} + \mu}. \tag{8.15.11}$$

Now we do the same trick for the bosonic summation over even integers n that we did above for the fermionic summation over odd integers m. There are poles at $iE_n = E_{\vec{p}+\vec{k}} - E_{\vec{p}}$ and $iE_n = iE_l - E_{\vec{q}-\vec{k}} + \mu$. We obtain

$$\Sigma_{\vec{q},l} = \sum_{\vec{p},\vec{k}} M_{\vec{k}}^2 \left(N_F(E_{\vec{p}} - \mu) - N_F(E_{\vec{p}+\vec{k}} - \mu) \right)$$

$$\times \frac{N_B(iE_l - E_{\vec{q}-\vec{k}} + \mu) - N_B(E_{\vec{p}+\vec{k}} - E_{\vec{p}})}{E_{\vec{p}} - E_{\vec{p}+\vec{k}} - E_{\vec{q}-\vec{k}} + iE_l + \mu}. \tag{8.15.12}$$

We can now use the relation $N_B(x + iE_l) = -N_F(x)$, which comes about because $e^{\beta E_l} = -1$ when l is an odd integer, and we use analytic continuation to substitute $iE_l \to E_{\vec{q}} - \mu + i\eta$. We therefore finally have

$$\Sigma_{\vec{q}} = \sum_{\vec{p},\vec{k}} M_{\vec{k}}^2 \left(N_F(E_{\vec{p}} - \mu) - N_F(E_{\vec{p}+\vec{k}} - \mu) \right)$$

$$\times \frac{N_F(-E_{\vec{q}-\vec{k}} + \mu) + N_B(E_{\vec{p}+\vec{k}} - E_{\vec{p}})}{E_{\vec{p}+\vec{k}} + E_{\vec{q}-\vec{k}} - E_{\vec{q}} - E_{\vec{p}} - i\eta}. \tag{8.15.13}$$

Connection to the quantum Boltzmann equation. The diagram shown in Figure 8.32 corresponds to the self-energy of an electron due to Coulomb scattering with another electron, in which the electrons with momenta $\vec{q}$ and $\vec{p}$ scatter into states $\vec{q} - \vec{k}$ and $\vec{p} + \vec{k}$. It is not very obvious from the form of (8.15.13) that this is the case, however.

We first note that the imaginary part of the self-energy gives the scattering rate, which leads to line broadening in the spectral function, as discussed in Section 8.9. The imaginary part is obtained here by using the Dirac formula (7.2.2) to obtain an energy-conserving δ-function, $\delta(E_{\vec{p}+\vec{k}} + E_{\vec{q}-\vec{k}} - E_{\vec{q}} - E_{\vec{p}})$.

We can then use several identities, namely

$$N_F(-x) = 1 - N_F(x), \tag{8.15.14}$$

$$(N_F(x) - N_F(y))N_B(y - x) = N_F(y)(1 - N_F(x)), \tag{8.15.15}$$

and

$$N_F(x)(1 - N_F(y)) - N_F(z)(1 - N_F(y)) + N_F(z)(1 - N_F(x))$$
$$= N_F(x)(1 - N_F(y))(1 - N_F(z) + N_F(y)N_F(z)(1 - N_F(x)) \tag{8.15.16}$$

to convert the occupation number factor in (8.15.13) to

$$N_F(E_{\vec{p}} - \mu)(1 - N_F(E_{\vec{q}-\vec{k}} - \mu))(1 - N_F(E_{\vec{p}+\vec{k}} - \mu))$$
$$+ N_F(E_{\vec{q}-\vec{k}} - \mu)N_F(E_{\vec{p}+\vec{q}} - \mu)(1 - N_F(E_{\vec{p}} - \mu)). \tag{8.15.17}$$

Comparing this to the quantum Boltzmann equation (4.8.16) for two-body scattering, derived in Section 4.8, we see that these two terms give the out-scattering rate for an electron in state $\vec{q}$, assuming it is occupied, and the in-scattering rate for the same state, assuming it is unoccupied, which is equivalent to the out-scattering rate of a *hole* in state $\vec{q}$. (The same result is obtained for the finite-temperature diagram method of Section 8.10; see Exercise 8.10.2.) Both of these terms contribute to the uncertainty of the state, which leads to line broadening.[3]

Fermi liquid theory. We can use the above result to find some interesting results for an electron gas at low temperature. This topic often goes under the name of Fermi liquid theory. There is nothing significant about whether we call this a Fermi "gas" or a Fermi "liquid," but the term "Fermi liquid" is typically used to refer to many-body effects of the Fermi system at low temperature. This is distinct from the "electron–hole liquid" that can occur in an excited semiconductor, which has equal numbers of electrons and holes in two different bands and has the properties of a true liquid, such as incompressibility and surface tension (see Rice *et al.* 1977).

As discussed above, the imaginary part of the self-energy in Figure 8.32 gives the scattering rate of an electron in the presence of a thermal electron bath. Without fully calculating this rate, we can show that it vanishes for particles at the Fermi level in the $T \to 0$ limit. The energy-conserving δ-function implies that if one particle has energy near the Fermi

---

[3] Another way to say this is that both terms contribute to the dephasing of the many-body wave function for fermions (see the discussion in Section 9.2). For bosons, the two terms have opposite sign, leading to the possibility of "enphasing," that is, onset of phase coherence below a critical temperature. See Snoke *et al.* (2012: appendix C).

level, then the other particle states involved in the scattering process also must be near the Fermi level. For example, if the particle in state $\vec{k}$ has energy $E_{\vec{k}} = E_F + \varepsilon$, then the second particle in state $\vec{p}$ must have energy below the Fermi level, but no less than $E_F - \varepsilon$, or else it cannot gain enough energy to go into a free state above the Fermi level. Assuming that the density of states is nearly constant near the Fermi level, the scattering rate is proportional to the number of these electrons times the number of empty states above the Fermi level they can scatter into, that is,

$$\int_{-\varepsilon}^{0} dE_1 \int_{0}^{\varepsilon + E_1} dE_2 = \frac{3}{2}\varepsilon^2. \tag{8.15.18}$$

The scattering state is therefore proportional to $(E_{\vec{p}} - \mu)^2$. A similar calculation can be done for holes below the Fermi level.

Because the imaginary self-energy of the particles vanishes at the Fermi surface, they have spectral functions that are δ-functions in energy; in other words, they are well-defined states with no smearing in energy, like particles in a noninteracting gas. The term "quasiparticles" is sometimes reserved for states like this with well-defined energy, but as we have seen throughout this book, all kinds of quasiparticle states exist with varying degrees of renormalization due to interactions, and there is no hard and fast distinction of how much spectral smearing disqualifies a state from being considered a good quasiparticle.

We can now use the spectral function relation (8.14.14) to calculate the momentum distribution of the electron gas. The result of a Dyson equation gives us the renormalized Green's function of the form (8.8.4). The renormalized spectral function is then

$$A^S(\vec{k}, E) = \frac{\Gamma_{\vec{k}}}{(E - E_{\vec{k}})^2 + \Gamma_{\vec{k}}^2}, \tag{8.15.19}$$

where the imaginary self-energy $\Gamma_{\vec{k}}$ depends on the energy $E_{\vec{k}}$, vanishing at the Fermi surface in the $T \to 0$ limit.

Equation (8.14.14) therefore gives a convolution with a spectral function of varying width. Figure 8.33 shows the result of this convolution for the electron system, assuming that the imaginary self-energy in the spectral function is constant for large $|E_{\vec{k}} - \mu|$ and decreases to zero at the Fermi surface as ε^2. As seen in this figure, there is still a vestige of the discontinuous jump in the occupation number that occurs for a Fermi gas at $T = 0$, but there is also a depletion of the ground state, with particles kicked to higher momentum, making the electrons appear to have an effective temperature, even though they are at $T = 0$. This will affect the heat capacity and other thermodynamics properties of the electrons. A similar effect of depletion of the ground state due to interactions occurs for bosons at $T = 0$, as discussed in Section 11.5.

Exercise 8.15.1   Prove the identities (8.15.14), (8.15.15), and (8.15.16), and therefore the conversion of the occupation number factor in (8.15.13) to (8.15.17).

Exercise 8.15.2   According to (4.8.15) in Section 4.8, the scattering rate for fermions is proportional to $\frac{1}{2}(U_D - U_E)^2$, where U_D is the direct interaction vertex and U_E is the exchange vertex. In the diagram shown in Figure 8.32, only the direct term proportional to U_D^2 occurs. Show that if you account for exchange (how many additional

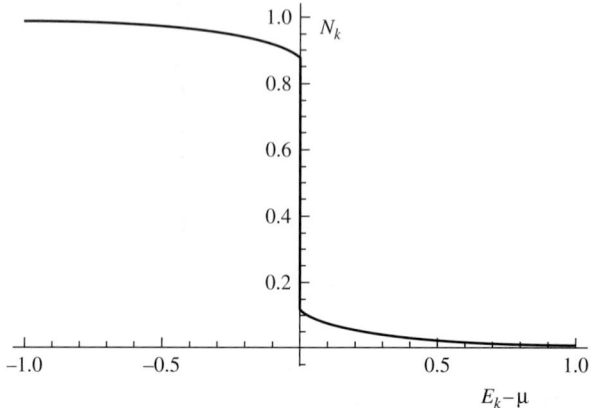

Fig. 8.33 Expectation value for the occupation number per $\vec{k}$-state, as a function of energy, for a Fermi system at zero temperature, when interactions are taken into account, for the spectral function discussed in the text.

diagrams are needed?), you get agreement of the Matusbara imaginary self-energy with the out-scattering rate of (4.8.15).

Exercise 8.15.3   (a) Write down the Matsubara sum for the electron self-energy diagram for phonon interaction shown in Figure 8.17, and resolve the sum using the method of residues given here.

(b) Using the identities of this section, show that you get the same answer as in Section 8.10 when using analytic continuation, in the limit of low electron density.

References

A.A. Abrikosov, L.P. Gorkov, and I.E. Dzyaloshinski, *Methods of Quantum Field Theory in Statistical Physics*, R. Silverman, trans. (Dover, 1975).

V. Alex, S. Finkbeiner, and K. Weber, "Temperature dependence of the indirect energy gap in crystalline silicon," *Journal of Applied Physics* **79**, 6953 (1996).

D. Ceperley and B.J. Adler, "Ground state of the electron gas by a stochastic method," *Physical Review Letters* **45**, 566 (1980).

M. Combescot, "On the generalized golden rule for transition probabilities," *Journal of Physics A* **34**, 6087 (2001).

A.L. Fetter and J.D. Walecka, *Quantum Theory of Many-Particle Systems* (McGraw-Hill, 1971).

H. Haug and S.W. Koch, *Quantum Theory of the Optical and Electronic Properties of Semiconductors*, 3rd ed. (World Scientific, 1990).

Y. Imry, "Casimir zero-point radiation pressure," *Physical Review Letters* **95**, 080404 (2005).

L.V. Keldysh, "Introduction," in *Electron–Hole Droplets in Semiconductors*, C.D. Jeffries and L.V. Keldysh, eds. (North-Holland, 1987).

C. Kittel, *Quantum Theory of Solids* (Wiley, 1963).

W. Kohn and P. Vashishta, *Theory of the Inhomogeneous Electron Gas* (Plenum, 1983).

G. Mahan, *Many-Body Physics*, 3rd ed. (Kluwer Academic, 2000).

F. Mandl and G. Shaw, *Quantum Field Theory* (Wiley, 1984).

R.D. Mattuck, *A Guide to Feynman Diagrams in the Many-Body Problem*, 2nd ed. (Dover, 1976).

J.P. Perdew and A. Zunger, "Self-interaction correction to density-functional approximations for many-electron systems," *Physical Review B* **23**, 5048 (1983).

T.M. Rice, J. Hensel, T. Phillips, and G.A. Thomas, in *Solid State Physics*, H. Ehrenreich, F. Seitz, and D. Turnbull, eds. (Academic Press 1977).

D.W. Snoke, A. Shields, and M. Cardona, "Phonon-absorption recombination luminescence of room temperature excitons in Cu_2O," *Physical Review B* **45**, 11693 (1992).

D. Snoke and J.D. Crawford, "Hysteresis in the Mott transition between plasma and insulating gas," *Physical Review E* **52**, 5796 (1995).

D.W. Snoke, G.-Q. Liu, and S. Girvin, "The basis of the second law of thermodynamics in quantum field theory," *Annals of Physics* **327**, 1825 (2012).

C.R. Smith, "Bound states in a Debye–Hückel potential," *Physical Review* **134**, A1235 (1964).

A. Yariv, *Quantum Electronics*, 3rd ed. (Wiley, 1989).

P.W. Yu, S. Chaudhuri, D.C. Reynolds, et al., "Temperature dependence of shop line photoluminescence in Ga As Al0.25 Ga 0.75As Multiple quantum well structures," *Solid State Communications* **54**, 159 (1985).

P.Y. Yu and M. Cardona, *Fundamentals of Semiconductors* (Springer, 1996).

9 Coherence and Correlation

Modern solid state physics has come to focus strongly on issues of coherence and decoherence. This topic relates to systems as diverse as nuclear magnetic resonance (NMR), ultrafast optics, and quantum computing.

There are actually three different classes of coherence systems that come under study. One type is a *coherent state* contructed of boson quasiparticles, discussed in Section 4.4. In this case, decoherence comes about by boson particles in the coherent state being kicked out into other states. This type of coherence can also include the case of waves that exist in a medium made of fermions, such as plasmons. In this case, although the ground state consists of fermions, when the system is properly quantized, the new quasiparticles are bosons, and we can talk of coherent states of these bosons.

A second type of coherence is the case of an ensemble of isolated oscillators coupled to a classical wave. This is the standard picture of NMR. In this case, decoherence is viewed as perturbations that cause the isolated oscillators to not be exactly identical. We have already discussed this model in Section 7.4 of Chapter 7. This is the type of coherence we will discuss first in this chapter.

A third type of coherence is the case of the wave function of a single electron, or quasiparticle, which extends through space. Decoherence in this case is viewed as scattering of this wave from various perturbations in the medium. We will discuss this type of coherence at the end of this chapter.

In each case, the common theme is that we keep track of a full wave function $\psi(\vec{r})$ which has two components, which we think of either as amplitude and phase, or as the real and imaginary components. A system is coherent if both components matter; it is incoherent if only the density, that is, the square of the amplitude, matters.

Initially much of the enthusiasm for quantum coherence was driven by the desire for a **quantum computer** which could potentially factor large numbers much faster than a digital computer (see Nielson and Chuang 2000). The focus on this objective has led to a much larger field which can be called **quantum information science**. More and more experiments now directly observe both amplitude and phase of quantum wave functions.

Coherence is not an all-or-nothing phenomenon – there are intermediate states between completely coherent and completely incoherent. To parametrize the state of a system that is not completely coherent, we introduce the concept of a **correlation function**, which will be the subject of the last half of this chapter.

9.1 Density Matrix Formalism

To begin our discussion of coherence, we first review the density matrix formalism of quantum mechanics. (For a review of density matrix formalism, see Cohen–Tannoudji *et al.* 1977: s. EIII.) The density operator can be defined as

$$\rho = |\psi\rangle\langle\psi|; \tag{9.1.1}$$

like any operator, it can be expressed as a matrix acting on a complete set of states, with elements $\langle\psi_n|\rho|\psi_m\rangle$. The density matrix is therefore

$$\rho_{mn} = \langle\psi_n|\rho|\psi_m\rangle = \langle\psi_n|\psi\rangle\langle\psi|\psi_m\rangle,$$
$$= c_m^* c_n, \tag{9.1.2}$$

where $|\psi_m\rangle$ and $|\psi_n\rangle$ are members of a complete set of states for the system, and c_m and c_n are the probability amplitudes for occupation of those states. The time evolution of the density matrix can be determined using the Schrödinger equation $(i\hbar)\partial/\partial t|\psi\rangle = H|\psi\rangle$,

$$\frac{\partial\rho(t)}{\partial t} = \left(\frac{\partial}{\partial t}|\psi\rangle\right)\langle\psi| + |\psi\rangle\left(\frac{\partial}{\partial t}\langle\psi|\right)$$
$$= \left(\frac{1}{i\hbar}H|\psi\rangle\right)\langle\psi| + |\psi\rangle\left(-\frac{1}{i\hbar}\langle\psi|H\right), \tag{9.1.3}$$

which is equivalent to

$$\boxed{\frac{\partial\rho(t)}{\partial t} = \frac{1}{i\hbar}[H, \rho(t)].} \tag{9.1.4}$$

This is known as the **Liouville** equation. Since the diagonal elements of the density matrix are the probabilities of occupation of the individual states, the density matrix obeys the normalization condition

$$\mathrm{Tr}\,\tilde{\rho} = \sum_n \rho_{nn} = \sum_n |c_n|^2 = 1. \tag{9.1.5}$$

The density matrix formalism gives us a natural way to account for coherent and incoherent populations. We distinguish between a system in a **pure** quantum state, which can be written as a linear superposition of other quantum states, and one which is in a **mixed** state, which consists of a random ensemble of pure states. In a pure state, there is uncertainty in the outcome of a measurement if the state is a linear superposition of other states, and this comes from the intrinsic uncertainty of quantum mechanics. A pure state can be called coherent since by a change of basis we could describe the system simply in terms of the amplitude and phase of a single quantum state. In a mixed state, in addition to the intrinsic uncertainty of quantum mechanics, there is randomness due to standard statistical uncertainty, such as thermal fluctuations or simple ignorance about the initial state of the system.

We suppose that there is some number of pure states that can be occupied with probability P_i, subject to the normalization $\sum_i P_i = 1$. The density operator in this case will be the sum of the density operators for all the separate pure states, weighted by the probability of being in each state:

$$\rho = \sum_i P_i \rho^{(i)} = \sum_i P_i |\psi_i\rangle \langle \psi_i|. \tag{9.1.6}$$

The probability of finding the system in state $|\psi_n\rangle$ is therefore

$$\sum_i P_i |\langle \psi_n | \psi_i \rangle|^2 = \sum_i P_i \langle \psi_n | \psi_i \rangle \langle \psi_i | \psi_n \rangle$$

$$= \langle \psi_n | \rho | \psi_n \rangle, \tag{9.1.7}$$

that is, the diagonal elements of the density operator, just as in the pure case. Since the density operator (9.1.6) is a linear superposition of pure-state operators, it still obeys the Liouville equation.

Suppose now that the system is in a pure state that is a linear superposition of states $|\psi_m\rangle$, $|\psi_n\rangle$, etc., with probability amplitudes c_n, c_m, respectively. The diagonal elements are $\rho_{nn} = |c_n|^2$, while the off-diagonal elements are $\rho_{nm} = c_m^* c_n$. Compare this to a statistical ensemble of states, each of which is a pure state $|\psi_n\rangle$, and the probability for each is $P_n = |c_n|^2$, where the probabilities are chosen to be exactly the same as in the pure linear superposition. If we look only at the diagonal elements of the density matrix, we cannot distinguish between these two cases. The off-diagonal elements in the statistical mixture, however, are

$$\rho_{mn} = \sum_i P_i \langle \psi_n | \psi_i \rangle \langle \psi_i | \psi_m \rangle$$

$$= \sum_i |c_i|^2 \delta_{ni} \delta_{im}$$

$$= |c_n|^2 \delta_{mn}. \tag{9.1.8}$$

That is, the off-diagonal elements are strictly zero. The off-diagonal elements of the density matrix are therefore good measures of the "pureness" of the system, which can also be called its coherence.

All of this formalism has been developed so far for the general wave function of any system. As mentioned at the end of Section 4.5, we can also generate a density matrix formalism for a many-body system in terms of creation and destruction operators. Suppose that we define a pure single-particle state as the superposition

$$|\psi\rangle = c_1 b_1^\dagger |0\rangle + c_2 b_2^\dagger |0\rangle + \cdots = \sum_i c_i b_i^\dagger |0\rangle, \tag{9.1.9}$$

where $|0\rangle$ is the vacuum (ground) state. Then if we define the density matrix element

$$\rho_{mn} = \langle \psi | b_m^\dagger b_n | \psi \rangle \tag{9.1.10}$$

we obtain

$$\rho_{mn} = \left(\langle 0| \sum_i c_i^* b_i \right) b_m^\dagger b_n \left(\sum_j c_j b_j^\dagger |0\rangle \right) = c_m^* c_n, \tag{9.1.11}$$

just the same as for definition (9.1.2).

In the case of a mixed state, we sum over all the density matrix elements weighted by the probability of their occupation, as before:

$$\rho_{mn} = \sum_i P_i \rho_{mn}^{(i)} = \sum_i P_i \langle \psi_i | b_m^\dagger b_n | \psi_i \rangle. \tag{9.1.12}$$

The time evolution of the density matrix in this formulation is

$$\frac{\partial \rho_{mn}(t)}{\partial t} = \frac{\partial}{\partial t} \langle \psi | b_m^\dagger b_n | \psi \rangle = \left(-\frac{1}{i\hbar} \langle \psi | H \right) b_m^\dagger b_n | \psi \rangle + \langle \psi | b_m^\dagger b_n \left(\frac{1}{i\hbar} H | \psi \rangle \right) \tag{9.1.13}$$

or,

$$\boxed{\frac{\partial \rho_{mn}(t)}{\partial t} = -\frac{1}{i\hbar} \langle \psi | [H, b_m^\dagger b_n] | \psi \rangle.} \tag{9.1.14}$$

The same approach works with bosonic operators as well. As discussed in Section 4.5, a coherent state $|\alpha_k\rangle$, which is a plane wave state with wave number k, can be viewed as a superposition of an infinite number of δ-function states in space. The probability amplitude for each δ-function state can be found using the spatial field operators,

$$\psi(r)|\alpha_k\rangle = \frac{e^{ikr}}{\sqrt{L}} A e^{i\theta} |\alpha_k\rangle \equiv c_r |\alpha_k\rangle, \tag{9.1.15}$$

and the off-diagonal elements of the density matrix are

$$\rho_{r,r'} = c_r^* c_{r'} = \langle \alpha_k | \psi^\dagger(r) \psi(r') | \alpha_k \rangle, \tag{9.1.16}$$

which, as discussed in Section 4.5, are nonzero for a coherent state. Incoherent light or sound can be described as a random statistical ensemble of different coherent states. The off-diagonal matrix elements will average to zero in this case, because they will be a sum of many plane waves with different wavelengths and phases. The density matrix of an incoherent field will therefore be the same as a statistical ensemble of Fock number states, with only diagonal matrix elements.

The common description in terms of density matrix formalism leads to interesting philosophical debates. We tend to view the electric field in a coherent state of photons, or the sound field in a coherent state of phonons, as a real physical entity, and the wave function of a single particle as a "useful fiction." The only difference in terms of the formalism, however, is whether we use fermion or boson field operators. Can we view a pure-state wave function of a single fermion as a physical entity in the same sense as a coherent state of electromagnetic field? We will see in Chapter 11 that bosons with mass can also form a coherent state; this is the basis of superfluidity and superconductivity. In that case, the wave function of the bosons is a macroscopic entity just as physically measurable as an electric field.

Exercise 9.1.1 Show that the density operator in the mixed case still obeys the same normalization rule,

$$\sum_n \rho_{nn} = 1. \tag{9.1.17}$$

Show that it still obeys the Liouville equation.

Exercise 9.1.2   Equation (9.1.14) seems to contradict (9.1.4), since it has a minus sign. Equation (9.1.14) is an equation for the density matrix elements, however, while (9.1.4) is for the time-dependent operator. Show that these two equations both give the same result for the time evolution of the element $\rho_{mn} = c_m^* c_n$ by explicit substitution, using $|\psi_m\rangle = b_m^\dagger|0\rangle$ and

$$H = \sum_i E_i b_i^\dagger b_i, \qquad |\psi\rangle = \sum_i c_i(t) b_i^\dagger |0\rangle. \qquad (9.1.18)$$

9.2 Magnetic Resonance: The Bloch Equations

The density matrix formalism is generic for all kinds of systems, but can be developed at length for the case of NMR. The same formalism is also used in the modern field of quantum information science, in which a **qubit** is defined as a system with two possible states.

We consider two spin states of a nucleus. In this case, the Hamiltonian is

$$H = -\gamma \vec{B} \cdot \vec{S} = -\gamma \frac{\hbar}{2}(B_x \sigma_x + B_y \sigma_y + B_z \sigma_z), \qquad (9.2.1)$$

where $\gamma = ge/2m$ is the gyromagnetic ratio and $\vec{S} = (\hbar/2)\vec{\sigma}$ is the spin. For a spin-$\frac{1}{2}$ system, the matrix operators for the spin are equal to

$$\sigma_z = \begin{pmatrix} 1 & 0 \\ 0 & -1 \end{pmatrix}, \qquad \sigma_x = \begin{pmatrix} 0 & 1 \\ 1 & 0 \end{pmatrix}, \qquad \sigma_y = \begin{pmatrix} 0 & -i \\ i & 0 \end{pmatrix}, \qquad (9.2.2)$$

(for a discussion of spin effects, see Appendix F). We suppose that there is a strong field B_0 in the z-direction, and a weak, oscillating field in the transverse direction. For convenience, we adopt the **rotating wave** picture, writing the transverse field as a clockwise circularly polarized wave, $B_1(\hat{x}\cos\omega t - \hat{y}\sin\omega t)$. Of course, a field polarized along one axis can be written as the sum of two circularly polarized waves rotating in opposite directions, so if the oscillating field is not rotating, we can use the solutions for both rotating waves.

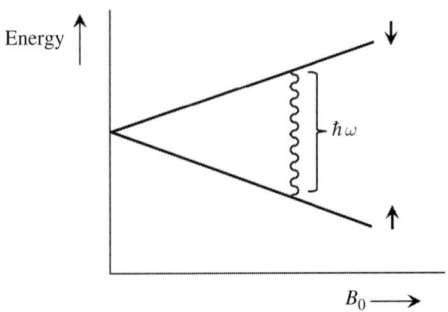

Fig. 9.1   Resonant excitation of two spin states split by magnetic field.

Using this rotating field for the transverse field, the Hamiltonian is then

$$H = -\gamma \frac{\hbar}{2} \begin{pmatrix} B_0 & B_1 e^{i\omega t} \\ B_1 e^{-i\omega t} & -B_0 \end{pmatrix}. \tag{9.2.3}$$

In other words, the magnetic field along z splits the spin states into two energy levels, with the spin-up state lowest, and a transverse magnetic field can induce transitions between the two levels. The resonant frequency (known as the **Larmor frequency**) is given by

$$\omega_0 = \gamma B_0, \tag{9.2.4}$$

and we define the **Rabi frequency**

$$\omega_R = \gamma B_1. \tag{9.2.5}$$

We then can write this Hamiltonian as

$$H = H_0 + V_{\text{int}}, \tag{9.2.6}$$

with

$$H_0 = \frac{\hbar\omega_0}{2} b_\downarrow^\dagger b_\downarrow - \frac{\hbar\omega_0}{2} b_\uparrow^\dagger b_\uparrow \tag{9.2.7}$$

and the interaction energy given by the off-diagonal terms in (9.2.3),

$$V_{\text{int}} = -\frac{\hbar\omega_R}{2} \left(\rho_{\downarrow\uparrow} e^{-i\omega t} + e^{i\omega t} \rho_{\uparrow\downarrow} \right), \tag{9.2.8}$$

where we have adopted the definition of ρ_{nm} for two states n and m from Section 9.1.

From (9.1.14), the time evolution of the density matrix is

$$\frac{\partial \rho_{mn}(t)}{\partial t} = \frac{i}{\hbar} \langle \psi | [H_0 + V_{\text{int}}, \rho_{mn}] | \psi \rangle. \tag{9.2.9}$$

This equation can be resolved using the general commutator relation,

$$[b_{m'}^\dagger b_{n'}, b_m^\dagger b_n] = b_{m'}^\dagger b_n \delta_{n',m} - b_m^\dagger b_{n'} \delta_{n,m'}. \tag{9.2.10}$$

We find for the H_0 term,

$$\langle \psi | [H_0, \rho_{mn}] | \psi \rangle = \frac{\hbar\omega_0}{2} \langle \psi | [b_\downarrow^\dagger b_\downarrow, b_m^\dagger b_n] | \psi \rangle - \frac{\hbar\omega_0}{2} \langle \psi | [b_\uparrow^\dagger b_\uparrow, b_m^\dagger b_n] | \psi \rangle$$

$$= \frac{\hbar\omega_0}{2} \left(\langle \psi | b_\downarrow^\dagger b_n | \psi \rangle \delta_{c,m} - \langle \psi | b_m^\dagger b_\downarrow | \psi \rangle \delta_{n,\downarrow} \right)$$

$$- \frac{\hbar\omega_0}{2} \left(\langle \psi | b_\uparrow^\dagger b_n | \psi \rangle \delta_{\uparrow,m} - \langle \psi | b_m^\dagger b_\uparrow | \psi \rangle \delta_{n,\uparrow} \right)$$

$$= \frac{\hbar\omega_0}{2} (\rho_{\downarrow n} \delta_{\downarrow,m} - \rho_{m\downarrow} \delta_{n,\downarrow}) - \frac{\hbar\omega_0}{2} (\rho_{\uparrow n} \delta_{\uparrow,m} - \rho_{m\uparrow} \delta_{n,\uparrow}). \tag{9.2.11}$$

Using the same commutator to act on the V_{int} term, we obtain

$$\frac{\partial \rho_{mn}(t)}{\partial t} = \frac{i}{\hbar} (E_m - E_n) \rho_{mn} \tag{9.2.12}$$

$$+ \frac{i}{\hbar} \frac{\hbar\omega_R}{2} [e^{-i\omega t} (\rho_{\downarrow n} \delta_{\uparrow,m} - \rho_{m\uparrow} \delta_{n,\downarrow}) + e^{i\omega t} (\rho_{\uparrow n} \delta_{\downarrow,m} - \rho_{m\downarrow} \delta_{n,\uparrow})].$$

Written out explicitly, this gives the evolution equations

$$\frac{\partial \rho_{\downarrow\downarrow}(t)}{\partial t} = -i\frac{\omega_R}{2}\left(e^{-i\omega t}\rho_{\downarrow\uparrow} - e^{i\omega t}\rho_{\uparrow\downarrow}\right) \tag{9.2.13}$$

$$\frac{\partial \rho_{\downarrow\uparrow}(t)}{\partial t} = i\omega_0\rho_{\downarrow\uparrow} - i\frac{\omega_R}{2}e^{i\omega t}\left(\rho_{\downarrow\downarrow} - \rho_{\uparrow\uparrow}\right),$$

with $\rho_{\uparrow\downarrow} = \rho_{\downarrow\uparrow}^*$ and $\dfrac{\partial \rho_{\uparrow\uparrow}}{\partial t} = -\dfrac{\partial \rho_{\downarrow\downarrow}}{\partial t}$.

We can present this in a simplified form by defining the unitless magnetization vector $\vec{m}$ with components

$$\begin{array}{rcl}
m_1 &=& \rho_{\downarrow\uparrow} + \rho_{\uparrow\downarrow} \\
m_2 &=& i(\rho_{\downarrow\uparrow} - \rho_{\uparrow\downarrow}) \\
m_3 &=& \rho_{\uparrow\uparrow} - \rho_{\downarrow\downarrow},
\end{array} \tag{9.2.14}$$

from the x, y, and z spin projections in (9.2.2). Then (9.2.13) becomes

$$\begin{array}{rcl}
\dfrac{\partial m_1}{\partial t} &=& \omega_0 m_2 + \omega_R \sin \omega t \, m_3 \\[2mm]
\dfrac{\partial m_2}{\partial t} &=& -\omega_0 m_1 + \omega_R \cos \omega t \, m_3 \\[2mm]
\dfrac{\partial m_3}{\partial t} &=& -\omega_R \sin \omega t \, m_1 - \omega_R \cos \omega t \, m_2.
\end{array} \tag{9.2.15}$$

These are the **Bloch equations**. They can be written more compactly as

$$\frac{\partial \vec{m}}{\partial t} = \gamma \vec{m} \times \vec{B}, \tag{9.2.16}$$

where $\vec{B} = B_1(\hat{x}\cos\omega t - \hat{y}\sin\omega t) + B_0\hat{z}$, and $\hat{u}_1$, $\hat{u}_2$, and $\hat{u}_3$ are unit vectors.

This is just as one would expect from the classical equation of motion for a magnetic moment in a magnetic field. The vector $\vec{m}$ thus gives the actual direction of the expectation value of the magnetic dipole in three dimensions, for a static $\vec{B}$-field in the $\hat{z}$-direction and an oscillating field in the transverse direction.

If there is no transverse field, the Bloch equations become simply

$$\frac{\partial m_1}{\partial t} = \omega_0 m_2$$

$$\frac{\partial m_2}{\partial t} = -\omega_0 m_1$$

$$\frac{\partial m_3}{\partial t} = 0; \tag{9.2.17}$$

in other words, in a static field the Bloch equations describe the precession of the magnetic moment around the z-axis with frequency ω_0. As is well known from classical mechanics, there is a natural rotation direction for the precession of a magnetic moment in a magnetic field (see, e.g., Melissinos 1966: p. 349).

This helps us to understand why we picked the rotating wave picture. In magnetic resonance, there is a natural precession due to the handedness of the magnetic field. As discussed above, an oscillating magnetic field along the x-direction corresponds to the sum of two rotating waves, one rotating in the same direction as the natural precession, and one rotating in the opposite direction. In principle, we should include the oppositely rotating wave in our analysis, which corresponds to negative frequency, $-\omega$, if we do not really have a rotating magnetic field (which is hardly ever the case in real experiments). It is common to ignore the negative-frequency term, however. The reason is that the system is resonant at $\omega = \omega_0$, and the negative-frequency term is typically far away from resonance. We will return to discuss this further below.

Rotating frame. We can switch to the rotating frame by choosing new coordinates m_1' and m_2' such that

$$m_1 = m_1' \cos \omega t + m_2' \sin \omega t$$
$$m_2 = -m_1' \sin \omega t + m_2' \cos \omega t$$
$$m_3 = m_3'. \tag{9.2.18}$$

We can transform the equations to the rotating frame by substituting these into (9.2.15) and equating $\sin \omega t$ and $\cos \omega t$ terms on each side of the equations. In the transformed coordinates, we then have

$$\begin{aligned}
\frac{\partial m_1'}{\partial t} &= \tilde{\omega} m_2' \\[2mm]
\frac{\partial m_2'}{\partial t} &= -\tilde{\omega} m_1' + \omega_R m_3' \\[2mm]
\frac{\partial m_3'}{\partial t} &= -\omega_R m_2',
\end{aligned} \tag{9.2.19}$$

where

$$\tilde{\omega} = \omega_0 - \omega \tag{9.2.20}$$

is the detuning of the driving field from the resonance.

Exercise 9.2.1   Prove the relation (9.2.10). Show that it is true for boson operators as well as fermion operators.

Exercise 9.2.2   Show that the Bloch equations (9.2.15) and (9.2.19) follow from (9.2.13) and definitions (9.2.14) and (9.2.18).

Bloch sphere representation. We can represent the motion of the magnetization vector on the **Bloch sphere**, shown in Figure 9.2. We define the **Bloch vector**, with $U_3 = -m_3$, so that the top of the sphere ($U_3 = 1$) represents the system completely in the higher-energy state, and the bottom ($U_3 = -1$) represents the system completely in the lower-energy state. Then the equations of motion for $\vec{U}$ are

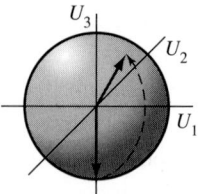

Fig. 9.2 The Bloch sphere. The vector pointing downward along the U_3 axis represents the initial, ground state. Excitation by a coherent field lifts the Bloch unit vector to another point on the sphere.

$$\frac{\partial U_1'}{\partial t} = \tilde{\omega} m_2'$$

$$\frac{\partial U_2'}{\partial t} = -\tilde{\omega} m_1' - \omega_R U_3'$$

$$\frac{\partial U_3'}{\partial t} = \omega_R U_2'. \tag{9.2.21}$$

Suppose that the system is initially in the lower state, $U_3' = -1$, and an electromagnetic field with frequency $\omega = \omega_0$ (i.e. $\tilde{\omega} = 0$) excites the system. Then the evolution obeys the equations

$$\frac{\partial U_1'}{\partial t} = 0$$

$$\frac{\partial U_2'}{\partial t} = -\omega_R U_3$$

$$\frac{\partial U_3'}{\partial t} = \omega_R U_2'. \tag{9.2.22}$$

The solution that satisfies the initial condition $U_3 = -1$ is

$$U_1' = 0$$
$$U_2' = \sin \omega_R t$$
$$U_3' = -\cos \omega_R t, \tag{9.2.23}$$

in other words, a simple rotation around the U_1' axis. The system oscillates from being entirely in the ground state, to a superposition of both states (along the U_2' axis) to entirely in the upper state, and then back down again. The frequency of the oscillation is proportional to the driving field, B_1. This is known as **Rabi oscillation**, or **Rabi flopping**. At first glance, Rabi oscillation is surprising since we are simply pumping the transition with an electromagnetic field composed of photons, and one would expect from Fermi's golden rule that this leads to transitions to the upper state, with spontaneous and stimulated emission back down to the ground state, leading to a steady-state population of both states. We cannot talk in terms of absorption and spontaneous emission obeying Fermi's golden rule, however, because Fermi's golden rule explicitly assumes an incoherent process, in which the transition is to a continuum of final states with different phases. Here the system is coherent.

In general, an electromagnetic pulse winds the Bloch vector around the U_1' axis by a fixed amount of phase rotation. If the system starts in the ground state, we can drive all of the electrons into the upper state by a pulse with $\omega_R t = \pi$, that is, pulse duration $t = \pi/\omega_R$, known as a "π-pulse;" if we drive it with a pulse with $\omega_R t = \pi/2$ (i.e., a "$\pi/2$-pulse"), we put the system into a 50% superposition of both states. The property of a pulse depends on the product of the strength of the driving field B_1 and the time duration. The product $B_1 t$ is known as the **area** of the pulse – as long as we keep this product constant, the pulse will have the same effect on the system.

Decay and dephasing. The above analysis is applies to the case of a single, isolated nucleus. Suppose that we do an experiment with many identical nuclei. In general, there will be processes that lead to the evolution of some of the nuclei becoming different from that of others over time. First, there will be relaxation of nuclei in the upper state down into the lower state, by spontaneous emission of photons. We label the lifetime for this process T_1, which is variously called the **energy relaxation time**, the **excitation lifetime**, or the **longitudinal relaxation time**. There also will be random processes by which the phase in the equatorial plane of the Bloch sphere becomes different for different nuclei. If each nucleus undergoes random jumps in phase, for example by random interactions with the environment, the average over all the nuclei of both U_1' and U_2' will tend toward zero. The lifetime for this process, T_2, is known variously as the **dephasing time**, the **decoherence time**, or the **transverse relaxation time**. A general rule is that $T_2 \leq T_1$, since processes that lead to energy relaxation also lead to dephasing.

A T_1 process corresponds to a loss of $\rho_{\downarrow\downarrow}$ with a gain of the same amount by $\rho_{\uparrow\uparrow}$. We therefore write

$$\frac{\partial \rho_{\downarrow\downarrow}}{\partial t} = -\frac{\rho_{\downarrow\downarrow}}{T_1} = -\frac{\partial \rho_{\uparrow\uparrow}}{\partial t}, \tag{9.2.24}$$

which implies

$$\frac{\partial U_3'}{\partial t} = \frac{\partial}{\partial t}(\rho_{\downarrow\downarrow} - \rho_{\uparrow\uparrow}) = -\frac{2}{T_1}\rho_{\downarrow\downarrow} \tag{9.2.25}$$

$$= -\frac{1}{T_1}(\rho_{\downarrow\downarrow} - \rho_{\uparrow\uparrow} + \rho_{\downarrow\downarrow} + \rho_{\uparrow\uparrow}) = -\frac{U_3' + 1}{T_1},$$

where we have used the conservation of probability, $\rho_{\downarrow\downarrow} + \rho_{\uparrow\uparrow} = 1$. Each of the off-diagonal components decays to zero as

$$\frac{\partial \rho_{\downarrow\uparrow}}{\partial t} = -\frac{\rho_{\downarrow\uparrow}}{T_2}, \tag{9.2.26}$$

leading both of the U_1' and U_2' components to decay. Adding these decay processes to our previous equations, the Bloch equations in the rotating frame are then

$$\frac{\partial U'_1}{\partial t} = -\frac{U'_1}{T_2} + \tilde{\omega} U'_2$$

$$\frac{\partial U'_2}{\partial t} = -\frac{U'_2}{T_2} - \tilde{\omega} U'_1 - \omega_R U'_3 \qquad (9.2.27)$$

$$\frac{\partial U'_3}{\partial t} = -\frac{U'_3 + 1}{T_1} + \omega_R U'_2.$$

Suppose now that the system begins in its ground state, $U'_3 = -1$, and we excite it with a weak pulse, at resonance with $\tilde{\omega} = 0$. If the duration of the pulse is short compared to T_1 and T_2, the system will be in a state with $U'_1 = 0$, $U'_2 \neq 0$, and $U'_3 \approx -1$. After the excitation pulse is finished, when $\omega_R = 0$, the equations of motion are

$$\frac{\partial U'_1}{\partial t} = -\frac{U'_1}{T_2} + \tilde{\omega} U'_2$$

$$\frac{\partial U'_2}{\partial t} = -\frac{U'_2}{T_2} - \tilde{\omega} U'_1$$

$$\frac{\partial U'_3}{\partial t} = -\frac{U'_3 + 1}{T_1}. \qquad (9.2.28)$$

In the non-rotating, laboratory frame, these equations of motion are

$$\frac{\partial U_1}{\partial t} = -\frac{U_1}{T_2} + \omega_0 U_2$$

$$\frac{\partial U_2}{\partial t} = -\frac{U_2}{T_2} - \omega_0 U_1$$

$$\frac{\partial U_3}{\partial t} = -\frac{U_3 + 1}{T_1}. \qquad (9.2.29)$$

The U_3 component decays with lifetime T_1, while the other two components are given by

$$U_1 = U_2(0) \sin \omega_0 t \, e^{-t/T_2}$$

$$U_2 = U_2(0) \cos \omega_0 t \, e^{-t/T_2}. \qquad (9.2.30)$$

The off-diagonal components of the density matrix, that is the coherence factors, oscillate with the intrinsic frequency of the energy gap. This is known as **free induction decay**, as illustrated in Figure 9.3. In terms of the Bloch vector, the state of the system spirals inward toward the origin in the U_1–U_2 plane.

There are actually two different sources of dephasing. When we discussed optical transitions in Section 8.4, we talked of **homogeneous** broadening, which we can visualize as random jumps in the phase of a single oscillator, such as by phonon scattering, and **inhomogeneous** broadening, which comes from the oscillators not being identical, but instead having a range of oscillator frequencies, that is energy gaps. The same two effects lead to dephasing here. Random jumps in phase of each individual nucleus will not be correlated,

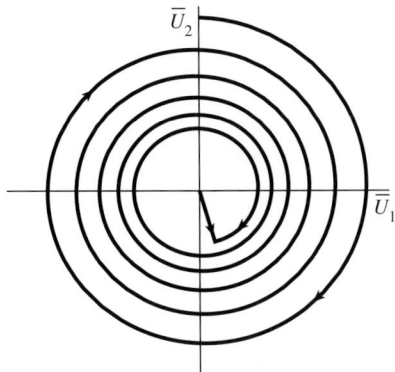

Free induction decay of the average Bloch vector of a set of oscillators.

and therefore the total amplitude in the U_1–U_2 plane after a long time will average to zero. If there is a range of energy gaps ω_0 for many different nuclei, perhaps because they are in some inhomogeneous matrix, the same effect will occur, because the Bloch vectors of the individual spins will rotate in the U_1–U_2 plane at different speeds. The dephasing time due to the effect of inhomogeneity can be called T_2'; the total dephasing rate is T_2^*, given by

$$\frac{1}{T_2^*} = \frac{1}{T_2} + \frac{1}{T_2'}. \tag{9.2.31}$$

We can distinguish between the decay of the phase due to inhomogeneous processes and the decay due to homogeneous processes by a clever experiment. Suppose that the system is initially in the ground state $U_3' = -1$ and at time $t = 0$ we excite it with a $\pi/2$-pulse, short in duration compared to T_1 and T_2. The Bloch vector in the rotating frame for all the spins will be along U_2'. After a time τ, the Bloch vectors of the individual spins will all have rotated to different points in the $U_1' - U_2'$ plane, as shown in Figure 9.4(a), since the solution of the equations of motion (9.2.28) in the rotating frame is

$$U_1' = U_2'(0) \sin \tilde{\omega} t \, e^{-t/T_2}$$
$$U_2' = U_2'(0) \cos \tilde{\omega} t \, e^{-t/T_2}, \tag{9.2.32}$$

where $\tilde{\omega}$ is different for each spin with a different ω_0. As discussed above, after a long time the average of all these vectors will be zero. If the system is excited by a π-pulse, however, all of the vectors will be flipped across the U_1' axis, as shown in Figure 9.4(b), since a π-pulse reverses the U_2' component while leaving the U_1' component the same. After that, they will continue to rotate in the same direction. This means that after a time τ they will all overlap on the U_2' axis again, in the opposite direction from the original state. If an experiment detects the magnitude of U_1', it will see an **echo pulse** when all the Bloch vectors line up. If a second π-pulse is sent into the system at time 3τ, another echo pulse will be created (Figure 9.5 shows the pulse sequence in time). The ratio of the two echo pulses will be equal to $e^{-2\tau/T_2}$, where T_2 is the homogeneous dephasing time due to

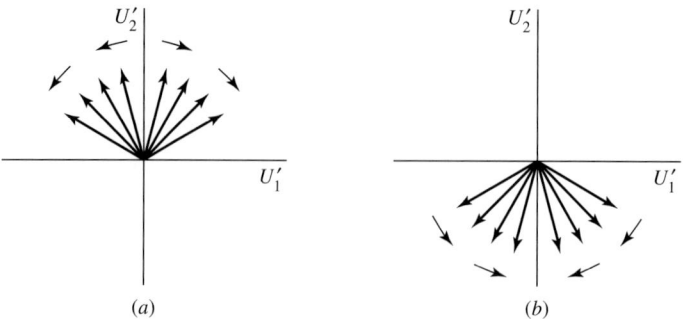

(a)   (b)

Fig. 9.4   (a) A set of two-level oscillators represented by different Bloch vectors in the $U'_1 - U'_2$ plane. When there is inhomogeneous broadening, ω_0 will different for different oscillators, and therefore some will rotate faster than others in the plane. (b) The same set of vectors after a π-pulse.

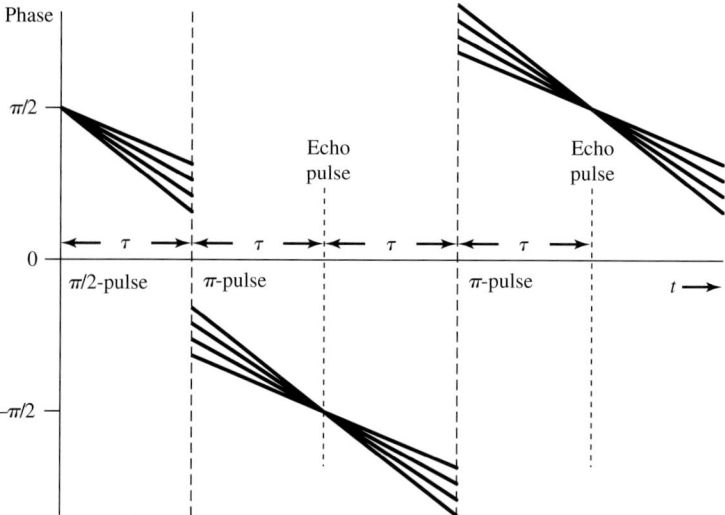

Fig. 9.5   The phase of several Bloch vectors with different angular frequencies $\tilde{\omega}$, relative to the U'_1 axis, for the pulse sequence discussed in the text for a T_2 measurement.

the intrinsic process acting on each oscillator. This gives a measurement of the intrinsic dephasing T_2 independent of T'_2.

Steady-state solution and magnetic susceptibility. At finite temperature, when $k_B T$ is comparable to the splitting between the states, we can approximate that U'_3 is controlled by the thermal occupation. Assuming U'_3 is a constant, the steady-state solution of (9.2.27) is found by setting the time derivatives to zero. We find for the magnetization $\vec{m}$,

$$m'_1 = \omega_R T_2 m'_3 \frac{\tilde{\omega} T_2}{1 + \tilde{\omega}^2 T_2^2}$$

$$m'_2 = \omega_R T_2 m'_3 \frac{1}{1 + \tilde{\omega}^2 T_2^2}. \tag{9.2.33}$$

The magnitude of the oscillation therefore has a resonance peak centered at $\tilde{\omega} = 0$, with a width determined by T_2. Note that this also tells us when we are justified in using the rotating-wave approximation. A field rotating with the opposite handedness corresponds to $\tilde{\omega} = 2\omega$. If this is much greater than the width $1/T_2$, then it will make very little contribution. Since a linearly polarized field is the sum of two oppositely rotating circularly polarized fields, we can therefore treat a linearly polarized field as a rotating wave.

We can rewrite these formulas using the complex susceptibility formalism of Chapter 7, by defining the **magnetic susceptibility** χ such that the induced magnetization field is given by

$$M = \chi H, \tag{9.2.34}$$

where $H = B/\mu_0$ is the applied field. Assuming that $k_B T \gg \gamma B_0$, we can write down the average magnetic moment $\vec{m}$ as just the difference between the thermal occupation of the levels:

$$m_3' = \frac{1}{Z}\left(f(\hbar\gamma B_0) - f(0)\right) = \frac{e^{-\hbar\gamma B_0/k_B T} - 1}{e^{-\hbar\gamma B_0/k_B T} + 1} \approx \frac{\hbar\gamma B_0}{2k_B T}, \tag{9.2.35}$$

which we can write as $m_3' = \chi_0 B_0/\mu_0$. We can now use this for m_3' in (9.2.33).

Setting $\omega_R = \gamma B_1$ and $\omega_0 = \gamma B_0$, we then can write for the transverse magnetization

$$m'(\omega) = m_1' + im_2' = \chi(\omega)B_1, \tag{9.2.36}$$

where

$$\chi(\omega) = \frac{\chi_0\omega_0}{\mu_0}\frac{(\omega_0 - \omega) + i\Gamma}{(\omega_0 - \omega)^2 + \Gamma^2}, \tag{9.2.37}$$

where $\Gamma = 1/T_2$. The imaginary term, proportional to $1/T_2$, is the dissipative term.

NMR (or MRI, magnetic resonance imaging, as it is now often called, since people fear the word "nuclear") has continued to have ever-increasing impact not only in medical and biological studies, but also chemical and materials research. There are two reasons. First, the resonance frequency is a signature for different nuclei, which allows identification of different chemicals. Second, T_1 and T_2 can depend sensitively on environmental parameters. NMR therefore can be a sensitive probe of chemical conditions. An extensive discussion of coherence in NMR can be found, for example, in Munowicz (1988) and Slichter (1989).

Exercise 9.2.3 (a) Prove that (9.2.33) is the steady-state solution to (9.2.15), in the limit $T_1 \gg T_2$. (b) Use a program like Mathematica to plot the magnitude of the Bloch vector as a function of $\tilde{\omega}$, for several values of T_2.

Exercise 9.2.4 Determine the Larmor frequency in hertz for a nucleus with spin-$\frac{1}{2}$ and atomic mass number of 50, in a magnetic field of 10 T.

9.3 Optical Bloch Equations

Most of the language of coherence and decoherence comes from the early work in NMR, and the formalism of the Bloch equation was developed for that work, as discussed in Section 9.2. The same formalism applies to any two-level system, however, and in particular coherent optics experiments, which have become a major topic of modern physics. Here we use the optics notation of Section 7.4. In Section 7.4, we solved for the response of a quantum harmonic oscillator under the assumption that the driving field was weak, so that the probability of occupation of the upper level was always small. Here we allow for the possibility that the system is strongly driven.

Consider a set of two-level quantum mechanical oscillators driven by a coherent electromagnetic wave, as shown in Figure 9.6. We write the Hamiltonian that includes the dipole electromagnetic interaction with the electrons as

$$H = H_0 + V_{\text{int}}, \tag{9.3.1}$$

with

$$H_0 = \sum_i \left(E_v b_{iv}^\dagger b_{iv} + E_c b_{ic}^\dagger b_{ic} \right), \tag{9.3.2}$$

where the index i labels the individual oscillators in the ensemble.

For the interaction energy, we use the electron–photon dipole interaction (5.2.25) derived in Section 5.2.2, keeping just two bands, c and v,

$$V_{\text{int}} = -iq \sum_{\vec{k},\vec{k}_1} \sqrt{\frac{\hbar\omega}{2\epsilon_0 V}} \left[\langle c|x|v \rangle \left(a_{\vec{k}} b_{c,\vec{k}_1+\vec{k}}^\dagger b_{v,\vec{k}_1} - a_{\vec{k}}^\dagger b_{c,\vec{k}_1-\vec{k}}^\dagger b_{v,\vec{k}_1} \right) \right.$$
$$\left. + \langle v|x|c \rangle \left(a_{\vec{k}} b_{v,\vec{k}_1+\vec{k}}^\dagger b_{c,\vec{k}_1} - a_{\vec{k}}^\dagger b_{v,\vec{k}_1-\vec{k}}^\dagger b_{c,\vec{k}_1} \right) \right], \tag{9.3.3}$$

neglecting the polarization. We can rewrite the k-dependent fermion operators in terms of localized atomic operators using the Fourier transform,

$$b_{n,\vec{k}_1}^\dagger = \frac{1}{\sqrt{N}} \sum_i e^{i\vec{k}_1 \cdot \vec{r}_i} b_{in}^\dagger. \tag{9.3.4}$$

Then, using the identity

$$\sum_{\vec{k}_1} e^{i\vec{k}_1 \cdot (\vec{r}_i - \vec{r}_j)} = N\delta_{i,j}, \tag{9.3.5}$$

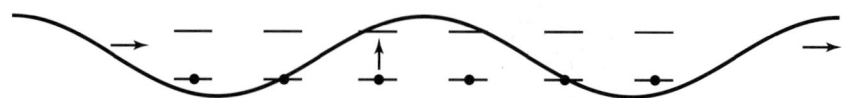

Fig. 9.6 An ensemble of two-level oscillators excited by a coherent wave.

where N is the number of oscillators, the interaction becomes

$$V_{int} = -iq \sum_{\vec{k}} \sqrt{\frac{\hbar\omega}{2\epsilon_0 V}} \sum_i \left[\langle c|x|v \rangle \left(e^{i\vec{k}\cdot\vec{r}_i} a_{\vec{k}} b_{ic}^\dagger b_{iv} - e^{-i\vec{k}\cdot\vec{r}_i} a_{\vec{k}}^\dagger b_{ic}^\dagger b_{iv} \right) \right.$$

$$\left. + \langle v|x|c \rangle \left(e^{i\vec{k}\cdot\vec{r}_i} a_{\vec{k}} b_{iv}^\dagger b_{ic} - e^{-i\vec{k}\cdot\vec{r}_i} a_{\vec{k}}^\dagger b_{iv}^\dagger b_{ic} \right) \right]. \quad (9.3.6)$$

For a coherent driving field, the boson operators can be replaced by

$$a_k \rightarrow -i\sqrt{\frac{\epsilon_0 V}{2\hbar\omega}} E_0 e^{-i\omega t}, \quad (9.3.7)$$

using (4.3.20) from Chapter 4.

We now make two approximations. First, we neglect the photon momentum, which is the same as assuming the electromagnetic field has long wavelength, so that the exponential factors with $\vec{k}$ can be dropped. Second, we make the rotating wave approximation, discussed in Section 9.2. This amounts to dropping the second term in the parentheses of the first line, and the first term of the second line, in (9.3.6). These correspond to transitions from the valence band v to the conduction band c by *emission* of a photon, and from the conduction band to the valence band by *absorption* of a photon, respectively. While energy conservation is not strictly required in computing transition rates, as we have seen in Chapter 8, we expect transitions like these to have large energy denominators that suppress their contribution. Keeping just the two other terms, we then have the simple form of the interaction of a two-level system with a coherent driving wave,

$$V_{int}^{(i)} = -\frac{q\langle c|x|v \rangle E_0}{2} \sum_i \left(e^{-i\omega t} b_{ic}^\dagger b_{iv} + e^{i\omega t} b_{iv}^\dagger b_{ic} \right). \quad (9.3.8)$$

We take the product $q\langle c|x|v \rangle$ to be real and positive, since the dipole moment $q\vec{x}$ from which the dipole interaction was derived in Section 5.2.2 is always in the direction of $\vec{E}$ in a polarizable medium.

As in Section 9.2, we define the resonance frequency $\omega_0 = (E_c - E_v)/\hbar$ and the Rabi frequency

$$\omega_R = \frac{e|\langle c|x|v \rangle|E_0}{\hbar}. \quad (9.3.9)$$

Since both H_0 and V_{int} are the sums of local terms for each location i, we can now drop the summation over i and just write the average Hamiltonian, as we did for each atom in Section 9.2:

$$H_0 = \frac{\hbar\omega_0}{2}(\rho_{cc} - \rho_{vv}) \quad (9.3.10)$$

$$V_{int} = -\frac{\hbar\omega_R}{2}(e^{-i\omega t}\rho_{cv} + e^{i\omega t}\rho_{vc}). \quad (9.3.11)$$

The interaction (9.3.11) has exactly the same form as (9.2.8). Therefore, we can use all of the same formalism for the Bloch equations and a Bloch vector. In particular, we define the Bloch vector

$$\begin{aligned} U_1 &= \rho_{cv} + \rho_{vc} \\ U_2 &= i(\rho_{cv} - \rho_{vc}) \\ U_3 &= \rho_{cc} - \rho_{vv}, \end{aligned} \qquad (9.3.12)$$

which allows us to use the same Bloch sphere representation, with equations of motion (9.2.27), derived in Section 9.2. The U_1–U_2 plane in this case does not correspond to x- and y-directions in real space, but rather to the real and imaginary parts of the density matrix in the complex plane. The Bloch vector will undergo Rabi flopping and free induction decay, and all the other effects discussed in Section 9.2.

In Section 7.4, we wrote the polarization operator for an ensemble of oscillators as (7.4.28),

$$\begin{aligned} \vec{P} &= \frac{Nq}{V} \sum_i \left(\langle c|\vec{x}|v\rangle b_{iv}^\dagger b_{ic} + \langle v|\vec{x}|c\rangle b_{ic}^\dagger b_{iv} \right) \\ &= 2\mathrm{Re}\, \frac{Nq}{V} \langle c|\vec{x}|v\rangle \rho_{cv}, \end{aligned} \qquad (9.3.13)$$

where in the second line we defined the average density matrix element

$$\rho_{cv} = \frac{1}{N} \sum_i b_{ic}^\dagger b_{iv}. \qquad (9.3.14)$$

In other words, the polarization is directly proportional to the average off-diagonal element of the density matrix ρ_{cv}. As discussed in Section 9.2, we can write this matrix element in terms of the Bloch vector components. In the rotating frame, we have

$$2\mathrm{Re}\,\rho_{cv} = U_1 = U_1' \cos \omega t + U_2' \sin \omega t. \qquad (9.3.15)$$

The free induction decay of the Bloch vector for an optical transition therefore corresponds to a real oscillating polarization in the medium, which can radiate electromagnetic field like an antenna. The system behaves essentially like a bell that has been struck – it rings for a while at its natural resonance until damping processes such as phonon emission and radiation remove the coherence of the oscillation.

Because the same Bloch equations apply, all of the same types of experiments can be done with optical systems as were described in Section 9.2 for magnetic resonance, including Rabi flopping and pulse-echo measurements. The main difference is that the energy scale, and therefore the frequency of the resonances, is much higher in optical transitions; energy gaps of eV correspond to frequencies $f \sim (1\ \mathrm{eV})/h \sim 4 \times 10^{15}$ Hz, which requires timing accuracy of the order of femtoseconds, while magnetic resonance experiments can be done at MHz or GHz frequencies. Also, in typical semiconductor systems, it is difficult to get high enough intensity of the pump light to get $\pi/2$ or π Rabi rotations. The optical Bloch equations can be modified to treat a continuum of states in semiconductors; the theory of the **semiconductor Bloch equations** is given, for example, by Haug and Koch (1990).

Exercise 9.3.1   Calculate the Rabi frequency for a two-band semiconductor with gap energy 1.5 eV, excited by a cw laser with power 500 mW and beam width 1 mm, for an oscillator strength approximately equal to unity.

For this laser characteristic, how long will a π-pulse be?

9.4 Quantum Coherent Effects

Whenever experiments can be done on time scales short compared to T_2, all kinds of fascinating effects can occur. These experiments can be done in any system, but we return to the case of optical transitions as the easiest case to demonstrate these.

As discussed in Section 9.3, an electromagnetic field driving an oscillation between two quantum states creates a time-varying polarization in the medium. This polarization then can be detected by a number of means.

Quantum beats. Suppose two states with different energies have both been excited from a common ground state, for example by a short pulse with spectral width wide enough to hit both resonances, as illustrated in Figure 9.8(a). As in the case of classical oscillators, the total amplitude of two oscillators at different frequencies will yield beats at the sum and difference frequencies:

$$\begin{aligned}
\rho_{\text{tot}} &= \rho_{12}(t) + \rho_{13}(t) \\
&= \rho_0 \sin \omega_{12} t + \rho_0 \sin \omega_{13} t \\
&= 2\rho_0 \sin \left(\frac{\omega_{12} + \omega_{13}}{2} t \right) \cos \left(\frac{\omega_{12} - \omega_{13}}{2} t \right).
\end{aligned} \tag{9.4.1}$$

At certain times, therefore, the total polarization will be zero. If there is decay of coherence of the states, then the cancellation will not be perfect, but there will still be oscillations in the polarization at the beat frequency.

Figure 9.7 shows an example of an experimental measurement of quantum beats, showing the intensity of the light radiated from the oscillating polarization, for two different values of the magnetic field, which is used to tune the energy splitting between two states, in this case exciton states in the semiconductor CdS. This is very different from what we would expect from the analysis of Section 5.2.1, which predicts a monotonic decay of the occupation of each of the upper states. Quantum beats are an intrinsically coherent effect, which occurs only on time scales short compared to the dephasing time of the states.

Quantum beats are a good method of determining very fine splittings between states. The method works even if the inhomogeneous broadening of the system is large compared to the splitting between the states, if the overall energy gap varies, but the splitting of the upper state is a constant, as illustrated in Figure 9.8(b). This often occurs when the splitting depends only on intrinsic parameters, such as the spin splitting in a magnetic field. In this case, the beat pattern will be the same for all the oscillators even though the absolute frequency of the oscillations is not.

Coherent control. Another fascinating experiment shows that it is possible to switch off a spontaneous emission process. In Section 5.2, we discussed light emission from an

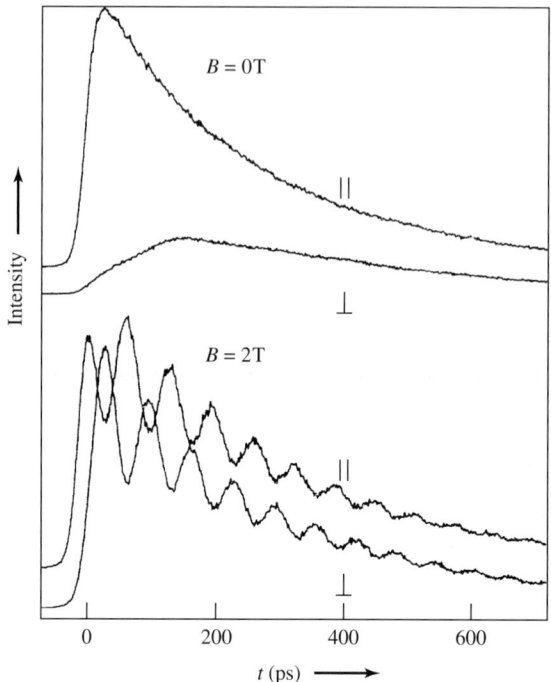

Fig. 9.7 Quantum beats of the light emission (perpendicular and parallel to the excitation polarization) from an exciton state in the semiconductor CdS, for a degenerate state and for a fixed splitting of the states as controlled by an applied magnetic field. From Stoltz *et al.* (1991).

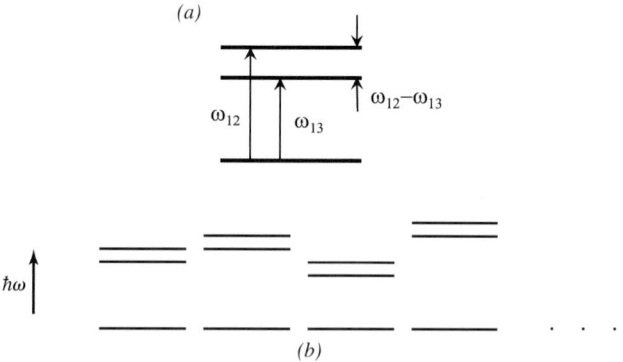

Fig. 9.8 (a) Excitation to two different states by a single laser pulse with broad spectral range. (b) An ensemble of three-level oscillators with varying fundamental gap but the same splitting of the upper levels.

excited state as an incoherent process. In this case, there is an intrinsic lifetime for an electron in an excited state to fall back down into a lower state. This is in general the picture of spontaneous emission according to Fermi's golden rule. On time scales short compared to the dephasing time, however, we must take into account coherent effects. This means that we can sometimes force the system back into its ground state faster than the intrinsic spontaneous emission time.

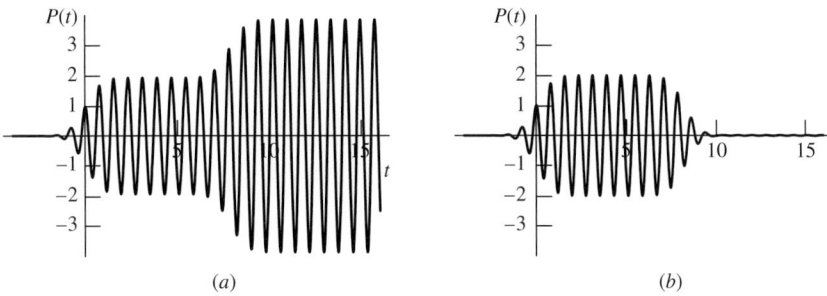

Fig. 9.9 Constructive or destructive interference of the response to two Gaussian pulses depending on the time delay. (a) Delay = 10 times the period of the oscillation. (b) Delay one half period longer.

Figure 9.9 shows the basic effect. Following the procedure of Section 7.4 for the general time dependence of a weakly excited oscillator driven by an electric field $E(t)$, we use (7.4.9) with the initial time $t_0 = -\infty$ in (7.4.4) for the polarization, but instead of using a steady-state driving field, we use an electric field $E(t)$ corresponding to the sequence of short pulses. This gives us the general formula

$$P(t) \propto e^{-i\omega_0 t} \int_{-\infty}^{t} E(t')e^{i\omega_0 t'}\, dt' + \text{c.c.} \qquad (9.4.2)$$

Since the total polarization depends on the integral of the electric field over time, if a first pulse excites the system, a second pulse with the correct phase can cancel it out, as shown in Figure 9.9. The effect is essentially the same as controlling the phase of when a push is given to a child on a swing set. If a push is given in phase with the existing swing, the child will swing higher. If a push is given at the wrong time, however, the child on the swing can be stopped short.

In terms of the Bloch sphere, we imagine that a pulse with very small area (which we define as $\eta\pi$, where $\eta \ll 1$), gives the system a Bloch vector that is slightly away from $U_3' = -1$, with a small U_2' component. As discussed in Section 9.2, this vector will then rotate around the U_3' axis with angular frequency ω_0. If we wait for an integer number of rotations around the U_3' axis, so that the Bloch vector is back in the $U_2' - U_3'$ plane, and then hit it with another, identical pulse, then the Bloch vector will be kicked up to an angle of $2\eta\pi$ relative to the U_3' axis. On the other hand, if we wait for a half-integer number of rotations, then the U_2' component will be negative, and a second pulse will rotate the Bloch vector in the U_2'–U_3' plane back down to $U_3' = -1$.

Figure 9.10 shows an experimental example. The reflectivity depends on the index of refraction, which depends on the real part of the polarization. As seen in this figure, for the correct phase of the second pulse, the system can be forced into its ground state much faster than expected for spontaneous emission.

Exercise 9.4.1   Use (9.4.2) to generate the response of a system to two pulses with Gaussian envelope functions, and create plots like those shown in Figure 9.9, using a program like Mathematica. To do this, you should first integrate the Gaussian pulse function

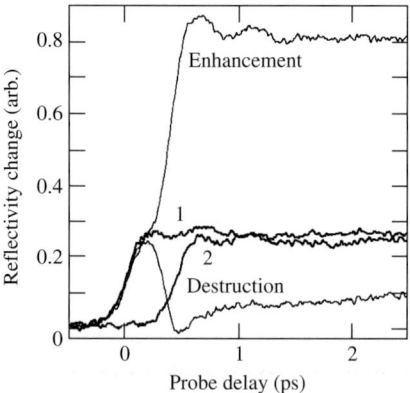

Fig. 9.10   Coherent control of the optical properties of GaAs. Curve 1 is the excitation that leads to enhancement, and Curve 2 leads to destruction of the initial state. From Heberle *et al.* (1996).

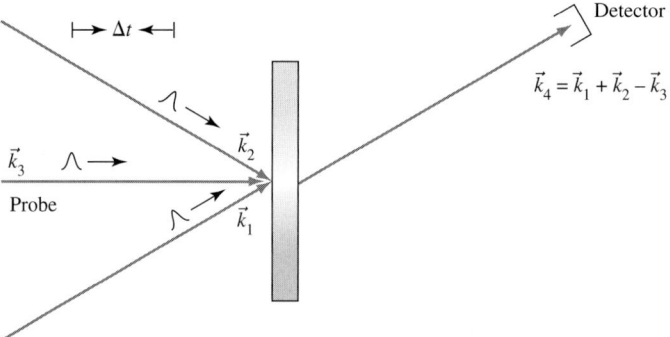

Fig. 9.11   A transient grating four-wave mixing experiment. The probe pulses arrive at the nonlinear medium a time delay Δt after the first two pulses.

$e^{-(t-t_0)^2/\tau^2}e^{-i\omega(t-t_0)}$, for $\omega = \omega_0$. Plot the response of the system to several different choices of time delay between the two pulses.

Time-dependent four-wave mixing. In Section 7.6.2, we discussed four-wave mixing as an example of a nonlinear optical effect, in which three waves in a medium mix to create a fourth wave. Based on the effect of coherent free induction decay discussed above, one or more of these waves can be polarization oscillations that persist in a medium after an initial excitation. The amplitude of such waves will decay with time constant T_2. Therefore, later laser pulses entering a medium will mix with pulses that came at earlier times.

Figure 9.11 gives an example of a four-wave mixing experiment, known as a **transient grating** measurement. The first two pulses arrive at the same time, creating an interference pattern in the medium. A third probe pulse then interacts with this pattern, which is effectively a diffraction grating. The efficiency of the diffraction will depend on the time delay between the original pulses and the probe, as the original polarization in the medium decays away. Anything that decreases the coherent polarization of the medium will lead to decay of the transient diffraction grating, and therefore the detected signal. In particular, diffusion of the excited carriers, if there are any, will wash the grating out, so that this method can be used to measure carrier diffusion constants.

Two-dimensional Fourier spectroscopy. Another variation of time-dependent four-wave mixing produces two-dimensional images of data that aid in understanding the correlations in a system. Suppose that instead of having the first two pulses arrive at the same time, they arrive separately, with time delay τ. Then the system is allowed to evolve for a time T, after which a third, probe pulse excites the system. Finally, after another time delay τ', a fourth pulse bypasses the sample altogether and overlaps the light emitted from the sample, leading to interference at the detector, in a heterodyne measurement. Figure 9.12 shows the sequence of the pulses.

Suppose that there are two resonances in the sample, one at frequency ω_A and the other at frequency ω_B. One can think of what happens in an experiment like this by imagining the action of two Bloch spheres, one for each resonance. When the first pulse arrives, with small area, the Bloch vector on both spheres will be rotated slightly away from $U'_3 = -1$, to obtain a U'_2 component. Afterwards, both Bloch vectors will rotate around the vertical axis, undergoing free induction decay like that shown in Figure 9.3, at different angular frequencies.

The second pulse then arrives after a time delay τ. When this second pulse hits the system, it again rotates the Bloch vectors around the U'_1 axis. The magnitude of the resulting Bloch vector will depend on the timing of this second pulse. If it hits when the Bloch vector has rotated to have a negative U'_2 component, then the action of the second pulse will rotate the Bloch vector down toward $U'_3 = -1$, as in the destructive interference of the coherent control experiment described above, leading to a smaller U'_2 component. If the second pulse arrives when the Bloch vector has a positive U'_2 component, it will rotate the Bloch vector to higher U'_2. Thus, the magnitude of the polarization oscillation after the second pulse will oscillate as a function of the interval τ. If there are two different resonances, each will oscillate at its own intrinsic frequency.

We now allow the system to evolve for some time T comparable to T_1 and T_2. When the third pulse arrives, there will again be a kick to rotate the Bloch vectors in their respective

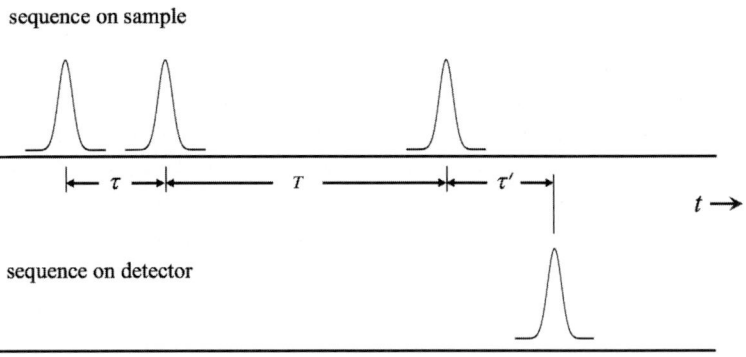

Fig. 9.12 Sequence of pulses used in a two-dimensional Fourier spectroscopy measurement.

spheres. The fourth pulse will sample the oscillation after this, as a function of the time delay τ'.

Thus, we have data for the amplitude of the final Bloch vector as a function of τ, T, and τ'. To view the data, we Fourier transform the data with respect to both τ and τ' to produce a two-dimensional Fourier transform as a function of frequencies ω and ω', which we can call the "pump" frequency and the "probe" frequency.

To interpret these two-dimensional data, consider first what will happen if only the third and fourth pulses (the "probe") are sent, and the first two pulses are not. In that case, the experiment will essentially be the same as the quantum beat experiment described above. The two resonances will both oscillate, and the Fourier transform as a function of ω' will have two peaks at ω_A and ω_B.

Now suppose that we also bring in the first two pulses, which combined make up the "pump." If the time interval T between the pump and probe pulses is long compared to T_1, then the situation will be no different from having no pump pulses at all; we will just see the spectrum of the two frequencies excited by the third pulse.

If the interval T is short enough, however, then when the third pulse arrives, the state of each oscillator will have memory of the action of the first two pulses. The amount of the perturbation from the pump pulses will depend on how much the original oscillation from the first pulse was canceled or enhanced by the second pulse. If the two types of oscillators A and B are completely isolated from each other, then we expect that oscillator A will not be affected by what happens to oscillator B during the pump sequence, and vice versa. We therefore expect that two isolated oscillators will produce two spots on the diagonal in the two-dimensional Fourier transform image, as shown in Figure 9.13(a). The probe measurement of each oscillator will be affected only by what happened to that oscillator during the pump sequence at earlier times.

Suppose now that electrons in the upper state B can fall down into the lower state A through some random process. Because the number of electrons in state B depends on the effect of the pump pulses on that state, the modulation of state B will affect the behavior of state A during the probe sequence. At low temperature, however, none of the electrons in state A will jump up to state B, so there will be no reverse effect. We therefore expect to see a pattern like that shown in Figure 9.13(b), with a spot off the diagonal. As temperature is raised, we expect to see a spot begin to appear on the opposite side of the diagonal corresponding to up-conversion.

Finally, let us suppose that the two states A and B are coupled together somehow, for example by a shared ground state, so that excitation of one resonance depletes the number of electrons left in the ground state that can be excited to the other. In this case, what happens to each during the pump sequence will affect the other, regardless of the temperature. We expect in this case to see two symmetric spots off the diagonal, as shown in Figure 9.13(c). We thus have a method to distinguish experimentally between two states that are coupled coherently and two that are not.

Two-dimensional spectral broadening. As discussed in Section 9.2, *inhomogeneous broadening* corresponds to a range of different resonance energies. This will lead to smearing along the diagonal of a two-dimensional image like those illustrated in Figure 9.13. On the other hand, *homogeneous* broadening corresponds to an intrinsic dephasing.

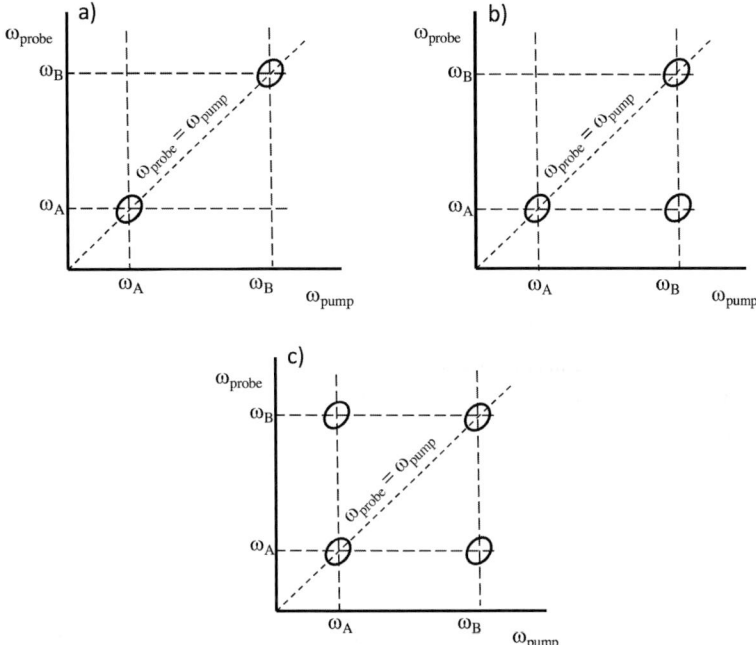

Fig. 9.13 (a) Generic two-dimensional Fourier transform image of a system with two uncoupled resonances. (b) Two-dimensional Fourier transform image of a system with incoherent down-conversion from a higher-energy state to a lower-energy state. (c) Two-dimensional Fourier transform image of a system with coherent coupling between two states with different energies.

A clever experiment allows us to distinguish inhomogeneous broadening and homogeneous broadening on a two-dimensional Fourier image.

As discussed in Section 7.6.2, the output signal of a four-wave mixing measurement requires phase matching, that is, the output beam $\vec{k}_4$ and the three input beams must be related by the relation $\vec{k}_4 = \pm \vec{k}_1 \pm \vec{k}_2 \pm \vec{k}_3$. The frequencies of the beams must also add up in the same way.

Since the amplitude of the oscillation after any pulse in the sequence of Figure 9.12 depends on the initial state of the Bloch vector when that pulse arrives, and that initial Bloch vector state depends on its time evolution left over from previous pulses, we can write the polarization of a resonance ω at the time of the fourth pulse as

$$P \propto \left(e^{i(\vec{k}_1 \cdot \vec{x} - \omega \tau_1) - \gamma \tau_1} + e^{-i(\vec{k}_1 \cdot \vec{x} - \omega \tau_1) - \gamma \tau_2} \right) \left(e^{i(\vec{k}_2 \cdot \vec{x} - \omega T) - \gamma T} + e^{-i(\vec{k}_2 \cdot \vec{x} - \omega T) - \gamma T} \right)$$
$$\times \left(e^{i(\vec{k}_3 \cdot \vec{x} - \omega \tau_2) - \gamma \tau_2} + e^{-i(\vec{k}_3 \cdot \vec{x} - \omega \tau_2) - \gamma \tau_2} \right), \tag{9.4.3}$$

where $\gamma = 1/T_2$ is the dephasing constant, and $\vec{k}_1$, $\vec{k}_2$, and $\vec{k}_3$ are the k-vectors of the three input pulses, which do not need to be the same. The modulation of each pulse by the preceding pulses leads to a large number of product terms, including

$$P \propto e^{i(\vec{k}_1+\vec{k}_2-\vec{k}_3)\cdot\vec{x}-\omega(\tau_1+T-\tau_2)}e^{-\gamma(\tau_1+T+\tau_2)} + \text{c.c.} \qquad (9.4.4)$$

As in any four-wave-mixing experiment, we can select the direction of observation $\vec{k}_4 = \vec{k}_1 + \vec{k}_2 - \vec{k}_3$, in which case the signal will be modulated by the time dependence $e^{-i\omega(\tau_1+T-\tau_2)}e^{-\gamma(\tau_1+T+\tau_2)}$.

The two-dimensional Fourier transform is then

$$F(\omega_1,\omega_2) \propto e^{-i\omega T}e^{-\gamma T}\int_0^\infty d\tau_1 \int_0^\infty d\tau_2\, e^{i\omega_1\tau_1}e^{i\omega_2\tau_2}e^{-i\omega(\tau_1-\tau_2)}e^{-\gamma(\tau_1+\tau_2)},$$

$$(9.4.5)$$

where the lower bounds of the integrals are zero because τ_1 and τ_2 are always positive. We switch variables to coordinates rotated by $45°$, $\tau = \tau_1 + \tau_2$ and $\tau' = \tau_1 - \tau_2$, and ignore the T dependence, to obtain

$$F(\omega_1,\omega_2) \propto \int_0^\infty d\tau \int_0^\infty d\tau'\, e^{i\omega_1(\tau+\tau')/2}e^{-i\omega_2(\tau-\tau')/2}e^{-i\omega\tau'}e^{-\gamma\tau}$$

$$= \int_0^\infty d\tau\, e^{i(\omega_1-\omega_2)\tau/2}e^{-\gamma\tau}\int_0^\infty d\tau'\, e^{i(\omega_1+\omega_2)\tau'/2}e^{-i\omega\tau'}.$$

$$(9.4.6)$$

Performing the integrals then gives

$$F(\omega_1,\omega_2) \propto \left(\frac{1}{(\omega_1-\omega_2)/2 + i\gamma}\right)\left(\frac{1}{(\omega_1+\omega_2-2\omega)+i\epsilon}\right), \qquad (9.4.7)$$

where we have added a small term $e^{-\epsilon\tau'}$ to the second integral to make it converge. When we square the terms to get the absolute magnitude, the first term will give us a Lorentzian of the form

$$\frac{1}{(\omega_1-\omega_2)^2 + 4\gamma^2}, \qquad (9.4.8)$$

and the second term will give us $\delta(\omega_1 + \omega_2 + 2\omega)$. If $\gamma \to 0$, this gives us a spot on the two-dimensional image at $\omega_1 = \omega_2 = \omega$. If γ is significant, we will see a Lorentzian broadening along the cross-diagonal, with a maximum on the diagonal at $\omega_1 = \omega_2$, with width given by the homogeneous broadening $\gamma = 1/T_2$. We thus see that we can read off of a two-dimensional image like those shown in Figure 9.13 the inhomogenous broadening as the width along the diagonal of a resonance spot, and the homogeneous broadening as the width in the cross-diagonal direction.

Time-resolved four-wave mixing measurement of T_2 is a complementary method to simply using a spectrometer to measure the line broadening of an optical transition. In Section 8.4, we saw that the homogeneous line width of a transition between two states is given by the out-scattering rate of the states. This is the same scattering rate that gives T_2. If T_2 is very long, then the homogeneous line broadening due to dephasing will be very small, making it difficult to deduce the dephasing rate from spectral line width measurements. In this case, however, the free induction decay will persist for a long time, making four-wave mixing measurements easier. If T_2 is short, four-wave mixing will be difficult because it will require very good time resolution, but the line broadening will be large, making it easy to deduce the dephasing time from spectral line width measurements.

This leads one to expect that the line width and the dephasing rate measured by four-wave-mixing should be inversely proportional. This is generally true, but can be incorrect when a continuum of states is excited (e.g., by an ultrafast pulse with considerable Heisenberg energy uncertainty), leading to interference between difference oscillator frequencies. Pulse-echo measurements, discussed in Section 9.2, can also be used to measure T_2 with optical transitions in semiconductors.

For a review of two-dimensional coherent spectroscopy, see Smallwood and Cundiff (2018).

9.5 Correlation Functions and Noise

As discussed at the beginning of this chapter, systems do not fall simply into two categories of "coherent" and "not coherent." In general, we want a quantitative way of characterizing systems that are partially coherent. There are two standard measurements we can make in this regard. First, if we create a coherent state in a system, we can measure how long the system takes to become incoherent. This is what we measure in a T_2 measurement, discussed in the previous sections. Second, given a system that is partially coherent, for example, a system driven by a coherent field but with noise, we can measure over what time periods or distances it can be considered coherent. The standard way of doing this is by measuring the **autocorrelation function**, also called simply the **correlation function**. We measure, essentially, whether the state of a system at one time has a definite phase relation with the state at a later time.

The concept of correlation is more general than the concept of coherence. Coherence can be defined as correlation of the off-diagonal terms of the density matrix. But other things can be correlated as well. The correlation function tells us generally how the state of a system at one time is related to the state at a later time. We will see that the correlation function is a useful measure of the thermal fluctuations in a system, since fluctuations destroy correlations. In this section, we discuss the correlation function for a classical field. We will generalize this to a quantum mechanical system in the next section.

We assume that some function $f(t)$ is measured over a time interval of duration T, and define the correlation function as

$$C_f(\tau) = \lim_{T \to \infty} \frac{1}{T} \int_{-T/2}^{T/2} f^*(t)f(t + \tau)dt. \tag{9.5.1}$$

If $f(t + \tau)$ is unrelated to $f(t)$, then the average of their product will be zero, since they will have opposite sign as often as the same sign, on average. If the function value at two different times is correlated, their average product will be nonzero. Sometimes functions are correlated for short times but lose correlation after a long time. Using the correlation function, we define the **coherence time**,

$$\tau_c = \int_0^\infty d\tau \left| \frac{C_f(\tau)}{C_f(0)} \right|^2. \tag{9.5.2}$$

If we view $f(t)$ as representing the electric field of an electromagnetic wave, we can also define the coherence *length*, $l_c = c\tau_c$.

If, for example, $f(t) = e^{-i\omega_0 t}$, then

$$C_f(\tau) = \lim_{T \to \infty} \frac{1}{T} \int_{-T/2}^{T/2} e^{i\omega_0 t} e^{-i\omega_0(t+\tau)} dt$$

$$= e^{-i\omega_0 \tau}. \tag{9.5.3}$$

The function remains finite for all τ up to $\tau \to \infty$. The coherence time and the coherence length are therefore infinite.

On the other hand, suppose that $f(t) = e^{-i(\omega_0 t + \theta(t))}$, where $\theta(t)$ is a phase that shifts randomly. We visualize the shift of the phase as a random walk, in which the phase jumps by some fixed amount $\Delta\theta_0$ in a time interval τ_s. The number of jumps in a time interval τ is τ/τ_s, and therefore the average of the square of the phase shift after a time interval τ is $\sigma^2 = (\Delta\theta_0)^2(|\tau|/\tau_s)$. The probability of a total phase shift $\Delta\theta(\tau) = \theta(t+\tau) - \theta(t)$ is given by the normal distribution

$$P(\Delta\theta) = \frac{e^{-\Delta\theta^2/2\sigma^2}}{\sqrt{2\pi}\sigma}. \tag{9.5.4}$$

The correlation function is

$$C_f(\tau) = \lim_{T \to \infty} \frac{1}{T} \int_{-T/2}^{T/2} e^{i(\omega_0 t + \theta(t))} e^{-i(\omega_0(t+\tau) + \theta(t+\tau))} dt$$

$$= \lim_{T \to \infty} \frac{1}{T} \int_{-T/2}^{T/2} e^{i\omega_0 \tau} e^{i(\theta(t) - \theta(t+\tau))} dt. \tag{9.5.5}$$

The integral over all time is the same as averaging over all $\Delta\theta$. We therefore write

$$C_f(\tau) = e^{i\omega_0 \tau} \int_{-\infty}^{\infty} d\Delta\theta \, e^{i\Delta\theta} P(\Delta\theta)$$

$$= e^{i\omega_0 \tau} e^{-\sigma^2/2}$$

$$= e^{i\omega_0 \tau} e^{-(\Delta\theta_0)^2|\tau|/2\tau_s}. \tag{9.5.6}$$

The coherence time is then

$$\tau_c = \int_0^{\infty} d\tau \, e^{-(\Delta\theta_0)^2|\tau|/\tau_s}$$

$$= \frac{\tau_s}{(\Delta\theta_0)^2}. \tag{9.5.7}$$

The correlation function can be expressed in another way. We write $f(t)$ in terms of its Fourier transform

$$f(t) = \frac{1}{2\pi} \int_{-\infty}^{\infty} d\omega \, F(\omega) e^{-i\omega t}, \tag{9.5.8}$$

where the Fourier transform is given by

$$F(\omega) = \int_{-T/2}^{T/2} dt \, f(t) e^{i\omega t}, \tag{9.5.9}$$

since the $f(t)$ is measured only during the interval of duration T, and therefore is zero outside this interval. Substituting the definition of $f(t)$ in terms of its Fourier transform into (9.5.1), we have

$$C_f(\tau) = \lim_{T \to \infty} \frac{1}{T} \int_{-T/2}^{T/2} dt \left(\frac{1}{2\pi} \int_{-\infty}^{\infty} F^*(\omega) e^{i\omega t} d\omega \right) \left(\frac{1}{2\pi} \int_{-\infty}^{\infty} F(\omega') e^{-i\omega'(t+\tau)} d\omega' \right).$$

$$(9.5.10)$$

In the limit $T \to \infty$, we can replace

$$\frac{1}{2\pi} \int_{-T/2}^{T/2} dt \, e^{i(\omega-\omega')t} \to \frac{1}{2\pi} \int_{-\infty}^{\infty} dt \, e^{i(\omega-\omega')t} = \delta(\omega - \omega'). \qquad (9.5.11)$$

This δ-function then eliminates one frequency integral, yielding

$$C_f(\tau) = \lim_{T \to \infty} \frac{1}{T} \frac{1}{2\pi} \int_{-\infty}^{\infty} d\omega \, F^*(\omega) F(\omega) e^{-i\omega\tau}.$$

$$= \frac{1}{2\pi} \int_{-\infty}^{\infty} d\omega \, S_f(\omega) e^{-i\omega\tau}, \qquad (9.5.12)$$

where

$$S_f(\omega) = \lim_{T \to \infty} \frac{1}{T} |F(\omega)|^2. \qquad (9.5.13)$$

If $f(t)$ is real, then $F^*(\omega) = F(-\omega)$ and

$$S_f(\omega) = \lim_{T \to \infty} \frac{1}{T} F(\omega) F(-\omega). \qquad (9.5.14)$$

If we set $\tau = 0$, then the correlation function gives simply the average power,

$$C_f(0) = \lim_{T \to \infty} \frac{1}{T} \int_{-T/2}^{T/2} dt \, |f(t)|^2$$

$$= \lim_{T \to \infty} \frac{1}{T} \int_{-T/2}^{T/2} dt \, P(t) = \bar{P}, \qquad (9.5.15)$$

where we set the power $P(t) = |f(t)|^2$, proportional to the square of the amplitude, absorbing any multiplicative constants into the definition of $f(t)$. According to (9.5.12), the average power is then equal to

$$\bar{P} = \frac{1}{2\pi} \int_{-\infty}^{\infty} d\omega \, S_f(\omega). \qquad (9.5.16)$$

$S_f(\omega) d\omega$ gives the average power in each frequency range $(\omega, \omega + d\omega)$. For this reason, $S_f(\omega)$ is called the **spectral density function**, or the power spectrum. Although a factor $1/T$ appears in definition (9.5.13), it does not vanish in the limit $T \to \infty$ because the magnitude of $F(\omega)$ increases with T; for example, for a single-frequency wave $f(t) = e^{-i\omega_0 t}$,

$$\bar{P} = \lim_{T \to \infty} \frac{1}{T} \int_{-\infty}^{\infty} d\omega \, \frac{1}{2\pi} \left(\int_{-T/2}^{T/2} dt \, e^{i\omega_0 t} e^{-i\omega t} \right) \left(\int_{-T/2}^{T/2} dt' \, e^{-i\omega_0 t} e^{i\omega t'} \right)$$

$$= \lim_{T \to \infty} \frac{1}{T} \int_{-\infty}^{\infty} d\omega \, \delta(\omega - \omega_0) \int_{-T/2}^{T/2} dt \, e^{i(\omega - \omega_0)t}$$

$$= \lim_{T \to \infty} \frac{1}{T} \int_{-T/2}^{T/2} dt$$

$$= 1, \tag{9.5.17}$$

where we have again used the limit (9.5.11) to obtain a δ-function, which eliminates the integration over ω.

According to (9.5.12), the correlation function is the Fourier transform of the spectral density function. We therefore have the Fourier pair,

$$
\boxed{
\begin{aligned}
C_f(\tau) &= \frac{1}{2\pi} \int_{-\infty}^{\infty} d\omega \, S_f(\omega) e^{-i\omega\tau} \\[2mm]
S_f(\omega) &= \int_{-\infty}^{\infty} d\tau \, C_f(\tau) e^{i\omega\tau}.
\end{aligned}
}
\tag{9.5.18}
$$

This is known as the **Wiener–Khintchine theorem** (the second name seems to have many variants in the literature, and so is apparently unspellable as well as unpronouncable).

Example: shot noise. Suppose that a signal $f(t)$ consists of random events at a series of times t_i. Each event is identical. The Fourier transform of each event is therefore identical except for a phase shift, as we see by a simple change of variables:

$$F_i(\omega) = \int_{-\infty}^{\infty} dt \, f(t - t_i) e^{i\omega t}$$

$$= e^{i\omega t_i} \int_{-\infty}^{\infty} dt' \, f(t') e^{i\omega t'}$$

$$= e^{i\omega t_i} F_0(\omega). \tag{9.5.19}$$

The total signal is then

$$F(\omega) = F_0(\omega) \sum_i e^{i\omega t_i}. \tag{9.5.20}$$

The spectral density function is then

$$S_f(\omega) = \lim_{T \to \infty} \frac{1}{T} |F(\omega)|^2 \left| \sum_i e^{i\omega t_i} \right|^2. \tag{9.5.21}$$

The square of the phase terms is just equal to N, the total number of events in time T, since the products of terms with different t_i cancel, on average. The spectral density function is then

$$S_f(\omega) = \bar{N} |F(\omega)|^2, \tag{9.5.22}$$

where $\bar{N} = N/T$.

Let us apply this to a current $I(t)$ that consists of a series of "shots" of charge e,

$$I(t) = \sum_i e\delta(t - t_i), \tag{9.5.23}$$

as we expect for the motion of single electrons. It is easy to show that the spectral density function is

$$S_I(\omega) = e\bar{N} = \text{constant}. \tag{9.5.24}$$

A flat spectrum, corresponding to δ-function shots in time, is known as **white noise**. If the power of the noise signal per frequency interval is compared to the total average current, a measurement of the shot noise gives a measurement of the charge of the particles. As mentioned in Section 2.9.4, shot noise has been used to verify the fractional charge of the quasiparticles in the fractional quantum Hall effect.

Line narrowing. As discussed in Section 4.4, a coherent state of bosons can be represented as a vector in the complex plane known as a phasor. As illustrated in Figure 9.14, the absorption or emission of a photon will lead to a random phase shift. The length of the original phasor is $\sqrt{N}$ while the length of the phasor for the single photon is 1. The change of phase due to adding or subtracting exactly one photon is found by adding a phasor with length 1 to the original phasor and averaging over all possible angles at which the unit-length phasor is added. Let us define θ' as the angle of the unit phasor relative to the original phasor. Then assuming $N \gg 1$, and therefore $\tan\theta \simeq \theta$, the mean-squared change in phase per single photon jump is

$$(\Delta\theta_0)^2 = \langle(\Delta\theta)^2\rangle \simeq \langle\tan^2\Delta\theta\rangle = \left\langle\left(\frac{\sin\theta'}{\sqrt{N}}\right)^2\right\rangle = \frac{1}{2N}. \tag{9.5.25}$$

Using the result (9.5.6) above for a random walk of the phase shift, we then have

$$\sigma^2 = (\Delta\theta_0)^2\frac{|\tau|}{\tau_s} = \frac{|\tau|}{2N\tau_s}, \tag{9.5.26}$$

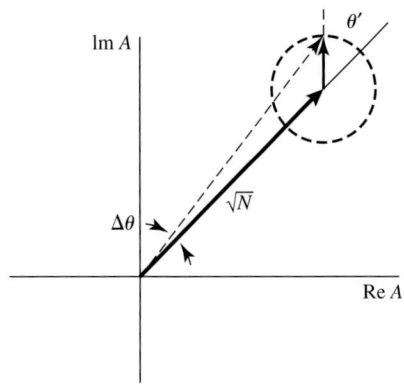

Fig. 9.14 Random phase shift of a phasor in the complex plane caused by absorption or emission of a photon.

and

$$C_A(\tau) = e^{i\omega_0\tau - |\tau|/4N\tau_s}. \tag{9.5.27}$$

Writing $\gamma = 1/4N\tau_s$, we can use the Wiener–Khintchine theorem to obtain the spectral density function,

$$
\begin{aligned}
S_A(\omega) &= \int_{-\infty}^{\infty} d\tau \; e^{-i\omega_0\tau - \gamma|\tau|} e^{i\omega\tau} \\
&= \int_{-\infty}^{0} d\tau \; e^{i(\omega-\omega_0)\tau + \gamma\tau} + \int_0^{\infty} d\tau \; e^{i(\omega-\omega_0)\tau - \gamma\tau} \\
&= \frac{2\gamma}{(\omega - \omega_0)^2 + \gamma^2}.
\end{aligned}
\tag{9.5.28}
$$

The spectral density is therefore a Lorentzian. This is the same form as (8.4.12), deduced from many-body theory in Section 8.4. As discussed in that section, line broadening can be viewed as a simple consequence of the uncertainty principle, in other words, of the properties of Fourier transforms – if the time spent in a single state is short, the spectral width will be large.

Notice that since the factor N appears in the denominator of γ, if N is large, the line will narrow. This is a ubiquitous effect for lasers and a standard test for whether a system is coherent.

Exercise 9.5.1   A laser pulse has electric field proportional to $e^{-i\omega_0 t} e^{-(t-t_0)^2/\tau_0^2}$, and is repeated every period T_0. Compute its correlation function, assuming $T_0 \gg \tau_0$, as well as its spectral density function (in arbitrary units), and its coherence length.

9.6 Correlations in Quantum Mechanics

The same type of analysis as done in the previous system can be extended to quantum mechanical systems. In this case, we must use quantum mechanical expectation values instead of classical averages. The quantum mechanical correlation function corresponds to the expectation value of the product of a measurement at one time, times the same measurement at a later time.

At time $t = 0$, the expectation value of a measurement of f on an initial state $|\psi_0\rangle$ is given by

$$\langle \psi_0|f|\psi_0\rangle. \tag{9.6.1}$$

Let the Hamiltonian of the system be H. Then at a later time, the state of the system is, according to the Schrödinger equation, $|\psi_t\rangle = e^{-iHt/\hbar}|\psi_0\rangle$, and the expectation value is

$$\langle \psi_t|f|\psi_t\rangle = \langle \psi_0|e^{iHt/\hbar} f e^{-iHt/\hbar}|\psi_0\rangle. \tag{9.6.2}$$

The operator that corresponds to a measurement after the system has evolved a time t is therefore

$$f(t) = e^{iHt/\hbar} f e^{-iHt/\hbar}. \tag{9.6.3}$$

The expectation value of the product of these two measurements can then be viewed as the inner product of the measurement $f(0)$ on the state $|\psi_0\rangle$ and the measurement on the same state at a later time τ:

$$C_f(\tau) = \left(\langle\psi_0|f^\dagger(\tau)\right)\left(f(0)|\psi_0\rangle\right) = \langle\psi_0|f^\dagger(\tau)f(0)|\psi_0\rangle. \tag{9.6.4}$$

Note that we want a single expectation value for a given state $|\psi_0\rangle$, not the product of two expectation values. If $|n\rangle$ is an eigenstate of the system, the expectation value $\langle n|f(\tau)|n\rangle$ will not be different from $\langle n|f(0)|n\rangle$, since an eigenstate by definition is invariant under a time shift. Thus, the product of the two expectation values will be a constant independent of τ, while the expectation value of the product can have very different dependence on τ, and can be nonzero even if the expectation value of a single measurement is zero.

Often we want to have a real observable. In that case, we take add it to its complex conjugate and divide by two,

$$C_f(\tau) = \frac{1}{2}\langle\psi_0|\left(f^\dagger(0)f(\tau) + f^\dagger(\tau)f(0)\right)|\psi_0\rangle. \tag{9.6.5}$$

We define a spectral density function as the Fourier transform of this function, just as we did for the classical Wiener–Khintchine theorem,

$$S_f(\omega) = \int_{-\infty}^{\infty} d\tau \, C_f(\tau)e^{i\omega\tau}. \tag{9.6.6}$$

We can rewrite this spectral function in terms of the eigenstates of the system. As with the Lehmann representation discussed in Section 8.14, we do not actually need to compute the true eigenstates of the system; we just assume that they exist, so that we can use them as a complete set of states.

If we assume that $|\psi_0\rangle$ is an eigenstate $|n\rangle$, then the spectral density function becomes

$$
\begin{aligned}
S_f(\omega) &= \int_{-\infty}^{\infty} d\tau \, \frac{1}{2}\left(\langle n|f^\dagger(0)f(\tau)|n\rangle + \langle n|f^\dagger(\tau)f(0)|n\rangle\right)e^{i\omega\tau} \\
&= \sum_m \int_{-\infty}^{\infty} d\tau \, \frac{1}{2}\left(\langle n|f^\dagger(0)|m\rangle\langle m|f(\tau)|n\rangle + \langle n|f^\dagger(\tau)|m\rangle\langle m|f(0)|n\rangle\right)e^{i\omega\tau} \\
&= \sum_m \frac{1}{2}|\langle m|f|n\rangle|^2 \int_{-\infty}^{\infty} d\tau \left(e^{i(E_m-E_n)\tau/\hbar} + e^{-i(E_m-E_n)\tau/\hbar}\right)e^{i\omega\tau},
\end{aligned}
$$

$$\tag{9.6.7}$$

where we have inserted a sum over the complete set of eigenstates $|m\rangle$. We then define $\omega_{mn} = (E_m - E_n)/\hbar$ and use the identity

$$\int_{-\infty}^{\infty} dt \, e^{i\omega t} = 2\pi\delta(\omega), \tag{9.6.8}$$

so that we finally obtain

$$\boxed{S_f(\omega) = \pi \sum_m |\langle m|f|n\rangle|^2 \left(\delta(\omega_{mn} + \omega) + \delta(\omega_{mn} - \omega)\right).} \tag{9.6.9}$$

Comparing this to (7.4.17), we see that the spectral density is proportional to the imaginary susceptibility, which in turn is proportional to the rate deduced from Fermi's golden rule for all transitions from the state $|n\rangle$ induced by the operator f. If we are interested in fluctuations of the energy, the matrix element is $|\langle m|H|n\rangle|^2$. Since the eigenstates are orthonormal, the only contribution to this for nonzero frequency will come from an interaction term not included in the original Hamiltonian. The energy spectrum of the fluctuations in a state $|n\rangle$ is then proportional to the sum of all possible spontaneous transitions out of that state.

Correlation functions and many-body theory. Comparing this formalism to that of Section 8.9, we see that there is a direct connection. The Green's functions of many-body theory introduced in Chapter 8 can be viewed as correlation functions of the quantum wave function amplitude $a_{\vec{k}}^{\dagger}$, multiplied by a Heaviside function to allow only retarded (causal) contributions:

$$G_{\vec{k}}(t) = -i\langle \text{vac}|a_{\vec{k}}(t)a_{\vec{k}}^{\dagger}(0)|\text{vac}\rangle\Theta(t). \tag{9.6.10}$$

We do not take the real part, as we are interested in the full, complex field.

The spectral function introduced in Section 8.9 is defined as

$$
\begin{aligned}
A(\vec{k}, \omega) &= -2\text{Im}\, G(\vec{k}, \omega) \\
&= iG(\vec{k}, \omega) - iG^*(\vec{k}, \omega) \\
&= \int_{-\infty}^{\infty} dt\, e^{i\omega t} \langle a_{\vec{k}}(t)a_{\vec{k}}^{\dagger}(0)\rangle\Theta(t) + \int_{-\infty}^{\infty} dt\, e^{-i\omega t} \langle a_{\vec{k}}(0)a_{\vec{k}}^{\dagger}(t)\rangle\Theta(t),
\end{aligned}
$$

where we have used the definition of the Fourier transform consistent with the Green's function definitions of Section 8.6. The average $\langle a_{\vec{k}}(0)a_{\vec{k}}^{\dagger}(t)\rangle$ is equal to $\langle a_{\vec{k}}(-t)a_{\vec{k}}^{\dagger}(0)\rangle$, and therefore after a change of variables from $t \rightarrow -t$ in the second integral, we have

$$
\begin{aligned}
A(\vec{k}, \omega) &= \int_{-\infty}^{\infty} dt\, e^{i\omega t} \langle a_{\vec{k}}(t)a_{\vec{k}}^{\dagger}(0)\rangle\Theta(t) + \int_{-\infty}^{\infty} dt\, e^{i\omega t} \langle a_{\vec{k}}(t)a_{\vec{k}}^{\dagger}(0)\rangle\Theta(-t) \\
&= \int_{-\infty}^{\infty} dt\, e^{i\omega t} \langle a_{\vec{k}}(t)a_{\vec{k}}^{\dagger}(0)\rangle. \tag{9.6.11}
\end{aligned}
$$

The spectral function that we introduced in Chapter 8 is therefore just equal to the spectral density function of the complex amplitude correlation function.

Density–density correlation function. By contrast, suppose that we are interested in the correlation function of the fluctuations of the total number of the particles $N = \sum N_k$; in other words:

$$
\begin{aligned}
C_N(t) &= \langle N(t)N(0)\rangle \\
&= \sum_{\vec{k},\vec{k}'} \langle a_{\vec{k}}^{\dagger}(t)a_{\vec{k}}(t)a_{\vec{k}'}^{\dagger}(0)a_{\vec{k}'}(0)\rangle. \tag{9.6.12}
\end{aligned}
$$

The correlation function for number fluctuations is closely related to the **density–density correlation function**. The particle density at a point $\vec{x}$ is given by

$$\rho(\vec{x}, t) = \psi^{\dagger}(\vec{x}, t)\psi(\vec{x}, t). \tag{9.6.13}$$

Substituting for the field operators in terms of the operators $a^{\dagger}_{\vec{k}}(t)$ and $a_{\vec{k}}(t)$ (see Section 4.5), this becomes

$$\rho(\vec{x}, t) = \frac{1}{V} \sum_{\vec{k}, \vec{k}'} e^{i(\vec{k} - \vec{k}') \cdot \vec{x}} \, a^{\dagger}_{\vec{k}'}(t) a_{\vec{k}}(t)$$

$$= \sum_{\vec{q}} e^{-i\vec{q} \cdot \vec{x}} \rho_{\vec{q}}(t), \qquad (9.6.14)$$

where

$$\rho_{\vec{q}}(t) = \frac{1}{V} \sum_{\vec{k}} a^{\dagger}_{\vec{k}+\vec{q}}(t) a_{\vec{k}}(t). \qquad (9.6.15)$$

We define the density–density correlation function, for fluctuations of density in both time and space, as

$$C_{\rho}(\vec{x}, t) = \frac{1}{V} \int d^3 x' \, \langle \rho(\vec{x}', t) \rho(\vec{x}' + \vec{x}, 0) \rangle, \qquad (9.6.16)$$

which after substituting in the definition of $\rho(\vec{x}, t)$ becomes

$$C_{\rho}(\vec{x}, t) = \sum_{\vec{q}} e^{i\vec{q} \cdot \vec{x}} \, \langle \rho_{\vec{q}}(t) \rho_{-\vec{q}}(0) \rangle. \qquad (9.6.17)$$

The quantity $\langle \rho_{\vec{q}}(t) \rho_{-\vec{q}}(0) \rangle$ is thus the Fourier component with wave vector $\vec{q}$ of the density–density correlation function. In the long wavelength limit $\vec{q} \to 0$, this is simply equal to

$$\frac{1}{V^2} \sum_{\vec{k}, \vec{k}'} \langle a^{\dagger}_{\vec{k}}(t) a_{\vec{k}}(t) a^{\dagger}_{\vec{k}'}(0) a_{\vec{k}'}(0) \rangle, \qquad (9.6.18)$$

which is just the same as the number correlation function (9.6.12) used above, with the number divided by the volume to give a density.

In the above, we have made the connection between the Green's function formalism of many-body theory and the formalism of correlation functions. The formalism also maps the opposite way. We can define Green's functions for any correlation function, including one for a classical field, and develop a Feynmann diagram method just as we did for the many-body theory of Chapter 8. For a good discussion of this, see Plischke and Bergerson (1989).

Exercise 9.6.1　　(a) The correlation function $\langle \text{vac} | a_{\vec{k}}(t) a^{\dagger}_{\vec{k}}(0) | \text{vac} \rangle$ used in (9.6.10) is not Hermitian. Show that if the correlation function is made Hermitian by the method used in (9.6.5), the spectral function will be symmetric with respect to $\omega = 0$.

　　(b) Find the spectral function for the case of a noninteracting boson gas, for a symmetrized correlation function.

Exercise 9.6.2　　Prove that (9.6.17) follows from putting the definition of $\rho(\vec{x}, t)$ into (9.6.16).

9.7 Particle–Particle Correlation

In Section 9.6, we discussed the density–density correlation function, which gives the spectrum of density fluctuations. A related, but different, correlation function is the **pair correlation function**,

$$K_{\vec{k}}(\tau) = \left(\langle \psi | a_{\vec{k}}^\dagger(0) a_{\vec{k}}^\dagger(\tau) \rangle \right) \left(a_{\vec{k}}(\tau) a_{\vec{k}}(0) | \psi \rangle \right). \tag{9.7.1}$$

This corresponds to the probability of starting in state $|\psi\rangle$ and detecting a particle at $t = 0$ and then a second particle at $t = \tau$.

This type of measurement can be done with photons using a Hanbury-Brown–Twiss (HBT) experiment[1] like that shown in Figure 9.15(a). The arrival time of one photon in a beam relative to the arrival time of another photon is recorded, and a histogram is made of all the arrival times. Using the commutation relations (8.6.13), the correlation function is

$$K_{\vec{k}}(\tau) = \langle \psi | a_{\vec{k}}^\dagger(0) a_{\vec{k}}^\dagger(\tau) a_{\vec{k}}(\tau) a_{\vec{k}}(0) | \psi \rangle$$

$$= \langle \psi | \left(a_{\vec{k}}^\dagger(\tau) a_{\vec{k}}(\tau) a_{\vec{k}}^\dagger(0) a_{\vec{k}}(0) - e^{-i\omega_{\vec{k}}\tau} a_{\vec{k}}^\dagger(\tau) a_{\vec{k}}(0) \right) | \psi \rangle. \tag{9.7.2}$$

By comparison, the correlation function (9.6.5) corresponds to an autocorrelation measurement, such as a Michelson interferometer, as illustrated in Figure 9.15(b), in which the measured intensity is proportional to

$$|\langle a_{\vec{k}}(0) + a_{\vec{k}}(\tau) \rangle|^2 = N_{\vec{k}}(0) + N_{\vec{k}}(\tau)$$

$$+ 2\langle a_{\vec{k}}^\dagger(0) a_{\vec{k}}(\tau) + a_{\vec{k}}^\dagger(\tau) a_{\vec{k}}(0) \rangle. \tag{9.7.3}$$

The zero-delay value of the correlation function (9.7.1) is related to the average number uncertainty by the following (we drop the subscript since we are concerned with only one $\vec{k}$ state):

$$K(0) = \langle (\hat{N} - 1)\hat{N} \rangle$$

$$= \langle \hat{N}^2 \rangle - \bar{N}. \tag{9.7.4}$$

Since the average number fluctuation is given by

$$(\Delta N)^2 = \langle (\hat{N} - \bar{N})^2 \rangle$$

$$= \langle \hat{N}^2 \rangle - 2\bar{N}\langle \hat{N} \rangle + \bar{N}^2$$

$$= \langle \hat{N}^2 \rangle - \bar{N}^2, \tag{9.7.5}$$

this implies

$$\boxed{K(0) = (\Delta N)^2 + \bar{N}^2 - \bar{N}.} \tag{9.7.6}$$

For a coherent state, which as discussed in Section 4.4 has a Poisson distribution of photon number, $(\Delta N)^2 = \bar{N}$, this implies

$$K(0) = \bar{N}^2. \tag{9.7.7}$$

---

[1] This is actually two names; the first author's last name was Hanbury Brown.

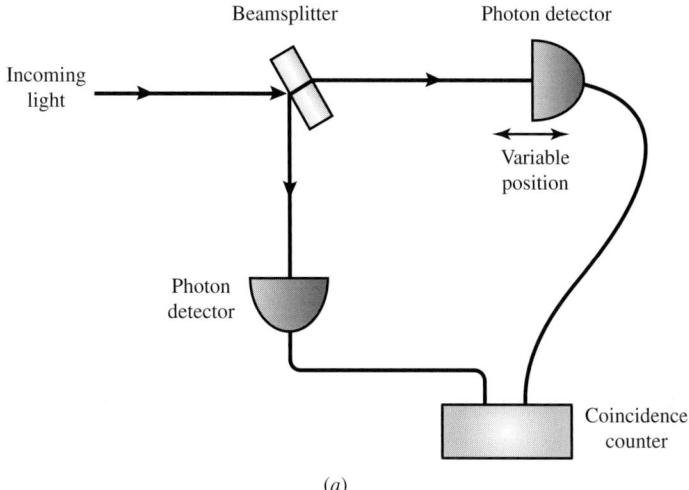

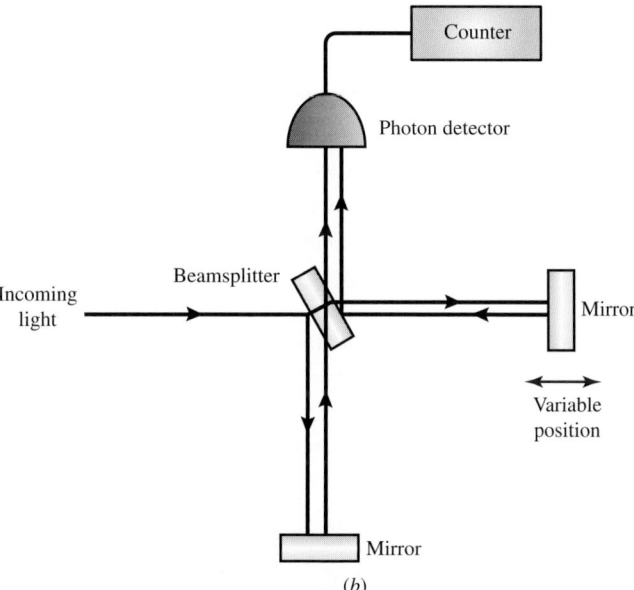

Fig. 9.15 (a) The HBT experiment for measuring photon–photon correlation. (b) The Michelson experiment for first-order correlation.

We could also have deduced this by noting that for a coherent state $a_{\vec{k}}|\alpha\rangle = \sqrt{\bar{N}}e^{i\omega_{\vec{k}}t}|\alpha\rangle$, where $\bar{N}$ is the average number of photons. Then in (9.7.1) all the phase factors cancel and we have simply

$$K(\tau) = \bar{N}^2, \tag{9.7.8}$$

independent of τ.

Photon bunching in thermal light. The number fluctuation of thermal light is different from that of coherent photons.

As already discussed in Section 4.10, the probability of N bosons in a mode with frequency ω is equal to

$$P(N) = \frac{1}{Z} e^{-\beta \hbar \omega N}, \tag{9.7.9}$$

where the partition function Z is

$$Z = \sum_{N=0}^{\infty} e^{-\beta \hbar \omega N} = \frac{1}{1 - e^{-\beta \hbar \omega}}. \tag{9.7.10}$$

The thermal average $\langle N^2 \rangle$ has already been computed in (4.10.17), as $\langle N^2 \rangle = \bar{N}(1 + 2\bar{N})$. Therefore, using (9.7.5), for bosons in a Planck distribution,

$$\boxed{(\Delta N)^2 = \bar{N}^2 + \bar{N}} \tag{9.7.11}$$

and

$$K(0) = 2\bar{N}^2. \tag{9.7.12}$$

At $\tau = \infty$, however, the particles are uncorrelated, and therefore $K(\infty) = \bar{N}^2$. In other words, it is twice as probable to find two bosons in a thermal distribution at the same point as it is to find them far away from each other. This is known in optics as **photon bunching** (it also occurs for *phonons* in a Planck distribution.) A photon–photon correlation measurement is therefore a way to distinguish a coherent state from a thermal state.

The elevated correlation function will only persist for a range of τ comparable to the coherence time of the distribution. In a Planck distribution with a continuum of modes of different frequency ω, a measurement that sums over many modes ω will be a sum of uncorrelated sources, and therefore will have a Poisson distribution. For a measurement that integrates a given spectral range $\Delta\omega$, the correlation time is just given by the uncertainty relation

$$\tau_{\text{coh}} = \frac{1}{\Delta\omega}. \tag{9.7.13}$$

Number-squeezed light. Suppose that instead of a coherent light source, or a thermal light source, we have a single-photon emitter, for example a quantum dot that emits exactly one photon in a Fock state on demand (such light sources are now widely available). Then $\bar{N} = 1$, and the number uncertainty $\Delta N = 0$, and the correlation function gives

$$K(0) = 0. \tag{9.7.14}$$

This is known as **number-squeezed** light. Figure 9.16 gives an example of the correlation function for a series of single photons, as measured with a HBT apparatus. A Fermi field, for example electrons, will also exhibit this property (see, e.g., Baym 1969: ch. 19).

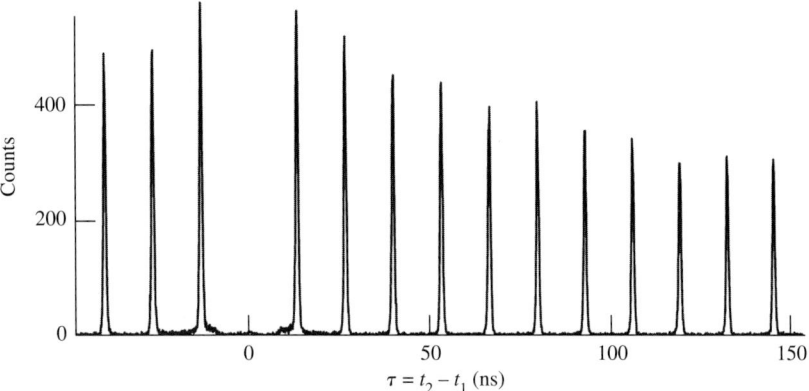

Fig. 9.16 The photon–photon correlation function for single photons emitted in a periodic sequence. The correlation function decreases at large τ because of fluctuations in the efficiency of the source. From Santori *et al.* (2002).

Number squeezing is impossible with a classical light field, even with random fluctuations. If we define the classical power $P(t) = |f(t)|^2$, then

$$(P(t) - P(t + \tau))^2 \geq 0, \tag{9.7.15}$$

since a square cannot be negative. The average of this value is

$$\overline{P^2(t)} - \overline{2P(t)P(t + \tau)} + \overline{P^2(t + \tau)} \geq 0, \tag{9.7.16}$$

but $\overline{P^2(t + \tau)} = \overline{P^2(t)} = K(0)$ and therefore

$$2K(0) - 2K(\tau) \geq 0 \tag{9.7.17}$$

or

$$K(\tau) \leq K(0) \tag{9.7.18}$$

for all τ, in contradiction with the result $K(0) = 0, K(\infty) = \bar{N}^2$, for single photons.

Figure 9.17 shows the predictions for the HBT measurement in the three cases of coherent, thermal, and number-squeezed waves.

Exercise 9.7.1   Show, using the same approach as for thermal bosons, that thermal fermions have $K(0) = 0$.

9.8 The Fluctuation–Dissipation Theorem

A very general theorem used in the study of correlations is the **fluctuation–dissipation theorem**. Like the Wiener–Khintchine theorem or the Kramers–Kronig relations, the fluctuation–dissipation theorem applies not only to the dielectric response function in

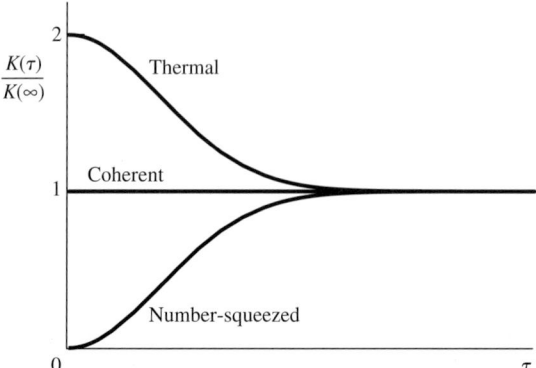

Fig. 9.17 Photon–photon correlation function for coherent, thermal (Planck) and number-squeezed fields.

optics, but to any response function. In general, if we consider an oscillator that can interact with an external system, we can on one hand determine the response of the oscillator to a driving force, or on the other hand we can determine the random fluctuations of the oscillator in the absence of a driving force. These are not independent of each other; the fluctuation–dissipation theorem gives us the relationship of the two.

Let us first determine the fluctuation spectrum. We define the interaction Hamiltonian

$$V_{\text{int}} = -xF, \tag{9.8.1}$$

where x is the generalized response of a system to a generalized driving force F; for example, as in our discussion of linear response theory in Section 7.4, x might be the polarization of a system and F the electric field that drives that polarization. The interaction Hamiltonian then has the form of a potential energy. In addition, we write

$$x = \chi_F F, \tag{9.8.2}$$

which defines the generalized susceptibility χ_F for the linear response of the system.

We use the definition of the correlation function in (9.6.5) of Section 9.6. For a state $|n\rangle$, this is

$$C_x(\tau) = \frac{1}{2} \langle n|x^\dagger(0)x(\tau) + x^\dagger(\tau)x(0)|n\rangle. \tag{9.8.3}$$

The corresponding spectral function is given by (9.6.9),

$$S_x(\omega) = \pi \sum_m |\langle m|x|n\rangle|^2 [\delta(\omega_{mn} + \omega) + \delta(\omega_{mn} - \omega)], \tag{9.8.4}$$

where $\omega_{mn} = (E_m - E_n)/\hbar$, and $|n\rangle$ and $|m\rangle$ are the eigenstates in the absence of the driving field; as in the Lehmann representation discussed in Section 8.14, we assume that these states exist even if we do not know what they are. We can then write the thermal average of the spectral function by weighting the probability of the occupation of the initial state $|n\rangle$:

$$\langle S_x(\omega)\rangle_T = \pi \frac{1}{Z} \sum_{m,n} e^{-E_n/k_B T} |\langle m|x|n\rangle|^2 [\delta(\omega_{mn} + \omega) + \delta(\omega_{mn} - \omega)]. \tag{9.8.5}$$

Because m and n are dummy variables, and x is Hermitian, we can equivalently write

$$\langle S_x(\omega)\rangle_T = \pi \frac{1}{Z} \sum_{m,n} \left(e^{-E_n/k_BT} + e^{-E_m/k_BT} \right) |\langle m|x|n\rangle|^2 \delta(\omega_{mn} - \omega)$$

$$= \pi \left(1 + e^{-\hbar\omega/k_BT} \right) \frac{1}{Z} \sum_{m,n} e^{-E_n/k_BT} |\langle m|x|n\rangle|^2 \delta(\omega_{mn} - \omega).$$

$$(9.8.6)$$

We now compute the response of the system to a driving force, namely the susceptibility. We can follow the procedure of Section 7.4 to deduce the susceptibility $\chi_F(\omega)$ for a driving force $F_0 \cos \omega t$, since the approach of that section is generally applicable to any two eigenstates coupled by a driving field. The analogous equation to (7.4.4) for the time evolution of the expectation value is

$$\langle x \rangle = \sum_m \langle \psi(t)|n\rangle \langle n|x|m\rangle \langle m|\psi(t)\rangle e^{-i\omega_{mn}t} + \text{c.c.},$$

$$(9.8.7)$$

where $|\psi(t)\rangle$ is the state the system evolves to after starting in state $|n\rangle$. Following the method of Section 7.4, we find to lowest order that $\langle \psi(t)|n\rangle \simeq 1$, while $\langle m|\psi(t)\rangle$ is obtained in the same way as (7.4.10), namely

$$\langle m|\psi(t)\rangle \simeq \frac{1}{2\hbar} \langle m|x|n\rangle F_0 \left(\frac{1}{\omega_{mn} + \omega - i\epsilon} e^{i(\omega_{mn}+\omega)t} + \frac{1}{\omega_{mn} - \omega - i\epsilon} e^{i(\omega_{mn}-\omega)t} \right).$$

$$(9.8.8)$$

Substituting this into (9.8.16), we have

$$\langle x \rangle \simeq \frac{1}{2\hbar} \sum_m |\langle m|x|n\rangle|^2 \left(\frac{F_0 e^{i\omega t}}{\omega_{mn} + \omega - i\epsilon} + \frac{F_0 e^{-i\omega t}}{\omega_{mn} - \omega - i\epsilon} \right) + \text{c.c.}$$

$$(9.8.9)$$

Averaging over all possible initial states weighted by their thermal probability, we have, for linear response,

$$\langle x \rangle_T = \frac{1}{2\hbar} \sum_{m,n} e^{-E_n/k_BT} |\langle m|x|n\rangle|^2 \left(\frac{F_0 e^{i\omega t}}{\omega_{mn} + \omega - i\epsilon} + \frac{F_0 e^{-i\omega t}}{\omega_{mn} - \omega - i\epsilon} \right) + \text{c.c.}$$

$$(9.8.10)$$

Grouping terms, and using the Dirac formula (7.2.2), we rewrite this as

$$\langle x \rangle_T = \frac{1}{\hbar} \sum_{m,n} e^{-E_n/k_BT} |\langle m|x|n\rangle|^2 F_0 \cos \omega t \left(\mathcal{P} \frac{1}{\omega_{mn} + \omega} + \mathcal{P} \frac{1}{\omega_{mn} - \omega} \right)$$

$$+ \frac{\pi}{\hbar} \sum_{m,n} e^{-E_n/k_BT} |\langle m|x|n\rangle|^2 F_0 \sin \omega t \left(\delta(\omega_{mn} - \omega) - \delta(\omega_{mn} + \omega) \right).$$

$$(9.8.11)$$

As discussed in Section 7.4, the term multiplying the $F_0 \sin \omega t$ term corresponds to the imaginary susceptibility, as defined for a classical oscillator, so that we can write

$$\langle \text{Im } \chi_F \rangle_T = \frac{\pi}{\hbar} \frac{1}{Z} \sum_{m,n} e^{-E_n/k_B T} |\langle m|x|n \rangle|^2 \left(\delta(\omega_{mn} - \omega) - \delta(\omega_{mn} + \omega) \right).$$

(9.8.12)

Since m and n are dummy variables, this can be rewritten as

$$\langle \text{Im } \chi_F \rangle_T = \frac{\pi}{\hbar} \frac{1}{Z} \sum_{m,n} \left(e^{-E_n/k_B T} - e^{-E_m/k_B T} \right) |\langle m|x|n \rangle|^2 \delta(\omega_{mn} - \omega)$$

$$= \frac{\pi}{\hbar} \left(1 - e^{-\hbar\omega/k_B T} \right) \frac{1}{Z} \sum_{m,n} e^{-E_n/k_B T} |\langle m|x|n \rangle|^2 \delta(\omega_{mn} - \omega).$$

(9.8.13)

Comparing this to (9.8.6), we see that the two are the same except for the thermal factors that depend only on ω. Computing the ratio, we are left with

$$\boxed{\langle S_x(\omega) \rangle_T = \hbar(2N(\omega) + 1)\langle \text{Im } \chi_F \rangle_T,}$$

(9.8.14)

where

$$N(\omega) = \frac{1}{e^{\hbar\omega/k_B T} - 1}$$

(9.8.15)

is the standard occupation number for a boson mode with frequency ω. Equation (9.8.14) is the quantum mechanical **fluctuation–dissipation theorem**. We have seen the $(2N(\omega) + 1)$ term appear before for the total rate of emission and absorption of excitations, such as in the Debye–Waller factor derived in Section 4.10. In the high-temperature limit, $N(\omega) \to k_B T/\hbar\omega \gg 1$, so the fluctuation–dissipation relation becomes

$$\boxed{\langle S_x(\omega) \rangle_T = \frac{2k_B T}{\omega} \langle \text{Im } \chi_F \rangle_T.}$$

(9.8.16)

As we saw in Section 7.4, the imaginary susceptibility corresponds to dissipation. We thus have connection between the dissipation under a driving force, given by the imaginary susceptibility, and the average amplitude of fluctuations, given by the spectral function.

Exercise 9.8.1 Show that the ratio of the thermal factors in (9.8.6) and (9.8.13) give the result (9.8.14).

Exercise 9.8.2 Show explicitly that the fluctuation–dissipation theorem applies to the case of the two-level oscillator driven by an electric field, discussed in Section 7.4, using the susceptibility defined in that section and the correlation function of the polarization P.

Onsager relations. The same approach can be generalized to the case of multiple forces in a system. In general, assuming linear response, we can write

$$\langle x_i \rangle_T = \sum_j \chi_{ij} \langle F_j \rangle_T, \qquad (9.8.17)$$

for some set of forces F_j proportional to the operators x_j.

We can then write for the response x_i to a force F_j, following the same logic that led us to (9.8.10),

$$
\begin{aligned}
\langle x_i \rangle_T = {} & \frac{1}{2\hbar} \sum_{m,n} e^{-E_n/k_B T} \langle n|x_i|m\rangle \langle m|x_j|n\rangle \left(\frac{F_j e^{i\omega t}}{\omega_{mn} + \omega - i\epsilon} + \frac{F_j e^{-i\omega t}}{\omega_{mn} - \omega - i\epsilon} \right) \\
& + \frac{1}{2\hbar} \sum_{m,n} e^{-E_n/k_B T} \langle m|x_i|n\rangle \langle n|x_j|m\rangle \left(\frac{F_j e^{-i\omega t}}{\omega_{mn} + \omega + i\epsilon} + \frac{F_j e^{i\omega t}}{\omega_{mn} - \omega + i\epsilon} \right),
\end{aligned}
$$

$$(9.8.18)$$

where we have written out the complex conjugate explicitly. For a conservative system with time-reversal symmetry, we must obtain the same expectation value if we reverse time. Following the same approach, we have for the time-reversed expectation value (note that we must also reverse the adiabatic turn-on invoked in Section 7.4 and use a factor $e^{-\epsilon t}$),

$$
\begin{aligned}
\langle x_i \rangle_T = {} & \frac{1}{2\hbar} \sum_{m,n} e^{-E_n/k_B T} \langle n|x_i|m\rangle \langle m|x_j|n\rangle \left(\frac{F_j e^{-i\omega t}}{\omega_{mn} + \omega + i\epsilon} + \frac{F_j e^{i\omega t}}{\omega_{mn} - \omega + i\epsilon} \right) \\
& + \frac{1}{2\hbar} \sum_{m,n} e^{-E_n/k_B T} \langle m|x_i|n\rangle \langle n|x_j|m\rangle \left(\frac{F_j e^{i\omega t}}{\omega_{mn} + \omega - i\epsilon} + \frac{F_j e^{-i\omega t}}{\omega_{mn} - \omega - i\epsilon} \right).
\end{aligned}
$$

$$(9.8.19)$$

On the other hand, the response x_j to a force F_i is found by switching i and j in (9.8.18):

$$
\begin{aligned}
\langle x_j \rangle_T = {} & \frac{1}{2\hbar} \sum_{m,n} e^{-E_n/k_B T} \langle n|x_j|m\rangle \langle m|x_i|n\rangle \left(\frac{F_i e^{i\omega t}}{\omega_{mn} + \omega - i\epsilon} + \frac{F_i e^{-i\omega t}}{\omega_{mn} - \omega - i\epsilon} \right) \\
& + \frac{1}{2\hbar} \sum_{m,n} e^{-E_n/k_B T} \langle m|x_j|n\rangle \langle n|x_i|m\rangle \left(\frac{F_i e^{-i\omega t}}{\omega_{mn} + \omega + i\epsilon} + \frac{F_i e^{i\omega t}}{\omega_{mn} - \omega + i\epsilon} \right).
\end{aligned}
$$

$$(9.8.20)$$

It is easy to see that (9.8.19) and (9.8.20) are the same except for the switch of F_j to F_i. We therefore have, for the linear susceptibility matrix,

$$\boxed{\chi_{ji} = \chi_{ij}.} \qquad (9.8.21)$$

This is the general form of the **Onsager relation**, which applies to any set of linear responses to forces with time-reversal invariance. For the case of a force that switches sign on time reversal, such as a magnetic field, we must reverse the sign of the response that occurs in (9.8.19), which then implies $\chi_{ji} = -\chi_{ij}$, as discussed in Section 5.10.

9.9 Current Fluctuations and the Nyquist Formula

As discussed in Section 9.8, the analysis of the fluctuation–dissipation theorem applies to all kinds of systems. In the context of electronic circuits, thermal fluctuations are often called **Johnson noise**.

A related theorem is known as the **Nyquist** formula. There are many versions of this formula, but the essence is that the conductivity is proportional to the current–current correlation function. We can see this easily as follows. We recall from Section 3.5.1 that the current density is given by

$$\vec{J} = \frac{\partial \vec{P}}{\partial t}, \tag{9.9.1}$$

where $\vec{P}$ is the polarization, and that

$$\vec{P} = \chi \epsilon_0 \vec{E} \tag{9.9.2}$$

in the linear response approximation. We define the conductivity σ by the relation

$$\vec{J} = \sigma \vec{E}. \tag{9.9.3}$$

Then if $\vec{E} = \vec{E}_0 e^{-i\omega t}$, we can equate

$$\vec{J} = -i\omega \chi \epsilon_0 \vec{E} = \sigma \vec{E} \tag{9.9.4}$$

and therefore

$$\sigma = -i\omega \chi \epsilon_0, \tag{9.9.5}$$

or in particular,

$$\text{Re } \sigma = \omega \text{ Im } \chi(\omega)\epsilon_0. \tag{9.9.6}$$

From the fluctuation–dissipation theorem, we can relate the imaginary susceptibility to the fluctuation spectrum. We first define the current–current fluctuation spectrum,

$$S_J(\omega) = \int_{-\infty}^{\infty} d\tau \, e^{-i\omega\tau} C_J(\tau)$$
$$= \omega^2 \int_{-\infty}^{\infty} d\tau \, e^{-i\omega\tau} C_P(\tau) = \omega^2 S_P(\omega), \tag{9.9.7}$$

since $\vec{J} = i\omega \vec{P}$. Substituting this into the fluctuation–dissipation theorem (9.8.14), with $x \to PV$, $F \to E$, and $\chi_F \to \epsilon_0 \chi V$, where V is the volume, we obtain the Nyquist formula

$$\boxed{\text{Re } \sigma = \frac{V}{\hbar\omega} S_J(\omega) \frac{1}{2N(\omega) + 1},} \tag{9.9.8}$$

or, in the classical limit,

$$\text{Re } \sigma = \frac{V}{2k_B T} S_J(\omega). \tag{9.9.9}$$

$$R = \frac{L}{\sigma A},$$

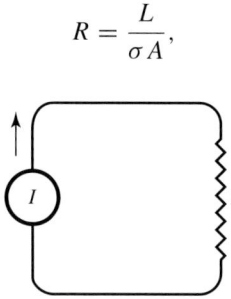

Model electrical circuit.

Again, this is expressing the same fact as the fluctuation–dissipation theorem, namely, that the response of a system to a driving field is controlled by the same physics as the fluctuations of the system in response to a random, thermal field.

Example: a simple electrical circuit. We can illustrate these concepts with a simple model of a classical electrical circuit, shown in Figure 9.18. The thermal fluctuations can be viewed as a random current source. The total current is $I = AJ$, and therefore $S_I(\omega) = A^2 S_J(\omega)$, where A is the cross-sectional area of the resistor, and the total resistance is

$$R = \frac{L}{\sigma A}, \tag{9.9.10}$$

where L is the length, and $V = AL$. The Nyquist formula (9.9.8) then gives

$$\frac{1}{R} = \frac{1}{\hbar\omega} \frac{1}{2N(\omega) + 1} S_I(\omega). \tag{9.9.11}$$

In the high-temperature limit, the Planck occupation number $N(\omega) \rightarrow k_B T / \hbar\omega$. Per definition (9.5.16), the spectral density has units of amplitude squared per unit frequency. Therefore, in the low-frequency limit $\omega \rightarrow 0$, we can rearrange to write

$$\frac{1}{2} R |I(f)|^2 = k_B T \Delta f, \tag{9.9.12}$$

where Δf is the frequency range of the measurement. The time-averaged power dissipation in the resistor is $\frac{1}{2} I^2 R$; we see, not surprisingly, that this is equal to the power expected for the Planck distribution for a one-dimensional system.

Exercise 9.9.1 Calculate the average power of Johnson noise in watts in the frequency range 100–110 MHz at $T = 300$ K, in a circuit with $R = 50\ \Omega$.

Exercise 9.9.2 Show that the result (9.9.12) is the same as expected for a one-dimensional Planck distribution, following the logic of Section 4.9.2 using a one-dimensional density of states.

9.10 The Kubo Formula and Many-Body Theory of Conductivity

We have already seen in Section 9.6 that correlation functions can be connected to the Green's functions of many-body theory. In this section, we will do an extended application of many-body correlation functions to the theory of metals. On one hand, it may seem that we could do this more easily using the quantum Boltzmann equation approach of Section 4.8. The Boltzmann equation approach, or Fermi's golden rule approach, will break down, however, when the wavelength of the particles becomes comparable to their mean free path as determined in the standard Boltzmann scattering theory. When that happens, we must use many-body theory.

Let us start from scratch, defining the Hamiltonian as we did for the linear response theory of Section 7.4. Recall that this linear response theory was then used in Section 9.8 to generate the fluctuation–dissipation theorem, to connect the susceptibility function to the correlation function for fluctuations. Here we will follow the same approach, but instead of using two-level oscillators as our model, we will treat free electrons. For more detail, see Doniach and Sondheimer (1998: ch. 5), and Mahan (2000: s. 3.8). One of the reasons for going through this derivation is to see what the correlation functions look like when we include spatial dependence (i.e., k-dependence).

In an electromagnetic field, the real velocity of the electrons is related to their total momentum $p = -i\hbar\nabla$ by the relation

$$m\vec{v} = \vec{p} - e\vec{A}, \tag{9.10.1}$$

where v is the velocity. The measured current density is therefore equal to

$$\vec{J} = \frac{e}{mV}\left(\vec{p} - e\vec{A}\right) \equiv \vec{j} - \frac{e^2}{mV}\vec{A}, \tag{9.10.2}$$

where the small $\vec{j}$ is the current derived from the quantum mechanical p operator, sometimes called the paramagnetic current. The quantum mechanical current is then defined in terms of the standard field operators as the real part of the expectation value,

$$\vec{J}(\vec{x}) = \text{Re}\,\Psi^{\dagger}(\vec{x})\left(-\frac{i\hbar e}{m}\nabla - \frac{e^2}{mV}\vec{A}\right)\Psi(\vec{x}), \tag{9.10.3}$$

where $\Psi(\vec{x})$ is the field operator defined in Section 4.6, (ignoring the Bloch function effects)

$$\Psi(\vec{x}) = \frac{1}{\sqrt{V}}\sum_{\vec{k}}e^{i\vec{k}\cdot\vec{x}}b_{\vec{k}}. \tag{9.10.4}$$

Substituting the definition of the field operators into the first terms, we obtain

$$\begin{aligned}\vec{j}(\vec{x}) &= -\frac{i\hbar e}{m}\left(\Psi^{\dagger}(\vec{x})\nabla\Psi(\vec{x}) - \nabla\Psi^{\dagger}(\vec{x})\Psi(\vec{x})\right) \\ &= \frac{\hbar e}{mV}\sum_{\vec{k},\vec{q}}e^{-i\vec{q}\cdot\vec{x}}\left(\vec{k} + \frac{\vec{q}}{2}\right)b^{\dagger}_{\vec{k}+\vec{q}}b_{\vec{k}},\end{aligned} \tag{9.10.5}$$

or

$$\vec{j}(\vec{x}) = \sum_{\vec{q}} e^{-i\vec{q}\cdot\vec{x}} \vec{j}_{\vec{q}}, \tag{9.10.6}$$

with

$$\vec{j}_{\vec{q}} = \frac{\hbar e}{mV} \sum_{\vec{k}} \left(\vec{k} + \frac{\vec{q}}{2}\right) b_{\vec{k}+\vec{q}}^{\dagger} b_{\vec{k}}. \tag{9.10.7}$$

In the long wavelength limit ($q \to 0$), this becomes simply

$$\vec{j}_0 = \frac{e}{mV} \sum_{\vec{k}} \hbar\vec{k}\, b_{\vec{k}}^{\dagger} b_{\vec{k}}, \tag{9.10.8}$$

that is, the total momentum $\vec{p}$ is just the sum of the momenta of the electrons.

The interaction Hamiltonian for electrons in the presence of electromagnetic field, which we have used many times (e.g., in Section 5.2), is $-(e/m)\vec{p}\cdot\vec{A}$. Using our definition of $\vec{j}$, this is the same as

$$V_{\text{int}} = -\int d^3x\, \vec{A}(\vec{x}) \cdot \vec{j}(\vec{x}). \tag{9.10.9}$$

Writing $\vec{A} = \vec{A}_0 e^{i(\vec{q}\cdot\vec{x}-\omega t)}$, this becomes

$$V_{\text{int}} = -V\vec{A}_0 \cdot \vec{j}_{\vec{q}}\, e^{-i\omega t}, \tag{9.10.10}$$

where we have used the identity (see Appendix C)

$$\frac{1}{(2\pi)^3} \int d^3x\, e^{i\vec{k}\cdot\vec{x}} = \delta^3(\vec{k}) \tag{9.10.11}$$

and the standard substitution

$$\sum_{\vec{k}} \to \frac{V}{(2\pi)^3} \int d^3k. \tag{9.10.12}$$

The Hamiltonian (9.10.10) is not Hermitian, but it will not matter, since we are just using complex wave notation to keep track of the phase.

Writing the initial state as $|\psi(t_0)\rangle$, the state of the system at time t is

$$|\psi(t)\rangle = |\psi(t_0)\rangle + \frac{1}{i\hbar} \int_{t_0}^{t} dt'\, V_{\text{int}}(t')|\psi(t_0)\rangle. \tag{9.10.13}$$

We assume that the current in the initial state is zero. The average current response at time t is therefore, to first order in V_{int},

$$\langle \vec{J}(\vec{x},t)\rangle = \langle\psi(t)|\vec{j}(\vec{x},t)|\psi(t)\rangle - \Psi^{\dagger}(\vec{x})\Psi(\vec{x})\frac{e^2}{mV}\vec{A}(\vec{x},t) \tag{9.10.14}$$

$$= \frac{1}{i\hbar} \int_{t_0}^{t} dt'\, \langle\psi(t_0)|[\vec{j}(\vec{x},t), V_{\text{int}}(t')]|\psi(t_0)\rangle - n(\vec{x})\frac{e^2}{m}\vec{A}(\vec{x},t),$$

where we have defined the local density $n(\vec{x}) = \langle\Psi^{\dagger}(\vec{x})\Psi(\vec{x})\rangle/V$. The commutator $[\vec{j}, V_{\text{int}}]$ arises from keeping the first-order time-dependent terms in both $|\psi(t)\rangle$ and $\langle\psi(t)|$.

As we did in Section 7.4, we can view the response of the system to a driving field as the current that occurs after a driving field has been turned on a long time earlier, so that transients have all died out. We therefore set $t_0 = -\infty$. Writing $J = \sigma E$, and $E = -\partial A/\partial t$ in the Coulomb gauge, and using (9.10.10) for V_{int}, we obtain

$$\sigma(\vec{x}) = \frac{V}{\hbar\omega} \int_{-\infty}^{t} dt' \; e^{-i\vec{q}\cdot\vec{x}} e^{i\omega(t-t')} \langle [j(\vec{x},t), j_{\vec{q}}(t')] \rangle - \frac{e^2 n(\vec{x})}{im\omega} \tag{9.10.15}$$

$$= \frac{V}{\hbar\omega} \sum_{\vec{q}'} e^{-i(\vec{q}+\vec{q}')\cdot\vec{x}} \int_{-\infty}^{t} dt' \; e^{i\omega(t-t')} \langle [j_{\vec{q}'}(t), j_{\vec{q}}(t')] \rangle - \frac{e^2 n(\vec{x})}{im\omega},$$

where we have dropped the vector notation for the current; that is, we restrict ourselves to $j = (\vec{j}\cdot\vec{A}_0)/|A_0|$, the magnitude of the current in the direction of the electric field. We then average over all $\vec{x}$; that is, we integrate $\int d^3x$ and divide by V, which gives a δ-function that eliminates the summation over $\vec{q}'$. For very large $t - t_0$, we can also take the upper bound as $t \to \infty$. We therefore finally have

$$\sigma = \frac{V}{\hbar\omega} \int_{-\infty}^{\infty} dt' \; e^{i\omega(t-t')} \langle [j_{-\vec{q}}(t), j_{\vec{q}}(t')] \rangle \Theta(t - t') - \frac{e^2 n}{im\omega}. \tag{9.10.16}$$

This is the **Kubo** formula. We thus have an equation for the linear response of the system in terms of a spectral density function, that is, the Fourier transform of the current–current correlation function, just as in the Nyquist theorem (which is simply another version of the fluctuation–dissipation theorem). If it were not for the second term on the right-hand side, this would just be another version of the Nyquist theorem, with the spatial q-dependence included.

The extra, imaginary term arises from the electromagnetic momentum term proportional to $\vec{A}$ in (9.10.2), which is necessary because we are concerned about a transport measurement, where the velocity $\vec{v}$ matters, not the dipole moment. The fact that the extra term depends on frequency as $1/\omega$ would seem to indicate that the conductivity approaches infinity in the DC limit. This is not the case, because the imaginary part of the first term will exactly cancel this term at $\omega = 0$, as we will see below.

In Section 9.6, we showed that the density–density correlation function can be expressed in terms of the many-body Green's functions of a system. We can do the same for the current–current correlation function. Substituting for $j_{\vec{q}}$ from (9.10.7), we have

$$\sigma = \frac{\hbar e^2}{m^2\omega V} \int_{-\infty}^{\infty} dt' \; e^{i\omega(t-t')} \Theta(t - t')$$

$$\times \sum_{\vec{k},\vec{k}'} (k'_x - \frac{1}{2}q'_x)(k_x + \frac{1}{2}q_x) \langle [b_{\vec{k}'-\vec{q}}^{\dagger}(t)b_{\vec{k}'}(t), b_{\vec{k}+\vec{q}}^{\dagger}(t')b_{\vec{k}}(t')] \rangle$$

$$- \frac{e^2 n}{im\omega}, \tag{9.10.17}$$

where we assume the direction of the electric field, and therefore the current, is along the $\hat{x}$ direction. The actions of the creation and destruction operators eliminate the summation variable $\vec{k}'$ since $\vec{k}' = \vec{k} + \vec{q}$. Then we have

$$\sigma = \frac{\hbar e^2}{m^2 \omega V} \sum_{\vec{k}} \left(k_x + \frac{1}{2} q_x \right)^2 \int_{-\infty}^{\infty} dt' e^{i\omega(t-t')} \Theta(t - t') \langle [b_{\vec{k}}^{\dagger}(t) b_{\vec{k}+\vec{q}}(t), b_{\vec{k}+\vec{q}}^{\dagger}(t') b_{\vec{k}}(t')] \rangle$$

$$- \frac{e^2 n}{im\omega}. \tag{9.10.18}$$

Using the definition of the fermion anticommutator in (8.6.13), the time integral in (9.10.18) is easily performed, to give

$$i \frac{N_{\vec{k}} - N_{\vec{k}+\vec{q}}}{\omega + \omega_{\vec{k}} - \omega_{\vec{k}+\vec{q}} + i\gamma}, \tag{9.10.19}$$

where $\omega_{\vec{k}} = E_{\vec{k}}/\hbar$ and $E_{\vec{k}}$ is the single-particle energy of state $\vec{k}$. In the case of a non-interacting gas, the imaginary term $i\gamma$ is infinitesimal, introduced in Section 8.6 in the definition of the Green's function. In the case of a renormalized Green's function for an interacting system, however, the imaginary term can be a finite self-energy, corresponding to i/τ, where τ is the characteristic scattering time.

The term (9.10.19) is equivalent to

$$i \frac{(\omega + \Delta\omega_q)\Delta N_q}{(\omega + \Delta\omega_q)^2 + \gamma^2} + \frac{\gamma \Delta N_q}{(\omega + \Delta\omega_q)^2 + \gamma^2}. \tag{9.10.20}$$

Let us focus on the first, imaginary term. Multiplying by the $1/\omega$ that appears in the prefactor of the spectral density function in (9.10.18), we have

$$\frac{1}{\omega} \frac{(\omega + \Delta\omega_q)\Delta N_q}{(\omega + \Delta\omega_q)^2 + \gamma^2}. \tag{9.10.21}$$

We must be careful how we evaluate this function since we want the DC limit $\omega \to 0$, the long wavelength limit $q \to 0$, which implies $\Delta\omega \to 0$, and also the weak scattering limit, $\gamma \ll \Delta\omega$, which implies $\gamma \to 0$. We first take the limit $\omega \to 0$, since the function has its maximum there, and take γ as negligible compared to $\Delta\omega$, which gives us

$$\frac{1}{\omega} \frac{\Delta N_q}{\Delta\omega_q}. \tag{9.10.22}$$

Finally, taking the long wavelength limit $q \to 0$, we obtain

$$\text{Im}\, \sigma = \frac{\hbar^2 e^2}{m^2 \omega V} \sum_{\vec{k}} k_x^2 \frac{\partial f}{\partial E_{\vec{k}}} + \frac{e^2 n}{m\omega}, \tag{9.10.23}$$

where we have treated the occupation number $N_{\vec{k}}$ as a continuous function $f(E)$. For a metal with a Fermi level, the derivative of the occupation number at the Fermi level is a δ-function. The first term therefore equals

$$-\frac{\hbar^2 e^2}{m^2 \omega V} \sum_{\vec{k}} k_x^2 \, \delta(E_F - E_{\vec{k}}) = -\frac{\hbar^2 e^2}{m^2 \omega V} \frac{k_F^2}{3} \mathcal{D}(E_F)$$

$$= -\frac{e^2 n}{m \omega}, \tag{9.10.24}$$

where we used the definitions of k_F and the density of states for a Fermi gas of Section 2.4.1. (We neglect spin; if we took into account spin, we would have to sum over spin as well as k in all the above.) This exactly cancels the second term, as we predicted above.

The real part of the conductivity is proportional to

$$\frac{1}{\omega} \frac{\gamma \, \Delta N_q}{(\omega + \Delta \omega_q)^2 + \gamma^2}. \tag{9.10.25}$$

In the limit $\gamma \to 0$, the term (9.10.25) is strongly peaked at $\omega = -\Delta \omega_q$; the Lorentzian $\gamma/((\omega + \Delta \omega_q)^2 + \gamma^2)$ approaches $\delta(\omega + \Delta \omega_q)$. Setting $\omega = -\Delta \omega_q$, this term then becomes, in the long wavelength limit $q \to 0$,

$$-\frac{\partial f}{\partial E_{\vec{k}}} \frac{1}{\gamma}. \tag{9.10.26}$$

It is easy to show, using the same integral as occurred in the imaginary part, that the conductivity is then

$$\operatorname{Re} \sigma = \frac{e^2 n \tau}{m}, \tag{9.10.27}$$

which is just the same as the result (5.8.6) from Section 5.8, that is, the Drude formula. This formula is valid when the dephasing time τ is long enough that γ is much less than the characteristic energy range of variation of the occupation number, $\Delta \omega_q$. This is the same condition of validity as (4.7.20), for Fermi's golden rule; in other words, the condition of validity for the relaxation time approximation to be valid, which we used in Section 5.8 in deriving the Drude formula.

In a sense, then, we have not learned anything from the Kubo formalism that we did not already know before from making the relaxation time approximation. The Kubo formalism, however, has allowed us to connect the conductivity to the more sophisticated Green's function formalism, which we can use to take into account many-body interactions. To see this, we note that the four-operator density-density correlation function in (9.10.18) can be written as

$$C_{\vec{q}}(t) = i \sum_{\vec{k}} \left(G_{\vec{k}+\vec{q}}^2(t) N_{\vec{k}} e^{i \omega_{\vec{k}} t} + G_{\vec{k}}^*(t) N_{\vec{k}+\vec{q}} e^{-i \omega_{\vec{k}+\vec{q}} t} \right), \tag{9.10.28}$$

where the $N_{\vec{k}}$ and $e^{i \omega_{\vec{k}} t}$ terms come from the un-contracted operators acting on the initial state. This corresponds to the spectral density function

$$S_{\vec{q}}(\omega) = i \sum_{\vec{k}} \left(N_{\vec{k}} G(\vec{k} + \vec{q}, \omega + \omega_{\vec{k}}) + N_{\vec{k}+\vec{q}} G^*(\vec{k}, \omega_{\vec{k}+\vec{q}} - \omega) \right), \tag{9.10.29}$$

which has the same form as the vertex correction from external legs used in the screening calculation of Section 8.11. The Lindhard formula (8.11.10) that arises from that analysis is the same as (9.10.19) above.

An important example of many-body corrections occurs at low temperature. As discussed in Section 5.8, we might expect that all metals become extremely good conductors in the low temperature limit, because τ from electron–phonon scattering approaches infinity as $T \to 0$. In fact, in two-dimensional systems with disorder, one can prove that all electron gases become insulating. This is known as Anderson localization. We will discuss this further in the next section.

Exercise 9.10.1 Prove, using the fermion anticommutator (8.6.13), that (9.10.19) is the result of the time integral in (9.10.18).

Exercise 9.10.2 Prove the formula (9.10.27) by taking the real part of the conductivity.

Exercise 9.10.3 (a) Show that the formula (9.10.23) also gives the same result (9.10.24) for a Maxwell–Boltzmann distribution $f(E) = f_0 e^{-E/k_B T}$.

9.11 Mesoscopic Effects

As discussed in the introduction to this chapter, there are several different classes of coherence. One example is a coherent state of bosons, which is effectively a classical wave. This type of coherence was discussed in Section 4.4. Another is the coherence of an ensemble of different quantum oscillators, as discussed in Sections 9.2–9.4.

Another type of coherence is the wave function of a single electron (or other quasiparticle). In Chapter 5, we discussed electron transport in terms of scattering events with probability determined by Fermi's golden rule. In other words, we ignored the phase of the electrons, keeping track only of the number of electrons in a given eigenstate. This is reasonable when there is a strong inelastic scattering rate, which means that phase is not a good observable. We distinguish here between *elastic* scattering, such as Rayleigh scattering from defects discussed in Section 5.3, and *inelastic* scattering, such as electron–phonon scattering discussed in Section 5.1. Elastic scattering does not intrinsically destroy phase coherence. In principle, one can write down a total Hamiltonian for the potential energy landscape felt by an electron, including all the elastic scattering centers, and solve this exactly for the eigenstates. These eigenstates would not be plane waves, and therefore if we tried to describe the system in terms of plane waves, we would have to use a perturbation theory and talk of random transitions between plane wave states, but the eigenstates of a system with static disorder are nevertheless time-invariant and lead to no loss of quantum information.

Inelastic scattering, in particular scattering of electrons with a Planck distribution of phonons, does destroy phase coherence. At low temperature, however, the number of phonons approaches zero, and therefore the coherence of the electron wave function can extend over significant distances.

The crossover between the regime of simple scattering and the regime of coherent wave propagation occurs when the thermal wavelength of the electrons becomes longer than the mean free path for inelastic scattering. The mean free path, defined in Section 5.6, is

$$l = \bar{v}\tau, \tag{9.11.1}$$

where $\bar{v}$ is the root-mean-square velocity and τ is the average scattering time by inelastic processes, determined by Fermi's golden rule. The condition $\lambda > l$ is actually the same as the condition for the breakdown of Fermi's golden rule, discussed in the previous section and in Section 4.7. The characteristic energy range of the system cannot be larger than $k_B T$; therefore the condition for the breakdown of Fermi's golden rule, from (4.7.20), is

$$\begin{aligned}
\tau &\sim \frac{\hbar}{\Delta E} \\
&\sim \frac{\hbar}{k_B T} \\
&\sim \frac{\hbar}{\frac{1}{2}mv^2}
\end{aligned} \tag{9.11.2}$$

or

$$v\tau \sim \frac{\hbar}{p} \sim \lambda. \tag{9.11.3}$$

If the system is small enough, the mean free path for inelastic scattering can be larger than the size of the system, in which case the condition for coherent effects to be important is $\lambda > L$, where L is the system size. At $T = 4$ K, the thermal de Broglie wavelength of an electron is around 100 nm, while the mean free path for inelastic scattering is comparable to or larger than this, using the typical time scale for electron–phonon scattering deduced in Section 5.1.4. If the length scale of the system under study is smaller than this, that is, of the order of hundreds of nanometers, then coherent effects will be important. Such a system is called *mesoscopic* – in the middle between microscopic and macroscopic. Fabrication of devices with hundred-nanometer length scales is now routine in solid state physics. Coherent effects are also important in larger systems when the temperature is lower. For example, at temperatures of 100 mK, easily obtainable in a modern lab, the mean free path for inelastic scattering can be hundreds of microns.

Anderson localization. As discussed in Section 9.10, at low temperature in the presence of disorder, an electron gas in two dimensions becomes an insulator, despite the fact that the electron–phonon scattering rate goes to zero and there is no energy gap. The scattering of the electrons from the disordered potential leads to confinement.

Consider a system with defects as shown in Figure 9.19. We can break all paths of the scattering electrons into two categories: those that start at a point $\bar{x}$ and do not return, and those that, by multiple scattering, return to the same point. Note that for every path that brings the electron back to the same point, there is another path that is exactly the same except it is time-reversed. These two paths will interfere constructively. Closed paths, in other words, paths that do not contribute to transport, have twice the relative weight in the electron transport calculation as open paths.

How important is the contribution from closed paths? To estimate this, let us think of the electron path as a tube, as illustrated in Figure 9.20, following the useful introduction of Hanke and Kopaev (1992). The length of the tube is given by the characteristic velocity times the time duration of the electron's motion. For an electron gas with a Fermi level, the

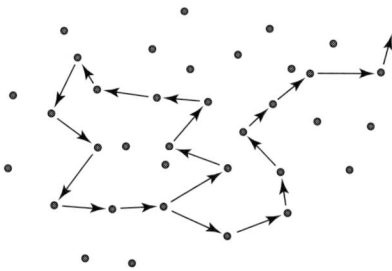

Fig. 9.19   Two types of paths in a system with disorder. Closed paths do not contribute to conductivity.

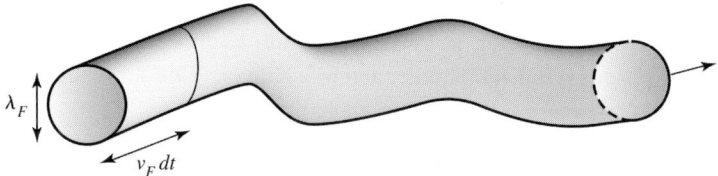

Fig. 9.20   The tube that gives the coherence volume of an electron path.

characteristic velocity is $v_F = p_F/m$, which depends only on the density of electrons, as discussed in Section 2.4.1.

If the particle were classical, the width of the tube would be zero – the electron would have an exactly defined path. Quantum mechanically, however, we cannot define the position of the electron on length scales less than its wavelength. We can therefore approximate the radius of the tube as given by the wavelength, which for a Fermi gas is $\lambda_F = h/p_F$, which also depends only on the electron density. The volume of the tube is therefore, in a d-dimensional system, in a time interval dt,

$$dV_{\text{coh}} = v_F dt \left(\frac{h}{mv_F} \right)^{d-1}. \tag{9.11.4}$$

The volume of this tube defines the region in which the wave nature of the electron is important – in which interference effects play a role.

On the other hand, we know from the considerations of Section 5.6 that a randomly scattering particle undergoes diffusion with a characteristic length $l = (Dt)^{1/2}$, where $D = \frac{1}{3}\bar{v}^2\tau = \frac{1}{3}\bar{v}l$. For elastic Rayleigh scattering, l depends only on the properties of the disorder potential and not on the temperature. The **diffusion volume** $V_{\text{diff}} = (Dt)^{d/2}$ is the total volume in d dimensions explored by the electron path within a time interval.

The number of times the tube crosses itself within an interval of time can therefore be estimated as the ratio of the volume of the tube to the total diffusion volume. This is equal to

$$N = \int_{\tau}^{\tau_{\text{in}}} \frac{dV_{\text{coh}}}{V_{\text{diff}}} = (\text{const.}) \int_{\tau}^{\tau_{\text{in}}} \frac{dt}{t^{d/2}}, \tag{9.11.5}$$

where the upper bound τ_{in} is the shortest inelastic scattering time (which destroys the phase coherence) and minimum time τ is the characteristic elastic scattering time.

In two dimensions, this integral is proportional to $\log(\tau_{in}/\tau)$. Therefore, when $\tau_{in} \to \infty$ as $T \to 0$, the number of closed loops increases, and therefore the conductivity decreases by an amount proportional to this number. Eventually the coherent volume is much larger than the diffusion volume and the electron cannot be said to diffuse at all; it is localized. This is known as **Anderson localization**. In this case, the integral is proportional to $\tau_{in}^{-1/2}$, which decreases to zero as $\tau_{in} \to \infty$. In three dimensions, there are many more ways for the tube to miss itself than in two dimensions. Detailed theory predicts that in three dimensions, Anderson localization will occur only for electrons with energies below a certain cutoff energy known as the **mobility edge**. See Abrahams (1979), the well-known "gang of four" paper.

A great deal of theoretical work has been aimed at understanding the conductor–insulator transition at low temperature. Useful review articles are (those by Hanke and Kopaev (1992) and Kramer and MacKinnon (1993). Localized states occur for any type of wave in a disordered system, including light.

Exercise 9.11.1   It is not hard, with modern computing power, to solve the time-dependent Schrödinger equation in one dimension using an iterative approach. We define the wave function $\psi(x,t)$ on a set of discrete points x separated by distance dx, and update $\psi(x,t)$ for a series of time steps dt according to the rule

$$\psi(x, t+dt) = \psi(x,t) + \frac{i}{\hbar} \left[\frac{\hbar^2}{2m} \frac{\psi(x-dx, t) - 2\psi(x,t) + \psi(x+dx, t)}{(dx)^2} \right.$$
$$\left. +V(x)\psi(x,t) \right] dt.$$

Use a program like Mathematica or MATLAB to define $\psi(x,t)$ on a array of several hundred x points. To simulate an infinite system, use the periodic boundary condition $\psi(L,t) = \psi(0,t)$, where $L = Ndx$ is the length of the x region. Start the simulation with $\psi(x,0) = e^{-x^2/l^2}$, where $l \ll L$, and $V = 0$, and pick a time interval dt small enough to see the evolution of the wave from a Gaussian peak to a constant spread through all space, as shown in Figure 9.21(a). For convenience, set $\hbar = m = 1$.

Next, let $V(x)$ be a random field, in which the value at each x is chosen randomly, independent of the value at any other x. Evolve the wave function in this potential. You should see that, consistent with the discussion of Anderson localization in this section, the wave function does not spread out evenly, but remains bunched near $x = 0$ at all later times, as shown in Figure 9.21(b).

Berry's phase. Much of the discussion of coherent effects in mesoscopic systems revolves around the concept of **Berry's phase**. This is a general concept which applies to all kinds of systems. Berry's original paper (Berry 1984) is an elegant and easy-to-read review. We can see how Berry's phase arises from a simple argument.

Consider a system with an explicitly time varying Hamiltonian $H(t)$. If $H(t)$ varies continuously, we expect that the eigenstates of the system will vary continuously. We write the eigenstates of $H(t)$ as $|\psi_n(t)\rangle$, defined by

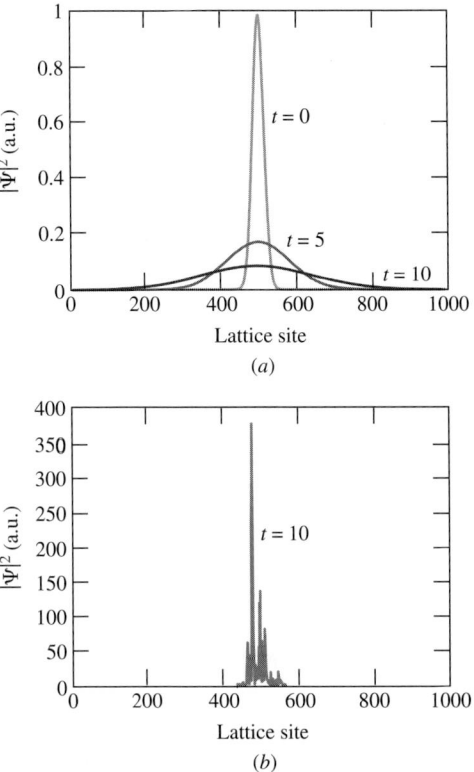

Fig. 9.21 Numerical solution of the one-dimensional Schrödinger equation, for (a) flat potential, and (b) random potential. Calculation by Z. Vörös.

$$H(t)|\psi_n(t)\rangle = E_n(t)|\psi(t)\rangle. \tag{9.11.6}$$

In defining the eigenstates this way, an arbitrary phase factor is left undetermined; we are only constrained to pick the phase factor so that the eigenstate function is continuous in time. In general, the actual state to which the system evolves in time may differ from the eigenstate we have defined by an overall phase factor. We write

$$|\psi(t)\rangle = |\psi_n(t)\rangle e^{i\phi(t)}. \tag{9.11.7}$$

Schrödinger equation's then gives us

$$H(t)|\psi(t)\rangle = i\hbar \frac{\partial}{\partial t}\left(|\psi_n(t)\rangle e^{i\phi(t)}\right)$$

$$\Rightarrow E_n(t)|\psi_n(t)\rangle e^{i\phi} = i\hbar e^{i\phi}\left(\frac{\partial}{\partial t}|\psi_n(t)\rangle + i|\psi_n(t)\rangle\frac{\partial\phi}{\partial t}\right). \tag{9.11.8}$$

Multiplying by $\langle\psi_n(t)|$, taking into account the normalization condition $\langle\psi_n(t)|\psi_n(t)\rangle = 1$, we obtain

$$\frac{\partial\phi}{\partial t} = i\langle\psi_n(t)|\frac{\partial}{\partial t}|\psi_n(t)\rangle - \frac{E_n(t)}{\hbar}. \tag{9.11.9}$$

The second term can be called the **dynamical** phase factor; it is just what we expect for the normal oscillation of the phase of an eigenstate, which for a time-independent Hamiltonian is simply given by $e^{-iE_n t/\hbar}$. The first term is an extra phase that only occurs if the eigenstates vary. This is Berry's phase.

Berry's phase is sometimes called the **geometrical** phase. Suppose that the time dependence of H occurs because the system is propagated along a path $\vec{R}(t)$ in space. (In general, $\vec{R}$ can be a set of parameters in some generalized parameter space and not necessarily a spatial position.) Then by the chain rule we have for Berry's phase

$$\frac{\partial \phi_B}{\partial t} = i \langle \psi_n(\vec{R}) | \nabla_{\vec{R}} | \psi_n(\vec{R}) \rangle \cdot \frac{\partial \vec{R}}{\partial t}, \tag{9.11.10}$$

and therefore the change in phase for some path is

$$\Delta \phi_B = i \int_{\vec{R}(0)}^{\vec{R}(t)} \langle \psi_n(\vec{R}) | \nabla_{\vec{R}} | \psi_n(\vec{R}) \rangle \cdot d\vec{R}. \tag{9.11.11}$$

In particular, around a closed loop,

$$\Delta \phi_B = i \oint \langle \psi_n(\vec{R}) | \nabla_{\vec{R}} | \psi_n(\vec{R}) \rangle \cdot d\vec{R}. \tag{9.11.12}$$

This integral, which does not depend on time at all, in general is nonzero. In other words, if a system moves continuously in a path and arrives back at its starting point, it will acquire a phase that depends only on the path chosen, in addition to the dynamical phase expected for the oscillation of the eigenstates in time.

A special case of Berry's phase is known as the **Aharonov–Bohm** effect. Consider the experiment shown in Figure 9.22(a). The magnetic field is nonzero inside a solenoid coil, in a localized region inside a metal ring, and is zero everywhere outside this region; in particular, it is zero everywhere in the metal ring. We might expect that the magnetic field has no effect on the electrons, since the electrons never feel a magnetic force. The vector $\vec{A}$-field does extend everywhere, however.

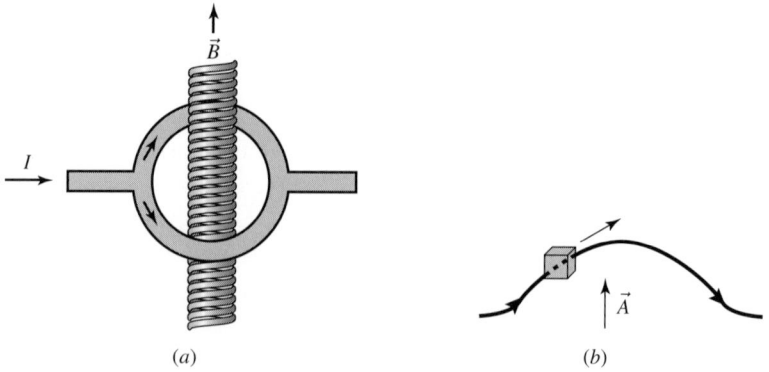

(a)   (b)

Fig. 9.22   (a) Physical layout for the Aharonov–Bohm effect. (b) The time evolution of a system following a path around the Aharonov–Bohm experiment.

Consider a system that consists of a box traveling in the ring, as illustrated in Figure 9.22(b). Within the box, the eigenstates are solutions of the Schrödinger equation

$$\left[\frac{1}{2m}\left(\frac{\hbar}{i}\nabla - q\vec{A}(\vec{r})\right)^2 + V(\vec{r})\right]\psi_n(\vec{r}) = E_n\psi_n(\vec{r}). \tag{9.11.13}$$

Assuming $\vec{A}$ is constant within the box centered at position $\vec{R}$, then if we know the solution for $\vec{A} = 0$, the solutions for finite $\vec{A}$ have the form

$$\psi_n^{(\vec{A})} = e^{iq\vec{A}\cdot(\vec{r}-\vec{R})/\hbar}\psi_n^{(0)}. \tag{9.11.14}$$

The Berry's phase matrix element is then found by integrating over the box,

$$\langle\psi_n|\nabla_{\vec{R}}|\psi_n\rangle = \int d^3r\,\psi_n^{(0)*}(\vec{r})\left(-\frac{iq}{\hbar}\vec{A} + \nabla_{\vec{R}}\right)\psi_n^{(0)}(\vec{r})$$

$$= -\frac{iq}{\hbar}\vec{A}, \tag{9.11.15}$$

where we have used the normalization condition $\langle\psi_n|\psi_n\rangle = 1$, and the second term vanishes by symmetry. The total phase change around the loop is then

$$\Delta\phi_B = \frac{q}{\hbar}\oint \vec{A}\cdot d\vec{R}$$

$$= \frac{q\Phi}{\hbar}, \tag{9.11.16}$$

where Φ is the total flux through the loop, regardless of whether the $\vec{B}$-field is nonzero anywhere in the path $\vec{R}$.

In other words, an electron moving around a path like this will experience a net phase change in coming back to its original starting point. Alternatively, if current is sent in one end of the loop and taken out at the other end, there will in general be interference between the two paths, with a phase difference that depends on the total flux going through the loop. This effect has been demonstrated experimentally with mesoscopic conducting wires, although not in cases where it could be proved the $\vec{B}$-field was strictly zero everywhere on the wires.

The Aharonov–Bohm effect was deduced long before it was seen to be a special case of Berry's phase. For the standard derivation, see Baym (1969: p.77). (The standard derivation has much in common with the effect of flux quantization in superconductors, which we will study in Section 11.10.) It surprised many people, by showing that the $\vec{A}$-field has a direct physical effect, even though we cannot assign an absolute value to it because of the flexibility of gauge invariance. Once again, we see that things that start out as "useful fictions" in theory, in this case the $\vec{A}$-field, can come to be seen as the fundamental realities instead of the things we first measured.

There is an equivalent effect with electric potential, in which a time-varying electric potential that is constant everywhere in space leads to a measurable phase shift, even though the electron feels zero electric field at all times. As shown by Snoke (2015: s. 10.4.3), the Aharonov–Bohm effect, which gives a phase shift $\Delta\phi = \vec{A}\cdot\vec{r}$, and the shift of phase due to a time-varying electric potential, $\Delta\phi = V(t)t$, are the actually same effect, just seen in different relativistic reference frames.

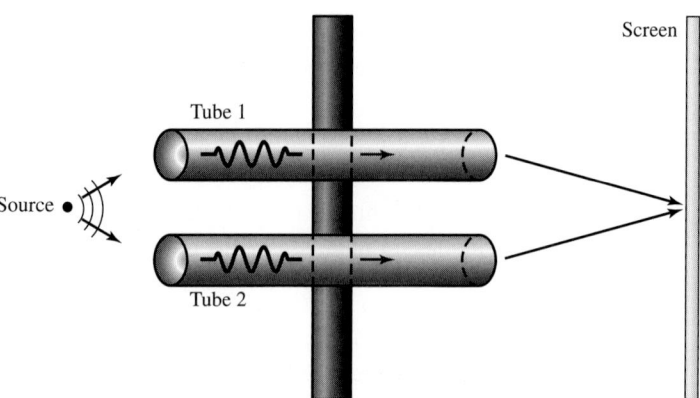

Fig. 9.23   An electron interference experiment.

Exercise 9.11.2   Imagine doing a "two-slit interference" experiment with electrons that go through two separate tubes (with no inelastic dephasing), as illustrated in Figure 9.23. The source emits wave packets that are well defined in time, so that for a certain period of time, we can say that the wave packet splits are definitely in the tubes. During this time a voltage V_1 is applied to tube 1, and a voltage V_2 is applied to tube 2. There is no gradient of voltage inside the tubes, so that there is no electric field felt by the electron during this time. The rest of the time there is no voltage on the tubes.

(a) After the electrons leave the ends of the tubes, they create an interference pattern on a screen. Describe how the interference pattern depends on V_1 and V_2.

(b) If the phase difference between the amplitudes arriving at the screen from the two tubes is Kz, where $K = 10^4$ cm^{-1}, by how many centimeters is the diffraction pattern shifted if $V_1 = 0$ and $V_2 = 10^{-6}$ V, and the time of exposure of the wave packet in the tubes to the voltages is $\Delta t = 10^{-9}$ s? (This exercise was originally suggested by Baym (1969).)

References

E. Abrahams, P.W. Anderson, D.C. Licciardello, and T.V. Ramakrishnan, "Scaling theory of localization: absence of quantum diffusion in two dimensions," *Physical Review Letters* **42**, 673 (1979).

G. Baym, *Lectures on Quantum Mechanics* (Benjamin Cummins, 1969).

M.V. Berry, "Quantal phase factors accompanying adiabatic changes," *Proceedings of the Royal Academy of London* **392**, 45 (1984).

C. Cohen–Tannoudji, B. Diu, and F. Laloë, *Quantum Mechanics* (Wiley, 1977).

S. Doniach and E.H. Sondheimer, *Green's Functions for Solid State Physicists* (Imperial College Press, 1998).

W. Hanke and Yu. V. Kopaev, *Electronic Phase Transitions* (North-Holland, 1992).

H. Haug and S.W. Koch, *Quantum Theory of the Optical and Electronic Properties of Semiconductors* (World Scientific, 1990).

A.P. Heberle A.P. Heberle, J.J. Baumberg et al., "Coherent control of exciton density and spin," *IEEE Journal of Selected Topics in Quantum Electronics* **2**, 769 (1996).

B. Kramer and A. MacKinnon, "Localization theory and experiment," *Reports on Progress in Physics* **56**, 1469 (1993).

G. Mahan, *Many-Body Physics*, 3rd ed. (Kluwer, 2000).

A.C. Melissinos, *Experiments in Modern Physics* (Academic Press, 1966).

M. Munowicz, *Coherence and NMR* (Wiley, 1988).

M.A. Nielsen and I.L. Chuang, *Quantum Computation and Quantum Information* (Cambridge University Press, 2000).

M. Plischke and B. Bergerson, *Equilibrium Statistical Physics* (Prentice Hall, 1989).

C. Santori, D. Fettal, J. Vučković, et al., "Indistinguishable photons from a single-photon device," *Nature* **419**, 594 (2002).

C.P. Slichter, *Principles of Magnetic Resonance*, 3rd ed. (Springer, 1989).

C.L. Smallwood and S.T. Cundiff, "Multidimensional coherent spectroscopy of semiconductors," *Laser and Photonics Reviews*, 1800171 (2018). *DOI*: 10.1002/lpor.201800171.

D.W. Snoke, *Electronics: A Physical Approach* (Pearson, 2015).

H. Stoltz, V. Langer, E. Schreiber, et al., "Picosecond quantum-Beat spectroscopy of bound excitons in CdS," *Physical Review Letters* **67**, 679 (1991).

10 Spin and Magnetic Systems

So far in this book, we have dealt with spin only in two straightforward ways. First, spin has simply been used to give a twofold degeneracy of the electron energy states, which can be split by a magnetic field. Second, spin has been invoked indirectly in enforcing Fermi statistics, which are connected with half-integer spin systems.

In modern solid state physics, issues of spin and magnetism are a major field of study. One reason is applied: Magnetic systems are used for recording in much of the computing industry. Another is more fundamental: Spin is a degree of freedom of solid systems to be manipulated in its own right. Most of what we have studied in earlier chapters has focused on the charge of the electronic carriers as the important degree of freedom to be manipulated, whether in transport measurements, optical transitions, or the Coulomb energy of electron–electron interactions. When we take into account spin–spin and spin–lattice interactions, however, then a whole new set of quasiparticles and quasiparticle interactions arises. Studies in this area are sometimes called **spintronics** to emphasize that information can be transported by spin degrees of freedom.

A major theme of this chapter is the theory of phase transitions, a broad topic of all condensed matter physics, which includes elements such as Ginzburg–Landau theory and renormalization group theory. We will see in Chapter 11 that we can also generalize Ginzburg–Landau theory to the theory of phase transitions that lead to spontaneous coherence. In general, the theory of phase transitions is a major topic of condensed matter physics. The basic theory of phase transitions is presented by Reif (1965) and Schwabl (2002); the Ginzburg–Landau theory is discussed in detail by Toledano and Toledano (1987), Mazenko (2003), and Chaikin and Lubensky (1995). Magnetic systems are the canonical example for this theory. Because of this, many of the concepts we study in this chapter, such as critical fluctuations, spontaneous symmetry breaking, Goldstone bosons, and domain walls, can be applied to other systems as well. For a general review of magnetic systems, see McCurrie (1994).

10.1 Overview of Magnetic Properties

In Chapter 7, we considered a single model for the optical or electromagnetic response of a solid, namely a charged dipole oscillator, and its quantum analog. Magnetic response has a much wider variety of types, with different models.

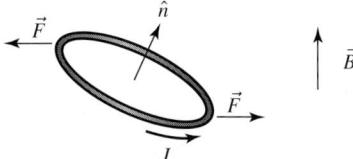

Fig. 10.1   Model of a current loop leading to paramagnetic response.

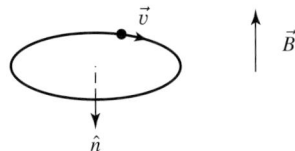

Fig. 10.2   The orbit of a free, positively charged particle leading to diamagnetic response.

A **paramagnetic** system has a response to a magnetic field that tends to *add to* the applied magnetic field. The classic model of paramagnetic response is a freely turning current loop, as shown in Figure 10.1. The current loop represents a magnetic dipole that can change its orientation. When the current loop is placed in a magnetic field, it feels a torque that tends to align it with the applied field. The magnetic field generated by the loop, according to the right-hand rule, will then point in the same direction as the applied field.

A **diamagnetic** system has a response to a magnetic field that tends to *cancel* the applied field. The model of diamagnetic response is a free electron in a conductor. As discussed in Section 2.9, a free electron in a magnetic field goes into a cyclotron orbit like that illustrated in Figure 10.2. The magnetic field generated by this orbit points in the direction opposite the applied field. This is actually just another case of the general rule that free electrons tend to screen out applied field – a perfect conductor perfectly excludes electric field, and a perfect conductor also perfectly excludes magnetic field.

All conductors have a diamagnetic response, but because of resistance that slows down electric current, most conductors have a very weak diamagnetic response. Superconductors, however, which we will discuss in Section 11.8, are nearly perfect conductors and therefore nearly perfectly exclude magnetic field.

A **ferromagnetic** material is similar to a paramagnetic material, but it can generate its own magnetic field even in the absence of an external applied field. The basic reason is that ferromagnetic materials contain magnetic dipoles that interact with each other and tend to line up. At high temperature, the magnetic dipoles may swing about randomly, but at low temperature they will favor lining up all in the same direction, generating a macroscopic magnetic field even if there is no external field. As illustrated in Figure 10.3(a), a large number of dipoles all aligned acts like a macroscopic current loop. Ferromagnetic systems have very complicated behavior, which we will discuss in the Sections 10.3 to 10.7.

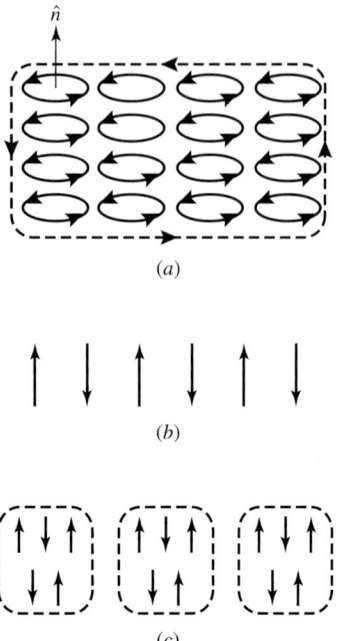

Fig. 10.3   (a) A large number of aligned dipoles, in a ferromagnetic system, acts like a macroscopic current loop, since currents internal to the system cancel out, while currents along the edge do not. (b) Antiferromagnetic response. (c) Ferrimagnetic response.

There are two other, more uncommon magnetic behaviors. One is **antiferromagnetism**. Antiferromagnetic materials have magnetic dipoles that interact, as in ferromagnetic materials, but with the opposite sign of the interaction – the lowest-energy state of two adjacent magnetic dipoles is when they are antialigned instead of aligned. At low temperature, an antiferromagnet will be strongly ordered but will have zero magnetization, as illustrated in Figure 10.3(b).

Another variation is **ferrimagnetism**. In a ferrimagnetic material, groups of dipoles interact such that some are antiparallel and some are parallel, with a net magnetization at low temperature, but less than what one would expect for all the dipoles aligned, as illustrated in Figure 10.3(c).

Spin–spin interactions lie at the root of ferromagnetism and these other types of systems. The actual details of the spin–spin interactions are nontrivial; we will return to discuss these in Section 10.8. For the moment, we will simply parametrize the spin–spin interaction by a single number J, which can be either positive (favoring aligned spins) or negative (favoring antialigned spins), as discussed in Section 10.3.

Magnetic dipole moment. The magnitude of magnetic response depends on the magnetic dipole moment induced by a magnetic field. At the microscopic level, the magnetic dipole moment is defined, in the MKS system, as

$$\vec{\mu} = \frac{e}{2m}(\vec{L} + 2\vec{S}), \tag{10.1.1}$$

where $\vec{L}$ is the orbital angular momentum and $\vec{S}$ is the spin angular momentum. Note that the magnetic dipole moment is not parallel to the total angular momentum $\vec{J} = \vec{L} + \vec{S}$. Another way of saying this is that the **gyromagnetic ratio**, that is, the ratio of magnetic dipole moment to angular momentum, is equal to $e/2m$ for a classical angular momentum and e/m for spin angular momentum.

The magnetic dipole moment due to spin can be written as

$$\vec{\mu} = g\mu_B \frac{\vec{\sigma}}{2}, \tag{10.1.2}$$

where the x, y, and z components of $\vec{\sigma}$ are given by the unitless Pauli spin matrices, listed in Section 1.13, and $\mu_B = e\hbar/2m = 5.79 \times 10^{-5} \text{eV/T}$ is the **Bohr magneton**. The unitless factor g is known as the **Landé g-factor**, which depends on both the orbital and spin angular momentum. For a bare electron, the g-factor is equal to 2 (with slight corrections due to relativistic field theory renormalization), but for electrons in a band in a solid, both the effective mass m and the g-factor can differ quite a bit from those of a free electron. We will discuss how to compute the g-factor of a band in Section 10.2. In general, the renormalization of the g-factor arises from the spin–orbit effect, which depends on the angular momentum of the atomic orbitals that contribute to an electron band; therefore materials with occupied atomic d-orbitals tend to have greater magnetic response because of their larger orbital angular momentum.

If there is a large ensemble of magnetic dipoles, we can talk of the **magnetization field**, that is, the magnetic field generated by the dipoles, which has the same units as the applied magnetic field $\vec{H}$. The magnetization field is simply the total magnetic dipole moment per unit volume,

$$\vec{M} = \frac{1}{V} \sum_i \vec{\mu}_i, \tag{10.1.3}$$

which, if all the dipoles are aligned, has magnitude simply equal to $M = n(g/2)\mu_B$, where $n = N/V$ is the number of spins per unit volume. The net magnetic field in a medium is then $\vec{B} = \mu_0(\vec{H} + \vec{M})$, where $\mu_0 = 4\pi \times 10^{-5}$ T-cm/A is the magnetic permeability of free space, the universal constant used in the MKS system ($\mu_0\text{H}$ has units of tesla). Note that we write the applied magnetic field as the non-italic H, to distinguish it from the Hamiltonian H. In general, a very useful conversion factor for magnetic field problems is one tesla $= 10^{-4}$ V-s/cm^2.

The energy of interaction of a magnetic dipole with an external magnetic field is then

$$H = -\mu_0(\vec{\mu} \cdot \vec{H}), \tag{10.1.4}$$

which has units of eV when we write the Bohr magneton in units of eV/T.

Anisotropy. One complication which we will not address in any depth in this chapter is the possibility of anisotropy. In general, magnetic systems may have an **easy axis** or an **easy plane**. In a system with an easy axis, the magnetic dipoles can lower their energy by aligning with the easy axis; in other words, there is an extra energy term of the form $-\Delta U \cos^2 \theta$, where θ is the angle of the dipole relative to the special axis (the c-axis, in the nomenclature of Chapter 3). For $\vec{\mu}$ nearly aligned with the c-axis, this can be approximated as $-\Delta U(2\cos\theta - 1)$. The term proportional to $\cos\theta$ implies that there will be a term in the Hamiltonian of the form

$$H_{An} = -2\frac{\Delta U}{|\vec{\mu}|}(\vec{\mu} \cdot \hat{z}). \tag{10.1.5}$$

There is thus an effective magnetic field, called the **anisotropy field**, in the system, with magnitude $H_{An} = \Delta U/(\mu_0|\vec{\mu}|)$. In a system with an easy *plane*, dipoles have *higher* energy when aligned with one axis, and therefore have lowest energy when aligned at any angle in the easy plane, that is perpendicular to the *c*-axis.

One can in general parametrize the degree of anisotropy of a magnetic system by the ratio $\kappa = H_{An}/M_s$, where M_s is the maximum magnetization field of the medium. If $\kappa \ll 1$, then the system is effectively isotropic. Such a material is known as a **soft** magnet. If $\kappa \sim 1$, then the system favors alignment along one axis, and is known as a **hard** magnet. For an extended discussion of anisotropy in magnets, see Bertotti (1998).

Exercise 10.1.1   Estimate the magnetization field $\vec{M}$ generated by a solid with 10^{23} atoms per cubic centimeter, all with a single spin aligned in the same direction, and a g-factor of 2. If this is a ferromagnet, what *B*-field does it generate at its surface?

10.2 Landé g-factor in Solids

As discussed at the end of Section 2.9.1, the Zeeman effect is altered in solids, due to the spin–orbit effect, leading in some cases to Zeeman splittings. The change of the g-factor in a solid can be deduced using the same second-order perturbation theory as used in Section 1.9.4. As in Section 5.2, we write

$$H\psi_n(\vec{r}) = \left(\frac{|\vec{p} - q\vec{A}|^2}{2m} + U(\vec{r})\right)\psi_n(\vec{r}) \tag{10.2.1}$$

to account for the vector potential $\vec{A}$ for an electromagnetic field. In the presence of a magnetic field, however, we cannot assume that the electron wave function $\psi(\vec{r})_n$ has the Bloch form $u_{n\vec{k}}(\vec{r})e^{i\vec{k}\cdot\vec{r}}$; the electrons will enter Landau levels, as discussed in Section 2.9.1, so that we cannot treat the states as proportional to pure plane waves. We can still approximate, however, as we have often done in this book,

$$\psi_n(\vec{r}) = u_{n0}(\vec{r})F_n(\vec{R}), \tag{10.2.2}$$

where $u_{n0}(\vec{r})$ is the cell function at zone center, and $F(\vec{R})$ is an envelope function that varies slowly over the length scale of a unit cell. Then $\nabla u_{n0}F = (\nabla u_{n0})F + u_{n0}(\nabla F)$. In this case, we can follow the approach of Section 1.9.4, to expand the Hamiltonian as

$$H\psi_n(\vec{r}) = \left(\frac{|\vec{p} + \vec{P} - q\vec{A}|^2}{2m} + U\right)\psi_n(\vec{r}), \tag{10.2.3}$$

where $\vec{p} = -i\hbar\nabla_r$ acts only on $\vec{r}$ in the cell function $u_{n0}(\vec{r})$, and $\vec{P} = -i\hbar\nabla_R$ acts only on $\vec{R}$ in the envelope wave function, and also on the long-range $\vec{R}$-dependence of $\vec{A}$. To first order in $\vec{A}$, this gives

$$H\psi_n(\vec{r}) = \left(\frac{p^2}{2m} + U - \frac{1}{m}\vec{p}\cdot(\vec{P} - q\vec{A}) \right) \psi_n(\vec{r}). \tag{10.2.4}$$

The first two terms are the energy of the electron in the absence of electromagnetic field, that is, the band energy at zone center. We therefore have the additional perturbation term in the presence of magnetic field, for $q = -e$,

$$H_1 = -\frac{1}{m}\vec{p}\cdot(\vec{P} + e\vec{A}). \tag{10.2.5}$$

Following second-order perturbation theory as in Section 1.9.4 for the cell functions, we have the interaction Hamiltonian

$$H_n = \frac{1}{m^2} \sum_{m\neq n} \frac{\langle u_{n0}|\vec{p}|u_{m0}\rangle \cdot (\vec{P} + e\vec{A})\langle u_{m0}|\vec{p}|u_{n0}\rangle \cdot (\vec{P} + e\vec{A})}{E_n - E_m}.$$

$$\tag{10.2.6}$$

To see how this can generate a Zeeman-like term, let us look at a specific example. We consider a static magnetic field in the z-direction, writing $\vec{A} = B_z x\hat{y}$, and we look at its effect on a band with s-symmetry at zone center, coupled to the six p-like states given in Section 1.13.1. For the split-off band functions given in (1.13.10), which are selected out by the spin–orbit interaction, the $\vec{p}$ operator acting on the orbital terms gives

$$\langle s, -\tfrac{1}{2}|p_x|\tfrac{1}{2}, -\tfrac{1}{2}\rangle\langle\tfrac{1}{2}, -\tfrac{1}{2}|p_y|s, -\tfrac{1}{2}\rangle = -\langle s, +\tfrac{1}{2}|p_x|\tfrac{1}{2}, +\tfrac{1}{2}\rangle\langle\tfrac{1}{2}, +\tfrac{1}{2}|p_y|s, +\tfrac{1}{2}\rangle,$$

$$\tag{10.2.7}$$

where the states $|s, \pm\tfrac{1}{2}\rangle$ are the s-like conduction band states. We abbreviate these terms as $\langle p_x\rangle\langle p_y\rangle\sigma_z$, where σ_z is the spin Pauli matrix. For the four $J = \tfrac{3}{2}$ states, the coupling from the s-states via p_x and p_y is zero, because there are two canceling transitions for each spin state. In addition, for both these terms,

$$\langle p_x\rangle\langle p_y\rangle = -\langle p_y\rangle\langle p_x\rangle. \tag{10.2.8}$$

We therefore have a term in (10.2.6) equal to

$$\frac{1}{m^2} \frac{\langle p_x\rangle\langle p_y\rangle\sigma_z}{E_s - E_p}[P_x, P_y + eB_z x] = -i\left(\frac{2\langle p_x\rangle\langle p_y\rangle}{m(E_s - E_p)} \right) \frac{e\hbar}{2m}\sigma_z B_z$$

$$\equiv \alpha_n\sigma_z B_z, \tag{10.2.9}$$

where the brackets are the commutator, which we resolve using $[P_x, x] = -i\hbar$. Terms involving $\langle p_z\rangle$, and the diagonal terms $|\langle p_x\rangle|^2$ and $|\langle p_y\rangle|^2$, also vanish for our choice of the direction of $\vec{B}$.

We therefore have a term that is proportional to $\vec{\sigma} \cdot \vec{B}$, of the same form as the Zeeman term derived in Appendix F. The term in parentheses in (10.2.9) is a unitless oscillator strength, and $e\hbar/2m$ is the Bohr magneton. The total g-factor of the electrons in a given band n will be the sum of the intrinsic vacuum g-factor plus this factor:

$$g_n = g_0 + \alpha_n. \tag{10.2.10}$$

In general, the effect of band coupling can give either positive or negative α_n; if it is negative, it can partially cancel out the intrinsic g-factor, or even make the total g-factor negative. It does not require the exact type of split-off band used in our example here, although this is a common band structure in III–V semiconductors such as GaAs. It can occur whenever there is any antisymmetry in the terms of the type in (10.2.7) and (10.2.8), which is produced here by the spin–orbit interaction.

Exercise 10.2.1   Prove the statements (10.2.7) and (10.2.8), and that the other terms vanish as stated, by explicitly evaluating the matrix elements, using the band functions in (1.13.9) and (1.13.10), and substituting $\Phi_x = x$, $\Phi_y = y$, $\Phi_z = z$, and $\Phi_s = 1$.

Show also that the term α_n is real for these states, because the factor i that appears in (10.2.9) cancels an i that occurs in the matrix elements.

10.3 The Ising Model

Ferromagnets are a fascinating example of surprising macroscopic behavior arising from simple underlying interactions. Like the effects of superfluidity and superconductivity discussed in Chapter 11, it is not clear whether anyone would ever have predicted ferromagnetism had we not first observed it experimentally. In each of these, as we will see, the quantum mechanics of exchange energy plays a major role.

The simplest model of ferromagnetic behavior is the **Ising** model. In this model, only nearest-neighbor interactions between magnetic dipoles on a lattice are taken into account, as illustrated in Figure 10.4, similar to the way the spring model of atomic bonds took into account only nearest-neighbor interactions in the model of phonons discussed in Section 3.1.2. The Hamiltonian of the system is written as

$$H = -\alpha \mathsf{H} \sum_i s_i - J \sum_{\langle i,j \rangle} s_i s_j, \tag{10.3.1}$$

where s_i is the z-component of the spin of a single dipole, which for convenience we assign to two values, ± 1. The first term in the Hamiltonian gives the energy of the total dipole moment of the collection of spins in the presence of an external magnetic field H. From the definitions in Section 10.1, the constant $\alpha = \mu_0(g/2)\mu_B$ for single spins, where g is the g-factor and μ_B is the Bohr magneton.

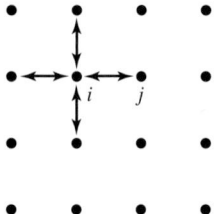

Fig. 10.4   The Ising model.

In the second term, the sum $\langle i, j \rangle$ represents a sum only over pairs i, j that are nearest neighbors, and J is a coupling constant. As we will discuss in Section 10.8, a lot of physics is swept into this single constant J, and calculating it from first principles is nontrivial. The basic effects of ferromagnetism, however, can be seen just by treating this as a constant.

10.3.1 Spontaneous Symmetry Breaking

In the absence of an external magnetic field H, it is easy to see that the ground state of the system, if $J > 0$, corresponds to all the spins aligned, that is, either all $s_i = 1$ or all $s_i = -1$. If the external magnetic field is strictly zero, it is not clear which of these two states the system will go into. This already raises the problem of **spontaneous symmetry breaking**. To minimize its energy, the system must go into one of these two states, but there is nothing to tell us which state it will choose.

The Ising model is one of the few exactly solvable models in the physics of phase transitions, and even so it can only be solved exactly in one or two dimensions. We will not discuss the exact solutions here; instead we will use the mean-field approximation, which will tell us the important physics.

Let us define m as the average value of the spin in the lattice (i.e., the magnetization). We can then rewrite the Hamiltonian, in the absence of an external magnetic field, as

$$H = -J \sum_{\langle i,j \rangle} (m + (s_i - m))(m + (s_j - m)), \tag{10.3.2}$$

where $(s_i - m)$ is the fluctuation of s_i away from the average value. Multiplying this out, we have

$$H = -J \sum_{\langle i,j \rangle} \left(m^2 + m(s_j - m) + m(s_i - m) + (s_i - m)(s_j - m) \right). \tag{10.3.3}$$

The mean-field approximation consists of dropping the last term, which is quadratic in the fluctuations, which will be small if the fluctuations are small compared to the average. We then have

$$H = -J \sum_{\langle i,j \rangle} \left(m^2 + m(s_i + s_j) - 2m^2 \right)$$
$$= J \sum_{\langle i,j \rangle} m^2 - J \sum_{\langle i,j \rangle} m(s_i + s_j). \tag{10.3.4}$$

For a given lattice, we can define the **coordination number** z as the number of nearest neighbors of any site. Then the mean-field Ising Hamiltonian can be written as

$$H = Jm^2 Nz - 2Jmz \sum_i s_i, \tag{10.3.5}$$

where N is the total number of sites.

To determine the temperature dependence of the system in classical thermodynamics, we want the partition function

$$Z = \sum_{\{s_i\}} e^{-\beta H}, \tag{10.3.6}$$

where the sum is over all possible spin configurations. Substituting in the mean-field Hamiltonian (10.3.5), we obtain

$$Z = \sum_{\{s_i\}} e^{-\beta J m^2 N z} \prod_i e^{2\beta J m z s_i}. \tag{10.3.7}$$

Since the sum over all possible configurations corresponds to a sum over each of the two possible spins at each site, we have

$$Z = e^{-\beta J m^2 N z} \prod_i \left(e^{2\beta J m z} + e^{-2\beta J m z} \right)$$

$$= e^{-\beta J m^2 N z} \prod_i 2 \cosh(2\beta J m z) \tag{10.3.8}$$

$$= \left(2 e^{-\beta J m^2 z} \cosh(2\beta J m z) \right)^N.$$

What we really want to know is the average spin at a given temperature. This is given by the weighted average

$$m = \frac{1}{N} \frac{\displaystyle\sum_{\{s_i\}} e^{-\beta H} \sum_i s_i}{\displaystyle\sum_{\{s_i\}} e^{-\beta H}}$$

$$= \frac{1}{N} \frac{\displaystyle\sum_{\{s_i\}} \prod_i e^{2\beta J m z s_i} \sum_i s_i}{\displaystyle\sum_{\{s_i\}} \prod_i e^{2\beta J m z s_i}}. \tag{10.3.9}$$

Defining $\nu = 2\beta J m z$, this is equivalent to

$$m = \frac{1}{N} \frac{\displaystyle\sum_{\{s_i\}} \frac{\partial}{\partial \nu} \prod_i e^{\nu s_i}}{\displaystyle\sum_{\{s_i\}} \prod_i e^{\nu s_i}}$$

$$= \frac{1}{N} \left(\frac{1}{Z'} \frac{\partial Z'}{\partial \nu} \right) = \frac{1}{N} \frac{\partial \ln Z'}{\partial \nu}, \tag{10.3.10}$$

where

$$Z' = \sum_{\{s_i\}} \prod_i e^{\nu s_i} = \cosh^N(2\beta J m z) \tag{10.3.11}$$

by the same logic as used to obtain (10.3.8). This implies

$$m = \frac{\partial \ln \cosh \nu}{\partial \nu}$$

$$= \tanh \nu. \tag{10.3.12}$$

To find the explicit T dependence, we recall the definition of ν, and define the parameter

$$T_c = 2Jz/k_B, \tag{10.3.13}$$

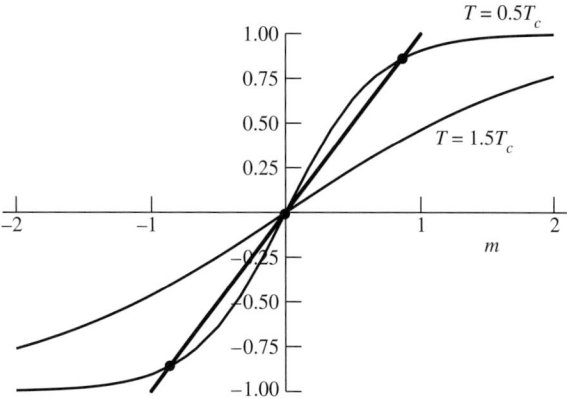

Fig. 10.5 Graphical solution of (10.3.14). The heavy solid is $y = m$.

so that we have

$$\boxed{m = \tanh\left(\frac{T_c}{T}m\right).}$$ (10.3.14)

The value of m must satisfy this equation. The solution can be found graphically, as shown in Figure 10.5. For $T > T_c$, there is only one solution, at $m = 0$. For $T < T_c$, there are three solutions. The lower the temperature, the closer m is to the values of ± 1.

The quantity m, which is the average spin in this system, is the **order parameter** of the phase transition. In general, in any phase transition, some order parameter, which is a measure of some macroscopic property of the system, goes from an average of zero, due to disorder and fluctuations, to a nonzero average value. The order has spontaneously increased.

Another way to see the behavior of m through the phase transition is to plot the Helmholtz free energy,

$$F = -\frac{1}{\beta}\ln Z,$$ (10.3.15)

which from thermodynamics gives the energy that a system will minimize in equilibrium.

Substituting in our mean-field solution (10.3.8), we have

$$F = -Nk_BT\left(\ln 2 - \frac{T_c}{2T}m^2 + \ln\cosh\left(\frac{T_c}{T}m\right)\right).$$ (10.3.16)

Suppose that m is small, that is T is near T_c. Then we can expand the cosh function as

$$\cosh\left(\frac{T_c}{T}m\right) = 1 + \frac{1}{2}\left(\frac{T_c}{T}\right)^2 m^2 + \frac{1}{4!}\left(\frac{T_c}{T}\right)^4 m^4 + \cdots$$ (10.3.17)

and we can expand

$$\ln(1 + x) = x - \frac{x^2}{2} + \cdots,$$ (10.3.18)

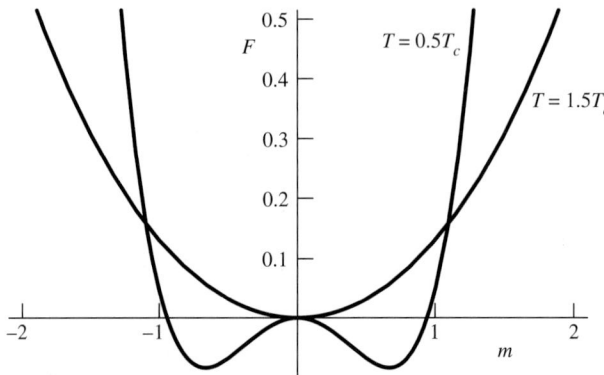

Fig. 10.6   Plot of the Helmholtz free energy (10.3.19) for the Ising model of a ferromagnet, for two temperatures, with $Nk_BT = 1$.

which yields

$$F = F(0) + Nk_BT \left(\frac{1}{2}m^2 \left(\frac{T_c}{T} \right) \left(\frac{T - T_c}{T} \right) + \frac{1}{12}m^4 \left(\frac{T_c}{T} \right)^4 + \cdots \right). \qquad (10.3.19)$$

When $T > T_c$, all the terms are positive, and there is a single minimum at $m = 0$. When $T < T_c$, however, the term proportional to m^2 is negative, which means that the system will have a double-minimum behavior as shown in Figure 10.6. This illustrates the effect of spontaneous symmetry breaking. Above T_c, the system is symmetric. Below T_c, the system is still symmetric, but the energy minimum is not at the symmetry point. In principle, the system could sit at the symmetry point $m = 0$. As we will see below, if there is some external magnetic field, no matter how tiny, then the system will fall down into one of the two minima, and the symmetry will be broken.

Exercise 10.3.1   The Ising model can easily be solved numerically, if you are familiar with a basic programming language. Create an array or matrix $s_i, i = 1, \ldots, N$, which can have value either 1 or -1. The initial values can be picked randomly. The array is updated by the following algorithm: (1) For each site in the array, calculate the interaction energy

$$E_i = -J \sum_j s_i s_j \qquad (10.3.20)$$

for each possible spin value $s_i = \pm 1$ based on the average of the nearest neighbors. (2) Set s_i to a new value, either spin $+1$ or -1, with probability

$$P(s_i = \pm 1) = \frac{e^{-\beta E_{\pm 1}}}{e^{-\beta E_{+1}} + e^{-\beta E_{-1}}}. \qquad (10.3.21)$$

(3) After updating the whole array, start over and continue to iterate the updating process until it converges to having the same average properties.

Show that for a one-dimensional system, the average value of m increases as T decreases, without any sharp transition, while for a two-dimensional system, there is a critical temperature at which the average spin jumps up, which depends on your choice of J.

10.3.2 External Magnetic Field: Hysteresis

Let us now consider the case of the mean-field Hamiltonian when the external field is nonzero:

$$
\begin{aligned}
H &= Jm^2 Nz - 2Jmz \sum_i s_i - \alpha H \sum_i s_i \\
&= Jm^2 Nz - (2Jmz + \alpha H) \sum_i s_i.
\end{aligned}
\tag{10.3.22}
$$

The only difference from before is an offset of the second term. We can therefore use the result (10.3.8), with the adjustment

$$
Z = \left(e^{-\beta Jm^2 z} \cosh(2\beta Jmz + \beta \alpha H) \right)^N
\tag{10.3.23}
$$

and

$$
F = -Nk_B T \left(\ln 2 - \frac{T_c}{2T} m^2 + \ln \cosh\left(\frac{T_c}{T} m + \frac{\alpha H}{k_B T} \right) \right),
\tag{10.3.24}
$$

which in the limit of small m, near T_c, is given by the Taylor expansion

$$
F = F(0) + Nk_B T \left(\frac{1}{2} m^2 \frac{T_c}{T} - \frac{1}{2}\left(\frac{T_c}{T} m + \frac{\alpha H}{k_B T} \right)^2 + \frac{1}{12}\left(\frac{T_c}{T} m + \frac{\alpha H}{k_B T} \right)^4 + \cdots \right).
\tag{10.3.25}
$$

To first order in H, this is

$$
\begin{aligned}
F = F(0) + Nk_B T \Bigg[&\frac{1}{2} m^2 \left(\frac{T_c}{T} \right)\left(\frac{T - T_c}{T} \right) + \frac{1}{12} m^4 \left(\frac{T_c}{T} \right)^4 \\
&- m\left(\frac{T_c}{T} \right)\frac{\alpha H}{k_B T} + \frac{1}{3} m^3 \left(\frac{T_c}{T} \right)^3 \frac{\alpha H}{k_B T} \Bigg]
\end{aligned}
\tag{10.3.26}
$$

and the slope is

$$
\frac{\partial F}{\partial m} =
\tag{10.3.27}
$$

$$
Nk_B T \left[m\left(\frac{T_c}{T} \right)\left(\frac{T - T_c}{T} \right) + \frac{1}{3} m^3 \left(\frac{T_c}{T} \right)^4 + \left(m^2 \left(\frac{T_c}{T} \right)^2 - 1 \right)\left(\frac{T_c}{T} \right)\frac{\alpha H}{k_B T} \right].
$$

Figure 10.7 shows a plot of the Helmholtz free energy for nonzero external H, for the two cases $T > T_c$ and $T < T_c$. At $m = 0$, the slope is nonzero, and equals, to lowest order,

$$
\frac{\partial F}{\partial m} = -N \frac{T_c}{T} \alpha H.
\tag{10.3.28}
$$

Even when T is below T_c, the symmetry is broken, and the system at $m = 0$ will evolve toward one of the two minima, even if the magnetic field is very weak.

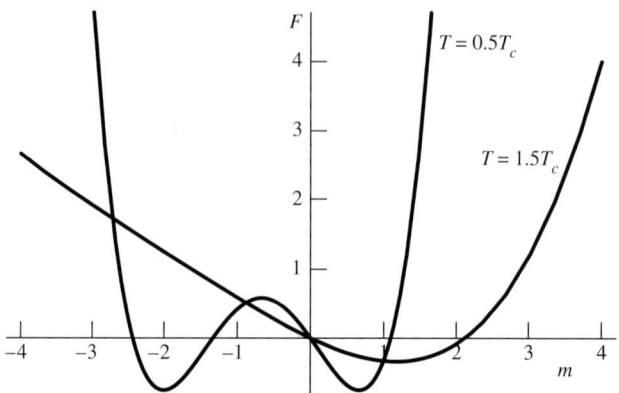

Fig. 10.7 Helmholtz free energy above and below T_c in the presence of an externally applied magnetic field, with $Nk_BT = 1$ and $\alpha H/k_BT = 1$.

This dependence on H implies **hysteresis** in the magnetic phase transition. Solving (10.3.27) for the points at which $\partial F/\partial m = 0$, we have

$$m\left(\frac{T - T_c}{T}\right) + \frac{1}{3}m^3\left(\frac{T_c}{T}\right)^3 + \left(m^2\left(\frac{T_c}{T}\right)^2 - 1\right)\frac{\alpha H}{k_BT} = 0. \qquad (10.3.29)$$

In general, there will be either one solution, when $T > T_c$, or three solutions of this cubic equation, two which are stable and one which is unstable. Figure 10.8(a) shows a plot of m_0 as a function of H when $T = 0.9T_c$. If the system starts on the bottom of the curve, it will continue in this state as H is increased until it reaches the upper critical field. At this point, it will jump to the upper curve. If the magnetic field is then reduced, the system will remain in the upper state until it jumps down to the lower state. Figure 10.8(b) shows an experimental hysteresis curve for comparison. Real hysteresis curves can vary greatly from the Ising model prediction, due to various complications such as anisotropy, formation of magnetic domains, and fluctuations (see Bertotti 1998). Even within the Ising model, which is greatly simplified, the upper critical field deduced here overestimates the point at which the system will jump to the other state, because fluctuations of the system can cause it to jump from a higher state to a lower one even if both are stable. We will discuss fluctuations in Section 10.4.

The theory of spontaneous symmetry breaking was first developed in solid state physics, but has since been adapted to all kinds of physical systems, including particle physics and early universe cosmology, such as the question of the asymmetry of matter and antimatter in the universe, discussed in Appendix F. Its use in early universe cosmology raises an interesting philosophical question. In the case of a solid state system, the symmetry is assumed to be slightly broken by some external perturbation, which is then amplified. In the case of the early universe, however, what is external to the system? In other words, if the universe is initially perfectly symmetric, then it would seem it could never become unsymmetric – the situation would be analogous to the Mexican hat potential in Figure 10.6 in which the system sits exactly at the unstable point $m = 0$ and no external perturbation

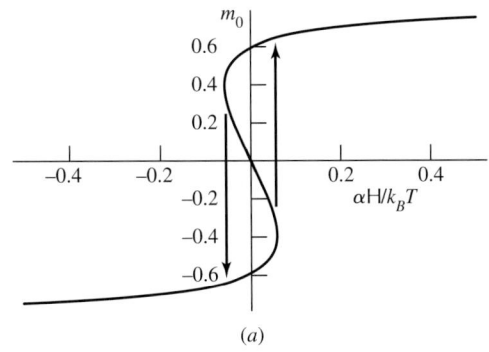

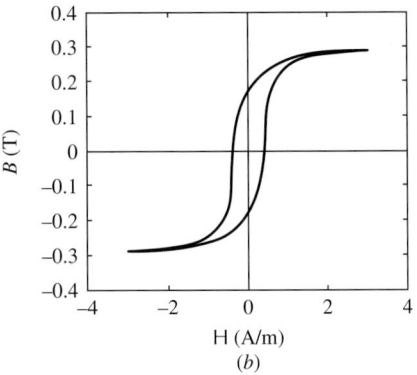

Fig. 10.8 (a) Hysteresis curve for the Ising ferromagnetic system discussed in the text, for $T = 0.9T_c$. (b) Hysteresis curve for a cobalt–iron–chromium–bismuth–silicon alloy. From Nowicki *et al.* (2018).

ever kicks it off. In other words, it would seem that either the universe even at its origin must have been slightly nonsymmetric, or that it received some external kick. This may not bother some people, but an overriding concern for many cosmologists is to have the laws of nature be entirely symmetric, following a sort of Platonic sense of "elegance" or "beauty" in the equations.

Exercise 10.3.2   Use a program like Mathematica to plot the solutions m_0 of (10.3.29) to generate a hysteresis curve like that shown in Figure 10.8. Plot the curve for a number of values of T_c/T and show how the system transforms as T moves through T_c.

10.4 Critical Exponents and Fluctuations

The exact value of the spontaneous spin alignment in the mean-field model can be determined by finding the point at which the derivative $\partial F/\partial m = 0$. From (10.3.29), when $H = 0$, the derivative is

$$\frac{\partial F}{\partial m} = Nk_B T \left(m \left(\frac{T_c}{T} \right) \left(\frac{T - T_c}{T} \right) + \frac{1}{3} m^3 \left(\frac{T_c}{T} \right)^4 \right), \tag{10.4.1}$$

which implies that at a stable point,

$$\left(\frac{T - T_c}{T} \right) + \frac{1}{3} m_0^2 \left(\frac{T_c}{T} \right)^3 = 0 \tag{10.4.2}$$

or

$$m_0 = \pm \sqrt{3} \left(\frac{T}{T_c} \right) \left(\frac{T_c - T}{T_c} \right)^{1/2}. \tag{10.4.3}$$

The exponent $\frac{1}{2}$ in (10.4.3) is called the **critical exponent** of the order parameter.

Another quantity for which we can define a critical exponent is the magnetic suscepti-bility, already introduced in Section 9.2,

$$\chi = \lim_{H \to 0} \frac{\partial M}{\partial H}, \tag{10.4.4}$$

where M is the magnetization field. From the basic definitions of Section 10.1, $M = n(g/2)\mu_B m$, with n the number of spins per unit volume. In the mean-field model, m is obtained by setting the slope (10.3.27) equal to zero, in the limit of small m and H. This yields

$$\frac{\partial F}{\partial m} = m \left(\frac{T - T_c}{T} \right) - \frac{\alpha H}{k_B T} = 0, \tag{10.4.5}$$

which implies

$$m = \frac{\alpha}{k_B (T - T_c)} H, \tag{10.4.6}$$

or

$$\chi = \left(\frac{g \mu_B}{2} \right)^2 \frac{\mu_0 n}{k_B (T - T_c)}. \tag{10.4.7}$$

This is known as the **Curie–Weiss** law, and gives a critical exponent for the susceptibility equal to -1. If $T_c = 0$ (i.e., if $J = 0$), then the system is paramagnetic and one has the simple Curie law, $\chi \propto 1/T$.

The mean-field Ising model of a ferromagnet therefore gives $m \propto |T - T_c|^\alpha$ and $\chi \propto |T - T_c|^{-\gamma}$, with $\alpha = \frac{1}{2}$ and $\gamma = 1$. Experiments with real ferromagnetic systems, however, give critical exponents of $\alpha = 0.3$–0.4 for the magnetization and $\gamma = 1.2$–1.4. The mean-field model is therefore close, but not quite right. What is the source of the discrepancy? The answer comes in large part from fluctuations, which we have ignored in the mean-field model. In the following, we will not go far enough to actually compute the correct critical exponents, but we will see the basic approach to treating fluctuations. The theory of critical phenomena is an active field of modern physics, treated in depth in books such as those by Chaikin and Lubensky (1995), Schwabl (2002), and Mazenko (2003).

Exercise 10.4.1 Show that near $T = T_c$, the solution (10.3.14) gives the same value of m_0 as (10.4.3).

Exercise 10.4.2   Compute the specific heat $C_v = \partial U / \partial T$, with $U = -\partial Z / \partial \beta = \partial (\beta F) / \partial \beta$, for the Ising model in the cases $T < T_c$ and $T > T_c$ and show that there is a discontinuity at T_c.

Critical fluctuations and Landau coarse graining theory. To discuss fluctuations quantitatively, we introduce the Landau **coarse graining** theory. In this approximation, we write the spin as the sum

$$s(\vec{x}) = m(\vec{x}) + \delta s, \tag{10.4.8}$$

where $m(\vec{x})$ gives the average spin over a large number of sites in a region, or grain, centered around $\vec{x}$, and δs is a small term that varies rapidly within a single grain. The size of the grain around $\vec{x}$ is not sharply defined; it is large compared to a single site but small compared to the whole system, similar to the long-wavelength, continuum approximation used for sound waves in Sections 3.4 and 4.2.

Treating the magnetization as a continuous function, the Hamiltonian becomes a functional of $m(\vec{x})$. Since δs is small, we can expand H as a Taylor series,

$$H[s(\vec{x})] = H[m(x)] + \left.\frac{\partial H}{\partial s}\right|_m \delta s + \left.\frac{\partial^2 H}{\partial s^2}\right|_m (\delta s)^2 + \cdots . \tag{10.4.9}$$

$$= H[m(x)] + H[\delta s].$$

Within a single grain, we assume m is constant, and define the local partition function in terms of all possible variations,

$$Z(\vec{x}) = \int \mathcal{D}(\delta s) \, e^{-\beta H[s]} = \int \mathcal{D}(\delta s) \, e^{-\beta (H[m] + H[\delta s])} \tag{10.4.10}$$

$$= e^{-\beta H[m]} \int \mathcal{D}(\delta s) \, e^{-\beta H[\delta s]},$$

in which $\mathcal{D}(\delta s)$ means that we integrate over all possible variations of δs. The Helmholtz free energy is then given by

$$F(\vec{x}) = -\frac{1}{\beta} \ln Z(\vec{x})$$

$$= -\frac{1}{\beta} \left(-\beta H[m(\vec{x})] + \mathcal{O}(\delta s) \right), \tag{10.4.11}$$

where the second term depends on the local fluctuations, which are small. Neglecting this allows us to equate the local free energy with the local Hamiltonian,

$$F(\vec{x}) \approx H[m(\vec{x})]. \tag{10.4.12}$$

This method is known in mathematics as the **saddle point** approximation. It amounts to saying that in the integral (10.4.10), the largest contribution to the partition function comes from very small deviations away from the mean value $m(\vec{x})$, and contributions from larger deviations drop away steeply, since they are exponentially suppressed.

We have already calculated the Helmholtz free energy of the Ising model in the mean-field approximation, in which we assumed that the system was spatially homogeneous. Assuming that the system is homogeneous inside a grain centered at $\vec{x}$, we can write, using (10.3.26),

$$\frac{1}{V}F(\vec{x}) = \frac{1}{2}am^2(\vec{x}) + \frac{1}{4}bm^4(\vec{x}) - h(\vec{x})m(\vec{x}), \tag{10.4.13}$$

where V is the volume of the system, $h(x)$ is proportional to the local magnetic field H, and a and b depend on T. Equating the coefficients with those in (10.3.26), we see that near $T = T_c$,

$$a \simeq (k_B Tn)(T - T_c)/T_c \qquad \text{and} \qquad b \simeq (k_B Tn)/3, \tag{10.4.14}$$

where $n = N/V$. Here we have kept terms up to fourth order in m, but if the magnetic field is small, we can keep only terms proportional to H that are linear in m.

If $m(\vec{x})$ has long-range fluctuations, however, then we must include an extra term that depends on the gradient of m. In (10.4.9), we expanded H in a Taylor series in terms of the small parameter δs. We can do the same in terms of a small variation dm with distance dx. The linear term must vanish, because when $B = 0$ the energy must be the same if the sign of m is reversed. Therefore, we add a term proportional to the square of the gradient, so that we have

$$\frac{1}{V}F(\vec{x}) = \frac{1}{2}am^2(\vec{x}) + \frac{1}{4}bm^4(\vec{x}) + \frac{1}{2}c\,|\nabla m(x)|^2 - h(\vec{x})m(\vec{x}), \tag{10.4.15}$$

where c is a constant which could in principle be determined knowing $\partial^2 H/\partial m^2$. This is the free energy per volume for a single grain. The total free energy is found by integrating over all space, in d dimensions,

$$F = \int d^d x \left(\frac{1}{2}am^2(x) + \frac{1}{4}bm^4(x) + \frac{1}{2}c\,|\nabla m(x)|^2 - h(x)m(x) \right). \tag{10.4.16}$$

To minimize the free energy, we can use a variational approach. Suppose we add a small change $\delta m(x)$ to the field $m(\vec{x})$. The change in energy is

$$\delta F = \int d^d x \left(am\delta m + bm^3\delta m + c\nabla m \cdot \nabla \delta m - h\delta m \right). \tag{10.4.17}$$

Integrating the term with the gradient of m by parts, assuming that $\delta m \to 0$ at a surface far away, we have

$$\delta F = \int d^d x\, \delta m \left(am + bm^3 - c\nabla^2 m - h \right). \tag{10.4.18}$$

We can therefore define the functional derivative

$$\frac{\delta F}{\delta m} = am + bm^3 - c\nabla^2 m - h, \tag{10.4.19}$$

which must vanish when F is minimized, that is

$$\boxed{am + bm^3 - c\nabla^2 m = h.} \tag{10.4.20}$$

This is known as a **Ginzburg–Landau** equation. If m is spatially homogeneous and $h = 0$, we have below T_c,

$$m_0 = \pm\sqrt{-\frac{a}{b}} = \pm\sqrt{3}\left(\frac{T_c - T}{T_c} \right)^{1/2}, \tag{10.4.21}$$

which is just the mean-field solution.

A point source. Suppose now that $h(\vec{x})$ is not zero, but is equal to $h_0\delta^d(\vec{x})$, in other words, a d-dimensional δ-function centered at the origin. Since the magnetic field is localized, we expect that far from the origin, $m(\vec{x}) = 0$. Above T_c, as we have seen, there is a single minimum in F at $m = 0$ in the mean-field model. Therefore, we assume m is small and drop the bm^3 term in (10.4.20). We then must solve the equation

$$-c\nabla^2 m + am = h_0\delta^d(\vec{x}) \tag{10.4.22}$$

subject to the boundary condition $m(\infty) \rightarrow 0$. Switching to frequency space using the Fourier transforms

$$m(\vec{x}) = \frac{1}{(2\pi)^d}\int d^d k\, e^{i\vec{k}\cdot\vec{x}} m(\vec{k})$$

$$h(\vec{x}) = \frac{1}{(2\pi)^d}\int d^d k\, e^{i\vec{k}\cdot\vec{x}} h(\vec{k}), \tag{10.4.23}$$

and using the relation

$$\delta^d(\vec{x}) = \frac{1}{(2\pi)^d}\int d^d k\, e^{i\vec{k}\cdot\vec{x}}, \tag{10.4.24}$$

equation (10.4.22) then becomes

$$(ck^2 + a)m(\vec{k}) = h_0. \tag{10.4.25}$$

In three dimensions, this has the solution

$$m(\vec{x}) = \frac{h_0}{c}\frac{1}{(2\pi)^3}\int d^3 k\, \frac{e^{i\vec{k}\cdot\vec{x}}}{k^2 + \xi^{-2}}$$

$$= \frac{h_0}{4\pi c}\frac{e^{-x/\xi}}{x}, \tag{10.4.26}$$

where $\xi^{-2}(T) = a(T)/c$. The net magnetization decays on a length scale of ξ. Since m is nonzero when the spins are correlated, and zero when they are uncorrelated, ξ is known as the **correlation length**. Because $a \simeq a_0(T - T_c)/T_c$, the correlation length diverges as $(T - T_c)^{-1/2}$ as T approaches T_c from above.

Below T_c, we can write the boundary value problem in terms of $m = m_0 + m'$, where m' is a small perturbation away from the equilibrium value m_0. Then the condition (10.4.22) becomes, to lowest order in m',

$$a(m_0 + m') + bm_0^3 + 3bm_0^2 m' - c\nabla^2 m' = h, \tag{10.4.27}$$

or since $m_0^2 = -a/b$,

$$-2am' - c\nabla^2 m' = h. \tag{10.4.28}$$

The boundary value problem for the case $h = h_0\delta^3(x)$ is just the same as above, but with $\xi^{-2}(T) = -2a(T)/c$. In other words, as T approaches T_c from below, the correlation length diverges as $(T_c - T)^{-1/2}$.

Near the critical temperature, then, a small local fluctuation in the magnetic field will lead to regions of correlated spin with very long correlation length. This effect is known as **critical fluctuations** and is ubiquitous for any phase transition. One way to think about how phase transitions occur is to think of domains of the new phase appearing due to fluctuations, which grow larger and larger as the system approaches the phase boundary, until they become the same size as the whole system, in which case the whole system has entered the new phase.

Spontaneous fluctuations. So far, we have posited a δ-function fluctuation imposed on the system by hand. What we really want is to determine the fluctuations that arise thermodynamically. To characterize the fluctuations, we define the spatial correlation function

$$C_m(\vec{x}) = \frac{1}{V} \int d^d x' \, \langle m(\vec{x}')m(\vec{x}' + \vec{x})\rangle, \tag{10.4.29}$$

where the brackets indicate a thermodynamic average. As in the case of the temporal correlation function discussed in Section 9.5, the spatial correlation function can be expressed in terms of a spectral density function, using the Wiener–Khintchine theorem, as the Fourier transform

$$C_m(\vec{x}) = \frac{1}{(2\pi)^d} \int d^d k \, S_m(\vec{k})e^{i\vec{k}\cdot\vec{x}}, \tag{10.4.30}$$

where

$$S_m(\vec{k}) = \frac{1}{V} \langle m(\vec{k})m(-\vec{k})\rangle. \tag{10.4.31}$$

As in Section 9.5, we assume $m(\vec{x})$ is real, and therefore $m^*(\vec{k}) = m(-\vec{k})$.

Because $m(\vec{k})$ is complex, we have to keep track of two numbers, which we can take as the amplitudes of $m(\vec{k})$ and $m^*(\vec{k})$. To perform the thermodynamic average explicitly, we integrate over all possible amplitudes of $m(\vec{k})$ and $m^*(\vec{k})$,

$$S_m(\vec{k}) = \left(\frac{1}{V}\right) \frac{\int \mathcal{D}m(\vec{k})\mathcal{D}m^*(\vec{k}) \, m(\vec{k})m^*(\vec{k})e^{-\beta H}}{\int \mathcal{D}m(\vec{k})\mathcal{D}m^*(\vec{k}) \, e^{-\beta H}}, \tag{10.4.32}$$

where, as above, $\mathcal{D}m$ means we integrate over all possible variations of m. Using the saddle point approximation discussed above, we can equate the local H with the local Helmholtz free energy F. Substituting the definitions (10.4.23) into the Landau free energy (10.4.16), and using the relation (10.4.24), we have for the free energy

$$F = \sum_{\vec{k}} \left(\frac{1}{2}(ck^2 + a)m(\vec{k})m(-\vec{k}) - h(-\vec{k})m(\vec{k})\right)$$

$$+ \sum_{\vec{k}_1,\vec{k}_2,\vec{k}_3} \frac{1}{4}bm(\vec{k}_1)m(\vec{k}_2)m(\vec{k}_3)m(-\vec{k}_1 - \vec{k}_2 - \vec{k}_3). \tag{10.4.33}$$

For $T > T_c$, we again assume that m is close to its minimum value of zero, and keep only terms up to quadratic in m (this is known as the **Gaussian** approximation). Therefore, we write the effective Hamiltonian as

$$H = \sum_{\vec{k}} \left(\frac{1}{2}(ck^2 + a)m(\vec{k})m(-\vec{k}) - h(-\vec{k})m(\vec{k}) \right)$$

$$= \frac{1}{2} \sum_{\vec{k}} \left(A_k m^* m - h^* m - hm^* \right), \tag{10.4.34}$$

where $A_k = ck^2 + a$. In the last line, we have symmetrized the terms using the fact that the sum extends over all $\vec{k}$, and that $m(-\vec{k}) = m^*(\vec{k})$ and $h(-\vec{k}) = h^*(\vec{k})$. The explicit thermal average in the spectral density function (10.4.32) is then equal to

$$S_m(\vec{k}) = \left(\frac{1}{V} \right) \frac{\int \mathcal{D}m(\vec{k})\mathcal{D}m^*(\vec{k}) \, \frac{1}{\beta^2} \frac{\partial}{\partial h(\vec{k})} \frac{\partial}{\partial h^*(\vec{k})} e^{-\beta H}}{\int \mathcal{D}m(\vec{k})\mathcal{D}m^*(\vec{k}) \, e^{-\beta H}}$$

$$= \frac{1}{V\beta^2} \frac{\partial}{\partial h(\vec{k})} \frac{\partial}{\partial h^*(\vec{k})} \ln \left[\int \mathcal{D}m(\vec{k})\mathcal{D}m^*(\vec{k}) \, e^{-\beta H} \right]$$

$$= \frac{1}{V\beta^2} \frac{\partial}{\partial h(\vec{k})} \frac{\partial}{\partial h^*(\vec{k})} \ln Z_{\vec{k}}, \tag{10.4.35}$$

where we have defined the partition function $Z_{\vec{k}}$ for fluctuations with wave vector $\vec{k}$ over the restricted sum that just includes variations of the amplitude of $m(\vec{k})$.

From (10.4.25), we know the value of m generated by a given $h(\vec{k})$, namely, $m(\vec{k}) = h(\vec{k})/A_k$. We can therefore write m as the sum of a term $m_0(\vec{k})$, which has a variable amplitude that does not depend on h, and this h-dependent term. Setting $m = m_0 + h/A_k$ in (10.4.34) gives

$$H = \frac{1}{2} \sum_{\vec{k}} \left(A_k m_0^* m_0 - \frac{h^* h}{A_k} \right). \tag{10.4.36}$$

The partition function is then

$$Z_{\vec{k}}(T, h) = Z_{\vec{k}}(T, 0) \exp \left[\frac{\beta}{2} \sum_{\vec{k}'} \frac{h^*(\vec{k}')h(\vec{k}')}{ck'^2 + a} \right]. \tag{10.4.37}$$

Since $\ln AB = \ln A + \ln B$, the spectral density function is therefore

$$S_m(\vec{k}) = \frac{1}{V\beta^2} \frac{\partial}{\partial h(\vec{k})} \frac{\partial}{\partial h^*(\vec{k})} \left(\frac{\beta}{2} \sum_{\vec{k}'} \frac{h^*(\vec{k}')h(\vec{k}')}{ck'^2 + a} \right)$$

$$= \frac{1}{V\beta} \frac{1}{ck^2 + a}. \tag{10.4.38}$$

(The factor of 2 in the denominator is canceled by the fact that for every $\vec{k}$ in the sum there are two terms, since we sum over both negative and positive $\vec{k}$, and $h^*(\vec{k}) = h(-\vec{k})$.)

Finally, the correlation function is, by the Wiener–Khintchine theorem (10.4.30),

$$C_m(\vec{x}) = \frac{1}{(2\pi)^d} \int d^d k \, \frac{1}{V\beta} \frac{1}{ck^2 + a} e^{i\vec{k}\cdot\vec{x}}. \tag{10.4.39}$$

The spontaneous correlation length has the same behavior as we found above for the field created by a fixed point source; in particular, the correlation length of spontaneous fluctuations diverges at the phase boundary. In three dimensions, this is equal to

$$C_m(\vec{x}) = \frac{1}{4\pi c V\beta} \frac{e^{-x/\xi}}{x}, \tag{10.4.40}$$

where $\xi^{-2} = a(T)/c$. In general, in d dimensions, the average of the square of the amplitude of the fluctuations, $C_m(0)$, analogous to the average power (9.5.15), can be written

$$C_m(0) = \frac{1}{c\beta\xi^{d-2}} \left(\frac{1}{(2\pi)^d} \int d^d y \frac{1}{y^2 + 1} \right) \sim \frac{1}{\xi^{d-2}}, \tag{10.4.41}$$

where the unitless integral in the parentheses has been obtained by changing variables to $\vec{y} = \vec{k}\xi$. We could do the same calculation for approaching T_c from below, as we did for a point source above, with the same result.

The magnitude of the spontaneous fluctuations, as a fraction of the total value of the field, is then

$$\frac{\sqrt{C_m(0)}}{m_0} \propto \frac{(T-T_c)^{\frac{d-2}{4}}}{(T-T_c)^{1/2}} \sim (T-T_c)^{\frac{d-4}{4}}, \tag{10.4.42}$$

where we have used the results $a \propto (T-T_c)$ and $m_0 \propto (T-T_c)^{1/2}$ from (10.4.14) and (10.4.3), respectively. This implies that for $d > 4$, the fluctuations are negligible, while for $d < 4$, not only the length scale, but the *magnitude* of the fluctuations diverges at the phase boundary. This means that our whole model will break down, because we have been assuming that the magnitude of the fluctuations is small.

As mentioned above, this behavior is not specific to magnetic systems. Critical fluctuations are ubiquitous in the physics of phase transitions. In each case, one defines an order parameter analogous to $m(\vec{x})$ and looks at fluctuations of that parameter.

Exercise 10.4.3   For $T < T_c$, we can write $m(\vec{x}) = m_0 + \delta m$, where m_0 is the mean-field solution of the Ising model. Show that for $h = 0$, Gaussian fluctuations around the ordered state m_0 lead to a divergent specific heat as T approaches T_c from below. (For definitions, see Exercise 10.3.2.)

10.5 Renormalization Group Methods

To treat the breakdown of the above approximations in the critical region near a phase boundary, a number of **renormalization group** methods have been developed. This approach has some similarities to the quantum mechanical renormalization which we discussed earlier in this book, although it deals with whole systems, not quasiparticles. It is called renormalization "group" theory because one renormalization succeeded by another renormalization is equivalent to a third, one of the characteristics of a group as defined in Chapter 6, but it has little else in common with symmetry group theory. A full treatment of renormalization group methods is beyond the scope of this book; for a more detailed treatment see Schwabl (2002), Mazenko (2003), and Chaikin and Lubensky (1995).

The basic idea of renormalization group theory is to remove the short wavelength fluctuations in a system by averaging over a suitable small collection of the interacting units. The average of this small collection then becomes a new, renormalized unit, which can interact with other renormalized units via a new, renormalized interaction. The new interaction is generally weaker than the original interaction, which allows a perturbation approach to be used.

Notice the similarity with quasiparticle renormalization discussed in previous chapters. In that approach, we defined the ground state of a system as the new vacuum and new quasiparticles as excitations of that vacuum. We could forget about all the original particles that went into the ground state. Then, if we chose, we could define a ground state for the new quasiparticles, and call that a new vacuum, and define new quasiparticles as excitations out of *that* vacuum. There is no limit to the number of times we can do this. In the same way, in the renormalization group method of phase transitions, once we have defined a new unit, we can forget about the underlying parts of which it is made. We can continue on again to define a higher-level renormalized unit, which consists of an average over the new, renormalized units. There is no limit to the number of times we can do this. Unlike the case of renormalization of quasiparticles, however, where the renormalization could lead to very different physics from the underlying particles, in the case of phase transitions we can often find limiting behavior, in which multiple renormalizations lead to the same behavior. This is what makes renormalization group theory powerful – there is no need to worry about the exact size of the collection over which you are averaging, because there is **scale-invariant** behavior that persists no matter how many times one does the renormalization.

To see the basic concepts, let us consider a one-dimensional chain of spins with periodic boundary conditions, as shown in Figure 10.9. The Hamiltonian for this system, in the absence of magnetic field, is

$$H = -J \sum_i s_i s_{i+1}.$$
(10.5.1)

The partition function is then the sum over all possible configurations of $s_i = \pm 1$,

$$Z = \sum_{\{s_i\}} e^{-\beta H}$$

$$= \sum_{\{s_i\}} e^{K \sum_i s_i s_{i+1}},$$
(10.5.2)

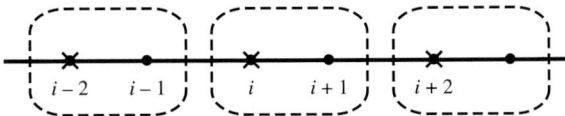

Fig. 10.9   Linear chain of spins. Blocks of two spins are chosen, in which one spin is removed, or **decimated**, and a new system is generated with interactions only between every second spin.

where we have defined the constant $K = \beta J$. Suppose now that we break the chain into blocks of two spins, as illustrated in Figure 10.9. We rewrite the partition function as a sum over each block, as follows:

$$Z = \sum_{\{s_i\}'} \sum_{s_i = \pm 1} e^{K \sum_i s_i(s_{i-1}+s_{i+1})}$$

$$= \sum_{\{s_i\}'} \prod_i 2 \cosh K(s_{i-1} + s_{i+1}), \tag{10.5.3}$$

where $\{s_i\}'$ means the sum over all configurations of only every second spin. We can rewrite the cosh term as an exponential,

$$2 \cosh K(s_{i-1} + s_{i+1}) = e^{C + K' s_{i-1} s_{i+1}}, \tag{10.5.4}$$

where the new constants C and K' are found by writing down the value of the cosh function in the two cases $s_{i+1} = s_{i-1}$ and $s_{i+1} = -s_{i-1}$:

$$2 = e^{C-K'}$$

$$2 \cosh 2K = e^{C+K'}, \tag{10.5.5}$$

which can be solved to give

$$K' = \frac{1}{2} \ln \cosh 2K$$

$$C = \ln 2 + K'. \tag{10.5.6}$$

The partition function then has the form

$$Z = \sum_{\{s_i\}'} e^{NC/2 + K' \sum s_{i-1} s_{i+1}}, \tag{10.5.7}$$

which involves only a sum over every second spin, with an additional constant added to the ground state energy that does not depend on the spin. The system has been *renormalized*, so that effectively we have a new system with half as many spins, with a new interaction constant and a new vacuum ground state energy. It is sometimes said the spin chain has been **decimated**, since half the spins have been removed from the partition function sum.

There is nothing, of course, to stop us from doing the same thing again, to renormalize this new system in blocks of two of the spins remaining in the sum. In general, we have for the nth renormalization the prescription

$$K^{(n)} = \frac{1}{2} \ln \cosh 2K^{(n-1)}$$

$$C = \ln 2 + K^{(n)}. \tag{10.5.8}$$

We can then ask what the limiting behavior of the system is if we keep on renormalizing the system. This corresponds to finding the K values that are fixed points, namely, values that have the property

$$K = \frac{1}{2} \ln \cosh 2K. \tag{10.5.9}$$

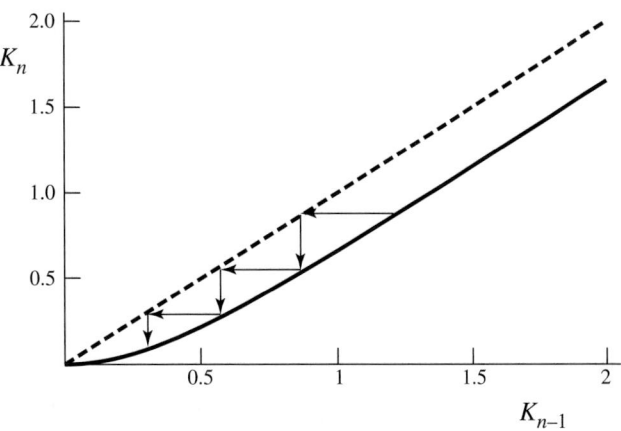

Fig. 10.10 Renormalization mapping for a one-dimensional Ising model.

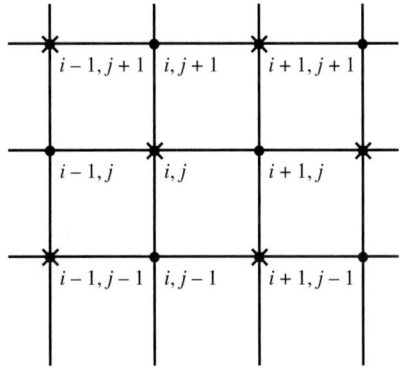

Fig. 10.11 Spins to be decimated in a two-dimensional Ising model.

There are two possible solutions: $K = 0$ and $K = \infty$. The second one, $K = \infty$, is unstable. As shown in Figure 10.10, any value of K less than infinity will converge toward $K = 0$ after successive renormalizations. Therefore, $K = 0$ is a stable fixed point and $K = \infty$ is an unstable fixed point. Since $K = J/k_B T$, this means that there are two possible physical solutions: either (1) infinite coupling constant, or $T = 0$, in which case the system is perfectly ordered; or (2) effectively zero interaction, or effectively infinite temperature, meaning a system with no order.

A one-dimensional system is in fact pathological, because it cannot become ordered except at $T = 0$. The same renormalization method can be extended to systems in higher dimensions, also. As shown in Figure 10.11, blocks of four spins can be defined similar to the way we defined blocks of two spins in the one-dimensional case. In two and higher dimensions, however, more complicated methods must be used, including numerical methods. In these cases, we still find that successive renormalizations lead the system to converge on stable fixed points, unless it starts already exactly at an unstable fixed point.

One of the great accomplishments of renormalization group theory has been the concept of **universality**. In the above example, we found that a one-dimensional system cannot undergo a phase transition except at $T = 0$. It did not matter what the interaction strength J was – this was renormalized away. All that enters the final result is the dimension of the system, in this case $d = 1$, and the assumption of nearest-neighbor interactions. Therefore, we can conclude that any system that is described by that type of Hamiltonian will have the same behavior. In the same way, in higher dimensions, critical exponents and other properties of the system near a phase transition can be found that depend only on the dimension and the form of the Hamiltonian, not on any specific interaction strength or parameter in the Hamiltonian. Broad classes of systems can therefore be grouped together under the same categories.

Closely related to the concept of universality is the idea of **scaling**. This general concept says that near a critical point, the only relevant length scale is the correlation length ξ, and all important properties of the system depend on this length and not on the microscopic length scales such as the lattice spacing in a crystal.

Exercise 10.5.1   What is the renormalization rule for the two-dimensional decimation procedure illustrated in Figure 10.11, analogous to (10.5.8)? How many constants need to be introduced?

10.6 Spin Waves and Goldstone Bosons

In the Ising model, we have been using, each spin feels a potential energy that is higher if it has the opposite spin of its neighbors. There is therefore a restoring force that causes the magnetization to move toward the average value. Since there is a restoring force for the spin, it should not be surprising that the system can support spin waves.

Recall that the Hamiltonian in the Ising model in the absence of external magnetic field is

$$H = -J \sum_{\langle i,j \rangle} s_i s_j, \tag{10.6.1}$$

where the brackets indicate that the second index is summed only over nearest neighbors of the first index. The problem with using this Hamiltonian for deducing wave behavior is that the values of the spin are discrete, $s_i = \pm 1$, while wave behavior involves oscillation of a continuous function. Therefore, although the Ising model can reproduce much of magnetic behavior, we need to turn to a more realistic model to discuss wave behavior.

The general spin–spin interaction Hamiltonian, often called the Heisenberg interaction, can be written as

$$H = -J \sum_{\langle i,j \rangle} \vec{\sigma}^{(i)} \cdot \vec{\sigma}^{(j)}, \tag{10.6.2}$$

where $\vec{\sigma}$ is given in terms of the standard Pauli matrices. The equation of motion in the interaction representation in quantum mechanics is given by

$$i\hbar\frac{\partial\sigma_a^{(i)}}{\partial t} = [\sigma_a^{(i)}, H].$$
(10.6.3)

Using the quantum mechanical identity $[A, BC] = B[A, C] + [A, B]C$ for any operators A, B, and C, we have for one component of the spin,

$$\begin{aligned}
[\sigma_a^{(i)}, H] &= -J\sum_{\langle i'j'\rangle}\sum_b[\sigma_a^{(i)}, \sigma_b^{(i')}\sigma_b^{(j')}] \\
&= -J\sum_{\langle i'j'\rangle}\sum_b\left(\sigma_b^{(i')}[\sigma_a^{(i)}, \sigma_b^{(j')}] + [\sigma_a^{(i)}, \sigma_b^{(i')}]\sigma_b^{(j')}\right) \\
&= -J\sum_{\langle j'\rangle}\sum_b\left(\sigma_b^{(j')}[\sigma_a^{(i)}, \sigma_b^{(i)}] + [\sigma_a^{(i)}, \sigma_b^{(i)}]\sigma_b^{(j')}\right),
\end{aligned}$$
(10.6.4)

where we have eliminated one sum, leaving only a sum over the nearest neighbors of i, because the spin operators for different sites commute. On the same site, the commutator of the two spin operators is given by the angular momentum commutation relation (see, e.g. Cohen-Tannoudji *et al.* 1977: 418)

$$[\sigma_a, \sigma_b] = 2i\epsilon_{abc}\sigma_c,$$
(10.6.5)

where ϵ_{abc} equals 1 for a cyclic ordering of the x, y, and z, it equals -1 for an anticyclic ordering, and it equals 0 otherwise. We therefore obtain

$$[\sigma_a^{(i)}, H] = -2iJ\sum_{\langle j'\rangle}\sum_b\left(\epsilon_{abc}\sigma_b^{(j')}\sigma_c^{(i)} + \epsilon_{abc}\sigma_c^{(i)}\sigma_b^{(j')}\right)$$

or

$$\boxed{\frac{\partial\vec{\sigma}^{(i)}}{\partial t} = \vec{\sigma}^{(i)} \times \frac{4J}{\hbar}\sum_{\langle j'\rangle}\vec{\sigma}^{(j')}.}$$
(10.6.6)

Essentially, this is the same as the classical equation of motion for a dipole moment in a magnetic field,

$$\frac{\partial\vec{m}}{\partial t} = \vec{m} \times \vec{B},$$
(10.6.7)

where the role of the magnetic field $\vec{B}$ is played by the average of the interaction with the nearest-neighbor spins.

As before, we can write the local spin as $\vec{\sigma} = \vec{m}_0 + \delta\vec{m}$, where $\vec{m}_0$ is the mean-field, average value of the spin, which we assume is along the $\hat{z}$-direction, and $\delta\vec{m}$ is a small perturbation. The x-component of $\vec{m}$ is therefore

$$\begin{aligned}
\frac{\partial m_x^{(i)}}{\partial t} &= 2J\sum_{\langle j'\rangle}\left[(\vec{m}_0 + \delta\vec{m}^{(i)}) \times (\vec{m}_0 + \delta\vec{m}^{(j')})\right]_x \\
&= 2Jm_0\sum_{\langle j'\rangle}\left(m_y^{(i)} - m_y^{(j')}\right),
\end{aligned}$$
(10.6.8)

where we have written the perturbation $\delta\vec{m}$ in terms of m_x and m_y, and similarly,

$$\frac{\partial m_y^{(i)}}{\partial t} = -2Jm_0 \sum_{\langle j' \rangle} \left(m_x^{(i)} - m_x^{(j)'} \right). \tag{10.6.9}$$

We define

$$m(\vec{r}_i) = m_x^{(i)} + i m_y^{(i)}, \tag{10.6.10}$$

which then gives us

$$i\frac{\partial m(\vec{r}_i)}{\partial t} = 2Jm_0 \sum_{\langle j' \rangle} (m(\vec{r}_i) - m(\vec{r}_{j'})). \tag{10.6.11}$$

For a cubic lattice with block a, we can write out explicitly the sum over nearest neighbors,

$$i\frac{\partial m(\vec{r}_i)}{\partial t} = -2Jm_0 \left(m(\vec{r}_i + a\hat{x}) - 2m(\vec{r}_i) + m(\vec{r} - a\hat{x}) \right)$$
$$-2Jm_0 \left(m(\vec{r}_i + a\hat{y}) - 2m(\vec{r}_i) + m(\vec{r} - a\hat{y}) \right). \tag{10.6.12}$$

Treating $m(\vec{r})$ as a continuous variable, this is equivalent to

$$\boxed{-i\frac{\partial m}{\partial t} = 2Jm_0a^2 \left(\frac{\partial^2 m}{\partial x^2} + \frac{\partial^2 m}{\partial y^2} \right).} \tag{10.6.13}$$

We have deduced a wave equation for spin waves, which when quantized are known as **magnons**. The value of m represents the magnitude of the spin in the x–y plane; the spin direction will oscillate around the z-axis. Equation (10.6.13) has the same form as the Schrödinger equation, in which a first time derivative is equal to the second spatial derivative. It implies that the dispersion will have the form $\omega \propto k^2$ in the long wavelength ($k \to 0$) limit, as seen in Figure 10.12.

Note that the spin waves will only exist when $m_0 \neq 0$, in other words, below T_c. This is actually a specific example of a very general principle, known as the **Goldstone** theorem: A broken symmetry in a continuous degree of freedom implies the existence of a new excitation mode that has zero frequency in the long wavelength limit. When this new excitation mode is quantized into quasiparticles, these must be bosons.

To see why this is so, consider Figure 10.13. The magnetization is indicated by the arrows in each case. By definition, if the symmetry is broken, there is a preferred direction for the magnetization, which is the order parameter of the transition. By definition also, this direction is arbitrary, since there was no energetic reason why it had to chose one direction or another; this is the essence of spontaneous symmetry breaking. The two cases (a) and (b) must therefore have the same energy. These two cases, however, can be viewed as two different snapshots in time of the infinite-wavelength limit of a spin wave. Figure 10.13(c) illustrates the spins in a long-wavelength, but not infinite-wavelength spin wave.

The same argument applies to any continuous broken symmetry. The order parameter has an arbitrary value in some parameter space. Therefore, variations of the order parameter in that space cost no energy.

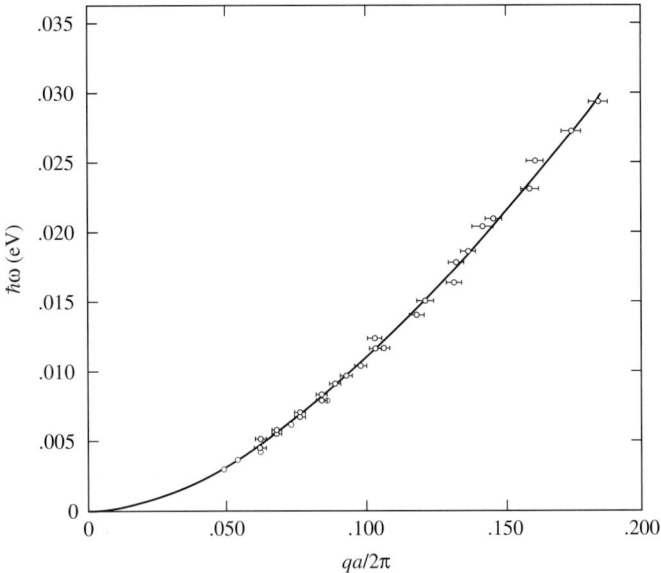

Fig. 10.12   The dispersion relation for magnons in an iron–silicon alloy, deduced from neutron scattering data. The dashed line is a fit with leading order proportional to k^2. From Shirane *et al.* (1965).

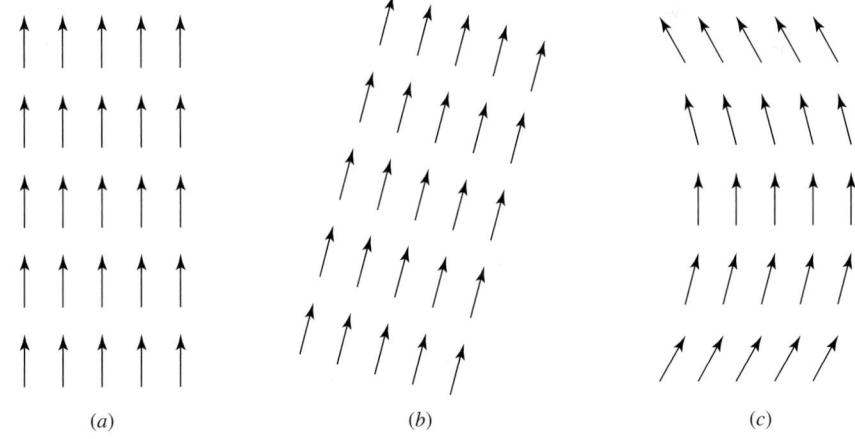

Fig. 10.13   (a) Spins aligned along $\hat{z}$. (b) Spins aligned along a different axis. (c) A long wavelength spin wave.

The Goldstone theorem has been used in high-energy particle physics to predict the existence of new types of particles. The same principle also applies to phonon modes in a solid. In a liquid, only longitudinal phonons exist. If there is a phase transition to a solid phase, new modes appear, with zero frequency at $\vec{k} = 0$, namely, transverse phonons. Transverse phonons can be viewed as the Goldstone bosons of the symmetry-breaking phase transition of the liquid–solid transition. A transverse phonon with zero frequency and infinite wavelength corresponds to a tilting of a crystal axis. Crystal axes do not exist in liquids, and therefore transverse phonons do not exist in liquids.

Exercise 10.6.1   Determine the speed of a magnon wave in the mean-field Ising model for a system at room temperature, with Curie temperature of 1000 K, and coordination number of 6.

10.7 Domains and Domain Walls

Consider a magnetic system such as we discussed in Section 10.3, with the Hamiltonian (10.3.1). We start with all the spins aligned, and then flip a small region of spins, as shown in Figure 10.14. In the absence of an external magnetic field, the energy penalty for doing this is given just by the number of sites at the boundary. In one dimension, this has very little energy cost. Since the energy cost of two adjacent sites with opposite spin is $2J$, then as seen in Figure 10.14(b), the total spin of the domain can be switched from $m = 1$ to $m = -1$ at the cost of just $4J$, or $4J/N$ per atom if there are N atoms in the system. This is demonstrates a very general result, which we have already seen in Section 10.5: in a one-dimensional system with short-range interactions, there can be no phase transition between an ordered and a disordered state. The energy cost for disorder is too small; in fact, the energy cost per unit cell vanishes in the limit $N \to \infty$.

In higher dimensions, there is an energy cost for domains of the wrong orientation, called the **surface energy**, which is proportional to the area of the surface between the different domains. For example, in the case of the two-dimensional domain shown in Figure 10.14(a), the energy cost is approximately equal to $4(L/a)2J$, where L is the length of a side and a is the lattice constant.

This model of domains assumes that the spin can be only up or down, as in the Ising model. If the spin can vary continuously, then a region of spins aligned in one direction can

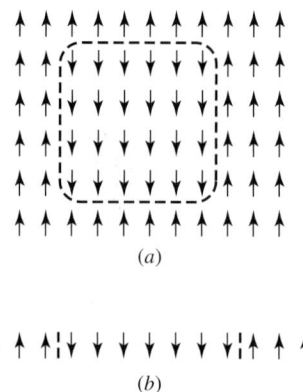

(a)

(b)

Fig. 10.14   A domain of misaligned spins, in the Ising model. (a) Two dimensions. (b) One dimension.

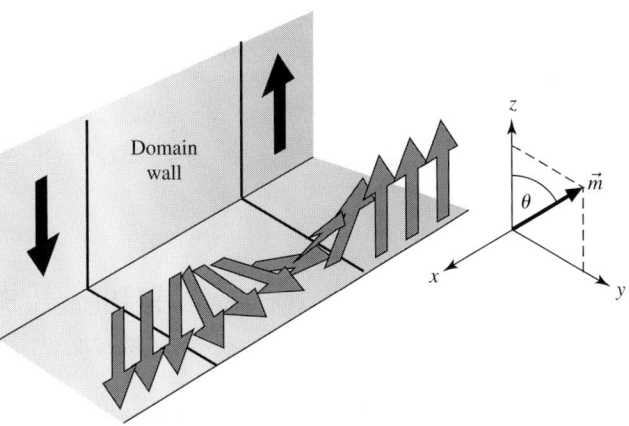

Fig. 10.15   A Bloch wall between two magnetic domains.

convert into a region with spins aligned in the opposite direction via a gradual transition, as in the spin waves discussed in Section 10.6. This is known as a **Bloch wall**, illustrated in Figure 10.15. If the energy cost for two adjacent antialigned spins is $2J$, then if the spin angle is allowed to vary continuously, the energy cost for two adjacent spins with relative angle θ, according to the Hamiltonian (10.6.2), is $2J(1 - \cos \theta)$. If a spin flip is spread over N sites, with equal relative angle between each site, then the relative angle between any two adjacent sites is π/N. The total energy cost of the wall is therefore $NJ(1 - \cos(\theta/N)) \approx NJ\frac{1}{2}(\pi/N)^2$, which decreases as $1/N$.

This would seem to imply that walls of infinite thickness are favored, since the energy is lowest for $N \rightarrow \infty$. If the system were perfectly isotropic, this would be true, and no domains would form. In most real systems, however, there is some small anisotropy energy, as discussed in Section 10.1. Spins that are not aligned in the preferred direction add an energy cost. Even if the anisotropy energy is very small, the total energy cost will add up to a large value, if the number of spins in the domain wall that are not aligned with the easy axis gets large. This means that in real systems, domain walls will have finite thickness. Figure 10.16 shows an example of pattern formation with well-defined domain walls due to anisotropy.

This discussion can be generalized to the theory of domains in many other systems, including domains of different crystal orientation in solids. In the absence of an external field that would favor one orientation of the crystal, there is no difference in energy for the material inside the domains. The only energy cost for having different domains comes from the mismatch of the orientation at the surfaces.

The existence of domains is directly related to the discussion of fluctuations in phase transitions, discussed in Section 10.4. Near a phase transition from a disordered to an ordered state, there will be fluctuations that create small ordered domains. In the absence of an external symmetry-breaking field, there is no reason for the order in one domain to line up with that in another domain. As the critical point is passed, the domains will grow

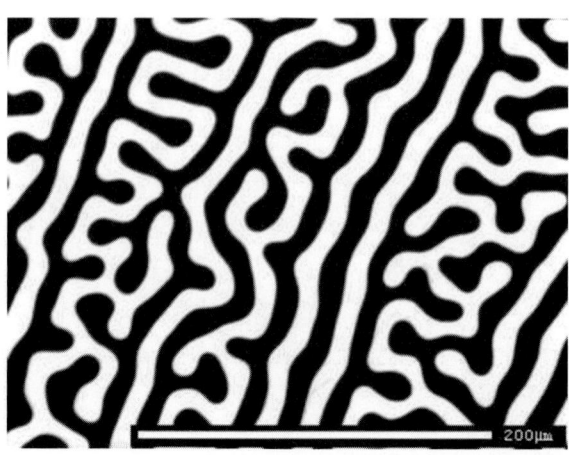

Fig. 10.16   Magnetic domains in a two-dimensional material (epitaxial garnet film) with easy axis perpendicular to the plane, with zero applied field. Courtesy of Jeffrey McCord, Kiel University.

larger, as fluctuations sometimes lead to two adjacent domains being aligned, which will have lower total free energy.

The *time scale* for fluctuations plays an important role. Suppose that a material is near a phase transition, with many domains of different orientation, and then is suddenly quenched to low temperature. Although the ground state of the system is to have the entire crystal aligned, at low temperature the time scale for fluctuations to bring this about is much too long for the different domains to line up. A standard process in metal working is to control the size of domains by **annealing** – raising the material to high temperature to allow fluctuations to increase the domain size – followed by rapid quenching to low temperature, which freezes in a particular domain size.

It may seem counterintuitive, but materials with many different domains, or grains, typically are stronger than single crystals. The reason is that single crystals have well-defined *cleave planes* – in certain directions relative to the crystal axes, the atomic layers can be easily slip relative to each other, as discussed in Section 5.12. Domains with different crystal orientations mean that there are no macroscopic cleave planes. On the other hand, if the domain size is too small, the material will be granular and can have decreased strength.

Besides different orientations, some crystals can have domains of different **polytypes**. These are regions with different crystal symmetry (represented by different groups, in the group theory of Chapter 6) with nearly the same free energy at room temperature.

The theory of domains in various materials is discussed at length by Mazenko (2003) and Bertotti (1998).

Exercise 10.7.1   (a) Compute the surface energy cost for a flat domain wall in three dimensions in the Ising model.

(b) Suppose that an Ising system at low temperature has nearly all spins aligned up. An external magnetic field is then applied, so that the lowest-energy state corresponds to all spins aligned down. To evolve to a state with all spins down, however, requires fluctuations that create domains large enough to overcome the surface energy cost. This is called **nucleation**.

Assuming a spherical domain with radius large enough to make using the surface energy of part (a) reasonable, calculate the total energy cost of a fluctuation that creates a magnetic domain of the opposite spin with radius R. What is the critical radius for the nucleated domain to be stable?

10.8 Spin–Spin Interaction

In all of the discussion of ferromagnetic systems in this chapter so far, we have assumed a spin–spin interaction of the form (10.3.1) or (10.6.2) that favors aligned spins, without discussing the details of this interaction. This is one of the beautiful aspects of the phase transition theory, that so many of the effects are universal and do not depend on the exact details of the interactions in the solid. Nevertheless, we would like to justify the existence of spin–spin interactions, and also be able to predict some things quantitatively such as the Curie temperature. It turns out that this is not so easy – the physics of the spin–spin interaction is quite subtle.

One might initially think that the interaction arises simply from the standard dipole–dipole interaction energy for two magnetic dipoles. A magnetic dipole generates a magnetic field, which can then be felt by another magnetic dipole. As discussed in Section 10.1, a magnetic dipole that is free to move tends to have a paramagnetic response to line up in such a way as to generate magnetic field in the same direction as the magnetic field it feels.

A close look at the numbers indicates that dipole–dipole magnetic interaction cannot be the main interaction in real ferromagnetic systems, however. The standard magnetic dipole–dipole interaction is given by (see, e.g., Cohen-Tannoudji *et al.* 1977, v. 2, s. B_{XI}):

$$H_{\text{int}} = \frac{\mu_0}{4\pi} \frac{\hbar^2 \gamma^2}{4r^3} \left(\vec{\sigma}^{(1)} \cdot \vec{\sigma}^{(2)} - 3(\vec{\sigma}^{(1)} \cdot \hat{n})(\vec{\sigma}^{(2)} \cdot \hat{n}) \right), \qquad (10.8.1)$$

where r is the distance between the two dipoles, $\hat{n}$ is the unit normal in the direction between the two dipoles, γ is the gyromagnetic ratio, and $\vec{\sigma}$ is given in terms of the Pauli spin matrices, with eigenvalues $s = \pm 1$. Note that when the dipoles are aligned along $\hat{n}$, the interaction energy is negative, as in the Ising Hamiltonian, but when they are side by side with dipole moment perpendicular to $\hat{n}$, the interaction energy is positive, favoring spin flip and antialignment. This is the result familiar to any child who has played with magnets, that magnetic dipoles tend to line up north pole to south pole, and it takes energy to force them to line up side to side.

The pure dipole–dipole interaction therefore does not give the right sign for ferromagnetism. Not only that, the strength of the interaction is far too weak to give the experimentally observed effects. Knowing the typical distance of atoms in a solid, one can estimate the magnitude of J for this interaction, where J has been defined in Section (10.3). It turns out that for realistic numbers, J deduced this way gives a Curie temperature much too small compared to experimental values for real ferromagnetic systems.

Exercise 10.8.1   Using the lattice constant of iron $r = 2.87$ Å, and assuming the spins are parallel and aligned along the same axis, estimate J for the interaction (10.8.1). Then estimate the interaction J for iron using $T_c = 2Jz/k_B$ as defined in Section 10.3, assuming $z = 8$ for a bcc crystal, and knowing the Curie point of iron of $770°$ C = 1043 K. How do the two values compare?

The magnetic fields generated by the spins are much too small to give the spin–spin interaction in solids. What else is left? Recall that in Section 8.12 we discussed *exchange* energy of electrons. When electrons are in the same spin state, they are indistinguishable, and therefore an extra term exists for the energy of interaction between the electrons due to the Coulomb interaction, which does not exist for the case of two electrons in different spin states, which are distinguishable. The difference in energy of these two cases leads to an effective spin–spin interaction. As seen in Section 8.12, for the case of the repulsive Coulomb interaction between electrons, the exchange term is negative; that is, exchange favors aligned spins.

A difference of energy for electrons with the same spin compared to electrons with different spins can in general be written as $E_0 - J\vec{\sigma}_1 \cdot \vec{\sigma}_2$, where E_0 and J are constants, because the case of the same spins corresponds to a spin triplet state and the case of different spins corresponds to a spin singlet state. Writing the total spin operator $\vec{S} = (\hbar/2)(\vec{\sigma}^{(1)} + \vec{\sigma}^{(2)}) \equiv \hbar\vec{s}/2$, we have

$$\vec{\sigma}^{(1)} \cdot \vec{\sigma}^{(2)} = \frac{1}{2}\left(|\vec{s}|^2 - |\vec{\sigma}^{(1)}|^2 - |\vec{\sigma}^{(2)}|^2\right). \tag{10.8.2}$$

Using the standard rules for angular momentum in quantum mechanics, the total spin $|\vec{S}|^2 = \hbar^2|\vec{s}|^2/4$ has eigenvalues $\hbar^2 s(s+1) = 2\hbar^2$ and 0, while the squares of the individual spins are equal to $\hbar^2(1/2)(1+1/2) = 3\hbar^2/4$. Therefore, $S^2 - S_1^2 - S_2^2$ has the values $+\hbar^2/4$ and $-3\hbar^2/4$, and the term $\vec{\sigma}^{(1)} \cdot \vec{\sigma}^{(2)}$ has the eigenvalues $+1$ and -3.

Exchange energy in the electron interactions is much more important than the direct magnetic spin–spin interaction. To get an idea of the order of magnitude, we can use (8.12.3) to estimate the exchange energy per particle for an average particle spacing of 2.87 Å, the lattice constant for iron. Assuming an effective mass equal to the vacuum electron mass, this corresponds to a Fermi momentum k_F approximately 1.1×10^8 cm^{-1}, or Fermi energy of 4.6 eV. By comparison, for this Fermi momentum, the exchange energy (8.12.3) is $\Sigma(0) = -2e^2 k_F/4\pi^2\epsilon_0 = -9.8$ eV, assuming the vacuum dielectric constant. Even if screening and the effective band masses are taken into account, one can see that the exchange energy will be significant.

Some astute readers may wonder whether this result, namely that exchange energy favors aligned spins, contradicts the well-known result of the Heitler–London model (see, e.g., Schiff 1968: s. 49) that the lowest energy state of a two-electron molecule such as H_2 is the spin singlet. If exchange energy favors aligned spins, why is that not the case in the case of molecular bonding?

The primary reason for the spin singlet having the lowest energy in the Heitler–London molecular bonding model is Pauli exclusion, not Coulomb exchange energy. The LCAO model, discussed in Section 1.1.2, gives a substantial energy splitting between the bonding and antibonding states. If both electrons are to occupy the bonding state, which has lower energy, they must have different spin, according to the Pauli exclusion rule. If they have the same spin, that is, are in a spin triplet state, then one of the electrons must be in the higher, antibonding state. This energy penalty can be substantial in molecular bonds. When the Coulomb interaction between the electrons is taken into account, exchange energy does reduce the overall energy for the same-spin electrons, but usually not enough to overcome the large bonding–antibonding energy difference.

This is the picture for two atoms. When a large number of atoms are bound together in a crystal, as discussed in Section 1.1, the overlap between the atomic wave functions gives rise to bands, with many electronic states with very small energy difference between them. In this case the energy cost for two electrons to be in different states according to Pauli exclusion is very small, and exchange energy can dominate.

Exchange energy underlies the first of **Hund's rules**, often learned in chemistry, namely that in an atomic or molecular system with degenerate orbitals, the electrons will occupy separate orbitals, with spin maximally aligned. Two aligned spins cannot occupy the same orbital because of Pauli exclusion. Two electrons of different spin in the same orbital will have strong Coulomb repulsion since they overlap in space. These two principles mean that electrons avoid occupying the same orbital if there are empty orbitals available. If they do occupy different, nearby orbitals, they can reduce their total energy by having the same spin, because of the exchange energy term.

10.8.1 Ferromagnetic Instability

As discussed in Section 1.11, covalent and ionic bonding favor bands that are either entirely full or entirely empty. In such a case, that is, in typical semiconductors or insulators, every band has an equal number of electrons with spin up and spin down, and there are no free states nearby into which an electron can move and flip its spin. There is therefore no way to have an excess of one spin over the other, and therefore we do not expect any ferromagnetic or other magnetic response.

In the case of metals, however, or in the case of semiconductors or insulators doped with impurities with an odd number of valence electrons, there are nearby empty states into which electrons can move, flipping their spin. In such cases, we expect ferromagnetism can arise under certain circumstances.

Consider a simple metal with a Fermi level, as we have discussed many times before. We normally assume that each energy state is occupied by two electrons with opposite spin. Suppose that we flip the spin of some electrons, though, so that there is an excess of one spin over the other. On one hand, this will raise the total kinetic energy, since Pauli exclusion demands that higher-energy states be occupied. On the other hand, the same-spin electrons feel a negative exchange energy. If this exchange energy is large enough, it could be enough to overcome the energy penalty of the extra kinetic energy, and the system could lose energy by spontaneously having an excess of one spin over the other.

From Section 2.4.1, the average kinetic energy for a single spin population n_s at $T = 0$ is given by

$$\bar{E}_{\text{kin}} = \frac{3}{5}E_F = \frac{3}{5}\left(\frac{3\pi^2\hbar^3}{\sqrt{2}m^{3/2}}\right)^{2/3} n_s^{2/3}. \tag{10.8.3}$$

On the other hand, as discussed in Section 8.12, for identical electrons we also must take into account the exchange energy. From Section 8.12, the unscreened exchange energy of a Fermi gas at $T = 0$ is equal to

$$\Sigma_{\text{exch}}(k) = -\frac{e^2 k_F}{4\pi^2\epsilon}\left(1 + \frac{1 - (k/k_F)^2}{2k/k_F}\ln\left|\frac{1 + k/k_F}{1 - k/k_F}\right|\right), \tag{10.8.4}$$

which has the average value

$$\bar{\Sigma}_{\text{exch}} = -\frac{e^2}{4\pi^2\epsilon}\frac{3k_F}{2} = -\frac{e^2}{4\pi^2\epsilon}\frac{3(6\pi^2)^{1/3}}{2}n_s^{1/3}. \tag{10.8.5}$$

Since the kinetic energy increases as $n_s^{2/3}$, and the magnitude of the exchange energy increases as $n_s^{1/3}$, this means that at low density the exchange energy will dominate. In this case, spontaneous symmetry breaking will occur, since the system can lose energy by increasing the fraction of electrons in one spin state relative to the other, thus increasing the contribution of exchange energy. This is known as a **Stoner instability**. Systems in which the kinetic energy increases slowly with increasing electron density are more likely to have this effect; since the effective mass appears in the denominator of (10.8.3), metallic solids with heavy band mass, that is nearly flat bands, are favored, as are materials with degenerate conduction bands. Iron and nickel are examples of metals believed to become ferromagnetic through this mechanism.

The unscreened exchange energy used above is, of course, an approximation. The screened Coulomb exchange interaction is given by generalizing (8.12.2) as

$$\Sigma_{\text{exch}}(k) = -\frac{1}{V}\sum_{\vec{p}} N_{s\vec{p}}\frac{e^2/\epsilon}{|\vec{k} - \vec{p}|^2 + \kappa^2}, \tag{10.8.6}$$

where $N_{s\vec{p}}$ is the occupation number for electrons of only one spin. The Thomas–Fermi screening relation (8.11.8) for an electron gas at $T = 0$ gives

$$\kappa^2 = \frac{3e^2 n_{\text{tot}}}{2\epsilon E_F} = \frac{e^2 m k_F}{\pi^2\epsilon\hbar^2}, \tag{10.8.7}$$

which is linear with k_F, while the typical value of $|\vec{k} - \vec{p}|^2$ is comparable to k_F^2, so that at low electron density, the interaction energy will be independent of the k-vector and depend only on the screening strength.

The screening does not depend on spin, to first order, and therefore we can take the interaction vertex as a constant that depends only on the total electron density. In this case, the exchange energy per particle for one spin population is given by

$$\bar{\Sigma}_{\text{exch}} = -\frac{e^2}{V\epsilon\kappa^2}\sum_{\vec{p}} N_{s\vec{p}} = -\frac{e^2}{\epsilon\kappa^2}n_s. \tag{10.8.8}$$

Figure 10.17(b) shows the total of the kinetic energy (10.8.3) minus the exchange energy (10.8.8) for an electron gas at $T = 0$, as a function of the fraction of the population in the spin-up state, $f = (n_\uparrow/n) = 1 - (n_\downarrow/n)$. Since (10.8.8) is an approximation, in Figures 10.17(a) and (c) the strength of the exchange energy is varied around this value (increasing exchange energy corresponds to decreasing effect of screening). In Figure 10.17(b), a classic spontaneous symmetry-breaking potential occurs as in the Ising model, which means the system will choose one ferromagnetic state. An important implication of this model is that for certain values of the interaction, the spins are not necessarily all aligned, even at $T = 0$, because of the kinetic energy penalty. This is consistent with experimental results from three-dimensional transition metals.

We can estimate the Curie temperature using the formalism developed in the beginning of this chapter. Let us assume that the kinetic energy is negligible, and the exchange energy vertex is independent of momentum, as discussed above. We can then write the Hamiltonian in a form similar to the Ising Hamiltonian (10.3.1),

$$H = -\frac{e^2}{V\epsilon\kappa^2} \sum_{\langle i,j\rangle} s_i s_j, \tag{10.8.9}$$

where the sum is over all pairs of electrons. Instead of nearest neighbors in real space, we can talk of nearest neighbors in k-space – essentially only those electrons that have momentum $\vec{p}$ within a radius of κ in k-space around a given electron's momentum $\vec{k}$ will contribute to the exchange energy, because the Coulomb exchange energy for other electrons is suppressed by the factor $1/(|\vec{k} - \vec{p}|^2 + \kappa^2)$.

The critical temperature (10.3.13) is $T_c = 2Jz/k_B$. The exchange energy constant is $J = e^2/V\epsilon\kappa^2$, while the effective number of nearest neighbors in k-space is the density of states times the volume in k-space, that is

$$z = \frac{V}{(2\pi)^3}\frac{4}{3}\pi\kappa^3, \tag{10.8.10}$$

which gives

$$T_c = \frac{e^2\kappa}{6\pi^2\epsilon k_B}. \tag{10.8.11}$$

For free electron density a few times 10^{22} and $\epsilon/\epsilon_0 \approx 10$, the value of κ is a few times 10^7 cm^{-1}, which implies T_c of the order of 1000 K. This is comparable to experimental values for the Curie temperature of standard ferromagnets.

We will see in Chapter 11 (in particular Section 11.3) that the superfluid and superconducting phase transitions for bosons are driven by exchange energy. Since the exchange energy is positive for bosons with repulsive interactions, the lowest-energy state of the system is that which minimizes the magnitude of the exchange energy, while for fermion systems the lowest energy state is that which maximizes the magnitude of the exchange energy. In each case, spontaneous symmetry breaking occurs, with dramatic macroscopic consequences.

Exercise 10.8.2   Show that (10.8.5) is correct by performing the average of the exchange energy (8.12.3) over all k.

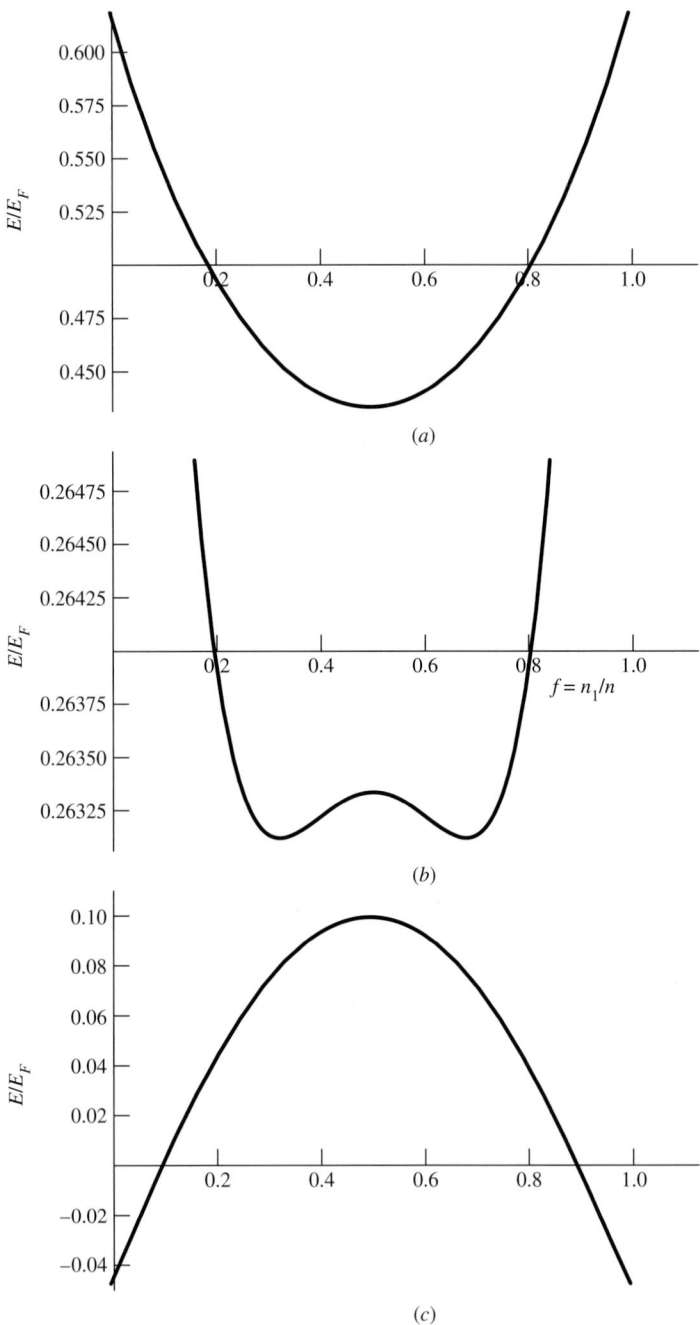

Total energy (kinetic + exchange) for an electron gas at $T = 0$ as a function of the fraction in one spin state. (a) Kinetic energy dominated: The exchange energy per particle is set to $0.5\,\bar{\Sigma}_{\text{exch}}$, where $\bar{\Sigma}_{\text{exch}}$ given by (10.8.8). (b) $\bar{\Sigma}_{\text{exch}}$ given by (10.8.8). (c) Exchange energy dominated: The exchange energy per particle is set to $1.5\,\bar{\Sigma}_{\text{exch}}$.

10.8.2 Localized States and RKKY Exchange Interaction

In the previous section, we looked at the case when there is substantial overlap of the wave functions of the electrons, leading to band formation with a Fermi level. In many magnetic materials, however, the magnetism arises from isolated, localized electrons with negligible wave function overlap, for example, in a material with dopant atoms, each of which has an extra unpaired electron.

If there is no overlap of the wave functions, and the distance between the localized electrons is large enough that Coulomb interaction is negligible, then it would seem at first that there can be no exchange interaction. In many systems, however, exchange interaction can come about due to the effect of an intermediate electron, which can be an electron in a metallic Fermi sea in the same medium as the localized states, or an **itinerant** electron hopping by thermal excitation from site to site. In this process, the intermediate electron has a spin-dependent interaction with one localized electron, then moves to interact with a second localized electron, keeping memory of the spin state of the first. Figure 10.18 shows the generalized Feynman diagram for this type of process, known as **indirect exchange**. Since the final states of the electrons are the same as the initial states, this is a self-energy diagram. The diagram in Figure 10.18(a) is a second-order Coulomb interaction, which can occur for electrons in any spin state. The diagram in Figure 10.18(b) can only occur if the two localized electrons have the same spin, that is, are identical. Although it looks complicated, as a higher-order diagram, remember that all we need is any spin-dependent term to give a spin–spin term in the Hamiltonian. Although this process may seem unlikely, we sum over all possible intermediate states $\vec{k}$, and therefore if there are a large number of intermediate electrons, the total contribution can be significant (though the number of intermediate electrons does not need to be comparable to the number of localized electrons).

One of the most well-known mechanisms like this, which can lead to ferromagnetic spin alignment, is known as the **RKKY** (Ruderman–Kittel–Kasuya–Yosida) interaction. In this model, the intermediate electron is assumed to be a free electron in a plane wave state in a Fermi sea near the localized states. The localized states cannot be treated as plane waves, and therefore we cannot use the standard vertex for Coulomb interaction given in

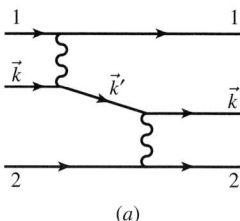 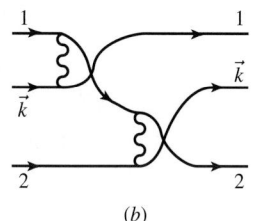

$$(a) \qquad\qquad\qquad (b)$$

Fig. 10.18 (a) Diagram for second-order Coulomb interaction in which an intermediate electron with momentum $\vec{k}$ scatters first with one localized electron and then another. (b) Exchange diagram for the same process, leading to an effective exchange between the two electrons in states 1 and 2.

Section 8.5. We must go back to deduce the vertex for scattering from a localized electron from scratch, following a method in many ways similar to the one for elastic scattering from an impurity in Section 5.3. Instead of scattering from a fixed region of constant potential energy, we envision scattering from a localized electron with a wave function spread out over some region of space.

The energy of interaction is given by the total Coulomb interaction energy, introduced in Section 5.5,

$$H_{\text{int}} = \frac{1}{2} \int d^3 r_1 d^3 r_2 \frac{e^2}{4\pi\epsilon |\vec{r}_1 - \vec{r}_2|} \Psi^\dagger(\vec{r}_1)\Psi^\dagger(\vec{r}_2)\Psi(\vec{r}_2)\Psi(\vec{r}_1), \qquad (10.8.12)$$

where the spatial field operators are related to the standard creation and destruction operators by

$$\Psi(\vec{r}) = \frac{1}{\sqrt{V}} \sum_{\vec{k}} e^{i\vec{k}\cdot\vec{r}} b_{\vec{k}}, \qquad (10.8.13)$$

(neglecting the Bloch cell function factors and spin indices). Recall that the action of a spatial field operator $\Psi^\dagger(\vec{r})$ is to create a particle at exactly one point $\vec{r}$. Therefore, creation of a localized electron state with wave function $\phi(\vec{r})$ can be written as a superposition of the creation of an electron at each point in space with probability amplitude given by $\phi(\vec{r})$:

$$\Psi^\dagger_{\text{loc}} = \int d^3 r \, \phi(\vec{r})\Psi^\dagger(\vec{r}). \qquad (10.8.14)$$

The interaction Hamiltonian is given by the Hamiltonian (10.8.12) in the presence of a localized electron, that is, with a localized electron created in both the initial and final state. Besides the integration over $\vec{r}_1$ and $\vec{r}_2$, we therefore also have integrations over $\vec{r}$ and $\vec{r}'$ and two extra spatial field operators for the creation of the localized electron in the initial and final states, giving terms like the following:

$$\left(\Psi(\vec{r}')\Psi^\dagger(\vec{r}_1)\Psi^\dagger(\vec{r}_2)\right)\left(\Psi(\vec{r}_2)\Psi(\vec{r}_1)\Psi^\dagger(\vec{r})\right). \qquad (10.8.15)$$

The three operators that appear in a single set of parentheses can be rewritten in normal order using the anticommutation relations of the field operators. Assuming that the electrons are identical (i.e., have the same spin), yields

$$\Psi(\vec{r}_2)\Psi(\vec{r}_1)\Psi^\dagger(\vec{r}) = \left(\Psi(\vec{r}_2)\delta(\vec{r} - \vec{r}_1) - \Psi(\vec{r}_1)\delta(\vec{r} - \vec{r}_2) + \Psi^\dagger(\vec{r})\Psi(\vec{r}_2)\Psi(\vec{r}_1)\right). \qquad (10.8.16)$$

The last term gives higher-order effects, namely the Coulomb interaction between free electrons and the effect of state filling on the localized state due to Pauli exclusion if there are many free electrons in the system. Dropping this term, and using the definition (10.8.13) of the spatial field operators, gives

$$H_{\text{int}} = \frac{1}{V} \sum_{\vec{k},\vec{k}'} b^\dagger_{\vec{k}'} b_{\vec{k}} \left(\int d^3 r_1 d^3 r_2 \, e^{-i\vec{k}'\cdot\vec{r}_2} \phi^*(\vec{r}_2) \frac{e^2}{4\pi\epsilon|\vec{r}_1 - \vec{r}_2|} e^{i\vec{k}\cdot\vec{r}_1} \phi_1(\vec{r}_1) \right.$$

$$\left. - \int d^3 r_1 d^3 r_2 \, e^{-i\vec{k}'\cdot\vec{r}_2} \phi^*(\vec{r}_1) \frac{e^2}{4\pi\epsilon|\vec{r}_1 - \vec{r}_2|} e^{i\vec{k}\cdot\vec{r}_1} \phi_1(\vec{r}_2) \right). \qquad (10.8.17)$$

The first term is the direct Coulomb interaction and the second term is the exchange interaction.

Let us suppose that all the localized electrons are in the same atomic or molecular orbital $\phi_{\text{loc}}(\vec{r})$ but centered different positions $\vec{R}$ in the material. In this case, just as in the case of elastic scattering discussed in Section 5.3, we can rewrite the wave function for a given localized electron as $\phi(\vec{r}) = \phi_{\text{loc}}(\vec{r} - \vec{R})$. The exchange term is then

$$
\begin{aligned}
H_{\text{int}} &= \\
&\frac{1}{V} \sum_{\vec{k}, \vec{k}'} b_{\vec{k}'}^{\dagger} b_{\vec{k}} \, e^{i(\vec{k} - \vec{k}') \cdot \vec{R}} \left(\int d^3 r_1 d^3 r_2 \, e^{i(\vec{k} \cdot \vec{r}_1 - \vec{k}' \cdot \vec{r}_2)} \phi_{\text{loc}}^*(\vec{r}_1) \frac{e^2}{4\pi\epsilon |\vec{r}_1 - \vec{r}_2|} \phi_{\text{loc}}(\vec{r}_2) \right) \\
&= \frac{1}{V} \sum_{\vec{k}, \vec{k}'} b_{\vec{k}'}^{\dagger} b_{\vec{k}} \, e^{i(\vec{k} - \vec{k}') \cdot \vec{R}} \, U(\vec{k}, \vec{k}'),
\end{aligned}
\tag{10.8.18}
$$

where $U(\vec{k}, \vec{k}')$ is a constant that depends on the form of the localized wave function but not on the position $\vec{R}$. This has the same form as the elastic scattering Hamiltonian (5.3.28). As with elastic scattering from an impurity, momentum is not conserved for the scattered electron going from state $\vec{k}$ to $\vec{k}'$, and so the interaction Hamiltonian has a sum over both momenta.

Summing over all possible initial electron states $\vec{k}$, we can therefore use second-order Rayleigh-Schrödinger perturbation theory to write the RKKY exchange energy corresponding to Figure 10.18(b), for one localized electron at $\vec{R}_1$ and another at $\vec{R}_2$, as

$$
J = \frac{1}{V^2} \sum_{|\vec{k}| < k_F} \sum_{|\vec{k}'| > k_F} e^{i(\vec{k} - \vec{k}') \cdot (\vec{R}_1 - \vec{R}_2)} \frac{|U(\vec{k}, \vec{k}')|^2}{E_k - E_{k'}},
\tag{10.8.19}
$$

where we require that the incoming electron belong to the Fermi sea of the metal, with magnitude of momentum less than k_F, while the scattered electron must go into an empty state above the Fermi sea. To first order, we can approximate that $|U(\vec{k}, \vec{k}')|^2 = U^2$ is roughly constant. Converting the sums to integrals, we then have

$$
J = \frac{U^2}{(2\pi)^4} \int_0^{k_F} k^2 dk \, d(\cos\theta) \int_{k_F}^{\infty} k'^2 dk' d(\cos\theta') \, \frac{e^{i(kR_{12}\cos\theta - k'R_{12}\cos\theta')}}{E_k - E_{k'}},
\tag{10.8.20}
$$

where $R_{12} = |\vec{R}_1 - \vec{R}_2|$. Performing the integrals over angle gives

$$
J = \frac{U^2}{(2\pi)^4} \int_{-k_F}^{k_F} k dk \, \frac{e^{ikR_{12}}}{R} \left(\int_{-\infty}^{-k_F} k' dk' + \int_{k_F}^{\infty} k' dk' \right) \frac{e^{ik'R_{12}}}{R} \frac{1}{E_k - E_{k'}}.
\tag{10.8.21}
$$

This can be simplified by noticing that the integrand is antisymmetric with respect to reversal of the dummy variables k and k'. Therefore, including the region $k' < k_F$ in the integration over k' does not change the final value of the integral, because integrating over both k and k' in that region gives zero by symmetry. We then have

$$
J = \frac{U^2}{R_{12}^2 (2\pi)^4} \int_{-k_F}^{k_F} k dk \, e^{ikR_{12}} \int_{-\infty}^{\infty} k' dk' \frac{e^{ik'R_{12}}}{E_k - E_{k'}}.
\tag{10.8.22}
$$

Assuming $E_k = \hbar^2 k^2 / 2m$, this can be resolved as

$$ J = \frac{mU^2 k_F}{2\hbar^2 (2\pi)^3 R_{12}^3} \left(\cos 2k_F R_{12} - \frac{\sin 2k_F R_{12}}{2k_F R_{12}} \right). \qquad (10.8.23) $$

At large R_{12}, the interaction energy oscillates as $\cos k_F R_{12}$. This comes about because we are essentially taking the Fourier transform of a function with a sharp cutoff at $k = k_F$. The same oscillatory behavior can be seen in the screening behavior of a Fermi sea, where it is known as **Friedel** oscillations (see, e.g., Ziman 1972: s. 5.5).

This term exists only if the two localized spins are the same. We therefore have a spin–spin interaction that can be either negative or positive, depending on the separation R_{12} of the localized states. This means that the RKKY interaction can in general give either ferromagnetic or antiferromagnetic behavior. This property has been used in various systems to engineer the magnetic response.

We have seen that electron–electron exchange is the main effect driving magnetic behavior. In general, quantitative calculations and experiments to determine the mechanisms of the exchange interactions leading to ferromagnetism and other magnetic states is still an active area of experimental and theoretical research. Mattis (2006), Maekawa (2006), and Du Trémolet de Lacheisserie *et al.* (2003) give general reviews; Chakravarty (1980) gives a comprehensive introduction.

Exercise 10.8.3   In deriving the interaction Hamiltonian for a free electron scattering from a localized electron, we showed that momentum conservation does not hold in the scattering process, and therefore an extra sum must be performed over all possible outgoing momenta. Show, following the same procedure as used to deduce (8.6.25), that energy is conserved at a vertex, by explicitly writing down the S-matrix for elastic scattering in time-dependent perturbation theory.

Exercise 10.8.4   Fill in the missing mathematical steps from (10.8.20) to (10.8.23). In particular, show the counterintuitive result that Pauli exclusion can be ignored for the final state of the scattered electron.

GMR effect. The RKKY interaction lies at the core of the most advanced method of reading out the orientation of magnetic field domains in magnetic hard drives. This is known as the **giant magnetoresistance** (GMR) effect, in which the spins of electrons affect the macroscopic electrical resistance.

An electron moving through a gas of other electrons with the same spin will see less scattering out of its original state than an electron moving through a gas of electrons of the opposite spin, because scattering into many of the possible final states of the electron is forbidden by Pauli exclusion, if those states are occupied by electrons with the same spin as the scattering electron. Since the scattering rate of the electrons determines the resistance of a conductor, as discussed in Section 5.8, the resistance of a conductor can be strongly dependent on the spins of the electrons.

Figure 10.19 shows a schematic of a GMR read head. The main element consists of two ferromagnetic layers separated by a very thin normal conductor (a few nanometers thick). In the presence of the thin conducting layer, the RKKY interaction can be tailored to make

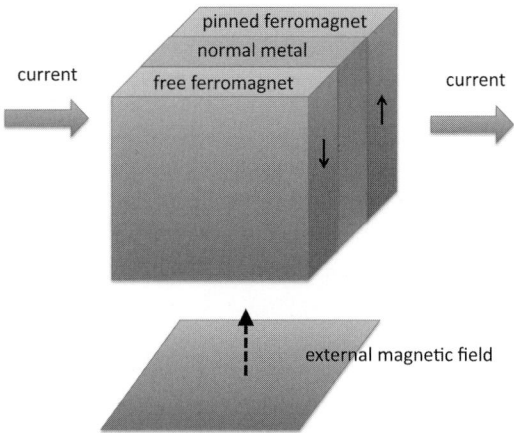

Fig. 10.19 A GMR magnetic read head, also known as a "spin valve." The small black arrows depict the unperturbed spin direction of the ferromagnetic layers. In the presence of an external magnetic field pointing upward, both layers will have their spins line up with the external field.

the spins in one ferromagnetic layer naturally line up opposite the spins in the adjacent layer. By carefully controlling the thickness of the metal layer to within a few nanometers, the distance between the electrons in the two ferromagnetic layers can be put at exactly the distance at which the RKKY interaction gives net energy savings by having electrons of opposite spin.

One of the ferromagnetic layers has spins pinned to always point in the same direction. (This can be done by doping the layer with impurities with spin that require a large energy cost to flip their spin.) In the presence of an external magnetic field in the same direction as the pinned layer, for example the magnetic field produced by a ferromagnetic domain on a hard drive, the second layer will flip its spin to align with the external magnetic field, which has a stronger effect than the RKKY interaction. In this case, both ferromagnetic layers will have their spins aligned in the same direction.

In this case, a current passing through both layers will see a lower resistance. We can see this using the equivalent parallel circuit shown in Figure 10.20, with the currents carried by the two different electron spins acting as two parallel conductors, and the two layers acting as two resistors in series. The resistances of the two layers are in series because we must sum up all the possible scattering processes in determining the overall resistance, and the electrons in each layer can scatter not only into their own layer but also into the adjacent ferromagnetic layer, because the conducting layer between them is so thin. When the spins in the two layers are opposite, the interlayer scattering process is greatly enhanced because there are many empty states with the same spin available in the other layer. This leads to a higher resistance due to the extra scattering process available. When the spins in the two layers are aligned, the scattering for one of the spins is greatly suppressed, leading to much lower resistance.

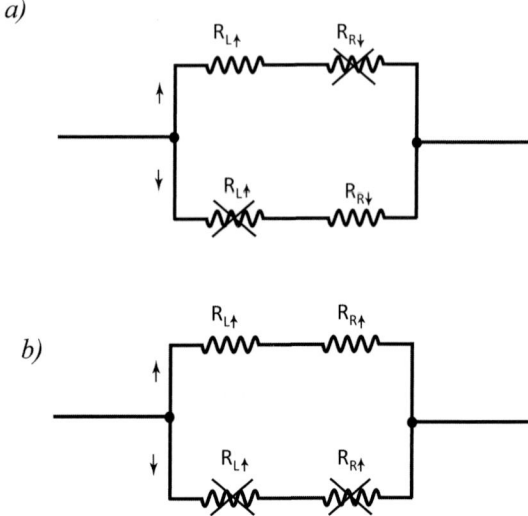

a)

b)

Fig. 10.20   Equivalent circuit for the GMR spin valve. (a) When the ferromagnetic domains are antialigned, and (b) when they are aligned. The $\times$ indicates high resistance.

An external circuit can be used to record the change of the resistance of the GMR spin valve as magnetic domains on the hard drive are moved under it. This gives a very sensitive readout of the magnetic field direction of the domains.

Exercise 10.8.5   It is worthwhile to consider to what degree it is valid to treat different scattering processes in a single resistor as a series circuit of resistors. Consider a standard resistor in which the electrons have two main scattering processes: Scattering with defects, and scattering with crystal vibrations. The scattering with defects has a scattering time τ_d given approximately by

$$\frac{1}{\tau_d} = n_D \sigma_D \bar{v},$$

where n_D is the density of defects, σ_D is the cross-sectional area of the defects, and $\bar{v}$ is the average velocity of the electrons. The scattering with crystal vibrations has a scattering time τ_p given approximately by (ignoring Pauli statistics)

$$\frac{1}{\tau_p} = n(T) \sigma_p \bar{v},$$

where $n(T)$ is the density of phonons according to the Planck equation, and σ_p is the cross-section for scattering of electrons with phonons. The total scattering rate is equal to the sum of the two scattering rates:

$$\frac{1}{\tau} = \frac{1}{\tau_D} + \frac{1}{\tau_p}.$$

The problem with this approach is that the average electron velocity $\bar{v}$ can depend on the voltage drop felt by the electrons, so we should write $\bar{v}(\Delta V)$. At low field, the average velocity $\bar{v}$ due to the thermal motion of the electrons is much higher than the extra velocity they get from acceleration due to the voltage drop, and so $\bar{v}$ can be taken as independent of ΔV. For very fast acceleration of the electrons in high electric field, however, their average velocity can be significantly affected. The effective cross-section for both defect scattering and phonon scattering can also depend on the electron velocity, which depends on the voltage drop. Show that accounting for these effects leads to voltage-dependent resistances and a breakdown of the series model for the resistance in the GMR effect.

10.8.3 Electron–Hole Exchange

The interaction between electrons and holes is governed by the same Coulomb interaction as between electrons. However, the processes involved in exchange are different. Figure 10.21(a) shows the direct Coulomb interaction between an electron and a hole, which looks the same as that between two electrons. We cannot make an exchange diagram by crossing the final two legs, however, because they are distinguishable. Instead, the electron–hole exchange diagram is that shown in Figure 10.21(b). In high-energy physics, with electrons and positrons, the equivalent process is known as **Bhabha** scattering. If the incoming electron and hole have the same spin, then they can annihilate each other in a virtual recombination process, in which a virtual photon is created and then excites another electron–hole pair, as shown in Figure 10.21(c). The outgoing spins must also be the same.

To determine the electron–hole exchange energy, we can follow the same process as Section 5.5, but explicitly account for spin. The interaction energy is written in terms of the electron Fermi field operators as

$$H = \frac{1}{2} \sum_{s,s'} \int d^3r \int d^3r' \, \frac{e^2}{4\pi\epsilon|\vec{r}_1 - \vec{r}_2|} \Psi_s^\dagger(\vec{r}_1)\Psi_{s'}^\dagger(\vec{r}_2)\Psi_{s'}(\vec{r}_2)\Psi_s(\vec{r}_1), \quad (10.8.24)$$

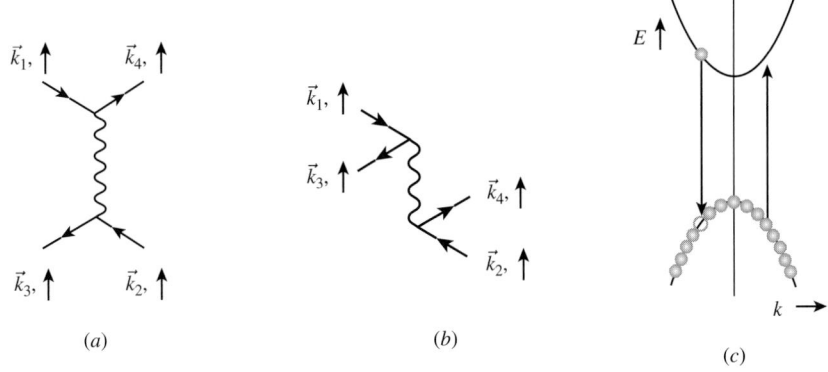

(a)   (b)   (c)

Fig. 10.21   (a) Direct Coulomb interaction of an electron and a hole. (b) Electron–hole exchange. (c) Electron–hole exchange process for simple conduction and valence bands.

where

$$\Psi_s(\vec{r}) = \frac{1}{\sqrt{V}} \sum_{n,\vec{k}} u_{n\vec{k}s}(\vec{r}) e^{i\vec{k}\cdot\vec{r}} b_{n\vec{k}s}, \tag{10.8.25}$$

in which we keep track of the spin s of the Bloch functions for an electron with band index n and momentum $\vec{k}$. Substituting these into (10.8.24) will give terms with four creation and destruction operators acting on momentum states. For a two-band semiconductor with conduction band c and valence band v, the term that corresponds to Figure 10.21(a) is

$$b^\dagger_{c\vec{k}_4 s} b^\dagger_{v\vec{k}_3 s'} b_{v\vec{k}_2 s'} b_{c\vec{k}_1 s}, \tag{10.8.26}$$

and the term that corresponds to Figure 10.21(b) is

$$b^\dagger_{v\vec{k}_3 s} b^\dagger_{c\vec{k}_4 s'} b_{v\vec{k}_2 s'} b_{c\vec{k}_1 s}. \tag{10.8.27}$$

In both terms, a creation operator acting on valence band state $\vec{k}$ acts as a destruction operator for a hole in that band, and vice versa.

For the electron–hole exchange interaction, we now drop the spin indices, and write

$$H_{eh} = \frac{1}{2V^2} \sum_{\vec{k}_1,\vec{k}_2,\vec{k}_3,\vec{k}_4} b^\dagger_{v\vec{k}_3} b^\dagger_{c\vec{k}_4} b_{v\vec{k}_2} b_{c\vec{k}_1} \int d^3 r_1 \int d^3 r_2 e^{-i(\vec{k}_3\cdot\vec{r}_1 + \vec{k}_4\cdot\vec{r}_2 - \vec{k}_2\cdot\vec{r}_2 - \vec{k}_1\cdot\vec{r}_1)}$$

$$\times u^*_{v\vec{k}_3}(\vec{r}_1) u^*_{c\vec{k}_4}(\vec{r}_2) u_{v\vec{k}_2}(\vec{r}_2) u_{c\vec{k}_1}(\vec{r}_1) \frac{e^2}{4\pi\epsilon|\vec{r}_1 - \vec{r}_2|}, \tag{10.8.28}$$

We make the long wavelength approximation that all k are small compared to the Brillouin zone, which means that the plane wave terms are nearly constant over a unit cell. For each $\vec{r}$, we write $\vec{r} = \vec{R} + \vec{x}$, where $\vec{R}$ is the position of a cell and $\vec{x}$ is the position inside a cell, and take the lowest order of the $\vec{k}\cdot\vec{p}$ expansion for the Bloch cell functions, to write

$$\frac{1}{V} \int d^3 r_1\, e^{-i(\vec{k}_3 - \vec{k}_1)\cdot\vec{r}_1}\, u^*_{v\vec{k}_3}(\vec{r}_1) u_{c\vec{k}_1}(\vec{r}_1)$$

$$\approx \frac{1}{N} \sum_{\vec{R}_1} e^{-i(\vec{k}_3 - \vec{k}_1)\cdot\vec{R}_1} \frac{1}{V_{\text{cell}}} \int_{\text{cell}} d^3 x_1\, u^*_{v0}(\vec{x}_1) u_{c0}(\vec{x}_1), \tag{10.8.29}$$

where N is the number of unit cells and V_{cell} is the volume of a unit cell. This yields the approximate result

$$H_{eh} \simeq \sum_{\vec{k}_1,\vec{k}_2,\vec{k}_3,\vec{k}_4} b^\dagger_{v\vec{k}_3} b^\dagger_{c\vec{k}_4} b_{v\vec{k}_2} b_{c\vec{k}_1} \frac{1}{N^2} \sum_{\vec{R}_1,\vec{R}_2} e^{-i(\vec{k}_3\cdot\vec{R}_1 + \vec{k}_4\cdot\vec{R}_2 - \vec{k}_2\cdot\vec{R}_2 - \vec{k}_1\cdot\vec{R}_1)}$$

$$\times \frac{1}{V^2_{\text{cell}}} \int_{\text{cell}} d^3 x_1\, u^*_{v0}(\vec{x}_1) u_{c0}(\vec{x}_1) \int_{\text{cell}} d^3 x_2\, u^*_{c0}(\vec{x}_2) u_{v0}(\vec{x}_2) \frac{e^2}{4\pi\epsilon|\vec{r}_1 - \vec{r}_2|}. \tag{10.8.30}$$

The denominator $|\vec{r}_1 - \vec{r}_2|$ must be treated with care. We break the sum over $\vec{R}_1$ and $\vec{R}_2$ into two parts, one with $\vec{R}_1 = \vec{R}_2$ (short range) and one with $\vec{R}_1 \neq \vec{R}_2$ (long range). For the short-range term, we have

$$H_{eh} = \sum_{\vec{k}_1,\vec{k}_2,\vec{k}_3,\vec{k}_4} b^\dagger_{v\vec{k}_3} b^\dagger_{c\vec{k}_4} b_{v\vec{k}_2} b_{c\vec{k}_1} \frac{1}{N^2} \sum_{\vec{R}_1} e^{-i(\vec{k}_3+\vec{k}_4-\vec{k}_2-\vec{k}_1)\cdot\vec{R}_1}$$

$$\times \frac{1}{V^2_{\text{cell}}} \int_{\text{cell}} d^3x_1 \, u^*_{c0}(\vec{x}_1)u_{v0}(\vec{x}_1) \int_{\text{cell}} d^3x_2 \, u^*_{v0}(\vec{x}_2)u_{v0}(\vec{x}_2)\frac{e^2}{4\pi\epsilon|\vec{x}_1-\vec{x}_2|}.$$

$$(10.8.31)$$

The sum over $\vec{R}_1$ gives a momentum-conserving δ-function that eliminates one sum over $\vec{k}$. Writing $N = V/V_{\text{cell}}$, the short-range exchange therefore has the form

$$\boxed{H^{(SR)}_{eh} = \frac{1}{V} \sum_{\vec{k}_1,\vec{k}_2,\vec{k}_3} V_{\text{cell}} J_{eh} \, b^\dagger_{v\vec{k}_3} b^\dagger_{c\vec{k}_4} b_{v\vec{k}_2} b_{c\vec{k}_1} \, \vec{\sigma}_e \cdot \vec{\sigma}_h,} \qquad (10.8.32)$$

with $\vec{k}_4 = \vec{k}_1 + \vec{k}_2 - \vec{k}_3$; the energy J_{eh} is an energy found by doing the integrals over $\vec{x}_1$ and $\vec{x}_2$ in (10.8.31). Here we have explicitly included the term $\vec{\sigma}_e \cdot \vec{\sigma}_h$, where $\vec{\sigma}_e$ and $\vec{\sigma}_h$ are the Pauli matrices for the electrons and holes, respectively, since this exchange term acts only on electrons and holes with the same spin. Typical values of Δ_{eh} for semiconductors are a few meV.

For the case of $\vec{R}_1 \neq \vec{R}_2$ in (10.8.30), we use the multipole expansion

$$\frac{1}{|\vec{r}_1 - \vec{r}_2|} \simeq \frac{1}{r} - \frac{\vec{r} \cdot (\vec{x}_1 - \vec{x}_2)}{r^3} + \frac{3|\vec{r} \cdot (\vec{x}_1 - \vec{x}_2)|^2 - r^2|\vec{x}_1 - \vec{x}_2|^2}{2r^5}, \qquad (10.8.33)$$

where $\vec{r} = \vec{R}_1 - \vec{R}_2$. Because of the orthogonality of the wave functions in different bands, only terms with products with both $\vec{x}_1$ and $\vec{x}_2$ will be nonzero. After integrating over $\vec{x}_1$ and $\vec{x}_2$, we thus have the term

$$\frac{-3|\vec{r} \cdot \vec{x}_{cv}|^2 + r^2|\vec{x}_{cv}|^2}{r^5}, \qquad (10.8.34)$$

where $\vec{x}_{cv}$ is the dipole matrix element between the bands. We can now convert the sums over cell positions to integrals using

$$\sum_{\vec{R}} \to \frac{1}{V_{\text{cell}}} \int d^3R, \qquad (10.8.35)$$

and switch variables to $\vec{r}$ and $\vec{r}_{cm} = (\vec{r}_1 + \vec{r}_2)/2$, as we did in Section 5.5. The integral over $\vec{r}_{cm}$ of the exponential factor gives a momentum-conserving δ-function, so that we have

$$H^{(LR)}_{eh} = \frac{1}{V} \sum_{\vec{k}_1,\vec{k}_2,\vec{k}_3} b^\dagger_{v\vec{k}_3} b^\dagger_{c\vec{k}_4} b_{v\vec{k}_2} b_{c\vec{k}_1} \int d^3r \frac{e^2}{4\pi\epsilon} \frac{r^2|\vec{x}_{cv}|^2 - 3|\vec{r} \cdot \vec{x}_{cv}|^2}{r^5}e^{i\Delta\vec{k}\cdot\vec{r}},$$

with $\vec{k}_4 = \vec{k}_1 + \vec{k}_2 - \vec{k}_3$ and $\Delta\vec{k} = \vec{k}_1 - \vec{k}_3 = \vec{k}_4 - \vec{k}_2$.

The integral over r, which is a three-dimensional Fourier transform, can be resolved by using the identity (Frahm 1983)

$$\frac{\partial^2}{\partial x^2}\frac{1}{r} = \frac{3x^2 - r^2}{r^5} - \frac{4\pi}{3}\delta^3(\vec{r}). \qquad (10.8.36)$$

We integrate by parts twice to obtain

$$\int d^3r \left(\frac{4\pi}{3} \delta^3(\vec{r}) + \frac{\partial^2}{\partial x^2} \frac{1}{r} \right) e^{i\vec{k}\cdot\vec{r}} = \frac{4\pi}{3} - k_x^2 \int d^3r \frac{e^{i\vec{k}\cdot\vec{r}}}{r}$$

$$= \frac{4\pi}{3} \left(\frac{k^2 - 3k_x^2}{k^2} \right). \tag{10.8.37}$$

The long-range exchange interaction is then

$$\boxed{H_{eh}^{(LR)} = \frac{e^2}{3\epsilon V} \sum_{\vec{k}_1,\vec{k}_2,\vec{k}_3} b_{v\vec{k}_3}^\dagger b_{c\vec{k}_4}^\dagger b_{v\vec{k}_2} b_{c\vec{k}_1} \left(\frac{|\Delta\vec{k}|^2|\vec{x}_{cv}|^2 - 3|\Delta\vec{k} \cdot \vec{x}_{cv}|^2}{|\Delta\vec{k}|^2} \right) \vec{\sigma}_e \cdot \vec{\sigma}_h,} \tag{10.8.38}$$

where we have again put in the spin dependence explicitly.

Exciton electron–hole exchange energy. To determine the exchange energy for an exciton, we use the Wannier exciton state, generated by the creation operator

$$|ex\rangle = c_K^\dagger|0\rangle = \sum_{\vec{k}} \phi(\vec{K}/2 - \vec{k}) b_{c,\vec{K}-\vec{k}}^\dagger b_{v,-\vec{k}}|0\rangle, \tag{10.8.39}$$

where c and v are the conduction and valence band labels, respectively, $\phi(\vec{k})$ is the momentum-space wave function of the relative electron–hole motion within the exciton, and $\vec{K}$ is the center-of-mass wave vector of the exciton. This form for a pair wave function will be derived in Section 11.6; it is easy to see for the case $\vec{K} = 0$ that it corresponds to a superposition of vertical transitions of electrons from the valence to the conduction band at the same $\vec{k}$.

The short-range exchange energy of an exciton with $\vec{K} = 0$ is

$$\langle ex\, |H_{eh}^{(SR)}|ex\rangle \simeq \left(\langle 0| \sum_{\vec{p}'} \phi^*(\vec{p}') b_{v\vec{p}'}^\dagger b_{c\vec{p}'} \right) \frac{V_{cell}J_{eh}}{V} \sum_{\vec{k},\vec{k}',\vec{q}} b_{v,\vec{k}'-\vec{q}}^\dagger b_{c,\vec{k}+\vec{q}}^\dagger b_{v,\vec{k}'} b_{c,\vec{k}}$$

$$\times \left(\sum_{\vec{p}} \phi(\vec{p}) b_{c\vec{p}}^\dagger b_{v\vec{p}}|0\rangle \right). \tag{10.8.40}$$

When all the creation and annihilation operators are resolved into constraints on the momentum vectors, this becomes

$$\langle ex|H_{eh}^{(SR)}|ex\rangle = \left(\sum_{\vec{p}'} \phi^*(\vec{p}') \right) \frac{V_{cell}J_{eh}}{V} \left(\sum_{\vec{p}} \phi(\vec{p}) \right) - \left(\sum_{\vec{p}} |\phi(\vec{p})|^2 \right) \frac{V_{cell}J_{eh}}{V}. \tag{10.8.41}$$

The second term is negligible since the wave function is normalized to unity, so this term is of order $1/V$ times the first term. We convert $\phi(\vec{k})$ to real space using the Fourier transform

$$\varphi(\vec{r}) = \frac{1}{\sqrt{V}} \sum_{\vec{k}} \phi(k) e^{i \vec{k} \cdot \vec{r}}, \tag{10.8.42}$$

so that we have

$$\langle ex|H_{eh}^{(SR)}|ex\rangle = V_{\text{cell}}|\varphi(0)|^2 J_{eh}. \tag{10.8.43}$$

The exciton wave function at $r = 0$ gives the probability of the electron and hole being at the same place; for the 1s-orbital Bohr wave function for Wannier excitons, $|\varphi(r)|^2 = (1/\pi a^3) e^{-r^2/a^2}$, where a is the exciton Bohr radius, and so $V_{\text{cell}}|\varphi(0)|^2 = V_{\text{cell}}/\pi a^3$. The short-range exciton exchange energy can therefore be viewed as proportional to the probability of the electron and hole being in the same unit cell.

The approach for the long-range exchange interaction is the same. Using the exciton wave function for $\vec{K} \neq 0$, we have

$$\langle ex\,|H_{eh}^{(LR)}|ex\rangle \simeq \left(\langle 0| \sum_{\vec{p}'} \phi^*(\vec{K}/2 - \vec{p}') b_{v,-\vec{p}'}^\dagger b_{c,\vec{K}-\vec{p}'} \right) \tag{10.8.44}$$

$$\times \frac{e^2}{3\epsilon V} \sum_{\vec{k},\vec{k}',\vec{q}} F(\vec{q}) b_{v,\vec{k}-\vec{q}}^\dagger b_{c,\vec{k}'+\vec{q}}^\dagger b_{v,\vec{k}'} b_{c,\vec{k}} \left(\sum_{\vec{p}} \phi(\vec{K}/2 - \vec{p}) b_{c,\vec{K}-\vec{p}}^\dagger b_{v,-\vec{p}} |0\rangle \right).$$

Matching the operators as we did for the first term of (10.8.41) gives us

$$\langle ex\,|H_{eh}^{(LR)}|ex\rangle = \left(\sum_{p'} \phi(\vec{K}/2 - \vec{p}) \right) \frac{e^2}{3\epsilon V} F(\vec{K}) \left(\sum_{p} \phi(\vec{K}/2 - \vec{p}) \right). \tag{10.8.45}$$

For small values of $\vec{K}$, this becomes

$$\langle ex\,|H_{eh}^{(LR)}|ex\rangle = \frac{e^2}{3\epsilon V} |\varphi(0)|^2 |\vec{x}_{cv}|^2 (1 - 3\cos^2\theta), \tag{10.8.46}$$

where θ is the angle between the exciton wave vector $\vec{K}$ and the polarization $\vec{x}_{cv}$. The long-range interaction leads to a splitting between excitons with $\vec{K}$ perpendicular to the polarization $\vec{x}_{cv}$, that is, transverse excitons, and ones with $\vec{K}$ parallel to the polarization $\vec{x}_{cv}$, that is, longitudinal excitons. This gives the transverse–longitudinal splitting invoked for exciton-polaritons in Section 7.5.2.

In summary, for simple conduction band and valence bands with two spin states, the short-range exchange interaction leads to a singlet–triplet splitting, in which the triplet excitons couple to the electromagnetic field via an allowed optical transition and the singlet exciton state does not couple to the electromagnetic field, and the long-range interaction leads to a further splitting of the triplet state into a transverse doublet and longitudinal singlet.

Exercise 10.8.6   In the semiconductor GaAs, the valence band states at zone center are split by spin–orbit interaction into four degenerate states (see Section 1.13.1),

$$| + \tfrac{3}{2}\rangle = -|1\rangle| \uparrow\rangle$$

$$| + \tfrac{1}{2}\rangle = -\frac{1}{\sqrt{3}}|1\rangle| \downarrow\rangle - \frac{\sqrt{2}}{\sqrt{3}}|0\rangle| \uparrow\rangle$$

$$| - \tfrac{1}{2}\rangle = \frac{1}{\sqrt{3}}| -1\rangle| \uparrow\rangle + \frac{\sqrt{2}}{\sqrt{3}}|0\rangle| \downarrow\rangle$$

$$| - \tfrac{3}{2}\rangle = | -1\rangle| \downarrow\rangle, \tag{10.8.47}$$

where the states $|1\rangle, |0\rangle$, and $| -1\rangle$ are the Bloch spatial cell functions with p-symmetry, and $| \downarrow\rangle$ and $| \uparrow\rangle$ are the pure spin states. The spin-$\tfrac{3}{2}$ states are the "heavy-hole" states and the spin-$\tfrac{1}{2}$ states are the "light-hole" states. The conduction band states are pure spin states. Determine the electron–hole exchange splitting of exciton states made from electrons and holes in these two bands.

10.9 Spin Flip and Spin Dephasing

So far, we have treated the equilibrium properties of systems with spin–spin interaction, and have showed that exchange gives a lower energy for aligned spins in many systems. We have not yet discussed how a system approaches an equilibrium spin configuration, however. If a system starts with equal numbers of up and down spins, and many of the electrons flip spin to enter a ferromagnetic state, where does the net extra angular momentum come from?

In general, to have a spin flip, there must be some term in the Hamiltonian that has products of the form $b^{\dagger}_{\uparrow,\vec{k}} b_{\downarrow,\vec{k}'}$. It is not easy, in general, to come up with such terms, since angular momentum is a conserved quantity. In this section, we will not do detailed calculations of spin flip processes, but will just sketch the mechanisms of how spin flip can come about at all. In general, when spin scattering processes are weak, spin can be quasi-conserved for periods of time long enough that spin-polarized electrons can travel macroscopic distances. This allows the possibility of **spintronics**, that is, manipulation of spin currents instead of charge currents. Recent advances in spintronics are reviewed by Maekawa (2006).

The Coulomb interaction manifestly does not change the total spin of an electron population, since it has terms of the form

$$b^{\dagger}_{\uparrow,\vec{k}+\Delta\vec{k}} b^{\dagger}_{\downarrow,\vec{k}-\Delta\vec{k}} b_{\downarrow,\vec{k}'} b_{\uparrow,\vec{k}}, \tag{10.9.1}$$

in which each spin that is removed by a destruction operator is replaced by a creation operator. If the electron population under consideration is coupled to another electron population, however, then spin can be transferred to the other population by Coulomb interaction. One possibility is that spin in the conduction-electron states is transferred to a population of holes via the electron–hole exchange interaction discussed in Section 10.8.3. An electron in the conduction band can flip its spin by flipping the spin of a hole in the valence band, since the diagram of Figure 10.21(b) allows the outgoing electron and hole to have either

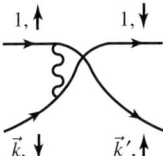

Fig. 10.22 Spin exchange with a localized electron.

spin, as long as it is the same for both. Spins in a free electron gas can also be flipped by interaction with localized electrons, such as we studied in Section 10.8.2. Figure 10.22 shows a process of spin flip via exchange with a localized electron. Again, the total angular momentum is conserved, but the spins of each have been flipped.

Spin flip by spin–orbit interaction. As discussed in Appendix F, relativistic corrections give a spin–orbit interaction, proportional to $(\nabla U \times \vec{p}) \cdot \vec{\sigma}$, where $U(\vec{r})$ is the potential felt by an electron in its orbital and $\vec{\sigma}$ is the Pauli spin operator. We have already used the spin–orbit interaction to derive its effect on single-particle band energies in Section 1.13.2. In general, we can modify the $k \cdot p$ theory presented in Section 1.9.4 to include the spin–orbit term, as follows:

$$\left(\frac{p^2}{2m} + U(\vec{r}) + \frac{\hbar}{4m^2c^2}(\nabla U \times \vec{p}) \cdot \vec{\sigma} \right) u_{n\vec{k}} e^{i\vec{k}\cdot\vec{r}} = E_n(\vec{k}) u_{n\vec{k}} e^{i\vec{k}\cdot\vec{r}}. \tag{10.9.2}$$

As in Section 1.9.4, we note that the $\vec{p}$ operator acts on the plane-wave phase factor to give a constant $\hbar\vec{k}$:

$$\left(\frac{1}{2m}|\vec{p} + \hbar\vec{k}|^2 + U(\vec{r}) + \frac{\hbar}{4m^2c^2}(\nabla U \times (\vec{p} + \hbar\vec{k})) \cdot \vec{\sigma} \right) u_{n\vec{k}} = E_n(\vec{k}) u_{n\vec{k}}. \tag{10.9.3}$$

The terms that do not depend on $\vec{k}$ just contribute to the band energy H_0; in particular, the term with $(\nabla U \times \vec{p}) \cdot \vec{\sigma}$ is the same as the $\vec{L} \cdot \vec{S}$ spin–orbit term discussed in Section 6.6, which leads to splitting of bands at $k = 0$ in some crystals. Taking the terms that are first order in $\vec{k}$, we have the effective k-dependent Hamiltonian

$$H = \frac{\hbar\vec{k}}{m} \cdot \left(\vec{p} + \frac{\hbar}{4mc^2} \vec{\sigma} \times \nabla U \right), \tag{10.9.4}$$

where we have reordered the operators using the fact that $\vec{k}$, $U(\vec{r})$, and $\vec{\sigma}$ commute. The term with $\vec{p}$ is just the standard $k \cdot p$ term introduced in Section 1.9.4. The remaining term is the k-dependent spin–orbit interaction,

$$\boxed{H_{\mathrm{so}} = \frac{\hbar\vec{k}}{m} \cdot \left(\frac{\hbar}{4mc^2} \vec{\sigma} \times \nabla U \right).} \tag{10.9.5}$$

Since H is a scalar, $\vec{\sigma} \times \nabla U$ must transform as a vector like $\vec{p}$.

If $|i\rangle$ is a pure spin eigenstate, the spin–orbit interaction will mix it with other states with opposite spin via the second-order perturbation according to

$$|i'\rangle = |i\rangle + \frac{\hbar^2}{4m^2c^2} \sum_{j\neq i} \frac{\vec{k}\cdot\langle j|\vec{\sigma}\times\nabla U|i\rangle}{E_i - E_j}|j\rangle, \tag{10.9.6}$$

where the sum over virtual intermediate states j includes all other electron bands and both spin states. The spin–orbit term can flip spin because the Pauli matrix has terms such as

$$\sigma_x = \begin{pmatrix} 0 & 1 \\ 1 & 0 \end{pmatrix}, \tag{10.9.7}$$

where off-diagonal elements correspond to the spin flip terms $b_{i\uparrow}^\dagger b_{i\downarrow} + b_{i\downarrow}^\dagger b_{i\uparrow}$. Specifically, for an original state with spin up, the new eigenstate (10.9.6) includes terms like the following:

$$k_z\langle j\downarrow|(\sigma_x\nabla_y U - \sigma_y\nabla_x U|i\uparrow\rangle = \frac{k_z}{V_{\text{cell}}}\int d^3r\, u_{j0}^*(\vec{r})(\nabla_y U(\vec{r}) - i\nabla_x U(\vec{r}))u_{i0}(\vec{r}). \tag{10.9.8}$$

For many crystal symmetries and bands, this term can be nonzero.

A surprising result of this analysis is that spin is not conserved in scattering when the spin–orbit term is taken into account, because spin is no longer a good quantum number for the eigenstates. We can write the perturbed eigenstates generally as

$$|i'\rangle = \alpha(\vec{k})|i\uparrow\rangle + \beta(\vec{k})|j\downarrow\rangle, \tag{10.9.9}$$

where the index j refers to the dominant band mixed in by the second-order perturbation summation (10.9.6). The factors $\alpha(\vec{k})$ and $\beta(\vec{k})$ will in general be different for different $\vec{k}$ states. Therefore, any process that leads to scattering to another $\vec{k}$ state, even if the matrix element is a scalar and does not itself have a spin flip term, will lead to a change in the fraction of the particle state that is spin up or spin down. For a scalar interaction such as Coulomb scattering or longitudinal phonon emission, the rate of such processes is proportional to

$$|\langle i'|i''\rangle|^2 = \left|\left(\alpha^*(\vec{k}')\langle i\uparrow| + \beta^*(\vec{k}')\langle j\downarrow|\right)\left(\alpha(\vec{k}'')|i\uparrow\rangle + \beta(\vec{k}'')|j\downarrow\rangle\right)\right|^2$$
$$= |\alpha^*(\vec{k}')\alpha(\vec{k}'') + \beta^*(\vec{k}')\beta(\vec{k}'')|^2. \tag{10.9.10}$$

This mechanism is known as an **Elliot–Yafet** mechanism.

Spin flip by hyperfine interaction. In systems in which an Elliot–Yafet phonon emission or absorption is forbidden, such as quantum-confined systems (e.g., quantum dots, discussed in Section 2.8) in which phonon emission is forbidden by energy and momentum conservation, the only remaining mechanism for spin flip is the hyperfine interaction of the electron spin with the nuclear spin. This term has the same form as the spin–spin interaction due to the magnetic dipole field generated by each spin, given in (10.8.1). The term

$$\vec{\sigma}^{(e)}\cdot\vec{\sigma}^{(\text{nuc})}, \tag{10.9.11}$$

for example, includes the term $\sigma_x^{(e)}\sigma_x^{(nuc)}$, which has four terms including, for example, $b_{e\uparrow}^\dagger b_{e\downarrow} b_{nuc\uparrow}^\dagger b_{nuc\downarrow}$. Spin can therefore be transferred from the electron to the fermions in the nuclei of the lattice. As discussed at the beginning of Section 10.8, however, this interaction is very weak, which corresponds to long spin flip times. In many systems, spin flip times can be microseconds or longer.

Spin dephasing. The spin flip processes discussed above are T_1 processes, in the language of coherent systems presented in Chapter 9. In many cases, we are also concerned about T_2 processes, that is, dephasing of spins without any net spin flip. Of course, all T_1 processes also act as T_2 processes, since any process that flips a spin also randomizes the phase.

One important spin dephasing process is the **Dyakonov–Perel** mechanism. In crystals without inversion symmetry, the spin degeneracy is broken for electronic states with finite k in some bands. The spin splitting term therefore acts like a magnetic field, and as in a magnetic field, the spin will precess.

The spin–orbit interaction (10.9.5) used above gives a k-dependent spin splitting. The first-order energy shift due to this term is proportional to

$$\langle \vec{k} \uparrow | (k_y \sigma_z \nabla_x U - k_x \sigma_z \nabla_y U)|\vec{k} \uparrow\rangle = \tag{10.9.12}$$

$$k_y \int d^3 r \, u_{i\vec{k}}^*(\vec{r})\nabla_x U(\vec{r})u_{i\vec{k}}(\vec{r}) - k_x \int d^3 r \, u_{i\vec{k}}^*(\vec{r})\nabla_y U(\vec{r})u_{i\vec{k}}(\vec{r})$$

for the spin-up state; the term for the spin-down state has the opposite sign. In a crystal with inversion symmetry, the gradient $\nabla U(\vec{r})$ is antisymmetric, and therefore this term will vanish. In crystals without inversion symmetry, however, this term can be nonzero. In this case, even in the absence of a magnetic field, there is a splitting of the spin degeneracy. This does not violate Kramer's rule, as discussed in Sections 1.13.2 and 6.9, which says

$$E_{\pm m_J}(-\vec{k}) = E_{\mp m_J}(\vec{k}). \tag{10.9.13}$$

States with the same $\vec{k}$ but opposite spin can have different energy, but two states with opposite spin and opposite $\vec{k}$ must have the same energy.

One example of this is the **Rashba effect** in a two-dimensional system, such as a surface or a quantum well. We assume that there is a gradient of the potential in the z-direction, normal to the two-dimensional plane. In this case, we can write the spin–orbit Hamiltonian (10.9.5) simply as

$$H_R = \frac{\hbar^2}{4m^2c^2}|\nabla U| \, \vec{k}\cdot(\vec{\sigma}\times\hat{z})$$

$$= \alpha_R(\sigma_y k_x - \sigma_x k_y), \tag{10.9.14}$$

where α_R is the Rashba constant, which includes the effective electric field E_0 due to the potential gradient. It is easy to solve for the eigenvectors and energies in this case; for $\vec{k}$ in the x-direction, the eigenvectors are $(i, 1)$ and $(-i, 1)$, which are the $\pm y$ spin eigenvectors. Adding the linear spin–orbit splitting due to the Rashba term to a standard quadratic dependence due to an effective band mass, we obtain the band structure shown in Figure 10.23. As seen in Figure 10.23(a), for equal energy, the spin direction rotates as the direction

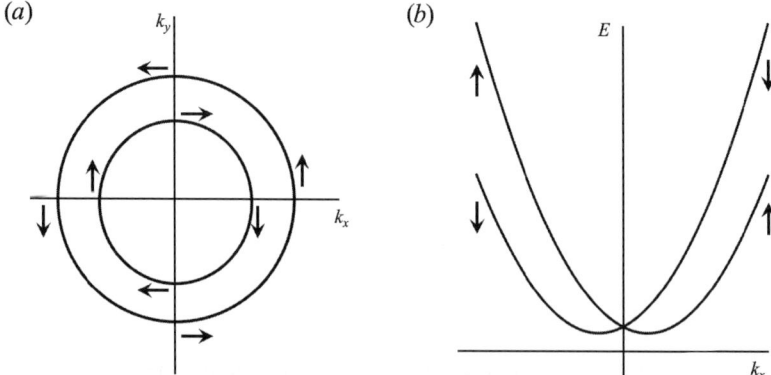

Fig. 10.23　Splitting of spin states due to the Rashba spin–orbit interaction. (a) Equal-energy lines in the two-dimensional plane in k-space. The arrows illustrate the eigenstate spin projection. (b) Energy as a function of k_x. The arrows illustrate the y-component of the spin for each band.

changes in k-space. At a given value of $\vec{k}$, there is a splitting between states with opposite spin.

A k-dependent spin splitting acts like a magnetic field, since it splits the energy of spin states. In general, as shown in Section 9.2, if the spin states are split, and the upper state is excited, the spin will oscillate between the two states; this is the quantum analog of a magnetic dipole precessing in a magnetic field. In the Dyakonov–Perel mechanism, an electron undergoes scattering to different k-states, and therefore feels random jumps in the effective magnetic field. This gives random jumps in the precession angle, that is, dephasing.

A counterintuitive property of this mechanism is that increasing scattering gives *decreasing* spin dephasing. This is a general effect of a fluctuating field, known as **motional narrowing** in the context of magnetic resonance. If we consider the evolution of a Bloch vector in the m_1–m_2 plane, using the Bloch sphere model of Chapter 9, a fluctuating field has the effect of random changes in the precession speed and direction. The Bloch vector therefore will undergo a type of random walk, or diffusion, in the m_1–m_2 plane. The more often the field fluctuates, the lower the diffusion constant will be; analogous to the transport diffusion constant discussed in Section 5.6, the phase diffusion constant will be proportional to $\bar{\omega}^2 \tau$, where $\bar{\omega}$ is the average angular velocity and τ is the average scattering time. Therefore, the shorter τ is, the slower the spin phase will diffuse away from its original value.

Exercise 10.9.1　Use the estimate of the spin–orbit interaction energy at the beginning of Section 1.13 to estimate the strength of the Rashba term. First, show that the Rashba energy is of order $(ka_{\text{eff}})\langle H_{SO}\rangle$, where $a = a_0/Z_{\text{eff}}$ is the effective Bohr radius of an atom and Z_{eff} is defined in Section 1.13. Then estimate the magnitude of this term for an electron with magnitude of its k-vector about one-tenth of the Brillouin zone, for a lattice constant $a = 5$ Å, and $Z_{\text{eff}} = 4$.

Exercise 10.9.2 Using the same order-of-magnitude approximation as the previous exercise, estimate the typical coefficient $\beta(\vec{k})$ that gives the fraction of opposite spin, according to (10.9.6). Assume that the nearby state mixed in is about 0.1 eV away, and take the other parameters from the previous exercise. Is there an appreciable fraction of opposite spin mixed in, for k around one-tenth of the Brillouin zone?

References

G. Bertotti, *Hysteresis in Magnetism* (Academic Press, 1998).

P.M. Chaikin and T.C. Lubensky, *Principles of Condensed Matter Physics* (Cambridge University Press, 1995).

A.S. Chakravarty, *Introduction to Magnetic Properties of Solids* (Wiley, 1980).

C. Cohen-Tannoudji, B. Diu, and F. Laloë, *Quantum Mechanics* (Wiley, 1979).

W. Du Trémolet de Lacheisserie, D. Gignous, and M. Schlenker, eds., *Magnetism: Fundamentals* (Kluwer, 2003).

C.P. Frahm, "Some novel delta-function identities," *American Journal of Physics* **51**, 826 (1983); J. Franklin, "Comment," *American Journal of Physics* **78**, 1225 (2010).

S. Maekawa, ed., *Concepts in Spin Electronics* (Oxford University Press, 2006).

D.C. Mattis, *Theory of Magnetism Made Simple* (World Scientific, 2006).

G. Mazenko, *Fluctuations, Order, and Defects* (Wiley, 2003).

R.A. McCurrie, *Ferromagnetic Materials: Structure and Properties* (Academic Press, 1994).

M. Nowicki, R. Szewczyk, T. Charubin, A. Marusenkov, A. Nosenko, and V. Kyrylchuk, "Modeling the hysteresis loop of ultra-high permeability amorphous alloy for space applications," *Materials* **11**, 2079 (2018).

F. Reif, *Fundamentals of Statistical and Thermal Physics* (McGraw-Hill, 1969).

F. Schwabl, *Statistical Mechanics*, W. Brewer, trans. (Springer, 2002).

L.I. Schiff, *Quantum Mechanics*, 3rd ed. (McGraw-Hill, 1968).

G. Shirane, R. Nathans, O. Steinsuoll, H.A. Alperin, and S.J. Pickard, "Measurement of the magnon dispersion relation of iron," *Physical Review Letters* **15**, 146 (1965).

J.C. Toledano and P. Toledano, *The Landau Theory of Phase Transitions* (World Scientific, 1987).

J.M. Ziman, *Principles of the Theory of Solids*, 2nd edition (Cambridge University Press).

11 Spontaneous Coherence in Matter

In the discussion of ferromagnets in Chapter 10, we introduced the basic concepts of phase transitions. In this chapter, we will discuss another type of phase transition, in which there is spontaneous symmetry breaking just as in other phase transitions, but the spontaneous symmetry breaking leads to macroscopic coherence. In other words, thermodynamics can cause matter to act as a wave on macroscopic scales. We may tend to think that the wave nature of matter only plays a role on microscopic length scales, but in the case of spontaneous coherence of matter, the proper description of a system with billions of particles is as a wave, and intrinsic wave properties such as the phase are directly related to macroscopic observables such as the current. Because of this, spontaneously coherent matter has sometimes bizarre properties unlike any other types of matter.

The primary experimental example of spontaneous coherence in solids is superconductivity. Before discussing superconductivity, however, we first review the theory of Bose–Einstein condensation, which is the paradigm for spontaneous coherence in matter. In recent years, this theory has been applied to solid state quasiparticles as well, such as polaritons and magnons, and we review some of these systems at the end of this chapter.

Many students of superconductivity miss the fact that superconductivity is essentially the same phenomenon as Bose–Einstein condensation. We typically think of Bose–Einstein condensation as occurring in a gaseous or liquid state, but as we will see, it involves the same underlying physics as superconductivity in solids. Both superfluids and superconductors are therefore often called **quantum liquids**. Good general texts on quantum liquids are those by Annett (2004) and Fujita and Godoy (1996); many of the details on the hydrodynamics are found in works by Noziéres and Pines (1990) and Lifshitz and Pitaevskii (1980).

As we have seen throughout this book, we often can describe the relevant properties of a system in terms of quasiparticles which are constructed from underlying degrees of freedom. In quantum liquids, these new quasiparticles play such an important role that some people reserve the term quasiparticle exclusively for these new states in superfluids and superconductors. As we have seen in earlier chapters, however, the quasiparticle concept is a general one for all of solid state physics, and is a general tool for all quantum field theories.

Consider a gas of particles, as illustrated in Figure 11.1. If the interactions between the particles are weak, we can treat this system theoretically in terms of the number of particles

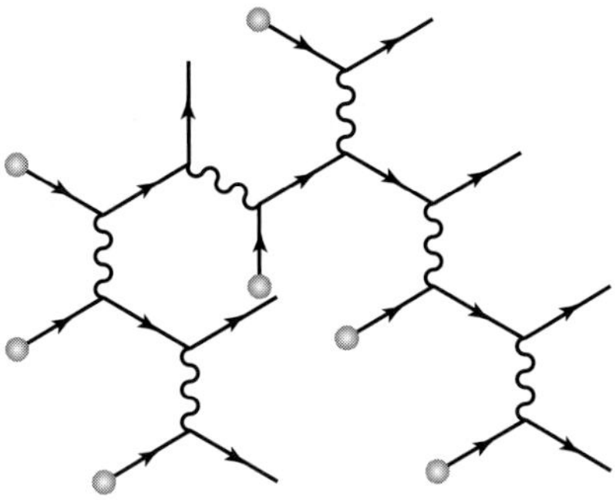

Fig. 11.1   A gas of interacting particles.

in each state, and account for the scattering transition rates from one state to another, using Fermi's golden rule as deduced in Chapter 4 with matrix elements like those deduced in Chapter 5. Suppose that we turn up the interaction strength, however. Now the original, unperturbed states are no longer good approximations of the eigenstates. We can account for the changes of the states using the many-body perturbation methods of Chapter 8. Each original state is mapped to a renormalized state.

Suppose that the interactions become even stronger. At some point, the interactions will be so strong that it does not make sense to view the new eigenstates as perturbations of the old ones. We must start with the full Hamiltonian and deduce the ground state and the eigenstates for excitations out of that ground state. These correspond to the quasiparticle states, which can be written in terms of the old particle states, but which cannot be mapped one-for-one to the old states in any meaningful way. The system has undergone a phase transition.

Phase transitions of electronic quasiparticles in solids are one step higher in this hierarchy. Suppose the original particles were electrons and nuclei in vacuum. A phase transition to a solid state leads to the appearance of new quasiparticles, namely free electrons and holes, phonons, and polaritons (photons in the medium). These new quasiparticles interact with each other with interaction Hamiltonians like those in Chapter 5. For weak interactions, we can describe the system in terms of transitions between states using Fermi's golden rule or the basic self-energy renormalization methods. If the interactions are strong, however, we solve for the ground state of the system including the interactions, and then solve for new quasiparticle states that are excitations of this ground state. This is the case with the superconducting phase transition; in the same way, magnons are quasiparticles of the ferromagnetic phase transition. In general, the new quasiparticles in this state themselves will also have interactions, which can lead to even higher-order renormalization effects.

As in Chapter 10, the Ginzburg–Landau theory of phase transitions plays a major role in our understanding of superfluids and superconductors. In general, once we have defined the proper order parameter for the quantum liquid phase transition, most of the standard theory of phase transitions applies.

11.1 Theory of the Ideal Bose Gas

We begin with the Hamiltonian for an ideal, spinless gas of bosons,

$$H = \sum_{\vec{k}} \frac{\hbar^2 k^2}{2m} a_{\vec{k}}^\dagger a_{\vec{k}}. \tag{11.1.1}$$

Here $a_{\vec{k}}^\dagger$ and $a_{\vec{k}}$ are the standard creation and annihilation operators. As we have already shown in Section 4.8.1, the distribution number of the particles is equal to

$$N_{\vec{k}} = \frac{1}{e^{\beta(\hbar^2 k^2/2m - \mu)} - 1}, \tag{11.1.2}$$

where μ is the chemical potential and $\beta = 1/k_B T$. In the normal state, the density of particles in the thermodynamic limit is simply the integral of N_k over all k:

$$n = \lim_{V \to \infty} \frac{N}{V} = \frac{1}{V} \sum_{\vec{k}} N_{\vec{k}}$$

$$= \frac{1}{(2\pi)^3} \int d^3 k \frac{1}{e^{\beta(\hbar^2 k^2/2m - \mu)} - 1}. \tag{11.1.3}$$

The integral (11.1.3) cannot account for all the particles in a system at all T and μ, since μ must have an upper bound of zero for the distribution function to remain defined at all energies. When $\mu = 0$, the integral becomes

$$n = 2.612 \frac{(m k_B T)^{3/2}}{(2\pi \hbar^2)^{3/2}}. \tag{11.1.4}$$

This does not mean that there is a real upper bound on the number of particles in the system. Formula (11.1.3) was generated by replacing the discrete states $\vec{k}$ with an integral over a continuous function. In this approximation, the density of states for a three-dimensional system, discussed in Section 1.8, is proportional to k. This implies zero states at zero kinetic energy, but of course we do not have zero states with zero energy; we have one state at zero energy. Therefore, we must treat this state separately. We write

$$n = \lim_{V \to \infty} \frac{N}{V} = n_0 + \frac{1}{(2\pi)^3} \int d^3 k \frac{1}{e^{\beta \hbar^2 k^2/2m} - 1}, \tag{11.1.5}$$

where

$$n_0 = \lim_{V \to \infty} \frac{N_0}{V}, \tag{11.1.6}$$

and N_0 is the number of particles in the ground state. This population is called the **condensate**, and can be a macroscopic fraction of the whole number of particles. In the thermodynamic limit, it is related to the chemical potential by the relation

$$\mu = -k_B T \ln \left(1 + \frac{1}{N_0} \right). \tag{11.1.7}$$

By inverting (11.1.4) we can define the critical temperature for the appearance of the condensate,

$$T_c = \frac{1.054\pi\hbar^2 n^{2/3}}{mk_B}. \tag{11.1.8}$$

Another way of writing this is in terms of the thermal DeBroglie wavelength, that is, the typical wavelength of a particle wave function at a given temperature, defined by $k_B T = p^2/2m = (2\pi\hbar)^2/2m\lambda_D^2$. The thermal wavelength at T_c is equal to

$$\lambda_D = 2.44 n^{-1/3}. \tag{11.1.9}$$

In other words, the phase transition occurs when the wavelength of the particles is comparable to the interparticle spacing. In the case of an interacting gas, this is the same point at which exchange effects become important, which are discussed in Section 11.3.

It follows from the above that the operators $a_0^\dagger$, a_0 asymptotically become c-numbers. To see this, consider the commutator

$$\left[\frac{a_0^\dagger}{\sqrt{V}}, \frac{a_0}{\sqrt{V}}\right] = \frac{1}{V}, \tag{11.1.10}$$

which asymptotically tends to zero. Therefore, the amplitudes $a_0^\dagger/\sqrt{V}, a_0/\sqrt{V}$ commute just like c-numbers. This is true for the operators for every k-state, but for the ground state, we also know that the condensate has the expectation value $\langle a_0^\dagger a_0\rangle/V = N_0/V = n_0$, which is nonzero in the thermodynamic limit. Since the product of the two operators is equal to n_0, and one is the complex conjugate of the other, we can then write

$$\frac{1}{\sqrt{V}}a_0^\dagger = \sqrt{n_0}e^{i\theta}, \qquad \frac{1}{\sqrt{V}}a_0 = \sqrt{n_0}e^{-i\theta}. \tag{11.1.11}$$

On the other hand, this leads to problems if we assume that the condensate is a Fock state with an exact number of particles. If we write

$$\langle a_0\rangle = \langle N_0|a_0|N_0\rangle \tag{11.1.12}$$

where the state $|N_0\rangle$ is a Fock state, then we obtain $\langle a_0\rangle = 0$.

How can this apparent contradiction be resolved? This has been the subject of a fair amount of controversy over the years. In general, one can note that in an interacting gas, the state of the system is never a pure Fock state, but is always a superposition of many Fock states

$$|\psi\rangle = \alpha|N_0, N_1, N_2, \ldots\rangle + \beta|N_0', N_1', N_2', \ldots\rangle + \cdots, \tag{11.1.13}$$

so that $\langle\psi|a_0|\psi\rangle \neq 0$. Therefore, as long as the condensate is not a pure Fock state, but is a superposition of states like this, it can have nonzero $\langle a_0\rangle$. Because we know that $\langle a_0/\sqrt{V}\rangle = \sqrt{n_0}e^{-i\theta}$, we can immediately write $\langle a_0\rangle = \sqrt{N_0}e^{-i\theta}$.

This means that we can effectively treat the condensate as a coherent state. These states were introduced in Section 4.4 as the eigenstates of the operator a_0. Even if the condensate is not exactly a coherent state, we can assume that the deviations are unimportant.

Recall from Section 4.4 that a coherent state with large amplitude acts as a macroscopic wave; we equated coherent states of phonons and photons with classical waves. The same occurs here. The condensate will act as a macroscopic classical wave, but with a complex amplitude. Notice that this is an example of spontaneous symmetry breaking, because the absolute phase θ of the condensate amplitude is arbitrary. The order parameter of the phase transition in this case is the wave function of the condensate.

Although the condensate in this picture is a superposition of many number states, it can nevertheless have a very accurately determined number. The coherent state (4.4.5) is

$$|\alpha_0\rangle = e^{-N_0/2} \sum_{N=0}^{\infty} \frac{N_0^{N/2}}{\sqrt{N!}} e^{i\theta N} |N\rangle. \tag{11.1.14}$$

Using Stirling's formula for $N!$ in the limit $|\alpha_0| = \sqrt{N_0} \to \infty$ gives

$$|\langle N|\alpha_0\rangle|^2 = \frac{e^{-(N_0-N)}}{\sqrt{2\pi N}} \left(\frac{N_0}{N}\right)^N, \tag{11.1.15}$$

which has a sharp peak at $N = N_0$, with width $\Delta N = \sqrt{N_0}$, as seen in Figure 11.2. As N_0 increases, this approaches a δ-function. In other words, although a coherent condensed state is not a definite number state, it looks very much like one. Number-phase uncertainty says that the number and phase of the condensate cannot both be measured perfectly, but the uncertainty $\Delta N/N \to 0$ as N_0 becomes large.

The coherence of the condensate can be described in terms of *off-diagonal long-range order*, discussed in Section 4.5. To see this, we write the spatial field operator $\psi(\vec{r})$

$$\psi(\vec{r}) = \frac{1}{\sqrt{V}} \sum_{\vec{k}} a_{\vec{k}} e^{i\vec{k}\cdot r} = \frac{1}{\sqrt{V}} a_0 + \frac{1}{\sqrt{V}} \sum_{\vec{k}\neq 0} a_{\vec{k}} e^{i\vec{k}\cdot r}. \tag{11.1.16}$$

Using (11.1.11), an off-diagonal element of the density matrix is then

$$\langle \psi^\dagger(\vec{r}')\psi(\vec{r})\rangle = \frac{N_0}{V} + \frac{1}{V}\sum_{\vec{k}\neq 0}\langle a_{\vec{k}}^\dagger a_{\vec{k}}\rangle e^{i\vec{k}\cdot(\vec{r}-\vec{r}')}, \tag{11.1.17}$$

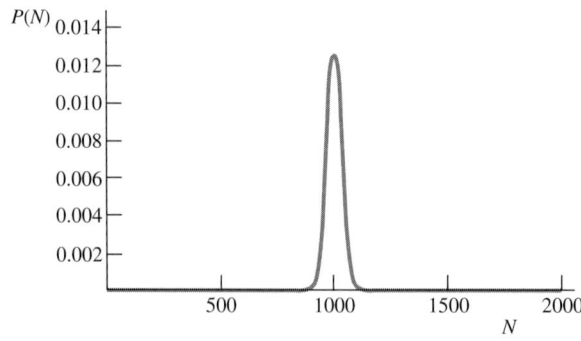

Fig. 11.2   Probability of a number N of particles in a coherent state when the amplitude of the wave function corresponds to an average of 1000 particles.

which remains finite at any distance $|\vec{r} - \vec{r}'|$ due to the first term in the right-hand side. In a normal system in thermal equilibrium, all off-diagonal elements of the density matrix are zero.

The above analysis relied on the interactions to make the system not have a pure Fock state for the condensate. What happens when we have a noninteracting, ideal gas at zero temperature? In that case, the system will be N_0 bosons in the ground state. As we will see in Section 11.3, in the case of an ideal gas, there is no symmetry-breaking mechanism, and therefore we cannot describe the condensate in terms of a definite phase. The ideal Bose gas is problematic in many ways; weak interactions, even very weak ones, are needed to make sense of many of the properties of condensates.

Exercise 11.1.1   Determine the critical *number* of particles for condensation at a fixed temperature in a three-dimensional harmonic potential with energy $U = \frac{1}{2}kr^2$. The quantum states in a harmonic potential are discrete, with energy $E = (n + \frac{3}{2})\hbar\omega$, where $\omega = \sqrt{k/m}$ and degeneracy $g = (n + 1)(n + 2)/2$, where $n = 0, 1, 2, \ldots$. Assume $k_B T \gg \hbar\omega$, so that one can treat the levels as a continuum of states with energy $E = n\hbar\omega$ and $g = (E/\hbar\omega)^2/2$.

11.2 The Bogoliubov Model

To model a weakly interacting Bose gas, we add a two-body interaction term to the Hamiltonian, as follows:

$$H = \sum_{\vec{k}} \frac{\hbar^2 \vec{k}^2}{2m} a_{\vec{k}}^{\dagger} a_{\vec{k}} + \frac{1}{2V} \sum_{\vec{p},\vec{q},\vec{k}} U(\vec{k}) a_{\vec{p}}^{\dagger} a_{\vec{q}}^{\dagger} a_{\vec{q}+\vec{k}} a_{\vec{p}-\vec{k}}. \tag{11.2.1}$$

The interaction term has the same form as the electron–electron interaction (5.5.7) deduced in Section 5.5; the form is the same for the Fourier transform of any short-range, two-body force.

Substituting the relations (11.1.11) in the Hamiltonian (11.2.1), we obtain

$$H = U(0)\frac{N_0^2}{2V} + \sum_{\vec{k} \neq 0} \frac{\hbar^2 \vec{k}^2}{2m} a_{\vec{k}}^{\dagger} a_{\vec{k}} + \sum_{\vec{k} \neq 0} U(0)\frac{N_0}{V} a_{\vec{k}}^{\dagger} a_{\vec{k}}$$

$$+ \sum_{\vec{k} \neq 0} U(\vec{k})\frac{N_0}{2V} \left(2a_{\vec{k}}^{\dagger} a_{\vec{k}} + a_{\vec{k}}^{\dagger} a_{-\vec{k}}^{\dagger} e^{-2i\theta} + a_{\vec{k}} a_{-\vec{k}} e^{2i\theta} \right)$$

$$+ \frac{\sqrt{N_0}}{V} \sum_{\vec{p},\vec{k} \neq 0} U(\vec{k}) \left(a_{\vec{p}}^{\dagger} a_{\vec{k}} a_{\vec{p}-\vec{k}} e^{i\theta} + a_{\vec{p}-\vec{k}}^{\dagger} a_{\vec{k}}^{\dagger} a_{\vec{p}} e^{-i\theta} \right)$$

$$+ \frac{1}{2V} \sum_{\vec{p},\vec{q},\vec{k} \neq 0} U(\vec{k}) \, a_{\vec{p}}^{\dagger} a_{\vec{q}}^{\dagger} a_{\vec{q}+\vec{k}} a_{\vec{p}-\vec{k}}, \tag{11.2.2}$$

where the sums do not include the condensate state. There are no terms with a single operator $a_{\vec{k}}$ or $a_{\vec{k}}^{\dagger}$ because momentum cannot be conserved in the interaction if three operators have $\vec{k} = 0$ and one does not.

We can rewrite the Hamiltonian in terms of the total number N by substituting

$$N_0 = N - \sum_{\vec{k} \neq 0} a_{\vec{k}}^{\dagger} a_{\vec{k}}. \tag{11.2.3}$$

If the number of particles in excited states is small compared to the number in the condensate, and the interactions between the excited particles are not too strong, we can drop all terms higher than quadratic order in the excited-states operators, to obtain

$$H = \frac{1}{2} L_0 N + \sum_{\vec{k} \neq 0} E_{\vec{k}} a_{\vec{k}}^{\dagger} a_{\vec{k}}$$
$$+ \sum_{\vec{k} \neq 0} \frac{L_{\vec{k}}}{2} \left(2 a_{\vec{k}}^{\dagger} a_{\vec{k}} + a_{\vec{k}}^{\dagger} a_{-\vec{k}}^{\dagger} e^{-2i\theta} + a_{\vec{k}} a_{-\vec{k}} e^{2i\theta} \right), \tag{11.2.4}$$

with

$$E_{\vec{k}} = \frac{\hbar^2 \vec{k}^2}{2m}, \qquad L_{\vec{k}} = U(\vec{k}) \frac{N}{V}. \tag{11.2.5}$$

This is the Bogoliubov Hamiltonian. We could set $\theta = 0$, since it is an arbitrary phase factor.

In Chapter 8, we dealt with systems in which we could not diagonalize the Hamiltonian, and instead developed methods for approximating the effects of interactions. In the present case, in which we assume the interactions are weak, we can diagonalize the interacting Hamiltonian. The Bogoliubov Hamiltonian is fairly special in this regard, because there are very few interacting systems that can be diagonalized analytically.

We can diagonalize the Hamiltonian by defining new operators. We write

$$a_{\vec{k}} = \frac{\xi_{\vec{k}} - A_{\vec{k}} \xi_{-\vec{k}}^{\dagger}}{\sqrt{1 - A_{\vec{k}}^2}} e^{-i\theta}, \qquad a_{\vec{k}}^{\dagger} = \frac{\xi_{\vec{k}}^{\dagger} - A_{\vec{k}} \xi_{-\vec{k}}}{\sqrt{1 - A_{\vec{k}}^2}} e^{i\theta}, \tag{11.2.6}$$

where $\xi_{\vec{k}}$ and $\xi_{\vec{k}}^{\dagger}$ are new bosonic quasiparticle operators, and $A_{\vec{k}}$ is defined so that the non-diagonal terms $a_{\vec{k}}^{\dagger} a_{-\vec{k}}^{\dagger}$ and $a_{\vec{k}} a_{-\vec{k}}$ vanish. Substituting these definitions into (11.2.4) and solving for $A_{\vec{k}}$ such that the non-diagonal terms vanish gives

$$A_{\vec{k}} = \frac{1}{L_{\vec{k}}} (E_{\vec{k}} + L_{\vec{k}} - E(\vec{k})), \tag{11.2.7}$$

with

$$E(\vec{k}) = \sqrt{E_{\vec{k}}^2 + 2 E_{\vec{k}} L_{\vec{k}}}. \tag{11.2.8}$$

If there are no interactions, then $L_{\vec{k}} = 0$ and $A_{\vec{k}} = 0$, and therefore $\xi_{\vec{k}} = a_{\vec{k}}$.

This transformation gives us the diagonal Hamiltonian

$$H = E_0 + \sum_{\vec{k}} E(\vec{k}) \xi_{\vec{k}}^{\dagger} \xi_{\vec{k}}, \tag{11.2.9}$$

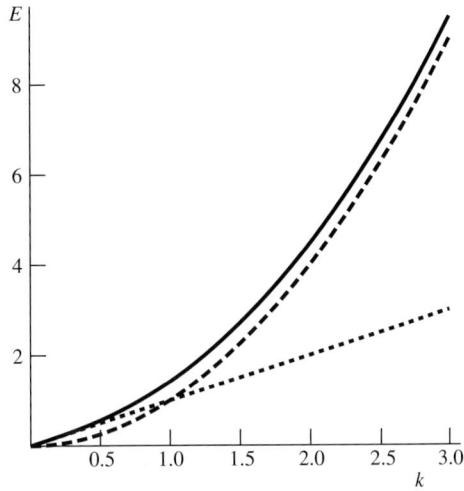

Fig. 11.3 Solid line: the Bogoliubov excitation spectrum. Short-dashed line: straight line with the same slope as the phonon-like part of the spectrum. Long-dashed line: $E = \hbar^2 k^2/2m$.

where

$$E_0 = \frac{1}{2}\frac{U(0)}{V}N^2 + \frac{1}{2}\sum_{\vec{k}}\left(E(\vec{k}) - E_{\vec{k}} - L_{\vec{k}}\right). \tag{11.2.10}$$

The sum over the $E(\vec{k}) - E_{\vec{k}} - L_{\vec{k}}$ will converge if $U(\vec{k}) \to 0$ as $\vec{k} \to \infty$. In the limit of weak interactions this sum is small compared to the first term.

The system therefore acts like a system with a new type of quasiparticle, which has the energy dispersion relation $E(\vec{k})$ given by (11.2.8). Figure 11.3 shows a typical plot of $E(\vec{k})$ for an interacting system. Near $\vec{k} = 0$, the dispersion is linear, like that of a phonon. At high energy, the dispersion is the same as the original kinetic energy of the particles in the absence of interactions, plus the potential energy term from interactions, L_k.

Exercise 11.2.1   Verify by direct substitution in the Hamiltonian (11.2.4) that the definition of $A_{\vec{k}}$ given in (11.2.6) will eliminate the off-diagonal terms.

Exercise 11.2.2   Show that the operators $\xi_{\vec{k}}^{\dagger}, \xi_{\vec{k}}$ obey boson commutation rules.

Self-energy. The ground state energy of a single particle is shifted upward relative to the noninteracting ground state energy by the chemical potential

$$\mu = \frac{\partial E_0}{\partial N} = \frac{U(0)}{V}N. \tag{11.2.11}$$

This extra energy per particle is not unique to the Bose-condensed system. Suppose that there is no condensate. Assuming no states have more than one particle, and that $U(\vec{k}) \approx U(0)$ is slowly varying with $\vec{k}$, the expectation value of the interaction energy term is

$$\frac{1}{2V} \sum_{\vec{p},\vec{q},\vec{k}} U(\vec{k}) \left\langle a_{\vec{p}}^{\dagger} a_{\vec{q}}^{\dagger} a_{\vec{q}+\vec{k}} a_{\vec{p}-\vec{k}} \right\rangle$$

$$= \frac{1}{2V} \sum_{\vec{p},\vec{q}} \left(U(0) \left\langle a_{\vec{p}}^{\dagger} a_{\vec{q}}^{\dagger} a_{\vec{q}} a_{\vec{p}} \right\rangle + U(\vec{p} - \vec{q}) \left\langle a_{\vec{p}}^{\dagger} a_{\vec{q}}^{\dagger} a_{\vec{p}} a_{\vec{q}} \right\rangle \right)$$

$$\approx \frac{U(0)}{V} \sum_{\vec{p},\vec{q}} \left\langle a_{\vec{p}}^{\dagger} a_{\vec{p}} a_{\vec{q}}^{\dagger} a_{\vec{q}} - a_{\vec{p}}^{\dagger} a_{\vec{p}} \delta_{\vec{q},\vec{p}} \right\rangle$$

$$= \frac{U(0)}{V} (N^2 - N) \simeq \frac{U(0)}{V} N^2, \tag{11.2.12}$$

and therefore

$$\mu = \frac{\partial E_0}{\partial N} = 2 \frac{U(0)}{V} N. \tag{11.2.13}$$

The shift of the energy per particle for the noncondensed case is twice that of the condensed case. This means that by becoming a condensate, the system saves a macroscopic energy. We will return to this in Section 11.3.

The energy shift is a straightforward first-order self-energy due to the particle–particle interaction. We see here several of the same effects that we saw in perturbation many-body theory in Chapter 8, such as energy renormalization, in which the single-particle ground state energy is shifted, and renormalization of the dispersion relation. There is no imaginary self-energy in this case because we have diagonalized to obtain exact eigenstates (under the approximation in which we discarded terms in (11.2.4)).

Note that if the interaction energy $U(\vec{k})$ is negative, the energy (11.2.8) can have an ill-defined region. For stability of the system in the Bogoliubov model, the interaction must be positive, that is a repulsive interaction between the particles. This does not mean that Bose–Einstein condensation cannot occur in a system with attractive interactions, however. If there is an attractive interaction, pair states of the bosons may arise. As long as the interactions between these pair states are repulsive, a condensate can exist.

This is another example of switching to a new quasiparticle picture of a system, as discussed at the beginning of this chapter. We have gone from an interacting system of one type of particle, to a new system with noninteracting particles. The new quasiparticles cannot be equated with the original particles, although in the limit of high kinetic energy they are nearly the same.

Exercise 11.2.3   Show, following a procedure similar to that of Section 5.5, that if the real-space interaction potential between two particles is proportional to a δ-function, $\delta(\vec{r}_1 - \vec{r}_2)$, then the interaction potential in k-space $U(\vec{k})$ is a constant.

11.3 The Stability of the Condensate: Analogy with Ferromagnets

In Section 11.2, we introduced the Bogoliubov model of a weakly interacting Bose gas and determined its energy spectrum. We have not yet seen what makes the condensate so special, however. We can gain insight into the special properties of the condensate by examining what happens if we try to break it apart.

We have already determined that the ground state energy of the condensate due to interactions is, to lowest order,

$$E_0 = \frac{1}{2}\frac{U(0)}{V}\langle a_0^\dagger a_0^\dagger a_0 a_0 \rangle = \frac{1}{2}\frac{U(0)}{V}N_0^2. \tag{11.3.1}$$

Suppose that instead of one condensate, in a single quantum state, we have two condensates in two degenerate states, with the same total number of particles $N_1 + N_2 = N_0$. We can always pick a second k-state so near to the ground state that its kinetic energy is negligible. In this case, assuming $U(\vec{k}_1 - \vec{k}_2) \simeq U(0)$, the ground state energy is equal to

$$
\begin{aligned}
E_0 &= \frac{1}{2}\frac{U(0)}{V}\langle a_1^\dagger a_1^\dagger a_1 a_1 + a_2^\dagger a_2^\dagger a_2 a_2 + a_1^\dagger a_1^\dagger a_1 a_2 + a_1^\dagger a_2^\dagger a_2 a_1 + a_2^\dagger a_1^\dagger a_1 a_2 + a_2^\dagger a_1^\dagger a_2 a_1\rangle \\
&= \frac{1}{2}\frac{U(0)}{V}\left(N_1^2 + N_2^2 + 4N_1 N_2\right) \\
&= \frac{1}{2}\frac{U(0)}{V}N_0^2 + \frac{U(0)}{V}N_1 N_2. \tag{11.3.2}
\end{aligned}
$$

Comparing this to (11.3.1), we see that there is an additional energy $U(0)N_1 N_2/V$ in this case, which is an exchange energy. To break up the condensate into two parts, we must pay a macroscopic energy penalty. In other words, the condensate will resist being broken up – it is a stable entity.

Another way to see the stability is by a kinetic argument. Recall from Section 4.7 that the scattering probability for bosons in Fermi's golden rule is proportional to $(1 + N_f)$, where N_f is the number of particles in the final state. Therefore, when there is a condensate, scattering processes into the condensate are enhanced by a macroscopic factor $(1 + N_0)$, while scattering out of the condensate into noncondensate states is not enhanced. If there are two condensates with slightly different numbers, then scattering will proceed preferentially from the one with the smaller number to the one with the larger number.

This analysis shows that the ideal gas is a pathological case. If there is no interaction between the particles, then there is no energy penalty for splitting the condensate endlessly. Alternatively, if there is no interaction, then there is no scattering matrix element with a final–states $(1 + N_0)$ factor to attract particles into the ground state.

If we go back to (11.2.2), we can estimate how the total energy, including exchange energy, depends on the fraction of the gas in the condensate. For constant $U(\vec{k}) = U_0$, at $T = 0$, the mean-field interaction energy of the system is given by the expectation value

$$\langle H \rangle = \frac{U_0}{2V}N_0^2 + \frac{U_0}{V}N_0\sum_{\vec{k}\neq 0}\langle a_{\vec{k}}^\dagger a_{\vec{k}}\rangle + \frac{U_0}{2V}\sum_{\vec{q},\vec{p}\neq 0}\langle a_{\vec{p}}^\dagger a_{\vec{q}}^\dagger a_{\vec{p}} a_{\vec{q}} + a_{\vec{p}}^\dagger a_{\vec{q}}^\dagger a_{\vec{q}} a_{\vec{p}}\rangle, \tag{11.3.3}$$

where only terms with equal number of creation and destruction operators acting on a given noncondensate state contribute, since the excited states are represented by a Fock state. Neglecting the effects of exchange among the excited state particles, that is, neglecting terms with $\vec{p} = \vec{q}$ as a small contribution to the whole sum, and writing $N_{ex} = N - N_0$, where N is the total number of particles, this becomes

$$\langle H \rangle = \frac{1}{2}\frac{U_0}{V}N_0^2 + \frac{U_0}{V}N_0 N_{ex} + \frac{U_0}{V}N_{ex}^2$$

$$= \frac{U_0}{V}\left(N^2 - NN_0 + \frac{1}{2}N_0^2\right). \tag{11.3.4}$$

As discussed above, putting particles in the condensate reduces the total energy of the system; when $N_0 = N$, the system has the lowest total energy. Actually, entropy considerations will deter this, because the condensate has negligible entropy. To account for the entropy, we write the free energy

$$F = \langle H \rangle - TS. \tag{11.3.5}$$

From (4.8.27), the entropy of a condensate in the limit $N \to \infty$ becomes vanishingly small. We can approximate the entropy of particles that leave the condensate by treating them as classical, distinguishable particles with negligible kinetic energy. From statistical mechanics, their entropy is given by the Sackur–Tetrode equation (see, e.g., Pathria and Beale 2011),

$$S = N_{ex}k_B\left(\ln\frac{V}{N_{ex}}\left(\frac{mk_BT}{2\pi\hbar^2}\right)^{\frac{3}{2}} + \frac{5}{2}\right), \tag{11.3.6}$$

which is linear with the number of particles in excited states, in the limit of low number. Adding this to the interaction energy (11.3.4), we can therefore write

$$\frac{F}{V} = A - U_0\frac{N}{V}|\psi_0|^2 + \frac{1}{2}U_0|\psi|^4 - Bk_BT(N - |\psi_0|^2), \tag{11.3.7}$$

where A and B are constants, and we have replaced the condensate number N_0 with the wave function ψ of the condensate, using $|\psi|^2 = N_0/V$.

The free energy (11.3.7) has the same generic form as the result for a ferromagnet in the Ising model below T_c, plotted in Figure 10.6. When we take into account the real and imaginary parts of the complex amplitude, we obtain a free energy surface like that shown in Figure 11.4. The symmetric center point at $\psi = 0$ is unstable, because increasing the number of particles in the condensate reduces the free energy. This occurs because exchange in an interacting bosonic system favors having particles in the same state, as we have seen. Eventually the entropy cost of adding particles to the low-entropy condensate state will prevent all the particles from entering it. In the two-dimensional surface, zero-energy variation of θ corresponds to the Goldstone mode, and oscillation of the magnitude in the radial direction is known as a **Higgs** mode.

As discussed at the end of Section 11.1, spontaneous symmetry breaking in this case means that the condensate will choose a phase angle θ in the complex plane. In other words, the system will have macroscopic coherence. Note that if there are no interactions, the minimum will be at $\langle \psi \rangle = 0$, and there will be nothing to drive the condensate to have a coherent complex amplitude.

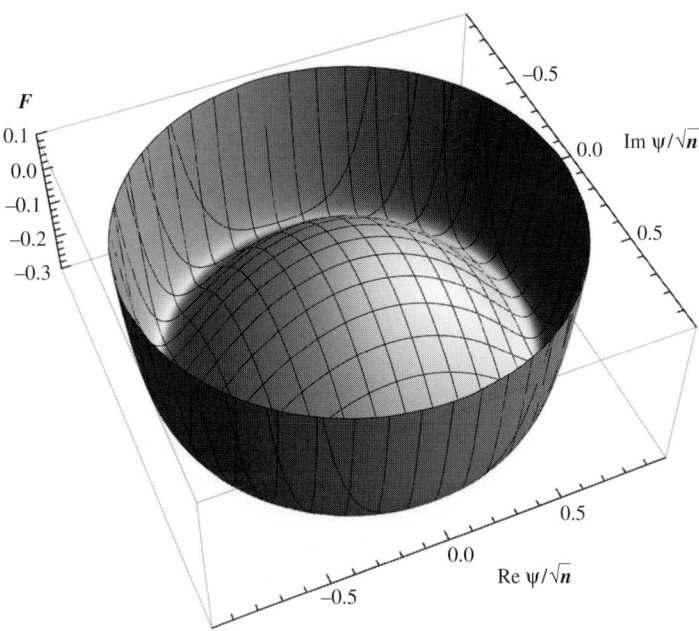

Fig. 11.4 The form of the free energy in the complex plane for a Bose condensate, for arbitrary units.

As we did for coupled spins in Section 10.3.1, we can analyze the fluctuations around equilibrium, using the coarse graining picture. Up to now, we have defined "a" phase of the condensate, θ, which we assigned to a $\vec{k} = 0$ state that extends through all space. The condensate need not be in a perfectly translationally invariant state, though. Any state that is macroscopically occupied will be a good condensate. When the system is out of equilibrium (e.g., a flowing condensate), or in a spatially inhomogeneous system (e.g., a condensate in a harmonic potential), we can still describe the condensate by a macroscopic wave function. To do this, we write the condensate wave function as

$$\psi(\vec{r}, t) = \sqrt{n_0(\vec{r}, t)}e^{i\theta(\vec{r},t)}, \qquad (11.3.8)$$

where the amplitude n_0 and the phase θ are both assumed to be constant over some fairly macroscopic distance, but allowed to vary over much longer distances.

Following the same argument as in Section 10.4, we can add a term to the free energy for spatial fluctuations of the condensate, to obtain

$$F = \int d^3x \left(a|\psi|^2 + \frac{1}{2}b|\psi|^4 + c|\nabla\psi|^2 \right), \qquad (11.3.9)$$

where a and b are constants. The order parameter in this case is complex, with two components. We account for these two components by treating ψ and ψ^* as independent. Minimizing the free energy with respect to variation of ψ^*, following a procedure just like that of Section 10.4, gives

$$a\psi + b|\psi|^2\psi - c\nabla^2\psi = 0. \qquad (11.3.10)$$

This is the Ginzburg–Landau equation for the condensate, analogous to (10.4.20) which we deduced in Section 10.4 for the Ising model of a ferromagnet. In fact, the Bose condensate wave function in the complex plane is completely analogous to a two-dimensional ferromagnet with spin oriented in a plane, known as the XY model.

Equation (11.3.10) has exactly the form of a Schrödinger equation for a particle in the condensate, with the potential energy given by $b|\psi|^2$, in other words, proportional to the local density. We can therefore identify the $\nabla^2\psi$ term as the kinetic energy, and rewrite the Ginzburg–Landau equation as

$$\left(U_0|\psi|^2 - \frac{\hbar^2}{2m}\nabla^2\right)\psi = E\psi. \tag{11.3.11}$$

The constant $-a$ can thus be seen as giving the energy per particle in the condensate, including all exchange effects, and the constant b as the potential energy felt by a particle due to the repulsion from all the other particles. We additionally recall that the energy E corresponds to the second time derivative in a time-dependent Schrödinger equation, and therefore write

$$\boxed{\left(U_0|\psi|^2 - \frac{\hbar^2}{2m}\nabla^2\right)\psi = i\hbar\frac{\partial}{\partial t}\psi.} \tag{11.3.12}$$

This is known as the **Gross–Pitaevskii** equation, or simply the **nonlinear Schrödinger equation**.

Converting the Ginzburg–Landau equation to a Schrödinger equation for particle behavior makes sense because, as discussed in Section 11.1, a condensate can be viewed as a classical wave. When a macroscopic number of particles all enter the same quantum mechanical state, we have simply the wave function of a single particle, with much larger amplitude.

Exercise 11.3.1   Show that linearizing the Gross–Pitaevskii equation leads to the Bogoliubov energy dispersion for excitations. To do this,

(a) Write the wave function of the condensate as $\psi_0 + \delta\psi$, where $\psi = \sqrt{n_0}e^{-i\omega_0 t}$ is the ground state wave function and $\delta\psi$ is a perturbation, and derive the linearized Gross–Pitaevskii equation by keeping terms no higher than linear in $\delta\psi$ or $\delta\psi^*$.

(b) Substitute into this equation the guess

$$\delta\psi = e^{-i\omega_0 t}(\alpha e^{i(kx-\omega t)} + \beta e^{-i(kx-\omega t)})$$

(the ground state phase, modulated by a general plane wave state), and write a 2×2 matrix equation for α and β. Solve this equation for the eigenvalues and eigenvectors and show that your solution gives you the Bogoliubov spectrum.

(c) Show that the solution in the limit $k \to 0$ corresponds to a density modulation of the condensate.

11.4 Bose Liquid Hydrodynamics

A system with a Bose–Einstein condensate is often called **superfluid**, because it has special properties. These arise fundamentally from the coherent wave nature of the condensate.

The proper quantum mechanical operator for the number current density given the probability density of a wave function is the symmetrized operator (see, e.g., Cohen-Tannoudji *et al.* 1977: s. 3.D.1.c.),

$$\vec{g} = \operatorname{Re}\left(\frac{1}{m}\psi^*(-i\hbar\nabla\psi)\right) = \frac{i\hbar}{2m}\left(\psi\nabla\psi^* - \psi^*\nabla\psi\right). \tag{11.4.1}$$

Subsituting in the field (11.3.8), this becomes

$$\begin{aligned}\vec{g} &= \frac{i\hbar}{2m}\left(\sqrt{n_0}e^{i\theta}(e^{-i\theta}\nabla(\sqrt{n_0}) - i\sqrt{n_0}e^{-i\theta}\nabla\theta)\right.\\ &\quad\left. - \sqrt{n_0}e^{-i\theta}(e^{i\theta}\nabla(\sqrt{n_0}) + i\sqrt{n_0}e^{i\theta}\nabla\theta)\right)\\ &= \frac{\hbar}{m}n_0\nabla\theta.\end{aligned} \tag{11.4.2}$$

In other words, the current is equal to the gradient of the phase. We see again here that the phase is a macroscopic observable, in contrast to the common notion that phase is only a mathematical construct. It plays the same role in relation to the superfluid current as electric potential plays in relation to electric field.

Standard vector calculus allows us to write down some relations based on this result. First, if the density n_0 is constant, which is the case if the temperature is constant, then we know that

$$\nabla\times\vec{g} = \frac{\hbar}{m}n_0\nabla\times(\nabla\theta) = 0, \tag{11.4.3}$$

since the curl of a gradient is zero. This implies that if the condensate density is the same everywhere, then the liquid is irrotational, since by Stokes' theorem, for any path that does not contain a pole,

$$\int_A (\nabla\times\vec{g})\cdot d\vec{A} = \oint_L \vec{g}\cdot d\vec{l} = 0, \tag{11.4.4}$$

where L is a closed loop and A is the area enclosed. More generally, if we allow for the possibility of a pole inside the loop, then we have

$$\begin{aligned}\oint_L \vec{g}\cdot d\vec{l} &= \frac{\hbar}{m}n_0\oint_L \nabla\theta\cdot d\vec{l}\\ &= \frac{\hbar}{m}n_0(\theta_2 - \theta_1),\end{aligned} \tag{11.4.5}$$

where θ_1 and θ_2 are the beginning and ending values of θ in going around the loop. If θ is single-valued, then we must have

$$\theta_2 - \theta_1 = 2\pi N, \tag{11.4.6}$$

where N is an integer. The circulation in any closed loop is therefore quantized:

$$\oint_L \vec{g} \cdot d\vec{l} = N \left(\frac{h}{m} n_0 \right). \tag{11.4.7}$$

We identify the number N as the number of **vortices**, that is poles in the superfluid current.

Exercise 11.4.1 The Navier–Stokes equation governing the flow of an incompressible fluid is (see, e.g., Chaikin and Lubensky 1995: 449)

$$\rho \frac{\partial \vec{v}}{\partial t} + \rho \vec{v} \cdot \nabla \vec{v} = -\nabla P + \eta \nabla^2 \vec{v}, \tag{11.4.8}$$

where ρ is the mass density, $\vec{v}$ is the fluid velocity, P is the pressure, and η is the viscosity. Show that the requirement $\nabla \times \vec{g} = 0$ implies that the viscosity term vanishes. Therefore, there can be no turbulence in an incompressible superfluid.

Exercise 11.4.2 (a) The Ginzburg–Landau equation can be written as

$$- U_0 n_0 \psi + U_0 |\psi|^2 \psi - \frac{\hbar^2}{2m} \nabla^2 \psi = 0, \tag{11.4.9}$$

where n_0 is the condensate density when $\nabla \psi = 0$. Show that in the case of a straight-line vortex with cylindrical symmetry, this can be written in unitless form as

$$\frac{1}{y} \frac{\partial}{\partial y} \left(y \frac{\partial f}{\partial y} \right) - \frac{f}{y^2} + f - f^3 = 0, \tag{11.4.10}$$

where $y = r/r_0$, with $r_0 = \hbar/\sqrt{2mU_0 n_0}$, and $\psi = \sqrt{n_0} e^{i\theta} f(r/r_0)$. To do this, you should assume that the condensate phase θ is equal to the azimuthal angle ϕ everywhere.

(b) Use a program like Mathematica to solve this differential equation numerically for y from 0 to 5, with the boundary conditions $f(0) = 0$ and $f(\infty) = 1$. The easiest way to do this is to use a "target shoot" method in which you solve the initial-value problem $y(0) = 0$, $y'(0) = a$, and pick values of the slope a until you hit the "target" of $y(5) \approx 1$. Because of the singularity in the equation at $y = 0$, the numerical solution will be better behaved if you pick in initial value of y close to, but not exactly equal to 0, for example $y = 0.01$. In Mathematica, to generate the solution of the second-order differential equation you can use the NDSolve function, and then use the Table function to generate values of $f(y)$ for specific y choices. You should be able to generate a plot like that shown in Figure 11.5.

Critical velocity. Suppose that the value of θ oscillates slowly in space, that is,

$$\psi(\vec{r}) = \sqrt{n_0} e^{i\theta(\vec{r})} = \sqrt{n_0} e^{i\vec{k}_0 \cdot \vec{r}}. \tag{11.4.11}$$

This is clearly the same as a superfluid in which the condensate is not in the ground state, but in a nonzero $\vec{k}$-state (i.e., a moving state). This corresponds to a moving condensate, with velocity

$$\vec{v}_s = \frac{\vec{g}}{n_0} = \frac{\hbar \vec{k}_0}{m}. \tag{11.4.12}$$

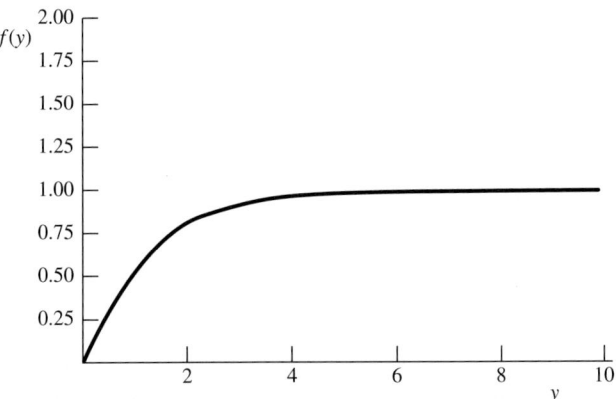

Fig. 11.5 Magnitude of $f = |\psi|/\sqrt{n_0}$ as a function of radial distance from the center of a vortex, from a solution of the Ginzburg–Landau equation.

In the condensate frame of reference, we define $k = 0$ as the condensate state, in which case the energy is given by

$$H = \sum_{\vec{k}} \frac{\hbar^2 |\vec{k} - \vec{k}_0|^2}{2m} a_{\vec{k}}^\dagger a_{\vec{k}} + \frac{1}{2V} \sum_{\vec{p},\vec{q},\vec{k}} U(\vec{k}) a_{\vec{p}}^\dagger a_{\vec{q}}^\dagger a_{\vec{q}+\vec{k}} a_{\vec{q}-\vec{k}} \tag{11.4.13}$$

$$= \left(\sum_{\vec{k}} \frac{\hbar^2 k^2}{2m} a_{\vec{k}}^\dagger a_{\vec{k}} + \frac{1}{2V} \sum_{\vec{p},\vec{q},\vec{k}} U(\vec{k}) a_{\vec{p}}^\dagger a_{\vec{q}}^\dagger a_{\vec{q}+\vec{k}} a_{\vec{q}-\vec{k}} \right) + N \frac{\hbar^2 k_0^2}{2m} - \sum_{\vec{k}} \frac{\hbar^2 \vec{k} \cdot \vec{k}_0}{m} a_{\vec{k}}^\dagger a_{\vec{k}}.$$

The term in parentheses is transformed by the same procedure as in Section 11.2 to a diagonal Hamiltonian in terms of the new quasiparticle operators $\xi_{\vec{k}}^\dagger, \xi_{\vec{k}}$. We can transform the last term as well, so that we obtain

$$H = E_0 + \sum_{\vec{k}} \left(E(\vec{k}) - \frac{\hbar^2 \vec{k} \cdot \vec{k}_0}{m} \right) \xi_{\vec{k}}^\dagger \xi_{\vec{k}} + N \frac{\hbar^2 k_0^2}{2m} \tag{11.4.14}$$

where $E(\vec{k})$ is given by (11.2.8). The total energy is increased by the kinetic energy of the moving condensate, but the excitation spectrum is also affected – the system will be unstable if $\vec{k}_0$ is large enough. As discussed in Section 11.2, the excitation energy $E(\vec{k})$ is linear near $\vec{k} = 0$, with the phonon-like dispersion

$$E(\vec{k}) \approx \hbar v_B k = \lim_{k \to 0} \sqrt{E_k^2 + 2 E_k L_k} = \hbar \sqrt{\frac{NU(0)}{mV}} k. \tag{11.4.15}$$

Therefore, when $v_s > v_B$, the system will be unstable to excitations with negative energy. Below this critical velocity, however, the system is stable. The condensate moves as a whole, without the random scattering processes discussed in Chapter 5. This is the essence of superfluidity, and as we will see, superconductivity.

Exercise 11.4.3 Show that

$$\sum_{\vec{k}} \frac{\hbar^2 \vec{k} \cdot \vec{k}_0}{m} a_{\vec{k}}^\dagger a_{\vec{k}} = \sum_{\vec{k}} \frac{\hbar^2 \vec{k} \cdot \vec{k}_0}{m} \xi_{\vec{k}}^\dagger \xi_{\vec{k}}, \tag{11.4.16}$$

as used in the transformation above. To do this, start by rewriting it as

$$\frac{1}{2} \sum_{\vec{k}} \frac{\hbar^2}{m} \left((\vec{k} \cdot \vec{k}_0) a_{\vec{k}}^\dagger a_{\vec{k}} - (\vec{k} \cdot \vec{k}_0) a_{-\vec{k}}^\dagger a_{-\vec{k}} \right)$$

$$= \frac{1}{2} \sum_{\vec{k}} \frac{\hbar^2}{m} (\vec{k} \cdot \vec{k}_0) \left(a_{\vec{k}}^\dagger a_{\vec{k}} - a_{-\vec{k}}^\dagger a_{-\vec{k}} \right), \tag{11.4.17}$$

and then use the transformation (11.2.6). You will need to use the fact that the $\xi_{\vec{k}}^\dagger, \xi_{\vec{k}}$ operators obey boson commutation relations, and to assume that $A_{-\vec{k}} = A_{\vec{k}}$.

Exercise 11.4.4 This analysis using the Bogoliubov model does not apply to liquid helium-4, because it is a strongly interacting system. Nevertheless, we can estimate the critical velocity for the superfluid by the following approximation:

(a) The Born approximation, also known as the s-wave scattering approximation, relates the scattering potential to the scattering cross-section as follows (e.g., Cohen-Tannoudji *et al.* 1977: s. VIII.B.4):

$$\sigma = \frac{m^2}{4\pi^2 \hbar^4} \left| \int d^3r \, e^{i\vec{k}\cdot\vec{r}} U(\vec{r}) \right|^2. \tag{11.4.18}$$

Assuming that the interaction is a contact potential $U(\vec{r}) = U_0 \delta(\vec{r})$, show that for bosons this implies $U_0 = 4\pi\hbar^2 a/m$, where a is the s-wave scattering length and U_0 is the potential $U(0)$ that appears in the Bogoliubov formulas.

(b) Assuming that the s-wave scattering length a is comparable to the helium atom size, and that the superfluid density is $10^{22} \mathrm{cm}^{-3}$, use the interaction potential of (a) to determine the critical velocity.

11.5 Superfluids versus Condensates

All of the above analysis has assumed a three-dimensional system. What happens in lower dimensions?

If we apply the analysis of Section 11.1 to the case of a two-dimensional system, we have the integral

$$n = \frac{1}{(2\pi)^2} \int d^2k \, \frac{1}{e^{\beta(k^2/2m - \mu)} - 1}, \tag{11.5.1}$$

in other words, the total number of particles in excited states is the sum over the occupation number of each excited state. Near $k = 0$, the occupation number is proportional to $1/k^2$, but the d^2k term is proportional to k, unlike the three-dimensional d^3k which is proportional to k^2. Therefore, there is no upper bound to the two-dimensional integral over excited states. By this argument there can be no Bose condensation in two dimensions. Researchers in the 1970s were surprised to find, however, that two-dimensional thin films of liquid helium still acted as a superfluid. What is wrong with this analysis?

One way to think about it is to note that when $\mu \to 0$ from below, the number of particles in the ground state is given by

$$N_0 \approx \frac{1}{-\mu/k_B T}. \tag{11.5.2}$$

In a two-dimensional system, μ never reaches zero. It can become extremely close to zero, however, which means that the number of particles N_0 can be extremely large, even though it does not become infinite. This means that the stimulated scattering factor $1+N_f$ discussed in Section 11.3 which leads to condensation is still a major factor in two-dimensional systems.

A two-dimensional system with finite size is not qualitatively different from a three-dimensional system of finite size. In each case, μ is not strictly zero, but has a finite negative value such that N_0 is very large, but not infinite. As the size of the system increases, the relative fraction of the particles in the $\vec{k} = 0$ state of a three-dimensional system will approach a constant, while in a two-dimensional system this fraction will approach zero. The fraction of particles in some range dk around $k = 0$ remains finite in a two-dimensional system, however. We can connect this to the spatial correlation function of the off-diagonal long-range order. A δ-function in k-space corresponds to an infinite coherence length. A narrow peak in the k-space spectral density function corresponds to a large, but not infinite, coherence length. Kosterlitz and Thouless (1973) showed that the correlation function of the wave function of a two-dimensional superfluid falls off as distance to some negative power. Locally, a two-dimensional superfluid acts like a condensate, but the phase correlation of the order parameter ψ is washed out over long distances.

It is helpful to distinguish between the **condensate fraction** and the **superfluid fraction**. These terms are not synonymous. In a three-dimensional system, the condensate fraction is given by the fraction of the particles in the $k = 0$ state, while the superfluid fraction is the fraction of the fluid that moves with the superfluid velocity v_s. This can be deduced as follows.

Let us adopt, again, the frame of reference in which the superfluid is at rest. In the laboratory frame, the quasiparticles of excitation, if there are any, have zero average velocity; therefore, in the superfluid frame, they have average velocity v_s. In this frame of reference, the entire momentum of the fluid is carried by the quasiparticles. The total current is

$$\vec{g} = \frac{1}{V} \sum_{\vec{k}} f(\vec{k}) \frac{\hbar \vec{k}}{m}, \tag{11.5.3}$$

where $f(\vec{k})$ is the thermal occupation number for the excitation quasiparticles, given by

$$f(\vec{k}) = \frac{1}{e^{\beta E'(\vec{k})} - 1}, \tag{11.5.4}$$

where the energy of the quasiparticles in the moving reference frame is

$$E'(\vec{k}) = E(\vec{k}) - \hbar^2 \vec{k} \cdot \vec{k}_0/m = E(\vec{k}) - \vec{p} \cdot \vec{v}_s, \tag{11.5.5}$$

as derived in Section 11.4. (Note that the number of quasiparticles is not conserved, and therefore they have a Planck distribution with zero chemical potential in equilibrium.)

Assuming that the superfluid velocity is small, we can then write

$$f(E'(\vec{k})) = f(E(\vec{k})) + \frac{\partial f}{\partial E}(E'(\vec{k}) - E(\vec{k})). \tag{11.5.6}$$

Using this in (11.5.3), we have

$$\vec{g} = \frac{1}{V} \sum_{\vec{k}} f(E(\vec{k})) \frac{\hbar \vec{k}}{m} - \frac{1}{V} \sum_{\vec{k}} \frac{\partial f}{\partial E}(\vec{p} \cdot \vec{v}_s) \frac{\hbar \vec{k}}{m}. \tag{11.5.7}$$

The first term vanishes, since the average velocity for the non-moving condensed system is zero. The second term has terms in the sum multiplied by $(p_x^2 \hat{x} + p_x p_y \hat{y} + p_x p_z \hat{z})$, assuming $\vec{v}_s$ is along $\hat{x}$. For an isotropic system, only the term proportional to p_x^2 is nonzero, equal to $\frac{1}{3} p^2$. This lets us rewrite the current, to first order, as

$$\vec{g} = -\frac{1}{3V} \sum_{\vec{p}} \frac{p^2}{m} \frac{\partial f}{\partial E} \vec{v}_s. \tag{11.5.8}$$

We define the *normal* density n_n as the density of the current moving with velocity v_s in this reference frame, that is at rest in the laboratory frame:

$$\vec{g} = n_n \vec{v}_s. \tag{11.5.9}$$

This implies

$$n_n = -\frac{1}{3V} \sum_{\vec{p}} \frac{p^2}{m} \frac{\partial f}{\partial E} = \frac{1}{3V} \sum_{\vec{p}} \frac{p^2}{m} \frac{\beta e^{\beta E(\vec{p})}}{(e^{\beta E(\vec{p})} - 1)^2}. \tag{11.5.10}$$

The superfluid density is then defined as

$$n_s = n - n_n. \tag{11.5.11}$$

On the other hand, let us calculate the *condensate* fraction, that is, the number of particles in the ground state. We define this as

$$N_0 = N - \sum_{\vec{k}} N_{\vec{k}}, \tag{11.5.12}$$

where N is the total number of particles and $N_{\vec{k}} = \langle a_{\vec{k}}^\dagger a_{\vec{k}} \rangle$. We can determine this in the Bogoliubov model using definitions (11.2.6)–(11.2.8), which yield

$$\langle a_{\vec{k}}^\dagger a_{\vec{k}} \rangle = \frac{\xi_{\vec{k}}^\dagger \xi_{\vec{k}} + A_{\vec{k}}^2 \xi_{-\vec{k}} \xi_{-\vec{k}}^\dagger}{1 - A_{\vec{k}}^2}, \tag{11.5.13}$$

where we have used the fact that terms with two creation or two destruction operators vanish in the expectation value. For an isotropic distribution, this is the same as

$$\langle a_{\vec{k}}^{\dagger} a_{\vec{k}} \rangle = \frac{f(k) + A_k^2(1 + f(k))}{1 - A_k^2}, \tag{11.5.14}$$

where, again, $f(k)$ is the occupation number of the quasiparticles. In the $T = 0$ limit, $f(k)$ is zero for all $\vec{k}$, and we have simply

$$N_0 = N - \sum_{\vec{k}} \frac{A_k^2}{1 - A_k^2}. \tag{11.5.15}$$

When $\vec{k} \to 0$, then $E_k \to 0$, proportional to k^2, and $E(\vec{k}) \to 0$, proportional to k. Therefore

$$\frac{A_k^2}{1 - A_k^2} \to \frac{L_k^2 - 2L_k E(\vec{k})}{L_k^2 - (L_k^2 - 2L_k E(\vec{k}))} \to \frac{L_k}{2E(\vec{k})} \sim \frac{1}{k}, \tag{11.5.16}$$

which is integrable in three dimensions. In the $k \to \infty$ limit, $E(\vec{k}) \to E_k + L_k - L_k^2/4E_k$, and therefore

$$\frac{A_k^2}{1 - A_k^2} \to \frac{(L_k^2/4E_k)^2}{1} \sim \frac{1}{k^4}. \tag{11.5.17}$$

The integral over finite-momentum states in (11.5.15) is therefore well behaved and nonzero.

Clearly, the superfluid fraction and the condensate fraction are not the same. When $T \to 0$, the formula (11.5.10) implies $n_s/n \to 1$, while the result (11.5.15) implies that even at $T = 0$ not all the particles will be in the $k = 0$ state; the ground state of the system has some of the particles kicked out of the $k = 0$ state due to the interactions. (Note that both of these formulas rely on the Bogoliubov model, which is valid only near $T = 0$, that is, when most of the particles are in the condensate. They will both break down near T_c.) One way to think of this is that at $T = 0$, all of the particles are in the ground state, so that 100% of the system is superfluid, but that state is not a plane wave, because the interactions smear it out in k-space.

A similar analysis applied to weakly interacting two-dimensional systems gives a nonzero value for the superfluid fraction, even while the condensate fraction, in an infinite system, is zero. The division of the system into a normal and a superfluid component is called the **two-fluid model**. A superfluid system acts as though it has two interpenetrating fluids, one which corresponds to the ground state of the system, and moves without drag, and one which consists of quasiparticle excitations out of this ground state, and which acts like a normal collection of scattering particles. This leads to the existence of **second sound**, which is an oscillation of the superfluid relative to the normal fluid. **First sound** is a simple density oscillation, which exists in normal fluids as well as superfluids; second sound is an oscillation of the fraction of the fluid not in the ground state, that is, in the excited quasiparticles. See Griffin (1993) for a discussion of second sound.

Exercise 11.5.1   Plot $k^2(A_k^2)/(1 - A_k^2)$ vs k assuming constant interaction vertex U, for various choices of U, to see what the distribution of excited particles looks like at $T = 0$.

11.6 Constructing Bosons from Fermions

All of normal matter is made of fermions, not pure bosons. How then can we have a Bose condensate of normal matter? The essence is fairly simple: When two fermions are bound together, their spins combine together according to the group-theoretic outer product rule $\frac{1}{2} \otimes \frac{1}{2} = 0 \oplus 1$. Two half-integer-spin fermions (or any even number) can join together to form an integer-spin boson. The same applies for any even number of fermions.

The canonical example is Cooper pairs of two electrons. We write the Hamiltonian for two fermions with opposite spin with a mutual interaction,

$$H = \sum_{\vec{k}} \frac{\hbar^2 k^2}{2m} b^\dagger_{\uparrow \vec{k}} b_{\uparrow \vec{k}} + \sum_{\vec{k}} \frac{\hbar^2 k^2}{2m} b^\dagger_{\downarrow \vec{k}} b_{\downarrow \vec{k}} + \frac{1}{2V} \sum_{\vec{k},\vec{k}',\vec{q}} U(\vec{q}) b^\dagger_{\downarrow \vec{k}} b^\dagger_{\uparrow \vec{k}'} b_{\uparrow,\vec{k}'+\vec{q}} b_{\downarrow,\vec{k}-\vec{q}}, \quad (11.6.1)$$

where $b^\dagger_{s,\vec{k}}$ and $b_{s,\vec{k}}$ are the fermion creation and destruction operators for a state with spin s and momentum $\vec{k}$. The role of spin here is to make the fermions distinguishable, another difference between the states would also give the same results, for example, if they were in different energy bands.

We assume that $U(\vec{k})$ is an attractive potential. In this case, we want to find the bound states of two particles. To find the pair eigenstates of this Hamiltonian, we guess that the state has the form

$$|\text{pair}\rangle = \sum_{\vec{k},\vec{k}'} A_{\vec{k},\vec{k}'} \, b^\dagger_{\downarrow \vec{k}} b^\dagger_{\uparrow \vec{k}'} |0\rangle, \quad (11.6.2)$$

where $|0\rangle$ is the vacuum state, and write the eigenvalue equation

$$H|\text{pair}\rangle = E|\text{pair}\rangle. \quad (11.6.3)$$

This becomes, after using the anticommutation relation $\{b_{s\vec{k}}, b^\dagger_{s'\vec{k}'}\} = \delta_{s,s'}\delta_{\vec{k},\vec{k}'}$, and switching the names of the dummy variables,

$$\sum_{\vec{k},\vec{k}'} \left(\left(\frac{\hbar^2 k^2}{2m} + \frac{\hbar^2 k'^2}{2m} - E \right) A_{\vec{k},\vec{k}'} + \frac{1}{2V} \sum_{\vec{q}} U(\vec{q}) A_{\vec{k}-\vec{q},\vec{k}'+\vec{q}} \right) b^\dagger_{\downarrow \vec{k}} b^\dagger_{\uparrow \vec{k}'} |0\rangle = 0. \quad (11.6.4)$$

For the pair state to be an eigenfunction, we assume that for every pair $b^\dagger_{\downarrow \vec{k}} b^\dagger_{\uparrow \vec{k}'}$, we have

$$\left(\frac{\hbar^2 k^2}{2m} + \frac{\hbar^2 k'^2}{2m} - E \right) A_{\vec{k},\vec{k}'} + \frac{1}{2V} \sum_{\vec{q}} U(\vec{q}) A_{\vec{k}-\vec{q},\vec{k}'+\vec{q}} = 0. \quad (11.6.5)$$

We multiply by $e^{-i\vec{k}\cdot\vec{r}_1}$ and $e^{-i\vec{k}'\cdot\vec{r}_2}$, and sum over $\vec{k}$ and $\vec{k}'$, to obtain the Fourier transform

$$\sum_{\vec{k},\vec{k}'} \left(\frac{\hbar^2 k^2}{2m} + \frac{\hbar^2 k'^2}{2m} - E \right) A_{\vec{k},\vec{k}'} e^{-i\vec{k}\cdot\vec{r}_1} e^{-i\vec{k}'\cdot\vec{r}_2}$$

$$+ \frac{1}{2V} \sum_{\vec{k},\vec{k}'} \sum_{\vec{q}} U(\vec{q}) A_{\vec{k}-\vec{q},\vec{k}'+\vec{q}} e^{-i\vec{k}\cdot\vec{r}_1} e^{-i\vec{k}'\cdot\vec{r}_2} = 0. \quad (11.6.6)$$

Defining $\vec{R} = (\vec{r}_1 + \vec{r}_2)/2$ and $\vec{r} = \vec{r}_1 - \vec{r}_2$, this becomes

$$\sum_{\vec{k},\vec{k}'} \left(-\frac{\hbar^2}{2M} \nabla_{\vec{R}}^2 - \frac{\hbar^2}{2m_r} \nabla_{\vec{r}}^2 - E \right) A_{\vec{k},\vec{k}'} e^{i(\vec{k}+\vec{k}')\cdot\vec{R}} e^{i(\vec{k}-\vec{k}')\cdot\vec{r}/2}$$

$$+ \frac{1}{2V} \sum_{\vec{k},\vec{k}'} \sum_{\vec{q}} U(\vec{q}) A_{\vec{k}-\vec{q},\vec{k}'+\vec{q}} \, e^{i(\vec{k}+\vec{k}')\cdot\vec{R}} e^{i(\vec{k}-\vec{k}')\cdot\vec{r}/2} = 0, \qquad (11.6.7)$$

where $M = 2m$ is the total mass of the pair and $m_r = \frac{1}{2}m$ is the reduced mass. We can then separate the center-of-mass motion $\vec{R}$ from the internal motion if $A_{\vec{k},\vec{k}'}$ has the form

$$A_{\vec{k},\vec{k}'} = \delta_{\vec{K},\vec{k}+\vec{k}'} \phi(\tfrac{1}{2}(\vec{k}-\vec{k}')), \qquad (11.6.8)$$

which transforms the eigenvalue equation into

$$\left(-\frac{\hbar^2}{2M} \nabla_{\vec{R}}^2 - \frac{\hbar^2}{2m_r} \nabla_{\vec{r}}^2 - E \right) e^{i\vec{K}\cdot\vec{R}} \sum_{\vec{q}'} \phi(\vec{q}') e^{i\vec{q}'\cdot\vec{r}}$$

$$+ \frac{1}{V} \sum_{\vec{q}} U(\vec{q}) e^{i\vec{q}\cdot\vec{r}} e^{i\vec{K}\cdot\vec{R}} \sum_{\vec{q}'} \phi(\vec{q}') e^{i\vec{q}'\cdot\vec{r}} = 0 \qquad (11.6.9)$$

or

$$\left(-\frac{\hbar^2}{2M} \nabla_{\vec{R}}^2 - E_K \right) e^{i\vec{K}\cdot\vec{R}} \phi_n(\vec{r}) + \left(-\frac{\hbar^2}{2m_r} \nabla_{\vec{r}}^2 + U(\vec{r}) - E_n \right) e^{i\vec{K}\cdot\vec{R}} \phi_n(\vec{r}) = 0,$$

$$(11.6.10)$$

where we have switched to the Fourier transform, in real space,

$$\phi_n(\vec{r}) = \frac{1}{\sqrt{V}} \sum_{\vec{q}} \phi_n(\vec{q}) e^{i\vec{q}\cdot\vec{r}}, \qquad (11.6.11)$$

and we define

$$U(\vec{r}) = \frac{1}{V} \sum_{\vec{q}} U(\vec{q}) e^{i\vec{q}\cdot\vec{r}} \qquad (11.6.12)$$

and $E = E_n + E_K$.

The eigenstate of a pair is therefore described by a plane wave that describes the center-of-mass motion with wave vector $\vec{K}$, times the wave function for the relative motion of the two particles, which is the solution to a simple Schrödinger equation,

$$-\frac{\hbar^2}{2m_r} \nabla_{\vec{r}}^2 \phi_n(\vec{r}) + U(\vec{r}) \phi_n(\vec{r}) = E_n \phi_n(\vec{r}). \qquad (11.6.13)$$

If $U(\vec{r})$ has a local minimum, then the energies E_n will be a discrete set of bound states; for example, if $U(\vec{r}) = -e^2/r$ then the bound states will be $E_n = -E_1/n^2$, the standard Bohr atom wave function.

We have deduced that the pair states are given by

$$|\text{pair}\rangle = \sum_{\vec{k}} \phi_n(\tfrac{1}{2}\vec{K} - \vec{k}) \, b^\dagger_{\downarrow,\vec{K}-\vec{k}} \, b^\dagger_{\uparrow,\vec{k}} |0\rangle. \qquad (11.6.14)$$

We can define a new operator for the creation of a pair with center-of-mass wave vector $\vec{K}$,

$$c^\dagger_{\vec{K},n} = \sum_{\vec{k}} \phi_n(\tfrac{1}{2}\vec{K} - \vec{k})\, b^\dagger_{\downarrow,\vec{K}-\vec{k}} b^\dagger_{\uparrow,\vec{k}}.$$

$$(11.6.15)$$

It is not hard to show that the commutation relations for this new operator are

$$\left[c_{\vec{K},n}, c_{\vec{K}',n}\right] = 0$$
$$\left[c^\dagger_{\vec{K},n}, c^\dagger_{\vec{K}',n}\right] = 0 \qquad\qquad (11.6.16)$$
$$\left[c_{\vec{K},n}, c^\dagger_{\vec{K}',n}\right] = \delta_{\vec{K},\vec{K}'} - \sum_{\vec{k}} \phi^*_n(\tfrac{1}{2}\vec{K} - \vec{k})\phi_n(\tfrac{1}{2}\vec{K}' - \vec{k})b^\dagger_{\downarrow,\vec{K}'-\vec{k}} b_{\downarrow,\vec{K}-\vec{k}}$$
$$- \sum_{\vec{k}} \phi^*_n(\vec{k} - \tfrac{1}{2}\vec{K})\phi_n(\vec{k} - \tfrac{1}{2}\vec{K}')b^\dagger_{\uparrow,\vec{K}'-\vec{k}} b_{\uparrow,\vec{K}-\vec{k}}.$$

If the spatial extent of the pair wave function is a, then the normalized k-space wave function will have $|\phi_n(0)|^2 \sim a^3/V$. Both terms after the $\delta_{\vec{K},\vec{K}'}$ are therefore of the order

$$|\phi_n(0)|^2 \sum_{\vec{k}} N_{\vec{k}} \sim na^3, \qquad\qquad (11.6.17)$$

where $n = N/V$ is the total particle density. In other words, the pairs will act as bosons, with commutation relation $[c_{\vec{K},n}, c^\dagger_{\vec{K}',n}] = \delta_{\vec{K},\vec{K}'}$, as long as the density is low enough, namely

$$na^3 \ll 1. \qquad\qquad (11.6.18)$$

If the individual fermions are charged particles such as electrons, then the system of pairs can undergo Bose–Einstein condensation and there will be a charged superfluid, known as a **superconductor**. The above formalism applies to any distinguishable fermions, however. We could equally well apply this process of **bosonization** to any two different types of particles. For example, as we have already studied in Section 2.3, electrons and holes can pair to form excitons. To apply the above formalism to excitons, we can simply switch the spin labels $\uparrow$ and $\downarrow$ to band labels (e.g., conduction and valence bands), and switch the creation operator with momentum $\vec{k}$ in (11.6.15) to a destruction operator with momentum $-\vec{k}$, which corresponds to the creation of a hole with momentum $\vec{k}$. Excitons are neutral quasi-bosons that can also in principle undergo Bose–Einstein condensation, as we will discuss in Section 11.13.

The density-dependent terms in (11.6.16) are typically important at the same densities at which interactions between the composite bosons are important. This leads to problems with treating the pairs as pure bosons. Instead of dropping the density-dependent terms, we could instead write

$$[c_{\vec{K},n}, c^\dagger_{\vec{K}',n}] = \delta_{\vec{K},\vec{K}'} - D_{\vec{K},\vec{K}'}, \qquad\qquad (11.6.19)$$

where $D_{\vec{K},\vec{K}'}$ is an operator, equal to the extra two terms in (11.6.16). It may seem that carrying this operator along will make the math intractable, but things are simplified by the following commutation relation, which can be proved,

$$[D_{\vec{K},\vec{K}'}, C_{\vec{K}''}^{\dagger}] = 2 \sum_{\vec{K}'''} \lambda_{\vec{K},\vec{K}',\vec{K}'',\vec{K}'''} \, C_{\vec{K}'''}^{\dagger}, \tag{11.6.20}$$

where $\lambda_{\vec{K},\vec{K}',\vec{K}'',\vec{K}'''}$ is a scalar. If we do not use the pure boson commutation relation, the Feynmann many-body formalism of Chapter 8 for pure bosons can be not be used, but an analogous diagrammatic expansion can be developed which treats the $D_{\vec{K},\vec{K}'}$ operator and the Hamiltonian term for V_{int} for the interaction between the composite bosons both as small perturbation terms on an equal footing (see Combescot and Betbeder–Matibet 2005). Another approach is to use an adjusted interaction energy that includes some of the effects of these terms (e.g., Laikhtman 2007).

Exercise 11.6.1   Show that (11.6.4) follows from (11.6.3), (11.6.2), and the fermion anticommutation relations.

Exercise 11.6.2   Prove the commutation relations (11.6.16), assuming that the wave function $\phi_n(\vec{k})$ is normalized and symmetric, that is $\phi_n(-\vec{k}) = \phi_n(\vec{k})$.

11.7 Cooper Pairing

We have seen that pairing can lead fermionic particles such as electrons to act as bosons. This pairing underlies the behavior of superconductors, as we shall see later in this chapter.

How can electrons pair up, however? Since they have the same charge, one would expect them to repel each other. Of course, at long range, the net repulsion of the negatively charged electrons is canceled by the positive background of the atomic nuclei. But for there to be a pairing, there must be some attractive interaction.

We have already seen in Sections 8.2 and 8.3 how the electron–phonon interaction can have a major effect on the electron states, leading to the polaron effect and the shift of band energies with temperature. These effects arose from the calculation of the self-energy,

$$\Delta^{(2)} = \sum_{m \neq i} \frac{|\langle m|V_{\mathrm{int}}|i\rangle|^2}{E_i - E_m}, \tag{11.7.1}$$

defined in Section 8.1. The same interaction also alters the interactions between electrons. We can use the same Rayleigh–Schrödinger theory to calculate the second-order transition matrix element, also defined in Section 8.1,

$$V_{if}^{(2)} = \sum_m \frac{\langle f|V_{\mathrm{int}}|m\rangle \langle m|V_{\mathrm{int}}|i\rangle}{E_i - E_m}. \tag{11.7.2}$$

As in Section 8.2, we use the Fröhlich electron–phonon interaction, neglecting the polarization vector for the direction of the electric field,

$$V_{\text{int}} = \frac{C_{\text{Fr}}}{\sqrt{V}} \sum_{\vec{k}} \sum_{\vec{k}_1} \frac{1}{k} \, i \left(a_{\vec{k}} b^{\dagger}_{n,\vec{k}_1+\vec{k}} b_{n\vec{k}_1} - a^{\dagger}_{\vec{k}} b^{\dagger}_{n,\vec{k}_1-\vec{k}} b_{n\vec{k}_1} \right), \tag{11.7.3}$$

where C_{Fr} is a constant that depends on the phonon frequency and the dielectric function. Since we are interested in the interaction of two electrons, we define the initial state $|i\rangle$ as a state with one electron in momentum state $\vec{p}$ and another electron in momentum state $\vec{p}'$, and we define the final state $|f\rangle$ of the two electrons as one that conserves total momentum, with one electron in state $\vec{p} + \vec{k}$ and the other with $\vec{p}' - \vec{k}$.

There are two possible intermediate states $|m\rangle$ that will couple $|i\rangle$ and $|f\rangle$ via the electron–phonon interaction: either a state in which the electron with momentum $\vec{p}$ has been kicked to $\vec{p} + \vec{k}$, by creation of a phonon with momentum $-\vec{k}$ (leaving the electron with $\vec{p}'$ untouched), or a state in which the electron with momentum $\vec{p}'$ has been kicked to $\vec{p}' - \vec{k}$, by creation of a phonon with momentum $\vec{k}$ (leaving the electron with $\vec{p}$ untouched). In the subsequent virtual transition to state $|f\rangle$, the created phonon is absorbed by the other electron, so that the final state has just the two electrons.

Assuming that neither the phonon nor electron states are highly occupied, the matrix element $\langle m|V|i\rangle$ in either case is

$$\langle m|V_{\text{int}}|i\rangle = i \frac{C_{\text{Fr}}}{\sqrt{V}} \frac{1}{k}. \tag{11.7.4}$$

The second-order matrix element is then

$$V^{(2)}_{if} = \frac{C^2_{\text{Fr}}}{V} \frac{1}{k^2} \left(\frac{1}{E(\vec{p}) - E(\vec{p} + \vec{k}) - \hbar\omega_{LO}} + \frac{1}{E(\vec{p}') - E(\vec{p}' - \vec{k}) - \hbar\omega_{LO}} \right). \tag{11.7.5}$$

By conservation of energy,

$$E(\vec{p}) + E(\vec{p}') = E(\vec{p} + \vec{k}) + E(\vec{p}' - \vec{k}), \tag{11.7.6}$$

and therefore

$$E(\vec{p}') - E(\vec{p}' - \vec{k}) = - \left(E(\vec{p}) - E(\vec{p} + \vec{k}) \right). \tag{11.7.7}$$

Thus

$$\begin{aligned} V^{(2)}_{if} &= \frac{C^2_{\text{Fr}}}{V} \frac{1}{k^2} \left(\frac{1}{E(\vec{p}) - E(\vec{p} + \vec{k}) - \hbar\omega_{LO}} - \frac{1}{E(\vec{p}) - E(\vec{p} + \vec{k}) + \hbar\omega_{LO}} \right) \\ &= \frac{C^2_{\text{Fr}}}{V} \frac{1}{k^2} \frac{2\hbar\omega_{LO}}{\left(E(\vec{p}) - E(\vec{p} + \vec{k}) \right)^2 - (\hbar\omega_{LO})^2}. \end{aligned} \tag{11.7.8}$$

The denominator of (11.7.8) is negative if $\Delta E = |E(\vec{p}) - E(\vec{p} + \vec{k})| < \hbar\omega_{LO}$. In other words, for low energy exchange, the interaction is attractive.

We could have gotten the same result using the Feynmann rules in Section 8.7 for the diagram shown in Figure 11.6, because the propagator for phonons corresponds to both

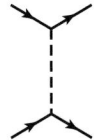

Fig. 11.6 Feynman diagram for the phonon-mediated Cooper pairing mechanism.

possibilities of absorption and emission of a phonon. This diagram looks just like the photon-mediated Coulomb interaction diagram, and in fact both are interactions mediated by a virtual, exchanged boson.

The total electron–electron interaction is found by adding the Coulomb interaction to the above phonon-mediated interaction. The total interaction is then, using the unscreened Coulomb interaction,

$$M_k = \frac{1}{V}\frac{e^2/\epsilon}{k^2} + \frac{1}{V}\frac{C_{\mathrm{Fr}}^2}{k^2}\frac{2\hbar\omega_{LO}}{(\Delta E)^2 - (\hbar\omega_{LO})^2}. \tag{11.7.9}$$

Looking up the vertex for the Fröhlich interaction (5.1.29), we have in the limit when $\epsilon(0) \gg \epsilon(\infty)$,

$$C_{\mathrm{Fr}} = \frac{e}{k\sqrt{\epsilon}}\frac{\hbar\omega_{LO}}{2}, \tag{11.7.10}$$

which yields

$$M_k = \frac{1}{V}\frac{e^2/\epsilon}{k^2}\left(1 + \frac{(\hbar\omega_{LO})^2}{(\Delta E)^2 - (\hbar\omega_{LO})^2}\right). \tag{11.7.11}$$

Recall that the Fröhlich interaction was deduced in Section 5.1.3 using the scalar electric potential generated by the oscillating dipole moment in a polar medium. Screening will therefore affect both the Coulomb interaction and the Fröhlich phonon-mediated interaction. When screening is taken into account, the total interaction becomes

$$M_k = \frac{1}{V}\frac{e^2/\epsilon}{k^2 + \kappa^2}\left(1 + \frac{(\hbar\omega_{LO})^2}{(\Delta E)^2 - (\hbar\omega_{LO})^2}\right), \tag{11.7.12}$$

which gives an overall attractive interaction between the electrons for $\Delta E < \hbar\omega_{LO}$. This is known as the **Bardeen–Pines** interaction.

This attractive interaction is what leads to Cooper pairing. The physical picture that is sometimes given is that of two people on a soft couch – when the cushions deform under their weight, they are pulled toward each other. As shown in Figure 11.7, the polaronic effect leads to a decrease of the energy near a free electron, due to positive charge in the background lattice moving slightly toward it. The overall energy will be minimized when the two electrons are near to each other.

Exercise 11.7.1　(a) What are the diagrams that you would write down in the Rayleigh–Schrödinger diagrammatic method that correspond to the two terms in (11.7.5)?

(b) What are the diagrams you would write down in the Feynman method for the screening of the phonon-mediated Cooper pairing interaction?

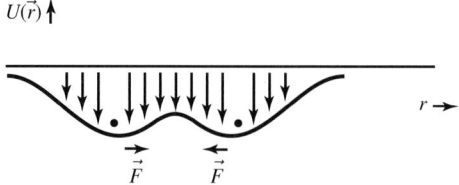

$U(\vec{r})$

$r \rightarrow$

$\vec{F}$ $\vec{F}$

Fig. 11.7   Cooper pairing via the polaron effect.

11.8 BCS Wave Function

This is a book on solid state physics, but the discussion in this chapter so far may seem to apply only to liquids and gases. Actually, all of the above theory lays the basis for the important solid state phenomenon of superconductivity. The essential physics of super-conductivity is Bose–Einstein condensation of Cooper pairs. This is often overlooked in introductory treatments of superconductivity. Most of the first part of this section, therefore, focuses on establishing the connection of Bose–Einstein condensation and superconduc-tivity. It turns out that condensate theory also can be used to describe optical condensates, as discussed in Section 11.13.

The theory of superconductivity continues to be a major subject of solid state physics, especially with the discovery of high-temperature superconductivity. For a more compre-hensive review of the basic theory, see, for example, Lifshitz and Pitaevskii (1980) or Leggett (2006).

In principle, superconductivity could occur as a Bose–Einstein condensate of electron pairs obeying the same theory we have described in Section 11.2. There are no known physical examples of this, however. In real superconductors, the electron density is much higher than the criterion (11.6.18) for the pairs to act as independent, localized bosons.

If the density is high, however, the pairing does not suddenly cease. The particles will still be correlated in pairs. Let us examine what happens if we increase the density. The Bose condensate consists of N pairs in the $\vec{K} = 0$ state. The creation operator for a particle with $\vec{K} = 0$ is

$$c_0^\dagger = \sum_{\vec{k}} \phi(\vec{k}) \, b_{\downarrow,-\vec{k}}^\dagger b_{\uparrow,\vec{k}}^\dagger, \tag{11.8.1}$$

and the condensate state is the coherent state

$$|\alpha\rangle = e^{-|\alpha|^2/2} \sum_N \frac{(\alpha c_0^\dagger)^N}{N!} |0\rangle, \tag{11.8.2}$$

with $\alpha = \sqrt{N_0} e^{i\theta}$.

At low density, putting more than one pair in the $\vec{K} = 0$ state just linearly increases the probability of finding a single particle in a given state. For example, suppose we put two pairs into the ground state, and we want to know the probability of finding a single particle in the $\vec{k} = 0$ state. The two-pair state is

$$|\Psi\rangle = \frac{1}{\sqrt{2!}} \left(\phi(0)b^\dagger_{\downarrow,0}b^\dagger_{\uparrow,0} + \sum_{\vec{k}\neq0} \phi(\vec{k})\, b^\dagger_{\downarrow,-\vec{k}}b^\dagger_{\uparrow,\vec{k}} \right) \tag{11.8.3}$$

$$\times \left(\phi(0)b^\dagger_{\downarrow,0}b^\dagger_{\uparrow,0} + \sum_{\vec{k}'\neq0} \phi(\vec{k}')\, b^\dagger_{\downarrow,-\vec{k}'}b^\dagger_{\uparrow,\vec{k}'} \right)|0\rangle.$$

The probability of finding a single particle in the $\vec{k} = 0$, spin-up state is $\langle\Psi|b^\dagger_{\uparrow,0}b_{\uparrow,0}|\Psi\rangle$. Since the particles are fermions, any terms with more than one creation operator acting on the same state vanish; therefore only cross terms of the type

$$b^\dagger_{\downarrow,0}b^\dagger_{\uparrow,0}b^\dagger_{\downarrow,-\vec{k}'}b^\dagger_{\uparrow,\vec{k}'}$$

survive in the product state. The probability of finding a single particle in the $\vec{k} = 0$, spin-up state is therefore

$$\langle\Psi|b^\dagger_{\uparrow,0}b_{\uparrow,0}|\Psi\rangle = 2|\phi(\vec{0})|^2 \sum_{\vec{k}\neq0} |\phi(\vec{k})|^2$$

$$= 2|\phi(\vec{0})|^2 \sim 2\frac{a^3}{V}, \tag{11.8.4}$$

where we have used the normalization of the wave function $\sum |\phi(\vec{k})|^2 = 1$, and where a is the spatial size of the relative motion of the pair state, introduced in Section 11.6. Putting two pairs in the $\vec{K} = 0$ state has just doubled the probability of finding a single particle in a given $\vec{k}$-state. In general, putting N pairs in the ground state gives, at low density,

$$\langle\Psi|b^\dagger_{\uparrow,0}b_{\uparrow,0}|\Psi\rangle \sim N\frac{a^3}{V}. \tag{11.8.5}$$

If we keep increasing the number of pairs in the ground state, however, the probability of finding a single particle in a given $\vec{k}$-state cannot keep increasing linearly. The probability of finding a fermion in any $\vec{k}$-state cannot exceed unity. Therefore, when $N > V/a^3$, the ground state cannot be simply the product of pair states (11.8.2). The function $\phi(\vec{k})$ must be altered, to keep a maximum of one fermion per $\vec{k}$-state. We don't yet know the exact form of $\phi(\vec{k})$, but we know that it must have a form that gives a probability of occupation of k-states like that shown in Figure 11.8. As the electron density increases, the pair wave function must evolve to keep the total probability per state less than or equal to unity. At high density, the pair wave function will resemble a Fermi sea, with all states filled below a density-dependent Fermi energy.

At high density, we can rewrite the ground state wave function (11.8.2) in a different form, which brings out directly the fact that no electron state has greater than unity occupation. As pointed out by Blatt (1964), the wave function (11.8.2) is equivalent to the Bardeen–Cooper–Schrieffer (BCS) ground state wave function, written as

$$\boxed{|\Psi\rangle = \prod_{\vec{k}} \left(u_k + v_k b^\dagger_{\downarrow,-\vec{k}}b^\dagger_{\uparrow,\vec{k}} \right)|0\rangle,} \tag{11.8.6}$$

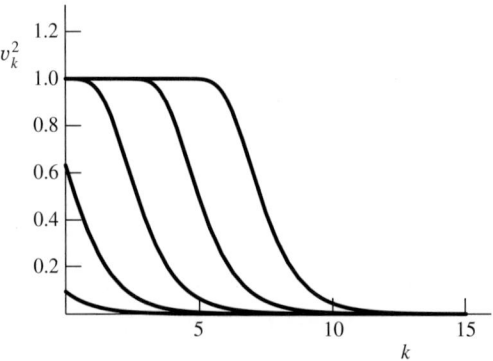

Fig. 11.8 The probability of finding a fermion as a function of k, for various pair densities.

where v_k is the probability amplitude for finding a pair occupied, and u_k is the probability amplitude for finding it unoccupied, with the normalization condition

$$u_k^2 + v_k^2 = 1. \tag{11.8.7}$$

The set of $\vec{k}$ in the product of (11.8.6) makes up a half-space, so that there is no double creation of pairs. The equivalence of (11.8.6) with (11.8.2) can be seen by rewriting (11.8.6) as

$$|\Psi\rangle = \prod_{\vec{k}'} u_{k'} \prod_{\vec{k}} \left(1 + \alpha\phi(\vec{k})b_{\downarrow,-\vec{k}}^\dagger b_{\uparrow,\vec{k}}^\dagger\right)|0\rangle, \tag{11.8.8}$$

where $\alpha\phi(\vec{k}) = v_k/u_k$, assuming u_k is never strictly equal to zero for any $\vec{k}$. In the expansion of the product over $\vec{k}$, the presence of the 1 means that we have every possible number of pairs of operators. The expansion give us

$$\prod_{\vec{k}} \left(1 + \alpha\phi(\vec{k})b_{\downarrow,-\vec{k}}^\dagger b_{\uparrow,\vec{k}}^\dagger\right) =$$

$$1 + \alpha\sum_{\vec{k}}\phi(\vec{k})b_{\downarrow,-\vec{k}}^\dagger b_{\uparrow,\vec{k}}^\dagger + \frac{1}{2}\left(\alpha\sum_{k}\phi(\vec{k})b_{\downarrow,-\vec{k}}^\dagger b_{\uparrow,\vec{k}}^\dagger\right)^2 + \cdots$$

$$= \sum_{N=0}^{\infty} \frac{(\alpha c_0^\dagger)^N}{N!}|0\rangle. \tag{11.8.9}$$

The higher-order terms are divided by $N!$ because there are $N!$ terms that have a sum over $\vec{k}$ to the Nth power, but these are all the same after changing the order of the operators. In the products of the sums in the expansion of (11.8.9), there are terms with more than one creation operator for the same state, which do not appear in the product (11.8.8), but these terms are zero because of Pauli exclusion.

The prefactor $C \equiv \prod u_k$ that appears in (11.8.8) is just a normalization constant, determined by the condition

$$\langle \Psi | \Psi \rangle = 1 \tag{11.8.10}$$

$$= \langle 0 | \sum_N \frac{1}{N!} \left(\alpha^* \sum_k \phi^*(\vec{k}) b_{\uparrow,\vec{k}} b_{\downarrow,-\vec{k}} \right)^N |C|^2 \sum_{N'} \frac{1}{N'!} \left(\alpha \sum_k \phi(\vec{k}) b^\dagger_{\downarrow,-\vec{k}} b^\dagger_{\uparrow,\vec{k}} \right)^{N'} |0\rangle.$$

There are $N!$ non-zero terms in each Nth-order term in the sum. The normalization condition becomes

$$\langle \Psi | \Psi \rangle = |C|^2 \langle 0 | \sum_N \left(1 + |\alpha|^2 \sum_{\vec{k}} |\phi(\vec{k})|^2 + \frac{1}{2} \left(|\alpha|^2 \sum_{\vec{k}} |\phi(\vec{k})|^2 \right)^2 + \cdots \right) |0\rangle$$

$$= |C|^2 e^{|\alpha|^2} = 1, \tag{11.8.11}$$

where we have used the normalization of the wave function $\sum |\phi(\vec{k})|^2 = 1$. This implies

$$|\Psi\rangle = e^{-|\alpha|^2/2} \sum_N \frac{(\alpha c_0^\dagger)^N}{N!} |0\rangle, \tag{11.8.12}$$

which is exactly the coherent state (11.8.2) we initially wrote for the condensate. The value of α is determined by

$$\prod_{\vec{k}} u_k = \prod_{\vec{k}} \sqrt{1 - v_k^2}$$

$$= \prod_{\vec{k}} \left(1 - \frac{1}{2} v_k^2 + \frac{1}{8} v_k^4 + \cdots \right)$$

$$\simeq 1 - \frac{1}{2} \sum_{\vec{k}} v_k^2 + \frac{1}{2} \left(\frac{1}{2} \sum_{\vec{k}} v_k^2 \right)^2 + \cdots$$

$$= \exp\left(-\frac{1}{2} \sum_{\vec{k}} v_k^2 \right) = \exp(-N_0/2), \tag{11.8.13}$$

where we use the normalization condition $\sum v_k^2 = N_0$, that is, the total probability in all states equals the number of pairs. This implies $|\alpha| = \sqrt{N_0}$, as we assumed in writing (11.8.2).

In Section 11.6, we found that the operators $c_{\vec{k}}^\dagger, c_{\vec{k}}$ do not exactly obey the boson commutation relations at high density. Even in the high-density limit, however, the operator c_0 has exactly the same action on the condensate state as the a_0 operator had for the boson condensate:

$$c_0 |\Psi\rangle = \left(\sum_{\vec{k}} \phi^*(\vec{k}) b_{\uparrow,\vec{k}} b_{\downarrow,-\vec{k}} \right) e^{-|\alpha|^2/2} \left(1 + \alpha \sum_{\vec{k}'} \phi(\vec{k}') b^\dagger_{\downarrow,-\vec{k}'} b^\dagger_{\uparrow,\vec{k}'} \right.$$

$$\left. + \frac{1}{2} \alpha^2 \sum_{\vec{k}',\vec{k}''} \phi(\vec{k}') \phi(\vec{k}'') b^\dagger_{\downarrow,-\vec{k}'} b^\dagger_{\uparrow,\vec{k}'} b^\dagger_{\downarrow,-\vec{k}''} b^\dagger_{\uparrow,\vec{k}''} + \cdots \right) |0\rangle$$

$$= e^{-|\alpha|^2/2} \left(\alpha \sum_{\vec{k}} |\phi(\vec{k})|^2 + \alpha^2 \sum_{\vec{k}} |\phi(\vec{k})|^2 \sum_{\vec{k}'} \phi(\vec{k}') b^\dagger_{\downarrow,-\vec{k}'} b^\dagger_{\uparrow,\vec{k}'} + \cdots \right) |0\rangle$$

$$= \alpha |\Psi\rangle, \tag{11.8.14}$$

where we have again used the normalization condition $\sum |\phi(\vec{k})|^2 = 1$. In both this case and in the weakly interacting condensate, the operator acts like a simple c-number. We can therefore make the same argument for the stability of the condensate that we made in Section 11.3. In other words, the superconductor has its unique properties for the same reason that a weakly interacting Bose–Einstein condensate does: To minimize exchange energy, the pairs enter a single macroscopic wave state.[1]

It is often not appreciated that the BCS state (11.8.6) is equivalent to a coherent state. The high density of electrons means that we cannot use the pair states (11.6.15) as pure bosons with $\phi(\vec{K})$ equal to the same pair wave function as used for isolated pairs of particles at low density, but the pair wave function $\phi(\vec{K})$ is renormalized so that the operator $c_{\vec{K}}$ still acts as a boson destruction operator on the many-particle ground state.

Exercise 11.8.1 Show explicitly that (11.8.5) is correct for an arbitrary number N of pairs.

Exercise 11.8.2 Show explicitly that the second-order term of the normalization condition (11.8.11) is correct, by showing that

$$\langle \Psi | \Psi \rangle = |C|^2 |\alpha|^4 \langle 0 | \frac{1}{2} \left(\sum_k \phi^*(\vec{k}) b_{\uparrow,\vec{k}} b_{\downarrow,-\vec{k}} \sum_{k'} \phi^*(\vec{k}') b_{\uparrow,\vec{k}'} b_{\downarrow,-\vec{k}'} \right)$$

$$\times \frac{1}{2} \left(\sum_p \phi(\vec{p}) b^\dagger_{\downarrow,-\vec{p}} b^\dagger_{\uparrow,\vec{p}} \sum_{p'} \phi(\vec{p}') b^\dagger_{\downarrow,-\vec{p}'} b^\dagger_{\uparrow,\vec{p}'} \right) |0\rangle \tag{11.8.15}$$

becomes

$$\langle \Psi | \Psi \rangle = |C|^2 |\alpha|^4 \frac{1}{4} \sum_{k,k',p,p'} \phi^*(\vec{k}) \phi^*(\vec{k}') \phi(\vec{p}) \phi(\vec{p}') \tag{11.8.16}$$

$$\times (\delta_{k,p'} \delta_{k'p} + \delta_{p,k} \delta_{k',p'} - \delta_{k,p} \delta_{k',p'} \delta_{p,p'} \delta_{k'p} - \delta_{k,p'} \delta_{k,p} \delta_{k',p'} \delta_{k',p}).$$

Show that the last two terms with four δ-functions are negligible compared to the first two terms.

11.9 Excitation Spectrum of a Superconductor

The BCS form of the ground state wave function leads to a natural way to find the excitation spectrum of the superconductor. An excited state will consist of at least one free, unpaired electron. We therefore define an excited state by starting with the ground state

---

[1] This is shown explicitly for a gas of composite bosons by Combescot and Snoke (2008).

wave function of the superconductor and putting one uncorrelated fermion in a definite momentum and spin state, as follows:

$$|\Psi\rangle = b^\dagger_{\uparrow,\vec{p}} \prod_{\vec{k}\neq\vec{p}} \left(u_k + v_k b^\dagger_{\downarrow,-\vec{k}} b^\dagger_{\uparrow,\vec{k}} \right) |0\rangle. \tag{11.9.1}$$

This is the same as acting on the ground state wave function (11.8.6) with the operator

$$\xi^\dagger_{\uparrow,\vec{p}} = \left(u_p b^\dagger_{\uparrow,\vec{p}} + v_p b_{\downarrow,-\vec{p}} \right). \tag{11.9.2}$$

Equivalently, an unpaired spin-down particle is created by the operator

$$\xi^\dagger_{\downarrow,\vec{p}} = \left(u_p b^\dagger_{\downarrow,\vec{p}} - v_p b_{\uparrow,-\vec{p}} \right). \tag{11.9.3}$$

These new operators are fermion creation operators which obey the fermion anticommutation rule $\{\xi_{s,\vec{k}}, \xi^\dagger_{s',\vec{k}'}\} = \delta_{s,s'}\delta_{\vec{k},\vec{k}'}$.

We now want to express the Hamiltonian of the system in terms of these new quasiparticle operators, taking the ground state as the new vacuum. We approximate the full Hamiltonian (11.6.1) by the *reduced* Hamiltonian

$$H = \sum_{\vec{k}} \frac{\hbar^2 k^2}{2m} \left(b^\dagger_{\uparrow,\vec{k}} b_{\uparrow,\vec{k}} + b^\dagger_{\downarrow,\vec{k}} b_{\downarrow,\vec{k}} \right) - \frac{U_0}{2V} \sum_{\vec{k},\vec{q}} b^\dagger_{\downarrow,\vec{k}} b^\dagger_{\uparrow,-\vec{k}} b_{\uparrow,-\vec{k}-\vec{q}} b_{\downarrow,\vec{k}+\vec{q}}, \tag{11.9.4}$$

where we have kept only terms that act on electrons with opposite momenta. Since the ground state consists of paired states, we expect that these terms will dominate. We have also assumed that the interaction $U(\vec{q})$ is not a strong function of $\vec{q}$, and replaced it with a constant $-U_0$. A negative constant corresponds to an attractive interaction.

The inverse transformation of the definitions (11.9.2) and (11.9.3) are

$$b^\dagger_{\uparrow,\vec{p}} = u_p \xi^\dagger_{\uparrow,\vec{p}} - v_p \xi_{\downarrow,-\vec{p}},$$
$$b^\dagger_{\downarrow,\vec{p}} = u_p \xi^\dagger_{\downarrow,\vec{p}} + v_p \xi_{\uparrow,-\vec{p}}. \tag{11.9.5}$$

As with the Bogoliubov model, we can transform the Hamiltonian into one written in terms of the quasiparticle operators. Substituting (11.9.5) into the reduced Hamiltonian, and switching the names of dummy summation variables, we obtain

$$H = 2\sum_{\vec{k}} \frac{\hbar^2 k^2}{2m} v_k^2 + \sum_{\vec{k}} \frac{\hbar^2 k^2}{2m} \left(u_k^2 - v_k^2 \right) \left(\xi^\dagger_{\uparrow,\vec{k}} \xi_{\uparrow,\vec{k}} + \xi^\dagger_{\downarrow,\vec{k}} \xi_{\downarrow,\vec{k}} \right)$$
$$+ 2\sum_{\vec{k}} \frac{\hbar^2 k^2}{2m} u_k v_k \left(\xi^\dagger_{\downarrow,\vec{k}} \xi^\dagger_{\uparrow,-\vec{k}} + \xi_{\uparrow,-\vec{k}} \xi_{\downarrow,\vec{k}} \right) \tag{11.9.6}$$
$$- \frac{U_0}{2V} \left| \sum_{\vec{k}} \left(u_k^2 \xi^\dagger_{\uparrow,\vec{k}} \xi^\dagger_{\downarrow,-\vec{k}} - v_k^2 \xi_{\downarrow,-\vec{k}} \xi_{\uparrow,\vec{k}} - u_k v_k \xi_{\downarrow,-\vec{k}} \xi^\dagger_{\downarrow,-\vec{k}} + u_k v_k \xi^\dagger_{\uparrow,\vec{k}} \xi_{\uparrow,\vec{k}} \right) \right|^2.$$

In terms of this Hamiltonian, the ground state consists of the absence of any quasiparticles, and the excited states consist of Fock number states of the new quasiparticles.

We can solve for u_k and v_k in the ground state by minimizing the expectation value of

$$E = \langle H - \mu N \rangle, \tag{11.9.7}$$

where we adopt the grand canonical ensemble by adding a term $(-\mu N)$, keeping the chemical potential fixed and allowing the number of particles to vary. Since the total particle number is

$$N = \sum_{\vec{k}} \left(b^{\dagger}_{\uparrow,\vec{k}} b_{\uparrow,\vec{k}} + b^{\dagger}_{\downarrow,\vec{k}} b_{\downarrow,\vec{k}} \right), \tag{11.9.8}$$

adding this term to (11.9.4) corresponds simply to shifting the definition of the kinetic energy to

$$E_k = \frac{\hbar^2 k^2}{2m} - \mu. \tag{11.9.9}$$

The thermal average of the expectation value is then

$$E = 2 \sum_{\vec{k}} E_k v_k^2 + \sum_{\vec{k}} E_k \left(u_k^2 - v_k^2 \right) \left(N_{\uparrow,\vec{k}} + N_{\downarrow,\vec{k}} \right)$$

$$- \frac{U_0}{2V} \left| \sum_{\vec{k}} u_k v_k \left(1 - N_{\uparrow,\vec{k}} - N_{\downarrow,\vec{k}} \right) \right|^2, \tag{11.9.10}$$

where $N_{s,\vec{k}} = \xi^{\dagger}_{s,\vec{k}} \xi_{s,\vec{k}}$ is the occupation number of the quasiparticles, equal to a Fermi–Dirac distribution in equilibrium at finite temperature. Since the expectation value is taken for a Fock state of the excitation quasiparticles, only terms in the Hamiltonian that have an equal number of creation and destruction operators contribute. The last term in (11.9.6) actually also contributes cross terms of the product of the sums of the form $\sum u_k^2 v_k^2 N_{\uparrow,\vec{k}} N_{\downarrow,-\vec{k}}$, but these are small compared to the square of the sum, and so we drop these.

Minimizing E with respect to u_k, recalling $v_k^2 = 1 - u_k^2$, we have the condition

$$\frac{\partial E}{\partial u_k} = -4 E_k u_k + 4 E_k u_k (N_{\uparrow,\vec{k}} + N_{\downarrow,\vec{k}})$$

$$- \frac{U_0}{V} \frac{v_k^2 - u_k^2}{v_k} \left(1 - N_{\uparrow,\vec{k}} - N_{\downarrow,\vec{k}} \right) \sum_{\vec{k}'} u_{k'} v_{k'} \left(1 - N_{\uparrow,\vec{k}'} - N_{\downarrow,\vec{k}'} \right)$$

$$= -\frac{2}{v_k} \left(1 - N_{\uparrow,\vec{k}} - N_{\downarrow,\vec{k}} \right) \left(2 E_k u_k v_k - \Delta(u_k^2 - v_k^2) \right) = 0, \tag{11.9.11}$$

where we have defined

$$\Delta = \frac{U_0}{2V} \sum_{\vec{k}} u_k v_k \left(1 - N_{\uparrow,\vec{k}} - N_{\downarrow,\vec{k}} \right). \tag{11.9.12}$$

To satisfy this, independent of the number of quasiparticles $N_{\uparrow,\vec{k}}$ and $N_{\downarrow,\vec{k}}$, we must have

$$2 E_k u_k v_k = \Delta(u_k^2 - v_k^2), \tag{11.9.13}$$

which can be solved for u_k and v_k, giving

$$u_k^2 = \frac{1}{2} \left(1 + \frac{E_k}{\sqrt{\Delta^2 + E_k^2}} \right),$$

$$v_k^2 = \frac{1}{2} \left(1 - \frac{E_k}{\sqrt{\Delta^2 + E_k^2}} \right). \qquad (11.9.14)$$

As expected, v_k has the form shown in Figure 11.8; namely, for constant Δ, it is nearly equal to unity well below the Fermi level, that is, when E_k is large and negative.

The energy per quasiparticle with momentum $\vec{k}$ is found as the derivative of the total energy (11.9.10) with respect to the number of quasiparticles at momentum $\vec{k}$, which, using our results (11.9.13) and (11.9.14), is

$$E(\vec{k}) = \frac{\partial E}{\partial N_{\uparrow, \vec{k}}} = E_k \left(u_k^2 - v_k^2 \right) + 2\Delta u_k v_k$$

$$= (E_k + \frac{\Delta^2}{E_k}) \left(u_k^2 - v_k^2 \right)$$

$$= \sqrt{\Delta^2 + E_k^2}. \qquad (11.9.15)$$

The quasiparticle spectrum therefore has a *gap*, which is equal to Δ, as shown in Figure 11.9.

Exercise 11.9.1   Prove that the operators (11.9.2) and (11.9.3) obey the fermion anticommutation rules.

Exercise 11.9.2   Show that substitution of the operators (11.9.5) into the Hamiltonian (11.9.4) gives the Hamiltonian (11.9.6).

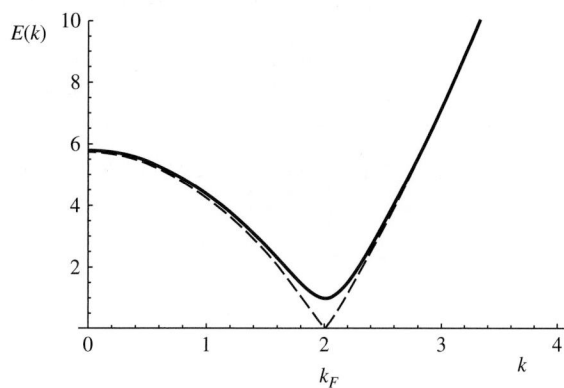

Fig. 11.9   Solid line: the excitation spectrum for a BCS superconductor. Dashed line: excitation spectrum for a noninteracting Fermi gas, consisting of the parabolic dispersion of the holes for momentum below k_F, and for free electrons above k_F.

11.9.1 Density of States and Tunneling Spectroscopy

The gap in the excitation spectrum leads to a van Hove singularity in the density of states of the quasiparticles (see Section 1.8). In the case of the electron density of states near a band minimum or maximum, the van Hove singularity does not lead to a divergence of the density of states in a three-dimensional system, but in the case of quasiparticles in a superconductor, the density of states does diverge. From Section 1.8, we have for an isotropic system,

$$\mathcal{D}(E)dE = \frac{V}{(2\pi)^3} 4\pi \, dE \, \frac{1}{|\nabla_{\vec{k}} E|} \, k^2(E). \tag{11.9.16}$$

Since the energy minimum of the quasiparticle spectrum occurs at finite k, this will diverge at $E_k = 0$, which corresponds to $k = k_F$.

Figure 11.10 illustrates the overall picture of quasiparticle excitations in a superconductor in terms of the density of states of the quasiparticles. Note that if we want to put energy into a superconductor without changing the total charge, we must remove an electron (creating a hole) below the Fermi level and create a free electron above the Fermi level. The overall excitation is therefore a bosonic two-fermion complex, like an exciton in a semiconductor. This is sometimes called **pair breaking** since the final state has two free, uncorrelated electrons at different momenta.

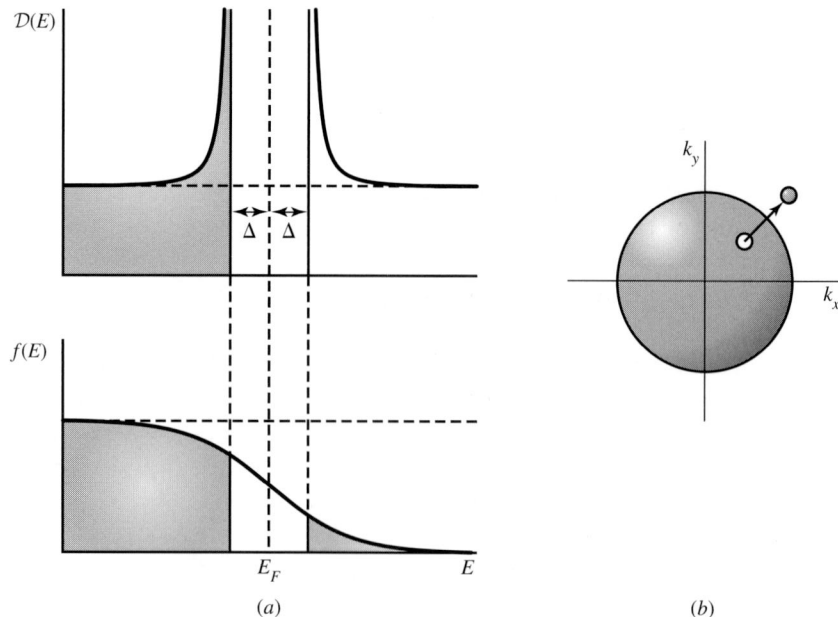

(a)   (b)

Fig. 11.10 (a) Semiconductor picture of the density of states of quasiparticles in a superconductor. Top: density of states of the quasiparticles. The ground state corresponds to a full "valence" band (no quasiholes) and empty "conduction" band (no quasielectrons). Bottom: Fermi-Dirac occupation number of the quasielectrons at finite temperature. (b) A typical excitation, which creates both a quasihole and a quasielectron.

The definition of the quasiparticle operator (11.9.2) implies that above the Fermi level, the quasiparticle is mostly the creation of a free electron (well above the Fermi level, when $u_p \to 1$ and $v_p \to 0$, we have $\xi^\dagger_{\uparrow,\vec{p}} = b^\dagger_{\uparrow,\vec{p}}$, which is exactly the creation of a free electron). We can call this a **quasielectron**. Below the Fermi level, the quasiparticle is mostly the creation of a hole (well below the Fermi level, when $u_p \to 0$ and $v_p \to 1$, we have $\xi^\dagger_{\uparrow,\vec{p}} = b_{\downarrow,-\vec{p}}$, which is the same as creation of a hole); we will call this a **quasihole**. The minimum energy for an excitation is therefore 2Δ, the energy to create two quasiparticles right at the Fermi level. This is quite different from a normal metal, in which the energy to move an electron from just below the Fermi level to just above it, leading to a hole below the Fermi level and a free electron above it, can be as small as we want.

As discussed in Section 2.5.1, when there is a gap, the chemical potential of the fermions is pinned at the center of the gap for a system with equal density of states in both bands. The number of quasielectrons above the gap must always equal the number of quasiholes below the gap, as illustrated in lower plot of Figure 11.10(a). At $T = 0$, there are no quasiparticles, while at finite temperature, the fermion equilibrium distribution $f(E) = 1/(e^{(E-\mu)/k_B T} + 1)$ has a tail in both bands, which corresponds to quasielectrons above the gap and quasiholes below the gap.

Note that the chemical potential of the quasielectrons has nothing to do with the chemical potential of the electrons that underlie the superconducting state. In the renormalized picture, the vacuum is the ground state of the system, which is the superconducting state with electron occupation number given by the BCS factor v_p^2, which has the smeared-out form shown in Figure 11.8, even at $T = 0$. With this now defined as the vacuum state, the number of quasiparticles is not conserved – it depends on the temperature. The chemical potential for the quasielectrons is always pinned at the midpoint of the gap in the quasiparticle spectrum.

The gap in the excitation spectrum of a superconductor is a conspicuous property of the superconducting state. The gap by itself does not cause the superconductivity, however. Many systems have energy gaps but are not superconductors – for example, every semiconductor – and gapless superconductors are possible (see, e.g., de Gennes 1966: s. 8–2).

We have not considered here other types of excitation besides the pair breaking, for example superfluid density fluctuations. Pair breaking excitations are the most important for transport measurements, because Cooper pairs cannot travel outside the superconductor, and therefore they must break apart if current is to leave the superconductor; conversely, current entering a superconductor enters as unpaired quasiparticles.

The semiconductor picture of the density of states allows us to explain experimental measurements of current at tunneling junctions of superconductors and normal metals, or junctions between two superconductors, which were important historically in confirming the BCS theory. Figures 11.11, 11.12, and 11.13 show examples of the current–voltage characteristics of tunneling junctions. We can assume that phase coherence is lost in the tunneling process, which is not necessarily the case in very thin junctions, as discussed below in Section 11.11. When the phase is not important, the current is simply

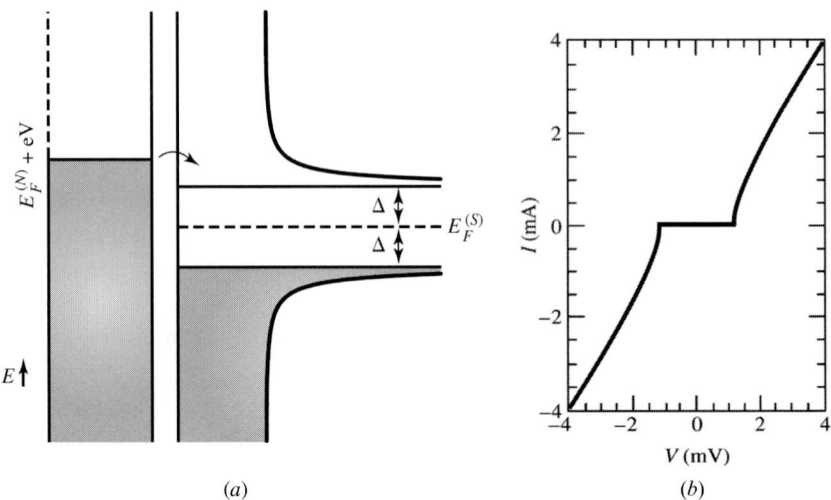

Fig. 11.11 (a) Effective density of states of a normal metal and a superconductor at different voltages, separated by an insulator. (b) Experimental current as a function of the voltage at $T = 0.38$ K for a junction of superconducting Nb and normal copper separated by insulator Al_2O_3. From Castellano *et al.* (1997).

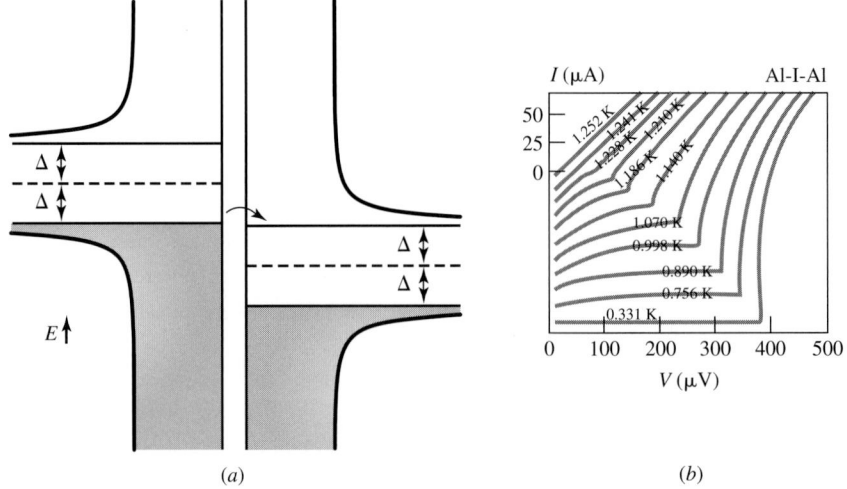

Fig. 11.12 (a) Effective density of states for a junction of two identical superconductors at different voltages, separated by an insulator. (b) Experimental current as a function of the voltage for a junction of two superconducting aluminum layers separated by insulator Al_2O_3. Successive curves at different temperatures are displaced for clarity. From Blackford and March (1968).

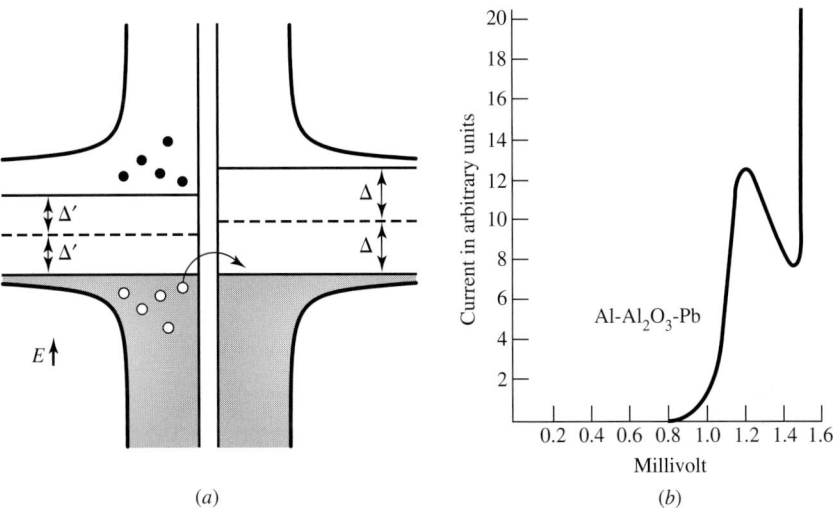

Fig. 11.13   (a) Density of states for a junction of two superconductors with different gaps. (b) Current as a function of the voltage at $T = 1.0$ K for a junction of superconducting aluminum and lead separated by insulator Al_2O_3. From Giaever (1960).

proportional to the final density of states available to the tunneling carriers, per Fermi's golden rule.

In the case of a normal metal–insulator–superconductor junction, assuming that the density of states of the normal metal is constant near the Fermi level, the current is simply proportional to the integral of the density of states of the superconductor, as shown in Figure 11.11. At zero temperature, no current can tunnel across until the Fermi level of the normal metal is above the effective gap energy 2Δ.

In the case of a superconductor–insulator–superconductor junction, there is a discontinuous jump when two singularities in the quasiparticle density of states cross. The low temperature data of Figure 11.12 show this jump clearly. At higher temperatures, thermally excited quasiparticles can tunnel even when the potential difference is less than 2Δ.

As shown in Figure 11.13, in the case of a junction of two superconductors with different gaps, at certain voltages one sees negative differential resistance – increasing the voltage leads to lower current. This is because at finite temperature, there is some occupation of the lower band of the narrower-gap superconductor by thermally excited quasiholes, which can tunnel into the lower band of the wider-gap superconductor. Fewer excited quasiparticles exist in the wider-gap superconductor. When the voltage is increased, most of the quasiholes at the top of the lower band in the narrower-gap superconductor will have energy inside the gap of the wider-gap superconductor, and therefore tunneling will be suppressed.

Exercise 11.9.3   Compute the integral of the density of states (11.9.16) for the quasiparticle energy spectrum (11.9.15), and plot it (setting $\mu = 1$ and $\Delta = 0.02$, and all other constants to unity). It should have the same form as the experimental results for the normal–insulator–superconductor junction plotted in Figure 11.11(b).

11.9.2 Temperature Dependence of the Gap

Equation (11.9.12) for the gap energy implies a temperature dependence for the gap energy, and this dependence is seen dramatically in Figure 11.12 – the gap shrinks as T is increased. Using our results (11.9.14) for u_k and v_k in (11.9.12), and assuming that the populations of quasiholes and quasielectrons are equal, we obtain

$$\Delta = \frac{U_0}{2V} \sum_{\vec{k}} \frac{\Delta}{2E(k)} \left(1 - 2\frac{1}{e^{E(k)/k_B T} + 1} \right)$$

$$= \frac{U_0}{2V} \sum_{\vec{k}} \frac{\Delta}{2E(k)} \tanh\left(E(k)/2k_B T \right). \tag{11.9.17}$$

The gap energy Δ in the numerator on both sides cancels, but is still present in the definition of $E(k) = \sqrt{\Delta^2 + E_k^2}$. Assuming that the density of k-states corresponding to E_k is roughly constant near $E_k = 0$, that is, approximately equal to the density of states at the surface of a Fermi sphere, the sum can be converted to an integral, so that we have

$$1 = \frac{U_0}{2V} \mathcal{D}(E_F) \int_0^{E_c} dE_k \frac{1}{2\sqrt{\Delta^2 + E_k^2}} \tanh\left(\sqrt{\Delta^2 + E_k^2}/2k_B T \right), \tag{11.9.18}$$

where the integration is up to some cutoff energy E_c, since we assume that only quasi-particles near the Fermi surface contribute. We thus have a self-consistency equation for Δ at a given T. Even without knowing the strength of the interaction U_0, this equation implies a universal shape for the curve $\Delta(T)/\Delta(0)$ for BCS superconductors. The form of $\Delta(T)/\Delta(0)$ computed numerically from (11.9.18), compared to experimental data, is shown in Figure 11.14. At large T, the $\tanh(E(k)/kBT)$ function decreases, which implies $\Delta(T)$ must also decrease to keep the integral constant. At some critical temperature T_c, the gap energy $\Delta(T)$ goes to zero, and the material reverts to the normal state.

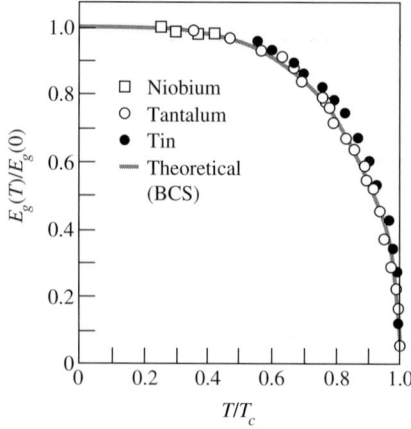

Fig. 11.14   Energy gap Δ as a function of T for various superconductors, as measured by tunneling experiments. From Townsend (1962).

By comparing the two cases of $T = 0$ and $T = T_c$, we can come up with another universal result. At T_c, the self-consistency equation is

$$\frac{4}{U_0 \mathcal{D}(E_F)/V} = \int_0^{E_c} dE_k \frac{1}{E_k} \tanh{(E_k/2k_B T_c)} . \tag{11.9.19}$$

When U_0 is small, the integral on the right-hand side must be large. In the limit of large E_c, the integral is approximately equal to $\ln(1.13 E_c/k_B T)$, which implies

$$k_B T_c = 1.13 E_c e^{-4V/U_0 \mathcal{D}(E_F)} . \tag{11.9.20}$$

On the other hand, at $T = 0$, the consistency equation is

$$\frac{4}{U_0 \mathcal{D}(E_F)/V} = \int_0^{E_c} dE_k \frac{1}{\sqrt{\Delta^2 + E_k^2}} . \tag{11.9.21}$$

This integral can be computed analytically, so that we have

$$\frac{4}{U_0 \mathcal{D}(E_F)/V} = \sinh^{-1}(E_c/\Delta). \tag{11.9.22}$$

Taking the limit of small U_0, as we did above for T_c, we then have

$$\Delta = \frac{E_c}{\sinh(4V/U_0 \mathcal{D}(E_F))} \simeq 2 E_c e^{-4V/U_0 \mathcal{D}(E_F)} . \tag{11.9.23}$$

Comparing (11.9.20) and (11.9.23), we obtain the approximate relation

$$2\Delta \simeq \frac{7}{2} k_B T_c. \tag{11.9.24}$$

This relation, deduced in the original work by Bardeen, Schrieffer, and Cooper, has been verified for a wide variety of standard superconductors, as shown in Figure 11.15.

Exercise 11.9.4   Compute numerically and plot the universal curve $\Delta(T)/\Delta(0)$ implied by (11.9.18).

Exercise 11.9.5   What is the critical temperature for Bose–Einstein condensation of a gas of electron pairs at a density of 10^{22} cm^{-3}, according to the ideal gas formula of Section 11.1, assuming that the electrons have the same effective mass as in vacuum? How does this T_c compare to typical superconductor T_c values of a few kelvin? This points out that the superconducting T_c discussed in this section is fundamentally related to pair breaking; as long as the pairs are stable, they are well below the critical temperature for Bose–Einstein condensation. It has been suggested that some materials might have the reverse: The T_c for pair formation could be above the temperature for condensation, in which case preformed Cooper pairs could exist without superconductivity.

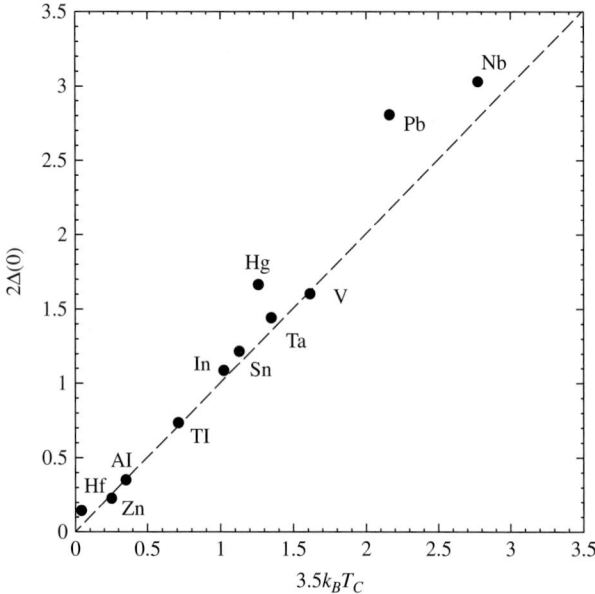

Fig. 11.15 Energy gap compared to $k_B T_c$ for several superconductors. Data from Mitrovic *et al.* (1984) and Poole *et al.* (1997).

11.10 Magnetic Effects of Superconductors

In Section 11.4, we showed how the hydrodynamics of a condensate are related to the phase of the wave function. For superconductors, since the particles are charged, we must alter these equations to take into account the effect of magnetic field.

Using the same approach as Section 11.4, we write the electrical current as the charge per particle times the particle current density,

$$\vec{J} = q\vec{g} = \text{Re}\left(\frac{q}{m}\psi^*(-i\hbar\nabla\psi)\right). \tag{11.10.1}$$

In the presence of magnetic field, the velocity is given by

$$\vec{v} = \frac{1}{m}\left(\vec{p} - q\vec{A}\right)$$
$$= \frac{1}{m}\left(-i\hbar\nabla - q\vec{A}\right), \tag{11.10.2}$$

where $\vec{A}$ is the vector potential field. Substituting $(-i\hbar\nabla - q\vec{A})$ for $-i\hbar\nabla$ in the definition of the current, with $\psi = \sqrt{n_0}e^{i\theta}$, we obtain

$$\vec{J} = \frac{q}{2m}\left(\psi^*\left(-i\hbar\nabla - q\vec{A}\right)\psi + \psi\left(i\hbar\nabla - q\vec{A}\right)\psi^*\right)$$

or

$$\vec{J} = \frac{q}{m}\left(\hbar n_0\nabla\theta - qn_0\vec{A}\right). \tag{11.10.3}$$

Taking the curl of both sides, and using the vector identity $\nabla \times \nabla\theta = 0$ for any scalar field θ, and the definition $\vec{B} = \nabla \times \vec{A}$, we have

$$\nabla \times \vec{J} = -\frac{n_0 q^2}{m}\vec{B}. \tag{11.10.4}$$

This is known as the **London** equation. This equation, combined with Maxwell's equations, tells us the interaction of magnetic field with a superconductor. From Maxwell's equations, we have, in steady state,

$$\nabla \times \vec{B} = \mu_0 \vec{J}. \tag{11.10.5}$$

Taking the curl of both sides, and using the vector identity $\nabla \times \nabla \times \vec{B} = \nabla(\nabla \cdot \vec{B}) - \nabla^2 \vec{B}$, and $\nabla \cdot \vec{B} = 0$ from Maxwell's equations, we obtain

$$-\nabla^2 \vec{B} = \mu_0 \nabla \times \vec{J}$$
$$= -\mu_0 \frac{n_0 q^2}{m}\vec{B}. \tag{11.10.6}$$

We guess a solution of the form

$$\vec{B}(\vec{x}) = \vec{B}_0 e^{-z/\lambda_L}, \tag{11.10.7}$$

and substitute this into (11.10.6). This implies

$$\frac{1}{\lambda_L^2} = \frac{\mu_0 n_0 q^2}{m}. \tag{11.10.8}$$

In other words, if the magnetic field is nonzero at the boundary of a superconductor, it will penetrate only a distance λ_L into the superconductor. This is known as the **London penetration depth.** This distance is typically very small. For $n_0 = 10^{20}$ cm^{-3}, λ_L is approximately equal to 0.5 μm. The exclusion of magnetic field from a superconductor is known as the **Meissner–Ochsenfeld** effect, or simply the Meissner effect.

The London penetration depth gives one intrinsic length scale for superconductors. Another is the coherence length, or correlation length, which gives the distance over which the phase θ of the wave function remains correlated. As discussed in Section 11.3, the Ginzburg–Landau equation for a Bose condensate is

$$\left(b|\psi|^2 - \frac{\hbar^2}{2m}\nabla^2\right)\psi = -a\psi. \tag{11.10.9}$$

In superconductors, because the particles are charged, the momentum term must be modified, giving

$$\left(b|\psi|^2 + \frac{1}{2m}\left(-i\hbar\nabla - q\vec{A}\right)^2\right)\psi = -a\psi, \tag{11.10.10}$$

where the charge and mass are $-2e$ and $2m_e$, respectively, because the charge carrier is a Cooper pair.

By the analysis of Section 10.4, the correlation length, in the absence of magnetic field, below T_c (where a is negative), is

$$\frac{1}{\xi^2} = -\frac{2a}{c},$$ (11.10.11)

with $c = \hbar^2/2m$.

These two equations, the London equation and the Ginzburg–Landau equation, with their associated natural length scales, are the primary equations in determining the general behavior of a superconductor in a magnetic field.

Exercise 11.10.1   Verify the calculation that the London penetration depth in a realistic superconductor is very small, using the electron density $n_0 = 10^{20}$ cm^{-3} and the mass and charge of an electron in vacuum (which is, of course, an approximation since electrons in a solid can have mass different from this.) Recall that the charge of the Cooper pairs is $-2e$, and find the exact value for λ_L in this case.

Exercise 11.10.2   What is the sheet current density K (in units of amperes per cm) that will be generated on the surface of a cylindrical superconductor with radius 2 cm, according to (11.10.4), assuming all the current flows within one London penetration depth of the surface? Assume that the sheet current effectively makes a solenoid that generates magnetic field in the opposite direction of the applied magnetic field.

11.10.1 Critical Field

As discussed above, superconductors exclude magnetic field over distances greater than the London penetration depth. This can only occur up to some maximum value of the field, however. Above a critical threshold of the field, the superconductor will revert to the normal state. We can see why by the following thermodynamic argument.

Consider a superconductor in an external field H which is excluding magnetic field from its whole volume – we neglect the field within the penetration depth as occurring in negligible volume. This implies a magnetization field M, created by a surface current circulating around the material, since

$$B = \mu H = \mu_0(H + M) = 0,$$ (11.10.12)

and therefore $M = -H$. The energy associated with this circulating current is (e.g., Lorrain and Corson 1970: s. 8.5.1)

$$W = \int d^3x \, \frac{1}{2}\mu_0(-M)^2 = \int d^3x \, \frac{1}{2}\mu_0 H^2.$$ (11.10.13)

This implies that work must be done to set up the circulating current in the superconductor that screens out the field, either by the external current that creates H, or mechanically by something that brings the superconductor from far away into the presence of the magnetic field, against an opposing force. (A dramatic demonstration of this can be seen in the levitation of a magnet over a superconductor due to the force from flux exclusion.)

There is therefore an energy penalty for being in the superconducting state due to the magnetization current. At some point, as the magnetic field is increased, the penalty for this magnetization energy can be greater than the thermodynamic free energy reduction from going into the condensate state from the normal state which we discussed in Section 11.3. At this point, the system will revert to the normal state.

The difference in free energy between the superconducting and normal state can be set equal to the Ginzburg–Landau free energy (entirely due to the superconducting wave function) plus this magnetization energy. The Ginzburg–Landau free energy for a superconductor is the same as (11.3.9) for a superfluid, with the momentum term generalized to include the effect of magnetic field on a charged particle. The difference between the free energy of the superconducting and normal state is therefore

$$F_{\text{sc}} - F_{\text{norm}} = \int d^3x \left(a|\psi|^2 + \frac{1}{2}b|\psi|^4 + \frac{1}{2m}\left|\left(-i\hbar\nabla - q\vec{A}\right)\psi\right|^2 + \frac{1}{2}\mu_0\text{H}^2 \right).$$

$$(11.10.14)$$

In the absence of magnetic field, and in the absence of spatial inhomogeneity, the mean-field solution of the Ginzburg–Landau equation, analogous to (10.4.21), is

$$|\psi|^2 = -\frac{a}{b}.$$

$$(11.10.15)$$

The critical field H_c, at which superconductivity disappears, can be found by setting $F_{\text{sc}} - F_{\text{norm}} = 0$. Assuming that the condensate amplitude does not depend strongly on the magnetic field, this gives us

$$0 = -\left(\frac{a^2}{b}\right) + \frac{1}{2}\left(\frac{a^2}{b}\right) + \frac{1}{2}\mu_0\text{H}_c^2$$

$$(11.10.16)$$

or

$$\text{H}_c = \sqrt{\frac{a^2}{\mu_0 b}}.$$

$$(11.10.17)$$

This critical field will depend on temperature. From the standard definitions of thermodynamics (e.g., Reif 1965: s. 11.4), the free energy is

$$F = E - TS,$$

$$(11.10.18)$$

where E is the internal energy, and from the first law of thermodynamics, the heat flow is

$$dQ = TdS = dE + \mu_0 VMd\text{H},$$

$$(11.10.19)$$

where $\mu_0 Md\text{H}$ is the work done to set up the magnetization field. From these two relations we deduce

$$dF = -\mu_0 VMd\text{H} - SdT.$$

$$(11.10.20)$$

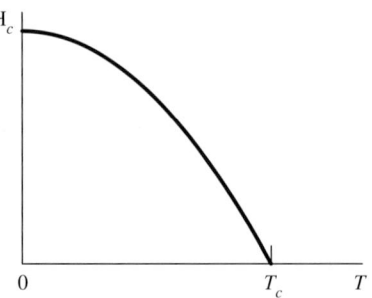

Fig. 11.16 Critical field of a superconductor as a function of temperature.

Suppose now we choose two values of the critical field H_c at two different temperatures. Since $F_{sc} = F_{norm}$ right at the critical field, we know that the difference between the free energy at each value of H_c for either state is the same,

$$dF_{sc} = dF_{norm}, \qquad (11.10.21)$$

or

$$-\mu_0 VM_{sc}dH_c - S_{sc}dT = -\mu_0 VM_{norm}dH_c - S_{norm}dT. \qquad (11.10.22)$$

Assuming that the magnetization of the normal state is negligible, this implies

$$S_{sc} - S_{norm} = -\mu_0 VM_{sc}\frac{dH_c}{dT}$$
$$= \mu_0 H_c\frac{dH_c}{dT}. \qquad (11.10.23)$$

When $T \to 0$, the third law of thermodynamics says that the entropy $S \to 0$ for both states, and therefore at $T = 0$ we must have $\partial H_c/\partial T = 0$. Figure 11.16 illustrates the phase boundary in the H–T plane. Near T_c, the parameter a is proportional to $T - T_c$ for the standard Ginzburg–Landau model, and therefore the critical field H_c has finite, negative slope at T_c.

The above also implies that there is a latent heat of the transition, equal to

$$Q = T(S_{norm} - S_{sc}) = -T\mu_0 VH_c\frac{\partial H_c}{\partial T}. \qquad (11.10.24)$$

Since $dH_c/dT < 0$, this is positive. In other words, to go from the superconducting state to the normal state at the same magnetic field H, heat must be put into the system. In the absence of magnetic field, there is no latent heat of the phase transition.

Critical current density. The existence of the upper critical field affects many of the applications for superconductors. When current is run through a superconductor, it generates a magnetic field. If the current is high enough, this magnetic field can exceed the critical magnetic field of the superconductor, causing it to revert to normal behavior. This implies that there is also a **critical current density**, above which the superconductor will revert to a normal metal. This is often the cause of catastrophic results, as the high current suddenly starts to generate ohmic heating due to the finite resistance of the conductor when it goes normal. This is known as a **quench**.

As discussed above, one of the implications of the London equation is that all of the current in a superconductor flows on its surface. Knowing that the current all flows within a London penetration depth of the surface, we can write down an equation for the critical current density of a superconductor. For simplicity, let us consider a wire with a circular cross-section. From basic electromagnetism (see, e.g., Griffiths 2013), the magnetic field at the surface of the wire, generated by the current flowing in the wire, has magnitude

$$B = \frac{\mu_0 I}{2\pi r},$$

(11.10.25)

where r is the radius of the wire. (This is easily deduced from the Maxwell equation $\nabla \times \vec{B} = \mu_0 \vec{J}$ for a DC system.) The critical current will occur when the magnetic field at the surface generated by the current equals the critical field B_c:

$$I_c = \frac{2\pi r B_c}{\mu_0}.$$

(11.10.26)

The area that this current flows through is the cross-section of the thin current sheet at the surface, equal to the circumference of the wire times the London penetration depth. Therefore, we can write the critical current density J_c,

$$J_c = \frac{I_c}{2\pi r \lambda_L} = \frac{B_c}{\mu_0 \lambda_L}.$$

(11.10.27)

This critical current density, $J_c = B_c/\mu_0\lambda_L$, is an intrinsic material property that depends only on microscopic parameters, not the geometry of the conductor.

Exercise 11.10.3   What is the value of the critical field if a is 50 meV and the condensate density is 10^{22} cm^{-3}? Is this obtainable by modern magnet systems?

Exercise 11.10.4   What is the critical current density, in amperes per cm^2, of a superconductor with critical field of 15 T and London penetration depth of 0.5 μm?

Exercise 11.10.5   Estimate the work needed to bring a superconductor with volume of 1 cm^3 from infinity (where the magnetic field is zero) to a location with constant magnetic field equal to 1 tesla, assuming all magnetic flux is excluded from the superconductor.

If a superconductor is placed in a uniform magnetic field, will it feel a force acting opposite to gravity to help levitate it? Give an argument for your answer.

11.10.2 Flux Quantization

Above, we deduced that the magnetic field inside a superconductor must be zero. But suppose we have a loop of superconducting wire, or a superconductor with a hole through it, or with some non-superconducting material going through it. Then there can be nonzero magnetic flux going through a closed path inside the superconductor, even though the magnetic field is zero inside the superconducting material itself. From Stokes' theorem, the path integral of $\vec{A}$ around a closed loop is equal to the total magnetic field flux through the loop:

$$\oint \vec{A} \cdot d\vec{l} = \int_A (\nabla \times \vec{A}) \cdot d\vec{a} = \int_A \vec{B} \cdot d\vec{a} \equiv \Phi.$$

(11.10.28)

This can be related to the total phase gradient around the loop. Because the current is zero deep inside a superconductor, from (11.10.3), we have

$$\frac{q}{m}\left(\hbar n_0 \nabla\theta - q n_0 \vec{A}\right) = 0. \tag{11.10.29}$$

Therefore, around any closed path in the bulk of the superconductor, we have

$$\hbar \oint \nabla\theta \cdot \vec{dl} = q \oint \vec{A} \cdot \vec{dl}. \tag{11.10.30}$$

The path integral of the gradient of phase θ around the loop is just equal to the total phase change around the loop. Since the phase has only one value at every point in space, this means that the total phase change around a loop must be equal to zero or equal to an integer multiple of 2π, since 2π radians is the same as zero radians. We therefore have

$$\hbar(2\pi N) = |q|\Phi, \tag{11.10.31}$$

where N is an integer, or

$$\Phi = \frac{2\pi\hbar}{|q|}N = \frac{h}{|q|}N. \tag{11.10.32}$$

The total flux through any closed path must therefore be either zero or an integer multiple of

$$\boxed{\Phi_0 = \frac{h}{|q|}.} \tag{11.10.33}$$

The quantity Φ_0 is known as a **flux quantum**. It is the same flux quantum as derived for free electrons in Landau orbits in Section 2.9. In the case of a superconductor, the magnitude of the charge q is equal to $2e$, because the relevant charged particle is the Cooper pair or electrons. As discussed in Section 2.9, the value of a flux quantum depends only on universal constants of nature, and not on any of the details of the superconducting material, such as what elements it is made out of. It is a very small value, equal to approximately 2.1×10^{-11} T-cm^2.

The result (11.10.32) says that magnetic field can go through the center of a super-conducting loop, or through a hole or a non-superconducting region in a superconductor, but that the flux cannot be just any value: It must be an integer number of flux quanta. The superconductor will generate surface current to create an opposing magnetic field to exactly cancel out any extra applied magnetic flux.

This surface current costs energy; equation (11.10.13) implies that the energy cost is proportional to the square of the magnetic flux excluded from the superconductor. Figure 11.17 shows how the energy cost depends on magnetic field in a superconductor. When there is an exact integer number of flux quanta going through the superconductor, there is no energy cost because there is no current in that case. When the flux is greater or less than an exact number of flux quanta, however, there must be surface current to generate field to cancel out the extra field. This surface current has energy cost proportional to the square of the amount of field to be canceled.

As seen in Figure 11.17, when the magnetic field is greater than $\frac{1}{2}\Phi_0$, the energy of the system is reduced, allowing one flux quantum to pass through the superconductor. When

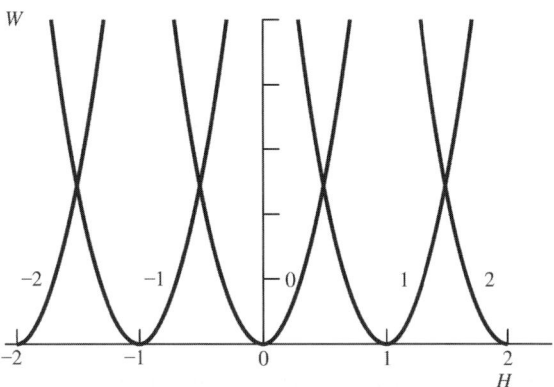

Fig. 11.17 Energy cost for surface current of a superconductor as a function of the applied magnetic field, for various numbers of flux quanta going through the superconductor, as labeled. Negative signs refer to flux in the opposite direction.

this happens, the surface current must change direction to generate magnetic field in the opposite direction, to cancel out the excess field. As the applied magnetic field is increased, this process will happen many times, as each time the field exceeds a half-integer number of flux quanta, the system will allow another flux quantum to penetrate, and the sign of the surface current will reverse. Oddly, right at each crossover point, the system will be in a quantum-mechanical superposition of the surface current going both clockwise and counterclockwise. Is this an example of a "Schrödinger's cat"? On one hand, it is a superposition of two states of a macroscopic number of particles. On the other hand, it is a superposition with just one bit of information, either circulation in one direction or the other.

Exercise 11.10.6   Confirm, using the values of the fundamental constants and converting units, the value of the superconductor flux quantum Φ_0, equal to 2.1×10^{-11} T-cm^2.

11.10.3 Type I and Type II Superconductors

In deducing the critical field (11.10.17), we assumed that the penetration depth of the magnetic field into the volume of the superconductor was negligible. What happens if the penetration depth is not negligible?

Consider the two cases shown in Figure 11.18. When $\xi \gg \lambda_L$, the magnetic field is almost entirely excluded from the surface region, while the condensate wave function amplitude is substantially lower than its value in the bulk of the material in this region. The free energy per volume for excluding magnetic field is $\frac{1}{2}\mu_0 H^2$, which is positive, while the Ginzburg–Landau free energy per volume for the superconductor condensate is negative. Therefore, the net energy per volume in the surface region in this case is larger than in the bulk, since the positive contribution of excluding magnetic field from this volume is about the same as the bulk, while the negative contribution of the condensate free energy is lower. In other words, there is a positive surface energy, an additional free energy cost per area of surface.

In the case when $\xi \ll \lambda_l$, the superconducting wave function has nearly its full bulk value over the entire surface region, while magnetic field is not excluded from this region.

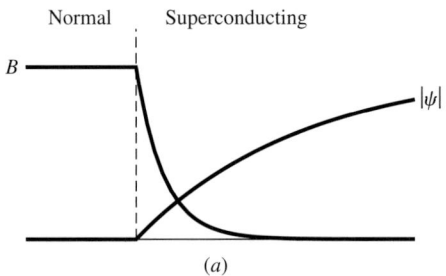

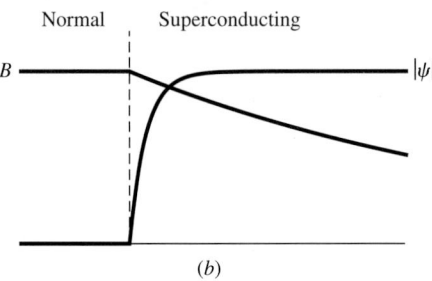

(a) Magnetic field and condensate wave function in the case $\xi \gg \lambda_l$ (Type I superconductor). (b) Magnetic field and condensate wave function in the case $\xi \ll \lambda_l$ (Type II superconductor).

Therefore, there is less free energy contribution from the exclusion of the magnetic field from this region than in the bulk, while the negative contribution of the condensate is about the same as in the bulk. Therefore, the free energy per volume is less than in the bulk, and so the surface energy is negative. In this case, the system will want to maximize the surface area of the interface between the superconducting and normal states.

There are therefore distinctly different behaviors depending on the ratio of the two parameters ξ and λ. We call a superconductor with positive surface energy a **Type I** superconductor, while a superconductor with negative surface energy is a **Type II** superconductor.

We can find the point of crossover by a simple argument. Consider a superconducting system near H_c. In this region of phase space, the amplitude ψ is small, and the Ginzburg–Landau equation (11.10.10) becomes, to lowest order in ψ,

$$\frac{1}{2m}\left(-i\hbar\nabla - q\vec{A}\right)^2 \psi = -a\psi. \qquad (11.10.34)$$

This is simply the equation for a charged particle in a magnetic field, which gives the cyclotron solution discussed in Section 2.9, with the lowest-energy eigenvalue

$$\frac{1}{2}\hbar\omega_c = \frac{\hbar qB}{2m} = \frac{\hbar q(\mu_0 H)}{2m}. \qquad (11.10.35)$$

The superconducting current in this case is circulating, with a wave function just like a single electron undergoing a circular orbit in a magnetic field. In other words, there is a vortex of superconducting current.

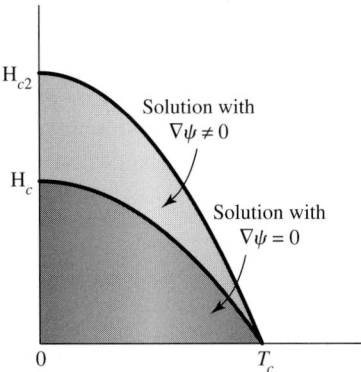

Phase diagram of a Type II superconductor.

Since $|a|$ is the energy per particle, we must have

$$|a| > \frac{1}{2}\hbar\omega_c, \tag{11.10.36}$$

which implies

$$H < \frac{2m|a|}{\mu_0\hbar q}. \tag{11.10.37}$$

In other words, there is a second critical field, which we can call H_{c2}, for solutions of the condensate wave function when ψ is small and $\nabla\psi \neq 0$.

Since the critical field H_c given by (11.10.17) for the case of a homogeneous superconductor also depends linearly on $|a|$, we can write H_{c2} in terms of H_c, as follows:

$$\frac{H_{c2}}{H_c} = \frac{2m}{\hbar q}\sqrt{\frac{b}{\mu_0}}. \tag{11.10.38}$$

Figure 11.19 illustrates the phase boundaries H_c and H_{c2} in the case of a Type II superconductor. Below the critical field H_c, there are solutions with $\nabla\theta$ of the condensate wave function equal to zero, while below the critical field H_{c2} there are solutions with $\nabla\theta \neq 0$. If $H_{c2} < H_c$, then everywhere there is a solution with $\nabla\theta \neq 0$, there is also a solution with $\nabla\theta = 0$. Since $\nabla\theta$ gives a positive kinetic energy term in the Ginzburg–Landau equation, we can assume that the $\nabla\theta = 0$ solution has lower energy per particle and is therefore favored. On the other hand, if $H_{c2} > H_c$, then there is a region where no $\nabla\theta = 0$ solution is allowed but a solution with $\nabla\theta$ (i.e., vortices), is allowed.

Using the definitions of the London penetration depth λ_L and the coherence length ξ, and the mean-field solution $n_0 = |a|/b$, the relation (11.10.38) is the same as

$$\frac{H_{c2}}{H_c} = \sqrt{2}\frac{\lambda_L}{\xi}. \tag{11.10.39}$$

Therefore, the two possible situations correspond to the same two situations discussed above, namely, when $H_{c2} < H_c$ the surface energy is positive, while if $H_{c2} > H_c$, the surface energy is negative. The crossover point is when $\lambda_L = \sqrt{2}\xi$. In the Type II case, there is a **mixed state** region in which normal metal and superconducting regions can coexist. The magnetic field penetrates the superconductor by creating vortices of the condensate surrounding filaments of nonzero magnetic field.

Flux lines in Type II superconductors. To find the magnetic field around a vortex, we can solve (11.10.6), which is a restatement of the London equation, for the case of cylindrical symmetry,

$$\nabla^2 B - \frac{B}{\lambda_L^2} = 0, \tag{11.10.40}$$

everywhere except at $r = 0$. The solution is

$$B(r) = CK_0(r/\lambda_L), \tag{11.10.41}$$

where $K_0(r/\lambda_L)$ is the modified Bessel function, proportional to e^{-r/λ_L} at large r, and C is a multiplicative constant. For small r, $K_0(r/\lambda_L)$ behaves as $-\ln(r/\lambda_L)$, which diverges at $r = 0$. The magnetic field does not really become infinite, because this solution breaks down for $r < \xi$, the coherence length. We therefore have the picture of a flux tube of radius ξ carrying almost all the magnetic field.

This implies that the current also drops to zero far from a vortex, since by the London equation, $\nabla \times \vec{J}$ is proportional to $\vec{B}$, and therefore, for the z-component of the cross product in cylindrical coordinates, at large r,

$$\frac{1}{r}\frac{\partial}{\partial r}(rJ) \propto e^{-r/\lambda_L}, \tag{11.10.42}$$

which has the solution $J(r) \propto (-e^{-r/\lambda_L}(1+r) + \text{const.})/r$. Therefore, the case of a vortex in a Type II superconductor is just the same as the case we looked at in Section 11.10.2; instead of a hole in the center of the superconductor that passes magnetic field, we have the core of the vortex which is non-superconducting, and allows magnetic field to pass through, and deep inside the superconductor, the magnetic field drops to zero. Therefore, the same argument applies, that the flux must be quantized into integer numbers of flux quanta.

It can be shown that the vortices repel each other. Therefore, the lowest-energy configuration is to have many flux lines each carrying one flux quantum. In typical superconductors, the conductivity is controlled by the motion of these vortices. Motion of a flux line gives rise to ohmic resistive loss.

Just like the dislocation lines discussed in Section 5.12, the vortex lines can become tangled or pinned, depending on the defects that exist in the superconductor. At low temperatures, the repulsion of the vortices leads to the appearance of an **Abrikosov lattice**, in which the vortices are ordered in a two-dimensional array, as shown in Figure 11.20. If the whole lattice is pinned to defects, then there is no motion of the flux lines, and the superconductor carries current with the least resistive loss. In other words, somewhat counterintuitively, defects are necessary for the best superconducting behavior. At higher temperatures, the Abrikosov lattice can undergo a phase transition to a vortex liquid analogous to standard crystal melting.

Fig. 11.20   Lattice of flux lines in NbSe$_2$, a Type II superconductor, observed with a scanning tunneling microscope. From Hess *et al.* (1989).

11.11 Josephson Junctions

The properties of superconductors lead to a strange effect. Consider a thin, insulating barrier between two superconducting layers, as shown in Figure 11.21, which is thin compared to the coherence length. Without knowing anything about the details of the barrier, we can write a set of coupled Schrödinger equations for the evolution of the wave function, in the absence of magnetic field,

$$i\hbar\frac{\partial\psi_1}{\partial t} = E_1\psi_1 + K\psi_2$$

$$i\hbar\frac{\partial\psi_2}{\partial t} = K^*\psi_1 + E_2\psi_2, \tag{11.11.1}$$

where K is some small coupling parameter. Setting $\psi_1 = \sqrt{n_1}e^{i\theta_1}$ and $\psi_2 = \sqrt{n_2}e^{i\theta_2}$, these equations become

$$\frac{1}{2\sqrt{n_1}}\frac{\partial n_1}{\partial t} + i\sqrt{n_1}\frac{\partial\theta_1}{\partial t} = -\frac{i}{\hbar}\left(E_1\sqrt{n_1} + K\sqrt{n_2}e^{i(\theta_1-\theta_2)}\right)$$

$$\frac{1}{2\sqrt{n_2}}\frac{\partial n_2}{\partial t} + i\sqrt{n_2}\frac{\partial\theta_2}{\partial t} = -\frac{i}{\hbar}\left(E_2\sqrt{n_2} + K\sqrt{n_1}e^{i(\theta_2-\theta_1)}\right). \tag{11.11.2}$$

Setting the real parts equal gives us

$$\frac{\partial n_1}{\partial t} = -\frac{\partial n_2}{\partial t} = \frac{2K}{\hbar}\sqrt{n_1 n_2}\sin(\theta_1 - \theta_2). \tag{11.11.3}$$

The rate of change of the charge density gives the current through the junction. Since the current is real, this implies that the coupling constant K must be real. For two identical superconductors, $n_1 \approx n_2 \equiv n_0$, and therefore we can replace $\sqrt{n_1 n_2}$ with simply n_0. Since we don't know K, to compare to the current formula (11.10.3) we can write

$$J = \frac{e\hbar n_0}{m\lambda_J}\sin(\theta_1 - \theta_2), \tag{11.11.4}$$

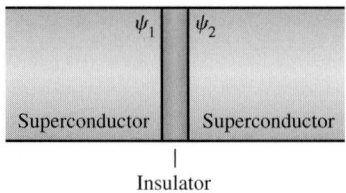

Fig. 11.21   A Josephson junction.

where λ_J is a characteristic length, which is another way of parametrizing the coupling.

Setting the imaginary parts of (11.11.2) equal gives

$$\sqrt{n_1}\frac{\partial\theta_1}{\partial t} = -\frac{E_1}{\hbar}\sqrt{n_1} - \frac{K}{\hbar}\sqrt{n_2}\cos(\theta_1 - \theta_2)$$

$$\sqrt{n_2}\frac{\partial\theta_2}{\partial t} = -\frac{E_2}{\hbar}\sqrt{n_2} - \frac{K}{\hbar}\sqrt{n_1}\cos(\theta_1 - \theta_2). \tag{11.11.5}$$

Subtracting these two equations gives

$$\frac{\partial}{\partial t}(\theta_1 - \theta_2) = -\frac{E_1 - E_2}{\hbar} = \frac{2eV}{\hbar}, \tag{11.11.6}$$

where V is the voltage applied across the junction. The charge $q = -2e$ is used since n_0 gives the number of pairs. For constant V, this implies

$$\theta_1 - \theta_2 = \theta_0 + \frac{2e}{\hbar}Vt. \tag{11.11.7}$$

Substituting into (11.11.4), we have

$$J = \frac{e\hbar n_0}{m\lambda_J}\sin\left(\theta_0 + \frac{2e}{\hbar}Vt\right). \tag{11.11.8}$$

In other words, a constant potential difference across the junction leads to an AC current with frequency proportional to the potential difference, similar to the Bloch oscillation effect discussed in Section 2.8.2. The power consumption for the superconducting AC current, $P = IV$, is zero, since V is constant and the average of the current, which oscillates as $\sin(\theta_1 - \theta_2)$, is zero. This is known as the **AC Josephson effect**. If $V = 0$, a current can still flow if $\theta_0 \neq 0$. This is known as the DC Josephson effect. The **DC Josephson effect** is just another way of saying that persistent currents can flow in a superconductor if the condensate is not in the ground state.

The AC Josephson effect can in principle be used as a very sensitive voltage standard, since the frequency of the oscillating current depends only on fundamental constants and the applied voltage.

SQUIDs. The property discussed in Section 11.10.2 that the surface current of the superconductor changes sign and passes through zero every time an additional flux quantum passes through it can be combined with the properties of Josephson junctions for a very sensitive measurement of magnetic field.

Consider a superconducting loop with two identical Josephson junctions, as illustrated in Figure 11.22. Each of the Josephson junctions can be modeled by the circuit shown in

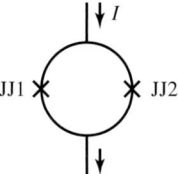

Fig. 11.22   A SQUID detector with two Josephson junctions.

Figure 11.23(a) with parallel capacitance and resistance, but for the moment we ignore these.

As discussed in Section 11.10.2, the total phase change around the superconducting loop gives the flux through the loop:

$$\oint \nabla \theta \cdot d\vec{l} = \frac{q}{\hbar} \oint \vec{A} \cdot d\vec{l} = 2\pi \frac{\Phi}{\Phi_0}, \tag{11.11.9}$$

where Φ is the total magnetic flux through the loop. (We will ignore any flux from self-inductance of the loop.) In this case, however, the total phase change does not have to be $2\pi N$; it is adjusted by the phase change across the Josephson junctions. If we define the phase differences for the two junctions as $\Delta\theta_1$ and $\Delta\theta_2$, each in the same direction from top to bottom, then taking the path integral around the loop adds one and subtracts the other:

$$2\pi N - \Delta\theta_1 + \Delta\theta_2 = 2\pi \frac{\Phi}{\Phi_0}. \tag{11.11.10}$$

The total current through the circuit is the sum of the two currents through the junctions, which we write using the form of the current (11.11.4),

$$I = I_c(\sin \Delta\theta_1 + \sin \Delta\theta_2). \tag{11.11.11}$$

Solving (11.11.10) for $\Delta\theta_2$ and substituting, we have

$$I = I_c\left(\sin \Delta\theta_1 + \sin \left(\Delta\theta_1 + 2\pi \Phi/\Phi_0\right) \right). \tag{11.11.12}$$

Using the trig identity $\sin A + \sin B = 2 \cos \frac{1}{2}(A - B) \sin \frac{1}{2}(A + B)$, this becomes

$$I = I_c \cos(\pi \Phi/\Phi_0) \sin(\Delta\theta_1 + \pi \Phi/\Phi_0). \tag{11.11.13}$$

The total coherent current therefore has a maximum amplitude that depends on the flux through the loop. The exact phase drop $\Delta\theta_1$ can be a complicated function of time, but when Φ/Φ_0 is equal to a half-integer number, no current will pass through. This is a coherent interference effect like the mesoscopic ring interference we studied in Section 9.11. This device is therefore known as a **SQUID** (superconducting quantum interference device).

If there is an external DC current source driving a current through the device, then when the coherent current is zero, the current must all be shunted through the parallel resistance, which we included in our model circuit in Figure 11.23(a). This will lead to a voltage drop across the resistance that can be measured by an external voltage detector; the voltage

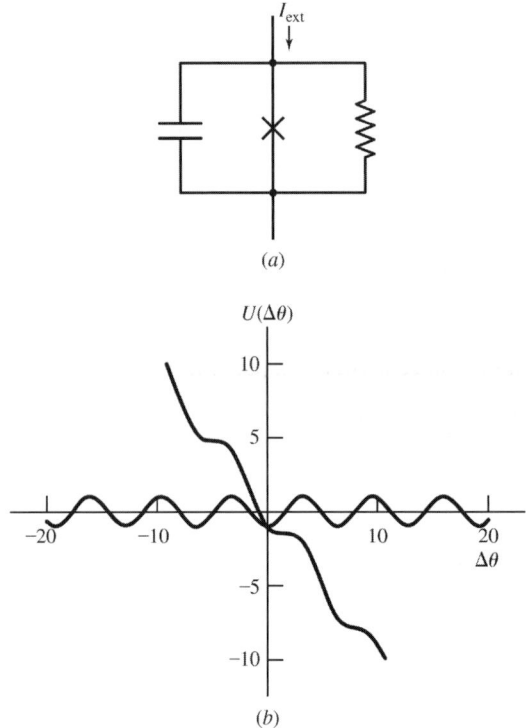

(a)

(b)

Fig. 11.23 (a) Equivalent circuit model of a real Josephson junction. (b) Effective potential $-I_{\text{ext}}(\Delta\theta) - I_c \cos(\Delta\theta)$ as a function of $\Delta\theta$ for this circuit for fixed DC current $I_{\text{ext}} = 0$ and $I_{\text{ext}} = I_c$.

drop is maximum when the coherent current is minimum. This means that the loop will be a very sensitive digital counter of the number of flux quanta. Since the value of a flux quantum is very small, $\Phi_0 = 2.1 \times 10^{-11}$ T-cm^2, SQUIDs are very sensitive magnetic field detectors. Often a metal truck driving down the street outside a building with a SQUID can be detected from the disturbance it creates in the magnetic field.

Note that the current depends on the total flux through the loop even if the B-field is zero everywhere on the wires themselves. In other words, the system responds to A-field, not B directly. This is another version of the **Aharonov–Bohm** effect (see Section 9.11). The basic principle is that whenever the system is sensitive to the phase of the wave function, it is sensitive to the A-field.

Exercise 11.11.1   Equation (11.11.13) was deduced neglecting the associated parallel resistance of the junctions, which shown in Figure 11.23(a). Rewrite this equation for the case in which each junction has a parallel resistance R, and determine the maximum voltage drop ΔV across the SQUID shown in Figure 11.22 for the case of constant input current I, as a function of the magnetic flux Φ through the loop.

Circuit model of a Josephson junction. Figure 11.23(a) shows a model of a realistic Josephson junction. The $\times$ represents the junction, which has the governing equations

$$I = I_c \sin(\Delta\theta), \qquad \Delta\dot{\theta} = \frac{2e}{\hbar}\Delta V, \qquad (11.11.14)$$

where we have rewritten (11.11.4) in terms of the total current, which of course depends on the cross-sectional area A of the junction. The multiplier $I_c = e\hbar n_0 A/m\lambda_J$ is often called the "critical current," but this should not be confused with the critical current density for breakdown of superconductivity, discussed in Section 11.10. It is, rather, the maximum possible coherent current through the junction.

In addition to the ideal junction, Figure 11.23(a) includes a parallel capacitance and a parallel resistance. The Josephson junction always has some intrinsic capacitance because it is an insulator between two conducting layers. There is also a resistance, which represents incoherent current through the junction. This occurs intrinsically because at finite temperature, some of the electrons are not paired in Cooper pairs. Often in real circuits a parallel resistance is added on purpose.

Suppose that a fixed DC current I_{ext} is driven through the junction. The total current through the parallel circuit is

$$I_{ext} = I_c \sin(\Delta\theta) + \frac{\Delta V}{R} + C\Delta\dot{V}. \tag{11.11.15}$$

We can rewrite this using the relation of ΔV and $\Delta\theta$ from (11.11.14):

$$I_{ext} - I_c \sin(\Delta\theta) - \frac{\hbar}{2eR}\Delta\dot{\theta} = \frac{\hbar C}{2e}\Delta\ddot{\theta}. \tag{11.11.16}$$

This has the same form as the force equation from classical mechanics,

$$F(x) - \gamma\dot{x} = m\ddot{x}, \tag{11.11.17}$$

for an object with mass m in the presence of an energy-conserving force and a drag force. The energy-conserving force can be written as the derivative of a potential energy $U(\Delta\theta)$,

$$I_{ext} - I_c \sin(\Delta\theta) = -\frac{\partial U}{\partial(\Delta\theta)} = -\frac{\partial}{\partial(\Delta\theta)}(-I_{ext}\Delta\theta - I_c \cos(\Delta\theta)). \tag{11.11.18}$$

Figure 11.23(b) shows this potential for zero and for finite bias current I.

For small $\Delta\theta$, one can approximate $\sin(\Delta\theta) \sim \Delta\theta$, so that the term $-I_c \sin(\Delta\theta) \sim -I_c(\Delta\theta)$ acts just like a spring force $F = -kx$, and (11.11.16) becomes exactly the equation of a resonant oscillator. This limit corresponds to the system sitting in one of the potential energy minima of $U(\Delta\theta)$ shown in Figure 11.23(b). In this limit, the Josephson junction acts as an inductor, since

$$\dot{I} = \cos(\Delta\theta)\Delta\dot{\theta} = \cos(\Delta\theta)\frac{2e}{\hbar}\Delta V \simeq \frac{2e}{\hbar}\Delta V, \tag{11.11.19}$$

which can be written as the inductor equation $\Delta V = L\dot{I}$, with $L = \hbar/2e$. The circuit in this limit therefore is a resonant LRC circuit.

If the bias current I_{ext} is made large enough, the phase will not oscillate about a fixed point, but instead will flow over the barrier and down the "washboard" potential shown in Figure 11.23(b). In this case, the oscillating term in (11.11.16) can be neglected and the effective force is nearly constant. As with the motion of an object at terminal velocity, in this limit the constant effective force leads to a constant average $\Delta\dot{\theta}$ in the presence of the drag force.

In between these two limits, the solution of (11.11.16) can be quite complicated, and requires numerical methods. This is because the Josephson junction is a *nonlinear* circuit element; the voltage drop across the junction is not in general proportional to the amplitude of the current.

Exercise 11.11.2 (a) Determine the natural oscillation frequency of the phase difference $\Delta\theta$ across the Josephson junction equivalent circuit with $I_{ext} = 0$, in the limit of low amplitude.

(b) At what value of I_{ext} will there be no local minima in the effective potential $U(\Delta\theta)$ at any value of $\Delta\theta$?

Exercise 11.11.3 Write down, but do not solve, the full equations for the dynamics of the phase jumps in the Josephson junctions of a SQUID in the presence of magnetic flux, using the model circuit given here. How are these equations altered if you include the self-inductance of the loop?

11.12 Spontaneous Optical Coherence: Lasing as a Phase Transition

In Section 7.4 and in Chapter 9, we discussed the effects of a coherent driving field on an ensemble of oscillators in a solid. In those discussions, we simply assumed the existence of a classical coherent field, which is equivalent to a coherent state of photons. How does one get such a state? There are two possible routes. At low frequency, a coherent electromagnetic field can be generated simply by a classical oscillation of a charged object. At optical frequencies, however, the frequency is comparable to the orbital frequencies of electrons around atoms, which means quantum mechanics must come into play. In general, the light emission from many atoms will be incoherent, since there is no correlation between isolated, noninteracting atoms. Under certain circumstances, however, spontaneous coherence can arise, as the dipole moments of the atoms become coupled through the electromagnetic field.

The coupling of the dipole moments in a medium and the electromagnetic field is given by Maxwell's wave equation, introduced in Section 3.5.1. We assume here that the electromagnetic wave has the simple form $E(\vec{k}, t) = E_0 e^{i(\vec{k}\cdot\vec{x} - \omega t)}$, which allows us to resolve the spatial second derivative that appears in Maxwell's wave equation as $\nabla^2 E = -k^2 E = -(\omega^2/c^2)E$. For an isotropic and spatially homogeneous system, Maxwell's wave equation is then

$$-\omega^2 E = \frac{\partial^2 E}{\partial t^2} + \frac{1}{\epsilon_0}\frac{\partial^2 P}{\partial t^2}. \tag{11.12.1}$$

We thus drop consideration of the spatial dependence of the electromagnetic field and write simply $E = E_0(t)e^{-i\omega t}$. For simplicity, we consider only one direction for the electric field, but we will allow for the possibility that the overall amplitude E_0 changes over time.

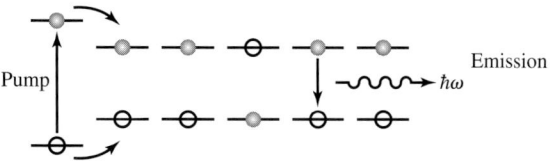

Fig. 11.24 Basic picture of a solid state laser.

The polarization of the medium P that appears in this equation can be connected to quantum mechanical states of an ensemble of two-level oscillators using the methods developed in the first few sections of Chapter 9. The standard model for a laser is a set of isolated two-level oscillators that can be pumped into their excited states by an incoherent external source (e.g., an electrical current, or an incoherent light), as illustrated in Figure 11.24. In Section 9.3, we showed that the Bloch equations (9.2.27) apply to an optically excited two-level oscillator with dephasing time T_2 and decay time T_1 for relaxation from the upper to the lower state. In the case when a coherent driving field is resonant with the energy gap between the two levels, these equations are

$$\frac{\partial U_1'}{\partial t} = -\frac{U_1'}{T_2}$$

$$\frac{\partial U_2'}{\partial t} = -\frac{U_2'}{T_2} - \omega_R U_3'$$

$$\frac{\partial U_3'}{\partial t} = -\frac{U_3' + 1}{T_1} + \omega_R U_2', \tag{11.12.2}$$

where, following the analysis of Section 9.3, U_1' and U_2' are the real and imaginary parts of the off-diagonal term of the density matrix in the rotating frame, $U_3' = \rho_{cc} - \rho_{vv}$ is the population inversion, and $\omega_R = e|\langle x \rangle|E_0/\hbar$ is the Rabi frequency. Let us add one extra term, which represents an incoherent pump that acts to promote electrons from the lower to the upper state, as in the process shown in Figure 11.24. This process will saturate due to Pauli exclusion when the upper level is fully occupied, that is, when $U_3' = 1$. We write

$$\frac{\partial U_1'}{\partial t} = -\frac{U_1'}{T}$$

$$\frac{\partial U_2'}{\partial t} = -\frac{U_2'}{T} - \omega_R U_3'$$

$$\frac{\partial U_3'}{\partial t} = -\frac{U_3' + 1}{T} + \omega_R U_2' + G(1 - U_3'), \tag{11.12.3}$$

where G is the rate of generating excited electrons in the upper state. Here we have also assumed that the main dephasing process is spontaneous emission by electrons in the upper state, so that we can set $T_2 = T_1 \equiv T$.

As discussed in Section 9.3, the polarization in the medium is proportional to the off-diagonal density matrix element, which is given by the Bloch vector in a two-level system. We write

$$P(t) = \frac{qN}{V} \langle c|x|v\rangle (U_1'(t) + iU_2'(t))e^{-i\omega t}, \tag{11.12.4}$$

where we have used the definition of the Bloch vector in the rotating frame: U_1' gives the component of the density matrix in phase with the driving field, and U_2' gives the component of the density matrix $90°$ out of phase with the driving field. To find a self-consistent solution for the electric field amplitude $E_0(t)$, we need to use this polarization in Maxwell's wave equation (11.12.1). Because the field amplitude E_0 appears in the Rabi frequency ω_R, the self-consistent solution can in general be a complicated function of time.

To simplify the solution, we assume that the time scale for change of the amplitude E_0 is long compared to all other time scales (this is known as the **slowly varying envelope approximation**). Taking the rate of change of E_0 as slow compared to the relaxation rate $1/T$ and the pumping rate G allows us to find the Bloch vector $\vec{U}'$ as the steady-state solution of (11.12.3) for a given value of E_0. Setting all time derivatives to zero yields

$$U_1' = 0, \quad U_2' = -\omega_R T \frac{GT - 1}{GT + 1 + \omega_R^2 T^2}, \quad U_3' = \frac{GT - 1}{GT + 1 + \omega_R^2 T^2}. \tag{11.12.5}$$

When the Rabi frequency is small, such as when the electric field is not too strong, we can rewrite the formula for U_2' as

$$U_2' = -\omega_R T \left(\frac{GT - 1}{GT + 1}\right) \frac{1}{1 + \omega_R^2 T^2/(GT + 1)}$$

$$\approx -\omega_R T \left(\frac{GT - 1}{GT + 1}\right) \left(1 - \frac{\omega_R^2 T^2}{GT + 1}\right). \tag{11.12.6}$$

Using the definition of the Rabi frequency gives us the polarization of the medium,

$$P(t) = -i\frac{e^2 \langle x\rangle^2 N}{\hbar V} E_0(t) T \left(\frac{GT - 1}{GT + 1}\right) \left(1 - e^2 \langle x\rangle^2 E_0^2(t) \frac{T^2}{GT + 1}\right) e^{-i\omega t}, \tag{11.12.7}$$

which can be written simply as

$$P(t) = -i(AE_0(t) - BE_0^3(t))e^{-i\omega t}, \tag{11.12.8}$$

where A and B are constants that depend on the properties of the medium and the incoherent pumping rate G. The constant A can be either positive or negative, depending on whether the generation rate G is larger than the relaxation rate $1/T$.

Substituting this into Maxwell's wave equation (11.12.1), we obtain

$$-\omega^2 E_0(t)e^{-i\omega t} = \left(1 - \frac{iA}{\epsilon_0}\right) \frac{\partial^2}{\partial t^2} E_0(t)e^{-i\omega t} + \frac{iB}{\epsilon_0} \frac{\partial^2}{\partial t^2} E_0^3(t)e^{-i\omega t}. \tag{11.12.9}$$

Since we assume that the rate of change of E_0 is slow, such that $(1/E_0)\partial E_0/\partial t \ll \omega$, we can approximate the time derivatives as

$$\frac{\partial^2}{\partial t^2} E_0(t)e^{-i\omega t} \approx \left(-2i\omega \frac{\partial E_0}{\partial t} - \omega^2 E_0\right) e^{-i\omega t}$$

$$\frac{\partial^2}{\partial t^2} E_0^3(t)e^{-i\omega t} \approx -\omega^2 E_0^3 e^{-i\omega t}, \tag{11.12.10}$$

so that (11.12.9) becomes

$$0 = \left(-2i\omega\frac{\partial E_0}{\partial t}\right) + \left(\frac{iA}{\epsilon_0}\right)\left(-2i\omega\frac{\partial E_0}{\partial t} - \omega^2 E_0\right) - \frac{iB}{\epsilon_0}\left(-\omega^2 E_0^3\right).$$

$$(11.12.11)$$

For typical doping densities in laser media, $A/\epsilon_0 \ll 1$. We can therefore drop the term in which A/ϵ_0 multiplies the time derivative of E_0 as small compared to the first term, and rewrite this equation as

$$\frac{\partial E_0}{\partial t} = \frac{\omega}{2\epsilon_0}(AE_0 - BE_0^3).$$

$$(11.12.12)$$

If A is negative, in other words, if the relaxation exceeds the excitation, then any coherent electric field amplitude will decay exponentially. If A is positive, in other words, if there is net *gain* in the system, then a small coherent electric field will grow exponentially. It will continue to grow until it reaches a steady-state value given by the solution of

$$-AE_0 + BE_0^3 = 0.$$

$$(11.12.13)$$

Note that this equation has the same form as the Ginzburg–Landau equation (10.4.20) in the case of a spatially homogeneous system, which gave us the phase transition behavior discussed in Sections 10.4 and 11.3. The appearance of a spontaneous coherent field involves the same type of spontaneous symmetry breaking as in a ferromagnet or a Bose condensate. Any fluctuation that gives rise to a small coherent field will be amplified until a macroscopic coherent field is established. We could also write down a Mexican hat function like that shown in Figure 11.4; in this case, since the electric is real, the angle in the complex plane is a bookkeeping tool that gives us the phase of the oscillation in time.

Note that this treatment of lasing did not require any discussion of mirrors. All that is needed is for the gain to outweigh the loss for one state. Using mirrors to reflect photons back through the medium is one way to decrease the loss for one state. We have, of course, introduced many simplifications in this model of lasing, as we did in the Ising model for ferromagnets, but the effect of spontaneous onset of optical coherence is a general one.

Exercise 11.12.1   What density of two-level oscillators is implied by the assumption made above, namely $A/\epsilon_0 \ll 1$, for ω in the visible optical range of the spectrum, oscillator strength of the order of unity, and spontaneous recombination time of a few picoseconds?

11.13 Excitonic Condensation

As discussed in Section 11.7, excitons, like Cooper pairs, are quasibosons and can therefore also, in principle, undergo Bose–Einstein condensation. Excitons are normally metastable, however, and can turn into photons that leave the system. Since a condensate is a coherent state, the light emitted by an excitonic condensate will also be coherent. An excitonic

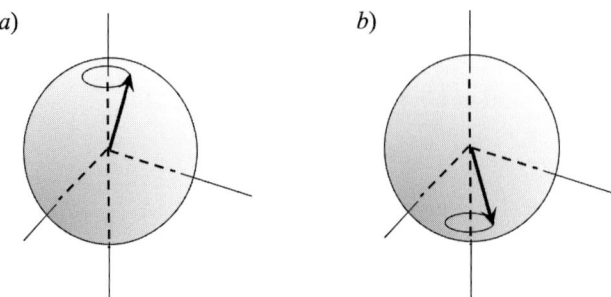

Fig. 11.25   The Bloch sphere representing the state of the oscillators in a medium for (a) lasing, and (b) excitonic (or polaritonic) Bose–Einstein condensation.

condensate therefore has much in common with a laser – energy is pumped in incoherently, putting electrons into excited states, and coherent light is emitted.

Unlike a laser, however, excitonic condensates do not require inversion. In Section 11.12, the spontaneous symmetry breaking was driven by the gain term A, which from (11.12.7) and (11.12.8) becomes positive when there is inversion, that is, when $GT > 1$, leading to optical gain.[2] In an excitonic condensate, symmetry breaking happens by the same mechanism as other condensates, namely thermalization into a coherent condensate state via particle–particle interactions. The repulsion of the excitons on each other plays the same role as the interaction term in the Hamiltonian (11.2.1), which allows the discussion of Section 11.3 to apply. Excitons can also be pushed to high density, in which case BCS theory applies; in fact, the earliest work on BEC–BCS crossover was developed for the case of excitons. (See Randeria 1995 for a review of this early work.)

Figure 11.25 shows the Bloch sphere representing the state of the polarizable medium in the case of a laser and an excitonic condensate. In each case, an ensemble of two-level oscillators oscillates in phase with the coherent electromagnetic field, but in the case of an excitonic condensate, there is no need for inversion; instead, the interaction between the particles creates the coherence. There are actually several different states of matter that involve coherence and inversion besides these two. Table 11.1 summarizes these different states. **Superfluorescence** is a process that has onset of coherence by exactly the same process as lasing, described in Section 11.12, but without any one particular emission transition selected out by mirrors; when inversion occurs in an ensemble of two-level oscillators, and there is very little dephasing, the system can spontaneously pick out a state into which to radiate coherently, often after a measurable time delay as a tiny fluctuation is amplified. Thus, there is not just spontaneous symmetry breaking in the appearance of coherence, but also in the radiative mode that is selected. **Superradiance** is similar, but the seed that is amplified is a coherent driving pulse, which can come from outside the system. Superradiance maps exactly to the classical case of coupled antennas, in which the coherent radiation from one drives the others, which then also radiate. Another variation is

---

[2] In systems with three or more levels, interference effects can be used to effectively turn off the absorption from lower to higher states, leading to "lasing without inversion" (Harris 1989; Scully *et al.* 1989). These systems still rely on optical gain for the spontaneous symmetry breaking.

Table 11.1 Types of coherence in solids			
	Inversion	Coherent medium	Spontaneous symmetry breaking
Lasing	yes	yes	yes
Superfluorescence	yes	yes	yes
Superradiance	yes	yes	no
Amplified spontaneous emission (ASE)	yes	no	no
Dielectric medium	no	yes	no
Excitonic BEC	no	yes	yes

a **photon condensate**, in which the photons are held in a high-Q optical cavity for a long enough time that they can thermalize and form a Bose–Einstein condensate. The mode in which coherence appears corresponds to the lowest-energy cavity mode, that is, the ground state of the photon gas (see, e.g., the article by Klaers and Weitz in Proukakis *et al.* 2017).

At the opposite limit, it is possible to have inversion without coherence. In **amplified spontaneous emission** (ASE), the dephasing is so strong that no coherence appears, but the inversion of the system still leads to optical gain. Finally, every dielectric medium has coherent oscillation of the medium as in a laser. An external coherent wave drives the polarizable oscillators, which then also oscillate in phase. Like an excitonic condensate, there is coherence but no inversion, but unlike an excitonic condensate, there is no spontaneous symmetry breaking.

11.13.1 Microcavity Polaritons

The most successful type of excitonic condensate is one in which the excitons are strongly coupled to photon states, to make polaritons. The mechanism of exciton-polariton formation is the same as discussed in Section 7.5.2, with the 2×2 matrix (7.5.42) for the coupling of the photon and excitons, but in this case the photon states are in an optical cavity, as shown in Figure 11.26(a), which leads to the dispersion relation of the photons $\hbar\omega = ((N\pi/L)^2 + k_\parallel^2)^{1/2}$, where L is the length of the cavity, N is an integer, and $k_\parallel$ is the wave number in a direction parallel to the planes of the mirrors. Photons are free to move in the two directions parallel to the mirrors, but confined to specific modes in the direction perpendicular to the mirrors.

As shown in Figure 11.26(b), when the exciton and cavity photon energies are resonant, the coupling of the photon and exciton leads to an anticrossing and formation of two new levels known as the upper and lower polariton. The system is in the limit of **strong coupling** when the line broadening of the polariton states (see Section 8.4) is significantly less than the Rabi splitting between the states, so that the polaritons can be viewed as nearly eigenstates.

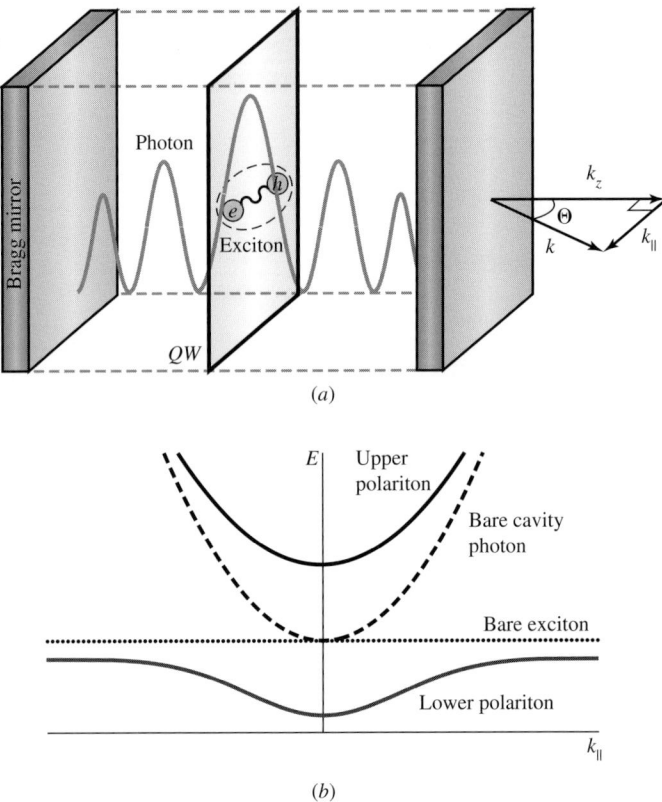

(a)

(b)

Fig. 11.26   (a) Microcavity structure with embedded quantum well at an antinode of the confined cavity mode. (b) Dashed line: dispersion of photons in a planar microcavity; dotted line: energy of excitons in a quantum well resonant with the cavity photon energy. Compared to the dispersion of the cavity photons, the exciton energy is essentially constant near $k_\parallel = 0$. Solid lines: upper and lower polariton modes formed by anticrossing of the photon and exciton states.

The effective mass of the lower polariton is given by the curvature of their dispersion at $k_\parallel$, as discussed in Section 2.2. For typical experimental structures, this effective mass can be as low as 10^{-4} times the vacuum electron mass, which of course is 1800 times less than a proton mass. The light mass of polaritons means that they can exhibit condensation effects at high temperature. As discussed in Section 11.1, the condition for quantum degeneracy is that the DeBroglie wavelength of the particles be comparable to the average distance between them. This implies that $T_c \sim 1/m$ in two dimensions, as well as in three dimensions. Several experimental groups have reported condensation of exciton-polaritons at room temperature. Since excitons are held together by Coulomb attraction between the electron and hole instead of by a phonon-mediated interaction, excitons can be much more tightly bound than Cooper pairs, and in many semiconductors and organic solids, excitons and polaritons are stable at room temperature. When the excitons are more tightly bound, however, they are generally also smaller, as discussed in Section 2.3, and therefore more sensitive to local potential fluctuations due to disorder.

A polariton condensate obeys a Ginzburg–Landau equation, or Gross–Pitaevskii equation, just as other types of condensate do. In this case, the order parameter, that is, the wave function of the condensate, is the electric field in the polarizable medium. To see this, we start with Maxwell's wave equation in a nonlinear isotropic medium. Substituting the third-order term in (7.6.1) of Section 7.6 into the relation (3.5.8) and the general Maxwell wave equation (3.5.4) in Section 3.5.1, the Maxwell wave equation for a third-order nonlinearity is

$$\nabla^2 E = \frac{n^2}{c^2}\frac{\partial^2 E}{\partial t^2} + 4\mu_0\chi^{(3)}\frac{\partial^2}{\partial t^2}|E|^2 E, \tag{11.13.1}$$

where $\chi^{(3)}$ is the nonlinear optical constant that corresponds to a photon–photon interaction. In the case of exciton-polaritons, this term comes from the exciton–exciton repulsion due to the exciton part included in the polariton states. We write the plane wave solution

$$E(\vec{x}, t) = \psi(\vec{x}, t)e^{-i\omega_0 t}, \tag{11.13.2}$$

where ψ is an envelope amplitude which may vary in time and space, and ω_0 is the frequency of the lowest polariton state. We write this envelope amplitude suggestively as ψ because we will see that it plays the same role as the wave function ψ in a matter wave.

Keeping only leading terms in frequency (the slowly varying envelope approximation), we have for the time derivative of E,

$$\frac{\partial^2 E}{\partial t^2} \simeq \left(-\omega_0^2\psi - 2i\omega_0\frac{\partial\psi}{\partial t}\right)e^{-i\omega_0 t}, \tag{11.13.3}$$

and for the time derivative of the nonlinear term

$$\frac{\partial^2}{\partial t^2}|E|^2 E \simeq -\omega_0^2|\psi|^2\psi e^{-i\omega_0 t}.$$

As discussed above, the standard polariton structure used a planar or nearly planar cavity to give one confined direction of the optical mode. We therefore distinguish between the component of momentum k_z in the direction of the cavity confinement and the momentum $k_\parallel$ in the two-dimensional plane perpendicular to this direction. We therefore write

$$\psi = \psi(\vec{x})e^{ik_z z}, \tag{11.13.4}$$

for $\vec{x}$ in the plane. The full Maxwell wave equation (11.13.1) then becomes

$$-k_z^2\psi + \nabla_\parallel^2\psi = (n/c)^2\left(-\omega_0^2\psi - 2i\omega_0\frac{\partial\psi}{\partial t}\right) - 4\mu_0\chi^{(3)}\omega_0^2|\psi|^2\psi. \tag{11.13.5}$$

From (7.5.13) in Section 7.5.2, the index of refraction taking into account the exciton polarization can be written as

$$n^2 = n_\infty^2\left(1 + \frac{\Omega^2}{\omega_{ex}^2 - \omega^2}\right), \tag{11.13.6}$$

where we have defined $\Omega^2 = q^2 N/\epsilon_\infty m V$, consistent with the discussion in Section 7.5.3, and $n_\infty^2 = \epsilon_\infty/\epsilon_0$. The frequency ω is equal to the ground state frequency ω_0 plus the additional amount ω' due to the oscillation of the in-plane envelope function ψ. Writing (11.13.6) as a Taylor series in ω', we have

$$n^2 \simeq n_\infty^2 \left(1 + \frac{\Omega^2}{\omega_{ex}^2 - \omega_0^2} - \frac{\Omega^2}{(\omega_{ex}^2 - \omega_0^2)^2}(-2\omega_0)\omega'\right). \tag{11.13.7}$$

As we have seen in Section 7.5.3 for polaritons, for a resonant system, $\omega_0 = \omega_{ex} - \Omega/2$, which gives

$$n^2 \simeq n_0^2 + n_\infty^2 \frac{2i\omega_0}{\omega_{ex}^2} \frac{\partial}{\partial t}, \tag{11.13.8}$$

where we have replaced ω' with the time derivative that generates it; n_0 is the index of refraction at ω_0.

We can now substitute this for n^2 in (11.13.5). The $n_0^2\omega_0^2$ term corresponds to the frequency of the ground state in the polariton band, and therefore cancels the k_z^2 term on the other side. Keeping only first derivatives in time, we have

$$\nabla_\parallel^2 \psi \simeq \frac{n_0^2}{c^2}(-2i\omega_0)\frac{\partial \psi}{\partial t} + \frac{n_\infty^2}{c^2}\frac{-2i\omega_0^2}{\omega_{ex}^2}\frac{\partial}{\partial t} - 4\mu_0\chi^{(3)}\omega_0^2|\psi|^2\psi$$

$$\simeq -4i\frac{n_0^2}{c^2}\omega_0\frac{\partial \psi}{\partial t} - 4\mu_0\chi^{(3)}\omega_0^2|\psi|^2\psi. \tag{11.13.9}$$

Defining $m = 2\hbar\omega_0(n_0^2/c^2)$, this becomes

$$i\hbar\frac{\partial \psi}{\partial t} = -\frac{\hbar^2}{2m}\nabla_\parallel^2\psi - \frac{2\mu_0\chi^{(3)}(\hbar\omega)^2}{m}|\psi|^2\psi, \tag{11.13.10}$$

which we can rewrite as

$$i\hbar\frac{\partial \psi}{\partial t} = -\frac{\hbar^2}{2m}\nabla_\parallel^2\psi + U|\psi|^2\psi. \tag{11.13.11}$$

This is the same as the Gross–Pitaevskii equation (11.3.12), or nonlinear Schrödinger equation, which we wrote down for a condensate in Section 11.3. Note that although the Maxwell wave equation is second order in the time derivative, this equation is first order in the time derivative, as in a typical Schrödinger equation.

The fact that that polariton condensate Gross–Pitaevskii equation can be generated from a nonlinear Maxwell wave equation in this system should not be cause to doubt the validity of the BEC description. Every condensate is described by a classical wave equation, because the Gross–Pitaevskii equation is a classical wave equation. In the case of polariton condensates, all of the effects of the Gross–Pitaevskii, or Ginzburg–Landau equation, for condensates can also be seen, such as quantized vorticity. Figure 11.27 shows an example of quantized vorticity of a polariton condensate in a ring, seen in an interference measurement. Because the polaritons inside the cavity couple directly into photons that leak through the mirrors, the phase coherence of the condensate can be measured directly using an optical interferometer to look at the light emitted from the cavity. As discussed in Section 11.4, the gradient of the phase corresponds to the flow of the condensate. In Figure 11.27(b), the fringe pattern of the interference pattern has been removed by Fourier transforming the image, erasing the frequency peak that corresponds to this

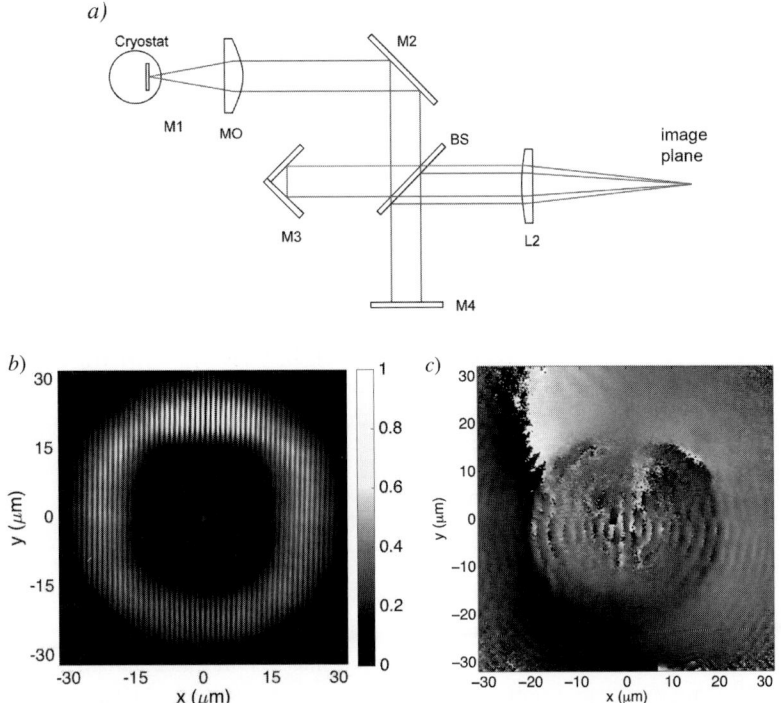

Fig. 11.27 (a) Interferometry system that flips one image across a spatial axis (mirror M3) to create a self-interferogram. (b) Interference pattern for an exciton-polariton condensate in a ring trap, recorded using the Michelson interferometer geometry shown in (a), showing that the condensate is coherent across the whole ring. (c) The phase map extracted from an image like that in (b), showing that there is quantized circulation in the ring. From Liu *et al.* (2015).

fringe pattern, and then Fourier transforming back to real space – this extracts the phase of the condensate, and is known as a **phase map**. The gray scale of this image gives the value of the phase from 0 to 2π. Since, as we saw in Section 11.4. a phase gradient corresponds to flow of a condensate, there is a net rotation of the condensate in the ring, which is subject to the constraint that the condensate wave function satisfies periodic boundary conditions. An interesting twist in this experiment is that the condensate actually undergoes only a π phase advance, not 2π, around the ring. This is allowed because a polariton condensate has two degenerate circular polarization components, corresponding to the two allowed circular polarizations of light in the optical cavity. Therefore, because linear polarization can always be made from a superposition of two circular polarizations, the condensate can have a linear polarization that rotates, or precesses, around the ring. A π phase advance due to the flow of the condensate, accompanied by a π rotation of the linear polarization direction, gives a total of 2π and allows the boundary conditions to be satisfied. This is known as a "half-quantized vortex," or a "half vortex."

Since the exciton-polariton condensate in a planar structure is two-dimensional, it cannot have infinitely long-range phase order, as discussed in Section 11.5, but it will still have many of the properties of condensates. Other condensate effects observed for exciton-polaritons include such a peak at the ground state in momentum space, Josephson oscillations, phase locking of two or more condensates, and a critical velocity for the onset of dissipation. For a survey of exciton-polariton condensate effects, see articles in Proukakis *et al.* (2017).

Exercise 11.13.1   Typical values of the nonlinear interaction potential for excitons are $U \sim 10^{-11}$ meV-cm^2. Convert this value to a photon–photon $\chi^{(3)}$ value in units of m^2/V^2, using the formulas of Section 4.4 to convert electric field amplitude E to photon number density, and taking a typical solid state microcavity length of 500 nm, the dielectric constant $\kappa = 13$ for a typical III–V semiconductor, and optical wavelength of 800 nm. You will first need to find the effective mass of the particles. How does your number compare to the $\chi^{(3)}$ values of other nonlinear optical media that you can look up?

11.13.2 Other Quasiparticle Condensates

In recent years, many quasiparticle systems have been studied that can undergo Bose–Einstein condensation and show at least some of its effects. All that is needed is to have bosonic particles with an effective mass, some mechanism for them to thermalize, and approximate number conservation over the period of time during which they thermalize and are observed. Besides polaritons and photons, other examples in which evidence of condensation has been seen include magnons, "triplons," and permanent excitons. (For a general review of various Bose condensate systems, see Proukakis *et al.* 2017.)

Although excitons are composed of an electron and hole, and therefore would seem to be unstable to decay, there are various ways to make excitons stable. Such a system can be engineered, for example, by creating two quantum wells near each other with a high barrier between them, and using doping or electric field to put free electrons into the system. A magnetic field can then be applied to make the electrons exactly half-fill the lowest Landau level in each quantum well with the same density, which also then corresponds to each level being half-filled with holes. If the barrier is thin enough, the electrons and holes will feel a Coulomb attraction to each other, leading to anticorrelation in which an electron in one layer lies opposite a hole in the other layer, as illustrated in Figure 11.28. The many-body state that describes this system will be the same as the BCS state (11.8.6), where instead of a spin index to distinguish the electrons, an index is used to label which of the two quantum wells the electron is in. Since this is the ground state of the system, there is no

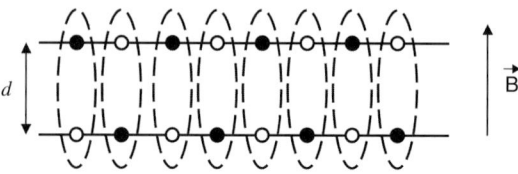

Fig. 11.28   Illustration of the bilayer system that produces a permanent set of electron–hole pairs in a BCS-type state.

recombination of the electrons and holes. Various measurements of the current transport from one well to the other, or along the wells, are consistent with a coherent state. For example, when all of the electrons and holes are fully paired up, there is a dramatic increase in the tunneling rate between the two wells, analogous to superradiance. For further details, see Eisenstein (2014).

References

J.F. Annett, *Superconductivity, Superfluids, and Condensates* (Oxford University Press, 2004).

B.L. Blackford and R.H. March, "Temperature dependence of the energy gap in superconducting Al–Al$_2$O$_3$–Al tunnel junctions," *Canadian Journal of Physics* **46**, 141 (1968).

J.M. Blatt, *Theory of Superconductivity* (Academic Press, 1964).

M.G. Castellano, R. Leoni, G. Torrioli *et al.*, "Superconductor–insulator–metal tunnel junctions for on-chip measurement of the temperature," *IEEE Transactions on Applied Superconductivity* **7**, 3251 (1997).

P.M. Chaikin and T.C. Lubensky, *Principles of Condensed Matter Physics* (Cambridge University Press, 1995).

C. Cohen-Tannoudji, B. Diu, and F. Laloë, *Quantum Mechanics* (Wiley, 1979).

M. Combescot and O. Betbeder-Matibet, "How composite bosons really interact," *European Physical Journal* B **48**, 469 (2005).

M. Combescot and D.W. Snoke, "Stability of a Bose–Einstein condensate revisited for composite bosons," *Physical Review B* **78**, 144303 (2008).

J. Eisenstein, "Exciton condensation in bilayer quantum Hall systems," *Annual Review of Condensed Matter Physics* **5**, 159 (2014).

S. Fujita and S. Godoy, *Quantum Statistical Theory of Superconductivity* (Plenum, 1996).

P. de Gennes, *Superconductivity of Metals and Alloys* (Benjamin Cummings, 1966).

I. Giaever, "Electron tunneling between two superconductors," *Physical Review Letters* **5**, 464 (1960).

A. Griffin, *Excitations in a Bose-Condensed Liquid* (Cambridge University Press, 1993).

R.J. Griffiths, *Introduction to Electrodynamics*, 4th ed. (Pearson, 2011).

S. Harris, "Lasers without inversion: interference of lifetime-broadened resonances," *Physical Review Letters* **62**, 1033 (1989).

H.F. Hess, R.B. Robinson, R.C. Dynes, J.M. Valles, and J.V. Waszczak, "Scanning–tunneling–microscope observation of Abrikosov flux lattice and the density of states near and inside a fluxoid," *Physical Review Letters* **62**, 214 (1989).

J.M. Kosterlitz and D.J. Thouless, *Journal of Physics C* **6**, 1181 (1973).

B. Laikhtman, "Are excitons really bosons?" *Journal of Physics: Condensed Matter* **19**, 295214 (2007).

A. Leggett, *Quantum Liquids* (Oxford University Press, 2006).

E.M. Lifshitz and L.P. Pitaevskii, *Statistical Physics: Part 2* (Pergamon Press, 1980).

G.-Q. Liu, D.W. Snoke, A.J. Daley, L.N. Pfeiffer, and K. West, "A new type of half-quantum circulation in a macroscopic polariton spinor ring condensate," *Proceedings of the National Academy of Sciences USA* **112**, 2676 (2015).

P. Lorrain and D.R. Corson, *Electromagnetic Fields and Waves*, 2nd ed. (Freeman, 1970).

B. Mitrovic, H.G. Zarate, and J.P. Carbotte, "The ratio $2\Delta_0/k_B T$ within Eliashberg theory," *Physical Review B* **29**, 184 (1984).

P. Noziéres and D. Pines, *The Theory of Quantum Liquids, Volume II: Superfluid Bose Liquids* (Addison-Wesley, 1990).

P.K. Pathria and P.D. Beale, *Statistical Mechanics*, 3rd ed. (Elsevier, 2011).

C.P. Poole, H.A. Farach, R.J. Creswick, and R. Prozorov, *Superconductivity*, 2nd ed. (Elsevier, 1997).

N. Proukakis, D.W. Snoke, and P.B. Littlewood, *Universal Themes of Bose–Einstein Condensation* (Cambridge University Press, 2017).

F. Reif *Fundamentals of Statistical and Thermal Physics* (McGraw-Hill, 1969).

M. Randeria, "Crossover from BCS theory to BEC," in *Bose-Einstein Condensation*, A. Griffin, D.W. Snoke, and S. Stringari, eds. (Cambridge University Press, 1995).

M.O. Scully, S.-Y. Zhu, and A. Gavrielides, "Degenerate quantum-beat laser: lasing without inversion and inversion without lasing," *Physical Review Letters* **62**, 2813 (1989).

P. Townsend and J. Sutton, "Investigation by electron tunneling of the superconducting energy gaps in Nb, Ta, Sn, and Pb," *Physical Review* **128**, 591 (1962).

Appendix A Review of Bra-Ket Notation

The bra-ket notation commonly used in quantum mechanics can be easily understood in terms of vector algebra. We write a vector as a "ket,"

$$|v\rangle = \begin{pmatrix} v_1 \\ v_2 \\ \cdot \\ \cdot \\ \cdot \\ v_n \end{pmatrix}. \tag{A.1}$$

The "bra" is the same vector transposed, and with the complex conjugate taken:

$$\langle v| = \begin{pmatrix} v_1^* & v_2^* & \cdot & \cdot & \cdot & v_n^* \end{pmatrix}. \tag{A.2}$$

The inner product of two vectors is then

$$\begin{aligned}
\langle v|u\rangle &= \begin{pmatrix} v_1^* & v_2^* & \cdot & \cdot & \cdot & v_n^* \end{pmatrix} \begin{pmatrix} u_1 \\ u_2 \\ \cdot \\ \cdot \\ \cdot \\ u_n \end{pmatrix} \\
&= v_1^* u_1 + v_2^* u_2 + \cdots v_n^* u_n.
\end{aligned} \tag{A.3}$$

For this to make sense, the two vectors must have the same dimension, in other words, span the same space. We see that any term in a bracket ("bra-ket") is a simple number, which commutes with all other terms.

The outer product is formed by writing a bra and ket in reverse order. For example,

$$|v\rangle\langle v| = \begin{pmatrix} v_1 \\ v_2 \\ \cdot \\ \cdot \\ \cdot \\ v_n \end{pmatrix} \begin{pmatrix} v_1^* & v_2^* & \cdot & \cdot & \cdot & v_n^* \end{pmatrix}$$

$$= \begin{pmatrix} v_1^* v_1 & v_2^* v_1 & \cdot & \cdot & \cdot & v_n^* v_1 \\ v_1^* v_2 & v_2^* v_2 & & & & v_n^* v_2 \\ & & \cdot & & & \cdot \\ \cdot & & & \cdot & & \\ \cdot & & & & \cdot & \\ v_1^* v_n & v_2^* v_n & \cdot & \cdot & \cdot & v_n^* v_n \end{pmatrix}. \tag{A.4}$$

A square matrix acting on vectors is known as an "operator" because when it multiplies a vector, it transforms that vector into another vector. The above matrix is known as the "projection operator" because it projects a vector onto another vector, giving a result which has length equal to the inner product of the two vectors, and which points in the direction of $|v\rangle$. This is written succinctly in bra-ket notation as

$$P_v |u\rangle = \big(|v\rangle \langle v| \big) |u\rangle = |v\rangle \langle v|u\rangle. \tag{A.5}$$

A continuous function $f(x)$ can be treated as a vector with an infinite number of components by writing

$$|f\rangle = \begin{pmatrix} f(x_1) \\ f(x_2) = f(x_1 + dx) \\ f(x_3) = f(x_1 + 2dx) \\ \cdot \\ \cdot \\ \cdot \\ f(x_n) \end{pmatrix}. \tag{A.6}$$

To get an integral instead of a sum for the inner product, the bra in this case must be adjusted by multiplying by dx:

$$\langle f| = \big(f^*(x_1) \quad f^*(x_2) \quad f^*(x_3) \quad \cdot \quad \cdot \quad \cdot \quad f^*(x_n) \big) dx. \tag{A.7}$$

The inner product is then

$$\langle f|g\rangle = \lim_{dx \to 0} \sum_i f^*(x_i) g(x_i) dx$$

$$= \int f^*(x) g(x) dx, \tag{A.8}$$

which is the standard definition of the inner product of two continuous functions in Schrödinger wave mechanics.

Appendix B Review of Fourier Series and Fourier Transforms

The Fourier *series* theorem says that any continuous function $f(x)$ defined in the interval $(0, 2L)$ can be written as

$$f(x) = \sum_{n=0}^{\infty} \left(a_n \cos \frac{2\pi nx}{a} + b_n \sin \frac{2\pi nx}{a} \right),$$ (B.1)

where the weight factors a_n and b_n are given by

$$a_n = \frac{2}{a} \int_0^a f(x) \cos \frac{2\pi nx}{a} dx$$

$$b_n = \frac{2}{a} \int_0^a f(x) \sin \frac{2\pi nx}{a} dx.$$ (B.2)

It is natural to use the Fourier series theorem for functions which are periodic with period a, but it can also be used for any function defined in a finite interval.

The same theorem can be written in complex notation, using $e^{i\theta} = \cos\theta + i\sin\theta$. Defining the frequency $k_n = n(2\pi/a)$, we have

$$f(x) = \sum_{n=-\infty}^{\infty} c_n e^{-ik_n x},$$ (B.3)

with

$$c_n = \frac{1}{a} \int_0^a f(x) e^{ik_n x} dx.$$ (B.4)

The values of c_n are connected to the values of a_n and b_n introduced above, by

$$c_n = \begin{cases} \frac{1}{2}(a_n + ib_n), & n > 0 \\ \frac{1}{2}(a_{-n} - ib_{-n}), & n < 0 \\ \frac{1}{2}a_0, & n = 0. \end{cases}$$ (B.5)

In the limit of $a \to \infty$, which applies to a nonperiodic function, the Fourier series theorem maps directly to the Fourier *transform* theorem, in which the sum over frequencies k_n becomes an integral:

$$f(x) = \frac{1}{2\pi} \int F(k) e^{-ikx} dk,$$ (B.6)

with the Fourier transform function having the role of the weight factor c_n,

$$F(k) = \int_{-\infty}^{\infty} f(x)e^{ikx}dx. \tag{B.7}$$

Equation (B.6) is often called the "inverse Fourier transform" since it starts with the Fourier transform and reverts it to the real-space function. Note that the normalization factor of 2π which occurs in (B.6) is not universal; the prefactor can be absorbed into the definition of $F(\vec{k})$, in which case a prefactor of $\frac{1}{2\pi}$ will occur in the Fourier transform (B.7) instead in the inverse Fourier transform.

In three dimensions, the Fourier transform theorem is generalized to

$$f(\vec{x}) = \frac{1}{(2\pi)^3} \int_{-\infty}^{\infty} F(\vec{k})e^{-i\vec{k}\cdot\vec{x}}d^3k, \tag{B.8}$$

$$F(\vec{k}) = \int_{-\infty}^{\infty} f(\vec{x})e^{i\vec{k}\cdot\vec{x}}d^3x. \tag{B.9}$$

Since the same information is contained in either the real-space function $f(\vec{x})$ or the frequency-space function $F(\vec{k})$, we can talk of switching descriptions between "real space" and "k-space." Since in quantum mechanics, the momentum is $\vec{p} = \hbar\vec{k}$, the latter is also often called "momentum space." Formulas (B.6) and (B.7) can equally well be written with the substitutions $x \rightarrow t$ and $k \rightarrow \omega$ to allow their use in the time domain.

Commonly used Fourier transform pairs include

$$f(x) = e^{-bx^2}$$
$$F(k) = \sqrt{\frac{\pi}{b}}e^{-k^2/4b} \tag{B.10}$$

and

$$f(x) = e^{-ik_0x}$$
$$F(k) = 2\pi\,\delta(k - k_0), \tag{B.11}$$

where $\delta(k)$ is the Dirac δ-function (see Appendix C). The Gaussian pair (B.10) has the property that a narrow Gaussian in real space has a broad Gaussian in k-space, and vice versa. This is the basis of the uncertainty Principle in quantum mechanics, when $f(x)$ is taken as a particle wave function. The Fourier pair (B.11) can be seen as a limiting case of the uncertainty Principle: When the width in real space is infinitely broad (a constant), the Fourier transform is infinitesimally narrow (the δ-function).

If the function $f(x)$ is a periodic function,

$$f(x) = \sum_{n=-\infty}^{\infty} f_P(x - an), \tag{B.12}$$

where $f_p(x)$ is nonzero only in the interval $(0, a)$, then its Fourier transform is

$$F(k) = \int_{-\infty}^{\infty} \sum_{n=-\infty}^{\infty} f_P(x - an)e^{ikx}dx$$

$$= \sum_{n=-\infty}^{\infty} \int_0^a f_P(x')e^{ik(x'+an)}dx'$$

$$= \left(\sum_{n=-\infty}^{\infty} e^{ikan} \right) \int_0^a f_P(x')e^{ikx'}dx'. \tag{B.13}$$

Using Equation (C.11) in Appendix C, switching variables $x \to k$ and $k_0 = 2\pi/L \to a$, the sum in parentheses can be equated to

$$\lim_{N\to\infty} \sum_{n=-N/2}^{N/2} e^{ikan} = \frac{2\pi}{a} \sum_{m=-\infty}^{\infty} \delta(k - 2\pi m/a). \tag{B.14}$$

Doing the inverse Fourier transform, we then have

$$f(x) = \frac{1}{2\pi} \int_{-\infty}^{\infty} F(k)e^{-ikx}dk$$

$$= \frac{1}{2\pi} \int_{-\infty}^{\infty} \frac{2\pi}{a} \sum_{m=-\infty}^{\infty} \delta(k - 2\pi m/a) \left(\int_0^a f_P(x')e^{ikx'}dx' \right) e^{-ikx}dk$$

$$= \sum_{m=-\infty}^{\infty} \left(\frac{1}{a} \int_0^a f_P(x')e^{ik_m x'}dx' \right) e^{-ik_m x}, \tag{B.15}$$

where $k_m = 2\pi m/L$. We thus see that we recover the Fourier series, given by formulas (B.3) and (B.4), as a special case of the more general Fourier transform. The term in parentheses, equal to c_n in (B.3), plays the same role as the structure factor introduced in Section 1.4.

Appendix C Delta-Function Identities

The easiest way to think of the Dirac δ-function is as a high and narrow peak, normalized to have area equal to 1. In fact, the Dirac δ-function can be defined as the limit of a Gaussian peak,

$$\delta(x - x_0) = \lim_{\sigma \to 0} \frac{1}{\sqrt{4\pi\sigma^2}} e^{-(x-x_0)^2/4\sigma^2} \tag{C.1}$$

or the limit of a Lorentzian peak,

$$\delta(x - x_0) = \lim_{\eta \to 0} \frac{1}{\pi} \frac{\eta}{(x - x_0)^2 + \eta^2}, \tag{C.2}$$

or

$$\delta(x - x_0) = \lim_{\epsilon \to 0} \frac{\sin(x - x_0)/\epsilon}{\pi(x - x_0)}. \tag{C.3}$$

It is easy to check that the integral over x of these functions is 1 in each case.

When a sharp peak is multiplied by a continuous function $f(x)$, the product will be non-negligible only at $x = x_0$, which leads to the property

$$\int_a^b \delta(x - x_0) f(x) dx = f(x_0) \Theta(x_0 - a) \Theta(b - x_0), \tag{C.4}$$

where $\Theta(x)$ is the Heaviside function, defined as

$$\Theta(x) = \begin{cases} 1, & x > 0 \\ \frac{1}{2}, & x = 0 \\ 0, & x < 0. \end{cases} \tag{C.5}$$

A three-dimensional δ-function can be written as the product

$$\delta(x - x_0)\delta(y - y_0)\delta(z - z_0) \equiv \delta(\vec{x} - \vec{x}_0). \tag{C.6}$$

The property (C.4) ensures that the δ-function is also the inverse Fourier transform (see Appendix B) of a plane wave e^{ikx_0},

$$\delta(x - x_0) = \frac{1}{2\pi} \int_{-\infty}^{\infty} dk \, e^{-ik(x-x_0)}, \tag{C.7}$$

or in three dimensions,

$$\delta(\vec{x} - \vec{x}_0) = \frac{1}{(2\pi)^3} \int_{-\infty}^{\infty} d^3k \, e^{-i\vec{k}\cdot(\vec{x}-\vec{x}_0)}. \tag{C.8}$$

The one-dimensional δ-function has units of inverse length; the three-dimensional δ-function has units of inverse volume. These formulas can, of course, be inverted to give a k-space δ-function, in terms of an integral over spatial coordinates, in which case the k-space δ-function has units of volume.

Periodic boundary conditions and discrete sums. The above apply when k varies continuously. Suppose that we impose the condition that the system is periodic with length L (e.g., the Born–von Karman boundary conditions discussed in Section 1.7). The boundary conditions in this case then will imply that only discrete values of k are allowed. In one dimension, we will have $k = n(2\pi/L)$, where L is the total length of the system and n is an integer that runs from $-\infty$ to $+\infty$. Then the range dk per state is $2\pi/L$, and we convert the integral (C.7) to a sum,

$$\int_{-\infty}^{\infty} dk \, e^{-ik(x-x_0)} = k_0 \sum_{n=-\infty}^{\infty} e^{-ink_0(x-x_0)}, \tag{C.9}$$

where $k_0 = 2\pi/L$. We can resolve this sum by using a finite range of n, from $-N/2$ to $N/2$, and then taking the limit $N \to \infty$, that is,

$$\lim_{N\to\infty} k_0 \sum_{n=-N/2}^{N/2} e^{ik_0 n(x-x_0)} = \lim_{N\to\infty} k_0 \frac{\sin[k_0(x-x_0)(N+1/2)]}{\sin[k_0(x-x_0)/2]}. \tag{C.10}$$

When N is large, this function will be peaked at values of $k_0(x-x_0) = 2\pi m$, where m is an integer. The integral of the function over x from $x_0 - \pi$ to $x_0 + \pi$ is equal to a constant, 2π. Since the function keeps a constant area as it grows more and more peaked, it also has the limit of being a δ-function centered at x_0. This is true for every value $x_0 + Lm$. Therefore, we can write

$$\lim_{N\to\infty} k_0 \sum_{n=-N/2}^{N/2} e^{ik_0 n(x-x_0)} = 2\pi \sum_{m=-\infty}^{\infty} \delta(x - x_0 - Lm). \tag{C.11}$$

In three dimensions, we then have

$$\frac{1}{V} \sum_{\vec{k}} e^{i\vec{k}\cdot(\vec{x}-\vec{x}_0)} = \sum_{\vec{X}} \delta(\vec{x} - \vec{x}_0 - \vec{X}), \tag{C.12}$$

where $\vec{X} = m_1 L_1 \hat{x} + m_2 L_2 \hat{y} + m_3 L_3 \hat{z}$, and L_1, L_2, and L_3 are the dimensions of the finite volume which is repeated in the periodic system. If $\vec{x} - \vec{x}_0$ is limited to stay within this volume, then we have simply

$$\frac{1}{V} \sum_{\vec{k}} e^{-i\vec{k}\cdot(\vec{x}-\vec{x}_0)} = \delta(\vec{x} - \vec{x}_0). \tag{C.13}$$

Momentum–space versions. The above formulas can be written for k as well as for x, by a simple change of variables. In three dimensions, we have, for $\vec{k}$ and $\vec{k}'$ inside the Brillouin zone,

$$\sum_{\vec{R}} e^{i(\vec{k}-\vec{k}')\cdot\vec{R}} = \frac{(2\pi)^3}{V_{\text{cell}}}\delta(\vec{k}-\vec{k}'), \tag{C.14}$$

where V_{cell} is the volume of the repeated cell.

Suppose that we restrict the sum over $\vec{R}$ to a finite range, with N total terms in the sum. Then it is easy to see that when $\vec{k}-\vec{k}'=0$, the sum is equal to N, since each term in the sum will be equal to 1. We can therefore normalize the sum by dividing by N, to obtain

$$\frac{1}{N}\sum_{\vec{R}} e^{-i(\vec{k}-\vec{k}')\cdot\vec{R}} = \delta_{\vec{k},\vec{k}'}, \tag{C.15}$$

where $\delta_{\vec{k},\vec{k}'}$ is the unitless **Kronecker** δ-function, defined as

$$\delta_{\vec{k},\vec{k}'} = \begin{cases} 1, & \vec{k}=\vec{k}' \\ 0, & \text{else.} \end{cases} \tag{C.16}$$

Alternatively, suppose we integrate over a continuous region of large but finite size. Then we have

$$\frac{1}{V}\int d^3r\, e^{-i(\vec{k}-\vec{k}')\cdot\vec{r}} = \delta_{\vec{k},\vec{k}'}. \tag{C.17}$$

This is easily seen by noticing that if $\vec{k}=\vec{k}'$, the exponential factor equals 1 and the integral gives simply the volume V of the region.

Appendix D Quantum Single Harmonic Oscillator

Suppose that the Hamiltonian of a single-particle system is

$$H = \frac{p^2}{2m} + \frac{kx^2}{2}. \tag{D.1}$$

If we define the operators

$$\hat{x} = \sqrt{\frac{M\omega_0}{\hbar}}x$$

$$\hat{p} = \sqrt{\frac{1}{M\hbar\omega_0}}p, \tag{D.2}$$

then the Hamiltonian takes on the simple form

$$H = \frac{1}{2}(\hat{x}^2 + \hat{p}^2)\hbar\omega_0. \tag{D.3}$$

The commutation relation $[x, p] = i\hbar$ implies the commutation relation between $\hat{x}$ and $\hat{p}$,

$$[\hat{x}, \hat{p}] = i. \tag{D.4}$$

The Hamiltonian is further simplified by defining the new operators,

$$a = \frac{1}{\sqrt{2}}(\hat{x} + i\hat{p})$$

$$a^\dagger = \frac{1}{\sqrt{2}}(\hat{x} - i\hat{p}). \tag{D.5}$$

Then we have

$$H = \left(a^\dagger a + \tfrac{1}{2}\right)\hbar\omega_0$$
$$= \left(\hat{N} + \tfrac{1}{2}\right)\hbar\omega_0, \tag{D.6}$$

where we have defined $\hat{N} = a^\dagger a$. The operators a and $a^\dagger$ have the commutation relation

$$[a, a^\dagger] = aa^\dagger - a^\dagger a = 1. \tag{D.7}$$

This implies the following commutation relations:

$$[\hat{N}, a] = a^\dagger aa - aa^\dagger a$$
$$= a^\dagger aa - (a^\dagger a + 1)a$$
$$= -a \tag{D.8}$$

$$[\hat{N}, a^\dagger] = a^\dagger a a^\dagger - a^\dagger a^\dagger a$$
$$= a^\dagger(a^\dagger a + 1) - a^\dagger a^\dagger a$$
$$= a^\dagger. \tag{D.9}$$

Let $|\phi_N\rangle$ be an eigenstate of $\hat{N}$ with eigenvalue N. Any state has a norm which is real and greater than or equal to zero:

$$\left(\langle\phi_N|a^\dagger\right)(a|\phi_N\rangle) \geq 0. \tag{D.10}$$

This implies

$$\langle\phi_N|a^\dagger a|\phi_N\rangle = \langle\phi_N|\hat{N}|\phi_N\rangle = N \geq 0. \tag{D.11}$$

Since the norm is real and non-negative, the eigenvalues of $\hat{N}$ must be real, and the lowest eigenvalue is 0. We also have

$$a^\dagger|\phi_N\rangle = (\hat{N}a^\dagger - a^\dagger\hat{N})|\phi_N\rangle$$
$$= (\hat{N} - N)a^\dagger|\phi_N\rangle$$
$$\Rightarrow \hat{N}a^\dagger|\phi_N\rangle = (N + 1)a^\dagger|\phi_N\rangle. \tag{D.12}$$

Thus, $a^\dagger|\phi_N\rangle$ is an eigenstate of $\hat{N}$ with eigenvalue $N + 1$, which we write as $\beta_N|\phi_{N+1}\rangle$, where β_N is a complex number. Assuming each of the eigenstates is normalized so that $\langle\phi_N|\phi_N\rangle = 1$, we obtain

$$\langle\phi_N|aa^\dagger|\phi_N\rangle = \langle\phi_N|(a^\dagger a + 1)|\phi_N\rangle$$
$$\langle\phi_{N+1}|\beta_N^*\beta_N|\phi_{N+1}\rangle = \langle\phi_N|(N + 1)|\phi_N\rangle = N + 1, \tag{D.13}$$

which implies $|\beta_N|^2 = N + 1$. Since an eigenstate multiplied by any phase factor $e^{i\theta}$ is also an eigenstate, we can always write

$$a^\dagger|\phi_N\rangle = \beta_N|\phi_N + 1\rangle = e^{i\theta_N}\sqrt{N + 1}|\phi_{N+1}\rangle$$
$$= \sqrt{N + 1}\left(e^{i\theta_N}|\phi_{N+1}\rangle\right) = \sqrt{N + 1}|\phi'_{N+1}\rangle, \tag{D.14}$$

where $|\phi'_{N+1}\rangle$ is a new definition of the eigenstate. Choosing this phase convention for every state, we can therefore write

$$a^\dagger|\phi_N\rangle = \sqrt{N + 1}|\phi_{N+1}\rangle. \tag{D.15}$$

Similarly,

$$a|\phi_N\rangle = (a\hat{N} - \hat{N}a)|\phi_N\rangle$$
$$= (N - \hat{N})a|\phi_N\rangle$$
$$\Rightarrow \hat{N}a|\phi_N\rangle = (N - 1)a|\phi_N\rangle. \tag{D.16}$$

Defining $a|\phi_N\rangle = \gamma_N|\phi_{N-1}\rangle$, we have

$$\langle\phi_N|a^\dagger a|\phi_N\rangle = \langle\phi_N|\hat{N}|\phi_N\rangle$$
$$\langle\phi_{N-1}|\gamma_N^*\gamma_N|\phi_{N-1}\rangle = \langle\phi_N|N|\phi_N\rangle = N, \tag{D.17}$$

which implies $|\gamma_N|^2 = N$, and by the same phase convention,

$$a|\phi_N\rangle = \sqrt{N}|\phi_{N-1}\rangle. \tag{D.18}$$

If N is a fraction between 0 and 1, then $\hat{N}|\phi_{N-1}\rangle = (N-1)|\phi_{N-1}\rangle$, which gives an eigenvalue less than zero, which is not allowed according to (D.11). Since such a state could be generated by applying a successively to any state with fractional n greater than 1, all states with fractional n are forbidden, and n must be an integer. We therefore have a ladder of eigenstates of $\hat{N}$ equal to all the non-negative integers. The $a^\dagger$ and a operators act as "creation" and "destruction" operators, or "raising" and "lowering" operators, which take one eigenstate to another. Note, however, that the definition (D.5) implies that these are *amplitude* operators, which measure the amount of excursion of the oscillator.

The Hamiltonian (D.6) therefore has eigenvalues $\hbar\omega_0(N + \frac{1}{2})$, for N an integer from 0 to infinity, or equivalently, eigenvalues $\hbar\omega_0(N - \frac{1}{2})$ for N ranging from 1 to infinity.

Appendix E Second-Order Perturbation Theory

In this appendix, we summarize the results of nondegenerate and degenerate perturbation theory. The approach is summarized in many textbooks on quantum mechanics (e.g., Sakurai 1995: ch. 5).

Suppose that there are N orthonormal eigenstates $|\psi_n\rangle$ of the Hamiltonian H_0 with the same energy, and a small term V added to the Hamiltonian. When this is added to H_0, the eigenstates become new states $|\psi_n'\rangle$ with new energies E_i':

$$(H_0 + V)|\psi_i'\rangle \equiv E_n'|\psi_n'\rangle. \tag{E.1}$$

In the lowest order of approximation, we assume that the eigenstates and energies are unchanged:

$$\begin{aligned} E_n^{(0)} &\simeq E_n \\ |\psi_n^{(0)}\rangle &\simeq |\psi_n\rangle. \end{aligned} \tag{E.2}$$

For the next order of approximation, we take the zero-order approximation for the eigenstates and find the energy

$$E_n^{(1)} \simeq \langle\psi_n|(H_0 + V)|\psi_n\rangle = E_n + \langle\psi_n|V|\psi_n\rangle \equiv E_n + \Delta_n. \tag{E.3}$$

For the first-order approximation of the eigenstates, we assume that they are adjusted by a small mixing in of other eigenstates:

$$|\psi_n^{(1)}\rangle = |\psi_n\rangle + \sum_{m \neq n} \alpha_m|\psi_m\rangle. \tag{E.4}$$

We then write

$$\langle\psi_m|H|\psi_n^{(1)}\rangle = \langle\psi_m|H|\psi_n\rangle + \sum_{m' \neq n} \alpha_{m'}\langle\psi_m|H|\psi_{m'}\rangle. \tag{E.5}$$

On the left side, we have, using the first-order approximation for the energy,

$$H|\psi_n^{(1)}\rangle \simeq (E_n + \Delta_n)\left(|\psi_n\rangle + \sum_{m' \neq n} \alpha_{m'}|\psi_{m'}\rangle\right). \tag{E.6}$$

For $m \neq n$, this gives

$$\langle \psi_m | H | \psi_n^{(1)} \rangle \simeq (E_n + \Delta_n) \left(\sum_{m' \neq n} \alpha_{m'} \langle \psi_m | \psi_{m'} \rangle \right)$$

$$= (E_n + \Delta_n)\alpha_m$$

$$\simeq E_n \alpha_m, \tag{E.7}$$

where in the last step we have dropped the term with two small parameters, Δ_n and α_m. On the right side of (E.5), we have

$$\langle \psi_m | (H_0 + V) | \psi_n \rangle + \sum_{m' \neq n} \alpha_{m'} \langle \psi_m | (H_0 + V) | \psi_{m'} \rangle$$

$$\simeq \langle \psi_m | V | \psi_n \rangle + \sum_{m' \neq n} \alpha_{m'} \langle \psi_m | H_0 | \psi_{m'} \rangle$$

$$= \langle \psi_m | V | \psi_n \rangle + \alpha_m E_m, \tag{E.8}$$

where in the second line we have also dropped terms with two small parameters, namely V and $\alpha_{m'}$. Equating (E.7) and (E.8), we obtain

$$E_n \alpha_m = \langle \psi_m | V | \psi_n \rangle + \alpha_m E_m, \tag{E.9}$$

or

$$\alpha_m = \frac{\langle \psi_m | V | \psi_n \rangle}{E_n - E_m}. \tag{E.10}$$

Then the first-order approximation of the state is

$$|\psi_n^{(1)}\rangle = |\psi_n\rangle + \sum_{m \neq n} \frac{\langle \psi_m | V | \psi_n \rangle}{E_n - E_m} |\psi_m\rangle. \tag{E.11}$$

The second-order energy is then found by using the first-order eigenstates, keeping terms with one or two small parameters but dropping those with three. In this case, we must also account for the normalization of the wave function, which has second-order corrections:

$$\langle \psi_n^{(1)} | \psi_n^{(1)} \rangle = \left(\langle \psi_n | + \sum_{m \neq n} \alpha_m^* \langle \psi_m | \right) \left(|\psi_n\rangle + \sum_{p \neq n} \alpha_p |\psi_p\rangle \right)$$

$$= 1 + \sum_{m \neq n} |\alpha_m|^2. \tag{E.12}$$

The normalization factor is

$$\frac{1}{\sqrt{1 + \sum_{m \neq n} |\alpha_m|^2}} \simeq 1 - \frac{1}{2} \sum_{m \neq n} |\alpha_m|^2. \tag{E.13}$$

The second-order energy is then

$$E_n^{(2)} \simeq \left(\langle \psi_n | + \sum_{m\neq n} \alpha_m^* \langle \psi_m | - \frac{1}{2} \sum_{m\neq n} |\alpha_m|^2 \langle \psi_n | \right) (H_0 + V)$$

$$\times \left(|\psi_n\rangle + \sum_{p\neq n} \alpha_p |\psi_p\rangle - \frac{1}{2} \sum_{p\neq n} |\alpha_p|^2 |\psi_n\rangle \right)$$

$$\simeq \langle \psi_n | (H_0 + V) | \psi_n \rangle + \sum_{m\neq n} \alpha_m^* \langle \psi_m | V | \psi_n \rangle + \sum_{p\neq n} \alpha_p \langle \psi_n | V | \psi_p \rangle$$

$$+ \sum_{m\neq n} E_m |\alpha_m|^2 - E_n \sum_{m\neq n} |\alpha_m|^2, \tag{E.14}$$

or

$$E_n^{(2)} = E_n + \langle \psi_n | V | \psi_n \rangle + \sum_{m\neq n} \frac{|\langle \psi_m | V | \psi_n \rangle|^2}{E_n - E_m}. \tag{E.15}$$

Degenerate-state perturbation theory. The above approach works as long as $E_n \neq E_m$. If there are several degenerate eigenstates, we must be more careful. In this case, we form the matrix elements $\langle \psi_k | V | \psi_l \rangle$, where $|\psi_k\rangle$ and $|l\rangle$ are states within the degenerate set with energy E_n, which we call $\{n\}$. We then diagonalize the matrix formed from these matrix elements, to obtain eigenstates $|\psi_i\rangle$ and $|\psi_j\rangle$ such that $\langle \psi_i | V | \psi_j \rangle = \Delta_i \delta_{ij}$. We write the perturbed state as

$$|\psi_i^{(1)}\rangle = |\psi_i\rangle + \sum_{\substack{j\in\{n\}\\ j\neq i}} \beta_j |\psi_j\rangle + \sum_{m\notin\{n\}} \alpha_m |\psi_m\rangle, \tag{E.16}$$

where, as before, α_m and β_j are assumed to be small.

To find α_m, we follow the same process as (E.5)–(E.10). All the first-order terms with β_j drop out because $\langle \psi_m | \psi_j \rangle = 0$ by assumption, leaving us with

$$\alpha_m = \frac{\langle \psi_m | V | \psi_i \rangle}{E_n - E_m}. \tag{E.17}$$

To find β_j, we can follow a similar approach. We construct, for $j \neq i$,

$$\langle \psi_j | H | \psi_i^{(1)} \rangle = \langle \psi_j | H | \psi_i \rangle + \sum_{\substack{j'\in\{n\}\\ j'\neq i}} \beta_{j'} \langle \psi_j | H | \psi_{j'} \rangle + \sum_{m'\neq n} \alpha_{m'} \langle \psi_j | H | \psi_{m'} \rangle. \tag{E.18}$$

On the left side, we have

$$\langle \psi_j | H | \psi_i^{(1)} \rangle \simeq (E_n + \Delta_i)\beta_j, \tag{E.19}$$

while on the right side, we have

$$\langle \psi_j | (H_0 + V) | \psi_i \rangle + \sum_{\substack{j'\in\{n\}\\ j'\neq i}} \beta_{j'} \langle \psi_j | (H_0 + V) | \psi_{j'} \rangle + \sum_{m'\notin\{n\}} \alpha_{m'} \langle \psi_j | (H_0 + V) | \psi_{m'} \rangle$$

$$= \beta_j (E_n + \Delta_j) + \sum_{m'\notin\{n\}} \alpha_{m'} \langle \psi_j | V | \psi_{m'} \rangle, \tag{E.20}$$

where we have used $\langle\psi_i|V|\psi_j\rangle = \Delta_i\delta_{ij}$. This yields

$$\beta_j = \sum_{m'\notin\{n\}} \alpha_{m'} \frac{\langle\psi_j|V|\psi_{m'}\rangle}{\Delta_i - \Delta_j}, \tag{E.21}$$

which is a bit ugly, but as we will see, it will not be needed in the calculation of the second-order energy shifts. This is a first-order term because the two small parameters in the numerator, $\alpha_{m'}$ and V, are divided by another small parameter linear with V. The denominator $\Delta_i - \Delta_j$ is assumed to be nonzero because we have removed the degeneracy by the diagonalization of V.

The normalization factor is then

$$\frac{1}{\sqrt{1 + \sum_{m\notin\{n\}} |\alpha_m|^2 + \sum_{j\in\{n\}} |\beta_j|^2}} \simeq 1 - \frac{1}{2}\sum_{m\notin\{n\}} |\alpha_m|^2 - \frac{1}{2}\sum_{j\in\{n\}} |\beta_j|^2, \tag{E.22}$$

and the energy is, to second order in small parameters,

$$\begin{aligned}
E_i^{(2)} \simeq & \left(\langle\psi_i| + \sum_{\substack{j\in\{n\}\\j\neq i}} \beta_j^*\langle\psi_j| + \sum_{m\notin\{n\}} \alpha_m^*\langle\psi_m| - \frac{1}{2}\sum_{m\notin\{n\}} |\alpha_m|^2\langle\psi_i| - \frac{1}{2}\sum_{\substack{j\in\{n\}\\j\neq i}} |\beta_j|^2\langle\psi_i| \right) \\
& \times (H_0 + V) \\
& \times \left(|\psi_i\rangle + \sum_{\substack{k\in\{n\}\\k\neq i}} \beta_k|\psi_k\rangle + \sum_{p\notin\{n\}} \alpha_p|\psi_p\rangle - \frac{1}{2}\sum_{p\notin\{n\}} |\alpha_p|^2|\psi_i\rangle - \frac{1}{2}\sum_{\substack{k\in\{n\}\\k\neq i}} |\beta_k|^2|\psi_i\rangle \right) \\
\simeq & \langle\psi_i|(H_0 + V)|\psi_i\rangle + \sum_{m\notin\{n\}} \alpha_m^*\langle\psi_m|V|\psi_i\rangle + \sum_{p\notin\{n\}} \alpha_p\langle\psi_i|V|\psi_p\rangle \\
& + \sum_{m\notin\{n\}} E_m|\alpha_m|^2 - E_n\sum_{m\notin\{n\}} |\alpha_m|^2 + \frac{E_n}{2}\sum_{\substack{j\in\{n\}\\j\neq i}} |\beta_j|^2 - \frac{E_n}{2}\sum_{\substack{k\in\{n\}\\k\neq i}} |\beta_k|^2,
\end{aligned} \tag{E.23}$$

which simplifies to

$$E_i^{(2)} = E_n + \langle\psi_i|V|\psi_i\rangle + \sum_{m\notin\{n\}} \frac{|\langle\psi_m|V|\psi_i\rangle|^2}{E_n - E_m}, \tag{E.24}$$

where we have again used $\langle\psi_i|V|\psi_j\rangle = \Delta_i\delta_{ij}$. The second-order energy in the degenerate case thus has a simple formula similar to (E.15).

Löwdin perturbation theory. An alternative approach gives the same energies to the same order of approximation, but can be computationally easier in some cases, especially in the case of degenerate states. For a review of Löwdin perturbation theory, see Loehr (1998: appendix A).

In this case, we write any exact perturbed state arising from a group of degenerate states $\{n\}$ as

$$|\psi\rangle = \sum_{j\in\{n\}} \gamma_j|\psi_j\rangle + \sum_{m\notin\{n\}} \alpha_m|\psi_m\rangle, \tag{E.25}$$

where we do *not* assume that the factors γ_j are small. The index j runs over all of the degenerate states corresponding to energy E_n.

We then form the equation

$$\langle\psi_j|H|\psi\rangle = E\langle\psi_j|\psi\rangle, \tag{E.26}$$

where E is the eigenvalue of the exact eigenstate $|\psi\rangle$ of the perturbed Hamiltonian. Substituting in the definition (E.25), we obtain

$$\gamma_j E_n + \sum_{i\in\{n\}} \gamma_i\langle\psi_j|V|\psi_i\rangle + \sum_{m\notin\{n\}} \alpha_m\langle\psi_j|V|\psi_m\rangle = \gamma_j E. \tag{E.27}$$

Since we do assume that α_m is small, we can do a perturbative expansion using it. We solve for α_m by forming

$$\langle\psi_m|H|\psi\rangle = E\langle\psi_m|\psi\rangle, \tag{E.28}$$

and again substitute the definition (E.25), to obtain

$$\alpha_m E_m + \sum_{i\in\{n\}} \gamma_i\langle\psi_m|V|\psi_i\rangle + \sum_{m'\notin\{n\}} \alpha_{m'}\langle\psi_m|V|\psi_{m'}\rangle = \alpha_m E. \tag{E.29}$$

For the first-order approximation, we drop the third term on the left side because it has two small parameters $\alpha_{m'}$. In the same way, as before, we approximate $E \simeq E_n$, because the correction to E is small. We then for α_m find

$$\alpha_m = \sum_{i\in\{n\}} \gamma_i \frac{\langle\psi_m|V|\psi_i\rangle}{E_n - E_m}. \tag{E.30}$$

Substituting this into (E.31), we obtain

$$\sum_{i\in\{n\}} \gamma_i \left(E_n\delta_{ij} + \langle\psi_j|V|\psi_i\rangle + \sum_{m\notin\{n\}} \frac{\langle\psi_j|V|\psi_m\rangle\langle\psi_m|V|\psi_i\rangle}{E_n - E_m} \right) = \gamma_j E.$$

$$\tag{E.31}$$

This is a matrix eigenvalue equation defined on the degenerate states with energy E_n, with matrix elements

$$H_{ij}^{(2)} = E_n\delta_{ij} + \langle\psi_j|V|\psi_i\rangle + \sum_{m\notin\{n\}} \frac{\langle\psi_j|V|\psi_m\rangle\langle\psi_m|V|\psi_i\rangle}{E_n - E_m}. \tag{E.32}$$

Diagonalizing this matrix will give the eigenvectors and energies of the perturbed states to second order in V.

This method gives the same perturbed energies as (E.24) for sufficiently small V, but the two methods can deviate from each other significantly when V is large.

References

John P. Loehr, *Physics of Strained Quantum Well Lasers* (Springer, 1998).

J.J. Sakurai, *Modern Quantum Mechanics*, revised edition (Addison-Wesley, 1995).

The theory of fermions is one of the great triumphs of twentieth-century physics. Most of the credit belongs to Paul Dirac, who started, like Einstein, with some simple assumptions and laid the foundations for the Pauli exclusion principle of chemistry, Fermi statistics in solids, and antimatter in particle physics.

We begin by reproducing Dirac's simple but elegant argument (Dirac 1947) to deduce the relativistic wave equation for particles with mass. The basic problem is that the Schrödinger equation,

$$ i\hbar \frac{\partial}{\partial t}\psi = H\psi, \tag{F.1} $$

is not relativistically invariant if H is given by the standard kinetic energy $p^2/2m = -\hbar^2\nabla^2/2m$. In relativity, the conserved quantity is $E^2 = (mc^2)^2 + (cp)^2$. The energy is squared in this relativistic invariant, however, while Dirac believed strongly that the equations for the particles should be linear in the time dependence. His argument was that if it is not so, then the probability of finding the particle is not conserved over time. Consider the equation,

$$ -\frac{1}{c^2}\frac{\partial^2}{\partial t^2}\psi = \frac{m^2c^2}{\hbar^2}\psi - \nabla^2\psi, \tag{F.2} $$

which is second order in the time derivative and has the standing wave solution

$$ \psi(x, t) = \psi_0 \cos \omega t \cos kx, \tag{F.3} $$

subject to the condition $(\hbar\omega)^2 = (mc^2)^2 + (c\hbar k)^2$. Equation (F.2) is known as the *Klein–Gordon equation*, and it satisfies the need for relativistic invariance, when we equate $E = \hbar\omega$ and $p = \hbar k$, but it has the odd property that the solution (F.3) "winks out" twice every cycle; that is, at certain times when $\cos \omega t = 0$, there is zero probability of finding the particle anywhere. We are familiar with this type of wave for bosons such as phonons and phonons; the measurable field is real, and complex notation is only used to keep track of the oscillations. By contrast, for the Schrödinger equation (F.1), the solution for a standing wave is the complex wave

$$ \psi(x, t) = \psi_0 e^{-i\omega t} \cos kx, \tag{F.4} $$

which has nonzero value of $|\psi|^2$ at all times. Dirac felt that normal particles with mass should not disappear at certain times, and therefore worked to find a linear relativistic equation.[1]

To obtain a linear equation, we can factor the relativistically invariant term $E^2 - (mc^2)^2 - (cp)^2$ into two linear terms as follows:

$$E^2 - (mc^2)^2 - |c\vec{p}|^2 = (E + \alpha_0 mc^2 + c\vec{\alpha} \cdot \vec{p})(E - \alpha_0 mc^2 - c\vec{\alpha} \cdot \vec{p}) = 0,$$

(F.5)

where the α_i are four new operators that commute with $\vec{p}$. Clearly, if either of the two factors is zero, then the full relativistic term is zero, satisfying relativistic invariance. Since the α_i all commute with the components of $\vec{p}$, they must describe some extra degree of freedom.

It turns out that this factorization is only possible if the operator $\vec{\alpha}$ has the anticommutation property

$$\{\alpha_i, \alpha_j\} = 2\delta_{ij}.$$

(F.6)

In order to have four linearly independent operators that anticommute, the new operator must be represented by a matrix with at least four rows and columns. There is not one unique choice for these matrices, and various theories have been developed for different representations. The standard choice, following Dirac, is the following:

$$\alpha_0 = \left(\begin{array}{c|c} E & 0 \\ \hline 0 & -E \end{array} \right), \quad \alpha_i = \left(\begin{array}{c|c} 0 & \sigma_i \\ \hline \sigma_i & 0 \end{array} \right),$$

(F.7)

where the σ_i are the standard 2×2 Pauli spin matrices

$$\sigma_x = \left(\begin{array}{cc} 0 & 1 \\ 1 & 0 \end{array} \right), \quad \sigma_y = \left(\begin{array}{cc} 0 & -i \\ i & 0 \end{array} \right), \quad \sigma_z = \left(\begin{array}{cc} 1 & 0 \\ 0 & -1 \end{array} \right),$$

(F.8)

and E represents the 2×2 identity matrix. These are the standard spin matrices used in quantum mechanics textbooks (e.g., Cohen-Tannoudji *et al.* 1977: s. VI.C.3.c).

We can therefore write a relativistically invariant wave equation for particles with mass as follows:

$$i\hbar\frac{\partial}{\partial t}|\psi\rangle = H|\psi\rangle = (\alpha_0 mc^2 + c\vec{\alpha} \cdot \vec{p})|\psi\rangle,$$

(F.9)

where $|\psi\rangle$ has four components. This is the **Dirac equation**. Notice that even when the momentum $p_i = -i\hbar\nabla$ gives zero contribution, there are both positive and negative energy solutions, or bands, with $\langle H \rangle = \pm mc^2$. This symmetry means it doesn't matter that we used (F.9) instead of $H = -\alpha_0 mc^2 - c\vec{\alpha} \cdot \vec{p}$, though both are equally valid according to (F.5). It does raise the interesting equation of what we mean by negative energy solutions, though. Dirac hypothesized that the negative-energy solutions are all filled with electrons, one per state according to the Pauli exclusion principle. Therefore, unless they are given enough energy to jump up to the positive-energy band (at least $2mc^2$,

---

[1] The recently discovered Higgs boson may be the first example of an elementary particle with mass that obeys the Klein–Gordon equation. Until its discovery, Dirac's expectation that all elementary bosons are massless had held up.

which is one million electron volts), they will have no effect on electrons in positive states. The energy of this **Dirac sea** does not matter because it is a constant; as we have done throughout this book, we are free to define the vacuum as the ground state of the system. The existence of the negative-energy sea led Dirac to predict the existence of **positrons** as holes in the Dirac sea just like holes in the valence band of a semiconductor. This was a tremendous success of theoretical physics driving new experiments, with a successful prediction.

As with all of the quasiparticles we have studied in this book, we are free to define the vacuum as the ground state of the system, and the particles as the excitations out of that. It is therefore common in particle physics to drop all discussion of the negative-energy Dirac sea and just consider processes by which electron–hole pairs are generated, such as, by coupling to photons via diagrams like that in Figure 8.13. Thus, although we may think of electrons as elementary particles, they can be seen as quasiparticles from a deeper underlying state, no different in nature from other quasiparticles.

The symmetry of the Dirac equations raises a philosophical problem. If the ground state of the universe is a filled Dirac sea with no excitations (free electrons or holes), we should expect equal amounts of matter and antimatter. If this were the case, however, we could not endure the energy released by all the recombination. The asymmetry favoring normal matter over antimatter has been proposed as an example of spontaneous symmetry breaking due to a phase transition in the early universe due to some unknown higher-order effect.

The Dirac equation and spin energies. In writing down (F.5) we had to introduce a new degree of freedom. We can see the connection of the extra degree of freedom with angular momentum by looking at the effect of a magnetic field on the particles. In the presence of a magnetic field, we modify the relativistically invariant term by the standard substitution $\vec{p} \rightarrow \vec{p} - q\vec{A}$, where q is the charge, and $\vec{A}$ is the vector potential. The invariant term (F.5) thus becomes

$$E^2 - (mc^2)^2 - |c\vec{\alpha} \cdot (\vec{p} - q\vec{A})|^2 = 0. \tag{F.10}$$

We would like to find the Hamiltonian that includes magnetic field in the nonrelativistic limit. To do this, we can adopt the strategy used in standard relativistic mechanics, in which we assume that the momentum terms are small compared to mc, and expand in a Taylor series, which allows us to write $E = \sqrt{(mc^2)^2 + (cp)^2} \simeq mc^2 + p^2/2m$. The last term of (F.10) can be simplified by using the properties of the α_i matrices, namely, the anticommutation rule (F.6) and $\alpha_x \alpha_y = i\sigma_z$, for cyclic permutations, as well as the definition $p = -i\hbar\nabla$, so that we obtain

$$E^2 = (mc^2)^2 + c^2|\vec{p} - q\vec{A}|^2 + \hbar c^2 \vec{\sigma} \cdot (\nabla \times \vec{A}). \tag{F.11}$$

Writing E as a Taylor series of the square root of the right side, and recalling $\vec{B} = \nabla \times \vec{A}$, we then have

$$E \simeq mc^2 + \frac{1}{2m}|\vec{p} - q\vec{A}|^2 + \frac{q\hbar}{2m}\vec{\sigma} \cdot \vec{B}. \tag{F.12}$$

The first two terms correspond to the standard nonrelativistic Hamiltonian for a particle in a magnetic field, while the last term is a new term that arises from the fact that we needed to introduce the α matrices for the relativistic wave equation. This term is the Zeeman spin splitting in magnetic field which we have used multiple times in this book, which corresponds to particles with magnetic moment of $\pm\hbar/2$. Thus, we obtain the standard picture of fermions as having half-integer spin.

If we neglect the $\vec{A}$-field but include a nonzero electrostatic field, we must adjust the energy by the potential energy $U = qV(\vec{r})$. We rewrite (F.9) as

$$E|\psi\rangle = \left(\alpha_0 mc^2 + c\vec{\alpha}\cdot\vec{p} + U(\vec{r})\right)|\psi\rangle. \tag{F.13}$$

The structure of the matrices given in (F.7) means that we can rewrite this in terms of two two-component states, $|\psi_1\rangle$ for positive-energy states and $|\psi_2\rangle$ for negative-energy states, as follows:

$$E|\psi_1\rangle = mc^2|\psi_1\rangle + c\vec{\sigma}\cdot\vec{p}|\psi_2\rangle + U(\vec{r})|\psi_1\rangle$$
$$E|\psi_2\rangle = -mc^2|\psi_2\rangle + c\vec{\sigma}\cdot\vec{p}|\psi_1\rangle + U(\vec{r})|\psi_2\rangle. \tag{F.14}$$

Because we are free to define energy relative to any zero, we can write this as

$$E'|\psi_1\rangle = (E - 2mc^2)|\psi_1\rangle = c\vec{\sigma}\cdot\vec{p}|\psi_2\rangle + U(\vec{r})|\psi_1\rangle$$
$$E'|\psi_2\rangle = c\vec{\sigma}\cdot\vec{p}|\psi_1\rangle + (U(\vec{r}) - 2mc^2)|\psi_2\rangle. \tag{F.15}$$

We can write the second equation as

$$|\psi_2\rangle = \frac{1}{2mc^2 + E' - U}c\vec{\sigma}\cdot\vec{p}|\psi_1\rangle, \tag{F.16}$$

which for small $E' - U$ can be approximated as

$$|\psi_2\rangle = \frac{1}{2mc}\left(1 - \frac{E'}{2mc^2} + \frac{U}{2mc^2}\right)\vec{\sigma}\cdot\vec{p}|\psi_1\rangle. \tag{F.17}$$

We can then substitute this into the first equation of (F.15) to obtain

$$E' = \frac{1}{2m}\vec{\sigma}\cdot\vec{p}\left(1 - \frac{E'}{2mc^2} + \frac{U}{2mc^2}\right)\vec{\sigma}\cdot\vec{p} + U(\vec{r}). \tag{F.18}$$

The momentum operator $\vec{p}$ and U do not commute; using the general relation $[A, f(B)] = [A, B]f'(B)$ when $[A, B]$ is a c-number, we have

$$\vec{p}\, U(\vec{r}) = U(\vec{r})\vec{p} - [\vec{p}, U(\vec{r})] = U(\vec{r})\vec{p} - i\hbar\nabla U. \tag{F.19}$$

We also use the identity for any two vectors $\vec{A}$ and $\vec{B}$,

$$(\vec{\sigma}\cdot\vec{A})\cdot(\vec{\sigma}\cdot\vec{B}) = \vec{A}\cdot\vec{B} + i\vec{\sigma}\cdot(\vec{A}\times\vec{B}), \tag{F.20}$$

to finally obtain

$$E' = \frac{p^2}{2m}\left(1 - \frac{E' - U}{2mc^2}\right) + U(\vec{r}) + \frac{1}{4m^2c^2}(i\hbar\nabla U)\cdot\vec{p} + \frac{\hbar}{4m^2c^2}\vec{\sigma}\cdot(\nabla U \times \vec{p}). \tag{F.21}$$

The first term gives the kinetic energy with a relativistic correction for the mass; the third term is relativistic correction for the potential energy. The last term is the standard spin–orbit energy, which we use throughout this book.

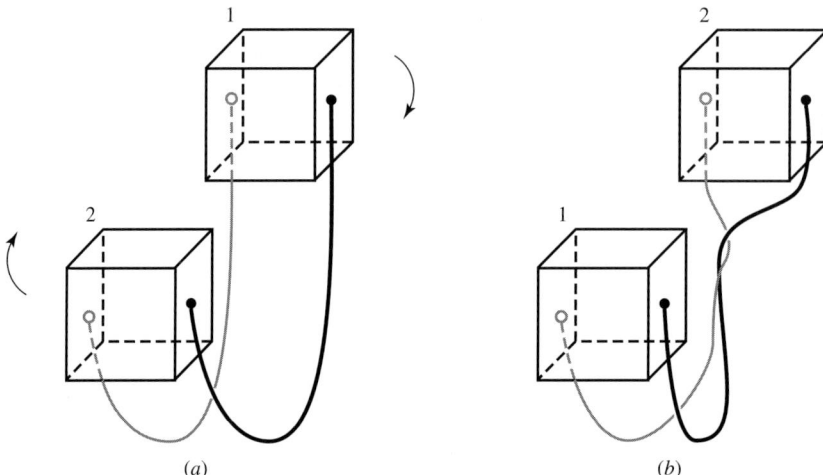

(a) (b)

Fig. F.1   (a) Two cubes attached by strings on their sides. An interchange is made by pure translation around each other, without rotating the cubes. (b) After the interchange, there is a twist in the strings, which can be removed by rotating one of the blocks 360°.

Spin-statistics connection. Having established that the quantization of the field requires particles of half-integer spin, we can then deduce that they must obey fermion statistics, namely the anticommutation relation $\{b_k, b_k^\dagger\} = 1$. Spin-statistics theorems have been an active theoretical field over the years. One argument is simply to note that if fermions did not obey this relation, that is, if they did not obey Pauli exclusion, then the negative-energy Dirac sea discussed above could not be filled up and would be unstable. A more direct argument comes from topological considerations. We first note that half-integer spin implies that a rotation of 2π yields a minus sign. The rotation operator for rotations about the z-axis is given by (see, e.g., Cohen-Tannoudji *et al.* 1977: s. B$_{IV}$.3.c.γ):

$$R_z(\theta) = e^{-i\theta L_z/\hbar}. \tag{F.22}$$

Simple substitution of $L_z = \hbar/2$ and $\theta = 2\pi$ gives $R_z = -1$. This is the result used in the character tables for double groups used in Chapter 6.

It can then be argued that interchanging two particles is topologically equivalent to a 2π rotation. Figure F.1 illustrates this. Imagine two blocks connected by strings, as shown in Figure F.1(a). If the positions of the two objects are interchanged by translating the blocks (e.g., taking block 1 to the right, then moving block 2 forward into its place, then moving block 1 back and to the left to take the place of block 2) then after the interchange, the strings will be twisted. To untwist the strings and restore the system back to its original state (but with block 1 and block 2 interchanged), one of the blocks must be rotated by 360°. The same thing can be demonstrated for an interchange via translation of one block up and over the other block, if the blocks are connected by strings on the front and back faces.

The change of sign on interchange of two particles implies that $\psi(r_1)\psi(r_2) = -\psi(r_2)\psi(r_1)$, which in turn, interpreting the ψ as spatial field operators, implies the anticommutation relation $\{b_k, b_k^\dagger\} = 1$.

References

C. Cohen-Tannoudji, B. Diu, and F. Laloë, *Quantum Mechanics* (Wiley, 1977).

P.A.M. Dirac, *The Principles of Quantum Mechanics*, 3rd ed. (Oxford University Press, 1947).

Index